Nanomaterials in Environmental Remediation

Environmental contamination poses a significant threat to ecosystems and human health. Nanomaterials, with their unique properties and versatility, have emerged as promising tools for environmental remediation. This book collates recent studies and new developments in effective environmental remediation using nanomaterials for cleaning pollutants in different environmental matrices. It explores the role of carbon and composite nanomaterials, bio nanomaterials, nanosheets, and phytonanotechnology. It also delves into photocatalytic applications of nanomaterials, using 3D-printed materials for the remediation of the environment, along with a deep analysis of the potential environmental risks, challenges, and the responsible use of nanomaterials. The authors also cover the prospective application and advantages of engineered nanomaterials, emphasizing the continuous growth and potential metamorphic impact of nanotechnology.

Features

- Provides a comprehensive review of the rapidly growing field of nanotechnology.
- Includes in-depth discussion on the detection of pollutants with the help of nanomaterials.
- Offers critical insight into the synthesis and characterization of nanomaterials.
- Covers a vast array of applications illustrating the wide use of different domains of nanotechnology along with their possible toxicological aspects.
- Presents the future of green nanomaterials as sustainable solutions in environmental remediation.

This book is a great resource for researchers, academicians, students, and professionals in environmental and sustainable engineering, nanotechnology, and environmental remediation.

Nanomaterials in Environmental Remediation

Developments and Challenges

Edited by
Neha Agarwal
Vijendra Singh Solanki
Neetu Singh
Maulin P. Shah

CRC Press
Taylor & Francis Group
Boca Raton London New York

CRC Press is an imprint of the
Taylor & Francis Group, an **informa** business

Designed cover image: Shutterstock

First edition published 2025
by CRC Press
2385 NW Executive Center Drive, Suite 320, Boca Raton FL 33431

and by CRC Press
4 Park Square, Milton Park, Abingdon, Oxon, OX14 4RN

CRC Press is an imprint of Taylor & Francis Group, LLC

ISBN: 978-1-032-77649-1 (hbk)
ISBN: 978-1-032-77650-7 (pbk)
ISBN: 978-1-003-48418-9 (ebk)

DOI: 10.1201/9781003484189

Typeset in Times
by SPi Technologies India Pvt Ltd (Straive)

Contents

Foreword

The realm of nanomaterials stands before the leading scientific innovations, opening a new age in the understanding of materials at the atomic and molecular levels. The compilation presented in this book provides a vast overview of the multifaceted world of nanomaterials along with their synthesis, properties, and a multifarious range of applications, particularly in the field of environmental remediation.

In "Nanomaterials in Environmental Remediation: Developments and Challenges" the journey begins with the fundamental characteristics of nanomaterials, offering a deep understanding of their design and the advanced techniques used to modify their properties for particular applications. This constructs an insightful understanding of the advanced nanomaterials studied and discussed in the following chapters.

The book then shows the transitions into particular themes, such as bioinspired green synthesis, sustainable nanomaterials, and their roles in reducing toxic organic pollutants. These comprehensive discussions highlight the significant role of nanomaterials in encouraging sustainability and environmental health, which shows their potential to change environmental remediation strategies.

Chapters that discuss functionalized nanomaterials, carbon nanomaterials, and bio-composite nanomaterials for wastewater remediation further elucidate the potency of these materials. The investigation of bio-nanomaterials and nanosheet composites presents innovative ideas for environmental remediation and water pollutant remediation.

As we look into the integration of 3D printed materials in remediation efforts, and photocatalytic activity of nanomaterials; these nanomaterials find potential in the field of environmental science. The analysis of the challenges, opportunities, and the assessment of potential environmental risks of using nanomaterials offer a balanced viewpoint, certifying a holistic comprehension of this field.

The final chapters on the current status and future prospects of engineered nanomaterials, along with the advancements in nanoremediation techniques for environmental remediation, hold a promise for a sustainable environment.

This book is an indispensable resource for scientists, scholars, researchers, technologists, practitioners, and students in the area of nanotechnology and environmental science, providing a detailed and insightful overview of the remarkable steps being made in the research and application of nanomaterials. This book will inspire further innovation and cultivate a great appreciation for the potential of nanomaterials in creating a sustainable environment.

Khac-Uan Do, PhD. Associate Professor
School of Chemistry and Life Sciences
Hanoi University of Science and Technology
Vietnam

Preface

Nanotechnology, a leading-edge frontier of science and technology, has vastly altered the prospects of materials science, nanoscience, engineering, and environmental science. The various applications of nanomaterials promise unmatched development in a variety of fields, particularly in the field of environmental remediation. This book, titled ***"Nanomaterials in Environmental Remediation: Developments and Challenges"***, provides a deep understanding regarding the exploration of the synthesis, properties, and applications of several nanomaterials, along with addressing the various environmental challenges.

The book has been structured into twenty chapters and each chapter provides an extensive overview of the current and future potential of nanomaterials in environmental remediation. The initial chapters deeply discuss the foundation of basic concepts and synthesis methods of various nanomaterials. Chapter 1 thoroughly investigates the basics and provides deep insights into the properties and distinct applications of nanomaterials. Chapter 2 discusses the introduction to advanced nanomaterials and nanotechnology.

Subsequent chapters delve into specific types of nanomaterials and their unique properties and applications in the remediation of the environment. These chapters cover a wide range of materials and their synthesis methods, from the biosynthesis of nanomaterials to the utilization of functionalized nanomaterials for the safety of the environment. For example, Chapters 3 and 4 study the various sustainable and green synthesis methods, while Chapters 5 to 14 investigate the potential applications such as the degradation of toxic organic pollutants and the remediation of wastewater by nanosheets, nanocomposites, bionanomaterials, carbon nanomaterials, along with the photocatalytic applications of nanomaterials in remediation of environment.

The book also provides an insightful discussion of the incorporation of nanotechnology with other advanced technologies. Chapter 15 emphasizes the usage of 3D-printed materials for the remediation of the environment, focussing on the intersection of nanotechnology and additive manufacturing. Moreover, Chapters 16 to 18 of this book also provide a deep analysis of the potential environmental risks and challenges associated with nanoremediation and the responsible use of nanomaterials.

Finally, the book concludes with advancements and innovations in nanoremediation techniques, along with the prospective application and advantages of engineered nanomaterials, emphasizing the continuous growth and the potential metamorphic impact of nanotechnology.

We hope this compendium serves as a valuable and potential resource for scientists, researchers, technologists, practitioners, and students working in the domain of environmental science and nanotechnology, inspiring future innovation and proffering to the sustainable management of our natural environment.

About the Editors

Dr. Neha Agarwal is an Assistant Professor in the Department of Chemistry, Navyug Kanya Mahavidylaya, University of Lucknow, India. She obtained her Ph.D. from the University of Lucknow after qualifying for UGC-CSIR-NET Chemical Science and GATE-Chemical Science. She has published many research papers and review articles on environmental remediation and nanomaterials in high-impact Scopus Indexed journals. She has published three patents, edited books with reputable publishers, and authored two textbooks for undergraduate students. Her research interests include the study of the environmental impacts of a vast variety of organic pollutants such as pharmaceuticals, dyes, and pesticides and finding economically viable treatment techniques to degrade these pollutants into simple and non-toxic substances in an environmentally friendly manner. Currently, she is working in the field of environmental remediation including wastewater remediation with a special focus on synthesis, characterization, and multifaceted applications of nanomaterials. She has been awarded the Best Researcher award by VDGOOD Professionals India.

Dr. Vijendra Singh Solanki is an Assistant Professor in the Department of Chemistry at the Institute of Science and Research, IPS Academy, Indore, India. He has an M. Phil and Ph.D. His PhD. research area was based on degradation studies of tough dyes from wastewater for a sustainable environment. He has published many research papers on dye degradation, nanoparticle synthesis, characterization, and applications, with a special focus on environmental applications. He has been granted many Indian patents, published many research articles in Scopus-indexed journals, and edited a book with a well-known publisher. His research interests include photocatalytic degradation of organic pollutants and wastewater remediation. Green chemistry, synthesis, and characterization of nanomaterials. He is an active member of many scientific organizations and has been invited many times for expert talks in the respective field.

Dr. Neetu Singh has a Ph.D. Degree in Physics and is working as an Assistant Professor (Physics) at Maharaja Bijli Pasi Post Graduate Degree College, Lucknow, Uttar Pradesh, India. Dr. Neetu is a reviewer of various reputed international journals and has reviewed many papers. She has more than 10 years of research experience, and her research area focuses on environmental science, nanomaterials, polymers, polymer nanocomposites, metal oxides, photocatalysis, and gas sensing. She has also published several research papers and book chapters in reputed international journals.

Dr. Maulin P. Shah has emerged as a prominent researcher with his dedication to research spanning over two decades. He completed his BSc in Microbiology from Gujarat University, Godhra (Gujarat), India, in 1999. He further completed his PhD from Sardar Patel University, Vallabh Vidyanagar (Gujarat), India in 2005. His expertise and efforts are committed to investigating the complex relation between pollution and microbial diversity of wastewater via both cultivation-dependent and independent analyses. Dr. Shah's prolific scholarly journey is underscored by a staggering publication record, comprising more than 250 research papers across esteemed national and international journals. Beyond his research articles, he holds the distinguished role of editor for over 150 internationally renowned books, published by esteemed publishers such as Elsevier, Springer, RSC, and CRC Press, Wiley, IWA, IOP, Bentham, Apple Academic Press. He actively serves on the editorial boards of top-rated journals. Alongside his advisory role includes as the Editorial Board Member of CLEAN-Soil, Air, Water- Wiley; Current Pollution Reports- Springer Nature; Journal of Biotechnology & Biotechnological Equipment-Taylor & Francis; Applied Water Science - Springer Nature; Frontiers in Bio engineering & Biotechnology; Archives of Microbiology-Springer; Green Technology, Resilience and Sustainability-Springer; Biomass Conversion & Bio refinery-Springer; Journal of Basic Microbiology - Wiley; Energy Nexus-Elsevier; e Prime - Elsevier; IET Nanobiotechnology-Wiley; Journal of Water Chemistry & Technology- Springer; Nanotechnology

Reviews - De Gruyter; Soil & Environmental Health - Elsevier; Environmental Chemistry & Ecotoxicology-Elsevier; Open Life Sciences- De Gruyter; Green Technology and Sustainability-Elsevier; Emerging Contaminants - Elsevier; All Life - Taylor & Francis; Current Indian Science - Bentham Science; South African Journal of Chemical Engineering - Elsevier; Industrial Biotechnology - Mary Ann Libert Inc. Publisher; Arab Journal of Basic and Applied Sciences-Taylor & Francis; Environmental Surfaces and Interfaces-KeAi; Microbial Cell Factories – BMC Springer; Environmental Geochemistry and Health - Springer.

Contributors

Agarwal, Neha
Department of Chemistry
Navyug Kanya Mahavidyalaya (University of Lucknow)
Lucknow, Uttar Pradesh, India

Agrahari, Saumya
Department of Chemistry
University of Allahabad
Prayagraj Uttar Pradesh

Anita
Department of Chemistry
Kishan Lal Public College
Rewari, Haryana, India

Babu, A. Ramesh
Department of Chemistry
SVA Government Degree College(M), Srikalahasti
Tirupati, Andhra Pradesh, India

Banerjee, Saikat
College of Engineering and Technology
Engineering Department
University of Technology and Applied Sciences-Salalah
Oman

Bhattacharya, Chitra
Department of Life Sciences
Atmiya University
Rajkot, Gujarat, India

Chaudhary, Anita
Department of Chemistry
Kishan Lal Public College
Rewari, Haryana, India

Dwivedi, Jaya
Department of Chemistry
Banasthali Vidyapith, Banasthali, Rajasthan, India

Gautam, Sakshi
Department of Chemistry
Banasthali University
Jaipur, Rajasthan, India

Geedkar, Deepika
School of Chemical Sciences
Devi Ahilya University
Indore, Madhya Pradesh, India

Ghanghas, Priyanka
Department of Chemistry
Banasthali Vidyapith
Newai, Rajasthan, India

Ghosh, Prabir
Department of Chemical Engineering
National Institute of Technology
Raipur, Chhattisgarh, India

Inwati, Gajendra Kumar
School of Chemical Sciences
Devi Ahilya University
Indore, Madhya Pradesh, India

Jagtap, Chandrashekhar Y
Central Ayurveda Research Institute
Jhansi, Uttar Pradesh, India

Jangid, Nirmala Kumari
Department of Chemistry
Banasthali University
Jaipur, Rajasthan, India

Kaur, Navjeet
Department of Chemistry
Lovely Professional University
Phagwara, Punjab, India

Korlam, Saivenkatesh
Department of Botany
SVA Government Degree College (M), Srikalahasti
Tirupati, Andhra Pradesh, India

Kumar, Ashok
School of Chemical Sciences
Devi Ahilya University
Indore, Madhya Pradesh, India

Kumar, Bhumika
Department of Pharmaceutics, School of Pharmacy
Sharda University
Greater Noida, Uttar Pradesh, India

Kumar, Manoj
Department of Hydrology
Indian Institute of Technology, Roorkee
Roorkee, Uttarakhand, India

Kumar, Nakul
Gandhinagar Institute of Science
Gandhinagar University, Khatraj
Gandhinagar, Gujarat, India

Kumar, Neha Rani
Dhemaji College
Dhemaji, Assam, India

Kumar, Pankaj
Department of Earth and Environmental Science
Parul Institute of Applied Sciences
Parul University
Vadodara, Gujarat, India

Kumar, Prasann
Department of Agronomy
School of Agriculture
Lovely Professional University
Phagwara, Punjab, India

Kumar, Vijay
Central Ayurveda Research Institute
Jhansi, Uttar Pradesh, India

Maurya, Rahul
Central Ayurveda Research Institute
Jhansi, Uttar Pradesh, India

Meka, Uma Reddy
College of Engineering and Technology
Engineering Department
University of Technology and Applied Sciences-Salalah
Oman

Miditana, Sankara Rao
Department of Chemistry
Government Degree College, Puttur
Tirupati, Andhra Pradesh, India

Mishra, Ashwini Kumar
Department of Pharmaceutics,
School of Pharmacy & Technology Management,
SVKM'S NMIMS Deemed-to-be University,
Shirpur, Maharashtra, India

Mitra, Amrit Krishna
Department of Chemistry
Government General Degree College, Singur
Singur, Hooghly, West Bengal, India

Nachiappan, Senthilnathan
Chemical Engineering, Engineering Department
University of Technology and Applied Sciences
Salalah Branch, Oman

Pandey, Bhawana
Department of Biotechnology & Microbiology
Bhilai Mahila Mahavidyalaya
Bhilai, Chhattisgarh, India

Patel, Prabhat Kumar
Civil Engineering Department
Institute of Engineering and Technology
Lucknow, Uttar Pradesh, India

Patra, Asha Rani
Department of Chemical Engineering
National Institute of Technology
Raipur, Chhattisgarh, India

Pattnaik, Amruta
Department of Chemical Engineering
National Institute of Technology
Raipur, Chhattisgarh, India

Pinar, Orkun
Department of Separation Science
LUT School of Engineering Science
LUT University
Sammonkatu, Mikkeli, Finland

Poonia, Anil Kumar
Department of Chemical Engineering
National Institute of Technology
Raipur, Chhattisgarh, India

Poonia, Kavita
Department of Chemistry
Banasthali Vidyapith
Newai, Rajasthan, India

Prasad, B.S. Naveen
College of Engineering and Technology
Engineering Department
University of Technology and Applied Sciences-Salalah
Oman

Rakesh, Namdeti
Chemical Engineering, Engineering Department
University of Technology and Applied Sciences
Salalah Branch, Oman

Rayjada, Nidhiba
Department of Life Science
Atmiya University
Rajkot, Gujarat, India

Rodríguez-Couto Susana
Department of Separation Science
LUT School of Engineering Science
LUT University
Sammonkatu, Mikkeli, Finland

Saha, Rama
Department of Chemistry
Rani Rashmoni Green University
Tarakeswar, Hooghly, West Bengal, India

Saini, Nisha
Department of Chemistry
Banasthali Vidyapith, Banasthali, Rajasthan, India

Shafi, Huma
Smt. Vidyawati Group of Institutions Goramachhiya
Jhansi, Uttar Pradesh, India

Sharma, Apurva
Department of Earth and Environmental Science
Parul Institute of Applied Sciences, Parul University
Vadodara, Gujarat, India

Sharma, Pratibha
School of Chemical Sciences
Devi Ahilya University
Indore, Madhya Pradesh, India

Singh, Asit
Civil Engineering Department
Institute of Engineering and Technology
Lucknow, Uttar Pradesh, India

Singh, Divya
Department of Chemistry
University of Allahabad
Prayagraj Uttar Pradesh

Singh, Hrishikesh Kumar
Civil Engineering Department
Institute of Engineering and Technology
Lucknow, Uttar Pradesh, India

Singh, Joginder
Department of Botany
Nagaland University
Nagaland, India

Singh, Rajesh
Central Ayurveda Research Institute
Jhansi, Uttar Pradesh, India

Singh, Snigdha
Department of Earth and Environmental Science
Parul Institute of Applied Sciences
Parul University
Vadodara, Gujarat, India

Sivamani, Selvaraju
College of Engineering and Technology
Engineering Department
University of Technology and Applied Sciences-Salalah
Oman

Solanki, Vijendra Singh
Department of Chemistry
ISR, IPS Academy
Indore, Madhya Pradesh, India

Srivastava, Anamika
Former Assistant Professor
Department of Chemistry
Banasthali Vidyapith
Rajasthan, 304022, India

Srivastava, Manish
Associate Professor
Department of Chemistry
University of Allahabad
Prayagraj, Uttar Pradesh, India

Tariq, Mohd
Department of Life Science
Parul Institute of Applied Sciences, Parul University
Vadodara, Gujarat, India

Yadav, Anjali
Department of Chemistry
JECRC University
Jaipur, Rajasthan, India

Yadav, Anshul
Department of Chemistry
Banasthali Vidyapith
Newai, Rajasthan, India

Yadav, Virendra Kumar
Marwadi University Research Center
Marwadi University
Rajkot, Gujarat, India

Yamee Shah
Department of Nutrition and Dietetics
Parul Institute of Applied Sciences
Parul University
Vadodara, Gujarat, India

1 Delving into Nanomaterials
Fundamentals, Synthetic Techniques, Properties, Diverse Applications, and Recent Advancements

Deepika Geedkar, Ashok Kumar, Gajendra Kumar Inwati, and Pratibha Sharma

1.1 INTRODUCTION

The term "nano" derives from the Greek word "nanos," signifying something exceptionally small, akin to a dwarf or minuscule entity. In scientific measurement, the prefix "nano" denotes a fraction of a unit, such as a nanometer representing one billionth of a meter or one-millionth of a millimeter. This prefix has permeated contemporary science, spanning diverse fields of knowledge and literature (Shinde et al., 2022; Alqahtani et al., 2022). NMs, characterized by dimensions ranging from 1 to 100 nanometers, constitute a fundamental domain of modern science. Within this scale, (nanoparticles) NPs are defined by materials possessing at least one nanoscale dimension. Generally, for a material to be considered an NM, it must exhibit a surface area greater than 60 m^2/cm^3 Baig et al., 2021; Khan et al., 2019; Vollath, 2013). These materials encompass a spectrum of categories delineated by size, structural configuration, pore diameters, and potential toxicity (Trotta and Andrea, 2019). At the nanoscale, NPs exhibit distinctive chemical, physical, and biological attributes that diverge significantly from their bulk equivalents, exemplifying unique optical, magnetic, and electrical properties (Altammar et al., 2023; Khan et al., 2022). Furthermore, the emergence of hybrid NMs, combining diverse components at the nanoscale, has opened new avenues for tailoring properties to specific applications (Godja and Munteanu, 2024).

NMs can be synthesized using two primary strategies: bottom-up approaches, which involve assembling materials from atomic or molecular precursors, and top-down approaches, which entail reducing bulk materials to the nanoscale (Baig et al., 2021; Mekuye and Abera, 2023; Sajid and Płotka-Wasylka, 2020). Top-down techniques include processes such as lithography, mechanical milling, and laser ablation, while bottom-up methods encompass chemical vapor deposition, sol-gel synthesis, and biological synthesis (Kumari et al., 2023; Nyabadza et al., 2023). Recent advancements in synthesis methodologies, such as template-assisted approaches and self-assembly techniques, have enabled precise control over NM properties, leading to enhanced functionality and performance (Poolakkandy and Menamparambath, 2020; Singamaneni et al., 2011).

The interdisciplinary realm of nanoscience delves into the intricacies of matter at the nanoscale, emphasizing the size-dependent properties of solid-state materials (Mekuye and Abera, 2023; Joudeh and Linke, 2022). Nanotechnology, on the other hand, encompasses the synthesis, engineering, and application of NMs, garnering significant interest for its innovative applications across various industries (Sim and Wong, 2021; Malik et al., 2023; Kumar, 2023; Iavicoli et al., 2017). Industries spanning from agriculture to biomedicine, electronics, and energy have embraced NMs for their transformative potential (Iavicoli et al., 2017; Manickam, et al., 2023; Jariwala et al., 2013;

DOI: 10.1201/9781003484189-1

Hussein, 2015). NMs offer not only technological advancements but also solutions to pressing global challenges such as climate change and epidemic diseases (Pokrajac, et al., 2021). For instance, NM-based green technologies embrace a perspective for mitigating the effects of climate change (Chausali, et al., 2023), while their applications in diagnostics and treatment have revolutionized disease control (Thwala, et al., 2023), evident in their role in combating epidemics like COVID-19 and monkeypox (Xu, et al., 2022; Xiao, et al., 2022).

Anticipating the future, NMs, nanoscience, and nanotechnology are positioned to make significant contributions to worldwide progress, highlighting the necessity for widespread awareness and comprehension of these materials (Mansurov, 2020; Tiza, et al., 2023). In prospect, NMs boast substantial potential for tackling urgent global issues and fostering innovation across various sectors. As research progresses, the scope of applications for nanotechnology is anticipated to broaden, heralding groundbreaking breakthroughs and revolutionary technologies. Despite extensive research conducted on NPs and materials in the past decades, a comprehensive synthesis of NMs classification, preparation methods, properties, and applications remains elusive. This chapter endeavors to bridge this disparity by providing a comprehensive overview of NMs, encompassing their classification, preparation methods, properties, applications, historical context, significance, challenges, ethical considerations, recent advancements, future directions, and societal implications. The structure of the chapter will guide the reader through a systematic exploration of these topics, facilitating a holistic understanding of the field of NMs.

1.2 CLASSIFICATION OF NMs

NMs exhibit diverse properties and applications, making their classification essential for understanding their behavior and harnessing their potential effectively. They can be categorized into distinct groups based on various factors such as origin, structure, dimensionality, pore size, and potential toxicity as depicted in Figure 1.1 (Mekuye and Abera, 2023).

1.2.1 Origin-Based Classification

Natural NMs: These encompass a broad spectrum of materials found in nature, ranging from biological entities like viruses and proteins to mineral-based substances such as clay and opals. Natural colloids like milk and blood, as well as organic structures like insect wings and spider silk, also fall under this category (Mekuye and Abera, 2023).

Artificial NMs: These are synthesized through controlled processes. Examples include carbon nanotubes, quantum dots (QDs), and metal NPs like gold and silver. Artificial NMs provide tailored properties suitable for applications in electronics, medicine, and environmental remediation (Baig, et al., 2021; Mekuye and Abera, 2023).

1.2.2 Structural Configuration/Composition-Based Classification

Organic NMs: This category includes liposomes, dendrimers, and micelles, which are widely employed in drug delivery systems and molecular sensing applications (Patra, et al., 2018). Organic NPs offer advantages such as biocompatibility and tunable surface chemistry (Kyriakides, et al., 2021). Emerging trends in this category include bioinspired NMs and advancements in self-assembly techniques (Lombardo, et al., 2020).

A chart classifies nanomaterials based on origin, structural configuration, number of dimensions, pore dimension, and potential toxicity.

Inorganic NMs: Comprising metal-based and metal oxide NPs, these materials exhibit unique optical, electrical, and catalytic properties (Alhalili, 2023). Examples include silver NPs for antimicrobial applications and titanium oxide NPs for photocatalysis (Chand, et al., 2020).

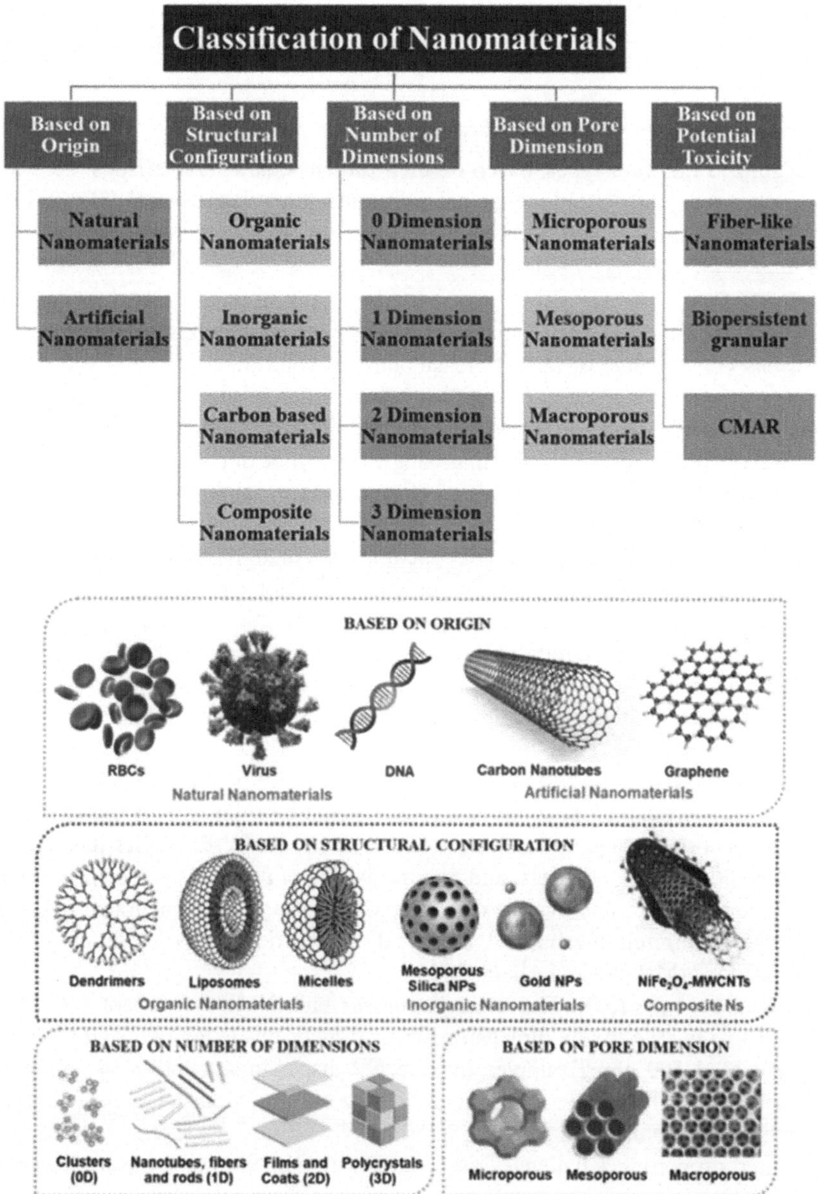

FIGURE 1.1 General classification of NMs.

Recent advancements in nanofabrication methods have facilitated meticulous manipulation of the structure and characteristics of inorganic NMs.

Carbon-Based NMs: This category encompasses carbon nanotubes, graphene, fullerenes, and carbon black, renowned for their outstanding strength, conductivity, and lightweight nature. Carbon-based NMs find applications in aerospace, automotive, and energy storage technologies (Harussani, et al., 2022). Standardization and characterization challenges persist, owing to the diverse forms and structures of carbon-based NMs.

Composite NMs: These amalgamated NPs with polymers, ceramics, or metals to enhance mechanical, thermal, or electrical properties. Composite NMs are utilized in diverse applications such as reinforced plastics, conductive coatings, and biomedical implants (Elbakyan and Zaporotskova, 2024). Advances in nanocomposite synthesis techniques have fostered the creation of innovative materials tailored for specific applications.

1.2.3 Dimension-Based Classification

NMs are categorized into four types, based on their dimensional characteristics: zero-dimensional (0D), one-dimensional (1D), two-dimensional (2D), and three-dimensional (3D), with each type possessing distinct properties and applications (Khan, et al., 2019; Mekuye and Abera, 2023; Byakodi, et al., 2022).

Zero-Dimensional NMs (0D): These materials occupy the nanoscale range in all three spatial dimensions (x, y, and z) or lack dimensionality beyond the nanometer scale (>10 nm). Prominent examples include quantum dots (QDs), fullerenes, and NPs (Khan, et al., 2019; Mekuye and Abera, 2023). They exhibit a wide array of attributes, ranging from amorphous to crystalline structures, and can be single crystalline or polycrystalline. Furthermore, they may exist in different dimensions and forms and can exhibit metallic or ceramic properties (Wang, et al., 2020).

One-Dimensional NMs (1D): This category encompasses materials with two dimensions (x, y) confined within the nanoscale range, while one dimension extends beyond the nanometer scale (>10 nm) (Khan, et al., 2019; Mekuye and Abera, 2023). Notable examples include nanofibers, nanotubes, nanorods, and thin films. These materials can exhibit diverse properties, such as crystallinity, chemical purity, and morphology. Additionally, they may function as stand-alone entities or be integrated into other materials, including metallic, ceramic, or polymeric matrices (Zhang, et al., 2019).

Two-Dimensional NMs (2D): NMs falling into this category possess plate-like shapes, with two dimensions extending beyond the nanometer scale, while one dimension (x) remains within the nanoscale range (1–100 nm) (Khan, et al., 2019). Examples include coatings, thin-film multilayers, nanosheets, and nanowalls. These materials can display a variety of chemical compositions and may be deposited onto substrates or integrated into surrounding matrices, exhibiting mechanical, electrical, and barrier properties suitable for numerous applications (Mekuye and Abera, 2023).

Three-Dimensional NMs (3D): This class comprises bulk NMs that do not exhibit confinement to the nanoscale in any dimension. Their dimensions extend beyond the nanometer scale or exceed 100 nm. Examples include NP dispersions, bundles of nanowires, and multi-nanolayers. These materials frequently integrate structural elements of zero, one, and two dimensions in proximity, fostering interfaces that contribute to their multifunctionality and tailored properties. Challenges associated with 3D NMs encompass scalability and reproducibility in manufacturing processes, owing to their intricate hierarchical structures (Mekuye and Abera, 2023; Manasa and Rout, 2024).

Each type of NM presents distinct opportunities and challenges, motivating continuous research and development endeavors focused on leveraging their unique properties for diverse applications in fields such as electronics, energy, healthcare, and environmental remediation.

1.2.3.1 Density of State in Various Dimensional NMs

Semiconductor particles known as quantum dots (QDs) typically possess diameters less than 10 nm and exhibit remarkable electronic and optical characteristics, owing to their nanoscale size. The phenomenon termed the quantum size effect describes the alteration of electronic properties as the size of the quantum dot diminishes, particularly noticeable when approaching the scale of the electron's wavelength. The scaling down of bulk crystalline materials to the nanoscale induces significant changes in their behavior, impacting applications like sensing, imaging, and energy conversion (Mekuye and Abera, 2023; Cotta, 2020).

The density of states $g(E)$ represents a fundamental conception in condensed matter physics, denoting the number of quantum states per unit energy interval within a system. It provides insight into how densely packed the quantum states are within the material's energy spectrum. This parameter is important in understanding the electronic properties and plays a fundamental role in assessing the effect of confinement on energy states (Mekuye and Abera, 2023; Mahmoud, et al., 2020).

The "particle in the box" problem is often utilized to elucidate the impact of confinement on electron energy levels, where an electron is confined within an infinitely deep potential well representing the dimensions of the nanostructure. In semiconductor materials, where carrier motion is confined to specific dimensions, the density of states serves as a significant parameter for predicting and interpreting electronic behaviour (Figure 1.2).

1.2.3.1.1 Density of State in 3D (Bulk) NMs

In three-dimensional NMs, electrons are characterized by three-dimensional wave functions extending throughout the material's volume. The computation of state density involves determining various parameters such as the volume of a single state, the volume of a sphere, and the number of field states. The density of states in 3D NMs follows a square root relationship with energy, indicating an increment in the density of available states, with increasing energy levels (Mekuye and Abera, 2023; Bokov, et al., 2021).

The volume of a single state,

$$V_{\text{Single state}} = \frac{\pi^3}{V} = \frac{\pi^3}{L^3} \tag{1.1}$$

The volume of the sphere,

$$V_{\text{Sphere}} = \frac{4}{3} \times \pi k^3 \tag{1.2}$$

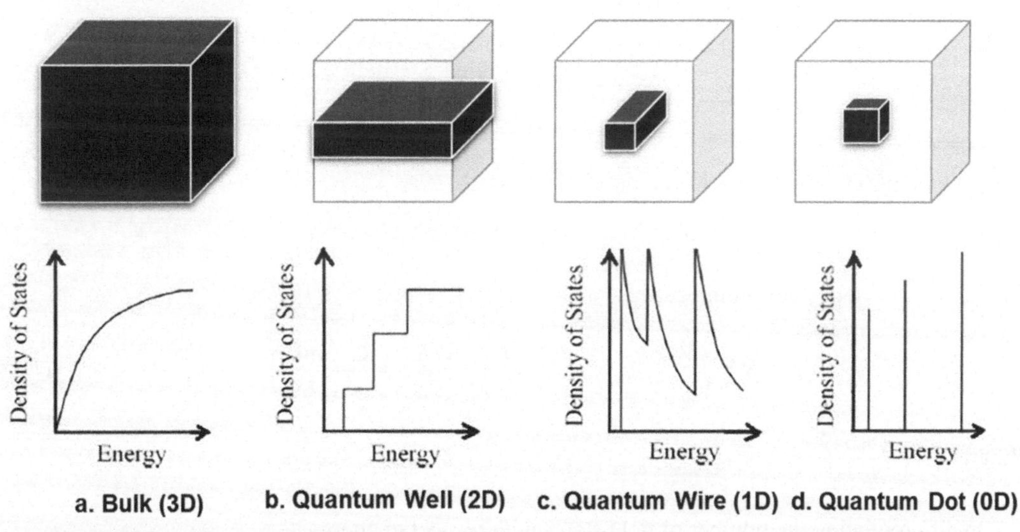

FIGURE 1.2 The impact of dimensional confinement on Density of States (DOS): (a) Bulk (3D) materials with a continuous DOS, (b) Quantum wells exhibiting a step-like DOS, (c) Quantum wires (1D) displaying a sawtooth-dependent DOS, and (d) Quantum dots (0D) characterized by discrete available states.

The number of field states in the cube,

$$N = \frac{V_{Sphere}}{V_{Single\ state}} \times 2 \times \left(\frac{1}{2} \times \frac{1}{2} \times \frac{1}{2}\right) = \frac{L^3 K^3}{3\pi^2} = \frac{L^3}{3\pi^2}\left(\frac{2mE}{\hbar^2}\right)^{3/2} \quad (1.3)$$

Where, $k = \left(\frac{2mE}{\hbar^2}\right)^{\frac{1}{2}}$

On differentiating the number of field states with respect to energy,

$$\frac{dN}{dE} = \frac{d}{dE}\left(\frac{L^3}{3\pi^2}\left(\frac{2mE}{\hbar^2}\right)^{\frac{3}{2}}\right) = \frac{L^3}{3\pi^2} \times \frac{3}{2} \times \left(\frac{2mE}{\hbar^2}\right)^{\frac{1}{2}} \times \left(\frac{2m}{\hbar^2}\right) = \frac{L^3}{3\pi^2} \times \frac{3}{2} \times \left(\frac{2mE}{\hbar^2}\right)^{\frac{1}{2}} \times \left(\frac{2m}{\hbar^2}\right)$$

$$\frac{dN}{dE} = \frac{L^3 m}{\pi^2 \hbar^2} \times \left(\frac{2mE}{\hbar^2}\right)^{\frac{1}{2}} = \frac{L^3 m}{\pi^2 \hbar^3}\sqrt{2mE} \quad (1.4)$$

We know, $g(E) = \frac{dN}{dE}$ divided by cell volume

$$g(E) = \frac{dN}{dt \times L^3} = \frac{m}{\pi^2 \hbar^3}\sqrt{2mE} = \frac{m^*}{\pi^2 \hbar^3}\sqrt{2mE} = \frac{m^*}{\pi^2 \hbar^3}\sqrt{2m(E - E_c)} \quad (1.5)$$

The expression $m \to m^*$ implies a direct correlation between the density of states and energy, implying that $g(E)$ is proportional to the square root of energy $\sqrt{E}$ (Mekuye and Abera, 2023).

1.2.3.1.2 Density of state in quantum wells (2D) NMs

In two-dimensional NMs, such as quantum wells, electrons are confined in one dimension while free to move within the plane of the material. Calculations for the state density involve determining the volumes of single states and spheres. The resulting density of states in 2D materials is independent of energy, indicating a constant density of available states across energy levels (Mekuye and Abera, 2023; Bokov, et al., 2021).

The volume of a single state,

$$V_{\text{Single state}} = \frac{\pi^2}{V} = \frac{\pi^2}{L^2} \quad (1.6)$$

The volume of the circle,

$$V_{\text{Circle}} = \pi k^2 \quad (1.7)$$

The number of field states in the cube,

$$N = \frac{V_{\text{Circle}}}{V_{\text{Single state}}} \times 2 \times \left(\frac{1}{2} \times \frac{1}{2}\right) = \frac{L^2 K^2}{2\pi} = \frac{L^2}{\pi}\left(\frac{mE}{\hbar^2}\right) \quad (1.8)$$

Where, $k = \left(\frac{2mE}{\hbar^2}\right)^{\frac{1}{2}}$

On differentiating the number of field states with respect to energy,

$$\frac{dN}{dE} = \frac{d}{dE}\left(\frac{L^2}{\pi}\left(\frac{mE}{\hbar^2}\right)\right) = \frac{L^2 m}{\pi \hbar^2} \quad (1.9)$$

We know, $g(E)=\frac{dN}{dE}$ is divided by cell volume,

$$g(E)=\frac{dN}{dt\times L^2}=\frac{L^2m}{\pi\hbar^2L^2}=\frac{m}{\pi\hbar^2}=\frac{m^*}{\pi\hbar^2} \tag{1.10}$$

As stated in Equation (1.10), the density of the state is independent of energy (Mekuye and Abera, 2023).

1.2.3.1.3 Density of state in quantum wires (1D) NMs

In one-dimensional NMs, such as quantum wires, electrons are confined in two dimensions while free to move along the length of the wire. Calculations for the density of states in these structures involve determining the volumes of single states and lines. The resulting density of states in 1D materials is inversely proportional to the square root of energy, indicating fewer available states at higher energy levels (Mekuye and Abera, 2023; Bokov, et al., 2021).

The volume of a single state,

$$V_{\text{Single state}}=\frac{\pi}{V}=\frac{\pi}{L} \tag{1.11}$$

The volume of the line,

$$V_{Line}=k \tag{1.12}$$

The number of field states in the cube,

$$N=\frac{V_{Line}}{V_{Single\ state}}\times 2\times\frac{1}{2}=\frac{Lk}{\pi}=\frac{L}{\pi}\left(\frac{2mE}{\hbar^2}\right)^{\frac{1}{2}}=\frac{L}{\pi\hbar}(2mE)^{\frac{1}{2}} \tag{1.13}$$

Where, $k=\left(\frac{2mE}{\hbar^2}\right)^{\frac{1}{2}}$

On differentiating the number of field states with respect to energy,

$$\frac{dN}{dE}=\frac{d}{dE}\left(\frac{L}{\pi\hbar}(2mE)^{\frac{1}{2}}\right)=\frac{L}{\pi\hbar}\times\frac{1}{2}\times(2mE)^{-\frac{1}{2}}\times(2m)=\frac{Lm}{\pi\hbar}\times(2mE)^{-\frac{1}{2}}=\frac{Lm}{\pi\hbar\sqrt{2mE}} \tag{1.14}$$

We know, $g(E)=\frac{dN}{dE}$ divided by cell volume,

$$g(E)=\frac{dN}{dt\times L}=\frac{Lm}{\pi\hbar\sqrt{2mE}\times L}=\frac{m}{\pi\hbar\sqrt{2mE}}=\frac{1}{\pi\hbar}\sqrt{\frac{m}{2E}}=\frac{1}{\pi\hbar}\sqrt{\frac{m^*}{2(E-E_C)}} \tag{1.15}$$

When the electron mass $m\rightarrow m*$

$$g(E)=\frac{1}{\pi\hbar}\sqrt{\frac{m^*}{2(E-E_C)}} \tag{1.16}$$

The Equation (1.16), elucidates that the density of a state exhibits an inverse relationship with energy, implying that $g(E)$ is proportional to $\sqrt{\frac{1}{E}}$ (Mekuye and Abera, 2023).

1.2.3.1.4 Density of state in quantum dots (0D) NMs

In zero-dimensional NMs, such as quantum dots, electrons are confined in all three dimensions, resulting in discrete energy levels where available states exist solely at specific energies. The density of states in quantum dots is represented by the delta Dirac function, indicating a discrete distribution of states (Mekuye and Abera, 2023; Bokov, et al., 2021).

Density of states,

$$g(E)=2\delta(E-E_C) \tag{1.17}$$

According to Equation (1.17), the density of state is characterized by the delta Dirac function, indicating a proportional relationship between energy (E) and the density of states (Mekuye and Abera, 2023).

The density of states in various dimensional NMs provides profound insights into their electronic behaviors and potential applications across diverse scientific fields.

1.2.4 Pore Dimension-Based Classification

Micro Porous Materials: Distinguished by narrow pores measuring less than 2 nm, microporous materials such as zeolites and activated carbon find applications in gas separation, purification, and storage, owing to their elevated surface area and selective adsorption capabilities. Progress in this domain encompasses advancements like the creation of nanoporous materials tailored for applications in energy storage and environmental remediation (Mekuye and Abera, 2023; Wu, et al., 2020; Mintova, et al., 2015).

Mesoporous Materials: With pore sizes ranging from above 2 nm to below 50 nm, materials such as silica-based mesoporous silicates and metal-organic frameworks present opportunities for controlled drug delivery, catalysis, and environmental remediation. Recent advancements in mesoporous materials encompass the development of stimuli-responsive drug delivery systems (Song, et al., 2017).

Macroporous Materials: Characterized by pores exceeding 50 nm in diameter, macroporous materials like aerogels and foams find applications in insulation, filtration, and tissue engineering, owing to their interconnected pore networks and high porosity (Wang, et al., 2017). Challenges in this category include achieving uniform pore size distribution and mechanical stability in macroporous materials.

1.2.5 Toxicity-Based Classification

Fiber-Like NPs: These include rigid structures like carbon nanotubes and metal oxides with fiber-like morphologies. While they offer exceptional mechanical properties, their resemblance to asbestos raises concerns about potential health hazards, necessitating strict safety protocols. Ongoing research focuses on mitigating the health risks associated with fiber-like NPs through surface modification and encapsulation techniques (Rashid, et al., 2024)

Persistent Granular NPs: Materials such as gold, silver, and metal oxides fall under this category (Mekuye and Abera, 2023). While they exhibit low toxicity compared to fiber-like NPs, their persistent nature may raise environmental concerns, especially in large quantities or under specific conditions. Strategies for the sustainable use and disposal of persistent granular NPs are under investigation (Mbanga, et al., 2023).

CMAR (Carcinogenic, Mutagenic, Asthmogenic, Reproductive Toxin) NPs: This group includes NPs containing elements like nickel, cadmium, chromium, and arsenic, which pose significant health risks owing to their carcinogenic, mutagenic, or respiratory toxicity

(Mekuye and Abera, 2023; Tchounwou, et al., 2012). Implementing precise protocols for the handling and disposal of CMAR NPs is imperative to effectively manage and minimize the associated risks within scientific and industrial contexts. Such procedures are vital not only for ensuring occupational safety but also for environmental protection and public health.

1.2.6 Hybrid NMs

Hybrid NMs combine different types of NMs to achieve synergistic properties. These combinations can lead to enhanced functionalities, such as improved mechanical strength, conductivity, or catalytic activity. For example, combining carbon nanotubes with metal NPs can create hybrid materials with excellent electrical conductivity and catalytic performance (Lopes, et al., 2022)

Challenges associated with hybrid NMs encompass the attainment of homogeneous dispersion, regulation of interfacial interactions, and preservation of stability across diverse conditions. Additionally, assessing the performance and potential ecological implications of hybrid NMs is critical for ensuring their safe and sustainable integration. Ongoing trends in hybrid NMs research involve the progression of multifaceted materials tailored for varied applications in energy, environmental remediation, and biomedical engineering.

1.3 PROPERTIES OF NMS

Nanometer-scale materials display a multitude of characteristics that set them apart from both individual atoms and bulk materials. These distinctions arise from factors such as surface charge/interaction, crystallography, composition, surface area, and nanoscale size effects. These unique attributes span a diverse array of domains, encompassing magnetic, optical, electrical, mechanical, chemical, and physical properties inherent to NMs (Mekuye and Abera, 2023; Barhoum and García-Betancourt, 2018).

The integrity and efficacy of NPs are intricately associated with their chemical or elemental composition. Particle size emerges as a fundamental metric for NP characterization, commonly evaluated through electron microscopy. For instance, gold NPs of different sizes exhibit distinct colors due to their surface plasmon resonance, which finds applications in biological imaging and sensing. (Mekuye and Abera, 2023; Lee, et al., 2018).

Additionally, the surface area-to-volume ratio profoundly influences NP performance and properties, often quantified using Brunauer–Emmett–Teller (BET) analysis (Teng, et al., 2006) This is exemplified in the catalytic activity of platinum NPs, which have a large surface area and are utilized in fuel cells for energy conversion.

The interactions between NPs and their environment are governed by their surface charge or overall charge. Zeta potential measurement, facilitated by a zeta potentiometer, serves as a common technique for assessing surface charges and the stability of dispersion in solution. For instance, positively charged NPs can adhere to negatively charged cell membranes, facilitating drug delivery in targeted therapy (Honary and Zahir, 2013).

1.3.1 Physical Properties of NMs

The melting temperature of bulk materials remains consistent regardless of size, while the melting point of NMs diminishes as particle size reduces, owing to the prevalence of unbounded surface atoms. This reduction is attributed to the amplified surface area-to-volume ratio at the nanoscale, resulting in heightened surface energy and a propensity for agglomeration. Remarkably, the size-dependent colors of gold and silver NMs impact the coloring of stained glass, demonstrating the significance of size control in material applications (Mekuye and Abera, 2023; Ray, 2010).

1.3.2 Magnetic Properties of NMs

At the nanoscale, the magnetic behavior of elements endures profound alterations attributable to the size-dependent properties of magnetic NPs. The nanostructuring bulk magnetic materials facilitate precise modulation of their magnetic hysteresis loops, thus affording the fabrication of soft or hard magnetic materials endowed with superior functionalities (Mekuye and Abera, 2023; Zhu, et al., 2018). This nuanced control over magnetic characteristics holds considerable promise across a spectrum of biomedical applications, encompassing targeted drug delivery, high-resolution magnetic resonance imaging (MRI), and innovative approaches like magnetic fluid hyperthermia for cancer therapy. For instance, superparamagnetic iron oxide NPs, functioning as contrast agents in MRI, not only facilitate the detection but also enable the delineation of tumour boundaries, thus enhancing diagnostic accuracy and treatment planning in oncology (Li, et al., 2013).

1.3.3 Optical Properties of NMs

NPs demonstrate distinctive optical properties, including localized surface plasmon resonance (LSPR), influenced by NP size. Reduction in NP size can induce shifts in emission light position, thereby impacting color and optical behavior. Quantum confinement effects in NMs enable tunable emission wavelengths and color changes, augmenting their optical versatility. Quantum dots, semiconductor NPs, are utilized in displays and lighting applications due to their size-tunable emission properties (Kelly, et al., 2003; Bera, et al., 2010).

1.3.4 Electrical Properties of NMs

Conductivity in ceramics may increase at the nanoscale, while electric resistance in metals may rise due to electron confinement effects. Nanostructured materials feature altered electron conduction pathways, often resulting in semiconductor behavior. Carbon nanotubes, for instance, may exhibit either conducting or semiconducting behavior based on their structure, offering unique opportunities for electrical applications. For example, carbon nanotubes are utilized in flexible electronics and conductive films (Lue, 2007; Cai and Wang, 2015).

1.3.5 Chemical Properties of NMs

The chemical attributes of NMs influence their reactivity, stability, and susceptibility to external elements such as moisture, heat, and light. These characteristics significantly determine the versatility of NMs across various fields, often showcasing superior catalytic prowess in comparison to bulk materials (Khan, et al., 2019). For example, NPs with catalytic properties, such as palladium NPs, find application in environmental cleanup endeavors, specifically in the breakdown of pollutants during wastewater treatment processes (Anjum, et al., 2019).

1.3.6 Mechanical Properties of NMs

Mechanical characteristics like elasticity, ductility, and strength hold considerable significance in assessing the applicability of NMs across diverse fields (Guo, et al., 2013). Nanostructured materials often display enhanced traits such as hardness, yield strength, and toughness relative to bulk counterparts, attributed to factors like reduced grain size and boundary deformation. Notably, graphene, an exemplary two-dimensional NM, stands renowned for its outstanding mechanical attributes, serving as a reinforcement component in composite materials (Kumar, et al., 2020).

Recent strides in NM advancement encompass the innovation of stimuli-responsive variants tailored for controlled drug delivery, alongside their integration into sophisticated sensors for environmental surveillance. However, challenges persist in the characterization and utilization of NMs, spanning the standardization of assessment techniques, scalability of synthesis methodologies, and the potential implications on environmental and human health. Moreover, ensuring the reproducibility and consistency of NM properties across multiple production batches remains a significant hurdle in their broad application. Addressing these complexities emphasizes the necessity for continued research and the conscientious deployment of NMs within scientific and technological realms.

1.4 SYNTHESIS OF NMS

The synthesis of NMs involves a diverse range of methods, broadly categorized into two main approaches: bottom-up and top-down methods (Mekuye and Abera, 2023; Sajid and Płotka-Wasylka, 2020) as depicted in Figure 1.3 and the complete classification of these methods in Figure 1.4.

1.4.1 Top-Down Methods

Top-down methodologies, alternatively termed deconstructive approaches, encompass the fragmentation of bulk materials into smaller entities until nano-sized dimensions are attained. This strategy affords meticulous regulation over particle size and morphology (Tour, 2014). Numerous methodologies are encompassed within the realm of top-down techniques, each contributing to the precise manipulation of material properties at the nanoscale (Figure 1.5).

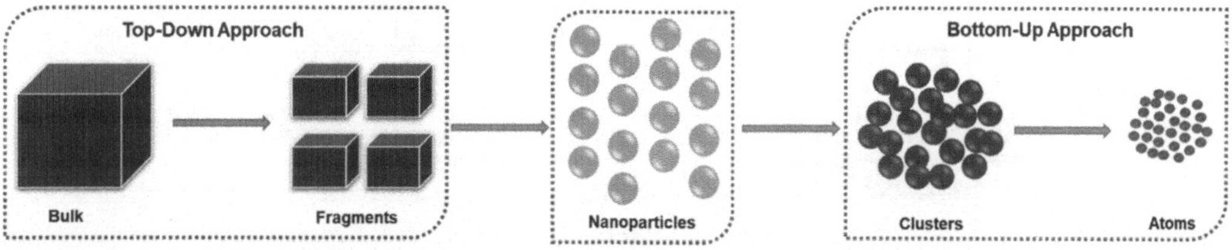

FIGURE 1.3 The top-down and bottom-up approach for the synthesis of NPs (NPs).

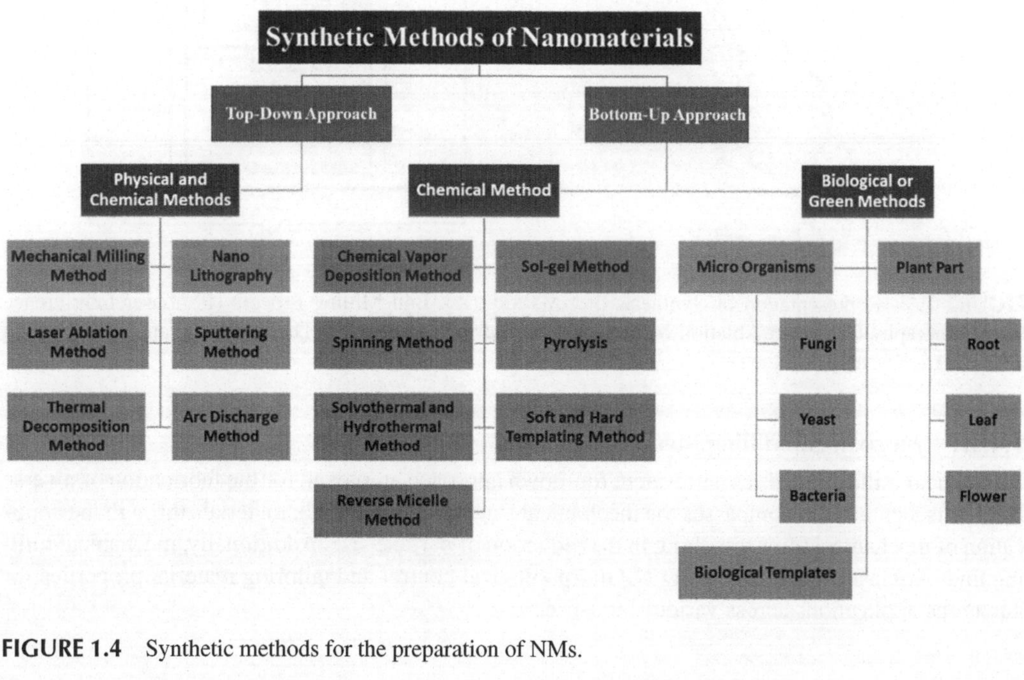

FIGURE 1.4 Synthetic methods for the preparation of NMs.

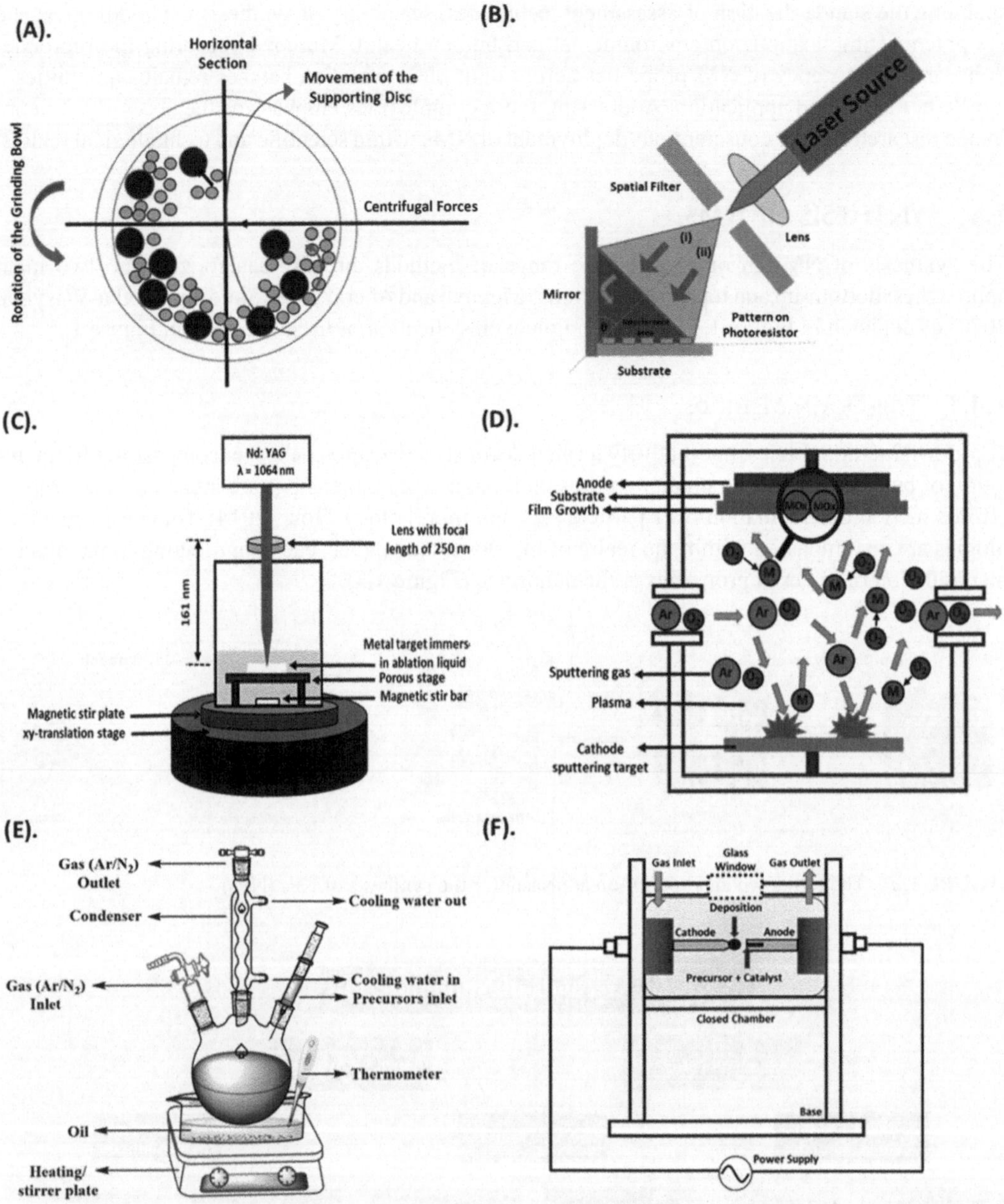

FIGURE 1.5 Representation of Synthesis of NMs via. (A). Ball Milling Process (B). Laser Interference Nanolithography (C). Laser Ablation Method (D). Sputtering Method (E). Thermal Decomposition Method and (F). Arc Discharge Method.

1.4.1.1 Mechanical Milling Method

Mechanical milling stands as a prevalent top-down approach employed for the fabrication of diverse NPs. This technique encompasses the mechanical fragmentation of bulk materials through the application of mechanical force, resulting in the reduction of particle size. Additionally, mechanical milling finds extensive use in achieving NP morphological control and tailoring material properties for numerous applications across various industries.

Process: In mechanical milling processes, bulk materials are introduced into a milling chamber alongside grinding media, like balls or rods. Subsequently, the milling chamber undergoes rotation, prompting collisions between the grinding media and the material, leading to size reduction (Baheti, et al., 2012).

Applications: Mechanical milling finds extensive utilization in the fabrication of oxide and carbide-strengthened alloys, wear-resistant coatings, and a diverse range of composite materials. Moreover, it serves as a crucial method for achieving tailored material properties in various industrial sectors, including aerospace, automotive, and biomedical engineering, etc. (Mekuye and Abera, 2023).

1.4.1.2 Nanolithography Method

Nanolithography denotes a method employed for the precise patterning and fabrication of nanoscale structures onto a substrate.

Process: Diverse techniques encompass nanolithography, including photolithography, electron beam lithography, and nanoimprint lithography. These methods entail the selective removal of material from a substrate to fashion desired shapes and intricate patterns with nanoscale precision (Wang and Gao, 2019).

Applications: Widely utilized in the semiconductor industry for fabricating integrated circuits and MEMS devices, as well as in nanoelectronics, nanophotonics, and nanobiotechnology.

1.4.1.3 Laser Ablation Method

The laser ablation technique involves striking a target material with a powerful laser beam to generate NPs, particularly effective for producing metal NPs.

Process: In laser ablation, the target material is irradiated with short pulses of laser light, causing rapid heating and vaporization. The resulting vapor condenses to form NPs (Abdullah, et al., 2021).

Applications: Used in catalysis, sensing, and biomedical imaging.

1.4.1.4 Sputtering Method

Physical vapor deposition techniques, such as sputtering, are employed to deposit thin films of material onto a substrate surface, facilitating various applications across industries.

Process: In a vacuum chamber, high-energy plasma is generated, leading to ion collisions with the target material and subsequent atom ejection. The ejected atoms then deposit to form a thin film on the substrate (Abdullah, et al., 2021).

Applications: This method is utilized in semiconductor fabrication, optics, electronics, and magnetic storage media production.

1.4.1.5 Thermal Decomposition Method

Thermal decomposition entails the decomposition of chemical compounds into simpler substances by applying heat. It is utilized in NP synthesis to decompose precursor molecules and facilitate NP formation.

Process: Precursor molecules are heated to high temperatures, causing them to decompose and form NPs, with reaction conditions carefully controlled to achieve the desired properties (Mahhouti, et al., 2019).

Applications: Employed in the synthesis of metal, metal oxide, and semiconductor NPs, with controlled properties.

1.4.1.6 Arc Discharge Method

The arc discharge technique is employed for the fabrication of carbon-based NMs like carbon nanotubes and fullerenes through the vaporization of carbon-containing substances in an electric arc discharge.

Process: A high-current electric arc is induced between two carbon electrodes within an inert gas environment, causing the vaporization of carbon atoms from the electrodes. Subsequently, these vaporized atoms condense and solidify into NMs (Arora and Sharma, 2014).

Applications: Carbon nanotubes produced via this method demonstrate outstanding mechanical, electrical, and thermal characteristics, rendering them suitable for a diverse array of applications.

1.4.2 Bottom-Up Methods

Bottom-up methods encompass the assembly or growth of NMs from atomic or molecular precursors. These techniques provide meticulous control over the dimensions, morphology, composition, and characteristics of the synthesized NMs. Below are detailed descriptions of several key bottom-up synthesis methods.

1.4.2.1 Chemical Vapor Deposition (CVD) Method

Chemical vapor deposition is a prevalent method for synthesizing thin films, coatings, and nanostructures. It involves the deposition of material from gaseous precursors onto a substrate surface through vapor-phase chemical reactions.

Process: Gaseous precursors are introduced into a reaction chamber, where they undergo chemical reactions induced by heat or plasma, leading to the deposition of solid-phase material on the substrate surface (Zaytseva and Neumann, 2016) (Figure 1.6).

Applications: Utilized in the production of thin films for integrated circuits, optical coatings, functional NMs, and more.

1.4.2.2 Sol-Gel Method

The sol-gel process is a wet chemical synthesis approach used to produce inorganic and hybrid organic-inorganic NMs from molecular precursors, offering versatility and control over composition, structure, and morphology.

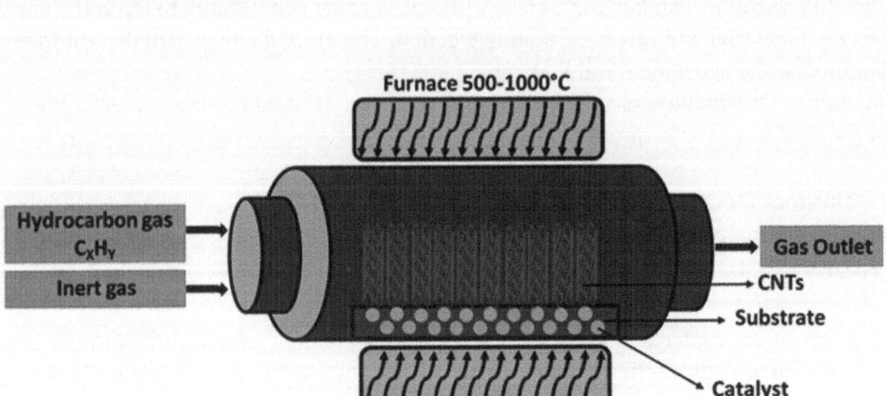

FIGURE 1.6 Representation of chemical vapor deposition method for the synthesis of CNTs.

Process: The process entails the hydrolysis and condensation of metal alkoxides or metal salts in a solution, resulting in the formation of a colloidal suspension (sol). This suspension subsequently undergoes gelation to create a three-dimensional network of interconnected NPs or a gel (Pierre, 2020) (Figure 1.7).

Applications: Employed in the production of NPs, thin films, coatings, and monoliths for catalysis, sensors, optics, biomaterials, and more.

1.4.2.3 Spinning Method

The spinning method, or spinning disc processing, is a versatile approach known for producing NPs and nanocomposites with controlled morphologies, offering scalability, simplicity, and flexibility.

Process: Precursor solutions are deposited onto a rotating disc substrate, leading to the development of thin films or NPs through centrifugal forces, with parameters modified to regulate dimensions, morphology, and composition (Wang and Gao, 2019) (Figure 1.8).

Applications: Utilized in the synthesis of magnetic NPs, quantum dots, thin films, functional coatings, and biomedical materials.

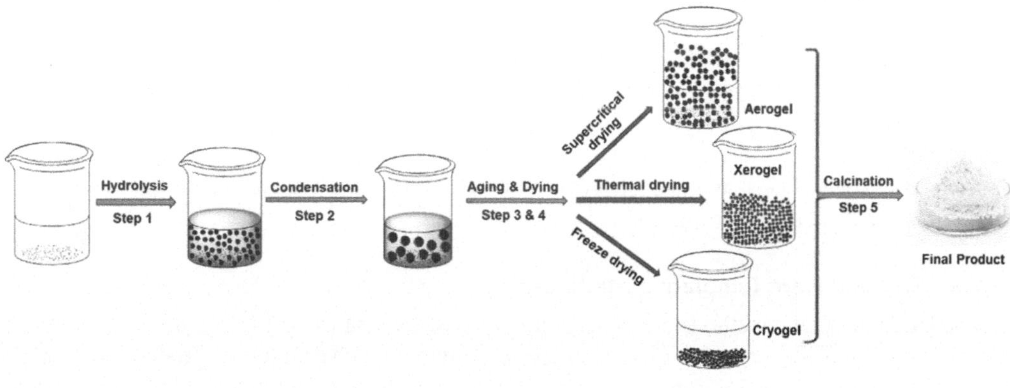

FIGURE 1.7 Schematic representation of different stages of the sol-gel process.

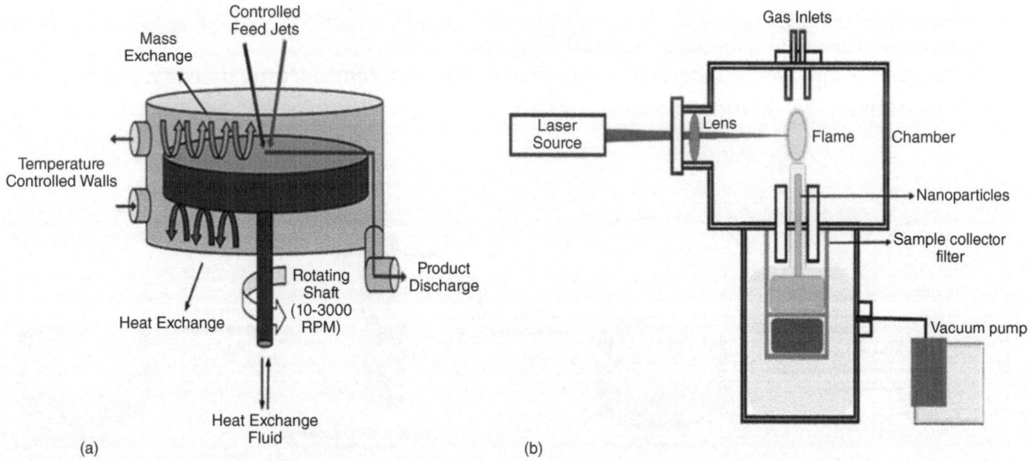

FIGURE 1.8 Schematic representation of a typical apparatus for (a). Spinning disk reactor and (b). Laser pyrolysis.

1.4.2.4 Pyrolysis Method

Pyrolysis, a thermal decomposition process, serves as a fundamental method for transforming organic precursors into various carbonaceous NMs, encompassing carbon nanotubes, graphene, carbon nanofibers, carbon dots, and carbon NPs. This versatile technique provides simplicity, scalability, and precise control over the structure and properties of the resultant materials.

Process: Organic precursors are subjected to heat treatment in a controlled environment, leading to decomposition and carbonization, with volatile components driven off, leaving behind carbon-rich residues that form NPs (Wang and Gao, 2019; Devi, et al., 2021).

Applications: This method finds application in a wide range of industries, spanning electronics, energy storage, composite materials, catalysis, environmental remediation, and biomedical engineering, showcasing its versatility across diverse fields.

1.4.2.5 Solvothermal and Hydrothermal Methods

Solvothermal and hydrothermal methods encompass variations of a heterogeneous reaction carried out in a solution, either aqueous or solvent-based, under elevated temperatures and pressures. These methods afford meticulous control over reaction conditions, leading to the synthesis of nanostructured materials with tailored properties.

Process: Hydrothermal synthesis embraces sealed aqueous solutions undergoing reactions above the boiling point of water under elevated pressures, while solvothermal synthesis uses organic solvents instead of water (Lai, et al., 2015) (Figure 1.9).

Applications: Utilized in the production of various nano geometries, including nanowires, nanorods, nanosheets, and nanospheres, for electronics, catalysis, energy storage, sensors, and biomedical engineering.

1.4.2.6 Soft and Hard Templating Methods

Soft and hard templating methods are widely employed in the synthesis of NMs, utilizing templates or scaffolds to direct the formation of nanostructures. This approach enables the production of nanoporous materials with precise control over pore sizes, shapes, and structures, addressing various application needs.

Process: Soft templating utilizes adaptable templates, including block copolymers or surfactants, whereas hard templating involves the use of rigid templates such as porous membranes or sacrificial templates (Yan, 2012) (Figure 1.10).

Applications: Employed in catalysis, separation, adsorption, sensing, drug delivery, and membrane technology, offering enhanced surface areas, high porosities, and tunable properties.

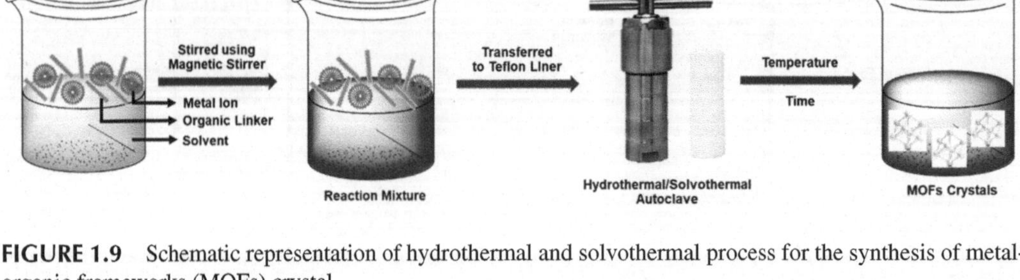

FIGURE 1.9 Schematic representation of hydrothermal and solvothermal process for the synthesis of metal-organic frameworks (MOFs) crystal.

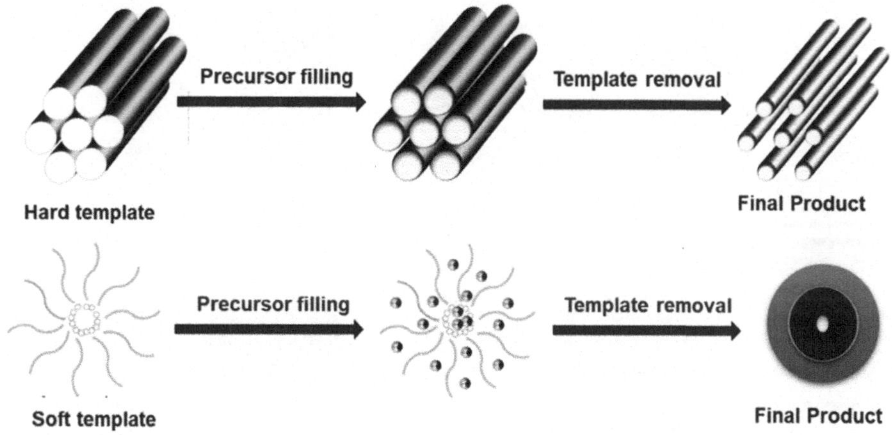

FIGURE 1.10 Representation of soft and hard templating method for the synthesis of NMs.

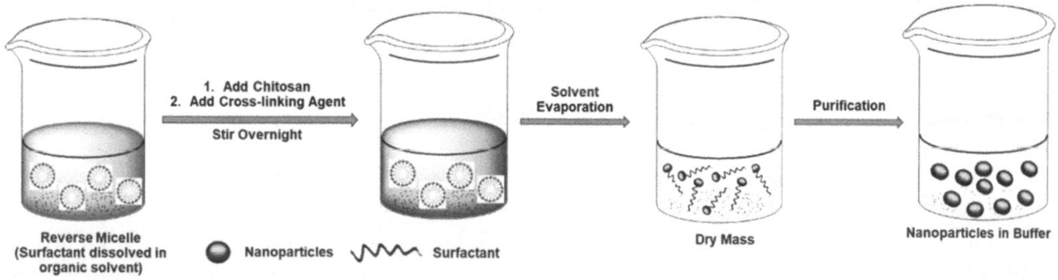

FIGURE 1.11 Preparation of Chitosan NPs using the reverse micellar method.

1.4.2.7 Reverse Micelle Method

The reverse micelle technique enables precise control over the production of NMs, ensuring they possess specific shapes and dimensions. This is achieved through the modulation of surfactant molecule self-assembly within a water-in-oil emulsion system, offering versatility and efficiency in NM synthesis.

Process: Surfactant molecules dissolved in an organic solvent form micelle around water droplets, with precursor ions or molecules introduced into the water cores of reverse micelles for chemical reactions or nucleation processes (Malik, et al., 2012) (Figure 1.11).

Applications: Employed in the synthesis of NPs, nanocrystals, and nanocomposites for catalysis, sensors, drug delivery, imaging, and nanotechnology.

1.4.2.8 Biosynthesis/Biological Method

Biological synthesis involves the use of biological organisms or components to produce NMs, utilizing the inherent properties of biological systems to control NP synthesis.

Process: Microorganisms, fungi, algae, or plant extracts are incubated with metal salts or precursors, facilitating the reduction and stabilization of NPs, with biological agents serving as reducing agents, capping agents, or templates (Gahlawat and Choudhury, 2019) (Figure 1.12).

Applications: In the realm of biomedical fields, encompassing drug administration, imaging techniques, diagnostic methods, and therapeutic interventions, as well as environmentally conscious approaches for synthesizing catalysts, and sensors, addressing environmental issues, and enhancing agricultural practices, NPs find extensive utility.

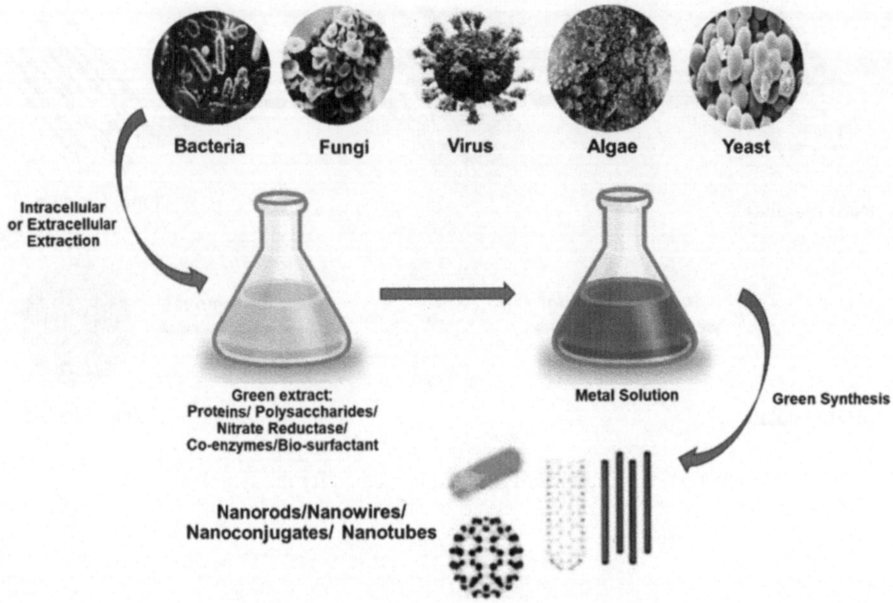

FIGURE 1.12 Biological approach for the synthesis of NPs.

1.5 APPLICATION OF NMS

NMs exhibit broad applications spanning diverse fields such as nanomedicine, chemical and cosmetic industries, materials science, food sciences, environmental and energy sectors, electronics and semiconductors, scientific tools, agriculture, and many more as depicted in Figure 1.13.

1.5.1 NANOMEDICINE

1.5.1.1 Nano Drugs

NPs, including liposomes, polymeric NPs, and dendrimers, are intricately designed to encapsulate pharmaceuticals, thereby targeting precise cells or tissues within the body. This sophisticated drug delivery mechanism not only enhances drug solubility and bioavailability but also mitigates systemic toxicity. By precisely delivering therapeutic agents to their intended destinations, these NPs hold promise for revolutionizing treatment efficacy, while minimizing adverse effects (Din, et al., 2017).

1.5.1.2 Medical Devices

Nanotechnology has revolutionized medical devices by enabling the advancement of materials with enhanced properties. For instance, nanocomposites increase the mechanical strength and biocompatibility of implants, while nanosensors provide real-time monitoring of physiological parameters (Lu, et al., 2023).

1.5.1.3 Tissue Engineering

In tissue engineering, NMs function as scaffolds that facilitate cell growth, proliferation, and differentiation. Nanofibrous scaffolds mimicking the extracellular matrix promote tissue regeneration, making them invaluable in regenerative medicine for repairing damaged tissues and organs (Menaa, et al., 2015).

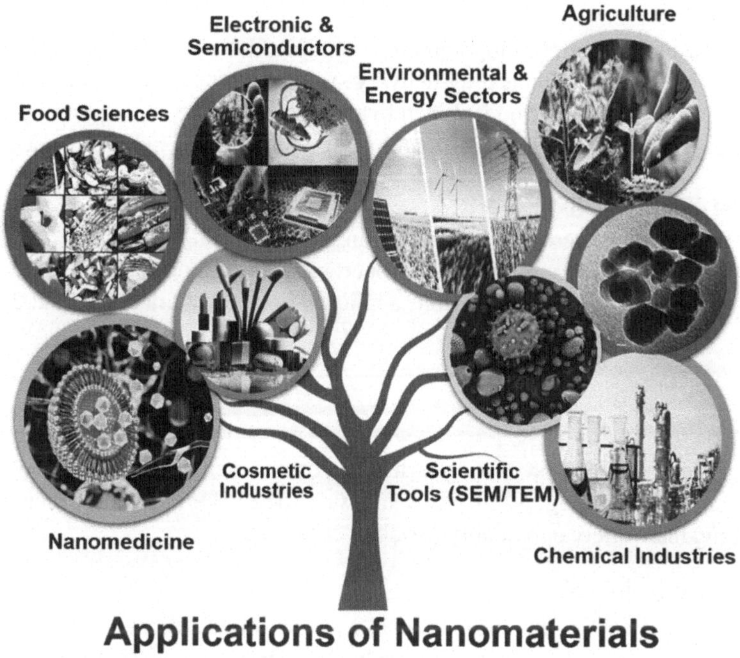

FIGURE 1.13 Representation of some applications of NMs.

1.5.2 Chemical and Cosmetic Fields

NMs find applications in cosmetics, paints, and coatings owing to their unique properties. For instance, NPs of titanium dioxide and zinc oxide are used in sunscreen formulations for their ability to block UV radiation effectively without leaving a white residue on the skin (Addae and Weiss, 2024).

1.5.3 Materials Science

1.5.3.1 NPs

NPs demonstrate distinctive attributes in contrast to bulk materials, rendering them versatile for a myriad of applications. Their unique properties open doors to innovative solutions across industries, from healthcare to electronics. For example, silver NPs, renowned for their antimicrobial prowess, play a significant role in preventing infections in wound dressings and medical textiles. Beyond this, NPs are paving the way for breakthroughs in drug delivery, environmental remediation, and advanced materials science, underscoring their immense potential to shape the future of technology and medicine (Rajski, et al., 2019).

1.5.3.2 Carbon Nanotubes

Carbon nanotubes exhibit outstanding mechanical, electrical, and thermal characteristics, positioning them as a cornerstone in reinforcing composite materials, crafting resilient lightweight components, and bolstering the efficiency of electronics and energy storage devices. Their remarkable properties not only enhance the structural integrity of materials but also contribute to advancements in various fields, from aerospace engineering to renewable energy. As researchers delve deeper into their applications, carbon nanotubes continue to revolutionize technology, promising even greater breakthroughs in the future (Ali, et al., 2023).

1.5.3.3 Biopolymers

Biodegradable and biocompatible biopolymers, including chitosan and polylactic acid find wide applications in various fields such as drug delivery systems, tissue engineering scaffolds, and food packaging materials. Their utilization is driven by their sustainability, compatibility with biological systems, and their ability to contribute to environmental preservation and health safety.

1.5.3.4 Paints and Coatings

NMs enhance the performance of paints and coatings by providing functionalities such as scratch resistance, UV protection, and antimicrobial properties. Nanocomposite coatings can improve the durability and longevity of surfaces in various applications (Udayakumar, et al., 2021).

1.5.4 Food Sciences

1.5.4.1 Processing

Nanotechnology improves food processing techniques by enhancing the efficiency of food preservation, sterilization, and packaging. NM-based sensors can detect spoilage and contamination in food products, ensuring their safety and quality (Sridhar, et al., 2021).

1.5.4.2 Nutraceutical Food

Nanocapsules and NPs enable the encapsulation and controlled release of bioactive compounds, vitamins, and nutrients in food products. This technology enhances the stability and bioavailability of nutritional supplements and functional foods (Katouzian and Jafari, 2016; Martínez-Ballesta et al., 2018).

1.5.5 Environment and Energy

1.5.5.1 Water and Air Purification

NMs are used in filtration membranes and adsorbents for removing pollutants and contaminants from water and air. Nanocomposite materials with high surface area and adsorption capacity improve the efficiency of purification processes (Ibrahim, et al., 2016).

1.5.5.2 Fuel Cells

Nanotechnology enhances the performance and efficiency of fuel cells by improving catalysts, membranes, and electrode materials (Qiao and Li, 2011). NP-based catalysts increase the reaction kinetics and reduce the cost of fuel cell systems.

1.5.5.3 Photovoltaic

NMs enable the development of high-efficiency solar cells and photovoltaic devices by enhancing light absorption, charge transport, and energy conversion efficiency (Yu, et al., 2012). Nanocomposite materials with tailored optical and electrical properties optimize solar energy harvesting.

1.5.6 Electronics and Semiconductors

1.5.6.1 Chips and Memory Storage

Nanotechnology drives advancements in semiconductor manufacturing, leading to smaller, faster, and more efficient electronic devices (Yu and Meyyappan, 2006). Nanoscale transistors, memory cells, and interconnects enable the production of high-performance integrated circuits with increased functionality and reduced power consumption.

1.5.6.2 Photonics and Optoelectronics

NMs enable the fabrication of optical components, sensors, and devices for telecommunications, imaging, and sensing applications. Nanophotonic structures, such as plasmonic nanoantennas and photonic crystals, manipulate light at the nanoscale for enhanced functionality and performance (Xavier, et al., 2018).

1.5.7 Scientific Tools

1.5.7.1 Atomic Force Microscopy

Nanotechnology facilitates the development of atomic force microscopes with high resolution and sensitivity, enabling imaging and manipulation of materials at the atomic scale. Nanoscale tips and cantilevers improve the spatial resolution and accuracy of AFM measurements.

1.5.7.2 Scanning Tunneling Microscope

NMs are utilized in scanning tunneling microscopes for imaging and analyzing surfaces with atomic precision. Atomically sharp tips and conductive substrates enable STM imaging of surface morphology, electronic structure, and molecular interactions at the nanoscale (Prezzi, et al., 2014).

1.5.8 Agriculture

1.5.8.1 Agricultural Applications

Nanotechnology offers innovative solutions to agricultural challenges, such as improving crop yield, nutrient uptake, and pest control. NMs, such as nanofertilizers, nanopesticides, and nanosensors, enhance plant growth, health, and resilience to environmental stressors (Ijaz, et al., 2023).

1.5.8.2 Precision Agriculture

Nanotechnology facilitates the advancement of precision agriculture practices by enabling continuous monitoring of soil conditions, plant health, and environmental factors. Nanoscale sensors and imaging technologies optimize resource utilization, minimize environmental impact, and increase agricultural productivity (Yadav, et al., 2023).

1.6 RECENT ADVANCEMENTS IN NANOTECHNOLOGY

In addition to the established applications of NMs, recent developments in nanotechnology have led to pioneering advancements in various fields as depicted in Figure 1.14.

1.6.1 Cancer Treatment Using NP-Based Therapies

NPs have become increasingly recognized as valuable assets in cancer management, providing targeted drug delivery, augmented therapeutic effectiveness, and minimized side effects. NP-based therapies, including liposomes and polymeric NPs, are designed to precisely transport anticancer drugs to tumor sites, thereby enhancing treatment outcomes and increasing patient survival rates (Bajpai, et al., 2021).

1.6.1.1 Environmental Remediation Using NMs for Groundwater Clean-up

NMs are increasingly utilized in environmental remediation efforts, particularly for groundwater cleanup. NPs possessing significant large surface area and reactivity, for instance, iron-based NPs, aid in eliminating pollutants such as heavy metals and organic contaminants from groundwater, fostering sustainable remediation practices and environmental preservation (Rajan, 2011).

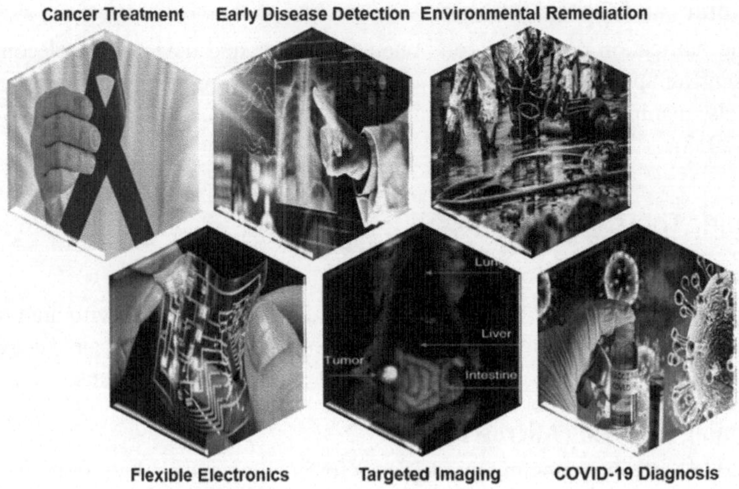

FIGURE 1.14 Recent advances in nanotechnology.

1.6.1.2 Flexible Electronics Utilizing NMs for Bendable and Stretchable Devices

NMs are essential in advancing flexible electronics, facilitating the production of bendable, stretchable, and wearable devices with improved mechanical properties and electrical performance. Nanocomposites and nanofabrication techniques are utilized in designing flexible substrates, conductive inks, and stretchable electrodes, opening avenues for next-generation electronic applications in robotics, healthcare, and consumer electronics (Wagholikar, et al., 2024).

1.6.2 Innovative Nanomedicine Approaches for Early Disease Detection and Targeted Imaging

Nanotechnology has stimulated the advancement of inventive nanomedicine approaches for the early identification of diseases and targeted imaging. NP-based contrast agents and biosensors enable highly sensitive and specific identification of disease indicators, facilitating prompt diagnosis and personalized treatment strategies. Furthermore, functionalized NPs can precisely target affected tissues or cells, augmenting the precision of imaging methods like magnetic resonance imaging (MRI) and positron emission tomography (PET) (Hu, et al., 2020; Pan, 2024).

1.7 CONCLUSION

In conclusion, this chapter has provided a comprehensive exploration of NMs, delving into their classification, synthesis methods, properties, and wide-ranging applications across diverse fields. By understanding the distinct properties and behaviors of NMs, we unlock their immense potential for innovation and advancement in various sectors. Through classification based on origin, structure, dimensions, and other factors, we gain insight into the vast landscape of NMs and their versatility. Synthesis methods, ranging from top-down approaches like lithography to bottom-up techniques such as chemical vapor deposition, offer diverse pathways for the development of nanoscale structures with tailored properties. The size-dependent properties of NMs, including physical, chemical, and mechanical attributes, underscore their significance in numerous applications. From agriculture to medicine to electrical engineering, NMs facilitate breakthroughs in drug delivery, tissue engineering, environmental remediation, electronics, and beyond. Furthermore, recent developments in nanotechnology, such as advancements in cancer treatment using NP-based therapies, environmental remediation utilizing NMs for groundwater cleanup, and the emergence of flexible electronics employing NMs for bendable and stretchable devices, exemplify the dynamic nature of nanotechnology's evolution. Additionally, innovative nanomedicine approaches for early disease detection and targeted imaging showcase the ongoing strides in leveraging NMs for improving healthcare outcomes. As we navigate the realm of nanotechnology, it becomes evident that NMs hold the key to addressing complex challenges and driving societal progress. Their ability to enhance functionality, improve efficiency, and enable novel solutions underscores their transformative potential. In essence, the study of NMs catalyzes innovation, propelling us toward a future where nanotechnology has a significant role in shaping our world. By harnessing the power of NMs and embracing recent advancements, we pave the way for groundbreaking discoveries, sustainable technologies, and a brighter tomorrow.

REFERENCES

Abdullah, H. I., Al-Amiery, A. A., Al-Baghdadi, S. B., 2021. The using of NMs as catalysts for photodegradations. *J. Phys. Conf. Ser.* 1853(1), 012052. https://doi.org/10.1088/1742-6596/1853/1/012052

Addae, A. J., Weiss, P. S., 2024. Standardizing the white cast potential of sunscreens with metal oxide ultraviolet filters. *Acc. Mater. Res.* 5(4), 392–399. https://doi.org/10.1021/accountsmr.4c00004

Alhalili, Z., 2023. Metal oxides NPs: General structural description, chemical, physical, and biological synthesis methods, role in pesticides and heavy metal removal through wastewater treatment. *Molecules*. 28(7), 3086. https://doi.org/10.3390%2F molecules28073086

Ali, A., Koloor, S. S. R., Alshehri, A. H., Arockiarajan, A. 2023. Carbon nanotube characteristics and enhancement effects on the mechanical features of polymer-based materials and structures: A review. *J. Mater. Res. Technol.* 24(5), 6495–6521. https://doi.org/10.1016/j.jmrt.2023.04.072

Alqahtani, M. S., Abbas, M., Abdulmuqeet, M., Alqahtani, A. S., Alshahrani, M. Y., Alsabaani, A., Ramalingam, M., 2022. Forecasting the post-pandemic effects of the SARS-CoV-2 virus using the bullwhip phenomenon alongside use of nanosensors for disease containment and cure. *Materials*. 15(14), 5078. https://doi.org/10.3390/ma15145078

Altammar, K. A., 2023. A review on NPs: Characteristics, synthesis, applications, and challenges. *Front. Microbiol.* 14, 1155622. https://doi.org/10.3389%2Ffmicb.2023.1155622

Anjum, M., Miandad, R., Waqas, M., Gehany, F., Barakat, M. A., 2019. Remediation of wastewater using various nano-materials. *Arab. J. Chem.* 12(8), 4897–4919. https://doi.org/10.1016/j.arabjc.2016.10.004

Arora, N., Sharma, N. N., 2014. Arc discharge synthesis of carbon nanotubes: Comprehensive review. *Diam. Relat. Mater.* 50, 135–150. https://doi.org/10.1016/j.diamond.2014.10.001

Baheti, V., Abbasi, R., Militky, J., 2012. Ball milling of jute fibre wastes to prepare nanocellulose. *World J. Eng.* 9(1), 45–50. https://doi.org/10.1260/1708-5284.9.1.45

Baig, N., Kammakakam, I., Falath, W., 2021. NMs: A review of synthesis methods, properties, recent progress, and challenges. *Mater. Adv.* 2(6), 1821–1871. https://doi.org/10.1039/D0MA00807A

Bajpai, S., Tiwary, S.K., Sonker, M., Joshi, A., Gupta, V., Kumar, Y., Shreyash, N., Biswas, S., 2021. Recent advances in NP-based cancer treatment: A review. *ACS Appl. Nano Mater.* 4(7), 6441–6470. https://doi.org/10.1021/acsanm.1c00779

Barhoum, A., García-Betancourt, M. L., 2018. Physicochemical characterization of NMs: Size, morphology, optical, magnetic, and electrical properties. *Emerg. Appl. NPs Arch. Nanostruct.* 10, 279–304. https://doi.org/10.1016/B978-0-323-51254-1.00010-5

Bera, D., Qian, L., Tseng, T. K., Holloway, P. H., 2010. Quantum dots and their multimodal applications: A review. *Materials*. 3(4), 2260–2345. https://doi.org/10.3390 %2Fma3042260

Bokov, D., Jalil, A.T., Chupradit, S., Suksatan, W., Ansari, M.J., Shewael, I.H., Valiev, G.H., Kianfar. E., 2021. NM by sol-gel method: Synthesis and application. *Adv. Mater. Sci. Eng.* 2021, 5102014. https://doi.org/10.1155/2021/5102014

Byakodi, M., Shrikrishna, N. S., Sharma, R., Bhansali, S., Mishra, Y., Kaushik, A., Gandhi, S., 2022. Emerging 0D, 1D, 2D, and 3D nanostructures for efficient point-of-care biosensing. *Biosens. Bioelectron.*: X. 12, 100284. https://doi.org/10.1016/j.biosx.2022.100284

Cai, L., Wang, C., 2015. Carbon nanotube flexible and stretchable electronics. *Nanoscale Res. Lett.* 10 (320), 1–21. https://doi.org/10.1186/s11671-015-1013-1

Chand, K., Cao, D., Fouad, D.E., Shah, A.H., Lakhan, M.N., Dayo, A.Q., Sagar, H.J., Zhu, K., Mohamed, A.M.A., 2020. Photocatalytic and antimicrobial activity of biosynthesized silver and titanium dioxide NPs: A comparative study. *J. Mol. Liq*. 316, 113821. https://doi.org/10.1016/j.molliq.2020.113821

Chausali, N., Saxena, J., Prasad, R., 2023. Nanotechnology as a sustainable approach for combating the environmental effects of climate change. *J. Agric. Food Res.* 12, 100541. https://doi.org/10.1016/j.jafr.2023.100541

Cotta, M. A., 2020. Quantum dots and their applications: What lies ahead? *ACS Appl. Nano Mater*. 3(6), 4920–4924. https://doi.org/10.1021/acsanm.0c01386

Devi, M., Rawat, S., Sharma, S., 2021. A comprehensive review of the pyrolysis process: From carbon NM synthesis to waste treatment. *Oxf. Open Mater. Sci*. 1(1), 014. https://doi.org/10.1093/oxfmat/itab014

Din, F. U., Aman, W., Ullah, I., Qureshi, O. S., Mustapha, O., Shafique, S., Zeb, A., 2017. Effective use of nanocarriers as drug delivery systems for the treatment of selected tumors. *Int. J. Nanomedicine*. 12, 7291–7309. https://doi.org/10.2147%2FIJN.S146315

Elbakyan, L., Zaporotskova, I., 2024. Composite NMs based on polymethylmethacrylate doped with carbon nanotubes and NPs: A review. *Polymers*. 16(9), 1242. https://doi.org/10.3390/polym16091242

Gahlawat, G., Choudhury, A. R., 2019. A review on the biosynthesis of metal and metal salt NPs by microbes. *RSC Adv*. 9(23), 12944–12967. https://doi.org/10.1039/ C8RA10483B

Godja, N. C., Munteanu, F. D., 2024. Hybrid NMs: A brief overview of versatile solutions for sensor technology in healthcare and environmental applications. *Biosensors*. 14(2), 67. https://doi.org/10.3390/bios14020067

Guo, D., Xie, G., Luo, J., 2013. Mechanical properties of NPs: Basics and applications. *J. Phys. D: Appl. Phys. 47*(1), 013001. https://doi.org/10.1088/0022-3727/47/1/013001

Harussani, M. M., Sapuan, S. M., Nadeem, G., Rafin, T., Kirubaanand, W., 2022. Recent applications of carbon-based composites in defence industry: A review. *Def. Technol*. 18(8), 1281–1300. https://doi.org/10.1016/j.dt.2022.03.006

Honary, S., Zahir, F., 2013. Effect of zeta potential on the properties of nano-drug delivery systems-a review (Part 1). *Trop. J. Pharm. Res*. 12(2), 255–264. https://doi.org/10.4314/tjpr.v12i2.19

Hu, B., Boakye-Yiadom K.O., Yu, W., Yuan, Z.W., Ho, W., Xu, X., Zhang, X.Q., 2020. Nanomedicine approaches for advanced diagnosis and treatment of atherosclerosis and related ischemic diseases. *Adv. Healthc. Mater*. 9(16), 2000336. https://doi.org/10.1002/adhm.202000336

Hussein, A. K., 2015. Applications of nanotechnology in renewable energies: A comprehensive overview and understanding. *Renew. Sustain. Energy Rev*. 42, 460–476. https://doi.org/10.1016/j.rser.2014.10.027

Iavicoli, I., Leso, V., Beezhold, D. H., Shvedova, A. A., 2017. Nanotechnology in agriculture: Opportunities, toxicological implications, and occupational risks. *Toxicol. Appl. Pharmacol*. 329, 96–111. https://doi.org/10.1016/j.taap.2017.05.025

Ibrahim, R. K., Hayyan, M., AlSaadi, M. A., Hayyan, A., Ibrahim, S., 2016. Environmental application of nanotechnology: Air, soil, and water. *Environ. Sci. Pollut. Res. Int*. 23(14), 13754–13788. https://doi.org/10.1007/s11356-016-6457-z

Ijaz, M., Khan, F., Ahmed, T., Noman, M., Zulfiqar, F., Rizwan, M., Chen, J., Siddique, K.H.M., Li, B., 2023. Nanobiotechnology to advance stress resilience in plants: Current opportunities and challenges. *Materials Today Bio*. 22, zy. https://doi.org/10.1016/j.mtbio.2023.100759

Jariwala, D., Sangwan, V. K., Lauhon, L. J., Marks, T. J., Hersam, M. C., 2013. Carbon NMs for electronics, optoelectronics, photovoltaics, and sensing. *Chem. Soc. Rev*. 42(7), 2824–2860. https://doi.org/10.1039/C2CS35335K

Joudeh, N., Linke, D., 2022. NP classification, physicochemical properties, characterization, and applications: A comprehensive review for biologists. *J. Nanobiotechnol*. 20(1), 262. https://doi.org/10.1186/s12951-022-01477-8

Katouzian, I., Jafari, S. M., 2016. Nano-encapsulation as a promising approach for targeted delivery and controlled release of vitamins. *Trends Food Sci. Technol*. 53, 34–48. http://doi.org/10.1016/j.tifs.2016.05.002

Kelly, K. L., Coronado, E., Zhao, L. L., Schatz, G. C., 2003. The optical properties of metal NPs: The influence of size, shape, and dielectric environment. *J. Phys. Chem. B*. 107(3), 668–677. https://doi.org/10.1021/jp026731y

Khan, I., Saeed, K., Khan, I., 2019. NPs: Properties, applications and toxicities. *Arab. J. Chem.*, 12(7), 908–931. https://doi.org/10.1016/j.arabjc.2017.05.011

Khan, Y., Sadia, H., Shah, S.Z.A., Khan, M.N., Shah, A.A., Ullah, N., Ullah, M.F., Bibi, H., Bafakeeh, O.T., Khedher, N.B., Eldin, S.M., Fadhi, B.M., Khan, M.I., 2022. Classification, synthetic, and characterization approaches to NPs, and their applications in various fields of nanotechnology: A review. *Catalysts.* 12(11), 1386. https://doi.org/10.3390/catal12111386

Kumar, A., Sharma, K., Dixit, A. R., 2020. Carbon nanotube-and graphene-reinforced multiphase polymeric composites: Review on their properties and applications. *J. Mater. Sci.* 55(7), 2682–2724. https://doi.org/10.1007/s10853-019-04196-y

Kumar, S., 2023. Smart and innovative nanotechnology applications for water purification. *Hybrid Adv.* 3, 100044. https://doi.org/10.1016/j.hybadv.2023.100044

Kumari, S., Raturi, S., Kulshrestha, S., Chauhan, K., Dhingra, S., András, K., Thu, K., Khargotra, R. Singh, T., 2023. A comprehensive review on various techniques used for synthesizing NPs. *J. Mater. Res. Technol.* 27, 1739–1763. https://doi.org/10.1016/j.jmrt.2023.09.291

Kyriakides, T.R., Raj, A., Tseng, T.H., Xiao, H., Nguyen, R., Mohammed, F.S., Halder, S., Xu, M., Wu, M.J., Bao, S., Sheu, W.C., 2021. Biocompatibility of NMs and their immunological properties. *Biomed. Mater.* 16(4), 042005. https://doi.org/10.1088/1748-605x/abe5fa

Lai, J., Niu, W., Luque, R., Xu, G., 2015. Solvothermal synthesis of metal nanocrystals and their applications. *Nano Today.* 10(2), 240–267. https://doi.org/10.1016/j.nantod.2015.03.001

Lee, J. H., Cho, H. Y., Choi, H. K., Lee, J. Y., Choi, J. W., 2018. Application of gold NP to plasmonic biosensors. *Int. J. Mol. Sci.* 19(7), 2021. https://doi.org/10.3390/ijms19072021

Li, L., Jiang, W., Luo, K., Song, H., Lan, F., Wu, Y., Gu, Z., 2013. Superparamagnetic iron oxide NPs as MRI contrast agents for non-invasive stem cell labeling and tracking. *Theranostics.* 3(8), 595. https://doi.org/10.7150%2Fthno.5366

Lombardo, D., Calandra, P., Pasqua, L., Magazù, S., 2020. Self-assembly of organic NMs and biomaterials: The bottom-up approach for functional nanostructures formation and advanced applications. *Materials.* 13(5), 1048. https://doi.org/10.3390/ma13051048

Lopes, R. C., Rocha, B. G., Maçoas, E. M., Marques, E. F., Martinho, J. M., 2022. Combining metal nanoclusters and carbon NMs: Opportunities and challenges in advanced nanohybrids. *Adv. Colloid Interface Sci.* 304, 102667. https://doi.org/10.1016/j.cis.2022.102667

Lu, T., Ji, S., Jin, W., Yang, Q., Luo, Q., Ren, T. L., 2023. Biocompatible and long-term monitoring strategies of wearable, ingestible and implantable biosensors: Reform the next generation healthcare. *Sensors.* 23(6), 2991. https://doi.org/10.3390/s23062991

Lue, J. T., 2007. Physical properties of NMs. *Encyclopedia of Nanoscience and Nanotechnology.* 10(1), 1–46. ISBN: 1-58883-058-6/$35.00

Mahhouti, Z., El Moussaoui, H., Mahfoud, T., Hamedoun, M., El Marssi, M., Lahmar, A., Benyoussef, A., 2019. Chemical synthesis and magnetic properties of monodisperse cobalt ferrite NPs. *J. Mater. Sci.: Mater. Electron.* 30(16), 14913–14922. https://doi.org/10.1007/s10854-019-01863-3

Mahmoud, C. B., Anelli, A., Csányi, G., Ceriotti, M., 2020. Learning the electronic density of o2f states in condensed matter. *Phys. Rev. B.* 102(23), 235130. https://doi.org/10.1103/PhysRevB.102.235130

Malik, M. A., Wani, M. Y., Hashim, M. A., 2012. Microemulsion method: A novel route to synthesize organic and inorganic NMs: 1st Nano Update. *Arab. J. Chem.* 5(4), 397–417. https://doi.org/10.1016/j.arabjc.2010.09.027

Malik, S., Muhammad, K., Waheed, Y., 2023. Nanotechnology: A revolution in modern industry. *Molecules.* 28(2), 661. https://doi.org/10.3390/molecules28020661

Manasa, G., Rout, C. S., 2024. Versatile MXenes as electrochemical sensors for heavy metal ions and phenolic moiety-containing industrial chemicals: Recent development and prospects. *Mater. Adv.* 5, 83–122. https://doi.org/10.1039/D3MA00362K

Manickam, R., Ravichandran, J., Chinnaiyan, D., Lakshminarasimhan, R.K., 2023. Nanotechgy-enabled biosensors: A review of fundamentals, design principles, materials, and applications. *Biosensors.* 13(1), 40. https://doi.org/10.3390/bios13010040

Mansurov, Z. A., 2020. Recent achievements and future challenges in nanoscience and nanotechnology. *Eurasian Chem.-Technol. J.* 22(4), 241–253. https://doi.org/10.18321/ ectj994

Martínez-Ballesta, M., Gil-Izquierdo, A., García-Viguera, C., Domínguez-Perles, R., 2018. NPs and controlled delivery for bioactive compounds: Outlining challenges for new "smart-foods" for health. *Foods.* 7(5), 72. https://doi.org/10.3390/foods7050072

Mbanga, O., Cukrowska, E., Gulumian, M., 2023. A comparative study of the biodurability and persistence of gold, silver, and titanium dioxide NPs using the continuous flow-through system. *Nanomater.* 13(10), 1653. https://doi.org/10.3390/nano13101653

Mekuye, B., Abera, B., 2023. NMs: An overview of synthesis, classification, characterization, and applications. *Nano Select.* 4(8), 486–501. https://doi.org/10.1002/nano.202300038

Menaa, F., Abdelghani, A., Menaa, B., 2015. Graphene NMs as biocompatible and conductive scaffolds for stem cells: Impact for tissue engineering and regenerative medicine. *J. Tissue Eng. Regen. Med.* 9(12), 1321–1338. https://doi.org/10.1002/term.1910

Mintova, S., Jaber, M., Valtchev, V., 2015. Nanosized microporous crystals: Emerging applications. *Chem. Soc. Rev.* 44(20), 7207–7233. https://doi.org/10.1039/C5CS00210A

Nyabadza, A., McCarthy, E., Makhesana, M., Heidarinassab, S., Plouze, A., Vazquez, M., Brabazon, D., 2023. A review of physical, chemical and biological synthesis methods of bimetallic NPs and applications in sensing, water treatment, biomedicine, catalysis and hydrogen storage. *Adv. Colloid and Interface Sci.* 321, 103010. https://doi.org/10.1016/j.cis.2023.103010

Pan, D., 2024. Nanomedicine and nanobiotechnology in India. *Wiley Interdiscip. Rev.: Nanomed. Nanobiotechnology*, 16(1), 1939. https://doi.org/10.1002/wnan.1939

Patra, J.K., Das, G., Fraceto, L.F., Campos, E.V.R., Rodriguez-Torres, M.D.P., Acosta-Torres, L.S., Diaz-Torres, L.A., 2018. Nano based drug delivery systems: Recent developments and future prospects. *J. Nanobiotechnology.*, 16(1), 71. https://doi.org/10.1186/s12951-018-0392-8

Pierre, A. C., 2020. Introduction to sol-gel processing. *Springer Nature.* https://doi.org/10.1007/978-3-030-38144-8

Pokrajac, L., Abbas, A., Chrzanowski, W., Dias, G.M., Eggleton, B.J., Maguire, S., Maine, E., 2021. Nanotechnology for a sustainable future: Addressing global challenges with the international network4sustainable nanotechnology. *ACS Nano.* 15(!2), 18608–18623. https://doi.org/10.1021/acsnano.1c10919

Poolakkandy, R. R., Menamparambath, M. M., 2020. Soft-template-assisted synthesis: A promising approach for the fabrication of transition metal oxides. *Nanoscale Adv.* 2(11), 5015–5045. https://doi.org/10.1039/D0NA00599A

Prezzi, D., Eom, D., Rim, K.T., Zhou, H., Lefenfeld, M., Xiao, S., Nuckolls, C., Heinz, T.F., Flynn, G.W., Hybertsen. M.S., 2014. Edge structures for nanoscale graphene islands on Co (0001) surfaces. *ACS Nano.* 8(6), 5765–5773. https://doi.org/10.1021/nn500583a

Qiao, Y., Li, C. M., 2011. Nanostructured catalysts in fuel cells. *J. Mater. Chem.* 21(12), 4027–4036. https://doi.org/10.1039/C0JM02871A

Rajan, C. S., 2011. Nanotechnology in groundwater remediation. *International J. Environ. Sci. Dev.* 2(3), 182. https://doi.org/10.7763/IJESD.2011.V2.121

Rajski, L., Juda, M., Los, A., Witun, E., Malm, A., 2019. Medical textiles with silver/nanosilver and their potential application for the prevention and control of healthcare-associated infections–mini-review. *Curr. Issues Pharm. Med. Sci.* 32(2), 104–107. https://doi.org/10.2478/cipms-2019-0020

Rashid, A. B., Haque, M., Islam, S. M., Labib, K. R. U., 2024. Nanotechnology-enhanced fiber-reinforced polymer composites: Recent advancements on processing techniques and applications. *Heliyon.* 10(!), 24692. https://doi.org/10.1016/j.heliyon.2024.e24692

Ray, P. C., 2010. Size and shape-dependent second-order nonlinear optical properties of NMs and their application in biological and chemical sensing. *Chem. Rev.* 110(9), 5332–5365. https://doi.org/10.1021/cr900335q

Sajid, M., Płotka-Wasylka, J., 2020. NPs: Synthesis, characteristics, and applications in analytical and other sciences. *Microchem. J.* 154, 104623. https://doi.org/10.1016/j.microc.2020.104623

Shinde, M. U., Patwekar, M., Patwekar, F., Bajaber, M. A., Medikeri, A., Mohammad, F. S., Jawaid, T., 2022. NMs: A potential hope for life sciences from bench to bedside. *J. Nanomater.* 22, 1–13. https://doi.org/10.1155/2022/5968131

Sim, S., Wong, N. K., 2021. Nanotechnology and its use in imaging and drug delivery. *Biomed. Rep.* 14(5), 42. https://doi.org/10.3892%2Fbr.2021.1418

Singamaneni, S., Bliznyuk, V.N., Binek, C., Tsymbal, E.Y., 2011. Magnetic NPs: Recent advances in synthesis, self-assembly and applications. *J. Mater. Chem.* 21(42), 16819–16845. https://doi.org/10.1039/C1JM11845E

Song, Y., Li, Y., Xu, Q., Liu, Z., 2017. Mesoporous silica NPs for stimuli-responsive controlled drug delivery: Advances,challenges,andoutlook.*Int.J.Nanomedicine.*12,87–110.https://doi.org/10.2147%2FIJN.S117495

Sridhar, A., Ponnuchamy, M., Kumar, P. S., Kapoor, A., 2021. Food preservation techniques and nanotechnology for increased shelf life of fruits, vegetables, beverages, and spices: A review. *Environ. Chem. Lett.* 19(2), 1715–1735. https://doi.org/10.1007%2Fs10311-020-01126-2

Tchounwou, P. B., Yedjou, C. G., Patlolla, A. K., Sutton, D. J., 2012. Heavy metal toxicity and the environment. *Molecular, Clinical and Environmental Toxicology.* 101, 133–164. https://doi.org/10.1007/978-3-7643-8340-4_6

Teng, X., Liang, X., Maksimuk, S., Yang, H., 2006. Synthesis of porous platinum NPs. *Small.* 2(2), 249–253. https://doi.org/10.1002/smll.200500244

Thwala, L. N., Ndlovu, S. C., Mpofu, K. T., Lugongolo, M. Y., Mthunzi-Kufa, P., 2023. Nanotechnology-based diagnostics for diseases prevalent in developing countries: Current advances in point-of-care tests. *Nanomater.* 13(7), 1247. https://doi.org/10.3390/nano13071247

Tiza, T.M., Kpur, G., Ogunleye, E., Sharma, S., Singh, S.K., Likassa, D.M., 2023. The potency of functionalized NMs for industrial applications. *Mater. Today: Proc.* https://doi.org/10.1016/j.matpr.2023.03.212

Tour, J. M., 2014. Top-down versus bottom-up fabrication of graphene-based electronics. *Chem. Mater. 26*(1), 163–171. https://doi.org/10.1021/cm402179h

Trotta, F., Andrea M., 2019. NMs: Classification and properties. Nanosponges: Synthesis and applications. *Nanosponges.* 1–26. https://doi.org/10.1002/ 9783527341009.ch1

Udayakumar, G. P., Muthusamy, S., Selvaganesh, B., Sivarajasekar, N., Rambabu, K., Banat, F., Show, P. L., 2021. Biopolymers and composites: Properties, characterization and their applications in food, medical and pharmaceutical industries. *J. Environ. Chem. Eng.* 9(4), 105322. https://doi.org/10.1016/j.jece.2021.105322

Vollath, D, 2013. *NMs: An introduction to synthesis, properties, and applications.* John Wiley & Sons. 1–386. 978-3-527-67186-1

Wagholikar, S., Mulay, P., Wagholikar, O., Mulay, V., 2024. NMs used in flexible electronics: Recent trends and future approaches. *Recent Pat. Eng.* 18(6), 27–38. https://doi.org/10.2174/1872212118666230519161338

Wang, B., Prinsen, P., Wang, H., Bai, Z., Wang, H., Luque, R., Xuan, J., 2017. Macroporous materials: Microfluidic fabrication, functionalization, and applications. *Chem. Soc. Rev.* 46(3), 855–914. https://doi.org/10.1039/C5CS00065C

Wang, S., Gao, L., 2019. Laser-driven NMs and laser-enabled nanofabrication for industrial applications. In *Industrial Applications of NMs.* 181–203. https://doi.org/10.1016/B978-0-12-815749-7.00007-4

Wang, Z., Hu, T., Liang, R., Wei, M., 2020. Application of zero-dimensional NMs in biosensing. *Front. Chem.* 8, 320. https://doi.org/10.3389%2F fchem.2020.00320

Wu, L., Li, Y., Fu, Z., Su, B. L., 2020. Hierarchically structured porous materials: Synthesis strategies and applications in energy storage. *Nati. Sci. Rev.* 7(11), 1667–1701. https://doi.org/10.1093/nsr/nwaa183

Xavier, J., Vincent, S., Meder, F., Vollmer, F., 2018. Advances in optoplasmonic sensors–combining optical nano/microcavities and photonic crystals with plasmonic nanostructures and NPs. *Nanophotonics.* 7(1), 1–38. https://doi.org/10.1515/nanoph-2017-0064

Xiao, M. F., Zeng, C., Li, S. H., Yuan, F. L., 2022. Applications of NMs in COVID-19 pandemic. *Rare Metals.* 41(1), 1–13. https://doi.org/10.1007%2Fs12598-021-01789-y

Xu, C., Lei, C., Hosseinpour, S., Ivanovski, S., Walsh, L. J., Khademhosseini, A., 2022. Nanotechnology for the management of COVID-19 during the pandemic and in the post-pandemic era. *Natl. Sci. Rev.* 9(10), 124. https://doi.org/10.1093%2Fnsr%2Fnwac124

Yadav, A., Yadav, K., Ahmad, R., Abd-Elsalam, K. A., 2023. Emerging frontiers in nanotechnology for precision agriculture: Advancements, hurdles, and prospects. *Agrochem.* 2(2), 220–256. https://doi.org/10.3390/agrochemicals2020016

Yan, H., 2012. Soft-templating synthesis of mesoporous graphitic carbon nitride with enhanced photocatalytic H2 evolution under visible light. *Chem. Comm.* 48(28), 3430–3432. https://doi.org/10.1039/C2CC00001F

Yu, B., Meyyappan, M., 2006. Nanotechnology: Role in emerging nanoelectronics. *Solid-state Electron.* 50(4), 536–544. https://doi.org/10.1016/j.sse.2006.03.028

Yu, R., Lin, Q., Leung, S. F., Fan, Z., 2012. NMs and nanostructures for efficient light absorption and photovoltaics. *Nano Energy.* 1(1), 57–72. https://doi.org/10.1016/j.nanoen.2011.10.002

Zaytseva, O., Neumann, G., 2016. Carbon NMs: Production, impact on plant development, agricultural and environmental applications. *Chem. Biol. Technol. Agric.* 3(17), 1–26. https://doi.org/10.1186/s40538-016-0070-8

Zhang, H., Marwat, M.A., Xie, B., Ashtar, M., Liu, K., Zhu, Y., Zhang, L., Fan, P., Samart, C., Ye, Z.G., 2019. Polymer matrix nanocomposites with 1D ceramic nanofillers for energy storage capacitor applications. *ACS Appl. Mater. Interfaces*. 12(1), 1–37. https://doi.org/10.1021/acsami.9b15005

Zhu, K., Ju, Y., Xu, J., Yang, Z., Gao, S., Hou, Y., 2018. Magnetic NMs: Chemical design, synthesis, and potential applications. *Acc. Chem. Res*. 51(2), 404–413. https://doi.org/10.1021/acs.accounts.7b00407

2 An Introduction to Advanced Nanomaterials and Nanotechnology

Anshul Yadav, Priyanka Ghanghas, Anita, and Kavita Poonia

2.1 INTRODUCTION

Advanced science and engineering, or nanotechnology, is concerned with the manipulation and study of materials and structures at the nanoscale, which is often defined as 1 to 100 nm. This tiny scale shows distinctive properties and behaviors of materials, giving engineers and scientists the capacity to alter and apply them for a range of purposes. Nanotechnology carries significant potential across various sectors, such as electronics, healthcare, materials science, and energy. It can transform technology and tackle some of the most critical global issues. This field has made it possible to create new materials, tools, and systems that are incredibly accurate and efficient (Hulla et al., 2015).

The industrial utilization of matter at the atomic, molecular, and supramolecular levels is referred to as nanotechnology (Bayda et al., 2019). The exact manipulation of atoms and molecules to manufacture structures at a macro scale was the core of the first and most commonly accepted definition of nanotechnology, now known as molecular nanotechnology (Ariga, 2021). A broader definition of nanotechnology was later defined by the National Nanotechnology Initiative. The technique of modifying materials with at least one dimension spanning from 1 to 100 nm is referred to as nanotechnology (Mostafavi et al., 2019).

A definition of nanotechnology based on the scale would encompass energy storage, surface science, organic chemistry, molecular biology, semiconductor physics, engineering, micro-manufacturing, and molecular engineering, to name only a few scientific fields (Madkour, 2019).

Consumer goods, nanomedicine, nanoelectronics, biomaterials, and energy generation are just a few of the new things that nanotechnology has the potential to produce (Shinde et al., 2022). On the other hand, there are a lot of worries associated with nanotechnology, just like with any new technology. These include concerns over nanoparticle (NP) toxicity and impacts on the environment, their potential to have a major impact on world economics, and the possibility of several catastrophic events (Timmermans et al., 2020). These worries have sparked discussions among advocacy groups and political entities over the need for particular nanotechnology legislation (Grunwald, 2020).

2.2 NANOTECHNOLOGY

New material discoveries and the advancement of industrial research have been accelerated by the first technological revolution, which took place in the late 1700s. Right now, the difficulties are in making tools and gadgets smaller, lighter, and more efficient, while maintaining functionality. Progress depends on the identification of novel, aesthetically pleasing materials and the capacity to precisely construct small structures. This development is not without its challenges, though. Among the most amazing approaches developed to deal with these issues is nanotechnology (Giron et al., 2019).

The Greek word "nano," which means "dwarf" or "very small," is where the term "nano" originates. Nanotechnology encompasses materials with incredibly tiny dimensions as a result (Nikalje, 2015). It

DOI: 10.1201/9781003484189-2

is, in essence, the fabrication of materials, parts, apparatuses, or systems at dimensions close to those of an atom or molecule. This cutting-edge technology includes nanoscale imaging, fabrication, modification, measurement, modeling, and miniaturization of matter. Its attributes, which make it appropriate for a variety of applications, include affordability, accuracy, environmental friendliness, high strength, low weight, and specificity (Asmatulu et al., 2010).

The definition of nanotechnology is divided into two parts. The first part deals with the production process at sizes between 1 and 100 nanometres (nm), and the second part concerns the special properties of materials at the nanoscale that make them useful in novel applications. The increased significance of surface area and interface area is the primary cause of this change in performance. According to this novel approach, one must think conceptually at the atomic level, taking into account how atoms interact to create structures, gadgets, and systems that are helpful (Raza & Raza, 2013).

Additionally, nanotechnology holds promise for the domains of biology, physics, pharmacology, and material science, all of which might have an impact on healthcare. Since the anticipated therapeutic advantages of nanotechnology are not completely understood, despite thirty years of exploration in healthcare research, the field is still in its infancy. Even so, the field is developing quickly and presents many chances for creative thinkers to use this cutting-edge technology to improve human health (Daniel & Astruc, 2004).

2.2.1 Historical Development of Nanotechnology

The initial concept of nanotechnology was proposed by Richard Feynman, a laureate of the Nobel Prize. The term "nanometer" was first introduced in 1925 by another Nobel Prize winner in Chemistry, Richard Zsigmondy. He specifically used the term "nanometer" to describe particle size and was the first to use a microscope to measure particle size. Interestingly, NPs and structures were utilized by the Romans in the 4th century CE, marking one of the earliest instances of nanotechnology. During the evolution of nanotechnology, a Japanese scientist named Iijima made significant contributions by developing carbon nanotubes (Bayda et al., 2019). The historical development of nanotechnology can be seen in Figure 2.1.

2.3 NANOMATERIALS (NMS)

NMs, once referred to as "Magic Bullets" by Paul Ehrlich, are among the most extensively studied materials of this era, leading to the emergence of a new scientific field, nanotechnology (Khan et al., 2017). NMs are chemical entities or substances that are engineered and utilized on an extremely small scale. The term "material" encompasses a vast array of components, collectively exhibiting a performance that is statistically averaged. Consequently, the behavior of NMs is influenced by unique interfacial effects and exhibits properties that are determined by their size and the limited quantity of components (Barhoum et al., 2022).

2.3.1 Classification of NMs

Many nations' economies are now benefiting from, and will continue to benefit from, the nanoscale manufacture of traditional goods. Numerous new product categories are expected to surface in the future, in addition to the variety of nanoscale items that have been documented so far. This has made the necessity for their classification all the more necessary. Gleiter made the first suggestion for the categorization of materials at the nanoscale in 2000 (Gleiter, 2000). Numerous terms, including nanostructured, nanometer-sized, ultrafine-grained, and so on, have been applied to these novel materials. Nanoscale materials can have a wide range of properties that are dependent on their size. They can be made up of one or more types of atoms or molecules. Nanoscale materials fill the energy state gap between tiny molecules and bulk materials within this size range. NMs can occur

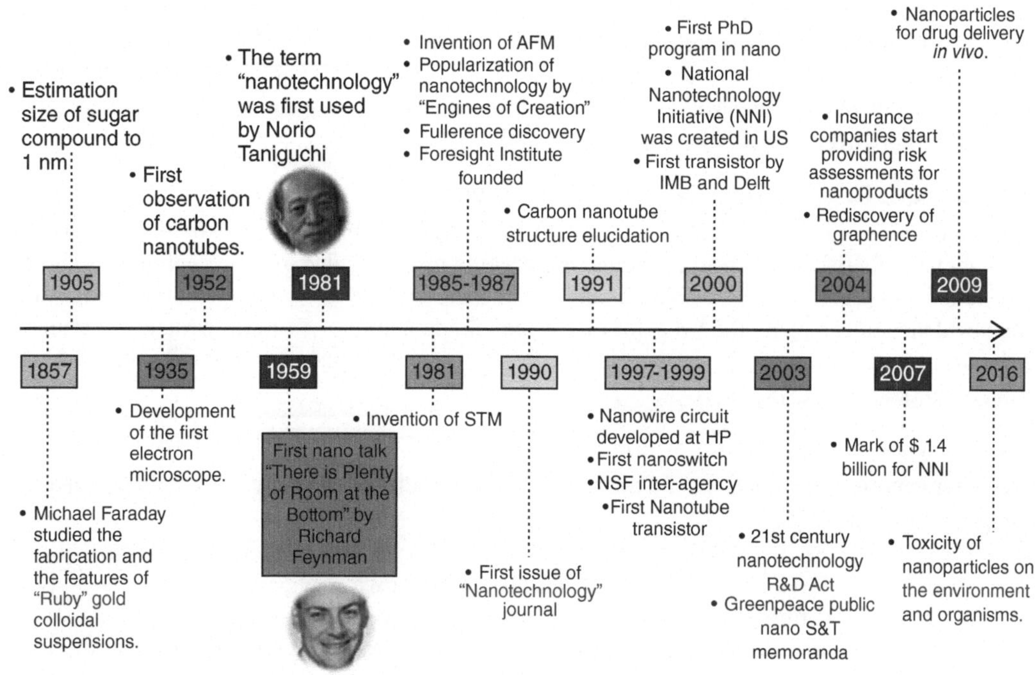

FIGURE 2.1 Timeline of NMs discovery in the modern era of nanotechnology (Barhoum et al., 2022).

naturally or can be produced using various methods, such as chemical, mechanical, physical, or biological processes, and they can exhibit a diverse range of structures (Smith & Nie, 2010). Nanoscale materials can be classified according to various specific criteria, including their source; their chemical makeup; the type of material they are based on; and their dimensional characteristics (Saleh, 2016).

2.3.1.1 NMs Based on Their Origin

Nanoscale materials can be bifurcated into two groups, depending on their source.

Natural NMs: These materials are nanoscale entities that either arise spontaneously from human action (e.g., NPs created during the combustion of diesel) or naturally occur in the environment (e.g., proteins, viruses). They may arise naturally or because of actions taken by humans. Nature is a readily available supply for the development of artificial surfaces, with excellent micro and nanoscale characteristics and patterns for industrial purposes. According to J. K. Patel and others, naturally occurring NMs may be found in all three of Earth's spheres: the hydrosphere, the atmosphere, and the lithosphere, which includes rocks, soils, magma, and lava at different phases of their evolutionary histories. Aside from human-caused activities, they are also found in the biosphere, which includes humans and other higher animals, as well as microbes (Sharma et al., 2015).

Synthetic NMs: These NMs can be created synthetically by physical, chemical, biological, or hybrid processes, or they can be purposefully manufactured utilizing certain techniques such as mechanical grinding, smoke, and engine exhaust. Natural materials, including fullerenes, micelle-like amphiphilic polyurethane particles, and dendrimers, as well as inorganic materials like elemental metals, metal oxides, metal salts, and aluminosilicates, are all included in the broad category of synthetic NMs. The rising manufacturing of engineered NMs and their subsequent release, along with their application in consumer goods and industrial equipment, have made a risk assessment strategy necessary. The main problem with created NPs is figuring out if the data available can accurately anticipate their behavior or if they operate differently in the environment than natural NMs. Currently, the production of tailored NMs uses a variety of sources that take prospective applications into account (Pawliszyn, 2012).

2.3.1.2 NMs Based on Their Chemical Composition

The use of NMs in nano remediation mainly poses several possible hazards to human health. The development of a risk assessment technique has become necessary due to the recent increase in the production and use of engineered NMs in consumer items and industrial equipment. Finding sufficient evidence to anticipate the behavior of created NPs in contrast to natural NMs is the primary challenge. Currently, a variety of sources that consider possible applications are utilized in the creation of tailored NMs (Wagner et al., 2014).

2.3.1.3 Material-Based Composition

The latest nanoscale materials can be categorized into four distinct groups, based on their material composition (Figure 2.2).

Carbon-Based NMs: Carbon-based NMs usually refer to a wide range of chemicals such as graphene (Gr), fullerenes (C60), carbon nanotubes (CNTs), carbon nanofibers, carbon black, and carbon onions. Except carbon black, several methods, including chemical vapor deposition (CVD), arc discharge, and laser ablation are used to create these carbon-based NMs (Jeevanandam et al., 2018).

Inorganic-Based NMs: Metal oxide and metal NPs are examples of inorganic-based NMs. The aforementioned NMs may be synthesized into metals, such as silver or gold NPs (Ag NPs), metal oxides, such as zinc oxide (ZnO) and titanium dioxide (TiO2) NPs, and semiconductors: silicon and ceramics (Jeevanandam et al., 2018).

Organic-Based NMs: Carbon-based and inorganic-based materials are not commonly used to create organic-based NMs. These organic NMs are transformed into desirable forms, including dendrimers, micelles, liposomes, and polymeric NPs, by using noncovalent interactions for the self-organization and design of molecules.

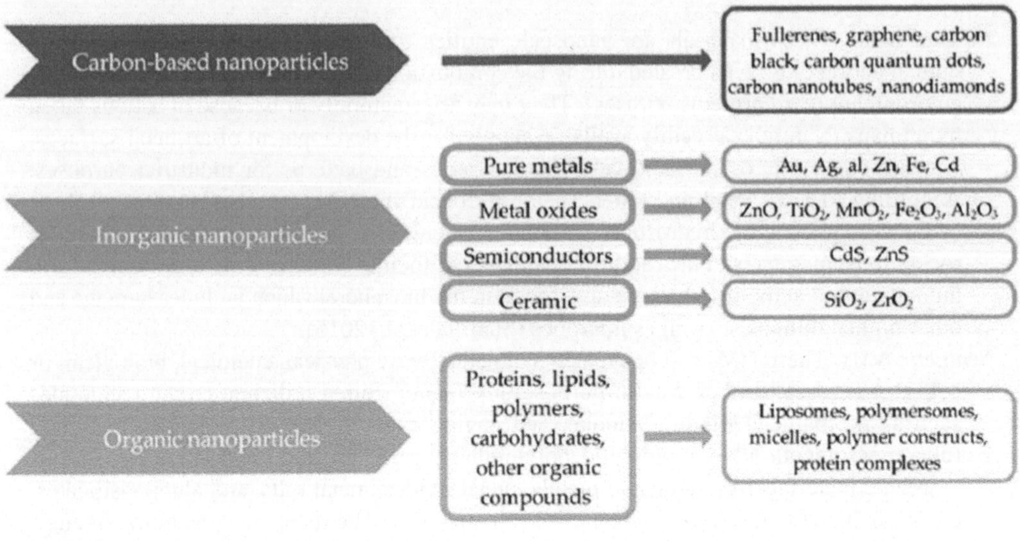

FIGURE 2.2 Classification of NMs, based on their chemical composition (Kučuk et al., 2023).

Composite-Based NMs: Multiphase NPs (NPs) with a single nanoscale phase are known as composite NMs. These can be used to join NPs (NPs) to one another or bigger, bulkier materials like hybrid nanofibers or even complex structures like metal-organic frameworks. Composites can consist of any kind of metal or polymer mixed with NPs based on carbon, metal, or organic elements. Different forms and structures of NMs are produced according to the attributes required for the desired use (Tiwari et al., 2012).

2.3.1.4 NMs Based on Their Dimension

Pokropivny and Skorokhod presented a unique system of categorization for NMs in 2007. The technique employed in this study classified the newly developed composite NMs into four distinct dimensions: zero, one, two, and three dimensions (Figure 2.3). The way electrons behave inside the NMs' dimensions has a significant impact on this categorization. Electrons, for instance, are restricted in a dimensionless region in 0-D NMs, whereas in 1-D NMs, electrons can travel down the x-axis, which is less than 100 nm. Similarly, electrons in 2D and 3D NMs migrate along the x, y, and z axes, respectively. The benefit of classifying NMs is based on the predictability of their characteristics (Tiwari et al., 2012).

Based on this concept, NMs can be categorized as follows:

Zero-dimensional (0-D): These are groups of atoms that have diameters between 2 and 100 nm and range in number from a few hundred to a few thousand. 0-D NMs are materials whose dimensions are at the nanoscale scale. In all spatial directions, the diameters of nanoclusters, which are typically spherical nanostructures, range from 1 to 100 nm. They are limited to one length, one breadth, and one height. They may have a crystalline or amorphous structure. In disciplines like electronics, engineering, and technology, 0-D NMs are crucial (Baláž & Baláž, 2008).

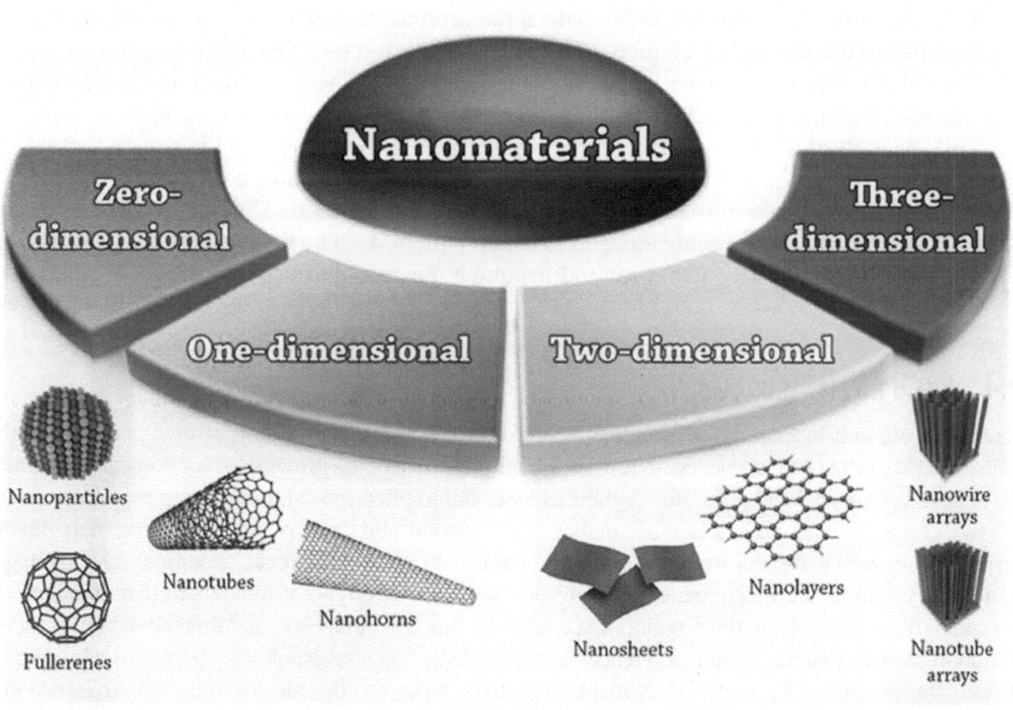

FIGURE 2.3 Dimension-based classification of NMs (Joudeh & Linke, 2022).

One-Dimensional (1-D): The second category of nanoscale materials, known as 1-D NMs, is designated for those materials that possess nanoscale dimensions identical in every direction except one. One-dimensional NMs, or 1-D NMs, are a type of NM where one of their exterior dimensions falls within the nanoscale. They have a single dimension, which could be length, breadth, or height. These materials often take the form of needle-like nanostructures, such as nanotubes, nanowires, nanofibers, and nanorods. Their diameter typically ranges from 1 to 100 nm, while their length can be significantly larger, classifying them as 1-D nanostructures. They can be either amorphous or crystalline in their structure. Nanofibers and nanowires are similar. One dimension in these structures is much larger than the other two, which are still in the nanoscale. Materials with a single dimension at the nanoscale are frequently referred to as thin films or surface coatings. For many years, they have been manufactured and applied in a wide range of industries, such as solar cell applications, computer memory chips, biological and chemical sensors, anti-reflective coatings on eyewear, electronics, information storage systems, and optical devices. Numerous techniques may be used to apply thin films, which can also be controlled to develop at the atomic level (Thomas et al., 2014).

Two-Dimensional (2-D): Inside this class of NMs, two dimensions are greater than the nanoscale, while only one dimension is included inside the nanoscale. It has only two measurements: width and length. 2-D nanostructures resemble flat plates in shape. Nanotubes, dendrimers, wires, fibers, films, layers, thin films, textured surfaces, and nano-coatings are a few examples of 2-D nanostructures. These two-dimensional NPs can have crystalline or amorphous forms. They may be utilized as a single layer or a multilayer structure and are made of a range of chemical compositions. The manufacturing procedures and the overall understanding of the properties of 2-D systems are not as advanced. To improve the required surface attributes—such as resistance to corrosion and wear and friction reduction—while preserving the material's intrinsic properties, 2-D systems are used to improve structural bulk materials (Saleh, 2016).

Three-Dimensional (3-D): Materials with three arbitrary dimensions that go beyond the nanoscale are referred to as three-dimensional (3-D) structures. All three dimensions are covered by these materials: length, breadth, and height. They have the quality of being nanocrystalline. NMs include particles, precipitates, colloids, fullerenes, quantum dots, and nanocrystals. Many well-known 3D systems exist, including metallic oxides, carbon black, TiO_2, ZnO, and natural NMs. Others, including quantum dots, fullerenes, and dendrimers, provide substantial difficulties in terms of production and property comprehension. Many nanocrystals grouped make up bulk NMs. Three-dimensional (3D) NMs include NP dispersion, nanowire and nanotube clusters, and numerous layers of NMs (West & Halas, 2003).

2.4 PROPERTIES OF NMS

Recently, research in material science has been focusing on the exploration of new materials with enhanced properties and innovative synthesis techniques to meet the growing technological demands. NMs have become a focal point due to their exceptional applications and intriguing properties.

The impact of size effects is what makes NMs special and fascinating. These size-dependent effects are essential for improving the spectroscopic, structural, chemical, electrical, electromagnetic, and thermodynamic characteristics of these systems. The type of movement that an electron may undertake depends on the amount of space that it has at its disposal, and this movement affects the characteristics of the material. Hence, a certain length scale, usually in the nanoscale range, defines the qualities of a material. A material's characteristics alter and become sensitive to both form and size when its physical size is lowered below this length scale. Our capacity to control atoms allows us to produce NMs that are suited for certain uses (Stucky & Mac Dougall, 1990).

When the size is reduced and the energy difference between electronic levels is greater than the thermal energy, dramatic changes in fundamental electrical and optical characteristics can be detected in any material. Small nanocrystals exhibit discrete electronic energy levels or finite density of states, in contrast to bulk materials because the electronic wave function is confined to the particle's physical dimensions. Due of this phenomenon, which is called quantum confinement, nanocrystals are sometimes referred to as quantum dots (Richards & Bönnemann, 2005). In addition, a considerable fraction of the atoms in nanocrystals are found on their surface, giving them a higher surface area. This factor might result in size impacts altering the chemical and physical characteristics of the nanocrystals; it is mostly dependent on the particle size.

2.4.1 Confinement Effect

The "dimensionality" of a system that is in the nanoscale range is related to quantum size effects. The dual nature of particles and waves, as well as the fundamentals of quantum physics, give birth to quantum phenomena. They take place when the size of the system is comparable to the de Broglie wavelengths of the circulating electrons, phonons, or excitons. Electrons display both particle and wave properties at the same time. They travel the whole area, in which they are in unrestricted motion like waves. The nanograin functions as if it were a kind of box, with certain properties either present or absent. The characteristics of the property are directly and exclusively dependent on the size of the grain below a threshold level. The confinement effect is the name given to this phenomenon. When influencing the physical and chemical features, such as charge transport methods and electronic structure, quantum size effects play a crucial role. For the investigation of these systems, optical and electro-tunneling spectroscopies are indispensable.

2.4.2 Surface Effect

In comparison to atoms in the bulk, atoms on surfaces contain fewer adjacent atoms. Surface atoms are not as stable as bulk atoms because of this decreased coordination and unmet bonding. A tiny particle has a high average binding energy per atom and a sizable percentage of atoms on its surface. Many attributes follow the same scaling equation when the surface-to-volume ratio inversely scales with size. Even fewer coordinated atoms can be found near edges and corners, where they can bond alien atoms and molecules more firmly. In tiny holes, the quantity of coordinated atoms is likewise constrained (De Rogatis et al., 2008).

Reducing size has an impact on more than just a material's physical characteristics and atomic structure. It becomes hard to precisely describe a surface layer if the structure of a particle's surface area is overemphasized concerning the particle's size. It is known that the free surface of a material may be modified to change its composition or crystal structure. In materials at the nanoscale, the volume of this surface layer becomes important. As a result, the outside area of a particle, where its composition or crystal structure varies from that of the inside, is known as the surface layer of NMs (Meyers et al., 2006).

2.4.3 Mechanical Properties

When it comes to mechanical properties, nanostructures outperform bulk materials. The superior mechanical characteristics of NPs are ascribed to their flawless structure. Internal structural defects, including dislocations, micro-twins, and impurity precipitations, are absent from these microscopic materials. Mechanical failures are frequently the result of more active NM flaws propagating to the surface during annealing, purifying the material. The NMs' perfect material structures are preserved as a result of the flaws migrating to the surface. When compared to bulk materials, the outer surface of nanostructures contains fewer or no defects. As a result, materials with lower defect counts have better mechanical qualities (Zhang et al., 2002).

Hardness is frequently seen as the main mechanical property in many NMs. This is especially true for nitrides, carbides, and borides used to create ultra-hard nanocomposite. These nanocomposites were made using special manufacturing techniques, namely the physical vapor deposition method and the plasma-induced chemical approach (Lu et al., 2004). NMs are engineered with superior mechanical characteristics for future use across macro, micro, and nanoscale applications. Compared to conventional copper, nanocrystalline copper exhibits three times the resistance and also offers enhanced flexibility. Electromechanical resonators at high frequencies are built using carbon nanotubes and nanowires. These can be used to manipulate NMs at the nanoscale scale as nanoprobes or nanotweezers. Isotropic elastic materials undergo tiny, continuous, and reversible deformations, which are included in the idea of elasticity. Three primary characteristics define an elastic material: it is homogeneous, isotropic, and uniformly dispersed across its volume; it deforms under tension and returns to its original shape when the load is released. Materials are usually polycrystalline, meaning that their grains have varying forms and orientations, which prevents them from being isotropic. However, since the dimensions of the materials are significantly larger than the average grain size, the assumptions of homogeneity and isotropy are often more or less met. As a result, the theory of elasticity is also applicable to polycrystalline materials (Juvé et al., 2010).

2.4.4 Structural Properties

The space between atoms in a material changes when the size of its particles decreases. This leads to an increase in the material's surface area and surface energy. When particles are reduced to nanoscale sizes, structural changes become visible. Cuboctahedral and doubly twinned decahedra are two examples of polyhedral structures that gold NPs (Au NPs) can take on. Through the expansion of crystalline structures, the previously described forms may be analyzed and understood while accounting for varied crystallographic orientations and energies of distinct crystallographic planes. Crystalline solids have distinct patterns and a long-range periodic atomic structure. On the other hand, this long-range periodic structure is absent from quasiperiodic crystals. An organized structure devoid of periodicity is called a quasiperiodic crystal. The shape, size, and morphological composition of nanostructured materials are their main features. Surfactants are chemicals that may be used to change the surface structure of various materials. Because NPs may take on a variety of forms, we can manipulate them to investigate their properties (Labastie & Calvo, 2007).

2.4.5 Thermal Properties

The space between atoms in a material changes by adjusting a material's size at the nanoscale, many of its properties may be customized. These customized nanostructures may be used to meet the needs of many applications. In contrast to macroscopic systems, where the particle count grows towards infinity, a nanosystem's thermodynamics is different. The fact that increased surface energy changes consistently with size is accommodated by the structure of thermodynamics. The usual experimental variations in the melting point of spherical gallium nitride NPs, which are correlated with the particle size, include the melting and other phase transition temperatures. The amplitude of atomic oscillations is enhanced, surface energy increases, and therefore, thermal vibration energy expands on the surface (Singh et al., 2017). These are the physical causes of it. It has been noted that when the particle size decreases, the specific heat increases, but the melting entropy and enthalpy both drop (Murty et al., 2013).

One intriguing use of NMs' thermal properties is the use of nanofluid to improve thermal transmission. Generally speaking, nanofluids are composite materials made up of particles that are scattered in a liquid media and have a size between 1 and 100 nm (Gorji & Ranjbar, 2017). Due to their improved thermal characteristics over their basic fluids, nanofluids are attracting more and more attention in both research and practical applications. NPs of metals, metal carbides, oxides, and nitrides, as well as nanofibers like single- and multi-wall carbon nanotubes, may all be utilized in nanofluids. Depending on the possible uses, they can be mixed with different base liquids such as oils, ethylene glycol, and water (Keblinski et al., 2005).

Several experimental studies have revealed that the most prominent feature of nanofluids is a significant increase in heat conductivity when compared to liquids without NPs (Berger, 2012). Nanofluid-based devices will facilitate the creation of real-time, minimally intrusive medical diagnostic systems to track astronaut health and support disease detection and therapy. As a result, scientists are having difficulty studying heat transmission in NMs theoretically. Many variables, including the small size, shape, and large contact area of the particles, can modify the thermal properties of NMs. As a result, compared to their bulk counterparts, NMs' thermal properties are far more diverse. Phonon transport inside the material is significantly altered when the material's size approaches the nanoscale when it approaches the wavelength and mean free path of phonons. Thermal characteristics naturally alter as a result of this shift in phonon confinement and the quantization of phonon transport.

2.4.6 Optical Properties

Electrical structure affects the optical properties, and variations in the zone structure affect the absorption and luminescence spectra. Their distinct characteristics, such as spectral breadth and location and sensitivity to light polarization, rely on both their surroundings and the inherent characteristics of the nanostructures, such as composition, structure, size, and form (Pattnaik, 2005).

Materials' optical characteristics are also impacted by size reduction. There are two possible explanations for the significant dependence of these features on particle size. One is caused by the higher energy level spacing brought on by the enhanced confinement structure. Surface plasmon resonance is connected to the other. Surface plasmon resonance is a process that determines the optical characteristics of metallic NPs (SPR). Strong absorption in the UV-visible region occurs when the particle's size is lowered below the electron mean free path or the separation an electron takes between scattering collisions with lattice centers. Surface plasmon absorption is the result of SPR's optical stimulation (Schmitt-Rink et al., 1987).

Mostly, the effects of quantum size are taken into account for semiconducting materials. A semiconducting material's bandgap increases when its particle size decreases because the inter-band transition moves to a higher frequency. Typically, semiconductors have an energy gap of a few electron volts, which grows quickly as particle size reduces. In the bandgap, the phenomenon of quantum confinement causes a shift toward the blue spectrum. In nanostructured semiconductors, the size of the particles greatly influences their optical properties. Therefore, it is easy to modify the optical characteristics of these materials by changing the particle size. Because of their higher carrier confinement and energy density states, nanostructured semiconducting materials are very effective and appropriate for use in laser systems. Surface plasmon resonance is produced when the size of the particles within metal nanostructures is smaller than the wavelength of the incoming radiation. It is evident from the debate that came before it that particle size has a big impact on the optical properties of materials. Advanced optical materials may be customized for a range of devices by varying their size at the nanoscale scale (Petracic, 2010).

2.4.7 Magnetic Properties

In the context of modern solid-state magnetism and nanotechnology, nano-magnetism is an exciting and captivating topic. Both great technological value and substantial scientific curiosity are attached to it. In several applications, such as magnetic recording and hard and soft magnets, ferromagnetic NPs may be superior to current materials. It is commonly known that a magnetic material's size has a major impact on its coercivity or resistance to demagnetization. Magnetic coercivity increases with particle size, reaching a peak at the single domain size as the particles get smaller and closer to the nanometer range. On the other hand, thermal factors cause the coercivity to decline again for very tiny particles, ultimately reaching zero at the superparamagnetic particle size. It's interesting to note that iron, a soft magnetic material with a coercivity of around 20 Oersted (Oe) at room temperature, can be induced to become "hard" with a coercivity of 540Oe (Mills & Bland, 2006). The phenomena of massive magnetoresistance (GMR) in magnetic multilayers is another intriguing example. Hard drive performance has been greatly enhanced by this, and in a few years, their capacity has increased by more than a hundredfold (Hiergeist et al., 1999).

The magnetic properties find use in many applications, including medication delivery (Sundaresan et al., 2006) and ferrofluids (Wu et al., 2011). NPs may have different magnetic characteristics than the comparable bulk material. The particles' decreased size results in an increase in surface area and a stronger magnetic interaction between nearby atoms. Different magnetic qualities follow from this.

It is possible to discover ferromagnetism at very tiny sizes. For clusters of little more than 100–200 atoms, the magnetic torques behave like individual atoms. The magnetic torque drops and approaches the bulk limit when the size approaches 700 atoms, most likely because of structural alterations or vibrations brought on by surface-induced spin density waves. One worldwide characteristic of NPs derived from nonmagnetic oxides is their ferromagnetism (Sato et al., 2007). Ferromagnetic particles become unstable when their size falls below a particular threshold. This instability results from both the high enough surface energy and the spontaneous polarization of domains. This property causes ferromagnetic materials to change to paramagnetic at the nanoscale. Superparamagnetism is the term used to describe this phenomenon because of the way the materials act in contrast to conventional paramagnetic materials. In bulk form, ferromagnetic materials usually consist of many magnetic domains. When these ferromagnetic materials are nanostructured, they only have one domain and are made up of small magnetic NPs. Due to thermal fluctuations, the domains of these various particles are randomly dispersed; nevertheless, when an external magnetic field is introduced, the domains align (Rosei, 2004).

2.5 SYNTHESIS STRATEGIES

Recently, in material science a lot of research investigations have focused on the development of NMs, which poses a big problem (Charitidis et al., 2014). This is a multidisciplinary area that includes materials science, engineering, biology, chemistry, and physics. New materials with specific features will certainly be developed as a result of the collaboration between researchers from these many fields. Strong industry-academic cooperation is essential for the development of nanomanufacturing to identify current demands and future issues and create solutions that the industrial sector can easily embrace. Precursor materials in their solid, liquid, or gaseous phases can be used in the production of NMs. Different types of NMs, such as buckyballs, clusters, colloids, powders, rods, thin films, tubes, wires, and so on, may be produced using a variety of techniques. We may mold synthetic materials in a way that makes them appropriate for particular purposes by employing a variety of techniques. To create novel NMs, several conventional production techniques for various materials have been improved, and numerous new techniques have been created (Wang & Xia, 2004).

Nanoscale materials may be produced using a variety of techniques. These techniques may be broadly divided into two groups: (a) the top-down approach and (b) the bottom-up approach. The "top-down" technique to manufacturing NMs involves shrinking bigger structures, whereas the "bottom-up" approach assembles smaller components to produce larger structures. These techniques further divide into a number of subclasses based on the operating procedure, reaction circumstances, and protocols employed. Nonetheless, a third strategy known as effective fabrication (computer simulations) has been proposed by certain academics (Iravani, 2011).

2.5.1 Bottom-Up Procedures

This methodology, also known as the building-up approach, is applied in reverse, generating NMs from relatively basic chemicals. Systems for sedimentation and reduction are two examples of this method. It includes procedures like spinning, sol-gel methods, green manufacturing, and biochemical synthesis (Blum et al., 2005).

Using this technique, little parts are combined to create complex structures. These constituents have distinct binding capabilities, commonly known as molecular recognition attributes that facilitate their appropriate self-organization. One essential component of bottom-up techniques is self-assembly (Dhingra et al., 2010).

By beginning at the atomic or molecular level and gradually increasing through nucleation and/or growth from solid, liquid, or gaseous precursors, bottom-up techniques mimic natural processes. Physical or chemical procedures are used to accomplish these results (Yaya et al., 2012). Microemulsions and other colloidal dispersions are excellent examples of the bottom-up method for creating NMs. Typically, products created using the bottom-up method exhibit superior purity, enhanced control over particle size, and surface chemistry. Despite being in its nascent stages of development, the bottom-up approach promises significant transformations of contemporary manufacturing techniques.

2.5.2 Top-Down Procedures

The concept of top-down design involves taking processes known from the macroscopic world and adapting them to perform similar tasks on a smaller scale. Large particles are progressively reduced to smaller ones. This approach is typically applied to solids or dispersed solids. Since ancient times, humans have created art and tools by manipulating materials. A prime example is stone sculpture, which involves creating visually appealing 3D objects from stone. This ancient craft involves shaping irregular pieces of natural stone by selectively removing parts to achieve the desired shape (Iravani, 2011).

This idea applies a reductionist approach. A big molecule is the starting point, and it is divided into smaller components. After that, these units are transformed into appropriate NMs. This strategy is demonstrated by several techniques, such as physical vapor deposition, grinding and milling, and other breakdown techniques (Yaya et al., 2012).

For its manufacturing, this approach often depends on physical processes, or a combination of physical, chemical, electrical, or thermal processes (Dhingra et al., 2010). Although this is the most well-established type of nanotechnology, top-down approaches are often thought to produce more waste (Rineesh et al., 2018).

2.6 APPLICATIONS

Platinum is known for its usage in energy storage, whereas NMs consisting of gold, silver, zinc, and other metals have attracted a lot of attention in the field of biological applications, due to their distinctive physicochemical features. Using silver NPs' optical characteristics as the functional component of several goods and sensors is gaining popularity. Unlike many dyes and pigments, the size and structure of silver NPs determine their hue, making them incredibly efficient at both scattering and absorbing light. Their expanding popularity stems from their application as antibacterial agents in textiles, wound dressings, medical equipment, and home appliances like washing machines and refrigerators (Chen et al., 2017). NMs (NMs) based on carbon have a wide range of applications in various fields. These include drug delivery, the immobilization of enzymes, the creation of biosensors, bioimaging, and the removal of pollutants (Li & Jin, 2013).

2.6.1 Catalysis

Catalysis applications have been pioneered by NPs. For many years, a wide range of substances and elements, including clays, silica, titanium dioxide, iron, and aluminum, have been used as catalysts at the nanoscale. Because of nanocatalysis's great activity, selectivity, and productivity, it has recently become a prominent scientific topic. Tiny metal NPs, with diameters between 1 and 10 nm, exhibit remarkable catalytic activity, sometimes even outperforming their matching metal complexes. Several important characteristics, including a high surface-to-volume ratio, surface geometric effect, electronic impact, and quantum size effect, are responsible for the high activity of nanocatalysts. Because metal NPs in suspension have facile product separation, simple recovery, and great recyclability, they are often utilized as efficient heterogeneous catalysts. This makes metal nanocatalysts ecologically beneficial. The size of metal NPs significantly influences the catalytic performance (e.g., conversion and selectivity) of nanocatalysts. Utilizing catalysts can lower the temperature required for a process, minimize waste from reagents, and improve reaction selectivity, potentially preventing undesirable side reactions and promoting environmentally friendly practices. Many items, including medications, fine chemicals, polymers, textiles, fuels, paints, lubricants, and many other necessary value-added products, could not be produced without catalysts (Tang et al., 2014). Consequently, the application of catalysts can lead to more economical, sustainable, and ecologically friendly production processes. In the field of chemistry, carbon nanotubes are used widely in photocatalytic processes, the partial oxidation process in fuel cells, and the creation of ammonia and methane.

2.6.2 Water Treatment

There is a lot of interest in the application of NPs in wastewater and water treatment. NMs have a huge surface area due to their tiny size, which increases their adsorption and reactivity capability. Various types of NMs are successful in eliminating heavy metals (Yan et al., 2015), organic contaminants (Liu et al., 2014), inorganic anions (Kalhapure et al., 2015), and microorganisms (Lu et al., 2016). Lately, the use of metal oxide NPs like TiO_2 for photocatalytic degradation has proven effective in breaking down contaminants in water and wastewater. Because of its strong photoactivity, chemical stability, ease of market availability, and non-toxicity, TiO2 has been the focus of a great deal of study (Bratovčić, 2019). Pollutants can gradually oxidize into low molecular weight intermediate products when they are exposed to light and a catalyst. These eventually become CO_2, H_2O, and anions, including NO_3^-, PO_4^{3-}, and Cl^- (Manikandan et al., 2018).

2.6.3 Sensors

The development of different NMs has made it possible to employ them in the production of effective electrochemical sensors for environmental monitoring, food safety, and medical diagnostics. According to the review of (Chen et al., 2018), various NMs have been synthesized for the electrochemical determination of common contaminants and additives, such as hydrazine (N_2H_4), malachite green (MG), bisphenol A (BPA), ascorbic acid (AA), caffeine, caffeic acid (CA), sulfite (SO_3^{2-}), and nitrite (NO_2^-), which are commonly found in food and beverages. Biosensors have been greatly influenced by nanotechnology, particularly in terms of their high selectivity and accuracy, as well as their capacity to reduce the size of sensor devices. Here, novel nanostructured biosensors for glucose monitoring have been developed using fluorescent NMs and nanostructures. People with diabetes frequently employ the electrochemical method of glucose measurement, usually in the form of a blood glucose monitor (Brainina et al., 2018). Moreover, carbon nanotubes may be used to develop molecular detection systems, such as chromatography applications, electrochemical detectors, gas sensors, and small molecule sensors. By adding NPs and other NMs to the macro electrode, one may improve the sensitivity and selectivity of the measurements, while lowering the detection threshold (Duan et al., 2018).

2.6.4 Energy Storage

Carbon nanotubes may be used to develop molecular detection systems, such as electrochemical detectors, gas sensors, small molecule sensors, and chromatography applications. We may increase the sensitivity and selectivity of the measurements and decrease the detection threshold by modifying the macro electrode with different NMs and NPs. Using platinum-based NMs significantly improves the number of open active sites and specific surface area, which significantly enhances the use of platinum atoms (Murali et al., 2018).

2.6.5 Nanomedicine

The characteristics of silver NPs, including their potent and broad-spectrum antibacterial activity, are extensively studied. A variety of applications for nano silver have been developed in consumer products, from the disinfection of medical equipment and household appliances to water treatment processes, as well as in the field of nanomedicine.

The potential applications of biocompatible gold NPs in nanomedicine, together with their fascinating size-dependent chemical, electrical, and optical properties, have spurred a major increase in research on these particles in recent years. Their pertinent uses in gene therapy, photodynamic treatment, photothermal therapy, drug delivery, bio labeling, biosensing, and other areas are revolutionizing the biomedical sector and generating a lot of attention from researchers (Qi & Gao, 2008). The non-toxic nature of gold NPs (AuNPs) and their vast surface area make them ideal for facile modification with target molecules in medicinal applications. They are therefore better than other kinds of NPs. Targeted drug delivery is the most successful type of therapy since it solely targets the cells or locations that are diseased, which minimizes medication side effects. Because medications may be given directly to malignant cells while sparing healthy cells, this method is especially helpful in the treatment of cancer. In addition, quantum dots are discussed in this context (Chertok et al., 2008). ZnO (Hutter & Maysinger, 2011) and Fe_3O_4 (Chen et al., 2008; Rasmussen et al., 2010) are useful for focused medication administration. AuNPs, or gold NPs, can be used for bio-imaging applications such as biosensing and labeling. They have been used for a very long time in cellular and molecular imaging as contrast agents. Ma and his colleagues created a novel sensor for Salmonella typhimurium colorimetric detection that makes use of the color-changing capabilities of gold NPs (Ma et al., 2017). Gold NPs are used in many different biosensing applications these days (Saha et al., 2012). Biosensing may be used to efficiently monitor infections (Phillips et al., 2008), identify bacteria (Alvarez-Puebla et al. 2007), and viruses (Li et al., 2010), detect pollutants (Li et al., 2010), and determine blood glucose levels (Phillips et al., 2008).

2.7 IMPACTS OF NMS ON THE ENVIRONMENT

Human and environmental exposure to NMs (NMs) is increasing due to their growing production and use in building, medical, and consumer goods, among other industrial processes. Humans are exposed to NPs through a variety of routes and sources, including ingestion of airborne particles, direct skin contact with consumer items, and food consumption. The increasing use of NMs in consumer items and other areas of our lives have inevitably resulted in the accumulation of NPs in the environment. Physical or chemical characteristics of these particles, as well as their surroundings, have a significant impact on their future. For example, when NPs disperse into the environment, their surface area and other physicochemical characteristics can have a major impact on their transformation and bioavailability. To assess the possible pollution consequences of toxicity and NPs in the environment, we currently need a thorough system. These buildings have an influence on the environment in some ways, including 1) raising the air, water, and soil pollution levels; 2) building up within the environmental system and causing both immediate and long-term consequences; and 3) altering the life-cycle of living things in the environment. Furthermore, the amounts of NPs exposed in environmental media are still unknown because there are no quantitative analytical techniques available to monitor these particles. The analysis of NMs in environmental samples frequently requires the integration of many different methods. These techniques for characterization include microscopy, chromatography, centrifugation, filtering, and spectroscopy, among others

(Batley et al., 2013). The main method for examining the properties of individual NMs is electron microscopy. When determining the size, shape, and aggregation condition of NMs smaller than 10 nm, two different techniques can be used: scanning electron microscopy and transmission electron microscopy. Surface runoff from soils, wastewater treatment effluents, and industrial discharge are three ways that NPs might enter the aquatic environment quickly (Mitrano et al., 2015).

2.7.1 Positive Impacts

When engineered NMs are released into the atmosphere, they are usually exposed to a lot more sunlight and ultraviolet radiation than when they are put into other settings (Slaoui et al., 2017). Exposure of this kind is probably going to increase the possibility of photochemical changes to NMs.

NMs are employed in aircraft as an alternative to traditional composites, aiding in reducing the weight of the aircraft and consequently saving thousands of tons of fuel. Furthermore, NMs are used in wind turbine blades to improve their strength and light weight, which raises the efficiency of energy conversion (Etim et al., 2018). To enhance chemical reactions and lower expenses and emissions at the same time, NMs are being used more and more in petroleum refining and automobile exhaust systems. Through the use of oxidation, filtration, and adsorption, NMs have proven to be more efficient than conventional approaches in the purification of air and water. The antibacterial properties of nano-silver particles are broad and effective against a variety of Gram-positive and Gram-negative bacteria, including antibiotic-resistant species (Banerjee et al., 2017).

Through the use of oxidation, filtration, and adsorption, NMs have proven to be more efficient than conventional approaches in the purification of air and water. On a variety of Gram-positive and Gram-negative bacteria, including antibiotic-resistant strains, nano silver particles demonstrate a broad antibacterial effect (de Francisco & García-Estepa, 2018). Significant uses of nanotechnology may be found in many different fields, most notably the agro-food sector. Primary production, food processing and packaging, and food supplements are its main uses (Pradhan et al., 2015). With its wide range of uses, nanotechnology is widely employed in many different sectors. For example, nano-sensors are used to monitor the health of livestock, such as pigs and sheep, to control soil cultivation conditions, to identify the presence of pesticides in agricultural produce, and many other applications (Milani et al., 2012). By improving the management and preservation of plant and animal production resources, applications of nanotechnology have the potential to completely transform agricultural operations. When it comes to benefits, nano-fertilizers outperform conventional ones. To increase the solubility of zinc fertilizers, for instance, zinc oxide NPs are included (Milani et al., 2015).

2.7.2 Negative Impacts

Unexpected interactions between novel NMs and biomolecules, cells, organs, and organisms are possible. As a result, exposure to these NMs by humans and the environment may have negative consequences. Attention should be paid to public and regulatory approval, which is an essential aspect of nanotechnology. The primary focus lies in the materials and products' impact on the environment and human health and safety. Certain materials could be quite useful in some situations, but their effects on the environment might make them inappropriate (Dekkers et al., 2016). The size, content, and clustering of these NPs are some of the elements that affect their toxicological effect. Recent research has brought attention to the detrimental impact that NPs in water or at the water-water interface have on aquatic life (Dekkers et al., 2016).

The environment and biological systems may be harmed by NPs. For example, if food crops are treated with a high concentration of nano silver, they may pose chemical dangers. In certain cases, free radicals produced by NMs can harm DNA in live tissue. Thus, it's imperative to evaluate nanotechnology in depth before increasing the usage of nano-agro materials (Ghosh et al., 2010).

Ag_2S NPs' effects on the environment have lately been investigated by scientists. Plant growth was seen to decrease in correlation with the overexpression of genes caused by the absorption of these NPs. There is a greater chance that these structures will go up the food chain because the majority of the Ag_2S NPs were discovered to concentrate in the leaves of the plants that were the subject of the investigation. Additionally, even at low doses of 0.25 mM, it was shown that certain NPs, such as TiO_2, might have genotoxic effects, harming DNA in plants like Nicotiana tabacum and Allium cepa at greater concentrations (Gao et al., 2018).

It has been demonstrated that the presence of NMs (NMs) has a spectrum of harmful effects on aquatic life, from low to high. According to toxicological research, NPs may affect bigger animals like fish and Daphnia, as well as single-celled aquatic species (Abou El-Nour et al., 2010). Several studies suggest that the capacity of silver NPs to release silver ions into cells is the cause of their deleterious effects. Similar degrees of cytotoxicity have been reported for both silver ions and silver NPs. Although the exact nature of the harmful process is yet unknown, it seems that nano-silver particles ionize inside of cells. Ion channels are triggered by this ionization, which also modifies the permeability of the cell membrane to sodium and potassium. Additionally, it interacts with mitochondria and, by producing reactive oxygen species, initiates the apoptotic pathway, which eventually results in cell death.

2.8 CONCLUDING REMARKS

The understanding, manipulation, and control of materials with dimensions approximately between 1 and 100 nm, where conventional physics is no longer applicable, is known as nanotechnology. Scientists view it as the cutting-edge technology of the twenty-first century. Materials that comprise particles in an unbound state or as aggregates or agglomerates natural, accidental, or engineered—are known as NMs. These materials consist of fundamental structural elements, fibers, particles, or other important constituents that are, in at least one dimension, less than 100 nm. Because of their distinct nanoscale features, NMs are different from their bulk counterparts and cannot be treated in the same way as tiny molecules or bulk materials. NMs' properties are determined by their chemistry, composition, size, and interactions with other materials. The two main techniques for creating NMs are called top-down and bottom-up approaches. In the former case, the bulk material is broken down into increasingly smaller dimensions, whereas in the latter case, the structure is built up from smaller clusters. The fields of nanoscience and nanotechnology have great promise for improving the quality of life and tackling several global issues of our day. The creation of novel and useful goods in a variety of fields is still greatly aided by the application of nanotechnology. On the other hand, NMs are becoming more and more prevalent in the environment. Humans are indirectly impacted by the dangers that NMs offer to bacteria, plants, and animals. Therefore, to completely comprehend the influence of nanotechnology on science, technology, and society, it is imperative to keep an eye on continuing research and breakthroughs in the field.

REFERENCES

Abou El-Nour, K. M., Eftaiha, A. A., Al-Warthan, A., & Ammar, R. A. (2010). Synthesis and applications of silver NPs. *Arabian Journal of Chemistry*, 3(3), 135–140. https://doi.org/10.1016/j.arabjc.2010.04.008

Alvarez-Puebla, R. A., dos Santos, D. S., & Aroca, R. F. (2007). SERS detection of environmental pollutants in humic acid: Gold NP composite materials. *Analyst*, 132(12), 1210–1214. https://doi.org/10.1039/B711361G

Ariga, K. (2021). Nanoarchitectonics: What's coming next after nanotechnology?. *Nanoscale Horizons*, 6(5), 364–378. https://doi.org/10.1039/D0NH00680G

Asmatulu, R., Asmatulu, E., & Zhang, B. (2010, September). Nanotechnology and nanoethics in engineering education. In *Proceedings of the 2010 Midwest Section Conference of the American Society for Engineering Education*. Lawrence, KS (pp. 1–11).

Baláž, P., & Baláž, P. (2008). *Mechanochemistry in minerals engineering* (pp. 257–296). Springer Berlin Heidelberg. https://doi.org/10.1007/978-3-540-74855-7_5

Banerjee, T., Shelby, T., & Santra, S. (2017). How can nanosensors detect bacterial contamination before it ever reaches the dinner table?. *Future Microbiology*, 12(2), 97–100. https://doi.org/10.2217/fmb-2016-0202

Barhoum, A., García-Betancourt, M.L., Jeevanandam, J., Hussien, E.A., Mekkawy, S.A., Mostafa, M., Omran, M.M., S. Abdalla, & Bechelany, M. (2022). Review on natural, incidental, bioinspired, and engineered NMs: History, definitions, classifications, synthesis, properties, market, toxicities, risks, and regulations. *NMs*, 12(2), 177. https://doi.org/10.3390/nano12020177

Batley, G. E., Kirby, J. K., & McLaughlin, M. J. (2013). Fate and risks of NMs in aquatic and terrestrial environments. *Accounts of Chemical Research*, 46(3), 854–862. https://doi.org/10.1021/ar2003368

Bayda, S., Adeel, M., Tuccinardi, T., Cordani, M., & Rizzolio, F. (2019). The history of nanoscience and nanotechnology: From chemical: Physical applications to nanomedicine. *Molecules*, 25(1), 112. https://doi.org/10.3390/molecules25010112

Berger, M. (2012). NASA and nanotechnology. 2012-10-23)[2012-12-01]. https://www.nanowerk.com/spotlight/spotid=27065.php#ixzz38CDvHTag

Blum, A. S., Soto, C. M., Wilson, C. D., Brower, T. L., Pollack, S. K., Schull, T. L., … & Ratna, B. R. (2005). An engineered virus as a scaffold for three-dimensional self-assembly on the nanoscale. *Small*, 1(7), 702–706. https://doi.org/10.1002/smll.200500021

Brainina, K., Stozhko, N., Bukharinova, M., & Vikulova, E. (2018). NMs: Electrochemical properties and application in sensors. *Physical Sciences Reviews*, 3(9), 20188050. https://doi.org/10.1515/psr-2018-8050

Bratovčić, A. (2019). Photocatalytic degradation of organic compounds in wastewaters. *Technologica Acta: Scientific/Professional Journal of Chemistry and Technology*, 11(2), 17–23. https://doi.org/10.5281/zenodo.2563022

Charitidis, C. A., Georgiou, P., Koklioti, M. A., Trompeta, A. F., & Markakis, V. (2014). Manufacturing NMs: From research to industry. *Manufacturing Review*, 1, 11. https://doi.org/10.1051/mfreview/2014009

Chen, F. H., Gao, Q., & Ni, J. Z. (2008). The grafting and release behavior of doxorubincin from Fe3O4@ SiO2 core–shell structure NPs via an acid cleaving amide bond: The potential for magnetic targeting drug delivery. *Nanotechnology*, 19(16), 165103. doi: 10.1088/0957-4484/19/16/165103

Chen, L., Hwang, E., & Zhang, J. (2018). Fluorescent nanobiosensors for sensing glucose. *Sensors*, 18(5), 1440. https://doi.org/10.3390/s18051440

Chen, M., Qin, X., & Zeng, G. (2017). Biodegradation of carbon nanotubes, graphene, and their derivatives. *Trends in Biotechnology*, 35(9), 836–846. https://doi.org/10.1016/j.tibtech.2016.12.001

Chertok, B., Moffat, B. A., David, A. E., Yu, F., Bergemann, C., Ross, B. D., & Yang, V. C. (2008). Iron oxide NPs as a drug delivery vehicle for MRI monitored magnetic targeting of brain tumors. *Biomaterials*, 29(4), 487–496. https://doi.org/10.1016/j.biomaterials.2007.08.050

Daniel, M. C., & Astruc, D. (2004). Gold NPs: assembly, supramolecular chemistry, quantum-size-related properties, and applications toward biology, catalysis, and nanotechnology. *Chemical Reviews*, 104(1), 293–346. https://doi.org/10.1021/cr030698

de Francisco, E. V., & García-Estepa, R. M. (2018). Nanotechnology in the agrofood industry. *Journal of Food Engineering*, 238, 1–11. https://doi.org/10.1016/j.jfoodeng.2018.05.024

De Rogatis, L., Montini, T., Gombac, V., Cargnell, M., & Fornasiero, P. (2008). *Stabilized Metal NPs Embedded into Porous Oxides: A Challenging Approach for Robust Catalysts* (pp. 71–123). Nova Science Publishers: New York.

Dekkers, S., Oomen, A. G., Bleeker, E. A., Vandebriel, R. J., Micheletti, C., Cabellos, J., … & Wijnhoven, S. W. (2016). Towards a nanospecific approach for risk assessment. *Regulatory Toxicology and Pharmacology*, 80, 46–59. https://doi.org/10.1016/j.yrtph.2016.05.037

Dhingra, R., Naidu, S., Upreti, G., & Sawhney, R. (2010). Sustainable nanotechnology: Through green methods and life-cycle thinking. *Sustainability*, 2(10), 3323–3338. https://doi.org/10.3390/su2103323

Duan, S., Du, Z., Fan, H., & Wang, R. (2018). Nanostructure optimization of platinum-based NMs for catalytic applications. *NMs*, 8(11), 949. https://doi.org/10.3390/nano8110949

Etim, U. J., Bai, P., & Yan, Z. (2018). Nanotechnology applications in petroleum refining. *Nanotechnology in Oil and Gas Industries: Principles and Applications*, 37–65. https://doi.org/10.1007/978-3-319-60630-9_2

Gao, M., Zhang, Z., Lv, M., Song, W., & Lv, Y. (2018). Toxic effects of NM-adsorbed cadmium on Daphnia magna. *Ecotoxicology and Environmental Safety*, 148, 261–268. https://doi.org/10.1016/j.ecoenv.2017.10.038

Ghosh, M., Bandyopadhyay, M., & Mukherjee, A. (2010). Genotoxicity of titanium dioxide (TiO2) NPs at two trophic levels: Plant and human lymphocytes. *Chemosphere*, 81(10), 1253–1262. https://doi.org/10.1016/j.chemosphere.2010.09.022

Giron, F., Pastó, A., Tasciotti, E., & Abraham, B. P. (2019). Nanotechnology in the treatment of inflammatory bowel disease. *Inflammatory Bowel Diseases*, 25(12), 1871–1880. https://doi.org/10.1093/ibd/izz205

Gleiter, H. (2000). Nanostructured materials: Basic concepts and microstructure. *Acta Materialia*, 48(1), 1–29. https://doi.org/10.1016/S1359-6454(99)00285-2

Gorji, T. B., & Ranjbar, A. A. (2017). A review on optical properties and application of nanofluids in direct absorption solar collectors (DASCs). *Renewable and Sustainable Energy Reviews*, 72, 10–32. https://doi.org/10.1016/j.rser.2017.01.015

Grunwald, A. (2020). Nanotechnology-a new field of ethical inquiry?. In *The Ethics of Nanotechnology, Geoengineering, and Clean Energy* (pp. 17–31). London: Routledge.

Hiergeist, R., Andrä, W., Buske, N., Hergt, R., Hilger, I., Richter, U., & Kaiser, W. (1999). Application of magnetite ferrofluids for hyperthermia. *Journal of Magnetism and Magnetic Materials*, 201(1–3), 420–422. https://doi.org/10.1016/S0304-8853(99)00145-6

Hulla, J. E., Sahu, S. C., & Hayes, A. W. (2015). Nanotechnology: History and future. *Human & Experimental Toxicology*, 34(12), 1318–1321. https://doi.org/10.1177/0960327115603588

Hutter, E., & Maysinger, D. (2011). Gold NPs and quantum dots for bioimaging. *Microscopy Research and Technique*, 74(7), 592–604. https://doi.org/10.1002/jemt.20928

Iravani, S. (2011). Green synthesis of metal NPs using plants. *Green Chemistry*, 13(10), 2638–2650. https://doi.org/10.1039/C1GC15386B

Jeevanandam, J., Barhoum, A., Chan, Y. S., Dufresne, A., & Danquah, M. K. (2018). Review on NPs and nanostructured materials: History, sources, toxicity and regulations. *Beilstein Journal of Nanotechnology*, 9(1), 1050–1074. https://doi.org/10.3762/bjnano.9.98

Joudeh, N., & Linke, D. (2022). NP classification, physicochemical properties, characterization, and applications: A comprehensive review for biologists. *Journal of Nanobiotechnology*, 20(1), 262. https://doi.org/10.1186/s12951-022-01477-8

Juvé, V., Crut, A., Maioli, P., Pellarin, M., Broyer, M., Del Fatti, N., & Vallée, F. (2010). Probing elasticity at the nanoscale: Terahertz acoustic vibration of small metal NPs. *Nano Letters*, 10(5), 1853–1858. https://doi.org/10.1021/nl100604r

Kalhapure, R. S., Sonawane, S. J., Sikwal, D. R., Jadhav, M., Rambharose, S., Mocktar, C., & Govender, T. (2015). Solid lipid NPs of clotrimazole silver complex: An efficient nano antibacterial against Staphylococcus aureus and MRSA. *Colloids and Surfaces B: Biointerfaces*, 136, 651–658. https://doi.org/10.1016/j.colsurfb.2015.10.003

Keblinski, P., Eastman, J. A., & Cahill, D. G. (2005). Nanofluids for thermal transport. *Materials Today*, 8(6), 36–44. https://doi.org/10.1016/S1369-7021(05)70936-6

Khan, M. N., Mobin, M., Abbas, Z. K., AlMutairi, K. A., & Siddiqui, Z. H. (2017). Role of NMs in plants under challenging environments. *Plant Physiology and Biochemistry*, 110, 194–209. https://doi.org/10.1016/j.plaphy.2016.05.038

Kučuk, N., Primožič, M., Knez, Ž., & Leitgeb, M. (2023). Sustainable biodegradable biopolymer-based NPs for healthcare applications. *International Journal of Molecular Sciences*, 24(4), 3188. https://doi.org/10.3390/ijms24043188

Labastie, P., & Calvo, F. (2007). Thermodynamics and solid: Liquid transitions. In *NMs and Nanochemistry* (pp. 55–87). Berlin, Heidelberg: Springer Berlin Heidelberg. https://doi.org/10.1007/978-3-540-72993-8_3

Li, F., Zhao, Q., Wang, C., Lu, X., Li, X. F., & Le, X. C. (2010). Detection of Escherichia coli O157: H7 using gold NP labeling and inductively coupled plasma mass spectrometry. *Analytical Chemistry*, 82(8), 3399–3403. https://doi.org/10.1021/ac100325f

Li, G., & Jin, R. (2013). Catalysis by gold NPs: Carbon-carbon coupling reactions. *Nanotechnology Reviews*, 2(5), 529–545. https://doi.org/10.1515/ntrev-2013-0020

Liu, F., Yang, J., Zuo, J., Ma, D., Gan, L., Xie, B., … & Yang, B. (2014). Graphene-supported nanoscale zero-valent iron: Removal of phosphorus from aqueous solution and mechanistic study. *Journal of Environmental Sciences*, 26(8), 1751–1762. https://doi.org/10.1016/j.jes.2014.06.016

Lu, H., Wang, J., Stoller, M., Wang, T., Bao, Y., & Hao, H. (2016). An overview of NMs for water and wastewater treatment. *Advances in Materials Science and Engineering*, 2016. https://doi.org/10.1155/2016/4964828

Lu, L., Shen, Y., Chen, X., Qian, L., & Lu, K. (2004). Ultrahigh strength and high electrical conductivity in copper. *Science*, 304(5669), 422–426. https://doi.org/10.1126/science.1092905

Ma, X., Song, L., Zhou, N., Xia, Y., & Wang, Z. (2017). A novel aptasensor for the colorimetric detection of S. typhimurium based on gold NPs. *International Journal of Food Microbiology*, 245, 1–5. https://doi.org/10.1016/j.ijfoodmicro.2016.12.024

Madkour, L. H. (2019). *Nanoelectronic Materials: Fundamentals and Applications* (Vol. 116). Springer.

Manikandan, V. S., Adhikari, B., & Chen, A. (2018). NM based electrochemical sensors for the safety and quality control of food and beverages. *Analyst*, 143(19), 4537–4554. https://doi.org/10.1039/C8AN00497H

Meyers, M. A., Mishra, A., & Benson, D. J. (2006). Mechanical properties of nanocrystalline materials. *Progress in Materials Science*, 51(4), 427–556. https://doi.org/10.1016/j.pmatsci.2005.08.003

Milani, N., Hettiarachchi, G. M., Kirby, J. K., Beak, D. G., Stacey, S. P., & McLaughlin, M. J. (2015). Fate of zinc oxide NPs coated onto macronutrient fertilizers in an alkaline calcareous soil. *PLoS One*, 10(5), e0126275. https://doi.org/10.1371/journal.pone.0126275

Milani, N., McLaughlin, M. J., Stacey, S. P., Kirby, J. K., Hettiarachchi, G. M., Beak, D. G., & Cornelis, G. (2012). Dissolution kinetics of macronutrient fertilizers coated with manufactured zinc oxide NPs. *Journal of Agricultural and Food Chemistry* 60(16), 3991–3998. https://doi.org/10.1021/jf205191y

Mills, D. L. L., & Bland, J. A. C. (2006). *Nanomagnetism: Ultrathin Films, Multilayers and Nanostructures*. Elsevier.

Mitrano, D. M., Motellier, S., Clavaguera, S., & Nowack, B. (2015). Review of NM aging and transformations through the life cycle of nano-enhanced products. *Environment International*, 77, 132–147. https://doi.org/10.1016/j.envint.2015.01.013

Mostafavi, E., Soltantabar, P., & Webster, T. J. (2019). Nanotechnology and picotechnology: A new arena for translational medicine. In *Biomaterials in Translational Medicine* (pp. 191–212). Academic Press. https://doi.org/10.1016/B978-0-12-813477-1.00009-8

Murali, K., Neelakandan, M. S., & Thomas, S. (2018). Biomedical applications of gold NPs. *JSM Nanotechnol Nanomed*, 6(1), 1064.

Murty, B. S., Shankar, P., Raj, B., Rath, B. B., & Murday, J. (2013). *Textbook of Nanoscience and Nanotechnology*. Springer Science & Business Media.

Nikalje, A. P. (2015). Nanotechnology and its applications in medicine. *Med chem*, 5(2), 081–089.

Pattnaik, P. (2005). Surface plasmon resonance: Applications in understanding receptor-ligand interaction. *Applied Biochemistry and Biotechnology*, 126, 79–92. https://doi.org/10.1385/ABAB:126:2:079

Pawliszyn, J. (2012). *Comprehensive Sampling and Sample Preparation: Analytical Techniques for Scientists*. Academic Press.

Petracic, O. (2010). Superparamagnetic NP ensembles. *Superlattices and Microstructures*, 47(5), 569–578. https://doi.org/10.1016/j.spmi.2010.01.009

Phillips, R. L., Miranda, O. R., You, C. C., Rotello, V. M., & Bunz, U. H. (2008). Rapid and efficient identification of bacteria using gold-NP–poly (para-phenyleneethynylene) constructs. *Angewandte Chemie International Edition*, 47(14), 2590–2594.

Pradhan, N., Singh, S., Ojha, N., Shrivastava, A., Barla, A., Rai, V., & Bose, S. (2015). Facets of nanotechnology as seen in food processing, packaging, and preservation industry. *BioMed Research International*, 2015. https://doi.org/10.1155/2015/365672

Qi, L., & Gao, X. (2008). Emerging application of quantum dots for drug delivery and therapy. *Expert Opinion on Drug Delivery*, 5(3), 263–267. https://doi.org/10.1517/17425247.5.3.263

Rasmussen, J. W., Martinez, E., Louka, P., & Wingett, D. G. (2010). Zinc oxide NPs for selective destruction of tumor cells and potential for drug delivery applications. *Expert Opinion on Drug Delivery*, 7(9), 1063–1077. https://doi.org/10.1517/17425247.2010.502560

Raza, H., & Raza, T. Z. (2013). Introducing nanoengineering and nanotechnology to the first year students through an interactive seminar course. *Journal of Nano Education*, 4(1–2), 41–46. https://doi.org/10.48550/arXiv.1303.6910

Richards, R., & Bönnemann, H. (2005). Synthetic approaches to metallic NMs. *Nanofabrication Towards Biomedical Applications: Techniques, Tools, Applications, and Impact*, 1–32. https://doi.org/10.1002/3527603476

Rineesh, N. R., Neelakandan, M. S., & Thomas, S. J. J. N. N. (2018). Applications of silver NPs for medicinal purpose. *JSM Nanotechnol Nanomed*, 6(1), 1063.

Rosei, F. (2004). Nanostructured surfaces: Challenges and frontiers in nanotechnology. *Journal of Physics: Condensed Matter*, 16(17), S1373. 10.1088/0953-8984/16/17/001

Saha, K., Agasti, S. S., Kim, C., Li, X., & Rotello, V. M. (2012). Gold NPs in chemical and biological sensing. *Chemical Reviews*, 112(5), 2739–2779. https://doi.org/10.1021/cr2001178

Saleh, T. A. (2016). NMs for pharmaceuticals determination. *Bioenergetics*, 5(1), 1000226.

Sato, K., Fukushima, T., & Katayama-Yoshida, H. (2007). Super-paramagnetic blocking phenomena and room-temperature ferromagnetism in wide band-gap dilute magnetic semiconductor (Ga, Mn) N. *Japanese Journal of Applied Physics*, 46(7L), L682. 10.1143/JJAP.46.L682

Schmitt-Rink, S. D. A. B. M., Miller, D. A. B., & Chemla, D. S. (1987). Theory of the linear and nonlinear optical properties of semiconductor microcrystallites. *Physical Review B*, 35(15), 8113. https://doi.org/10.1103/PhysRevB.35.8113

Sharma, V. K., Filip, J., Zboril, R., & Varma, R. S. (2015). Natural inorganic NPs–formation, fate, and toxicity in the environment. Chemical Society Reviews, 44(23), 8410–8423. https://doi.org/10.1039/C5CS00236B

Shinde, M. U., Mohsina, P., Faheem, P., Bajaber, M. A., Medikeri, A., Mohammad, F. S., Mukim, M., Soni, S., Mallick, J., & Jawaid, T. NMs: A potential hope for life sciences from bench to bedside. *Journal of NMs* 2022 (June 30, 2022): 1–13. https://doi.org/10.1155/2022/5968131

Singh, M., Lara, S. O., & Tlali, S. (2017). Effects of size and shape on the specific heat, melting entropy and enthalpy of NMs. *Journal of Taibah University for Science*, 11(6), 922–929. https://doi.org/10.1016/j.jtusci.2016.09.011

Slaoui, A., Lincot, D., Guillemoles, J. F., & Escoubas, L. (2017). NMs for Photovoltaic Conversion. https://doi.org/10.1002/9783527696109

Smith, A. M., & Nie, S. (2010). Semiconductor nanocrystals: Structure, properties, and band gap engineering. *Accounts of Chemical Research*, 43(2), 190–200. https://doi.org/10.1021/ar9001069

Stucky, G. D., & Mac Dougall, J. E. (1990). Quantum confinement and host/guest chemistry: Probing a new dimension. *Science*, 247(4943), 669–678. https://doi.org/10.1126/science.247.4943.669

Sundaresan, A., Bhargavi, R., Rangarajan, N., Siddesh, U., & Rao, C. N. R. (2006). Ferromagnetism as a universal feature of NPs of the otherwise nonmagnetic oxides. *Physical Review B*, 74(16), 161306. https://doi.org/10.1103/PhysRevB.74.161306

Tang, W. W., Zeng, G. M., Gong, J. L., Liang, J., Xu, P., Zhang, C., & Huang, B. B. (2014). Impact of humic/fulvic acid on the removal of heavy metals from aqueous solutions using NMs: A review. *Science of the Total Environment*, 468, 1014–1027. https://doi.org/10.1016/j.scitotenv.2013.09.044

Thomas, S., Rafiei, S., Maghsoodlou, S., & Afzali, A. (2014). *Foundations of Nanotechnology, Volume Two: Nanoelements Formation and Interaction*. CRC Press.

Timmermans, J., Zhao, Y., & van den Hoven, J. (2020). Ethics and Naeopharmacy: Value Sensitive Design of New Drugs. In *The Ethical Challenges of Emerging Medical Technologies* (pp. 139–153). Routledge.

Tiwari, J. N., Tiwari, R. N., & Kim, K. S. (2012). Zero-dimensional, one-dimensional, two-dimensional and three-dimensional nanostructured materials for advanced electrochemical energy devices. *Progress in Materials Science*, 57(4), 724–803. https://doi.org/10.1016/j.pmatsci.2011.08.003

Wagner, S., Gondikas, A., Neubauer, E., Hofmann, T., & von der Kammer, F. (2014). Spot the difference: Engineered and natural NPs in the environment: Release, behavior, and fate. *Angewandte Chemie International Edition*, 53(46), 12398–12419. https://doi.org/10.1002/anie.201405050

Wang, Y., & Xia, Y. (2004). Bottom-up and top-down approaches to the synthesis of monodispersed spherical colloids of low melting-point metals. *Nano Letters*, 4(10), 2047–2050. https://doi.org/10.1021/nl048689j

West, J. L., & Halas, N. J. (2003). Engineered NMs for biophotonics applications: Improving sensing, imaging, and therapeutics. *Annual Review of Biomedical Engineering*, 5(1), 285–292. https://doi.org/10.1146/annurev.bioeng.5.011303.120723

Wu, H., Liu, G., Wang, X., Zhang, J., Chen, Y., Shi, J., … & Yang, S. (2011). Solvothermal synthesis of cobalt ferrite NPs loaded on multiwalled carbon nanotubes for magnetic resonance imaging and drug delivery. *Acta Biomaterialia*, 7(9), 3496–3504. https://doi.org/10.1016/j.actbio.2011.05.031

Yan, J., Han, L., Gao, W., Xue, S., & Chen, M. (2015). Biochar supported nanoscale zerovalent iron composite used as persulfate activator for removing trichloroethylene. *Bioresource Technology*, 175, 269–274. https://doi.org/10.1016/j.biortech.2014.10.103

Yaya, A., Agyei-Tuffour, B., Dodoo-Arhin, D., Nyankson, E., Annan, E., Konadu, D. S., Sinayobye, E., Baryeh, E. A., & Ewels, C. P. Layered NMs: A review. *Global Journal of Engineering Design and Technology* 2 (2012): 32–41.

Zhang, S., Deen, S. U. N., & Yongqing, F. U. (2002). Superhard nanocomposite coatings. *Journal of Materials Sciences and Technology*, 18(06), 485.

3 Bioinspired Green Synthesis of NMs

Sakshi Gautam, Anjali Yadav, Nirmala Kumari Jangid, Manish Srivastava, and Navjeet Kaur

3.1 INTRODUCTION

Emerging as a multidisciplinary area, nanotechnology links the fields of material science, chemistry, biology, physics, pharmacy, engineering, and medicine. The distinctive morphological and physicochemical characteristics of NPs have drawn attention to them. They also have optical, magnetic, thermal, and mechanical properties. Broadly speaking, there are two primary types of NPs: inorganic, which primarily consists of metals and semiconductors (TiO_2), and organic, also referred to as carbon-based NMs. Numerous techniques have been put forth for the preparation of NPs, which can be summarised as top-down and bottom-up strategies. Machining, etching, grinding, and milling are examples of top-down techniques that break down bulk material. On the other hand, bottom-up techniques like sol-gel and chemical reduction processes require assembling tiny building blocks to produce larger structures, a process known as NP growth. Chemical, physical, and biological techniques can be employed to create NPs (Figure 3.1). To synthesize NPs, a precursor material must

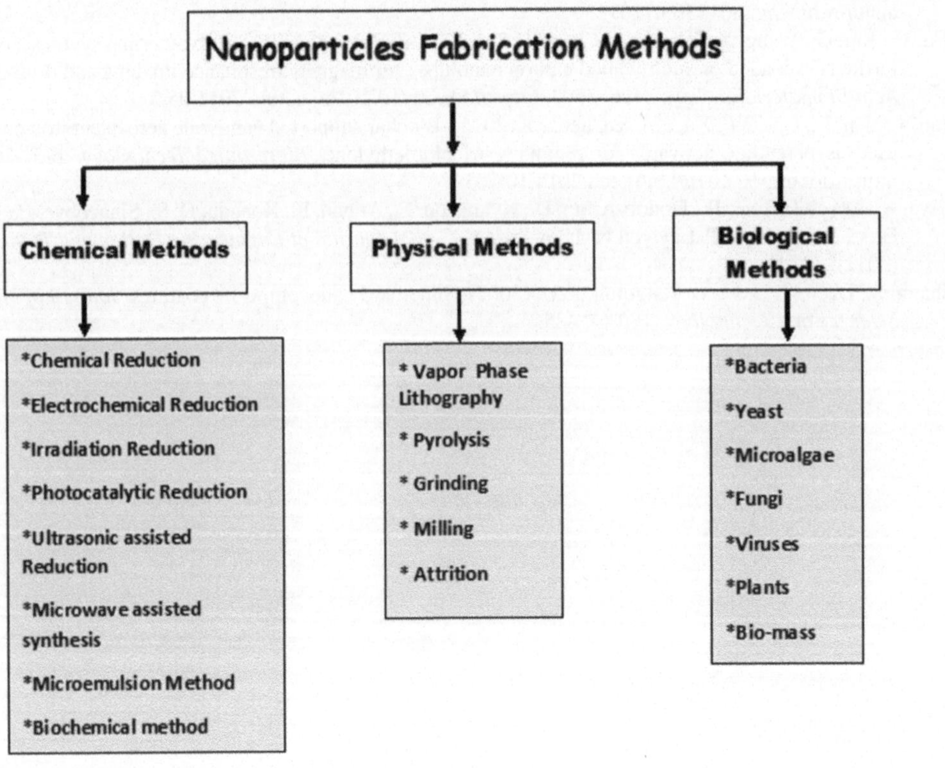

FIGURE 3.1 Different fabrication techniques for NPs.

DOI: 10.1201/9781003484189-3

interact with reducing agents such as hydrazine hydrate and sodium borohydride. Most of these reducing agents are hazardous chemicals that pose a risk to both human and the environment. The primary physical techniques utilized in the production of NPs are lithography, crushing, grinding, milling, pyrolysis, and physical vapour deposition. On the other hand, the chemosynthesis of NPs involves chemical reduction, sol gel synthesis, chemical vapour deposition, microwave-assisted reduction, photocatalytic reduction, and electrolysis synthesis (Koul et al., 2021).

This situation has necessitated study into biological methodologies, such as the development of simple, ecologically friendly, and green-reducing agents for NP manufacturing, resulting in the convergence of green chemistry, environmental remediation, and nanotechnology. Researchers have documented the utilization of microorganisms, including plant materials, including fruit, leaves, and seeds (Govindasamy et al., 2022, Awad et al., 2021), bacteria, algae, and fungi (Rajivgandhi et al., 2022; Khanna et al., 2019), bio- and agro-waste (Srivastava et al., 2021; Zamani et al., 2019), as more environmentally friendly methods of synthesising diverse NPs.

Comparing biosynthesized NPs to those made by physical and chemical means, it is frequently seen that the former exhibit greater stability, less toxicity, and less proliferation. This is the case as certain biological materials, especially plants, have bioactive components that have a stabilizing and capping action. Green synthesis is a method for producing NMs that is safe, economical, ecologically benign, and clean. Several active chemicals and precursors, including metal salt, determine the final shape and size of the NP. Green synthesis also offers advantages for NMs, such as natural reducing and stabilizing capabilities, antibacterial characteristics, and more (Sivaraj et al., 2020).

3.2 GREEN SYNTHESIS OF NMS

Green synthesis is a branch of green chemistry, which has emerged from the understanding that sustainable practices were necessary for chemical manufacture. An efficient and ecologically friendly method of synthesizing NPs ideally should make use of renewable energy sources, minimize waste discharge, and optimize energy usage, as per the principles of green chemistry. The goal of green chemistry is to create safer chemical products and processes by minimizing the use and production of unsafe substances. Thus, a number of fundamental ideas of green chemistry are applied in green synthesis, including waste reduction or elimination, the use of renewable feedstock, the use of non-toxic solvents, and the minimal usage of derivatives (Figure 3.2). Therefore, it can be said that in order to produce the active bio-components, the green synthesis method must employ

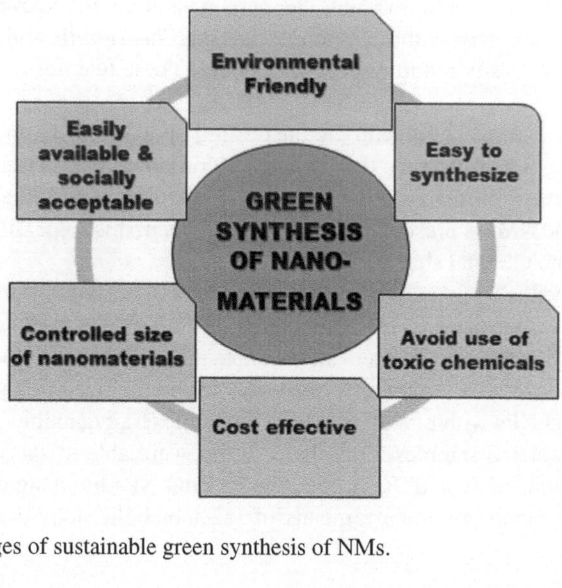

FIGURE 3.2 Advantages of sustainable green synthesis of NMs.

waste materials, microbes, plants, and have minimal heating requirements in addition to benign solvents (Bolade et al., 2019). Numerous studies have demonstrated the effectiveness of water, often referred to as the universal solvent, in reducing, capping, and stabilizing biomolecules that have been extracted from plants, particularly polyphenols. Vegetative wastes, as well as flowers, leaves, stems, roots, gums, seeds, fruits, and fruit peels have been found to contain these bioactive polyphenols (Yew et al., 2020). The main challenge in the chemical fabrication of NPs is agglomeration, which has been significantly reduced by green approach techniques. To ensure stability and evenly distributed particle size, the majority of produced NPs via a greener approach are typically stored in aqueous form at a certain pH (Bolade et al., 2019). Using chemical surface coatings and biocompatible reagents, which react less when in contact with fluids in the environment, can also lead to better stabilization. When NPs are retained in powder form, they must be kept under vacuum; otherwise, moisture from the environment causes agglomeration. The green synthesis of NPs utilizing various techniques is covered in the following subsections.

3.3 MICROBE BASED SYNTHESIS

Microbe-mediated green synthesis has been proposed as a substitute process for designing and developing NPs (Khalil et al., 2022; Ahmed et al., 2020). Microbes can be used safely and affordably to prepare metallic NPs such as silver, gold, nickel, copper, zinc, palladium, and titanium. Microbes can be used for both intracellular and extracellular NP synthesis (Koul et al., 2021). The culture filtrate is collected by centrifugation and added to an aqueous metallic salt solution for extracellular production. The colored shift of the mixed solution indicates the state of NP preparation. For example, changes in color from bright yellow to dark brown reflect the development of Ag NPs (Hossain et al., 2019; Ibrahim et al., 2020). After growing microorganisms under optimal circumstances and incubation with a metal ion solution, the biomass is thoroughly cleansed with sterile water before intracellular synthesis. As previously indicated, the shift in color signifies the formation of NPs. After that, NPs are collected using centrifugation, washing, and ultrasonication (Koul et al., 2021). Following is a sampling of metallic NPs that can be made using microbes.

3.3.1 Bacteria

Beneficial bacteria for the eco-friendly synthesis of NPs belong to a large family of unicellular organisms with cell walls but no organized nucleus or organelles. Many bacteria strains occur naturally in the body and provide little to no risk to the person handling them, even though some strains are dangerous. In addition, many of these strains—like *Bacillus subtilis* and *Escherichia coli*—are simple to culture and have easily modifiable genetic codes. These features make it possible for bacteria to synthesise NPs.

After the bacteria have grown aerobically to the desired optical density, the growth medium containing the cells and the precursor for NPs is mixed to produce NMs. High-speed centrifugation ($\geq$10,000 rpm) is performed on the media, following an incubation period and a discernible color shift. A suspension of the NMs is present in the supernatant from this spin. Different bacterial strains and precursors determine the final shape and size of the NP.

Actinomycetes are a valuable resource for the environmentally friendly synthesis of NMs (Hassan et al., 2019). They may synthesize NPs by extracellular or intracellular methods. Numerous secreted secondary metabolites give them an inherent surface area and size. Several documented studies used *actinomycetes* to synthesize NPs. *Streptomyces* species were employed in a recent study to synthesize exagon-shaped ZnO NPs with potential uses in medicine (Rajivgandhi et al., 2022). This study found that out of five isolated samples, only three strains were able to successfully generate ZnO NPs. These *actinomycetes* facilitated the production of ZnO NPs by using enzymes to reduce the particles from micro to nanosize and by means of the metabolic activities of the actinomycete isolates.

For the intracellular synthesis of AuNPs, various actinomycetes, such as Rhodococcus sp. and Thermomonospora sp., were employed. By employing Rhodococcus sp. for IC synthesis, uniformly sized AuNPs (5–15 nm) were produced. Several enzymes found in mycelium surfaces and cytoplasmic membranes considerably reduced the Au^{3+} ions. After Verticillium biomass was exposed to Ag^{+} ionic solutions, intracellular reductions and the subsequent synthesis of Ag-NPs were the ultimate outcomes (Khalil et al., 2022).

3.3.2 Yeast (Live and Extract)

Numerous investigations have demonstrated that yeasts can be effectively used to synthesise NPs. Yeasts are microorganisms that belong to the single-cell eukaryotic class and have developed from multicellular ancestors. Yeasts come in over 1500 varieties, several of which are widely used in the manufacturing of metallic NPs. Yeasts have large surfaces that allow them to collect and accumulate large amounts of hazardous metals from their environment.

Yeasts adapt to toxic metals through a variety of detoxification processes, including as bioprecipitation, extracellular sequestration, biosorption, and chelation. These techniques, through which yeast cells have evolved, are used to produce NPs and increase their durability, resulting in variations in the size and properties of the particles. Numerous investigations have been carried out to look at the use of yeast in the synthesis of metallic NPs.

Sivaraj et al. (2020) reported the successful synthesis of silver chloride NPs, utilizing proteins extracted from commercial yeast. To produce the NPs, the scientists first added precursor solutions to commercial yeast extracts and allowed them to incubate for an entire day. Following the incubation period, they collected the solution and sterilely filtered it to produce a solution that included only the NPs. The researchers also discovered that the reduction of silver chloride into NPs was solely driven by the primary amine of numerous proteins contained in the yeast cell wall. Furthermore, it was revealed that this specific NP has beneficial anti-mycobacterial capabilities (Sivaraj et al., 2020).

M. Rasouli has demonstrated that the yeast *Nematospora coryli* can absorb the oxyanions of selenium and convert them intracellularly into zero-valent selenium (Se^0) NPs, which have a spherical shape and range in size from 50 to 250 nm (Rasouli, 2019). The outcomes validated the biological functions of the Se^0 NPs produced by yeast. This approach was also economical and didn‘t involve the use of any dangerous materials.

Using the yeast strain *Candida utilis* NCIM 3469, a recent study showed how to extracellularly biofabricate 20–80 nm-sized AgNPs (Waghmare et al., (2015). Additionally, they discovered that these biofabricated AgNPs exhibited strong antibacterial qualities against disease-causing strains of *P. aeruginosa*, *E. coli*, and *S. aureus* (Bhupendra Koul et al., 2021). Similarly, a yeast strain of *Candida lusitaniae* obtained from a termite's stomach was utilized to biosynthesize Ag NPs of 2–10 nm in size with high antibacterial activity against *K. pneumonia* and *S. aureus* (Koul et al., 2021).

As a result, using yeast as a biotemplate to synthesize diverse NPs has emerged as an essential, economical, and environmentally beneficial technique with a wide range of potential uses.

3.3.3 Algae

"Phyconanotechnology," or the synthesis of NPs based on microalgae, has been a fast-developing topic in recent years. Many photoautotrophic microbes are members of the microalgae family, which includes proteins, pigments, and secondary metabolites (Khanna et al., 2019). Due to their ability to reduce a variety of metal ions, including silver, gold, cadmium, and others, microalgae can be used as an effective bionanofactory. It is possible to create metallic NPs from the dried biomass of both living and dead microalgae.

In a recent work, the microalgae strain *Phaeodactylum tricornutum* and its supernatant were used to biosynthesize titanium NPs with an average particle diameter of 49.7 nm (Caliskan et al., 2022).

The microalgae-synthesized NPs had potential medical uses, such as imaging techniques, drug delivery systems, immune system studies, cancer treatments, hyperthermia, and biosensors because of their biogenic, cytotoxic, antistatic, and antibacterial properties.

Gold nanoparticles have drawn a lot of interest due to their plasmonic qualities in the management of cancer. Cystoseira baccata, a brown macroalgae, was utilized by González-Ballesteros et al. for the green, efficient, one-step AuNP synthesis (Khalil et al., 2022).

3.3.4 Fungi

Another biological technique for synthesizing NPs is biofabrication, which involves the use of fungi. In comparison to bacterial cells, fungi can produce more NPs, due to their higher potential to withstand metal concentrations and their strong cell wall binding potential with metal ions (Khalil et al., 2022). Fungi have a few added characteristics that aid in the synthesis of metallic NPs when compared to other microbes like bacteria.

Fungi produce a higher production of NPs than bacteria because they secrete more enzymes and proteins per unit mass. Numerous studies indicate that certain fungi produce synthesized particles of smaller size because of their large intracellular metal absorption volumes (Pandiarajan, 2021). On the other hand, the biosynthesis of metallic NPs might be significantly influenced by the culture conditions.

Clarance et al. (2020) reported that the endophytic fungus *Fusarium solani* ATLOY-8 produced AuNPs in the 40–45 nm size range from the plant *Chonemorpha fragrans*. The biosynthesized AuNPs showed good antitumor efficaciousness and were stable against the HeLa and MCF-7 cell lines (Clarance et al., 2020). The fungus *Cladosporium sp*., which was isolated from the medicinal plant *Commiphora wightii*, was the source of the spherical AuNPs with a size of 5–10 nm that Munawer et al. (2020) reported producing through extracellular green synthesis. The results demonstrated remarkable antiproliferative activity against the MCF-7 breast cancer cell line (Munawer et al., 2020).

The extracellular manufacturing of silver NPs produced by Trichoderma sp. fungus—which were isolated from *Bertholletia excelsa* plant seeds—was reported by Ramos and colleagues in 2020. They concluded that the biofabricated AgNPs had better antibacterial efficacy against bacteria that are Gram-negative (Ramos et al., 2020). According to reports, mouse mayo blast carcinoma C_2 C_{12} cells cultured for 24, 48, and 72 hours were treated with AuNPs biosynthesized from the fungus (*Penicillium brevicompactum*) cell filtrate (Khalil et al., 2022).

3.3.5 Virus

Viruses are coated with capsid proteins with nanoscale structures, which give metallic ions the right conditions to interact with the virus machinery. Genetic engineering can be used to modify the capsid protein to create NMs for targeted drug delivery and cancer treatment (Gahlawat and Choudhury, 2019). There are numerous cases of viruses being effectively used to produce NPs through biosynthesis.

Plant viruses are not dangerous to humans or animals, are easily propagated, and are stable. As a result, plant viruses can be employed to produce NPs that can be used by humans as well as animals.

In contrast, Czapar and colleagues (2016) produced hollow nanotubes with a polyanionic inner surface using the tobacco mosaic virus (TMV) with the goal of using them for the targeted delivery of the anticancer drug phenanthriplatin. This work effectively demonstrated the one-step process for phenanthriplatin loading and distribution utilizing TMV nanotubes using a mouse model (Koul et al., 2021). In a related investigation, Le and coworkers in 2017 used potato virus X-derived nanocarriers. For delivering doxorubicin to specific cancer patients, they created elongated, lamentous nanocarriers. When compared to spherical NPs, these virus-mediated nanocarriers demonstrated a higher tumor penetrating power (Koul et al., 2021). In a different study, Chen et al. (2018) reported the hepatitis E virus NP (HEVNP) generation using the hepatitis E viral capsule. In this work, HEVNP was combined with the cell-specific ligand LXY30 and a labeled near-infrared (NIR) fluorescent dye. Chen and colleagues (2018) have reported that these biosynthesized HEVNPs have the potential to serve as versatile delivery vehicles for tissue targeting, tumor imaging, and drug delivery (Koul et al., 2021).

Thangavelu et al. (2020) used the plant pathogenic squash leaf curl China virus to synthesize Au-Ag composite semiconductor NPs in their recent work. The virus was used to create the Au-Ag NMs, using it as a nanobiotemplate (32 nm). Using the "lab on a chip" technology and Keithley's picoammeter, the electrical conductivity of the Au-Ag NM was examined. The hybrid NM Au-Ag was shown to have good biocompatibility and outstanding semiconductive qualities, making it suitable for use in biomedical applications. The aforementioned reports demonstrate the successful use of viruses as a template for NP synthesis; however, there are still some drawbacks to virus-mediated NP synthesis, such as its large-scale application, requirement for a host for protein expression, biocompatibility, and lack of well-defined NP synthesis methods (Thangavelu et al., 2020).

3.4 PLANT MEDIATED SYNTHESIS

Plant-mediated synthesis is the most efficient green synthesis method for NPs because of its high yield. Since NPs are highly stable when synthesised in various plant species, there is a large yield of NPs produced. The stability of NMs is caused by the many metabolites and biochemical species found in plants (Gahlawat and Choudhury, 2019). Consequently, plant-mediated synthesis of NPs is an affordable and environmentally benign process. NPs produced by plants can be synthesised by three different processes: extracellular, intracellular, and phytochemical (Figure 3.3). More NPs are produced by the extracellular approach than by the intracellular method, which is the most efficient.

This could be due to the plant extract's high concentration of phytochemicals, which are essential to the synthesis process. The cellular enzymes found in various plant tissues are responsible for the crucial role that synthesis plays in the intracellular approach. The plant's cell wall must subsequently be broken by these enzymes, which require a series of complex steps to attract the synthesised NPs.

3.4.1 Leaf Extract

Aqueous leaf extracts of *Lantana camara* were recently used to sustainably generate yttrium oxide (Y_2O_3) NPs ranging in size from 20 to 45 nm (Govindasamy et al., 2022). When subjected to a 250 W high-pressure Xenon lamp, the synthesized Y_2O_3 NPs showed photocatalytic activity in the breakdown of Rhodamine B. Furthermore, the produced Y_2O_3 NPs have shown antibacterial efficacy

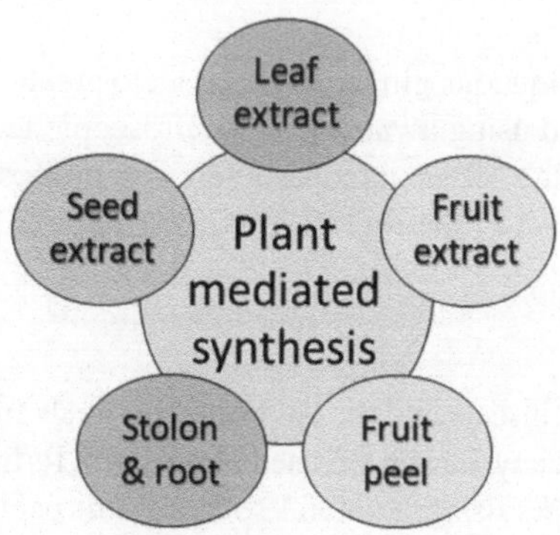

FIGURE 3.3 Different natural resources used for the green synthesis of NMs.

against both Gram-positive and Gram-negative bacteria, including *Bacillus subtilis* (MTCC 5981) and *E. coli* (MTCC 732).

These NPs can also be used to treat cervical cancer in humans because they have cytotoxic effects on HeLa cell lines. The findings demonstrated the high surface reactivity of these Y_2O_3 NPs, which made them promising candidates for targeted drug delivery in cancer therapy. As a result, these green NPs were well-suited for usage in a variety of medicinal applications.

Reetika Singh et al. (2021) synthesized silver nanoparticles by using an extract from the leaves of *Carissa carandas L.* as a capping and reducing agent (Singh et al., 2021).

3.4.2 Fruit Extract

Indian gooseberry (*Phyllanthus emblica*) fruit extract was used by Renuka and colleagues (Renuka et al., 2020) to synthesize silver nanoparticles. The synthetic nanoparticles had good size and shape.

In 2020, Ying Chun Liu and colleagues employed a plant-derived reducing agent, Amomum longiligulare (*A. longiligulare*), to manufacture zinc oxide nanoparticles by the co-precipitation method. This methodology was considered green biosynthetic (Liu et al., 2020).

Van-Dat Doan et al.'s green method for creating colloidal gold nanoparticles (AuNPs) employs Litsea cubeba (LC) fruit extract as a reducing and stabilizing agent at the same time (Doan et al., 2020).

Cornelian cherry was used by Elham Rostamizadeh and coworkers (Rostamizadeh et al., 2020) to manufacture Fe_2O_3 nanoparticles.

Renitta and coworkers (Renitta et al., 2022) biosynthesized TiO_2 nanoparticles using plant extract from *Cynodon dactylon*. Antibacterial activity against human infections will be evaluated for both the plant extract and the herb *Cynodon dactylon*'s titanium-based nanoparticles.

3.4.3 Seed Extract

Trigonella foenum-graecum seed (fenugreek) extract was used by Manal A. Awad et al. (2021) to synthesize silver nanoparticles with ultra stable characteristics and small-scale sizes Awad et al. (2021).

Titanium dioxide nanoparticles were produced by Sneha Susan Mathew et al. (2021) using seed extract of *Cuminum cyminum* (Mathew et al., 2021).

Shams Tabrez et al. (2022) produce zinc oxide nanoparticles using pumpkin seed extract and evaluate their anticancer potential using various molecular biology tests in human breast cancer (MDA-MB-231) cell lines (Tabrez et al., 2022).

The synthesised copper nanoparticles by Farah M. Ibrahim et al. (2023) using avocado seed extract may offer an inexpensive, simple, and safe way to adsorb these heavy metals (Ibrahim et al., 2023).

3.4.4 Stolon and Root

Morteza Yadi et al. (2022) used aqueous ginger root extract to create Au and Ag-based nanoparticles (NPs), which were then analyzed using a variety of spectroscopic techniques (Yadi et al., 2022).

A green method for producing silver nanoparticles was proposed by Muhammad Riaz et al. (2020) utilizing an aqueous extract of the root of *Angilica glauca Edgew*. (Riaz et al., 2020).

3.4.5 Fruit Peel

Fruit peels protect the flesh from insects and the environment. Peels from fruits and vegetables, which are generally seen as waste in society, have been used as a natural fertilizer. Nonetheless, such agricultural waste was used to create NPs using a natural source. Fruit peel extract has been the subject of extensive investigation when it comes to NP production. Using citrus fruit peels, which are typically discarded as waste or a byproduct of the juice production process, has been shown to effectively reduce waste and make efficient use of available resources. The peels were centrifuged and filtered via a membrane after being extracted with ultrapure water at high temperatures for 80 minutes.

Ayse Baran and colleagues conducted an environmentally friendly synthesis of silver nanoparticles from *Ananas comosus* fruit peels (Baran et al., 2021). Salem S. Salem (2022) used leftover orange peel to biosynthesize selenium nanoparticles (Se-NPs) using an environmentally benign and sustainable process, with the goal of using these nanoparticles as antibacterial and antibiofilm agents (Salem et al., 2022). In order to create biocompatible gold nanoparticles, Chiravoot Pechyen et al. (2021) used aqueous fruit peel extract from *Ananas comosus* (pineapple) and *Passiflora edulis* (passion fruit) (Pechyen et al., 2021).

A straightforward technique for producing gold (Au NPs) and silver nanoparticles (AgNPs) using an aqueous extract of leftover *Passiflora edulis* peel (PEP) as stabilizing and reducing agents was presented by T. My-Thao Nguyen and his colleagues in 2021 (Nguyen et al., 2021).

3.5 SYNTHESIS OF NPS USING BIOMASS AND ORGANIC WASTES

One of the main issues facing modern civilization with regard to disposal is biomass, which comprises wastes from a variety of sources, including municipal trash, forest wastes and litter, microbiological biomass, agricultural wastes, wastes from poultry, wastes from breweries, wastes from food businesses, and so on. Bioactive compounds (including glycoproteins, carbohydrates, polyphenols, and polysaccharides, etc.) that are abundant in this biomass may be used as capping and reducing agents. When compared to alternative synthesis pathways, using biomass as a raw material for NM production has a number of advantages. Serving as a raw material for the synthesis of NMs, waste biomass lowers associated costs, increasing the process's profitability. Consequently, it aids in solving the issue of disposing of garbage as it converts it into money.

Numerous investigations have focused on the production of iron-based NMs from waste biomass (Srivastava et al., 2021). α-FeOOH NPs with a particle size of 12.8 nm were synthesized by Ghanbariasad et al. (2019) using secretory chemicals obtained from *Chlorella vulgaris* cells in an ambient setting (Ghanbariasad et al., 2019). Therefore, it is vital to further research the potential commercial applications of lignocellulosic biomass and organic wastes from the juice and food industries as reagents for the preparation of NMs (Zamani et al., 2019).

3.6 CONCLUSION

NPs are relevant in a variety of fields due to their unique properties. Research on NMs is still ongoing, and its area and applications are growing. Traditional NM synthesis technologies affect the environment because they use harmful reagents and require a lot of energy. The use of these technologies is both irresponsible and naïve, given their grave effects on the environment during the current rise in climate change. Green synthesis techniques are just as successful as traditional synthesis techniques, with little to no impact on the environment or the individuals who work on them. Green methods of synthesizing NMs, like using waste biomass, plant extracts, and microbes, are becoming more and more popular. It is observed that various bioactive reagents in the extraction solution serve as capping and reducing agents concurrently, minimizing the use of hazardous chemicals needed for the chemical synthesis. Furthermore, the constant synthesis of bioactive compounds during photosynthesis, as well as the environmentally benign, non-toxic, easy, and free availability of plant material, may reduce the overall cost of NM synthesis, boosting their value and economy. Conversely, organic wastes produced by many sectors can be used as starting points for the production of NMs. We have discussed more environmentally friendly methods of creating NMs in this chapter, including using biomass, microbes, and green plants. In conclusion, this chapter has described different green nanotechnology processes. Future research and development of superior NMs to treat contaminated solutions and wastewater in practical applications with the lowest possible risk of a toxicological effect on the environment could benefit from these methods as a resource.

REFERENCES

Ahmed, T., Shahid, M., Noman, M., Niazi, M. B. K., Mahmood, F., Manzoor, I., Zhang, Y. et al. (2020). "Silver NPs synthesized by using Bacillus cereus SZT1 ameliorated the damage of bacterial leaf blight pathogen in rice." *Pathogens* 9(3), 160.

Awad, M. A., Hendi, A. A., Ortashi, K. M., Alzahrani, B., Soliman, D., Alanazi, A., … and Alomar, T. S. (2021). Biogenic synthesis of silver nanoparticles using Trigonella foenum-graecum seed extract: Characterization, photocatalytic and antibacterial activities. *Sensors and Actuators A: Physical*, 323, 112670.

Baran, A., Keskin, C., Baran, M. F., Huseynova, I., Khalilov, R., Eftekhari, A., … and Kavak, D. E. (2021). Ecofriendly synthesis of silver nanoparticles using ananas comosus fruit peels: Anticancer and antimicrobial activities. *Bioinorganic Chemistry and Applications*, 2021(1), 2058149.

Bolade, O. P., Williams, A. B., and Benson, N. U. (2019). "Green synthesis of iron-based NMs for environmental remediation: A review." *Environmental Nanotechnology, Monitoring & Management*, 13, 100279.

Caliskan, G., Mutaf, T., Agba, H. C., and Elibol, M. (2022). "Green synthesis and characterization of titanium NPs using microalga, Phaeodactylum tricornutum." *Geomicrobiology Journal*, 39(1), 83–96.

Chen, C. C., Marie, S., Mo, B., Holland, C. R. (2018). "Surface functionalization of hepatitis E virus nanoparticles using chemical conjugation methods". *Journal of Visualized Experiments*, 11 (135): 57020.

Clarance, P., Luvankar, B., Sales, J., Khusro, A., Paul Agastian, J.-C., Tack, M. M., Khulaifi, A. L. et al. (2020). "Green synthesis and characterization of gold NPs using endophytic fungi Fusarium solani and its in-vitro anticancer and biomedical applications." *Saudi Journal of Biological Sciences* 27 (2), 706–712.

Doan, V. D., Thieu, A. T., Nguyen, T. D., Nguyen, V. C., Cao, X. T., Nguyen, T. L. H., and Le, V. T. (2020). Biosynthesis of gold nanoparticles using Litsea cubeba fruit extract for catalytic reduction of 4-nitrophenol. *Journal of Nanomaterials*, 2020(1), 4548790.

Gahlawat, G., and Choudhury, A. R. (2019). "A review on the biosynthesis of metal and metal salt NPs by microbes." *RSC Advances* 9(23), 12944–12967.

Ghanbariasad, A., Taghizadeh, S.-M., Show, P. L., Nomanbhay, S., Berenjian, A., Ghasemi, Y., and Ebrahiminezhad, A. (2019). "Controlled synthesis of iron oxyhydroxide (FeOOH) NPs using secretory compounds from Chlorella vulgaris microalgae." *Bioengineered* 10(1), 390–396.

Govindasamy, R., Govindarasu, M., Alharthi, S. S., Mani, P., Bernaurdshaw, N., Gomathi, T., Ansari, M. A. et al. (2022). "Sustainable green synthesis of yttrium oxide (Y2O3) NPs using Lantana camara leaf extracts: Physicochemical characterization, photocatalytic degradation, antibacterial, and anticancer potency." *NMs* 12(14), 2393.

Hassan, S. El-D., Fouda, A., Radwan, A. A., Salem, S. S., Barghoth, M. G., Awad, M. A., Abdo, A. M., and El-Gamal, M. S. (2019). "Endophytic actinomycetes Streptomyces spp mediated biosynthesis of copper oxide NPs as a promising tool for biotechnological applications." *JBIC Journal of Biological Inorganic Chemistry*, 24, 377–393.

Hossain, A., Hong, X., Ibrahim, E., Li, B., Sun, G., Meng, Y., Wang, Y., and An, Q. (2019). "Green synthesis of silver NPs with culture supernatant of a bacterium Pseudomonas rhodesiae and their antibacterial activity against soft rot pathogen Dickeya dadantii." *Molecules*, 24(12), 2303.

Ibrahim, F. M., Najeeb, D. A., and ThamerSadeq, H. (2023). Green preparation of Cu nanoparticles of the avocado seed extract as an adsorbent surface. *Materials Science for Energy Technologies*, *6*, 130–136.

Ibrahim, E., Zhang, M., Zhang, Y., Hossain, A., Qiu, W., Chen, Y., Wang, Y., Wenge, W., Sun, G., and Li, B. (2020). "Green-synthesization of silver NPs using endophytic bacteria isolated from garlic and its antifungal activity against wheat Fusarium head blight pathogen Fusarium graminearum." *NMs*, 10(2), 219.

Khalil, A. T., Ovais, M., Iqbal, J., Ali, A., Ayaz, M., Abbas, M., … & Devkota, H. P. (2022, November). Microbes-mediated synthesis strategies of metal nanoparticles and their potential role in cancer therapeutics. In *Seminars in Cancer Biology* (Vol. 86, pp. 693–705). Academic Press.

Khanna, P., Kaur, A., and Goyal, D. (2019). "Algae-based metallic NPs: Synthesis, characterization and applications." *Journal of Microbiological Methods*, 163, 105656.

Koul, B., Poonia, A. K., Yadav, D., and Jin, J. O. (2021). "Microbe-mediated biosynthesis of NPs: Applications and future prospects." *Biomolecules*, 11(6), 886.

Liu, Y. C., Li, J., Ahn, J., Pu, J., Rupa, E. J., Huo, Y., and Yang, D. C. (2020). Biosynthesis of zinc oxide nanoparticles by one-pot green synthesis using fruit extract of Amomum longiligulare and its activity as a photocatalyst. *Optik*, 218, 165245.

Mathew, S. S., Sunny, N. E., and Shanmugam, V. (2021). Green synthesis of anatase titanium dioxide nanoparticles using Cuminum cyminum seed extract; effect on Mung bean (Vigna radiata) seed germination. *Inorganic Chemistry Communications*, 126, 108485.

Munawer, U., Raghavendra, V. B., Ningaraju, S., Krishna, K. L., Ghosh, A. R., Melappa, G., and Pugazhendhi, A. (2020). "Biofabrication of gold NPs mediated by the endophytic Cladosporium species: Photodegradation, in vitro anticancer activity and in vivo antitumor studies." *International Journal of Pharmaceutics*, 588, 119729.

Nguyen, T. M. T., Nguyen, T. A. T., Tuong-Van Pham, N., Ly, Q. V., Tran, T. T. Q., Thach, T. D., ... & Nguyen, T. D. (2021). Biosynthesis of metallic nanoparticles from waste Passiflora edulis peels for their antibacterial effect and catalytic activity. *Arabian Journal of Chemistry*, 14(4), 103096.

Pandiarajan, J. (2021). "Diverse manifolds of biogenic NPs in synthesis, characterization, and applications." In: Saglam, N., Korkusuz, F., Prasad, R. (eds.) *Nanotechnology Applications in Health and Environmental Sciences*, 1–28. Nanotechnology in the Life Sciences. Springer, Cham. https://doi.org/10.1007/978-3-030-64410-9_1

Pechyen, C., Ponsanti, K., Tangnorawich, B., and Ngernyuang, N. (2021). Waste fruit peel–Mediated green synthesis of biocompatible gold nanoparticles. *Journal of Materials Research and Technology*, 14, 2982–2991.

Rajivgandhi, G., Gnanamangai, B. M., Prabha, T. H., Poornima, S., Maruthupandy, M., Alharbi, N. S., Kadaikunnan, S., & Li, W.-J. (2022). "Biosynthesized zinc oxide NPs (ZnO NPs) using actinomycetes enhance the anti-bacterial efficacy against K.Pneumoniae." *Journal of King Saud University-Science*, 34(1), 101731.

Ramos, M. M., Morais, E. dos S., Sena, I. da S., Lima, A. L., de Oliveira, F. R., de Freitas, C. M., Fernandes, C. P., de Carvalho, J. C. T., and Ferreira, I. M. (2020). "Silver NP from whole cells of the fungi Trichoderma spp. isolated from Brazilian Amazon." *Biotechnology Letters*, 42, 833–843.

Rasouli, M. (2019). "Biosynthesis of selenium NPs using yeast Nematospora coryli and examination of their anti-candida and anti-oxidant activities." *IET Nanobiotechnology*, 13(2), 214–218.

Renitta, R. E., Jebaseeli, T. J., Dhanaraj, A., and Paul, S. (2022). Biological synthesis and characterization of titanium dioxide nanoparticle from Cynodon dactylon. *Journal of Achievements in Materials and Manufacturing Engineering*, 113(1): 31–41. https://doi.org/10.5604/01.3001.0016.0952

Renuka, R., Devi, K. R., Sivakami, M., Thilagavathi, T., Uthrakumar, R., and Kaviyarasu, K. (2020). Biosynthesis of silver nanoparticles using Phyllanthus emblica fruit extract for antimicrobial application. *Biocatalysis and Agricultural Biotechnology*, 24, 101567.

Riaz, M., Altaf, M., Ayaz, M., Sherkheli, M. A., and Islam, A. (2020). Antibacterial and antioxidant potential of biosynthesized silver nanoparticles using aqueous root extract of Angilica glauca. *Inorganic and Nano-Metal Chemistry*, 51(10), 1379–1385.

Rostamizadeh, E., Iranbakhsh, A., Majd, A., Arbabian, S., and Mehregan, I. (2020). Green synthesis of Fe 2 O 3 nanoparticles using fruit extract of Cornus mas L. and its growth-promoting roles in Barley. *Journal of Nanostructure in Chemistry*, 10, 125–130.

Salem, S. S., Badawy, M. S. E., Al-Askar, A. A., Arishi, A. A., Elkady, F. M., and Hashem, A. H. (2022). Green biosynthesis of selenium nanoparticles using orange peel waste: Characterization, antibacterial and antibiofilm activities against multidrug-resistant bacteria. *Life*, 12(6), 893.

Singh, R., Hano, C., Nath, G., and Sharma, B. (2021). Green biosynthesis of silver nanoparticles using leaf extract of Carissa carandas L. and their antioxidant and antimicrobial activity against human pathogenic bacteria. *Biomolecules*, 11(2), 299.

Sivaraj, A., Kumar, V., Sunder, R., Parthasarathy, K., and Kasivelu, G. (2020). "Commercial yeast extracts mediated green synthesis of silver chloride NPs and their anti-mycobacterial activity." *Journal of Cluster Science* 31, 287–291.

Srivastava, M., Srivastava, N., Mohd, S., Mishra, P. K., Amir, S., Gupta, V. K., and Malhotra, B. D. (2021). "Bioinspired synthesis of iron-based NMs for application in biofuels production: A new in-sight." *Renewable and Sustainable Energy Reviews*, 147, 111206.

Tabrez, S., Khan, A. U., Hoque, M., Suhail, M., Khan, M. I., and Zughaibi, T. A. (2022). Biosynthesis of ZnO NPs from pumpkin seeds' extract and elucidation of its anticancer potential against breast cancer. *Nanotechnology Reviews*, 11(1), 2714–2725.

Thangavelu, R. M., Ganapathy, R., Ramasamy, P., and Krishnan, K. (2020). "Fabrication of virus metal hybrid NMs: An ideal reference for bio semiconductor." *Arabian Journal of Chemistry* 13(1), 2750–2765.

Waghmare, S. R., Mulla, M. N., Marathe, S. N., Kailas, D. S. (2015). "Ecofriendly production of silver nanoparticles using Candida utilis and its mechanistic action against pathogenic microorganisms". 3 Biotech 5, 33–38. https://doi.org/10.1007/s13205-014-0196-y

Yadi, M., Azizi, M., Dianat-Moghadam, H., Akbarzadeh, A., Abyadeh, M., and Milani, M. (2022). Antibacterial activity of green gold and silver nanoparticles using ginger root extract. *Bioprocess and Biosystems Engineering*, 45(12), 1905–1917.

Yew, Y. P., Shameli, K., Miyake, M., Khairudin, N. B. B. A., Mohamad, S. E. B., Naiki, T., and Lee, K. X. (2020). "Green biosynthesis of superparamagnetic magnetite Fe3O4 NPs and biomedical applications in targeted anticancer drug delivery system: A review." *Arabian Journal of Chemistry*, 13(1), 2287–2308.

Zamani, A., Marjani, A. P., and Mousavi, Z. (2019). "Agricultural waste biomass-assisted nanostructures: Synthesis and application." *Green Processing and Synthesis*, 8(1), 421–429.

4 Sustainable Nanomaterials
Synthesis and Characterization

Amruta Pattnaik, Asha Rani Patra, Anil Kumar Poonia, and Prabir Ghosh

4.1 INTRODUCTION

The rapid progress of nanotechnological synthesis techniques has opened in a new age of discovery, unveiling a plethora of functional NMs with adjustable properties derived from non-renewable resources. These nanostructures boast better physical and chemical attributes than their bulk counterparts, offering promising alternatives to crucial societal concerns spanning health, renewable energy, information technology, and environmental sustainability as well. However, the future viability of these technologies faces scrutiny because of their dependence on non-renewable materials and energy-intensive fabricating processes. Other than that, there has been little emphasis paid to examining the potential effects of non-sustainable NMs on human health, climate change, and environmental deterioration. To achieve United Nation's Sustainable Development Goals, collaborative research initiatives are imperative to create sustainable NMs made with renewable resources (Y. Zhang et al., 2023). This entails maximizing performance, manufacturing processes, and production costs while adhering to sustainable principles. Integrating sustainable resources into nanotechnology is a critical step towards creating a sustainable society in the twenty-first century. Sustainable NM design attempts to save resources for current and future generations by using natural resources and low-energy production technologies. Furthermore, adhering to sustainable practices aims to reduce waste creation, while pushing for a circular economy that promotes long-term sustainable growth.

Establishing design standards that encourage alternatives to traditional, non-renewable production technologies is critical in paving the way for sustainable NMs. Optimization efforts should include renewable material selection, mechanical and thermal properties, biocompatibility, hazard testing, and environmental effect evaluation. By adopting these concepts and novel methodologies, the quest for sustainable NMs promises to usher in a new era of environmentally aware innovation and technological growth.

4.2 SYNTHESIS TECHNIQUES

Nanoparticles (NPs) are synthesized using a variety of processes, which are divided into bottom-up and top-down approaches. Figure 4.1 shows a simplified illustration of the procedure. The bottom-up or constructive technique involves the assembly of material from atoms to clusters to NPs. The most prevalent bottom-up processes for NP creation are sol-gel, spinning, chemical vapor deposition (CVD), pyrolysis, and biosynthesis. The top-down or destructive technique involves reducing a bulk substance to nanometric-scale particles. Mechanical milling, nanolithography, laser ablation, sputtering, and thermal breakdown are among the most common NP production processes.

4.2.1 BALL MILLING

Mechanical attrition produces nanostructures by decomposing coarser grained structures by plastic deformation in contrast to cluster creation. A notable example includes the synthesis of elemental

DOI: 10.1201/9781003484189-4

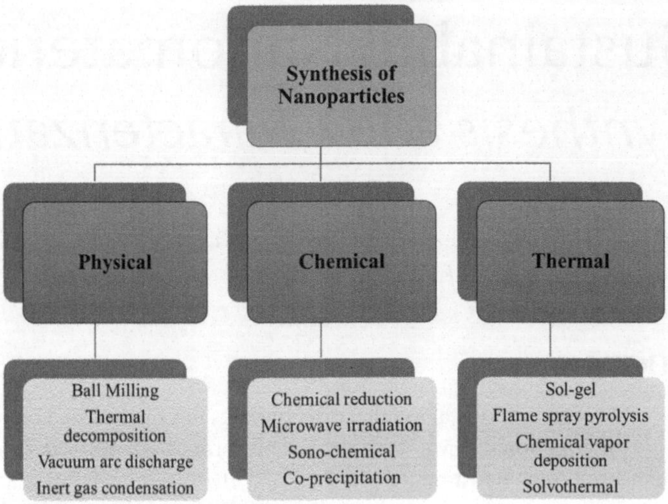

FIGURE 4.1 Classification of synthesis techniques.

powders of aluminium and beta-silicon carbide utilizing a high-energy ball mill. More recently, comparable processes have been used to fabricate ceramic/ceramic nanocomposite materials. The ball milling and rod milling procedures, which are part of the mechanical alloying approach, have received a lot of attention for their capacity to produce sophisticated materials. Notably, mechanical alloying is a unique method that may be carried out at room temperature. It includes both high-energy mills, such as centrifugal and vibratory kinds, and low-energy tumble mills (Arshad and Naseer, 2017).

While high-energy ball milling is a well-established technique, worries about contamination, particularly with iron, have deemed it "dirty." However, innovations like adding tungsten carbide components and using inert atmospheres or high vacuum techniques have reduced impurity levels to acceptable levels. Despite these advancements, difficulties, such as poor surface area, extremely polydisperse size distribution and partly amorphous powder, remain. Furthermore, the resultant powders are highly reactive with oxygen, hydrogen, and nitrogen. Nonetheless, mechanical alloying allows for the creation of alloys that are not possible to produce using traditional methods. For example, mechanical alloying using ball milling methods makes it possible to generate an aluminium-tantalum alloy, which was previously impossible due to the significant difference in melting temperatures (Mr. Sharad Kamble et al., 2023). This demonstrates the revolutionary potential of mechanical alloying in pushing the limits of material synthesis beyond the confines of previous approaches.

4.2.2 Thermal Decomposition

Thermal decomposition is an endothermic process in which chemical breakdown happens when heat is applied, resulting in the breaking of chemical bonds within the substance (Ealias and Saravanakumar, 2017). The decomposition temperature is the precise temperature at which an element undergoes chemical breakdown. In the context of NP production, metal breakdown at specified temperatures triggers a chemical process that results in the generation of NPs. Heat acts as a catalyst for this chemical breakdown, breaking chemical bonds within the molecule and accelerating the creation of NPs by metal decomposition at a certain temperature (Yadav et al. 2023).

4.2.3 Inert Gas Condensation

Gas condensation was one of the pioneering procedures used to create nanocrystalline metals and alloys. This process involves vaporizing a metallic or inorganic material using thermal evaporation sources such as Joule-heated refractory crucibles or electron beam evaporation devices in an environment that generally ranges from 1 to 50 mbar (Tissue and Yuan, 2003). Gas phase collisions at high residual gas pressures result in the creation of ultrafine particles, generally approximately 100 nm, as evaporated atoms contact with remaining gas molecules. Vaporization sources range from resistive heating to high-energy electron beams, low-energy electron beams, and inductive heating. Clusters develop near the source due to homogenous nucleation in the gas phase and increase as atoms are incorporated. The configuration usually consists of an ultra-high vacuum (UHV) system with an evaporation source, a cluster collecting device, such as a liquid nitrogen-filled cold finger scraper assembly, and a compaction device. During heating, atoms condense in the supersaturation zone near the heating device, and the NPs are collected by a scraper in the shape of a metal plate. Refractory metal crucibles composed of tungsten (W), tantalum (Ta), or molybdenum (Mo) are typically used for evaporation; however, electron beam evaporation may be required if metals react with the crucibles.

Despite its early use, gas condensation has limitations such as source-precursor incompatibility, restricted temperature ranges, and varying evaporation rates inside alloys. Alternative sources, such as sputtering or laser evaporation, have evolved throughout time and offer benefits in specific applications. Sputtering, a non-thermal phenomenon, is the physical ejection of surface atoms by momentum transfer from intense bombardment species of atomic or molecular size. Magnetron sputtering, in particular, restricts the majority of the plasma around the target region, providing benefits over diode and triode sputtering. Various energy sources, including sputtering, electron beam heating, and plasma techniques, have been used effectively to create clusters of ultrafine particles in low-pressure conditions. Sputtering, for example, has been used to create a wide range of clusters, including those made of silver (Ag), iron (Fe), and silicon (F. Hasany et al., 2013).

4.2.4 Microwave Irradiation

Microwave (MW) technology has developed as a powerful alternative energy source for rapid chemical transformations, notably in NM creation. Traditional thermal techniques use the conduction of blackbody radiation to drive reactions, which results in strong heat gradients and nonuniform reaction conditions. In contrast, MW heating delivers quick and consistent heating, which overcomes the limitations of traditional methods. This synthesis has various benefits, including improved reaction kinetics, faster starting heating, and higher reaction rates. This leads to cleaner reaction products, faster consumption of starting materials, and greater yields. Furthermore, MW approaches are relevant in a variety of settings, including enzymatic and biological systems. MW-assisted reduction of noble metal salts with aqueous sugar solutions allows for bulk and shape-controlled synthesis of noble nanostructures, including prisms, cubes, and hexagons. Adjusting the sugar content allows for easily regulating the size of the resultant NPs. Furthermore, MW technology enables the crosslinking of poly (vinyl alcohol) (PVA) with metallic and bimetallic systems, as well as carbon nanotubes. This method extends to the creation of biodegradable composite films with noble nanometals, allowing for metal ornamentation and alignment of carbon nanotubes (Varma, 2012).

Also, MW-assisted synthesis allows for shape-controlled bulk synthesis of a variety of nanorods, including Ag and Fe nanorods in polyethylene glycol (Nadagouda and Varma, 2008). Plus, optimized pathways for the creation of tantalum oxide NPs have been discovered by employing ethyl glycol-mediated techniques, resulting in a cleaner and more effective synthesis approach. Ultimately, MW-assisted synthesis serves as a diverse and successful method for the quick and regulated creation of NMs of a variety of applications.

4.2.5 Sonochemical Synthesis

Sonochemical synthesis is a sophisticated approach that provides an ecologically benign pathway for the manufacture of novel NMs and composites, with significant benefits in customising specific features. This method uses ultrasonic radiation, with frequencies ranging from 20 kHz to 10 MHz, to trigger or change whole chemical processes (Alam et al., 2024). The phenomenon of cavitation, caused by acoustic waves in a liquid media, is critical in producing the chemical effects of ultrasonication. During cavitation, gas bubbles compress and rarefy, causing rapid temperature and pressure fluctuations within the bubble. This dynamic environment changes the structures of NPs, allowing for exact control over their properties. It has several significant advantages. For instance, it speeds reactions by sonic cavitation, allowing for quick product creation, while also encouraging green chemistry principles by functioning under moderate, ecologically favorable settings. This technology is environmentally friendly since it reduces energy usage and eliminates the need for harmful chemicals. Furthermore, sonochemistry provides the freedom to modify NP size by adjusting the ultrasonic frequency and intensity.

NMs are widely used in environmental, biological, electrical, and sensor applications, demonstrating their flexibility. Metal oxide nanostructures stand out among the many types of NMs due to their remarkable physical and chemical characteristics, making them ideal for the construction of efficient electronics (Pattnaik et al., 2023). However, there is an increasing need to synthesize metal oxide NMs using environmentally friendly precursors and techniques, with a wide range of sizes and morphologies, to ensure optimal utilization in a variety of applications, including supercapacitors (SCs). In the context of SC electrodes, this research digs into the phases of mono, binary, and ternary metal oxides, analyzing how the synthesis affects their performance.

4.2.6 Co-Precipitation

Co-precipitation approaches are a popular method for NP (NP) production, using reactions that allow for concurrent nucleation, growth, and/or agglomeration processes. Under supersaturated circumstances, these reactions produce insoluble compounds, allowing for the development of large numbers of microscopic particles. Secondary reactions during co-precipitation can modify particle size, shape, and aggregation. This synthesis method's main benefits are its simplicity and quickness. It is also known as a solvent displacement technique, which happens during a wet chemical process. Examples of polymer solvents include ethanol, acetone, hexane, and non-solvent polymers, with polymer phases ranging from synthetic to natural (Yadav et al., 2023). To ensure that the polymer-solvent propagates quickly into the polymer's non-solvent phase, the polymer solution is normally added last to the combination. NP production is caused by interfacial tension between the two phases.

4.2.7 Spray Pyrolysis

Spray pyrolysis is the most recent innovation in flame aerosol technology. it is also a one-step combustion process in which the precursor, usually in liquid form and with a much higher combustion enthalpy (>50% of total energy of combustion), is frequently dissolved in an organic solvent (Dhand et al., 2015). Key technological components of FSP include a self-sustaining flame, the use of liquid feeds and less volatile precursors, demonstrated scalability, high-temperature flames, and significant temperature gradients. This approach is one of the most widely used for manufacturing complex and functional NPs. Sokolowski et al. originally used this technique to synthesize Al_2O_3 NPs by combusting an ultrasonically dispersed spray of aluminium acetylacetonate in a benzene-ethanol solvent combination (Sokolowski et al., 1977). It has since been extensively used to build numerous oxide and non-oxide ceramic NPs, such as CeO_2, SiO_2, Al_2O_3, MgO, C-Cu, $BaCO_3$, CaO, $CaCO_3$, and $CaSO_4$, among others. (Channei et al., 2013) synthesized Fe-doped CeO_2 NPs for photocatalysis applications and discovered that adding Fe improved photocatalytic efficiency. Similarly, (Siriwong et al., 2012) synthesized Pt-doped Zn_2TiO_4 NPs by flame-assisted spray pyrolysis, highlighting the process's promising potential for producing high-quality, size-controlled NPs.

4.2.8 Sol-Gel Synthesis

Sol-gel synthesis is a popular approach for producing NPs (NPs), which involves the development of a sol (a colloidal solution of solids floating in a liquid phase) followed by gelation, to generate a solid macromolecule immersed in a solvent. This bottom-up technique is popular because it is simple and versatile, providing a route for synthesizing a wide spectrum of NPs. Metal oxides and chlorides are typically used as precursors in the sol-gel method. Initially, the precursor is distributed in a host liquid by processes such as shaking, stirring, or sonication, yielding a system with both liquid and solid phases (Mann et al., 1997). NPs are recovered via phase separation procedures, such as sedimentation, filtering, or centrifugation, which are then dried to remove moisture. The sol-gel method's essential processes are the hydrolysis and condensation of metal alkoxides or metal oxide solutions. Hydrolysis produces a colloidal solution called a sol, which condenses to create a semisolid phase known as a gel. Following high-temperature drying, the required NPs form more easily. The sol-gel method's key advantages are simplicity, cheap cost, uniformity, and excellent purity of the resultant NPs (Liu et al., 2011).

4.2.9 Chemical Vapor Deposition

Chemical vapor deposition refers to the formation of a thin layer of gaseous reactants on a substrate. The deposition occurs in a reaction chamber at ambient temperature by mixing gas molecules. When a heated substrate comes into touch with the combined gas, it causes a chemical reaction. This reaction results in a thin coating of product on the substrate surface, which is recovered and utilized. In CVD, the controlling factor is the substrate temperature. The advantages of CVD include very pure, homogeneous, rigid, and robust NPs. The downsides of CVD include the need for specialized equipment and the presence of very hazardous gaseous byproducts (Ealias and Saravanakumar, 2017). It is also utilized to synthesize inorganic NPs on the surface of appropriate substrates. It usually happens in three phases (Mudshinge et al., 2011). First, a volatile precursor is injected into a reaction chamber holding the substrate via a carrier gas. The vapors adsorb on the surface of the substrate, resulting in certain intermediate products. In the last process, the intermediate products are decomposed to provide solid grains and NPs.

4.2.10 Solvothermal Synthesis

Solvothermal synthesis is a widely used chemical process that takes place in a solvent medium under precise temperature and pressure conditions. It is commonly used to create catalysts with core-shell structures. Catalysts produced using this approach have excellent crystallinity, tiny particle size, and uniform dispersion. Compared to other synthetic approaches, catalysts made using hydrothermal/solvothermal synthesis in a closed environment are purer, effectively preventing crystal defects in NM development and providing an ideal crystal growth environment. Key control factors are reaction temperature, precursor volume, reaction duration, and solution pH. These characteristics affect the material's crystallinity, shape, and, ultimately, the catalyst's activity (Morales-Narváez et al., 2017). As an instance, (Niu et al., 2018) reduced the reaction time by three hours which caused a change from yolk-shell to core-shell structure in Pd-Fe_3O_4-MOF. Furthermore, the precursor volume influences reaction nucleation; an excess of precursor may cause the active core to aggregate, lowering catalytic efficiency. Despite their benefits, these procedures may not produce high-quality encapsulated structures due to the presence of stabilizers, surfactants, and organic solvents. To improve the interaction between transition-metal active cores and shells, researchers used hydrothermal/solvothermal techniques coupled with calcination phases to create high-quality encapsulated catalysts (Yu et al., 2024).

4.3 CHARACTERIZATION TECHNIQUES

The features of NPs dictate their possible uses. As a result, many approaches and techniques are employed to analyse and characterize the physicochemical features of NPs. Table 4.1 displays some of the characterization approaches discussed in this paper, as well as the traits and features that each methodology may resolve.

4.3.1 XPS

The advent of NMs has revolutionized various fields, from electronics to biomedicine, owing to their unique properties. However, harnessing their full potential necessitates a profound understanding of surface interactions and modifications. Herein lies the significance of surface-sensitive techniques like X-ray photoelectron spectroscopy (XPS), which offers unparalleled insights into NM surface chemistry. It stands out as a pivotal technique for probing the intricate interplay between NMs and their surrounding environments. XPS stands out for elemental analysis in the surface layers of materials. Its ability to detect surface modifications is crucial for verifying the successful functionalization of NMs. Unlike conventional methods reliant on ultrahigh vacuum conditions, XPS transcends these limitations, enabling the exploration of liquid-NP interfaces (Jayawardena et al., 2021). Through its sensitivity and depth profiling abilities, XPS spearheads improvements in NM characterization, offering profound knowledge into surface interactions, functionalization, and reactions.

The chemical state noticed via XPS may fluctuate with particle concentration on a substrate, as evidenced by the study of ceria particles (Baer et al., 2013). Unbalanced distribution of particles on a silicon wafer resulted in differing chemical states (Ce^{3+} and Ce^{4+}), highlighting the importance of uniform deposition for proper analysis. By discerning distinct oxidation states, XPS facilitates the recognition of surface ligands. Exemplifying its significance, XPS has been instrumental in characterizing sulfur-containing ligands on PbS quantum dots, disclosing diverse binding motifs, as well as surface passivation states (Morris-Cohen et al., 2013). Recent strides in near ambient pressure show that XPS (NAP-XPS) facilitates in situ measurements and solution cell analyses at mbar pressures. NAP-XPS conquers the constraints of traditional vacuum conditions, such as peak shifting and reduced signal intensity, by offering opportunities to study surface dynamics and real-time interactions. Considering assumptions of NP homogeneity and ligand distribution, XPS gives accurate estimation of ligand shell thickness. Exemplifying this, Torelli and colleagues (Torelli et al., 2015) established an integrated experimental and computational approach to determine ligand shell thickness around gold NPs (AuNPs), accounting for scattering effects from both the Au core and ligand shell. Amstad and friends (Amstad, Textor, and Reimhult 2011) addressed the requirement of various techniques for NM analysis, highlighting the utilization of surface-sensitive methods. In their research on Fe metal-core oxide-shell NPs, XPS proved to be the primary tool for sustaining consistency and detecting surface alterations and contamination.

TABLE 4.1
Some Properties and Their Characterization Techniques

Property	Technique
Mass	Thermogravimetric analysis
Temperature	Differential thermal analysis
Heat flux	Differential scanning calorimetry
Mechanical	Thermomechanical analysis
Radioactive decay	Emanation thermal analysis
Volatiles	Evolved gas analysis

XPS is crucial to assess surface composition, chemical state, and the existence of functional groups and contaminants. Beyond that, it enables monitoring NP size, electrical properties, and the distribution of acid sites on zeolite surfaces. Various algorithms, like MultiQuantTM and QUASES (Baer et al., 2016) can derive coating thickness from XPS data by taking into account parameters such as particle shape and surface nanostructure. For particles exceeding 30-40 nm with curved surfaces, coating thickness is determined by particle shape but retains independence of size. For tiny particles, successful analysis requires data on both size and shape. Despite its wide adoption and value, traditional XPS setups need ultrahigh vacuum to examine particles hosted on substrates. Understanding the effect of angle on XPS spectra permits the extraction of a variety of information on surface characteristics and nanostructure. From elemental evaluation to ligand shell quantification, XPS reveals unparalleled insights into surface modifications and interactions, developing our understanding of NM surface chemistry.

4.3.2 EDX

Energy Dispersive X-ray Spectroscopy (EDX) is a potent analytical approach that works in conjunction with SEM to determine the elemental makeup of materials. It works by identifying the distinct X-rays generated when a sample is attacked with high-energy electrons from the SEM. Each element generates distinct X-ray fingerprints, enabling the accurate identification and quantification of components in the sample. EDX gives useful information on the chemical composition, distribution, and concentration of elements in micro- to nanoscale areas of a sample. This information is vital for different sectors, such as materials science, geology, forensics, and biological research, allowing researchers to better understand the characteristics and behaviour of materials at the atomic level (Notte and Goetze, 2014).

The essential premise of EDX is the excitation of electrons within the atoms of a specimen's surface by an electron beam from a scanning electron microscope (SEM). As these excited electrons return to their ground state, they produce distinctive X-rays. A detector collects these X-rays, producing a picture that reveals important information about the elemental makeup of the analysed region. EDS can detect elements ranging from carbon to uranium at concentrations as low as 1.0 wt.% (Chareev et al., 2018). When paired along with SEM, the analysis area can be effectively selected according to the magnification level.

A basic EDX system consists of three major components: an X-ray detector, a pulse processor that measures voltage pulses corresponding to X-ray energy, and a computer for data interpretation. The detector receives outgoing X-rays and converts them into voltage pulses proportionate to their energy. The computer then measures and plots these pulses as a histogram, displaying an X-ray energy spectrum and identifying the elements in the material. Notably, because X-ray interactions with electrons do not result in volume loss in the sample, SEM examination using EDX is considered non-destructive, allowing for repeated investigations of the same materials.

4.3.3 XRD

X-ray diffraction (XRD) is a well-known and non-destructive method for determining NM crystalline structure, chemical content, and physical characteristics (Mokhena et al., 2020). XRD may provide useful information on structures, phases, preferred crystallographic orientations, average crystallite/grain size, and so on. XRD can precisely analyse structure and conformation, while also mapping the location of surface ligands on NMs (Sakthivel et al., 2020). Nanocrystals, as well as atomically accurate nanoclusters with well-defined chemical formulas and ligand compositions, are ideal NMs for studying the overall structure of the ligand shell. XRD is mostly utilized for the surface investigation of NPs. Size may be determined using XRD spectroscopy and the Scherrer equation. The sharp XRD peaks are useful for determining the size of NPs. However, for non-crystalline

NPs, the XRD peaks are wide, making size determination more difficult than with TEM. XRD cannot identify the size of extremely tiny NPs.

XRD offers insight into the structures on a particle's surface, as well as its grain size. In addition, the constituents of the unknown NM may be identified via comparing the location and strength of the strong peaks in the sample's XRD pattern to the reference patterns from the International Center for Diffraction Data. Also, comparing a specific type of synthesised NM to current legitimate papers would be quite beneficial. As an illustration, the XRD distribution of the synthesized NPs by (Vaseghi et al., 2019) depicts the presence of all three metals at the intense peak sites. Beyond that, by evaluating XRD results using X'pert High Score plus software, they discovered that the peak locations were assigned to Cr_2O_3 and CuO for the position A, NiO, Cr_2O_3, Cu and Cr for position B and Cr_2O_3 and NiO for position C. The average crystallite size was also estimated independently for positions A, B, and C, using the Scherrer formula.

4.3.4 SEM and TEM

SEM is a proven electron microscopy method for spotting surfaces that may also be used to characterize nanoscale materials. It is a surface-imaging technique that generates images of a material by scanning it with a focused electron beam. The incident electron reacts with the sample, producing signals that represent direct images and the atomic composition of the scanned surface. The incident electrons emit elastically backscattered electrons, inelastically scattered electrons and distinctive X-rays from the atoms on the surface of the sample. The primary mode in SEM is secondary electron detection, which creates the sample picture by collecting the secondary electrons with a specific detector (Johal, 2012). Optical microscopes work with visible light sources and lenses or mirrors to magnify pictures at the micrometer scale. On the contrary, electron microscopes are the most widely utilized tools for giving immediate images and chemical information on nanoscale materials. Since electrons have a far shorter de Broglie wavelength, these can capture images at substantially better resolution than light microscopes (Barhoum and Luisa García-Betancourt, 2018).

Scanning Electron Microscopy (SEM) is a vital technique in nanotechnology, providing unmatched insight into the complex world of NPs. SEM's ability to create high-contrast 3D pictures of NPs at a stunning resolution limit of 1.2 nm allows researchers to investigate the microscopic surface properties of materials in unprecedented depth. What distinguishes SEM from other electron microscopy methods is its ability to capture a large region of the material while delivering detailed morphological and topological information. SEM is an important tool in environmental cleanup and energy storage because it allows for the investigation of material morphology. SEM allows researchers to understand the intricate structures underlying composite materials by investigating the ordering, configuration, interaction, crystallinity, and composition of diverse nano systems. SEM provides a variety of instruments for analyzing the composition and crystallinity of these materials, including Energy Dispersive X-ray Spectroscopy and Backscattered Electron Imaging (Shaheen et al., 2023). As nanotechnology evolves, SEM remains vital, providing unmatched insights into the tiny world and fostering innovation across numerous areas.

Transmission Electron Microscopy (TEM) provides a detailed look at the tiny world by sending electrons through a sample and seeing how they scatter. This contact with the sample atoms produces signals including information about surface topography, composition, and electrical conductivity. Unlike other techniques, TEM penetrates the material, making it excellent for studying interior structures. However, its picture production process is extremely complex, necessitating powerful magnetic fields to concentrate the electron beam and rigorously eliminating aberrations in the electron lenses to assure peak performance. TEM also reveals material diffraction patterns, which are caused by the interaction of the incident beam with the crystal lattice, as described by Bragg's law. As a result, by analyzing the diffraction patterns and the zone axis, researchers may readily determine the crystallographic orientation of the materials under investigation (Mayeen et al., 2018).

4.3.5 AFM

Atomic Force Microscopy (AFM) has several modes of operation, each customized to unique imaging requirements. The key distinction is how the tip interacts with the sample surface: contact, intermittent, and non-contact modes. In contact mode AFM, the tip remains in constant contact with the sample surface during scanning, similar to a profilometer's stylus (Prasad et al., 2024). This tight engagement, assisted by a very sensitive cantilever, allows for high-resolution imaging of both horizontal and vertical surface characteristics. As the tip moves across the surface, changes in vertical force generated by height differences are sensed via feedback control, allowing the tip to properly follow the surface contour. However, this constant contact might cause lateral stresses on the sample surface due to shear forces at the tip location, possibly damaging soft or biological materials. Understanding and managing these pressures is critical for proper image interpretation. To solve this issue, force modulation microscopy emerged, which maintains a constant force across the tip and the sample by modulating the tip's vertical movement. This modulation causes cantilever deflection, which changes depending on local forces and mechanical properties of the sample, providing vital information on surface features.

To summarize, the type of AFM mode used is determined by the unique imaging needs and the nature of the material under examination. While contact mode gives great resolution, it necessitates careful consideration of lateral forces, whereas force modulation microscopy provides a more nuanced knowledge of sample mechanics by controlling force variation.

4.3.6 FTIR

Infrared (IR) spectroscopy is a key tool for studying molecule structures at the microscopic level. IR spectroscopy, which probes the chemical composition and bonding patterns of materials, gives essential insights into molecular architecture. Fourier transform infrared spectroscopy (FTIR) is a key technique in this discipline, allowing for the capture of IR spectra via Fourier transformation of data from interferometers. FTIR allows for the identification of individual chemical components or atomic groups that absorb IR radiation at different frequencies, hence defining molecular structures. Its applications go beyond chemical analysis to include the determination of tacticity, crystallinity, and molecular strain. Vibrational movements of chemically bonded components resonate in the infrared spectrum, allowing for interaction with impinging IR radiation. In an IR experiment, the intensity of IR radiation prior to and following interaction with a sample is measured as a function of frequency to produce an IR spectrum. This spectrum reveals the identities, locations, and concentrations of chemical bonds found in the sample.

FTIR spectra of phosphor produced via the combustion technique were presented by (Salame et al., 2018) in their chapter. These spectra show a degradation phase and absorption over various areas, clarifying the structural relationship between them. Particularly, the breakdown of inorganic groups is visible, with extra peaks ascribed to OH and CH absorption. The research also displays distinctive metal-oxygen vibrations inside the area, clarifying the b-alumina structural phase. In conclusion, FTIR spectroscopy is a strong technique for clarifying molecular structures and unraveling complex material compositions, providing a better knowledge of chemical interactions and molecular dynamics.

4.3.7 TGA

Thermogravimetric Analysis (TGA) is a versatile approach to analysis aimed at offering insight into the composition of materials, stability, and thermal behaviour. By assessing shifts in weight as a function of temperature or time, TGA delivers insight into the complex dynamics of materials undergoing degradation, oxidation, or dehydration. It has an extensive list of applications, from nanotechnology to the study of polymers, delivering important details for industries and researchers alike. At its core, it operates by heating a sample under controlled conditions across multiple atmospheres, notably nitrogen, helium, air, and vacuum. Modern TGA systems are highly sensitive, with the ability to identify minute mass changes down to the level of ligand shells in NMs. The resultant thermogram, which plots % weight loss against temperature, serves as a road map for understanding material behavior under thermal stress.

In the domain of nanotechnology, TGA has emerged as an essential tool for forecasting the thermal stability of NPs. Researchers leverage its skills to evaluate surface changes, calculate ligand densities, and test NP purity. For instance, Ramanujan et al. (Kayal and Ramanujan 2010) showcased the utility of TGA in quantifying polyvinyl alcohol (PVA) bound to Fe oxide NPs, shedding light on surface functionalization. Mansfield et al. (Mansfield et al., 2014) assessed the purity of silica NPs, underlining the role of TGA in NP characterization. Despite its multiple significance, TGA incorporates some challenges, especially when dealing with complex organic NMs or multi-component systems. It might be difficult to distinguish between thermal degradation processes involving particular components. However, strategic measures, such as using derivative thermograms (DTG) and adding control samples, improve TGA efficacy. Additionally, preheating samples to eliminate trapped solvents and contaminants reduces possible skewing of results, assuring reliable characterization. (Z. Zhang et al., 2018) exemplifies TGA's competence in quantifying ligand density on NMs. The researchers created polystyrene-grafted cellulose nanocrystals (PS-g-CNCs), using two distinct approaches: surface-initiated atom transfer radical polymerization and surface-initiated activator regenerated by electron transfer. The researchers employed TGA analysis to determine the thermal breakdown profiles of PS-g-CNCs, illustrating peculiar decomposition phases. In an investigation, thermal degradation analysis employing TGA indicated distinct decomposition phases for chitin, chitin xerogel, bentonite, and their composites (Perez et al., 2020). Bentonite displayed two stages of decomposition due to water removal. Chitin and its derivatives exhibited three loss steps, encompassing water evaporation, depolymerization of acetylated groups, and pyrolysis. Interestingly, composites resulted in non-proportional loss in weight, with inclining bentonite concentration, indicating additional factors contribute to their thermal behavior. Thus, TGA gives valuable insights into the intricate degradation processes of the materials, resulting in a deeper understanding of the material's composition and stability. The TGA curves from Pan's work (Pan et al., 2022) illustrate distinct thermal activity for diverse materials in a nitrogen atmosphere. Graphene oxide (GO) depicts a minor initial weight loss due to adsorbed water, followed by a rapid decline in mass around 200°C, due to the pyrolysis of oxygen-containing functional groups. This method releases gases such as carbon monoxide, carbon dioxide, and water vapor, leading to intense gas release and frequent thermal expansion.

In simple terms, Thermogravimetric Analysis (TGA) is a key tool for determining the composition, stability, and thermal properties of materials in numerous kinds of usages. Its relevance in nanotechnology, polymer science, and materials analysis accentuates its importance in both research and industry. Regardless of challenges imposed by complex systems, strategic techniques strengthen the effectiveness of TGA, enabling precise and meaningful analysis. As technology advances, TGA stays at the forefront of testing techniques, solving the perplexing problems of material science with accuracy and clarity.

4.3.8 DSC

Differential Scanning Calorimetry (DSC) is a fundamental technique for understanding the thermal characteristics of substances, establishing the correlation between temperature and specific physical attributes. DSC enables the exact assessment of energy absorption or release during heating or cooling operations by monitoring variations in physical characteristics over time and at different temperatures. This versatility spans an array of applications, including the measurement of medicinal nanosolids, the characterisation of nanostructured lipid carriers, and the investigation of colloidal NPs.

In DSC, the heat flux and power-compensated mechanisms are the two primary approaches for detecting heat flow (Abraham et al., 2018). Heat flux DSC entails heating a sample and reference pan at linear rates within a thermoelectric disk-enclosed furnace. The variation in temperature is caused by the sample's heat capacity and measured to determine heat flow. In contrast, power-compensated DSC applies separate furnaces with independent heaters to maintain sample and reference temperatures exactly the same, allowing for faster equilibration. This method is particularly beneficial for fast calorimetry, which allows for in-depth research of structure development. Fast scanning calorimetry, a variant of power-compensated DSC, has gained recognition for its capacity to analyze materials that are far from thermodynamic equilibrium. With sensors positioned in a gas environment at the bottom of a tube, fast scanning calorimetry enables rapid temperature fluctuations to explore structure developments in polymers, metals, pharmaceuticals, and many more. The combination of analog electronic devices managed via computer software facilitates precise temperature control and data collection, which is essential for kinetics studies and thorough material characterization.

In addition to conventional DSC techniques, an abundance of new methods has been introduced that boost the properties of material like infrared-heated DSC, microelectromechanical DSC, modulated-temperature DSC, gas-modulated DSC, and more. Every strategy offers distinctive insights into the thermal behavior of nanostructures, making it important for researchers in a variety of domains. (Jennings et al., 2024) evaluated the thermal traits of whey protein employing DSC. The research investigation (Li et al., 2021) focused on the thermal excitation threshold and heat discharge of thermite in a DSC thermal analysis experiment performed at a linear heating rate of 15°C per minute. The exothermic temperature of the thermite system decreased as the particle size of Bi_2O_3 increased. In the meantime, when Bi_2O_3 particle size expands, it consequently increases the thermal excitation threshold. The analysis of the reactants indicated that all three Bi_2O_3 particle sizes had an impact on the composition of the thermite system. Initial exothermic reactions tend to be solid-solid (s-s), solid-liquid (s-l), and liquid-liquid (l-l) thermite interactions. These results provide light on the delicate link between Bi_2O_3 particle size, thermal behaviour, and reaction kinetics in thermite systems.

DSC has multiple uses in nanoscience, posing as an important approach for studying the thermal behaviour of NMs. From investigating glass transition temperatures of macromolecules in nanophases to identifying ion-chelating nanocarriers and self-assembled supramolecular nanostructures, DSC delivers indispensable information into the behavior of NMs. Finally, it emerges as a powerful tool for investigating the thermal trait possessed by the materials in various disciplines. Its capacity to detect heat flow with precision and versatility makes it paramount for research as well as industry. From fundamental principles to sophisticated methodologies, DSC is expanding the frontiers of material characterization, paving the way for breakthroughs in nanoscience and beyond.

4.3.9 ESR

Electron spin resonance (ESR), also referred to as electron paramagnetic resonance (EPR) spectroscopy, is an important method for investigating the magnetic characteristics of materials by measuring the unpaired electrons in a sample. In a high magnetic field, unpaired electrons with spin orientations (+ ½ and – ½) align alternatively, resulting in discrete energy levels known as the Zeeman effect. By exposing the sample with microwave pulses, ESR stimulates electron spins till a resonance state is reached by photon absorption or emission.

ESR spectroscopy is useful for assessing the ferromagnetic and antiferromagnetic characteristics of NPs (NPs). For example, researchers used ESR to characterize iron oxide NPs generated by *Ficus carica*, demonstrating changes in magnetic properties depending on environmental circumstances (Joudeh and Linke, 2022). Furthermore, ESR revealed a magnetic anisotropy in the signal, demonstrating varying magnetic behaviours at various temperatures. Similarly, ESR was used to analyse the magnetic properties of Se NPs produced by anaerobic granular sludge, revealing the presence of

Fe (III) atoms embedded in the Se NM, improving its overall magnetic characteristics and demonstrating ferromagnetic behaviour (Dixit et al., 2021). In a nutshell, ESR spectroscopy provides essential insights into materials' magnetic properties, giving light on their behavior under changing environmental circumstances and offering a more in-depth knowledge of them.

4.3.10 DMA

Dynamic mechanical analysis (DMA), sometimes known as dynamic mechanical thermal analysis (DMTA) when temperature response is addressed, is an analytical approach for determining the viscoelastic characteristics of materials, particularly polymers and composites. DMA assesses how materials respond to dynamically changing stress or strain applied to the sample. Storage modulus, loss modulus, and damping factor are key characteristics analysed in DMA for composite materials; they fluctuate with time, temperature, and frequency. The storage modulus characterizes a material's stiffness and elastic behavior, and are similar to, but not identical to, Young's modulus. In contrast, the loss modulus indicates a material's viscous response and thermal energy dissipation, which is impacted by molecular arrangements, heterogeneities, movements, and phase transitions.

The link between storage modulus and loss modulus may be shown with the example of a bouncing ball, which dissipates and stores energy. The damping factor (tan δ) is an important metric produced from DMA that represents the ratio of the loss modulus to the storage modulus. The complex modulus (E^*) describes a material's resistance to deformation by combining the storage and loss moduli. A larger damping factor in fiber-reinforced polymer composites suggests stronger interfacial adhesion between the fiber and matrix components (Bettaieb et al., 2015). This reduces energy loss owing to limited molecular mobility, resulting in a better storage-to-loss modulus ratio. Thus, materials with strong interfacial bonding have a greater damping factor, which contributes to improved mechanical characteristics.

4.3.11 UTM

The Universal Testing Machine (UTM) is flexible equipment used for a variety of tasks, most notably assessing the mechanical characteristics of materials. It may be used to determine factors such as tensile strength, compression resistance, and material thickness, making it appropriate for characterizing materials such as pineapple leaf fiber (Sufian & Othman, 2021). Due to their ability to accommodate a wide range of substrates, UTMs subject materials to external pressures and strains. Tensile testing, a frequent UTM application, evaluates the mechanical characteristics of materials by applying tension until failure occurs. This approach is good for bulk sample characterization and yields useful information about material behavior.

However, because the specimens generated are tiny in size, tensile testing may not be appropriate for micro- and nanostructured materials. Tensile strength tests are difficult to conduct effectively due to the complexities of these materials. Despite this restriction, UTM is still a useful technique for determining the mechanical characteristics of a variety of materials.

4.3.12 BET

The BET method, created by Brunauer, Emmett, and Teller, is a well-known methodology for determining the surface area of NPs. This approach uses adsorption and desorption principles to determine a material's surface area. In BET analysis, a partial vacuum is generated to promote adsorption between the sample and liquid nitrogen (N_2). The sample is chilled with liquid N_2 to allow observable adsorption, since the contact between the solid and gaseous phases is modest. Once adsorption monolayers have formed, the sample is withdrawn from the N_2 environment and heated to liberate the adsorbed N_2 (desorption), which is then measured (Khan et al., 2019).

For example, the BET technique was used to determine the surface area of CeO_2 NPs synthesized from *Eucalyptus globulus* leaf extract, resulting in a surface area of 40.96 m^2/g. This figure was far greater than that of commercial CeO_2 NPs, which had a surface area of 8.5 m^2/g. Similarly, BET methods were used to determine the surface area of SiO_2 NPs from rice husk, CuO NPs from *Leucaena leucocephala* leaf extract, and Ag NPs from *Acanthospermum hispidum* leaf extract, which had surface areas of 7.15, 47.54 and 9.91 m^2/g, respectively (Joudeh and Linke, 2022).

4.3.13 Zeta Potential

The zeta potential is critical for stabilizing NPs (NPs) in solution and preventing aggregation, either by electrostatic repulsion, steric hindrance, or a combination of both. The surface charge of NPs is an important component determining their stability, and while there is no direct analytical way to estimate this charge, the zeta potential is a useful metric. It depicts the charge at the surface of the electrical double layer that surrounds the particles in solution and is affected by a variety of elements, including surface chemistry, particle concentration, size, pH, temperature, solvent, and ionic strength. NPs are considered stable in solution when these values lie within a certain range, which generally spans from greater than +25 mV to less than -25 mV (Mudalige et al., 2018). However, this range can vary based on particle characteristics and environmental conditions.

In this regard, in the research conducted by (Ansar et al., 2017) on thiolated poly(acrylic acid) modified gold NPs, the zeta potential declined from 42 to 0 mV when the pH of the medium was lowered from 10 to 2, owing to surface carboxylate group neutralization. As the electrostatic repulsion between particles lessened, NPs aggregation occurred. To measure the zeta potential, the scientists used a Malvern Nano ZS device that included a HeNe laser and a scattering detector. More recently, (Go et al., 2017) examined the zeta potentials of NPs in food matrices to evaluate their interactions with food components. This demonstrates the significance of knowing zeta potential in a variety of applications.

4.4 CONCLUSION AND FUTURE PROSPECTS

Over the last few decades, the search for sustainable NMs has encouraged tremendous advances in nanotechnology and nanoscience. Semiconductor nanostructures, with their increased surface area, high surface-to-volume ratio, and unique physical and chemical characteristics resulting from size quantization, have enormous potential for a variety of applications. This chapter digs into NMs and sustainability, specifically how nanofabrication techniques enable the creation of customized nanostructures and functionality. The bottoms-up and top-down methods of nanofabrication are thoroughly examined, highlighting their critical significance in generating sustainable NMs. Furthermore, numerous characterization technologies for determining sustainability are described, emphasizing their relevance in evaluating NM performance and environmental effect. Looking ahead, the future of nanotechnology and nanoscience is dependent on the continuous development of sustainable low-dimensional NMs. For example, addressing problems in large-area chemical vapor deposition manufacturing of 2D NMs is critical for next-generation devices. Furthermore, researching processing approaches that are compatible with present silicon-based technology is critical. Semiconductor nanocomposites, on the other hand, are ready to enter commercial markets quickly, with applications ranging from sensors to coatings, energy storage, and conversion.

REFERENCES

Abraham, Jiji, Arif P. Mohammed, M. P. Ajith Kumar, Soney C. George, and Sabu Thomas. 2018. Thermoanalytical Techniques of NMs. Book: *Characterization of NMs: Advances and Key Technologies*. Elsevier Ltd. https://doi.org/10.1016/B978-0-08-101973-3.00008-0

Alam, Shahid, Qamar Zaman Khan, Abdelaziz Gassoumi, Muhammad Ishaq Khan, Muhammad Zahir Iqbal, and Zubair Ahmad. 2024. Innovating Synthesis Approaches in Advancing Electrochemical Efficiency: A

Journey into Hydrothermal and Sonochemical Realms. *Journal of Energy Storage* 78: 109821. https://doi.org/10.1016/j.est.2023.109821

Amstad, Esther, Marcus Textor, and Erik Reimhult. 2011. Stabilization and Functionalization of Iron Oxide NPs for Biomedical Applications. *Nanoscale*. 3 (7): 2819–2843. https://doi.org/10.1039/c1nr10173k

Ansar, Siyam M., Benjamin Fellows, Patrick Mispireta, O. Thompson Mefford, and Christopher L. Kitchens. 2017. PH Triggered Recovery and Reuse of Thiolated Poly(Acrylic Acid) Functionalized Gold NPs with Applications in Colloidal Catalysis. *Langmuir*. 33 (31): 7642–7648. https://doi.org/10.1021/acs.langmuir.7b00870

Arshad, Shumaila, and Ayesha Naseer. 2017. NPs: A Review. *IAJPS, CSK Publications*. 03 (11): 1389–1395. https://doi.org/10.5281/zenodo.208175

Baer, Donald R., Mark H. Engelhard, Grant E. Johnson, Julia Laskin, Jinfeng Lai, Karl Mueller, and Prabhakaran Munusamy. 2013. Surface Characterization of NMs and NPs: Important Needs and Challenging Opportunities. *Journal of Vacuum Science & Technology A: Vacuum, Surfaces, and Films*. 31 (5). https://doi.org/10.1116/1.4818423

Baer, Donald R., Yung-Cheng Wang, and David G. Castner. 2016. Use of XPS to Quantify Thickness of Coatings on NPs. *Microscopy Today*. 24(2): 40–45. https://doi.org/10.1017/s1551929516000109

Barhoum, Ahmed, and M. Luisa García-Betancourt. 2018. *Physicochemical Characterization of NMs: Size, Morphology, Optical, Magnetic, and Electrical Properties. Emerging Applications of NPs and Architectural Nanostructures: Current Prospects and Future Trends*. Elsevier Inc. https://doi.org/10.1016/B978-0-323-51254-1.00010-5

Bettaieb, Fedia, Ramzi Khiari, Alain Dufresne, Mohamed Farouk Mhenni, and Mohamed Naceur Belgacem. 2015. Mechanical and Thermal Properties of Posidonia Oceanica Cellulose Nanocrystal Reinforced Polymer. *Carbohydrate Polymers*. 123: 99–104. https://doi.org/10.1016/j.carbpol.2015.01.026

Channei, Duangdao, Burapat Inceesungvorn, Natda Wetchakun, Sukon Phanichphant, Auppatham Nakaruk, Pramod Koshy, and Charles C. Sorrell. 2013. Photocatalytic Activity under Visible Light of Fe-Doped CeO_2 NPs Synthesized by Flame Spray Pyrolysis. *Ceramics International*. 39 (3): 3129–3134. https://doi.org/10.1016/j.ceramint.2012.09.093

Chareev, Dmitriy, Yevgeniy Ovchenkov, Larisa Shvanskaya, Andrey Kovalskii, Mahmoud Abdel-Hafiez, Dan J. Trainer, Eric M. Lechner, Maria Iavarone, Olga Volkova, and Alexander Vasiliev. 2018. Single Crystal Growth, Transport and Scanning Tunneling Microscopy and Spectroscopy of $FeSe_{1-X}S_x$. *CrystEngComm*. 20(17): 2449–2454. https://doi.org/10.1039/c8ce00074c

Dhand, Chetna, Neeraj Dwivedi, Xian Jun Loh, Alice Ng Jie Ying, Navin Kumar Verma, Roger W. Beuerman, Rajamani Lakshminarayanan, and Seeram Ramakrishna. 2015. Methods and Strategies for the Synthesis of Diverse NPs and Their Applications: A Comprehensive Overview. *RSC Advances*. 5 (127): 105003–105037. https://doi.org/10.1039/c5ra19388e

Dixit, Rewati, Anirudh Gupta, Norbert Jordan, Shengqiang Zhou, Dieter Schild, Stephan Weiss, Emmanuel Guillon, Rohan Jain, and Piet Lens. 2021. Emerging Trends in Biotechnology for Sustainable Development and Pollution Prevention: Magnetic properties of biogenic selenium NMs. *Environmental Science and Pollution Research*. 40264–40274. https://doi.org/10.1007/s11356-020-11683-2

Ealias, Anu Mary, and M. P. Saravanakumar. 2017. A Review on the Classification, Characterisation, Synthesis of NPs and Their Application. *IOP Conference Series: Materials Science and Engineering*. 263 (3). https://doi.org/10.1088/1757-899X/263/3/032019

Hasany, Syed Farhan, Ahmed, Isteaque, Rajan, Jose, and Rehman, Allah 2013. Systematic Review of the Preparation Techniques of Iron Oxide Magnetic NPs. *Nanoscience and Nanotechnology*. 2 (6): 148–158. https://doi.org/10.5923/j.nn.20120206.01

Go, Mi Ran, Song Hwa Bae, Hyeon Jin Kim, Jin Yu, and Soo Jin Choi. 2017. Interactions between Food Additive Silica NPs and Food Matrices. *Frontiers in Microbiology*. 8: 1–12. https://doi.org/10.3389/fmicb.2017.01013

Jayawardena, H. Surangi N., Sajani H. Liyanage, Kavini Rathnayake, Unnati Patel, and Mingdi Yan. 2021. Analytical Methods for Characterization of NM Surfaces. *Analytical Chemistry*. 93(4): 1889–1911. https://doi.org/10.1021/acs.analchem.0c05208

Jennings, Charity C., McCall Freidenberger, Shawn A. Christensen, Joy Conlin, Olivia Freidenberger, and Jason D. Kenealey. 2024. Thermal Characterization and Separation of Whey Proteins by Differential Scanning Calorimetry. *Food Chemistry*. 441: 138347. https://doi.org/10.1016/J.FOODCHEM.2023.138347

Johal, Malkiat S. 2012. *Understanding NMs*. (1st ed.), CRC Press, Boca Raton, pages 328 https://doi.org/10.1201/b11545

Joudeh, Nadeem, and Dirk Linke. 2022. NP Classification, Physicochemical Properties, Characterization, and Applications: A Comprehensive Review for Biologists. *Journal of Nanobiotechnology*. 20 (1): 1–29. https://doi.org/10.1186/s12951-022-01477-8

Kayal, Sibnath, and Raju Vijayaraghavan Ramanujan. 2010. Doxorubicin Loaded PVA Coated Iron Oxide NPs for Targeted Drug Delivery. *Materials Science and Engineering: C*. 30 (3): 484–490. https://doi.org/10.1016/J.MSEC.2010.01.006

Khan, Ibrahim, Khalid Saeed, and Idrees Khan. 2019. NPs: Properties, Applications and Toxicities. *Arabian Journal of Chemistry*. 12(7): 908–931. https://doi.org/10.1016/j.arabjc.2017.05.011

Li, Shi, Tao Guo, Miao Yao, Jiaxing Song, Wen Ding, Yiming Mao, and Jialin Chen. 2021. Effect of Bismuth Oxide Particles Size on the Thermal Excitation and Combustion Properties of Thermite Systems. *ChemistryOpen*. 10(4): 464–470. https://doi.org/10.1002/open.202000358

Liu, Xianghong, Jun Zhang, Liwei Wang, Taili Yang, Xianzhi Guo, Shihua Wu, and Shurong Wang. 2011. 3D Hierarchically Porous ZnO Structures and Their Functionalization by Au NPs for Gas Sensors. *Journal of Materials Chemistry*. 21(2): 349–356. https://doi.org/10.1039/c0jm01800g

Mann, Stephen, Sandra L. Burkett, Sean A. Davis, Christabel E. Fowler, Neil H. Mendelson, Stephen D. Sims, Dominic Walsh, and Nicola T. Whilton. 1997. Sol-Gel Synthesis of Organized Matter. *Chemistry of Materials*. 9 (11): 2300–2310. https://doi.org/10.1021/cm970274u

Mansfield, Elisabeth, Katherine M. Tyner, Christopher M. Poling, and Jenifer L. Blacklock. 2014. Determination of NP Surface Coatings and NP Purity Using Microscale Thermogravimetric Analysis. *Analytical Chemistry*. 86 (3): 1478–1484. https://doi.org/10.1021/ac402888v

Mayeen, Anshida, Leyana K. Shaji, Anju K. Nair, and Nandakumar Kalarikkal. 2018. *Morphological Characterization of NMs. Characterization of NMs: Advances and Key Technologies*. Elsevier Ltd. https://doi.org/10.1016/B978-0-08-101973-3.00012-2

Mokhena, Teboho C., Maya. J. John, M. A. Sibeko, V. C. Agbakoba, M. J. Mochane, Asanda Mtibe, Thabang Hendrica Mokhothu, et al. 2020. NMs: Types, Synthesis and Characterization. *NMs in Biofuels Research*. 115–141. https://doi.org/10.1007/978-981-13-9333-4_5

Morales-Narváez, Eden, Lívia Florio Sgobbi, Sergio Antonio Spinola Machado, and Arben Merkoçi. 2017. Graphene-Encapsulated Materials: Synthesis, Applications and Trends. *Progress in Materials Science*. 86: 1–24. https://doi.org/10.1016/j.pmatsci.2017.01.001

Morris-Cohen, Adam J., Michał Malicki, Mark D. Peterson, John W.J. Slavin, and Emily A. Weiss. 2013. Chemical, Structural, and Quantitative Analysis of the Ligand Shells of Colloidal Quantum Dots. *Chemistry of Materials*. 25 (8): 1155–1165. https://doi.org/10.1021/cm302108j

Sharad Kamble, Kaveri Bhosale, Mahesh Mohite, and Swapnali Navale. 2023. Methods of Preparation of NPs. *International Journal of Advanced Research in Science, Communication and Technology*. 7 (4): 121–127. https://doi.org/10.48175/ijarsct-9485

Mudalige, Thilak, Haiou Qu, Desiree Van Haute, Siyam M. Ansar, Angel Paredes, and Taylor Ingle. 2018. *Characterization of NMs: Tools and Challenges*. Elsevier Inc. NMs for Food Applications. 313–353. https://doi.org/10.1016/B978-0-12-814130-4.00011-7

Mudshinge, Sagar R., Amol B. Deore, Sachin Patil, and Chetan M. Bhalgat. 2011. NPs: Emerging Carriers for Drug Delivery. *Saudi Pharmaceutical Journal*. 19 (3): 129–141. https://doi.org/10.1016/j.jsps.2011.04.001

Nadagouda, Mallikarjuna N., and Rajender S. Varma. 2008. Microwave-Assisted Shape-Controlled Bulk Synthesis of Ag and Fe Nanorods in Poly(Ethylene Glycol) Solutions. *Crystal Growth and Design*. 8 (1): 291–295. https://doi.org/10.1021/cg070473i

Niu, Hongyun, Yang Zheng, Saihua Wang, Lixia Zhao, Shipeng Yang, and Yaqi Cai. 2018. Continuous Generation of Hydroxyl Radicals for Highly Efficient Elimination of Chlorophenols and Phenols Catalyzed by Heterogeneous Fenton-like Catalysts Yolk/Shell Pd@Fe_3O_4@metal Organic Frameworks. *Journal of Hazardous Materials*. 346: 174–183. https://doi.org/10.1016/j.jhazmat.2017.12.027

Notte, John, and Bernhard Goetze. 2014. Imaging with the Helium Ion Microscope. Surface Analysis and Techniques in Biology. Book: *Surface Analysis and Techniques in Biology*. 171–194. https://doi.org/10.1007/978-3-319-01360-2_7

Pan, Yubo, Yali Zhang, Yaoguo Huang, Yun Jia, and Linlin Chen. 2022. Enhanced Photocatalytic Oxidation Degradability for Real Cyanide Wastewater by Designing Photocatalyst GO/TiO2/ZSM-5: Performance and Mechanism Research. Chemical Engineering Journal. 428: 131257. https://doi.org/10.1016/j.cej.2021.131257

Pattnaik, Amruta, Anil Kumar Poonia, and Prabir Ghosh. 2023. Bi-Tailored Compounds for Photocatalytic Environmental Applications: Current Trends, Advancements, Challenges and Future Perspectives. *Sustainable Materials and Technologies*. 38: e00769. https://doi.org/10.1016/j.susmat.2023.e00769

Perez, Jonás José, María Emilia Villanueva, Laura Sánchez, Romina Ollier, Vera Alvarez, and Guillermo Javier Copello. 2020. Low Cost and Regenerable Composites Based on Chitin/Bentonite for the Adsorption Potential Emerging Pollutants. *Applied Clay Science*. 194: 105703. https://doi.org/10.1016/J.CLAY.2020.105703

Prasad, Rai Dhirendra Rai Surendra Prasad Rai Bishvendra Prasad, Saurabh R. Prasad Shashi Bhushan Singha, Anand Dev Singha Rai Jitendra Prasad Pramod Sinha Sanjay Saxena Anil Kumar Vaidya Shivanand B Teli Umapati Rai Saxena, Avinash Harale, M B Deshmukh, M.N. Padvi, and G. J. Navathe. 2024. A Review on Modern Characterization Techniques for Analysis of NMs and Biomaterials. *ES Energy & Environment*. 23, 1087, https://doi.org/10.30919/esee1087

Sakthivel, Naga Arjun, Masoud Shabaninezhad, Luca Sementa, Bokwon Yoon, Mauro Stener, Robert L. Whetten, Guda Ramakrishna, Alessandro Fortunelli, Uzi Landman, and Amala Dass. 2020. The Missing Link: Au_{191}(SPh- tBu)$_{66}$ Janus NP with Molecular and Bulk-Metal-like Properties. *Journal of the American Chemical Society*. 142 (37): 15799–15814. https://doi.org/10.1021/jacs.0c05685

Salame, Paresh H., Vijay B. Pawade, and Bharat A. Bhanvase. 2018. *Characterization Tools and Techniques for NMs*. NMs for Green Energy. Elsevier Inc. https://doi.org/10.1016/B978-0-12-813731-4.00003-5

Shaheen, Iqra, Adnan Khalil, Rahila Shaheen, and M. Bilal Tahir. 2023. A Review on NMs: Types, Synthesis, Characterization Techniques, Properties and Applications. *Innovation in Science and Technology*. 2(1): 56–62. https://doi.org/10.56397/ist.2023.01.04

Siriwong, C., N. Wetchakun, B. Inceesungvorn, D. Channei, T. Samerjai, and S. Phanichphant. 2012. Doped-Metal Oxide NPs for Use as Photocatalysts. *Progress in Crystal Growth and Characterization of Materials*. 58 (2–3): 145–163. https://doi.org/10.1016/j.pcrysgrow.2012.02.004

Sokolowski, M., A. Sokolowska, A. Michalski, and B. Gokieli. 1977. The 'in-Flame-Reaction' Method for Al_2O_3 Aerosol Formation. *Journal of Aerosol Science*. 8 (4). https://doi.org/10.1016/0021-8502(77)90041-6

Sufian, Amirul Azan Mohd, and Siti Amira Othman. 2021. The Characterization of Pineapple Leaves Paper by using Scanning Electron Microscope (SEM) and Universal Testing Machine (UTM): A Review. *Asian Journal of Fundamental and Applied Sciences*. 1, no. (4): 58–60. https://myjms.mohe.gov.my/index.php/ajfas/article/view/11882

Tissue, B. M., and H. B. Yuan. 2003. Structure, Particle Size, and Annealing of Gas Phase-Condensed Eu3$^+$:Y_2O_3 Nanophosphors. *Journal of Solid State Chemistry*. 171 (1–2): 12–18. https://doi.org/10.1016/S0022-4596(02)00140-8

Torelli, Marco D., Rebecca A. Putans, Yizheng Tan, Samuel E. Lohse, Catherine J. Murphy, and Robert J. Hamers. 2015. Quantitative Determination of Ligand Densities on NMs by X-Ray Photoelectron Spectroscopy. *ACS Applied Materials and Interfaces*. 7 (3): 1720–1725. https://doi.org/10.1021/am507300x

Varma, Rajender S. 2012. Greener Approach to NMs and Their Sustainable Applications. *Current Opinion in Chemical Engineering*. 1(2): 123–128. https://doi.org/10.1016/j.coche.2011.12.002

Vaseghi, Zahra, Omid Tavakoli, and Ali Nematollahzadeh. 2019. New Insights into Mechanistic Aspects and Structure of Polycrystalline Cu/Cr/Ni Metal Oxide Nanoclusters Synthesized Using Eryngium Campestre and Froriepia Subpinnata. *Korean Journal of Chemical Engineering*. 36 (3): 489–499. https://doi.org/10.1007/s11814-018-0216-4

Yadav, Ritu Kamal, Shivani Jangeer, Ronit Parashar, Garima Sharma, Priyadarshi Meena, Mahesh Kumar, and Dev Dutt Patel. 2023. A Critical Review on NPs Synthesis: Physical. *Chemical and Biological Perspectives. IJCRT*. 11 (10): 834–848. https://ijcrt.org/papers/IJCRT2310318.pdf

Yu, Ting, Hong Chen, Tong Hu, Jing Feng, Wenle Xing, Lin Tang, and Wangwang Tang. 2024. Recent Advances in the Applications of Encapsulated Transition-Metal NPs in Advanced Oxidation Processes for Degradation of Organic Pollutants: A Critical Review. *Applied Catalysis B: Environmental*. 342: 123401. https://doi.org/10.1016/j.apcatb.2023.123401

Zhang, Yuhang, Kingsley Poon, Gweneth Sofia P. Masonsong, Yogambha Ramaswamy, and Gurvinder Singh. 2023. Sustainable NMs for Biomedical Applications. *Pharmaceutics*. 15 (3): 1–20. https://doi.org/10.3390/pharmaceutics15030922

Zhang, Zhen, Gilles Sèbe, Xiaosong Wang, and Kam C. Tam. 2018. Gold NPs Stabilized by Poly (4-Vinylpyridine) Grafted Cellulose Nanocrystals as Efficient and Recyclable Catalysts. *Carbohydrate Polymers*. 182, 61–68. https://doi.org/10.1016/j.carbpol.2017.10.094

5 Nanomaterials in the Degradation of Toxic Organic Pollutants

Prabhat Kumar Patel, Hrishikesh Kumar Singh, and Asit Singh

5.1 INTRODUCTION

Most of Earth's surface water is inappropriate for human intake, yet water is the most plentiful asset on Earth. Due to escalating costs of potable water, various environmental and ecological factors, and the rapid expansion of individuals worldwide (Patel, Pandey, and Uppaluri 2024b, 2023c), approximately 1.1 billion individuals lack access to an adequate supply of drinking water and are experiencing water scarcity. The primary challenge that the world is now dealing with is the contamination of surface water caused by industrial byproducts, runoff from agriculture, and improper waste dumping. According to a recent UN study, the supply of fresh water that has been cleansed will grow into a worldwide concern in the coming decades. When both inorganic and organic matter are present in water, it gets polluted and inappropriate for human intake or any other use (Patel, Pandey, and Uppaluri 2024a). According to a recent World Health Organization research, each year over 4 billion households experience different health problems as a result of waterborne infections, and approximately 1.7 million households lose their lives as a result of consuming unsafe water (Thirunavukkarasu, Nithya, and Sivashankar 2020). Filtration, chelation, reverse osmosis, chemical coagulation and precipitation, solvent extraction, and electrochemical operation have been a few traditional methods being used to mitigate hazardous pollutants cations from wastewater streams (Patel, Pandey, and Uppaluri 2023a; Patel et al. 2022). Nevertheless, these methods are often thought to be less effective in meeting the water quality criteria (Kumar, Mohan, and Uppaluri 2023; Patel, Pandey, and Uppaluri 2023b). Slowness, microbial toxicity, and the existence of pollutants that do not biodegrade are some of the negative aspects of the biological remediation process. While physical processes such as filtration might be useful in the removal of these contaminants these processes can release a lot of sludge, which is both harmful to the environment and difficult to properly dispose of (Bali, Çatalkaya, and Şengül 2003). For the wastewater remediation process, it is necessary to design technologies that are both dependable and environmentally friendly (Thirunavukkarasu and Nithya, 2019). Nanotechnology is regarded as an emergent solution that offers a substitute and potentially feasible method to effectively mitigate contaminants with minimal operating costs (Baruah, Najam Khan, and Dutta 2016). NMs, which range in size from 1-100 nm, have great promise for addressing a wide range of worldwide environmental issues. Several types of nano-materials may be found, such as quantum dots, particles, films, nanotubes, and nanowires. These nanoparticles (NPs) are very adaptable, surface-active, environmentally safe, etc., and may be used to remediate a variety of contaminants found in effluents.

The development of nano-sorbents and their efficacy in removing contaminants from wastewater has been supported in many literatures. It is anticipated that the produced nanosorbents will have highly capable surface sorption sites that are chemically active (Thirunavukkarasu, Nithya, and Sivashankar 2020). They can be divided into several categories, such as silica, composites, graphene, metal oxides, or carbon. Conversely, several nano-sized photocatalysts were created to break down the organic contaminants in the wastewater. These can be divided into three categories: electrocatalysts, Fenton-based catalysts, and catalysts with antimicrobial properties. Additionally, the

DOI: 10.1201/9781003484189-5

pressure-driven mechanism employed by the nanomembranes helps to eliminate pollutants and enhance water quality (Thirunavukkarasu, Nithya, and Sivashankar 2020). Although there are various membrane filtration methods, the most effective for industrial wastewater treatment is nanofiltration (NF), due to its minimal pores, cost-effectiveness, and simplicity of operation. To eliminate these organic contaminants from wastewater, the current chapter will highlight the four types of NMs and their possible roles in the process.

5.2 ORGANIC CONTAMINANTS PRESENT IN WASTEWATER

Organic pollutants are substances that have been utilized in ordinary life but become polluting when disposed of as trash in watercourses. These substances include colors, insecticides, medicines, and other harmful compounds. Many textile, leather, and paper industries utilize dyes as coloring agents; nevertheless, after usage, these dyes provide a significant challenge for these businesses needing to eradicate them from watercourses (Mohamed, Al-Harby, and Almarshed 2020). The discarded pigments have been dumped into nearby sources of water, such as ponds and rivers. The dyes ruin fish and other aquatic creatures immediately when they enter rivers, and then they go up the food chain and start harming other living things. Since the majority of the dyes are carcinogenic and transgenic, they have been classified as lethal. These pigments have a strong propensity to contaminate surroundings and disrupt ecosystems, since they are not biodegradable and remain in the surroundings for extended periods. In addition to colors, there are additional organic pollutants that are created to eradicate insects, weeds, and other entities that cause illness, such as herbicides, and insecticides (Scholz et al. 2012). These chemicals are used to boost agricultural productivity and quality, but in reality, they are lethal, destroying weeds and other unwanted plants while also posing a serious risk to anybody coming into contact with them. It has been shown that even a tiny amount of exposure to them may induce cancer and other illnesses that eventually result in mortality. Certain organic substances, such as pharmaceutical medications, are designed to prolong human life but, if disposed of improperly, may spread several illnesses. These pharmaceuticals are used to treat human ailments, but they eventually become dangerous. Critical organs, including the kidney, liver, brain, and central nervous system, are all immediately impacted. Other organic compounds are used in different chemical sectors to synthesize chemicals, and the environment is being contaminated by the byproducts that are emitted. Certain leftovers are volatile gases, whereas others are fluids. Prolonged exposure to these hazardous organic pollutants may be fatal. Due to their extreme toxicity, these organic pollutants, which include aromatic hydrocarbons like toluene, aromatic amines, phenol, etc., must be removed from water (J. G. Yu et al. 2014).

5.2.1 Dyes

When applied to various materials, such as textiles, papers, cosmetics, leather, etc., dyes provide color. You may also find dyes in paints, inks, and even as a laboratory strain. According to estimates, the yearly output of commercial pigments that are used in a variety of sectors is projected to exceed 7×10^4 tonnes per year. Furthermore, it is predicted that around 100 tonnes per year have been thrown as trash into water systems. Since the majority of these dyes are extremely hazardous, the presence of such a massive quantity of dye effluents in water streams may cause catastrophic impacts to all living organisms. Such harmful colors cause skin irritation, respiratory problems, and cancer if exposed to them over an extended period. Consequently, it is crucial to purify water that has been polluted with dyes. It is possible to categorize dyes according to their chemical structure, charge, color, and the applications they are used for. Cationic, anionic, and non-ionic are the terms used to describe them, based on their charges.

Cationic dyes: All basic dyes, such as xanthene, triphenylmethane, and azo dyes, are considered cationic dyes. When these pigments have been dissolved, they take on a positive charge. Malachite green, basic fuchsin safranin, basic yellow 28, methylene blue, basic red 29, crystal violet, basic blue 159, and so on are some of the most widely used cationic dyes.

These pigments are used to color paper, nylon, and polyester polymers as well as to prepare ink. These dyes are extremely apparent and generate vivid color, even in minimal quantities. These colorants have a strong malignant and transmissible effect on human organs, including the liver, the renal system, the nervous system, and the genital tract. These colors are infectious to aquatic organisms, in addition to being poisonous to humans. The substantial amount of such pigments in seawater prevents light from penetrating and interferes with aqua plants' capacity to photosynthesize. Because they include aromatic compounds that are either not sustainable or persistent in water for extended periods, as well as lethal metal ions, azo dyes are potentially hazardous.

Anionic dyes: Reactive dyes, acid dyes, and direct dyes are three different groups into which anionic dyes may be differentiated. In an aqueous solution, these pigments take on a negative charge. Direct Red 23, Congo red, Methyl orange, Reactive Black 5, and other colors are examples of anionic dyes. Every dye has a specific utility based on its composition and capacity for binding. These dyes have been employed to color materials, including paper, wool, silk, and leather. Reactive dyes are less biodegradable and easy to dissolve in water. Reactive dye-colored effluent is difficult to remove since it quickly travels through the purification operation. Its presence pollutes the environment in several ways.

Non-ionic dyes: Dispersed dyes have been classified as non-ionic dyes because they have been primarily utilized to produce colored polyamide, polyester, plastic, and other materials. Common non-ionic dye examples are Direct Green 97 and others. When colored substances come into contact with living organisms, this may lead to sensitization and allergies in the body. Furthermore, these colors are quite toxic (Yagub et al. 2014).

5.2.1.1 Toxic Effects of Dyes

Basic dyes have been prominent even at extremely low concentrations due to their strong color strength. The majority of complicated dyes include as their basis, chromium, which is hazardous. Various aquatic organisms may be affected by pigments because of the metals, aromatics, etc. Dyes may reduce light absorption, which can impact photosynthesis. Additionally, dyes have been shown to be teratogenic, mutagenic, or malignant in a variety of fishes and microbial species. Furthermore, humans may be severely impacted as dyes can impair their renal system, genital system, brain, liver, and central nervous system. Analogously, dyes based on anthraquinone have the highest resistance to deterioration and maintain their color for extended periods in wastewater. Water-soluble reactive dyes dissolve in 5–10% of the dye bath, producing brightly colored effluent that poses a major environmental risk (Yagub et al. 2014). Furthermore, it is crucial to get rid of reactive dyes as they are poorly biodegradable and chemically stable, making it probable that they will pass through traditional treatment facilities untreated.

5.2.2 Pesticides/Insecticides

Chemicals known as pesticides are made specifically to be employed in farms to shield crops from insects and weeds. Simply, pesticides are poisons that are specifically made to kill bugs or stop weed development. The problem with these pesticides is that they are dangerous and do not break down naturally. Once they get into the food chain, they kill not only the targeted species but also the non-targeted creatures, whether explicitly via food or subtly through water or soil. The environment is destroyed by pesticides in a similar manner. The existence of species, such as birds, aquatic creatures, and mammals, is threatened by these hazardous chemicals. Pesticides are becoming a major worry for the environment's survival. The environment is affected by pesticide pollution over an extended period. A pesticide such as atrazine disrupts hormone function in both humans and animals. Eliminating pesticides from drinking water is crucial for the survival of people and other animals, since certain pesticides are so toxic that minimal quantities may cause mortality (Firozjaee et al. 2018).

5.2.2.1 Toxic Effects of Pesticides

Pesticides can sneak their way into the water via a variety of means, including drift, runoff, leaching through the soil, and direct application, in some situations. Water tainted with pesticides is very dangerous for aquatic life. It may have an impact on aquatic vegetation, lower the amount of dissolved oxygen in the water, and alter fish species' metabolism. Gardening pesticides have been discovered in surface waterways and waterbodies, including lakes, ponds, and streams (Mahmood et al. 2016). Pesticides used on land end up in aquatic habitats where they affect fishes and other non-target species. In addition to being poisonous by themselves, some insecticides also have negative reactions, with stressors, such as toxic algal blooms. A decline in the number of fish species is noted as a result of excessive pesticide application (Scholz et al. 2012). Three approaches exist for aquatic animals to be exposed to pesticides (Helfrich, 2009).

- Orally: Ingress by drinking tainted water;
- Dermally:Immediate absorption via skin;
- Breathing: Absorption via gills.

Water plants offer around 80% of the dissolved oxygen, which is essential for aquatic life to survive. Ultimately, salmon asphyxiate. A decline in fish production has been caused by herbicide-killed aquatic vegetation, which substantially reduces O_2 levels (Helfrich, 2009). Due to surface runoff from farms and spray drift pollution, pesticide concentrations in surface waterways are often substantially greater than in the subsurface. However, inadequate disposition, unintentional spills and leaks, and seepage of polluted surface water are the ways that pesticides enter groundwater. Pesticides are making their way into lakes, ponds, and rivers, causing serious damage to aquatic ecosystems. Atrazine is not only hazardous to some species of fish, but it also has an indirect impact on the immune systems of certain categories of amphibians. The most significant threats to amphibians come from overexploitation, habitat loss, pesticide contamination of surface waters, and other human activities. Carbaryl has been shown to be harmful to several amphibian species, while the herbicide glyphosate is known to produce a significant death rate among tadpoles and young frogs. It has been shown that low levels of malathion alter the quantity and make-up of plankton and periphyton populations, which in turn impacts frog/tadpole development (Relyea and Hoverman 2008).

Furthermore, endosulfan and chlorpyrifos seriously harm amphibians as well. 10% of male frogs grown in atrazine-contaminated water turned into females. Genetically male frogs phenotypically developed ovaries within their testes. They also acquired the propensity to produce viable eggs and mate with other males. Herbicide spraying in weedy fish nurseries gradually lowers the level of protection that juvenile fish need to hide from predators, which lowers the reproductive capacity of aquatic life forms (Helfrich, 2009).

5.2.3 Pharmaceutical Drugs

Chemicals known as pharmaceutical medications are employed to safeguard the lives of both humans and animals. However, when these medications and their byproducts are discharged in huge quantities into water bodies, they endanger the lives of all living things. Because these treatments are specifically intended to breach the cell membrane, they can target important organs, which may lead to a variety of complications if they are exposed for an extended period. The ingestion of medications and chemicals is what causes them to be deadly. The drug production sector and hospitals are the main contributors to contamination. Additional sources include several research endeavors, as well as improper disposal of unused medications from residential settings. Hospitals are thought to be the main sources of pharmaceutical contamination, according to several sources. Hospital trash includes pathogens that cause illness, medications that are unused or dispersed, associated chemicals, and certain hazardous radioactive elements that are utilized in devices. Their careless discharge into waterbodies slows down the organic contaminants' breakdown process and has a negative impact on aquatic life. Pharmaceutical waste mostly consists of commonly used medications such as antibiotics, analgesics, antipyretics, and even certain antidepressants, as well as chemical compounds used in chemo. Plenty of studies have shown their high adverse effects and potential to induce fatal illnesses. Pharmaceuticals that contain significant levels of carcinogens include estrone, ethinylestrone, and estradiol. The renowned antibiotic tetracycline has several serious adverse effects on people. It is quite concerning that antibiotics such as sulphonamides and chloramphenicol are present. The industries prioritized their research and manufacturing of their synthetic analogs to meet the urgent demand for pharmaceutical items via quick production. This pharmaceutical manufacturing race generates a lot of trash, which eventually has an impact on the environment (Cincinelli et al. 2015).

5.2.4 Other Organic Contaminants

Volatile organic compounds, polyaromatic hydrocarbons, phenol, toluene, aromatic amines, humic acid, and other substances are among the additional organic pollutants. The ecosystem is negatively impacted by these organic contaminants. The brain and nervous system are affected right away by volatile organic chemicals like toluene. Continuous interaction with these pollutants can damage the lungs, liver, and kidneys. The primary application of aromatic compounds in industry is for the synthesis of chemicals. These polyaromatic hydrocarbons are also harmful because they affect the kidneys, liver, and eyes, as well as induce nausea, vomiting, cataracts, and other health problems. Due to the extreme toxicity of these chemicals, elimination is required.

5.3 REMOVAL OF ORGANIC CONTAMINANTS USING NMS

The remarkable adsorption capacity, enormous surface area, and controllable particle size of NMs have made them promising new adsorbents. In addition, they exhibit electromagnetic qualities, which expands the range of scientific domains, including chemistry, physics, and material science, in which they may be employed. They are employed in the production of premium catalysts, sensors, and digital gadgets. Because of their distinctive structural characteristics, high porosity, and excellent pollutant binding capabilities, NMs offer several benefits (Yang et al. 2015). Because of these characteristics, they work well as adsorbents and are employed for eliminating a variety of contaminants from water. The desorption mechanism can be employed also to renew NMs once they have been depleted. They also exhibit strong reactivity and catalytic potential. The susceptibility of NMs to either inorganic or organic modifications increases their specificity toward particular species. They are used as adsorbents in the purification of water because of their many distinctive qualities. NMs may be grouped according to their primary components and material type. Since carbon is the primary component of carbon-based NMs, such as carbon nanotubes and graphene, they fall under this category. They are a suitable option for wastewater treatment because of their small size, distinctive hollow shape, and high surface area with many adsorption sites. Certain metal-based NMs, such as iron oxide, titanium dioxide, zinc oxide, and silver and gold NPs, contain metal as their primary component. Due to their charge, conductivity, electrical, and magnetic characteristics, metal and metal oxide-based NPs are excellent choices for adsorbents. The primary component of metalloids such as silica NPs is silica. Because chitosan is non-toxic and biodegradable, it has become a popular biosorbent (Lu et al. 2016). Different kinds of NMs have been employed to eliminate organic pollutants from water. As shown in Figure 5.1, this study discusses four different kinds of NMs for the removal of organic contaminants from wastewater: those based on silica, metal oxides, carbon nanotubes, and chitosan.

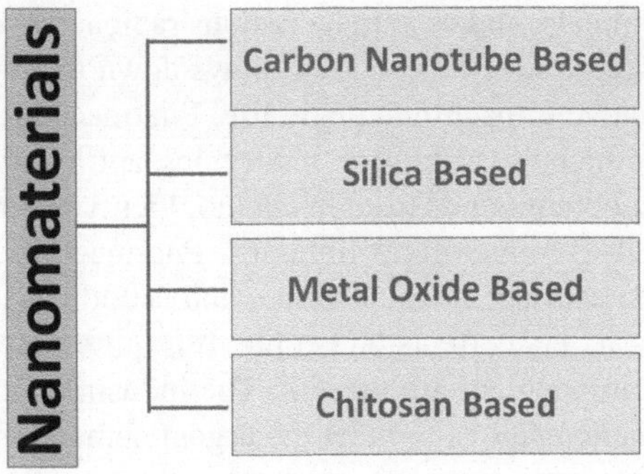

FIGURE 5.1 Classification of NMs.

5.3.1 Carbon Nanotube-Based NMs for Organic Pollutant Removal

A research group (Álvarez-Torrellas et al. 2016), analyzed the adsorptive characteristics of tetracycline using carbonaceous multiwall carbon nanotubes (Carbonaceous-MWCT). The authors concluded that the sorbent possessed a higher pollutant uptake capacity (192.7 mg g^{-1}). Another research group (Panić et al. 2015) noted an optimum sorption capacity at pH 9 and contact period of 60 min for Thiamethoxam adsorption using oxidized-MWCT. Also, another research group (Osman Duman et al. 2019) reported the maximum adsorption capacity of diquat dibromide (58.3 mg g^{-1}) at an optimal pH of 6.5 for the oxidized MWCT. Another research group (Seyed Arabi et al. 2019) targeted the removal of congo red azo dye using ZnO-MWCT. The authors inferred a maximum adsorption capacity of 249.5 mg g^{-1}. For the amino-MWCNT, a maximum capacity of 178.5 mg g^{-1} has been reported for Methylene Blue cationic dye (Ahamad et al. 2019). Also, for the pTSA-doped polyaniline@grapheneoxide-MWCNT, the maximum adsorption-based uptake of 60 mg g^{-1} for Congo Red adsorption has been reported (Ansari et al. 2017). Another group of researchers investigated Acid Orange 7 adsorption using Peroxymonosulfate activated by carbon nanotube and reported maximum adsorption capacity at optimum pH 6 (Chen et al. 2016). For the multi-walled carbon nanotubes, the optimum adsorption capacity of organophosphorus pesticide malathion was achieved at an optimal pH of 7 (Dehghani et al. 2017). The diquat dibromide adsorption on hydroxyl (OH)-functionalized SWCNT has been reported at a temperature of 20 °C (Dichiara, Harlander, and Rogers 2015). Other researchers targeted Methylene Blue adsorption using MWCNT-k-carrageenan-Fe_3O_4 nanocomposites and inferred a maximum adsorption capacity of 0.124 μmol g^{-1} at an optimum pH of 6.5 (O. Duman et al. 2017). For MWCNTs with a Carboxylic-to-Amine functional group, the optimum adsorption capacity of Methyl Orange was achieved at an optimal pH of 9 (Sadegh et al. 2019). The Methyl Orange dye adsorption on Al-doped carbon nanotubes has been reported at pH 4.5 and optimum adsorption capacity of 69.7 mg g^{-1} has been achieved (Kang et al. 2017).

A research group (Zhou et al. 2019) analyzed the adsorptive characteristics of oxidized multiwalled carbon nanotube for the mitigation of Perfloxacin. The authors found that the adsorbent possessed higher pollutant uptake characteristics of 61.1 mg g^{-1}. For the polypyrrole/CNTs-$CoFe_2O_4$ composite, maximum capacities of 137.00, 116.06, and 132.15 mg g^{-1} have been reported for Methyl Blue (MB), Methyl Orange (MO), and Acid Fuchsin (AF) dye (Li et al. 2017). Also, for the single, double, and multi-walled carbon nanotubes, the maximum adsorption-based uptake of 554.0, 507.0, and 391.0 mg g^{-1} respectively for Oxytetracycline adsorption have been reported (Ncibi and Sillanpää 2015). Another research group (Agarwal et al. 2016), reported maximum adsorption capacities of 123.3 and 276.2 mg g^{-1} for Methylene Blue and Bromothymol Blue respectively using polyvinyl alcohol adsorbent. Also, For the MWCNTs decorated on the calcined eggshell adsorbent, a maximum capacity of 136.99 mg g^{-1} has been reported for Congo Red dye (Seyahmazegi, Mohammad-Rezaei, and Razmi 2016). For the oxidized multi-walled carbon nanotube, the maximum adsorption capacity of Ciprofloxacin (20 mg g^{-1}) has been obtained at an optimal pH of 4 (F. Yu et al. 2016). Also, for the Granular MWCNTs, the maximum adsorption-based uptake of

369.5, 284.2, and 203.1 μmol g^{-1} respectively for Carbamazepine, Tetracycline, and Diclofenac sodium adsorption have been reported (Shan et al. 2016). A research group (Wu et al. 2016), analyzed the adsorptive characteristics of phenol using Halloysite/carbon nanocomposites. The authors found that the adsorbent possessed higher pollutant uptake characteristics (5.23 mg g^{-1}).

5.3.2 Silica-Based NMs for Organic Pollutant Removal

Other researchers targeted Methylene Orange dye adsorption using Silica NPs and inferred a maximum adsorption capacity was achieved at an optimum pH of 2 (Chaudhary, Sood, and Mehta 2016). For Monodispersed Mesoporous Silica NPs, the optimum adsorption capacities of 23.3, 34.2, 20.4, 20.1, and 14.7 mg g^{-1} respectively for Rhodamine B and Methylene Blue, Methyl Violet, Malachite Green, and Basic Fuchsin dye adsorption were achieved (Shinde et al. 2017). The adsorptive removal of Ciprofloxacin with Silica NPs has been studied from aqueous solutions (Nassar, Ahmed, and Raya 2019). For the chosen adsorbent, the highest pollutant uptake of 24.1 mg g^{-1} has been achieved for Ciprofloxacin adsorption at an optimum pH of 7.

In a relevant prior literatures, the authors prepared a solution of Methylene Blue and thereby studied the adsorptive characteristics of Modified nano-SiO_2 (Salimi et al. 2017). At the optimal process condition, the adsorption capacity of the adsorbent was inferred to be 9.5 mg g^{-1}. Other researchers studied the adsorption of Rhodamine 6G using Amphi-Functional Mesoporous Silica NPs (Shinde et al. 2017). The authors reported an adsorption capacity of 9.9 mmol g^{-1} at an optimal contact duration of 6 h. Adsorption characteristics of Xylenol Orange were examined (Zhang et al. 2016). In the article, the authors synthesized functionalized silica NPs and found that the highest adsorption capacity of 9 mg g^{-1} was achieved at an optimal pH of 3.

5.3.3 Metal Oxide-Based NMs for Organic Pollutant Removal

A research group (Chawla et al. 2017), analyzed the adsorptive characteristics of Congo Red using ZnO_2 NPs. The authors evaluated that the sorbent possessed a higher pollutant uptake capacity (208 mg g^{-1}) at an optimum pH of 7. Another research group (Heidarizad and Şengör 2016), stated an optimum sorption capacity and maximum adsorption capacity of 833 mg g^{-1} for Methylene Blue dye adsorption using MgO/GO nanocomposites. Also, another research group (Kalantary et al. 2016) reported the maximum adsorption capacity of Malathion (43.3 mg g^{-1}) for magnetic graphene oxide nanocomposite. Another research group (Muthukumaran, Sivakumar, and Thirumarimurugan 2016) targeted the removal of Crystal Violet from aqueous solution using novel Surfactant-Fe_3O_4 magnetic NP. The authors inferred a maximum adsorption capacity of 166.7 mg g^{-1}. For the novel Fe_3O_4– Trisodium citrate Nanocomposite, a maximum capacity of 435 mg g^{-1} has been reported for Malachite Green dye at an optimum pH of 4 (Alqadami et al. 2016). Also, for the Fe_3O_4/MgO NPs, the maximum adsorption-based uptake of 37.9 mg g^{-1} for Amaranth dye adsorption has been reported at an optimal pH of 9 (Salem, Ahmed, and El-Shahat 2016). Another group of researchers investigated the mitigation of Congo Red from an aqueous solution using synthesized quercetin-modified α-Fe_2O_3 NPs and reported a maximum adsorption capacity of 427.4 mg g^{-1} (Satheesh et al. 2016). For the k-carrageenan-Fe_3O_4@SiO2 siliceous shells, the optimum adsorption capacity of Metoprolol (447 mg g^{-1}) was achieved (Soares et al. 2016).

5.3.4 Chitosan-Based NMs for Organic Pollutant Removal

In a relevant prior article, the authors prepared a solution of Congo Red dye to study the adsorptive characteristics of Congo Red dye (Mohamed, Al-Harby, and Almarshed 2020). At the optimal process condition, the adsorption capacity of the adsorbent was inferred to be 63.05 mg g^{-1}. Other researchers studied the adsorption of Methylene Blue using H_2SO_4 crosslinked magnetic chitosan nanocomposite beads (Rahmi, Ismaturrahmi, and Mustafa 2019). The authors reported an

TABLE 5.1
A Summary of the Literature Reported Organic Pollutant Adsorption Characteristics of Various NMs

Adsorbents	Organic Pollutants	Optimum Contact Duration (min)	Optimum pH	Adsorption Capacity (mg g–¹)	References
Carbonaceous-MWCT	Tetracycline	–	7	192.7	Álvarez-Torrellas et al. (2016)
Oxidized-MWCT	Thiamethoxam	60	9	–	Panić et al. (2015)
Oxidized-MWCT	Diquat Dibromide	300	6.5	58.3	Osman Duman et al. (2019)
ZnO-MWCT	Congo Red	50	7	249.5	Seyed Arabi et al. (2019)
Mmino-MWCNT	Methylene Blue	60	8	178.5	Ahamad et al. (2019)
pTSA doped polyaniline@ grapheneoxide-MWCNT	Congo Red	300	5	60.0	Ansari et al. (2017)
Peroxymonosulfate activated by carbon nanotube	Acid Orange 7	–	6	–	Chen et al. (2016)
MWCTs	Malathion	30	7	–	Dehghani et al. (2017)
Hydroxyl (OH)-functionalized SWCNT	Diquat Dibromide	–	–	–	Dichiara, Harlander, and Rogers (2015)
MWCNT-k-carrageenan-Fe_3O_4 nanocomposites	Diquat Dibromide	60	6.5	0.124 μmol g^{-1}	O. Duman et al. (2017)
MWCNTs with Carboxylic-to-Amine functional group	Methyl Orange	50	9	96.0	Sadegh et al. (2019)
Al-doped carbon nanotubes	Methyl Orange	15	4.5	69.7	Kang et al. (2017)
Polypyrrole/CNTs-$CoFe_2O_4$ composite	Methyl Blue Methyl Orange Acid Fuchsin	30	3	137.0 116.0 132.1	Li et al. (2017)
SWCNT DWCNT MWCNT	Oxytetracycline	30	7	554.0 507.0 391.0	Ncibi and Sillanpää (2015)
MWCNTs decorated on the calcined eggshell	Oxytetracycline	40	9	136.9	Seyahmazegi, Mohammad-Rezaei, and Razmi (2016)
Granular MWCNTs	Carbamazepine Tetracycline Diclofenac Sodium	72 h	6	369.5 284.2 203.1 (μmol g^{-1})	Shan et al. (2016)
Halloysite/carbon nanocomposites	Phenol	–	–	5.23	Wu et al. (2016)
Oxidized-MWCNT	Ciprofloxacin	240	4	20.0	F. Yu et al. (2016)
Oxidized-MWCNT	Pefloxacin	240	6	61.1	Zhou et al. (2019)
Polyvinyl Alcohol	Methylene Blue Bromothymol Blue	10	6	123.3 276.2	Agarwal et al. (2016)
Silica NPs	Ciprofloxacin	24 h	7	24.1	Nassar, Ahmed, and Raya (2019)
Surfactant Functionalized Silica NPs	Methyl Orange	–	2	–	Chaudhary, Sood, and Mehta (2016)

(continued)

TABLE 5.1 (Continued)

Adsorbents	Organic Pollutants	Optimum Contact Duration (min)	Optimum pH	Adsorption Capacity (mg g^{-1})	References
Monodispersed Mesoporous Silica NPs	Rhodamine B Methylene Blue Methyl Violet Malachite Green Basic Fuchsin	6	7	23.3 34.2 20.4 20.1 14.7	Shinde et al. (2017)
Modified nano-SiO_2	Methylene Blue	20	6	9.5	Salimi et al. (2017)
Amphi-Functional Mesoporous Silica NPs	Rhodamine 6G	6 h	–	9.9 µmol g^{-1}	Shinde et al. (2017)
Functionalized Silica NPs	Xylenol Orange	180	3	9	Zhang et al. (2016)
ZnO_2 NPs	Congo Red	10	7	208.0	Chawla et al. (2017)
MgO/GO nanocomposites	Methylene Blue	20	11	833.0	Heidarizad and Şengör (2016)
Magnetic graphene oxide nanocomposite	Malathion	45	7	43.3	Kalantary et al. (2016)
Surfactant-Fe_3O_4 magnetic NP	Crystal Violet	10	6	166.7	Muthukumaran, Sivakumar, and Thirumarimurugan (2016)
Fe_3O_4– Trisodium citrate Nanocomposite	Malachite Green	40	4	435	Alqadami et al. (2016)
Fe_3O_4/MgO NPs	Amaranth	60	9	37.9	Salem, Ahmed, and El-Shahat (2016)
Quercetin-modified α-Fe_2O_3 NPs	Congo Red	140	5.4	427.4	Satheesh et al. (2016)
k-carrageenan-Fe_3O_4@SiO2 siliceous shells	Metoprolol	15	7	447	Soares et al. (2016)
Trimellitic anhydride isothiocyanate-cross-linked chitosan hydrogels	Congo Red	–	4	63.05	Mohamed, Al-Harby, and Almarshed (2020)
H_2SO_4 crosslinked magnetic chitosan nanocomposite beads	Methylene Blue	25	12	20.4	Rahmi, Ismaturrahmi, and Mustafa (2019)
Magnetic chitosan NPs	Tetracycline	–	5	78.1	Raeiatbin and Açıkel (2017)
Nano chitosan	Naphthalene	90	5	–	Tahvildari and Mojahedi (2016)
Magnetic chitosan nanocomposite	Crystal Violet	140	7	333.3	Massoudinejad, Rasoulzadeh, and Ghaderpoori (2019)
Magnetic amine-functionalized chitosan	Diclofenac	60	4.5	469.5	Liang et al. (2019)

adsorption capacity of 20.4 mg g^{-1}. Adsorption characteristics of Tetracycline were examined (Raeiatbin and Açıkel 2017). In the article, the authors synthesized magnetic chitosan NPs and found that the highest adsorption capacity of Tetracycline (78.1 mg g^{-1}) has been achieved at a pH of 5 after 90 min contact duration. Another research group (Tahvildari and Mojahedi 2016) targeted the removal of Naphthalene using Nano chitosan. The authors inferred an optimum adsorption at pH of 5 after 90 min contact duration.

For the magnetic chitosan nanocomposite, a maximum adsorption capacity of 333.3 mg g^{-1} has been reported for Crystal Violet (Massoudinejad, Rasoulzadeh, and Ghaderpoori 2019). Also, for the magnetic amine-functionalized chitosan, the maximum adsorption-based metal uptake of 469.5 mg g^{-1} for Diclofenac adsorption has been reported (Liang et al. 2019). A quantitative summary of the critical findings reported among the best available prior literature with respect to organic pollutants' adsorption attributes has been presented in **Table 5.1**.

5.4 MECHANISMS INVOLVED IN THE REMOVAL OF ORGANIC POLLUTANTS

The extent to which organic contaminants adhere to the surface of a sorbent is contingent upon the characteristics of both the sorbate (dyes, pesticides, pharmaceutical compounds, etc.) and the sorbent itself (silica NPs, chitosan NPs, metal oxide NMs, carbon nanotubes, etc.). The vast surface area and excellent sorption capability of these sorbents make them ideal for the mitigation of organic contaminants from wastewater and water streams. Both polar and non-polar organic substances interact and are sorbed on the sorbent's surface by a variety of mechanisms that maintain the two molecules' bond. Numerous probable reactions can occur between organic substances and the sorbent. These include van der Waals effects, electrostatic forces, hydrophobic bonds, and hydrogen bonds. The sorption process is mostly driven by these reactive forces, which also help in the eradication of biological contaminants in wastewater. Occasionally, the interaction of multiple forces is contingent upon the nature of the material. Every conceivable force acting on the sorbate and sorbent is being covered extensively here. The organic substances are desorbed with the help of an appropriate eluent after sorption. (Pan and Xing 2008).

5.4.1 Hydrophobic Interactions

The specific type of connection may occur when both the sorbate and the sorbent are hydrophobic in their pure form. In the majority of instances, this sort of contact happens when carbon nanotubes and organic toxins come together. Some chemical contaminants, such as Bisphenol-A (BPA) (Gupta et al. 2017) and 17α-ethinylestradiol (EE2), and carbon nanotubes fail to mix with water. Carbon nanotubes rely on hydrophobic interactions to eliminate these impurities. Through hydrophobic interactions, carbon nanotubes may also aid in the removal of perfluorinated compounds (PFCs) that exist in water.

5.4.2 π-π Stacking Interactions

When π-electrons have been identified in both the sorbent and the sorbate, π-π stacking contacts occur. Benzene compounds and polyaromatic hydrocarbons are examples of aromatic organic contaminants with π-electrons. Colorants like malachite green, congo red, bromothymol blue, methylene blue, etc., include aromatic ring structures with π-electrons, along with functional groups (Agarwal et al. 2016). Their π-electrons exhibit π-π reactions with sorbents such as carbon nanotubes, which also consist of π-electrons. Consequently, π-π stacking connections mediate the sorption that occurs when these organic pollutants come into proximity to carbon nanotubes. Sorbent and conjugated organic contaminants may sometimes combine in two distinct manners. Since hydrogen bonds and π-π stacking interactions are both involved in the sorption of methylene blue dye on oxidized MWCNTs, the substituent in both situations behaves differently from the primary molecule. The hydrogen bonds allow the nitrogen-containing group of the dye component to bind to the carboxyl group of the oxidized MWCNTs, while the π-π stacking connections include the aromatic ring of the methylene blue dye and the rolled graphene sheets of the MWCNTs, which also contain π-electrons (Agarwal et al. 2016).

5.4.3 Van der Waal's Forces

To hold the molecules together, weak forces like van der Waals' forces have been utilized. This contact occurs between molecules or atoms and relies on proximity. These interactions have no dependence on chemical electronic bonding like covalent or ionic bonds. Because of their relative vulnerability, they are more prone to disruption. As the proximity between interacting molecules broadens, the van der Waals force rapidly diminishes. The enhancement of these forces through hydrogen bonding can result in increased efficacy in the reduction of organic pollutants, including phenol and aniline compounds (Tan et al. 2015).

5.4.4 Hydrogen Bonding

It occurs between the sorbent and the sorbate when either of them possesses functionality of the H-F, N-H, or O-H variety or at least one possesses an electronegative substituent that may form a hydrogen bond through these functional groups. This kind of contact aids in the eradication of a variety of organic contaminants, such as phenolic substances, and aniline, and their substitutes found in dyes, prescribed drugs, pesticides, etc., that can create hydrogen bonds. Hydrogen bonding allows the oxidized or functionalized hydroxyl or carbonyl groups on the sorbent to bind to these organic pollutants (Saxena, Saxena, and Lochab 2020).

5.4.5 Electrostatic Interactions

Sometimes, when both the sorbate and the material exhibit charges, they combine electrostatically. Functional groups are added to adsorbents in order to impose charges on them. These groups are frequently composed of oxygen. Some polar or ionic organic contaminants, such as anionic dyes, cationic dyes, and ionic functionalized aromatic pollutants, may be removed from CNTs through the addition of hydroxyl, carbonyl, or amine groups (Gupta et al. 2013).

5.5 METHODS FOR CHARACTERIZATION OF NMS

Characterization techniques have become indispensable tools for understanding the complex nature of NMs, which have implications for numerous scientific and commercial activities. With the aid of high-tech imaging methods, such as transmission electron microscopy and scanning electron microscopy, researchers can visualize NMs, which then enables them to describe their sizes, shape, and surface characteristics. In addition, spectroscopic methods like X-ray diffraction, Fourier transform infrared spectroscopy, and Raman spectroscopy can help researchers uncover the NPs' crystal structure, functional groups, and molecular bonds. With all the above listed measurement techniques and physical and thermal analysis methodologies, such as surface area analysis and thermal gravimetric analysis, the researcher gains the most comprehensive understanding of NPs. It is also noteworthy to mention that the comprehensive insight into the NPs helped to achieve remarkable advancements in the field of nanoscience and nanotechnology. (Lin et al. 2014).

5.5.1 Thermogravimetric Measurement

Thermogravimetric analysis may be used to analyze the mass change of a product under a controlled environment. In NMs, The understanding of phenomena such as decomposition, sorption, desorption, oxidation, reduction, and sublimation in situations involving the loss of gaseous or volatile products during chemical reactions is facilitated by this kind of measurement. The device is made up of a thermostat controller, heater, sample holding, recorder, and electronic microbalance. The microbalance is used to record the changes related to the mass of the specimen being measured. The crucible houses the samples under investigation. Heaters are built with a linear heating rate throughout the investigation. Temperature is measured using thermocouples, and as the temperature rises throughout the analysis, the thermostat controller regulates the heating rate (Ziegler-Borowska et al. 2016).

5.5.2 Zeta Potential

The "stern" layer is a thin layer in an ionic solution that is formed by charged particles with oppositely charged ions. The outer portion of the solution has a diffusive layer that contains ions that are loosely linked. Electric dual layers are produced by these two layers. Ions migrate in the diffused layer due to the motion of a charged particle, rather than persisting in the bulk dispersion on the outer layer. Zeta potential, also known as shear surface electric potential, is typically measured in a sample solution by the rate of motion of charged elements toward the electrode in the existence of an exterior electric field. Zeta potential is measured by electrophoretic light scattering. This monitors a charged particle's liquid state velocity. However, the most significant issue with this approach is that it has an electroosmotic impact, which results in a decrease in accuracy (Sapsford et al. 2011).

5.5.3 Infrared Spectroscopy (IR)

By measuring the vibrational amplitudes of the bonding molecules, Fourier transforms infrared spectroscopy may assess the occurrence of functional groups. When the vibrational activation energy of molecules is in the range of 10^{13}–10^{14} Hz, the infrared spectra can be utilized to measure and describe the changes in the vibrational patterns of the functional groups in NPs (Riding et al. 2012).

5.5.4 X-Ray Diffraction (XRD)

Diffraction techniques are employed to ascertain the average size and structure of compounds. Alterations in crystal structure, phase measurement, and identification, the shape and size of crystallites, the deformation of the lattice, the size, regularity, and direction of noncrystalline materials, and other things, are examples of this type of knowledge. The crystal that needs to be examined is positioned on the goniometer, and then it is subjected to X-rays while being rotated progressively. The two-dimensional pictures are transformed into three-dimensional ones by use of the Fourier transform (FT). Inadequate resolution or mistakes could happen if the crystals are undersized (Senthil Kumar, Grace Pavithra, and Naushad 2019).

5.5.5 Mass Spectrometry (MS)

Numerous physiochemical properties of NMs, including composition, structure, and mass, may be ascertained by a variety of MS techniques. MS works by distinguishing charged atoms with various masses depending on their ratio of mass to charge. It provides excellent sensitivity and accuracy in monitoring. These days, environmental research, pharmacological examinations, and nanotechnology frequently utilize MS for assessment and characterization (Lavigne et al. 2013).

5.5.6 Thermogravimetric Measurement

Using predetermined parameters, thermogravimetric analysis (TGA) primarily examines the mass change of a sample. Thus, certain thermal processes such as sorption, vaporization, oxidation, sublimation, desorption, and reduction have been primarily highlighted by TGA. NMs, polymers, polymer nanocomposites, fibers, paints, coatings, films, and gaseous or volatile byproducts of these chemical processes may also be evaluated using TGA. TGA may be employed to predict how stable samples will be at different temperatures and also to study how chemical processes work in different contexts. The impact of ambient temperatures on sample mass is widely recognized. It should be noted that not all forms of temperature change may affect the weight change of samples. However, TGA cannot be employed to analyze the melting and crystallization characteristics of samples (Loganathan et al. 2017).

5.5.7 Scanning Electron Microscopy (SEM)

The adaptable characterizing instrument known as the scanning electron microscope (SEM) facilitates the micro-to-nanoscale inspection and investigation of both inorganic and organic heterogeneous materials. SEM can acquire surface pictures that resemble three dimensions from a vast array of materials. Even though scanning electron microscopy (SEM) is most often associated with imaging, it has several additional applications in microanalysis. A scanning electron microscope (SEM) utilizes a focused electron beam, either static or swept over an outer layer, to examine a sample under a microscope. Secondary electrons, backscatter electrons, distinctive X-rays, and other photons of varying energies may emerge from the surface of the material as a result of the electron's contact with it. Today, scanning electron microscopes include detectors specifically designed to pick up on each of these signals. Such signals are collected from a particular spot on the material and utilized to study its chemistry, crystallization, geography, and other properties. Scanning electron microscopy (SEM) necessitates an electron beam, which may be generated through either thermionic emission or field emission. A sharp, cold cathode filament is employed to produce electrons in field emission, which are then directed toward a sample by applying an amplified current. On the other hand, Thermionic emission necessitates a high-current application to a conducting filament. When it has been coupled with an anode placed in front of an electron cannon, it generates a stream of electrons that are directed toward the exterior of the sample (Senthil Kumar, Grace Pavithra, and Naushad 2019).

5.6 OBSTACLES POSED BY NPS IN EFFLUENT TREATMENT PROCESSES

There have been several problems with the commonly employed approaches in water technology. However, nanotechnology and NMs have rendered numerous novel avenues. NMs are very flexible and have their unique physical, chemical, and surface qualities (Liang et al. 2019). In the field of photo-catalysis, sorption, and membrane filtration, there has been an interesting transition in recent years from experimental to implementation on an international basis. Yet, to enhance the effectiveness of the procedures and reach the objective of long-term sustainability, systematic and ongoing changes are nevertheless essential. Furthermore, there has been a lack of evidence about the extensive research that has been conducted on the destiny of these nanocrystals in the surrounding environment, as well as their transit/persistency behavior. The US Environmental Protection Agency (EPA) emphasized on elucidating the overall elimination pathway and the potential production of byproducts during treatment. Furthermore, it is uncertain how these enhanced NPs respond to the various other pollutants. Even though studies have demonstrated that toxicity may be minimized through alternative synthesizing techniques, the recipients may still be seriously threatened.

5.7 CONCLUSION

Considering how quickly emerging countries are growing and becoming more mechanized, it has become essential to provide sustainable effluent processing facilities so that everyone has access to high-quality water. Nanotechnology stands forth as a promising option in the field of water management. Utilizing an extensive range of NPs with distinct surface-related and physicochemical characteristics, the treatment methods enhanced by NMs demonstrated the successful elimination of all potential contaminants from the wastewater. There have been limited efforts to enhance process effectiveness by the integration of biological functions. According to the prior literature, these techniques for eradication of organic contaminants from effluents have been very productive, safe for the environment, energy-efficient, and inexpensive.

5.7.1 Possible Future Prospects

Although NMs have contributed to substantial strides in effluent treatment, there have been no live or real-time surveillance technologies that could facilitate the continual evaluation of the strengthened NM in the treated effluent stream. The establishment of these real-time systems to forecast the behavior of the NPs during the treatment procedure may thus aid in a thorough comprehension of the eradication procedure and the secure disposal of the NMs. Furthermore, it is crucial to conduct a thorough investigation into the implementation of NMs in sewage treatment, contaminated water, and industrial wastewater sources while the process is ongoing. This research could yield valuable insights that might contribute to the further development of the wastewater treatment facility. The vast majority of academic scientists conducted their studies on the effectiveness of the means of elimination, the sorptive capacity, and the possibilities for rejuvenation. In contrast, comprehensive studies on the secure disposal of pollutant-bound NMs are scarce; thus, further innovative efforts are required to ensure their effective utilization. It is also critical to establish global, standard rules that outline the potential risks and uses of these NMs. In summary, academicians may be especially curious in the following areas: risk assessment, material toxicity, real-time monitoring, mechanism explanation, the possibility of recycling, fouling concerns, continual flow processes, the absence of commercialization, and the development of next-generation adaptive NMs.

REFERENCES

Agarwal, Shilpi, Hamidreza Sadegh, Majid Monajjemi, Abdel Salam Hamdy, Gomaa A.M. Ali, Amir O.H. Memar, Ramin Shahryari-Ghoshekandi, Inderjeet Tyagi, and Vinod Kumar Gupta. 2016. "Efficient Removal of Toxic Bromothymol Blue and Methylene Blue from Wastewater by Polyvinyl Alcohol." *Journal of Molecular Liquids* 218 (June). Elsevier B.V.: 191–97. doi:10.1016/j.molliq.2016.02.060

Ahamad, Tansir, Mu Naushad, Gaber E. Eldesoky, Sameerah I. Al-Saeedi, Ayman Nafady, Nada S. Al-Kadhi, Ala'a H. Al-Muhtaseb, Azmat Ali Khan, and Aslam Khan. 2019. "Effective and Fast Adsorptive Removal of Toxic Cationic Dye (MB) from Aqueous Medium Using Amino-Functionalized Magnetic Multiwall Carbon Nanotubes." *Journal of Molecular Liquids* 282 (May). Elsevier B.V.: 154–61. doi:10.1016/j.molliq.2019.02.128

Alqadami, Ayoub Abdullah, Mu Naushad, Mohammad Abulhassan Abdalla, Mohammad Rizwan Khan, and Zeid Abdullah Alothman. 2016. "Adsorptive Removal of Toxic Dye Using Fe3O4–TSC Nanocomposite: Equilibrium, Kinetic, and Thermodynamic Studies." *Journal of Chemical and Engineering Data* 61 (11). American Chemical Society: 3806–13. doi: 10.1021/acs.jced.6b00446

Álvarez-Torrellas, Silvia, Araceli Rodríguez, Gabriel Ovejero, and Juan García. 2016. "Comparative Adsorption Performance of Ibuprofen and Tetracycline from Aqueous Solution by Carbonaceous Materials." *Chemical Engineering Journal* 283 (January). Elsevier: 936–47. doi:10.1016/j.cej.2015.08.023

Ansari, Mohammad Omaish, Rajeev Kumar, Sajid Ali Ansari, Shahid Pervez Ansari, M. A. Barakat, Ahmed Alshahrie, and Moo Hwan Cho. 2017. "Anion Selective PTSA Doped Polyaniline@graphene Oxide-Multiwalled Carbon Nanotube Composite for Cr(VI) and Congo Red Adsorption." *Journal of Colloid and Interface Science* 496 (June). Academic Press Inc.: 407–15. doi:10.1016/j.jcis.2017.02.034

Bali, Ulusoy, Ebru Ç Çatalkaya, and Füsun Şengül. 2003. "Photochemical Degradation and Mineralization of Phenol: A Comparative Study." In *Journal of Environmental Science and Health - Part A Toxic/Hazardous Substances and Environmental Engineering*, 38:2259–75. doi:10.1081/ESE-120023373

Baruah, Sunandan, Muhammad Najam Khan, and Joydeep Dutta. 2016. "Perspectives and Applications of Nanotechnology in Water Treatment." *Environmental Chemistry Letters*. Springer Verlag. doi:10.1007/s10311-015-0542-2

Chaudhary, Savita, Aastha Sood, and S. K. Mehta. 2016. "Comparative Study of Dye Adsorption on Silica Nanoparticles: Effects of Surface Functionalization and Other Operational Parameters." *Journal of Nanoscience and Nanotechnology* 16 (7). American Scientific Publishers: 7433–43. doi:10.1166/jnn.2016.11128

Chawla, Sneha, Himani Uppal, Mohit Yadav, Nupur Bahadur, and Nahar Singh. 2017. "Zinc Peroxide Nanomaterial as an Adsorbent for Removal of Congo Red Dye from Waste Water." *Ecotoxicology and Environmental Safety* 135 (January). Academic Press: 68–74. doi:10.1016/j.ecoenv.2016.09.017

Chen, Jiabin, Liming Zhang, Tianyin Huang, Wenwei Li, Ying Wang, and Zhongming Wang. 2016. "Decolorization of Azo Dye by Peroxymonosulfate Activated by Carbon Nanotube: Radical versus Non-Radical Mechanism." *Journal of Hazardous Materials* 320 (December). Elsevier B.V.: 571–80. doi:10.1016/j.jhazmat.2016.07.038

Cincinelli, Alessandra, Tania Martellini, Ester Coppini, Donatella Fibbi, and Athanasios Katsoyiannis. 2015. "Nanotechnologies for Removal of Pharmaceuticals and Personal Care Products from Water and Wastewater: A Review." *Journal of Nanoscience and Nanotechnology*. American Scientific Publishers. doi:10.1166/jnn.2015.10036

Dehghani, Mohammad Hadi, Zahra Shariati Niasar, Mohammad Reza Mehrnia, Mansoreh Shayeghi, Mohammad A. Al-Ghouti, Behzad Heibati, Gordon McKay, and Kaan Yetilmezsoy. 2017. "Optimizing the Removal of Organophosphorus Pesticide Malathion from Water Using Multi-Walled Carbon Nanotubes." *Chemical Engineering Journal* 310 (February). Elsevier B.V.: 22–32. doi:10.1016/j.cej.2016.10.057

Dichiara, Anthony B., Samuel F. Harlander, and Reginald E. Rogers. 2015. "Fixed Bed Adsorption of Diquat Dibromide from Aqueous Solution Using Carbon Nanotubes." *RSC Advances* 5 (76). Royal Society of Chemistry: 61508–12. doi:10.1039/c5ra11167f

Duman, Osman, Sibel Tunç, Tülin Gürkan Polat, and Bahar Kancı Bozoğlan. 2017. "Corrigendum to 'Synthesis of Magnetic Oxidized Multiwalled Carbon Nanotube-κ-Carrageenan-Fe3O4 Nanocomposite Adsorbent and Its Application in Cationic Methylene Blue Dye Adsorption' [Carbohydrate Polymers, 147 (2016) 79–88] (S0144861716303526) (10.1016/j.Carbpol.2016.03.099)." *Carbohydrate Polymers*. Elsevier Ltd. doi:10.1016/j.carbpol.2017.02.002

Duman, Osman, Ceren Özcan, Tülin Gürkan Polat, and Sibel Tunç. 2019. "Carbon Nanotube-Based Magnetic and Non-Magnetic Adsorbents for the High-Efficiency Removal of Diquat Dibromide Herbicide from Water: OMWCNT, OMWCNT-Fe3O4 and OMWCNT-K-Carrageenan-Fe3O4 Nanocomposites." *Environmental Pollution* 244 (January). Elsevier Ltd: 723–32. doi:10.1016/j.envpol.2018.10.071

Firozjaee, Tahere Taghizade, Reza Nabi, T Taghizade Firozjaee, N Mehrdadi, M Baghdadi, and G R Nabi Bidhendi. 2018. "Application of Nanotechnology in Pesticides Removal from Aqueous Solutions: Review." *International Journal of Nanoscience* 14. https://www.researchgate.net/publication/329706104

Gupta, Vinod Kumar, Shilpi Agarwal, Hamidreza Sadegh, Gomaa A.M. Ali, Arvind Kumar Bharti, and Abdel Salam Hamdy Makhlouf. 2017. "Facile Route Synthesis of Novel Graphene Oxide-β-Cyclodextrin Nanocomposite and Its Application as Adsorbent for Removal of Toxic Bisphenol A from the Aqueous Phase." *Journal of Molecular Liquids* 237 (July). Elsevier B.V.: 466–72. doi:10.1016/j.molliq.2017.04.113

Gupta, Vinod Kumar, Rajeev Kumar, Arunima Nayak, Tawfik A. Saleh, and M. A. Barakat. 2013. "Adsorptive Removal of Dyes from Aqueous Solution onto Carbon Nanotubes: A Review." *Advances in Colloid and Interface Science*. Elsevier B.V. doi:10.1016/j.cis.2013.03.003

Heidarizad, Mahdi, and S. Sevinç Şengör. 2016. "Synthesis of Graphene Oxide/Magnesium Oxide Nanocomposites with High-Rate Adsorption of Methylene Blue." *Journal of Molecular Liquids* 224 (December). Elsevier B.V.: 607–17. doi:10.1016/j.molliq.2016.09.049

Helfrich, Louis A. 2009 "Pesticides and Aquatic Animals: A Guide to Reducing Impacts on Aquatic Systems 2 2." www.ext.vt.edu

Kalantary, Roshanak Rezaei, Ali Azari, Ali Esrafili, Kamyar Yaghmaeian, Masoud Moradi, and Kiomars Sharafi. 2016. "The Survey of Malathion Removal Using Magnetic Graphene Oxide Nanocomposite as a Novel Adsorbent: Thermodynamics, Isotherms, and Kinetic Study." *Desalination and Water Treatment* 57 (58). Taylor and Francis Inc.: 28460–73. doi:10.1080/19443994.2016.1178605

Kang, Dongjuan, Xiaolin Yu, Maofa Ge, Feng Xiao, and Hui Xu. 2017. "Novel Al-Doped Carbon Nanotubes with Adsorption and Coagulation Promotion for Organic Pollutant Removal." *Journal of Environmental Sciences (China)* 54 (April). Chinese Academy of Sciences: 1–12. doi:10.1016/j.jes.2016.04.022

Kumar, Prabhat, Lalit Mohan, and Ramagopal V S Uppaluri. 2023. "Cyclic Desorption Based Efficacy of Polyvinyl Alcohol-Chitosan Variant Resins for Multi Heavy-Metal Removal." *International Journal of Biological Macromolecules* 242 (P1). Elsevier B.V.: 124812. doi:10.1016/j.ijbiomac.2023.124812

Lavigne, Jean Philippe, Paula Espinal, Catherine Dunyach-Remy, Nourredine Messad, Alix Pantel, and Albert Sotto. 2013. "Mass Spectrometry: A Revolution in Clinical Microbiology?" *Clinical Chemistry and Laboratory Medicine*. doi:10.1515/cclm-2012-0291

Li, Xiaoli, Haijun Lu, Yun Zhang, and Fu He. 2017. "Efficient Removal of Organic Pollutants from Aqueous Media Using Newly Synthesized Polypyrrole/CNTs-CoFe2O4 Magnetic Nanocomposites." *Chemical Engineering Journal* 316. Elsevier B.V.: 893–902. doi:10.1016/j.cej.2017.02.037

Liang, Xue Xue, A. M. Omer, Zhao hong Hu, Yang guang Wang, Di Yu, and Xiao kun Ouyang. 2019. "Efficient Adsorption of Diclofenac Sodium from Aqueous Solutions Using Magnetic Amine-Functionalized Chitosan." *Chemosphere* 217 (February). Elsevier Ltd: 270–78. doi:10.1016/j.chemosphere.2018.11.023

Lin, Ping Chang, Stephen Lin, Paul C. Wang, and Rajagopalan Sridhar. 2014. "Techniques for Physicochemical Characterization of Nanomaterials." *Biotechnology Advances*. Elsevier Inc. doi:10.1016/j.biotechadv.2013.11.006

Loganathan, Sravanthi, Ravi Babu Valapa, Raghvendra Kumar Mishra, G. Pugazhenthi, and Sabu Thomas. 2017. "Thermogravimetric Analysis for Characterization of Nanomaterials." In *Thermal and Rheological Measurement Techniques for Nanomaterials Characterization*, 3:67–108. Elsevier. doi:10.1016/B978-0-323-46139-9.00004-9

Lu, Haijiao, Jingkang Wang, Marco Stoller, Ting Wang, Ying Bao, and Hongxun Hao. 2016. "An Overview of Nanomaterials for Water and Wastewater Treatment." *Advances in Materials Science and Engineering*. Hindawi Publishing Corporation. doi:10.1155/2016/4964828

Mahmood, Isra, Sameen Ruqia Imadi, Kanwal Shazadi, Alvina Gul, and Khalid Rehman Hakeem. 2016. "Effects of Pesticides on Environment." In *Plant, Soil and Microbes: Volume 1: Implications in Crop Science*, 253–69. Springer International Publishing. doi:10.1007/978-3-319-27455-3_13

Massoudinejad, Mohamadreza, Hassan Rasoulzadeh, and Mansour Ghaderpoori. 2019. "Magnetic Chitosan Nanocomposite: Fabrication, Properties, and Optimization for Adsorptive Removal of Crystal Violet from Aqueous Solutions." *Carbohydrate Polymers* 206 (February). Elsevier Ltd: 844–53. doi:10.1016/j.carbpol.2018.11.048

Mohamed, Nadia A., Nouf F. Al-Harby, and Mawaheb S. Almarshed. 2020. "Enhancement of Adsorption of Congo Red Dye onto Novel Antimicrobial Trimellitic Anhydride Isothiocyanate-Cross-Linked Chitosan Hydrogels." *Polymer Bulletin* 77 (12). Springer Science and Business Media Deutschland GmbH: 6135–60. doi:10.1007/s00289-019-03058-6

Muthukumaran, Chandrasekaran, Vaiyazhipalayam Murugaiyan Sivakumar, and Marimuthu Thirumarimurugan. 2016. "Adsorption Isotherms and Kinetic Studies of Crystal Violet Dye Removal from Aqueous Solution Using Surfactant Modified Magnetic Nanoadsorbent." *Journal of the Taiwan Institute of Chemical Engineers* 63 (June). Taiwan Institute of Chemical Engineers: 354–62. doi:10.1016/j.jtice.2016.03.034

Nassar, Mostafa Y., Ibrahim S. Ahmed, and Marwa A. Raya. 2019. "A Facile and Tunable Approach for Synthesis of Pure Silica Nanostructures from Rice Husk for the Removal of Ciprofloxacin Drug from Polluted Aqueous Solutions." *Journal of Molecular Liquids* 282 (May). Elsevier B.V.: 251–63. doi:10.1016/j.molliq.2019.03.017

Ncibi, Mohamed Chaker, and Mika Sillanpää. 2015. "Optimized Removal of Antibiotic Drugs from Aqueous Solutions Using Single, Double and Multi-Walled Carbon Nanotubes." *Journal of Hazardous Materials* 298 (November). Elsevier: 102–10. doi:10.1016/j.jhazmat.2015.05.025

Pan, Bo, and Baoshan Xing. 2008. "Adsorption Mechanisms of Organic Chemicals on Carbon Nanotubes." *Environmental Science and Technology*. doi:10.1021/es801777n

Panić, Sanja, Dušan Rakić, Valéria Guzsvány, Erne Kiss, Goran Boskovic, Zoltán Kónya, and Ákos Kukovecz. 2015. "Optimization of Thiamethoxam Adsorption Parameters Using Multi-Walled Carbon Nanotubes by Means of Fractional Factorial Design." *Chemosphere* 141. Elsevier Ltd: 87–93. doi:10.1016/j.chemosphere.2015.06.042

Patel, Prabhat Kumar, Srinu Nagireddi, Ramagopal V.S. Uppaluri, and Lalit Mohan Pandey. 2022. "Batch Adsorption Characteristics of Dowex Marathon MSA Commercial Resin for Au(III) Removal from Synthetic Electroless Plating Solutions." *Materials Today: Proceedings* 68 (January). Elsevier Ltd: 824–29. doi:10.1016/j.matpr.2022.06.258

Patel, Prabhat Kumar, Lalit Mohan Pandey, and Ramagopal Uppaluri. 2023a. "Multi-Metal Adsorption and Cyclic Desorption Characteristics of Zn+2 and Cu+2 Constituting Multi-Component Synthetic Wastewater System Using Commercial Resins." In *Sustainable Environment*, edited by Deepmoni Deka, Subrata Kumar Majumder, and Mihir Kumar Purkait, 3–27. Singapore: Springer Nature Singapore.

Patel, Prabhat Kumar, Lalit Mohan Pandey, and Ramagopal V.S. Uppaluri. 2023b. "Adsorptive Removal of Zn, Fe, and Pb from Zn Dominant Simulated Industrial Wastewater Solution Using Polyvinyl Alcohol

Grafted Chitosan Variant Resins." *Chemical Engineering Journal* 459 (January). Elsevier B.V.: 141563. doi:10.1016/j.cej.2023.141563

Patel, Prabhat Kumar, Lalit Mohan Pandey, and Ramagopal V.S. Uppaluri. 2023c. "Synthesized Carboxymethyl-Chitosan Variant Composites for Cyclic Adsorption- Desorption Based Removal of Fe, Pb, and Cu." *Chemosphere*. Elsevier Ltd, 139780. doi:10.1016/j.chemosphere.2023.139780

Patel, Prabhat Kumar, Lalit Mohan Pandey, and Ramagopal V.S. Uppaluri. 2024a. "Cyclic Adsorption and Desorption Characteristics of Citric Acid-Chitosan Variant Resins for Pb, Fe, and Zn Removal from Simulated Mining and Agricultural Wastewater System." *Journal of Polymers and the Environment*. Springer. doi:10.1007/s10924-024-03343-0

Patel, Prabhat Kumar, Lalit Mohan Pandey, and Ramagopal V.S. Uppaluri. 2024b. "Highly Effective Removal of Multi-Heavy Metals from Simulated Industrial Effluent through an Adsorption Process Employing Carboxymethyl-Chitosan Composites." *Environmental Research* 240 (January). Academic Press Inc. doi:10.1016/j.envres.2023.117502

Raeiatbin, Parinaz, and Yeşim Sağ Açıkel. 2017. "Removal of Tetracycline by Magnetic Chitosan Nanoparticles from Medical Wastewaters." *Desalination and Water Treatment* 73 (April). Desalination Publications: 380–88. doi:10.5004/dwt.2017.20421

Rahmi, Ismaturrahmi, and Irfan Mustafa. 2019. "Methylene Blue Removal from Water Using H2SO4 Crosslinked Magnetic Chitosan Nanocomposite Beads." *Microchemical Journal* 144 (January). Elsevier Inc.: 397–402. doi:10.1016/j.microc.2018.09.032

Relyea, Rick A., and Jason T. Hoverman. 2008. "Interactive Effects of Predators and a Pesticide on Aquatic Communities." *Oikos* 117 (11): 1647–58. doi:10.1111/j.1600-0706.2008.16933.x

Riding, Matthew J., Francis L. Martin, Júlio Trevisan, Valon Llabjani, Imran I. Patel, Kevin C. Jones, and Kirk T. Semple. 2012. "Concentration-Dependent Effects of Carbon Nanoparticles in Gram-Negative Bacteria Determined by Infrared Spectroscopy with Multivariate Analysis." *Environmental Pollution* 163 (April): 226–34. doi:10.1016/j.envpol.2011.12.027

Sadegh, Hamidreza, Gomaa Abdelgawad Mohammed Ali, Shilpi Agarwal, and Vinod Kumar Gupta. 2019. "Surface Modification of MWCNTs with Carboxylic-to-Amine and Their Superb Adsorption Performance." *International Journal of Environmental Research* 13 (3). Springer International Publishing: 523–31. doi:10.1007/s41742-019-00193-w

Salem, Abdel Naby M., M. A. Ahmed, and M. F. El-Shahat. 2016. "Selective Adsorption of Amaranth Dye on Fe3O4/MgO Nanoparticles." *Journal of Molecular Liquids* 219 (July). Elsevier B.V.: 780–88. doi:10.1016/j.molliq.2016.03.084

Salimi, Farhad, Keivan Tahmasobi, Changiz Karami, and Alireza Jahangiri. 2017. "Preparation of Modified Nano-SiO2 by Bismuth and Iron as a Novel Remover of Methylene Blue from Water Solution." *Journal of the Mexican Chemical Society* 61 (3). Sociedad Quimica de Mexico, A.C. doi:10.29356/jmcs.v61i3.351

Sapsford, Kim E., Katherine M. Tyner, Benita J. Dair, Jeffrey R. Deschamps, and Igor L. Medintz. 2011. "Analyzing Nanomaterial Bioconjugates: A Review of Current and Emerging Purification and Characterization Techniques." *Analytical Chemistry* doi:10.1021/ac200853a

Satheesh, R., K. Vignesh, M. Rajarajan, A. Suganthi, Srimala Sreekantan, Misook Kang, and Byeong Sub Kwak. 2016. "Removal of Congo Red from Water Using Quercetin Modified α-Fe2O3 Nanoparticles as Effective Nanoadsorbent." *Materials Chemistry and Physics* 180 (September). Elsevier Ltd: 53–65. doi:10.1016/j.matchemphys.2016.05.029

Saxena, Reena, Megha Saxena, and Amit Lochab. 2020. "Recent Progress in Nanomaterials for Adsorptive Removal of Organic Contaminants from Wastewater." *ChemistrySelect*. Wiley-Blackwell. doi:10.1002/slct.201903542

Scholz, Nathaniel L., Erica Fleishman, Larry Brown, Inge Werner, Michael L. Johnson, Marjorie L. Brooks, Carys L. Mitchelmore, and Daniel Schlenk. 2012. "A Perspective on Modern Pesticides, Pelagic Fish Declines, and Unknown Ecological Resilience in Highly Managed Ecosystems." *BioScience* doi:10.1525/bio.2012.62.4.13

Senthil Kumar, P., K. Grace Pavithra, and Mu Naushad. 2019. "Characterization Techniques for Nanomaterials." In *Nanomaterials for Solar Cell Applications*, 97–124. Elsevier. doi:10.1016/B978-0-12-813337-8.00004-7

Seyahmazegi, Esmat Naji, Rahim Mohammad-Rezaei, and Habib Razmi. 2016. "Multiwall Carbon Nanotubes Decorated on Calcined Eggshell Waste as a Novel Nano-Sorbent: Application for Anionic Dye Congo Red Removal." *Chemical Engineering Research and Design* 109 (May). Institution of Chemical Engineers: 824–34. doi:10.1016/j.cherd.2016.04.001

Seyed Arabi, Seyed Masoud, Rahman Shakibaei Lalehloo, Mohamad Reza Talei Bavil Olyai, Gomaa A.M. Ali, and Hamidreza Sadegh. 2019. "Removal of Congo Red Azo Dye from Aqueous Solution by ZnO Nanoparticles Loaded on Multiwall Carbon Nanotubes." *Physica E: Low-Dimensional Systems and Nanostructures* 106 (February). Elsevier B.V.: 150–55. doi:10.1016/j.physe.2018.10.030

Shan, Danna, Shubo Deng, Tianning Zhao, Gang Yu, Judith Winglee, and Mark R. Wiesner. 2016. "Preparation of Regenerable Granular Carbon Nanotubes by a Simple Heating-Filtration Method for Efficient Removal of Typical Pharmaceuticals." *Chemical Engineering Journal* 294 (June). Elsevier B.V.: 353–61. doi:10.1016/j.cej.2016.02.118

Shinde, Pravin, Sayam Sen Gupta, Baljeet Singh, Vivek Polshettiwar, and Bhagavatula L.V. Prasad. 2017. "Amphi-Functional Mesoporous Silica Nanoparticles for Dye Separation." *Journal of Materials Chemistry A* 5 (28). Royal Society of Chemistry: 14914–21. doi:10.1039/c7ta03904b

Soares, Sofia F., Tânia R. Simões, Maria António, Tito Trindade, and Ana L. Daniel-da-Silva. 2016. "Hybrid Nanoadsorbents for the Magnetically Assisted Removal of Metoprolol from Water." *Chemical Engineering Journal* 302 (October). Elsevier B.V.: 560–69. doi:10.1016/j.cej.2016.05.079

Tahvildari, Kambiz, and Hadi Mojahedi. 2016. "Investigation of Poly Aromatic Hydrocarbons Adsorption Using Chitosan and Its Synthetic Derivatives," 1. www.sciencejournal.in

Tan, Kok Bing, Mohammadtaghi Vakili, Bahman Amini Horri, Phaik Eong Poh, Ahmad Zuhairi Abdullah, and Babak Salamatinia. 2015. "Adsorption of Dyes by Nanomaterials: Recent Developments and Adsorption Mechanisms." *Separation and Purification Technology*. Elsevier B.V. doi:10.1016/j.seppur.2015.07.009

Thirunavukkarasu, Arunachalam, and Rajarathinam Nithya. 2019 "Response Surface Optimization of Cu(II) Biosorption onto Candida Tropicalis Immobilized Strontium Alginate Beads by Box-Behnken Experimental Design A BSTRACT." www.vinanie.com/jebr

Thirunavukkarasu, Arunachalam, Rajarathinam Nithya, and Raja Sivashankar. 2020. "A Review on the Role of Nanomaterials in the Removal of Organic Pollutants from Wastewater." *Reviews in Environmental Science and Biotechnology*. Springer Science and Business Media B.V. doi:10.1007/s11157-020-09548-8

Wu, Xueping, Cun Liu, Huijie Qi, Xianlong Zhang, Jianjian Dai, Qingxin Zhang, Liang Zhang, Yucheng Wu, and Xinhua Peng. 2016. "Synthesis and Adsorption Properties of Halloysite/Carbon Nanocomposites and Halloysite-Derived Carbon Nanotubes." *Applied Clay Science* 119 (January). Elsevier Ltd: 284–93. doi:10.1016/j.clay.2015.10.029

Yagub, Mustafa T., Tushar Kanti Sen, Sharmeen Afroze, and H. M. Ang. 2014. "Dye and Its Removal from Aqueous Solution by Adsorption: A Review." *Advances in Colloid and Interface Science*. Elsevier. doi:10.1016/j.cis.2014.04.002

Yang, Quanquan, Guangcai Chen, Jianfeng Zhang, and Helian Li. 2015. "Adsorption of Sulfamethazine by Multi-Walled Carbon Nanotubes: Effects of Aqueous Solution Chemistry." *RSC Advances* 5 (32). Royal Society of Chemistry: 25541–49. doi:10.1039/c4ra15056b

Yu, Fei, Sainan Sun, Sheng Han, Jie Zheng, and Jie Ma. 2016. "Adsorption Removal of Ciprofloxacin by Multi-Walled Carbon Nanotubes with Different Oxygen Contents from Aqueous Solutions." *Chemical Engineering Journal* 285 (February). Elsevier: 588–95. doi:10.1016/j.cej.2015.10.039

Yu, Jin Gang, Xiu Hui Zhao, Hua Yang, Xiao Hong Chen, Qiaoqin Yang, Lin Yan Yu, Jian Hui Jiang, and Xiao Qing Chen. 2014. "Aqueous Adsorption and Removal of Organic Contaminants by Carbon Nanotubes." *Science of the Total Environment*. Elsevier. doi:10.1016/j.scitotenv.2014.02.129

Zhang, Libo, Gengwei Zhang, Shixing Wang, Jinhui Peng, and Wei Cui. 2016. "Cation-Functionalized Silica Nanoparticle as an Adsorbent to Selectively Adsorb Anionic Dye from Aqueous Solutions." *Environmental Progress and Sustainable Energy* 35 (4). John Wiley and Sons Inc.: 1070–77. doi:10.1002/ep.12333

Zhou, Yaoyu, Yangzhuo He, Yujia Xiang, Sijun Meng, Xiaocheng Liu, Jiangfang Yu, Jian Yang, Jiachao Zhang, Pufeng Qin, and Lin Luo. 2019. "Single and Simultaneous Adsorption of Pefloxacin and Cu(II) Ions from Aqueous Solutions by Oxidized Multiwalled Carbon Nanotube." *Science of the Total Environment* 646 (January). Elsevier B.V.: 29–36. doi:10.1016/j.scitotenv.2018.07.267

Ziegler-Borowska, Marta, Dorota Chełminiak, Halina Kaczmarek, and Anna Kaczmarek-Kędziera. 2016. "Effect of Side Substituents on Thermal Stability of the Modified Chitosan and Its Nanocomposites with Magnetite." *Journal of Thermal Analysis and Calorimetry* 124 (3). Springer Netherlands: 1267–80. doi:10.1007/s10973-016-5260-x

6 Advantages of Nanomaterials for Environmental Remediation

Priyanka Ghanghas, Anita, Anshul Yadav, and Kavita Poonia

6.1 INTRODUCTION

The world stands on the brink of a significant environmental crisis that threatens our well-being and prosperity. Environmental degradation is an ever-worsening reality, with issues mounting globally, demanding urgent action. It is imperative that we adopt a new prospect and proactive approach, tackling impending disasters with innovative theory, strategies, and utmost seriousness. Pollution, whether in the air, water, or soil, takes millions of years to dissipate, with industrial and vehicular emissions being major contributors (Adeleye et al., 2016). Airborne pollutants such as NO(x), SO_2, and toxic organic compounds, including highly reactive persistent organic pollutants (POPs) like dioxins and polycyclic aromatic hydrocarbons (PAHs), pose significant risks to human health (Mancebo & Wang 2015). Inhalation of high levels of carbon monoxide (CO) can result in immediate poisoning, while heavy metals like lead (Pb) can cause acute or chronic poisoning upon ingestion (Roy 2021). These pollutants contribute to a range of health issues, including pulmonary diseases like chronic obstructive pulmonary disease (COPD), asthma, and bronchiolitis, as well as cancers, cardiovascular diseases, central nervous system dysfunctions, and skin disorders (Manisalidis et al., 2020).

Traditional methods such as flocculation, sand filtration, condensation, froth flotation, and activated carbon adsorption have been used for pollution control, but they have limitations, such as high energy requirements, inability to remove metal ions, and production of non-reusable chemicals. Nanotechnology emerges as a promising alternative to overcome these challenges.

Nanotechnology offers the potential to revolutionize environmental remediation with cleaner, more sustainable technologies that offer significant health and environmental benefits (Ahmed et al., 2021). Nanotechnology methods hold great promise for addressing pollution management and mitigation challenges while augmenting conventional environmental cleanup techniques. By reducing energy consumption in manufacturing, facilitating recycling of products, and developing environmentally friendly materials, nanotechnology can contribute to environmental preservation (Kaur & Roy 2021). However, it's essential to consider the potential environmental and health risks associated with nanotechnology (Khan, 2020).

In this chapter, we discuss the importance of various NMs in that they have the ability to remove pollutants, highlighting their potential contributions to addressing sustainability challenges while also acknowledging the need for careful assessment of their impacts on the environment and human health.

6.2 DIFFERENT TYPES OF POLLUTANTS

Pollutants are substances that pose harm to the environment upon their release. Exposure to pollutants can have detrimental effects on both humans and other organisms. These substances enter the environment through various natural and anthropogenic pathways (Seewagen, 2020). Here, we discuss different types of pollutants, including dyes, heavy metals, and pesticides(Figure 6.1).

DOI: 10.1201/9781003484189-6

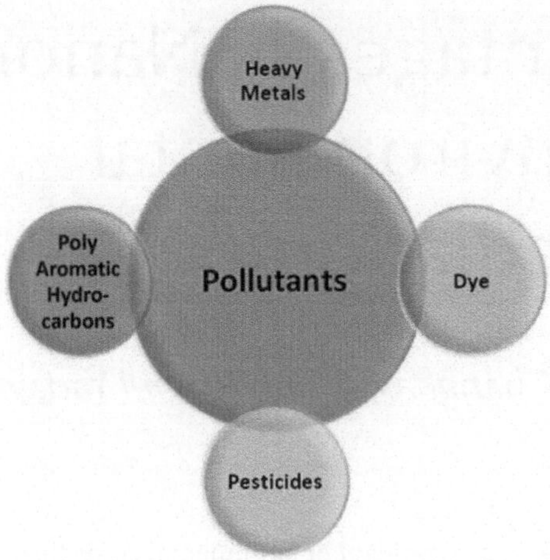

FIGURE 6.1 Different types of pollutants.

6.2.1 Dyes

Dyes are synthetic organic chemicals extensively used in various industries, notably textiles. Consequently, they have become common pollutants in industrial wastewater. Discharge of synthetic dyes into water bodies reduces light penetration, affecting the photosynthetic efficiency of aquatic plants and affects the food chain. Moreover, dyes contribute to biochemical oxygen demand (BOD), reducing dissolved oxygen levels crucial for aquatic life. Many dyes are persistent organic pollutants (POPs), posing significant environmental concerns due to their xenobiotic nature (Pereira & Alves, 2012). Some dyes, such as Azure-B, Disperse Orange 1, and Disperse Red 1, exhibit mutagenic properties, while others like Basic Red 9, Sudan I, and Crystal Violet are carcinogenic (Lellis et al., 2019).

6.2.2 Heavy Metals

Heavy metal pollution has become a global concern, due to its adverse impacts on the environment (Briffa et al., 2020). Various factors, including industrial activities, improper waste management, and agricultural practices, contribute to the release of heavy metals into soil, water, and air. Common heavy metals producing pollution include chromium, iron, titanium, cadmium, mercury, and lead. These metals are characterized by their high atomic weight and density. Anthropogenic activities such as mining, metallurgy, and industrial processes significantly contribute to heavy metal pollution, leading to detrimental effects on ecosystems and human health (Gautam et al., 2016).

6.2.3 Pesticides

Pesticides have been used in agriculture for decades to protect crops and livestock from pests and enhance productivity (Hassaan & Nemr, 2020). However, these chemicals pose risks to food safety, the environment, and human health. Pesticides are designed to eradicate unwanted organisms but can inadvertently harm non-target species and disrupt ecosystems. The World Health Organization (WHO) estimates that around 1 million people suffer from acute pesticide poisoning annually (Zhen et al., 2019). Pesticides can undergo processes such as biomagnification and bioconcentration, leading to their accumulation in organisms and subsequent health risks. DDT (dichlorodiphenyltrichloroethane) is a notorious example of a pesticide that accumulates in adipose tissues and poses risks to organisms along the food chain (Jia et al., 2020). Inhibition or inability to reproduce, suppression of the immune system or hormonal system disturbances on both animal and human, fish tumors or death, effects on intergenerational impacts of teratogens, cell damage, DNA, and significant muck on fish gills and (indicative of poor fish health) crusts, and other physiological impacts, including the weakening of eggshells, are some of the problems caused by pesticide pollutants (Hwang et al., 2018).

6.2.4 Polyaromatic Hydrocarbons (PAHs)

PAHs are persistent environmental contaminants known for their toxic, mutagenic, and carcinogenic properties. With increasing demand for petroleum products, emissions of PAHs have risen significantly. Sources include combustion of organic materials such as coal and firewood in oxygen-deficient conditions (Pandey et al., 2017). PAHs exhibit high thermal stability and resistance to decomposition. The melting and boiling temperatures of these aromatic compounds are quite high. Poly aromatic hydrocarbons with a lower molecular weight are typically water soluble, but their hydrophobicity and aqueous insolubility increase with each additional ring. (Masih et al., 2010). The US Environmental Protection Agency has identified several PAHs, including anthracene($C_6H_4CH)_2$, naphthalene, fluoranthene($C_{16}H_{10}$), phenanthrene (C_4H_{10}), pyrene($C_{16}H_{10}$), benzoanthracene($C_{18}H_{12}$), chrysene($C_{18}H_{12}$), and benzo fluoranthene ($C_{20}H_{12}$) as pollutants requiring stringent monitoring to maintain low concentrations.

6.3 ENVIRONMENTAL AND HEALTH CONSEQUENCES OF POLLUTANTS

In recent decades, advancements in analytical techniques have enabled the detection and analysis of trace pollutants. However, this has also led to the recognition of emerging environmental concerns. Incomplete toxicological information and the lack of data on long-term or low-level exposure to many chemicals obscure the environmental challenges we face. Identifying sustainable and enduring techniques to address both emerging and known pollutants, along with a thorough examination of their consequences, is crucial for the well-being of future generations.

6.3.1 Environmental Consequences

6.3.1.1 Aquatic Environment

Cultural eutrophication is a significant source of underwater pollution caused by environmental pollutants (Tilley et al., 2005). Increased food production and population growth have led to the discharge of phosphorus-rich and nitrogen compounds into water bodies. This has resulted in the proliferation of poisonous algal blooms, detrimental to endemic aquatic flora and fauna, and disruption of biogeochemical cycles of persistent organic contamination in the environment. Eutrophication has restricted water usage for fisheries, industries, and potable consumption due to decreased oxygen levels and elevated concentrations of toxic compounds (Khan & Mohammadm, 2014). Persistent organic pollutants can undergo biomagnifications, a process wherein poorly metabolizable, hydrophobic substances accumulate and magnify in concentration as they move up the food chain. This accumulation can lead to a significant reduction in the populations of native species crucial for ecosystem stability. Anthropogenic pollutants pose a threat to marine ecosystems, with elevated levels of carbon dioxide and trace metals disrupting acid-base balance in organisms, a phenomenon known as ocean acidification (Ivanina & Sokolova, 2015). This imbalance affects the survival and performance of marine organisms, leading to untimely calcification of coral reefs, zooplankton, and shellfish (Billé et al., 2013). Aquatic toxicology studies reveal the presence of mixed toxicities, including heavy metals and persistent organic pollutants (POPs), such as perfluoroalkylated acids (PFAS), on a global scale (Groffen et al., 2021). Water pollution remains a significant concern, since water is one of the most essential and vulnerable components of the environment.

6.3.1.2 Terrestrial

Industries are major contributors to environmental pollution, with unauthorized dumping of pollutants posing a threat to terrestrial ecosystems (Phale et al., 2019). Aromatic hydrocarbons, plasticizers, and xenobiotic compounds are among the pollutants with mutagenic and carcinogenic properties. Accumulation of these toxic compounds affects agricultural land productivity and seed dormancy. Municipal sewage discharge and extensive fertilizer use lead to groundwater vulnerability, due to leaching of pollutants such as heavy metals, pesticides, and fertilizers. Nutrient secretion on land can also result in water nutrient enrichment through processes like eutrophication. Instances of forest fires in areas contaminated with radioactive pollutants highlight the increasing risk posed by pollution (Álvarez et al., 2017). Increased deforestation due to various pollutants contributes to higher soil erosion rates, displacing sediment and nutrient-rich topsoil, which can lead to loose soil and increase the occurrence of flash floods.

6.3.1.3 Aerial Environment

Air pollution poses a significant threat to human civilization and ranks as the sixth leading cause of death globally. Aerial pollutants, such as carbon dioxide, fuel soot particles, methane, halocarbons, ozone, and nitrous oxides, contribute to major environmental challenges, including global warming. Global warming leads to ocean acidification, rising sea levels, increased frequency of tropical storms, loss of agricultural land, habitat destruction, and disrupted water supplies (Jacobson, 2009). Pollutants like nitrogenous and sulfur oxides, produced by combustion, interact with the atmosphere, leading to acid deposition in the form of rain, fog, or snow. This phenomenon destabilizes ecosystems, threatens food supplies, and has socioeconomic implications. In developing countries like India, urban and peri-urban areas experience substantial increases in air pollution levels, affecting nearby agricultural areas and crop production rates. High concentrations of particulate matter can induce emission and physical/chemical changes in the atmosphere, impacting air quality (Orru et al., 2017).

6.3.2 Health Consequences

Air pollutants such as oxides, particulate matter, and ground-level ozone pose significant health risks. Carbon monoxide (CO), for instance, reduces oxygen delivery to organs, impairs vision, and can be fatal at high concentrations. Sulfur oxides are reported to be carcinogenic over prolonged exposure (Kelishadi & Poursafa, 2010). These pollutants contribute to respiratory problems such as coughing, difficulty breathing, decreased lung function, asthma exacerbation, and bronchial inflammation. Children are particularly vulnerable, experiencing increased rates of infant mortality, perinatal effects, allergies, cardiovascular diseases, mental disorders, and endothelial dysfunction.

6.3.3 Traffic-Related Air Pollution

A mix of air pollutants from traffic sources can generate reactive oxygen species, leading to oxidative stress and an increased risk of cancer development (Ismail et al., 2019). Oxidative stress and impaired cardiovascular circulation can cause inflammation in organs, potentially leading to mental health issues. Nitrogen oxides and fine particulate matter can cross the blood-brain barrier, modulating vasoregulation and triggering neural inflammation, potentially leading to neurological disorders (Roberts et al., 2019). Emerging pollutants entering aquatic bodies, such as pesticides, synthetic fertilizers, dyes, detergents, and heavy metals, pose additional health risks. Eutrophic waters can cause skin and eye irritation, gastric inflammation, vomiting, and pose a risk of Blue Baby syndrome in infants due to high nitrogen levels. Heavy metals and per- and polyfluoroalkyl substances (PFAS) found in drinking water have endocrine-disrupting properties, leading to neurological and reproductive adverse effects, cancer, and obesity (Quinete and Hauser-Davis, 2021). Contaminants can also cause diseases such as diarrhea, cholera, typhoid, malaria, and congenital disorders.

6.4 TRADITIONAL METHODS OF REMOVING POLLUTANTS AND THEIR SIDE EFFECTS

Traditional methods such as coagulation, adsorption, and advanced oxidation processes, along with electrochemical and biological processes, have been widely employed for water treatment prior to the advent of nano-biotechnology.

Coagulation and flocculation, alongside adsorption, are indeed traditional methods used for treating polluted water. These methods play crucial roles in the removal of various contaminants from wastewater, enhancing its quality before further treatment or discharge. Coagulation involves the addition of coagulants like ferric chloride, which destabilize colloidal materials and small particles in the water, causing them to clump together into larger flocs that can be easily settled out. Flocculation then promotes the aggregation of these flocs into even larger particles, facilitating their removal from the water (Iwuozor 2019). This physiochemical approach is particularly effective for treating industrial wastewater containing metal pollutants and removing soluble compounds, suspended solids, and organic materials. Additionally, the coagulation-flocculation process has been combined with ozonation for treating waste from sanitary landfill leachates (Poznyak et al., 2008), further enhancing its effectiveness. Moreover, this process is known to remove nutrient pollutants like phosphorus and nitrogen water sources. In comparison, adsorption is a surface phenomenon that involves the adhesion of pollutants onto adsorbent materials. Membrane adsorption, for instance, utilizes membranes with high surface areas to adsorb contaminants from flowing water. Chromium (VI), zinc(II), and lead(II)are known to be successfully removed by polymer membrane-basedadsorptionmethodsandmembrane-basedadsorptionmethods, especially polypyrrole (Khulbe & Matsuura, 2018). Activated carbon as listed by the USEPA (Environmental Protection Agency), is one of the most widely used adsorbents, known for its effectiveness in removing organic dyes and discharges from pharmaceutical companies (Zambianchi et al., 2017). Zeolite adsorption and biosorption are both effective methods for the removal of pollutants from water sources, each offering unique advantages.

- *Zeolite Adsorption*: Natural zeolites are highly effective adsorbents, due to their high selectivity for heavy metals such as cadmium, copper, and zinc. They possess a high surface area and large pores, providing ample sites for adsorption. Additionally, zeolites are thermally stable, chemically inert, and relatively inexpensive. Their surface functional groups can be easily modified to enhance their selectivity for specific pollutants. Zeolites are widely used in water treatment, due to their excellent adsorption capacity and versatility.
- *Biosorption*: Biosorption involves the use of natural or engineered adsorbents, such as certain types of microorganisms or biomaterials, to remove pollutants from water. One key advantage of biosorption is its low cost, as many of the materials used are readily available and inexpensive. Biosorbents also exhibit high selectivity for pollutants and can be regenerated for multiple uses, reducing the overall environmental impact of the treatment process (Chaukura et al., 2016). Biosorption is particularly effective for the removal of secondary and tertiary effluent pollutants, complementing other water treatment methods.

Advanced oxidation processes (AOPs) offer effective solutions for the complete degradation of organic pollutants, achieving mineralization and producing less toxic byproducts (Lindholm-Lehto et al., 2015). These processes utilize various methods and combinations, including:

- *Ozonation*: Ozonation involves the use of ozone (O_3) to oxidize and degrade pollutants. It is particularly effective for the removal of a wide range of organic contaminants, including pesticides and aromatic compounds.
- *Fenton Reaction*: The Fenton reaction utilizes hydrogen peroxide (H_2O_2) and ferrous ions (Fe^{2+}) to produce highly reactive hydroxyl radicals (•OH), which oxidize organic compounds. Fenton-based processes can achieve significant reductions in pollutant concentrations and chemical oxygen demand.

- *UV Irradiation*: UV irradiation, often combined with hydrogen peroxide, can degrade pollutants through photolysis, generating highly reactive radicals that oxidize organic compounds. This process is effective for the degradation of pesticides and other organic pollutants.
- *Peroxide Handling*: Hydrogen peroxide (H_2O_2) can be used alone or in combination with other oxidants to degrade organic pollutants. Its use can enhance the efficiency of oxidation processes and contribute to the removal of various contaminants.
- *Photo-Fenton*: Photo-Fenton combines UV irradiation with the Fenton reaction, utilizing sunlight to activate the process. This method is particularly effective for treating recurring organic compounds and can achieve high removal efficiencies.
- *NaClO and Hydrogen Peroxide*: A combination of sodium hypochlorite (NaClO) and hydrogen peroxide (H_2O_2) can be used to remove toxic cyanide pollutants. This combination forms singlet oxygen and hydroxide ions, which react with and degrade the cyanide compounds.
- *Photolysis*: Photolysis involves the degradation of pollutants by light (photons) in the presence of hydrogen peroxide. This process can be used to degrade pesticides, especially when assisted by solvents like trichloromethane and carbon tetrachloride (Oliveira et al., 2014).

Heterogeneous photocatalysis using semiconductors, such as titanium dioxide (TiO_2), zinc oxide (ZnO), ferric oxide (Fe_2O_3), copper sulfide (CuS), and cadmium sulfide (CdS), is an effective method for degrading pollutants into biodegradable compounds and inorganic ions. These semiconductors act as catalysts in the presence of light, initiating oxidation reactions that break down organic pollutants. These processes offer several benefits, including low cost, easily crystallizable forms, phytochemical stability, and non-toxic behavior.

Ozonation is another effective method for pollutant removal, particularly for pesticides belonging to classes such as organochlorine, organophosphorus, and triazine. Ozonation oxidizes and degrades these compounds, leading to reductions in chemical oxygen demand and significant removal of aromatic compounds. Electrochemical processes, including electrocoagulation and electrochemical advanced oxidation processes (EAOPs), are traditional methods used for pollutant removal. Electrocoagulation involves the use of a sacrificial metal anode to dose water, providing cations without increasing the salinity of the solution. It offers advantages such as rapid sedimentation rates, reduced sludge production, and lower requirements for coagulant ions (Elazzouzi et al., 2017). EAOPs enable the complete mineralization of pollutants through the generation of hydroxyl ions under optimal conditions of current and catalyst. This approach promises complete automation of the process on a large scale, offering an environmentally friendly solution for pollutant removal. Electrokinetic remediation is a potential technique for the separation and remediation of areas contaminated with organic and inorganic pollutants (Hassan et al., 2015). It utilizes mechanisms such as electromigration, electroosmosis, and electrophoresis, with or without electrolysis, to remove pollutants from low permeable matrices. Electro Fenton is an electrochemical-based oxidation method that involves the indirect generation of hydroxyl radicals through electrochemical oxidation. Both hydrogen peroxide and ferrous ions can be electro generated under in situ conditions, facilitating the oxidation of organic pollutants. Biological methods for pollutant removal are cost-effective and environmentally friendly. These methods utilize microorganisms to degrade pollutants; they can be combined with other processes such as wet air oxidation and ozonation to enhance efficiency and remove toxic and refractory pollutants effectively (Jeworski & Heinzle, 2000).

6.5 ROLE OF NANOTECHNOLOGY AND NMS FOR REMEDIATION OF POLLUTANTS

Indeed, nanotechnology represents a groundbreaking field that involves manipulating materials at the scale of atoms and molecules, which is significantly smaller than what we typically encounter in our everyday lives. At this scale, conventional principles of physics and chemistry can behave quite differently, leading to unique properties and behaviors of materials. For example, carbon nanotubes are one of the most well-known NMs and exhibit extraordinary strength despite being incredibly lightweight. They are about a hundred times stronger than steel, yet six times lighter (Janas & Koziol, 2016). This exceptional strength-to-weight ratio makes them incredibly valuable for various applications across multiple industries.

Environmental remediation processes utilize a variety of methods to eliminate pollutants, including surface assimilation, absorption, synthetic reactions, catalysis using light, and filtering (Jiang et al., 2018). Nano materials are particularly advantageous in these processes due to their high surface-area-to-volume ratio, which often leads to greater biogeological and chemical reactivity compared to macroscale materials. The improved characteristics and efficacy of nanoscale materials make them well-suited for environmental remediation tasks (Guerra et al., 2018). They are being increasingly applied in various fields, including science, environmental protection, industry, and medicine, to address pollution and contamination challenges more effectively.

6.6 TYPES OF NANOPARTICLES (NPS) AND THEIR CHARACTERIZATION METHODS

NPs can be classified based on various criteria, including their origin, 3D structure, and material type.

6.6.1 Basis of Origin

NPs can originate from natural processes or be engineered through synthetic means. Natural NPs are formed through biological processes in nature or as a result of human activities (Hochella et al., 2015). On the other hand, synthetic NPs are intentionally produced through various methods such as mechanical processes, chemical reactions, biological synthesis, or physical techniques. These engineered NPs can be created individually or through a combination of different processes to achieve specific properties and functionalities (Wagner et al., 2014).

6.6.2 Basis of 3D Structure

NPs can be classified based on their structure into four categories: zero-dimensional (0D), one-dimensional (1D), two-dimensional (2D), and three-dimensional (3D). In zero-dimensional NPs, the electrons are confined within a dimensionless space, meaning they cannot move along any axis. Examples of zero-dimensional NPs include quantum dots (Jeevanandam et al., 2018). One-dimensional NPs are thin filaments with sizes typically ranging from 1 to 100 nanometers. Carbon nanotubes are a common example of one-dimensional NPs. Two-dimensional NPs include materials with structures that extend in two dimensions, such as nanosheets. Carbon nanotubes can also be considered two-dimensional when flattened. Three-dimensional NPs encompass structures that extend in three dimensions. Examples of three-dimensional NPs include dendrimers, quantum dots, and fullerenes (such as carbon-60) (Bhatia, 2016).

6.6.3 Based on Material Type

NMs can be further classified based on the material they are composed of, including inorganic, organic, carbon-containing, or composite substances.

- *Inorganic NMs*: These are composed of metals or metal oxides such as gold, silver, titanium dioxide, zinc oxide, or silicon-based compounds (Hasan, 2015). Inorganic NMs are widely used in various applications due to their unique properties.

- *Organic NMs*: These NMs are derived from organic matter, excluding carbon-based or inorganic-based sources. They can exist in various forms such as micelles, liposomes, and polymeric particles. Organic NMs find applications in drug delivery, cosmetics, and biotechnology.
- *Composite NPs*: Composite NPs are composed of multiple phases, where one phase can combine with other similar or dissimilar NPs to form complex structures. For example, a gold NP may be combined with a ceramic or polymer phase. These composite NPs exhibit unique properties derived from the combination of different materials.
- *Carbon-Based NPs*: Carbon-based NPs include fullerenes, nanotubes, and nanofibers. They are generally found in two-dimensional morphological structures and possess exceptional mechanical, electrical, and thermal properties. Carbon-based NPs are used in various fields including electronics, materials science, and biomedical applications (Nagore et al., 2021).

6.6.4 Characterization

Various techniques (Figure 6.2) are used for characterization, including UV-visible spectroscopy, X-ray diffraction (XRD), Fourier transform infrared (FTIR) spectroscopy, atomic force microscopy (AFM), scanning electron microscopy (SEM), transmission electron microscopy (TEM), and dynamic light scattering. These techniques help determine factors such as size, morphology, surface charge, and crystalline phase of NPs (Roy & Bharadvaja, 2021).

6.7 DIFFERENT APPLICATIONS OF NPS

The utilization of NPs and NMs has witnessed remarkable growth globally, spanning from medical applications to sensing tools, revolutionizing various techniques and processes.

6.7.1 Medical Applications

- NPs, especially silver NPs synthesized through green methods, exhibit antibiotic activities. They have been employed in medical implants such as silicone artificial heart valves, catheters, bone prosthetics, dental sets, and bandages. These particles possess multifactorial antimicrobial properties and can alter bacterial growth rates both

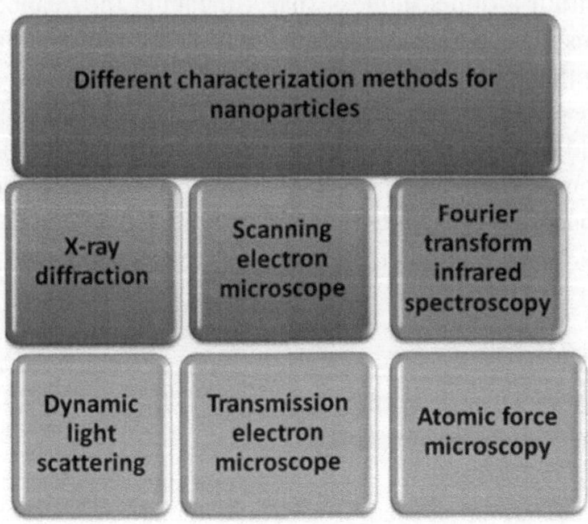

FIGURE 6.2 Different characterization methods for NPs.

extracellularly and intracellularly, inhibiting cytological transport and signaling. (Wang et al., 2013) Additionally, they have antifungal properties, capable of degrading cell walls effectively.

- Green-synthesized biodegradable NPs enable sustained drug release over extended periods without infection risks, facilitating therapeutic interventions.
- Gold nanoclusters have been utilized for bioimaging of carcinogenic cells in in vivo studies. These fluorescent NPs, when clustered, exhibit fluorescence activity, aiding in the visualization of cancerous cell lines such as hepatocarcinoma and leukemia (Huang et al., 2014).
- Carbon NPs, with negligible cytotoxicity and high biocompatibility, have been utilized for cell staining and imaging. They exhibit blue photoluminescence and are compatible with hydrophilic nature, enabling green-fluorescent imaging of human cells.

6.7.2 Therapeutic Applications

- NPs synthesized with antineoplastic properties are being researched extensively due to their potential for reduced side effects compared to conventional therapies. These NPs effectively target tumorigenic cell lines and can be used alone or in combination with high-dose drugs to optimize therapy outcomes (Tourinho et al., 2012).
- Targeted therapies employing a combination of anticarcinogenic NMs and carriers like polymeric micelles, dendrimers, and liposomes are being developed for improved efficacy and reduced toxicity.
- Inorganic metallic NPs, including metals and metal oxides, hold promise for future applications in cell imaging, sensor development, and drug delivery systems, contributing to advancements in personalized medicine and targeted therapies (Sanaeimehr et al., 2018).

6.7.3 NMs in Remediation of Soil

Soil contamination and degradation remain significant environmental challenges, posing a threat to agricultural production and food safety worldwide (Bakshi and Abhilash, 2020). Degraded soil reservoirs have a profound impact on agricultural productivity, exacerbating the need for immediate remediation efforts. Heavy metals, herbicides, and persistent organic pollutants (POPs) are among the substances that contaminate soil, further exacerbating the issue. Soil contaminated with these chemicals increases the risk of food chain contamination through biomagnification processes. The urgency to increase food production to meet the needs of a growing population, coupled with the challenge of preventing further land degradation, significantly impacts agricultural production. The immobilization or adsorption procedure is one of the most commonly utilized treatment techniques for addressing soil contamination (Bhadouria et al., 2020). This approach offers several benefits, including effectiveness, low cost, and environmental friendliness, particularly in removing metal pollutants from soils. These advantages make it anobviouschoice for soil remediation efforts. Carbon-based NMs, such as carbon nanotubes, metal oxides like ferric oxide and titanium oxide, and various nanocomposites, have been utilized to immobilize soil pollutants effectively (Arif et al., 2020). For instance, ferric oxide NPs demonstrate remarkable potential in absorbing and immobilizing heavy metals, such as cadmium (Cd) and arsenic (As), from various media samples. Oily sewage poses a significant challenge to aquatic ecosystems (Peng et al., 2020). Iron NPs have shown success in removing Total Petroleum Hydrocarbons (TPHs) from water, achieving enhanced removal rates of up to 88.34% (Murgueitio et al., 2018). Therefore, nano-technology-based treatments offer the potential to produce high-performance treated water with reduced impurities and toxic substances, along with the removal of heavy metals. Nanoscience can also be combined with biological treatments to mitigate the hazards posed by environmental pollutants. For example, organic dyes can be efficiently decolorized, using industrial processes supported by NPs (Sharma et al., 2015). Gold and silver NPs serve as highly effective catalysts in dye degradation, with the basic process involving two steps: the accumulation of electrons on the particle surface, followed by the diffusion of dye for a reduction reaction (Bastús et al., 2014).

Magnetic NPs have a wide range of applications in adsorption and catalytic pollution treatment (Ajmal et al., 2018). Polyacrylamide-modified magnetite NMs, for instance, can be utilized to address soil erosion and arsenate leaching (Zheng et al., 2020). On-site immobilization of pollutants has gained attention as a feasible and cost-effective method for remediating polluted soils. Various modifications of NMs have been explored to identify suitable materials that offer low cost, high efficiency, improved stability, minimal environmental impact, and maximum reusability. Nano-hydroxyapatite particles have been effective in reducing the exchangeable percentage of metals in polluted sediments and soils, successfully immobilizing metal concentrations in pore water (Wang et al., 2014). Research has demonstrated that the use of sodium carboxymethyl cellulose-stabilized nano zero-valent iron (nZVI) can remove up to 80 percent of soil-bound chromium (VI). Abiotic reductive reactions, particularly those involving redox-sensitive elements as pollutants, can directly impact their mobility and toxicity in soil, either alone or in combination with immobilization techniques. For example, adding nZVI to soil contaminated with pyrene creates strong reducing conditions, facilitating the reduction and removal of pyrene. NMs are also employed in advanced oxidation processes (AOPs) for the breakdown of organic contaminants in polluted soil (Usman et al., 2020). These processes involve the generation of highly reactive radicals, such as hydroxyl radicals and sulfate radicals, by combining hydrogen peroxide or persulfate with soluble iron (II) for the efficient oxidative breakdown of organic contaminants. However, the acidic pH required to prevent iron (II) precipitation poses challenges, including the expense of acidification and negative impacts on soil quality and microorganisms. As a result, solid iron phases have been proposed as an alternative to soluble Fe (II) to enable chemical oxidation without altering pH. Iron particles, when used with chelating ligands, can enhance the oxidation process (Fu et al., 2018). Green NPs offer significant potential for addressing environmental pollutants and can be utilized for remediating localized environments contaminated with toxic metal ions. These materials hold promise for cleaning up waste sites and remediating contaminated soil. Ferrous NPs can be applied to disinfect heavy metal-contaminated water and soil, serving as effective and eco-friendly fertilizers. These nano fertilizers can synchronize plant nutrient uptake while reducing groundwater contamination since they are immobilized and do not transform into permanent chemical or gaseous forms that are inaccessible to growing plants (Prasad et al., 2010). Magnetite (Fe_3O_4) and greigite (Fe_3S_4) with siliceous material produced using bacteria have been utilized in optical coatings for solar energy applications (Hussain et al., 2016).

6.7.4 NMs in Remediation of Water Pollution

Wastewater discharge from industrial and untreated residential sludge, as well as commercial factories and synthetic contaminants, pose significant threats to aquatic resources, harming both humans and ecosystems. Organic compounds, heavy metal ions, and oils are among the primary pollutants in water streams, rendering them unfit for utilization (Lu et al., 2016). With the improvement of nanotechnology, a large range of novel NM-based technologies has come out to address this issue. NMs, acting as porous systems, offer a large surface area for reactions at a much lower relative weight compared to traditional materials like activated carbon. They can effectively remove contaminants from water streams. Various NM-based adsorbents with different parameters and shapes, including 0-dimensional NPs, 1-dimensional nanofibers and 2-dimensional nanosheets, nanotubes, and 3-dimensional nanoflowers, have been studied for their efficacy in this regard (Homaeigohar, 2020). In terms of composition, there is a broad diversity of NMs that can assist in cleansing aquatic bodies, including both carbon-based and non-carbon-based materials. The segregation process depends on the contaminant's chemical/physical interaction with the surface of the NM. NMs can serve as a major component in the construction of permeable segregation units, acting as supreme filtration membranes or as thin film additions to enhance the hydrophilicity and thermomechanical characteristics of the membrane (Bansal et al., 2015). Materials such as zinc and titanium oxides, ceramics, nanowires, polymeric membranes, carbon nanotubes, and submicron particles are used in various remediation methods, such as lysis, filtration, adsorption, and oxidation (Tyagi et al., 2012). Dye pollutants are particularly concerning due to their carcinogenic nature. Dyes have been used in textiles, paintings, pigments, and other products for thousands of years, with approximately 0.1 million different synthetic dyes manufactured on an industrial scale today. It is estimated that roughly 1.6 million tonnes of dyes are used annually, with fifteen percent of the total being wasted during usage (Tan et al., 2015). Crocein orange G and acid green 25 exhibit high adsorption capability when using chitosan sugar-coated magnetite NPs. Notably, the adsorbent NPs can be easily recovered using a magnetic field (Chang and Chen, 2005). Building on this concept, Fe_3O_4/activated carbon NPs have been developed, capable of separating 138 mg/g and 166.6 mg/g of dyes (methylene blue and brilliant green) (Joshi & Garg, 2019). In another approach for removing methylene blue dye from water, electrospun polyether sulfonenanofibers containing vanadium NPs are employed. These nanofibers, with a low isoelectric point, develop a large, highly hydroxylated surface area in an alkaline medium, facilitating the adsorption of cationic methylene blue molecules (Homaeigohar et al., 2016).

However, adsorption methods have the limitation of not being able to completely eliminate or neutralize dye contaminants. Instead, they merely collect and gather dye constituents by moving them to different molecular phases. This characteristic may pose challenges, especially since dye-based sewage is difficult to remove and tends to be adsorbed onto membranes that are rarely reused. A variety of advanced oxidation processes that generate hydroxyl radicals (•OH) show promise for removing color from textile effluents. The hydroxyl radical is highly reactive and oxidizes refractory organic contaminants due to its unpaired electrons. Among these processes, photocatalysis, which involves the increasing rate of photoreaction, has emerged as one of the most studied advanced oxidative remediation methods and is considered a practical destructive technique for organic dyes and pesticides, due to its cost-effectiveness (Li et al., 2019). Metal oxides with semiconductor properties, such as zinc oxide and titanium oxide, have demonstrated impressive photodecomposition capabilities when used to degrade dye contaminants. However, despite the significant potential of semiconductors for decomposition in the presence of light, industries using visible light photocatalysts often face challenges with stability or efficiency when exposed to light. Semiconductors with energy gaps falling within the visible light spectrum, belonging to groups II–VI, have been proposed as promising alternatives to address these issues (Ghadimi et al., 2020).

6.7.5 NMs in Remediation of Air Pollution

Nanotechnology offers several methods for cleaning up air pollution. One approach involves the use of nano-catalysts with a better surface area to accelerate chemical reactions that convert dangerous vapors from industrial and automobiles facilities into harmless gases. This process helps mitigate the impact of gaseous pollutants on air quality (Jain & Singh, 2012). Additionally, filters play a crucial role in combating air pollution by trapping particles and pollutants. Nanotechnology has improved filter design by enabling the creation of small absorbent structures that allow gas to pass through while trapping molecules inside. These filters utilize three mechanisms to treat toxins: surface contact with the inertial force, filter structure applied as gas flows through, and electric charge generated by the particle and filter materials. This multi-faceted approach enhances the efficiency of air purification systems, contributing to cleaner and healthier air quality (Miaomiao et al., 2016). The primary disadvantage of traditional filters is the pressure drop they cause, leading to high energy consumption. Nano-filters offer a more efficient alternative, featuring pores ranging from 1 to 10 nm, which effectively remove various organic and microorganisms pollutants. Carbon nanotube-based membranes exhibit superior ability in isolating CO_2 from other gases. These membranes can capture gases at a rate 100 times fastened than current gas isolation techniques, making them suitable for large-scale applications. Carbon-based nanotube membranes address the reverse relationship between gas isolation efficiency and the volume flowing through conventional layers (Chaturvedi et al., 2012). Nanomembrane technology holds promise for separating and purifying gases and pollutant vapors across various industries, thereby preventing their release into the environment (Zhang et al., 2015). Nanosensors represent cutting-edge devices capable of detecting and responding to detectable physical changes on a nanometer scale. These sensors utilize NPs, which are small 3D circular substances, to provide details when a toxin is present in a given space. This technology enables rapid detection and response to environmental pollutants, enhancing environmental monitoring and protection efforts (Peng et al., 2014). In the commercial sector, the risk of obnoxious and unsafe spillage poses significant challenges to safety and environmental protection. Sensors deployed primarily at commercial levels play a crucial role in detecting even small amounts of these gases. This necessitates advancements in sensor technology to create faster and more accurate sensors. Biosensors, comprising single or multimembrane nanotubes, are capable of capturing dangerous gas molecules. A new type of nanosensor known as smart dust has emerged. Indeed, the development of ultralight nano computer sensors represents a significant advancement in sensor technology. These silicon-based ultrafine particles possess the capability to wirelessly transmit collected data to a central server, enabling real-time monitoring and analysis. With a data transportation rate of about 1 kbps or more, these sensors can efficiently relay crucial information.

Moreover, the utilization of solar power allows these nano sensors to remain suspended in the air for extended periods, providing continuous monitoring of environmental conditions. Smart dust, as these nano sensors are often referred to, can relay a wealth of information, including climate conditions, humidity levels, and the concentration of pollutants, over distances of up to 20 kilometers. They enable continuous monitoring of air pollution in a given environment, providing valuable data for managing environmental quality and mitigating pollution-related risks.

6.8 CONCLUSION AND FUTURE PROSPECTS

NPs indeed offer a revolutionary approach to environmental remediation, particularly when combined with biotechnology. As human industrialization and resource consumption continue to strain the environment, there is an urgent need for effective pollutant removal methods with minimal complexity. Standard methods like coagulation, Fenton reaction, adsorption, and advanced oxidation processes have been effective to some extent, but they often come with drawbacks such as high cost, limited effectiveness, and potential environmental impacts. NPs can be tailored to target specific pollutants in various environments, such as sewage spills and landmines, providing a more targeted and efficient approach to remediation. Additionally, advancements in nanotechnology have led to the improvement of NPs with reduced toxicity during production, making them safer for both the environment and human health. Moving forward, further research and development in NP technology hold the promise of decreased production costs and increased compatibility with existing remediation processes. By harnessing the power of NPs in combination with biotechnology, we can pave the way for more effective and environmentally friendly solutions to environmental pollution.

REFERENCES

Álvarez, X., Valero, E., Santos, R. M., Varandas, S. G. P., Fernandes, L. S., Pacheco, F. A. L., 2017. Anthropogenic nutrients and eutrophication in multiple land use water- sheds: Best management practices and policies for the protection of water resources. *Land Use Policy*.69, 1–11. https://doi.org/10.1016/j.landusepol.2017.08.028

Adeleye, S. A., Conway, J. R., Garner, K., Huang, Y., Su, Y., Keller, A. A., 2016. Engineered nanomaterials for water treatment and remediation: costs, benefits, and applicability. *J. Chem. Eng*.286, 640–662. https://doi.org/10.1016/j.cej.2015.10.105

Ahmed, M. H., Roy, A., Wahab, M.et al., 2021. Applications of nanomaterials in agrifood and pharmaceutical industry. *J. Nanomater*. ArticleID 1472096, 10 pages. https://doi.org/10.1155/2021/1472096

Ajmal, Z., Muhmood, A.,Usman, M.et al., 2018. Phosphate removal from aqueous solution using iron oxides: Adsorption, desorption and regeneration characteristics. *J.Colloid Interface Sci.*528, 145–155. https://doi.org/10.1016/j.jcis.2018.05.084

Arif, Z., Sethy, N. K., Kumari, L., Mishra, P. K., Upadhyay, S. N., 2020. Recent advances in functionalized poly- mer-based composite photocatalysts for wastewater treatment. *Nano-Materials as Photocatalysts for Degradation of Environmental Pollutants*. 39–64, Elsevier, Amsterdam, Netherlands. https://doi.org/10.1016/B978-0-12-818598-8.00003-1

Bakshi, M., Abhilash, P. C., 2020. *Nanotechnology for Soil Remediation: Revitalizing the Tarnished Resource in Nano- Materials as Photocatalysts for Degradation of Environmental Pollutants*. 345–370, Elsevier, Amsterdam, Netherlands. https://doi.org/10.1016/B978-0-12-818598-8.00017-1

Bansal, P., Chaudhary, G. R., Mehta, S. K., 2015. Comparative study of catalytic activity of ZrO_2 nanoparticles for sono-catalytic and photocatalytic degradation of cationic and anionic dyes. *J. Chem. Eng.*280, 475–485. https://doi.org/10.1016/j.cej.2015.06.039

Bastús, G. N., Merkoçi, F., Piella, J., Puntes, V., 2014. Synthesis of highly monodisperse citrate-stabilized silver nanoparticles of up to 200 nm: Kinetic control and catalytic properties. *Chem.Mater.*26(9), 2836–2846. https://doi.org/10.1021/cm500316k

Bhadouria, R., Mishra, D., Singh, V. K.et al., 2020. Nanocatalyst types and their potential impacts in agroecosystems: An overview. *Nano-Materials as Photocatalysts for Degradation of Environmental Pollutants*. 323–344, Elsevier, Amsterdam, Netherlands. https://doi.org/10.1016/B978-0-12-818598-8.00016-X

Bhatia, S., 2016. Nanoparticles types, classification, characterization, fabrication methods and drug delivery applications. *Natural Polymer Drug Delivery Systems*. 33–39, Springer, Berlin, Germany.

Billé, R., Kelly, R., Biastoch, A.et al., 2013. Taking action against Ocean Acidification: A review of management and policy options. *Environ. Manag.*52,761–779.

Briffa, J., Sinagra, E., Blundell, R., 2020. Heavy metal pollution in the environment and their toxicological effects on humans. *Heliyon.*6, 1–26

Chang, C. Y., Chen, D. H., 2005. Adsorption kinetics and thermodynamics of acid dyes on a carboxymethylated chitosan-conjugated magnetic nano-adsorbent. *Macromol.Biosci.*5, 254–261. https://doi.org/10.1002/mabi.200400153

Chaturvedi, S., Dave, P. N., Shah, N. K., 2012. Applications of nano-catalyst in new era. *J. Saudi Chem. Soc.*16(3), 307–325. https://doi.org/10.1016/j.jscs.2011.01.015

Chaukura, N., Gwenzi, W., Tavengwa, N., Manyuchi, M., 2016. Biosorbents for the removal of synthetic organics and emerging pollutants: Opportunities and challenges for developing countries. *Environmental Development.*19, 84–89. https://doi.org/10.1016/j.envdev.2016.05.002

Elazzouzi, M., Haboubi, K., Salahdine, E. M., 2017. Electrocoagulation flocculation as a low-cost process for pollutants removal from urban wastewater. *Chem.Eng.Res.Des.*117, 614–626. https://doi.org/10.1016/j.cherd.2016.11.011

Fu, H., Yang, Y., Zhu R.et al., 2018. Superior adsorption of phosphate by ferrihydrite-coated and lanthanum-decorated magnetite. *J. Colloid Interface Sci.*530, 704–713. https://doi.org/10.1016/j.jcis.2018.07.025

Gautam, P. K., Gautam, R. K., Chattopadhyaya, M. C., Banerjee, S., Chattopadhyaya, M. C., Pandey, J. D., 2016. Heavy metals in the environment: Fate, transport, toxicity and remediation technologies. *Nova Science Publishers.*60, 101–130.

Ghadimi, M., Zangenehtabar, S., Homaeigohar, S., 2020. An overview of the water remediation potential of nanomaterials and their ecotoxicological impacts. *Water Switzerland.*12(4) 1150. https://doi.org/10.3390/w12041150

Groffen, T.et al., 2021. Preliminary study on the distribution of metals and persistent organic pollutants POPs, including perfluoroalkylated acids PFAS, in the aquatic environment near Morogoro, Tanzania, and the potential health risks for humans. *Environ. Res. Anthropology and Aesthetics*. 192, Article ID110299. https://doi.org/10.1016/j.envres.2020.110299

Guerra, F., Attia, M., Whitehead, D. C., Alexis, F., 2018. Nanotechnology for environmental remediation: Materials and applications. *Molecules.*23(7), 1760. https://doi.org/10.3390/molecules23071760

Hasan, S., 2015. A review on nanoparticles: Their synthesis and types. *Res. J. Recent. Sci.*4, 1–32277, 2502.

Hassaan, A. M., Nemr, E. A., 2020. Pesticides pollution: Classifications, human health impact, extraction and treatment techniques. *Egypt. J. Aquat. Res.*46(3), 207–220. https://doi.org/10.1016/j.ejar.2020.08.007

Hassan, I., Mohamedelhassan, E., Yanful, E. K., 2015. Solar powered electrokinetic remediation of Cu polluted soil using a novel anode configuration. *Electrochim. Acta* 181, 58–67. https://doi.org/10.1016/j.electacta.2015.02.216

Hochella, F. M., Spencer, M. G., Jones, K. L., 2015. Nanotechnology: Nature's gift or scientists' brainchild. *Environ. Sci.: Nano.*2, 114–119.

Homaeigohar, S., 2020. The nanosized dye adsorbents for water treatment. *Nanomater Basel, Switzerland.*10(2), 1–43. https://doi.org/10.3390/nano10020295

Homaeigohar, S., Zillohu, A. U., Abdelaziz, R., Hedayati, M. K., Elbahri, M., 2016. A novel nanohybridnanofibrous adsorbent for water purification from dye pollutants. *Materials.*9(10), 848. https://doi.org/10.3390/ma9100848

Huang, H., Xu, Y., Tang, C. J.et al., 2014. Facile and green synthesis of photoluminescent carbon nanoparticles for cellular imaging. *New. J. Chem.*38(2), 784–789. https://doi.org/10.1039/C3NJ01185B

Hussain, I., Singh, N. B., Singh, A., Singh, H., Singh, S. C., 2016. Green synthesis of nanoparticles and its potential application. *Biotechnol. Lett.*38, 545–560.

Hwang, J. I., Zimmerman, A. R., Kim, J. E., 2018. Bio- concentration factor-based management of soil pesticide residues: Endosulfan uptake by carrot and potato plants. *Sci. Total Environ.*627, 514–522. https://doi.org/10.1016/j.scitotenv.2018.01.208

Ismail, N. I., Jalaludin, J., Bakar, S. A., Hisamuddin, N. H., Suhaimi, N. F., 2019. Association of traffic related air pollution TRAP with DNA damage and respiratory health symptoms among primary school children in Selangor. *Asian J. Atmos. Environ.*13, 106–116.

Ivanina, V. A., Sokolova, I. M., 2015. Interactive effects of metal pollution and ocean acidification on physiology of marine organisms. *Curr. Zool.*61(4), 653–668. https://doi.org/10.1093/czoolo/61.4.653

Iwuozor, O. K., 2019. Prospects and challenges of using coagu- lation-flocculation method in the treatment of effluents. *Adv. J. Chem. A.*2(2), 105–127.

Jacobson, Z. M., 2009. Review of solutions to global warming, air pollution, and energy security. *Energy Environ. Sci.*2,148–173.

Jain, C.K., Singh, D. R., 2012. Technological options for the removal of arsenic with special reference to South East Asia. *Environ. Manag.*107, 1–18. https://doi.org/10.1016/j.jenvman.2012.04.016

Janas, D., Koziol, K. K., 2016. Carbon nanotube fibers and films: Synthesis, applications and perspectives of the direct- spinning method. *Nanoscale.*8(47), 19475– 19490. https://doi.org/10.1039/C6NR07549E

Jeevanandam, J., Barhoum, A.et al., 2018. Review on nanoparticles and nano- structured materials: History, sources, toxicity and regulations. *Beilstein J. Nanotechnol.*9, 1050–1074. https://doi.org/10.3762/bjnano.9.98

Jeworski, M., Heinzle, E., 2000. Combined chemical-biological treatment of wastewater containing refractory pollutants. *Biotechnol. Annu. Rev.*6, 163–196. https://doi.org/10.1016/S1387-2656(00)06022-1

Jia, Z. Q.et al., 2020. Acute toxicity, bioconcentration, elimination, action mode and detoxification metabolism of broflanilide in zebrafish, Danio rerio. *J. Hazard. Mater*. 394, Article ID122521. https://doi.org/10.1016/j.jhazmat.2020.122521

Jiang, D., Zeng, G., Huang, D.et al., 2018. Remediation of contaminated soils by enhanced nanoscale zero valent iron. *Environ. Res.*163, 217–227. https://doi.org/10.1016/j.envres.2018.01.030

Joshi, S., Garg, V. K., 2019. Applications of Fe_3O_4@AC nanoparticles for dye removal from simulated wastewater. *Chemosphere.*236, https://doi.org/10.1016/j.chemosphere.2019.07.011

Kaur, S., Roy, A., 2021. Bioremediation of heavy metals from wastewater using nanomaterials. *Environ. Dev. Sustain.*.23, 9617–9640.

Kelishadi, R., Poursafa, P., 2010. Air pollution and non-respiratory health hazards for children. *Arch. Med. Res.*6(4), 483–495. https://doi.org/10.5114/aoms.2010.14458

Khan, H. S., 2020. Green nanotechnology for the environment and sustainable development. *Green Materials for Wastewater Treatment.*13, 13–46.

Khan, M. N., Mohammadm, F., 2014. Eutrophication: Challenges and Solutions BT. *Eutrophication: Causes, Consequences and Control.* Ansari S. and S. Gill, Eds., Springer, Dordrecht, Netherlands. 1–15.

Khulbe, K. C., Matsuura, T., 2018. Removal of heavy metals and pollutants by membrane adsorption techniques. *Appl. Water Sci.*8(19), 1–30. https://doi.org/10.1007/s13201-018-0661-6

Lellis, B.et al., 2019. Effects of textile dyes on health and the environment and bioremediation potential of living organisms. *BIORI.*3(2), 275–290. https://doi.org/10.1016/j.biori.2019.09.001

Li, Q., Zhao, T., Li, M.et al., 2019. One-step construction of Pickering emulsion via commercial TiO_2 nanoparticles for photocatalytic dye degradation. *Appl. Catal. B: Environ.*249, 1–8. https://doi.org/10.1016/j.apcatb.2019.02.057

Lindholm-Lehto, C. P., Juha, S. K., Heidi, S. J. A., Sirpa, H. H., 2015. Refractory organic pollutants and toxicity in pulp and paper mill wastewaters. *Environ. Sci. &Pollut.*22, 6473–6499.

Lu, H., Wang, J., Stoller, M., Wang, T., Bao, Y., Hao, H., 2016. An overview of nanomaterials for water and wastewater treatment. *Adv. Mater. Sci. Eng.*2016(10), Article ID 4964828.

Mancebo, S. E., Wang, S. Q., 2015. Recognizing the impact of ambient air pollution on skin health. *J. Eur. Acad. Dermatol. Venereol.*29, 2326–2332. https://doi.org/10.1111/jdv.13250

Manisalidis, I., Stavropoulou, E., Stavropoulos, A., Bezirtzoglou, E., 2020. Environmental and health impacts of air pollution: A review. *Public Health Front.*8. https://doi.org/10.3389/fpubh.2020.00014

Masih, J., Masih, A., Kulshrestha, A., Singhvi, R., Taneja, A., 2010. Characteristics of polycyclic aromatic hydrocarbons in indoor and outdoor atmosphere in the North central part of India. *J. Hazard. Mater.*177(1-3), 190–198. https://doi.org/10.1016/j.jhazmat.2009.12.017

Miaomiao, Zhu, Han, J., Wang, F., Shao, W., Xiong, R., Zhang, Q., Pan, H., Yong, Y., Samal, S. K., Zhang, F., Huang, C., 2016. Electrospun nanofibers membranes for effective air filtration. *Macromol.Mater. Eng.*302(1). https://doi.org/10.1002/mame.201600353

Murgueitio, E., Cumbal, L., Abril, M., Izquierdo, A., Debut, A., Tinoco, O., 2018. Green synthesis of iron nanoparticles: Application on the removal of petroleum oil from contaminated water and soils. *J. Nanotechnol.* 2018(1), Article ID4184769. https://doi.org/10.1155/2018/4184769

Nagore, P. K., Ghotekar, S.et al., 2021. Structural properties and antimicrobial activities of polyalthialongifolia leaf extract-mediated CuOnano- particles. *BioNanoScience.*11, 579–589.

Oliveira, A.G.et al., 2014. Degradation of the pesticide chlorpyrifos in aqueous solutions with UV/H_2O_2: Optimization and effect of interfering anions. *J. Adv. Oxid. Technol.*17, 133–138. https://doi.org/10.1515/jaots-2014-0119

Orru, H., Ebi, K. L., Forsberg, B., 2017. The interplay of climate change and air pollution on health. *Curr. Environ. Health Rep.*4, 504–513.

Pandey, R., Masood, F., Singh, H. P., Batish, D., 2017. Polycyclic aromatic hydrocarbons as environmental pol- lutants: A review. *Int. J. Eng. Sci. Technol.*6, 1361–1369.

Peng, F., Su, Y., Zhong, Y., Fan, C., Lee, S.T., He, Y., 2014. Silicon nanomaterials platform for bioimaging, biosensing, and cancer therapy. *Acc. Chem. Res.*47(2), 612–623. https://doi.org/10.1021/ar400221g

Peng, Yunyan, Yuan Liu, Jinyue Dai, Lijun Cao, Xiaoqing Liu, 2020. A sustainable strategy for remediation of oily sewage: Clean and safe. *Sep. Purif. Technol.* 240, Article ID 116592. https://doi.org/10.1016/j.seppur.2020.116592

Pereira, L., Alves, M., 2012. *Dyes environmental impact and remediation in Environmental Protection Strategies for Sustainable Development.* Springer, Berlin Germany. 111–162.

Phale, S. P., Sharma, A., Gautam, K., 2019. Microbial degra- dation of xenobiotics like aromatic pollutants from the terrestrial environments. *Pharmaceuticals and Personal Care Products: Waste Management and Treatment Technology.* 259–278. https://doi.org/10.1016/B978-0-12-816189-0.00011-1

Poznyak, T., Bautista, G. L., ChÁırez, I., Córdova, R. I., Rıos, L. E., 2008. Decomposition of toxic pollutants in landfill leachate by ozone after coagulation treatment. *J. Hazard. Mater.*152(3), 1108–1114. https://doi.org/10.1016/j.jhazmat.2007.07.098

Prasad, R., Jain, V. K., Varma, A., 2010. Role of nanomaterials in symbiotic fungus growth enhancement. *Curr. Sci.*99(9), 1189–1191.

Quinete, N., Hauser-Davis, R. A., 2021. Drinking water pollutants may affect the immune system: Concerns regarding COVID-19 health effects. *Environ. Sci. Pollut. Res.*28, 1235–1246.

Roberts, S.et al., 2019. Exploration of NO_2 and PM2.5 air pollution and mental health problems using high-resolution data in London-based children from a UK longitudinal cohort study. *Psychiatry Res.*272, 8–17. https://doi.org/10.1016/j.psychres.2018.12.050

Roy, A., 2021. Plant derived silver nanoparticles and their therapeutic applications. *Curr.Pharm. Biotechnol.*22(14), 1834–1847. https://doi.org/10.2174/1389201021666201027155708

Roy, A., Bharadvaja, N., 2021. Efficient removal of heavy metals from artificial wastewater using biochar. Environ. Nanotechnol. *Monit. Manag.* 16, Article ID 100602. https://doi.org/10.1016/j.enmm.2021.100602

Sanaeimehr, Z., Javadi, I., Namvar, F., 2018. Antiangiogenic and antiapoptotic effects of green-synthesized zinc oxide nanoparticles using Sargassummuticum algae extraction. *Cancer Nanotechnol.*9(3), 1–16.

Seewagen, C. L., 2020. The threat of global mercury pollution to bird migration: Potential mechanisms and current evidence. *Ecotoxicol.*29, 1254–1267.

Sharma, K., Singh, G., Singh, G., Kumar, M., Bhalla, V., 2015. Silver nanoparticles: Facile synthesis and their catalytic application for the degradation of dyes. *RSC Adv.*5(33), 25781–25788.

Tan, B. K., Vakili, M.et al., 2015. Adsorption of dyes by nanomaterials: Recent developments and adsorption mechanisms. *Sep. Purif. Technol.*150, 229–242. https://doi.org/10.1016/j.seppur.2015.07.009

Tilley, J. S., Orlova, E. V., Gilbert, R. J. C., Andrew, P. W., Saibil, H. R., 2005. Structural basis of pore formation by the bacterial toxin pneumolysin. *Cell.*121(2), 247–256. https://doi.org/10.1016/j.cell.2005.02.033

Tourinho, Paula S. et al., 2012. Metal-based nano- particles in soil: Fate, behavior, and effects on soil invertebrates. *Environ. Toxicol. Chem.*31(8), 1679–1692. https://doi.org/10.1002/etc.1880

Tyagi, K.P., Singh, R., Vats, S., Kumar, D., Shruti, T. 2012. *Nanomaterials use in wastewater treatment in Proceedings of the International Conference on Nanotechnology and Chemical Engineering 21,* Bangkuk, Thailand.

Usman, M., Farooq, M., Wakeel, A. et al., 2020. Nanotechnology in agriculture: Current status, challenges and future opportunities. *Sci. Total Environ.* 721, Article ID137778. https://doi.org/10.1016/j.scitotenv.2020.137778

Wagner, S.et al., 2014. Spot the difference: Engineered and natural nanoparticles in the environment release, behavior, and fate. *Angew.Chem.*53(46), 12398–12419. https://doi.org/10.1002/anie.201405050

Wang, G. J., Jang, S. S., Wong, D. S. H., Shieh, S. S., Wu, C. W., 2013. Soft-sensor development with adaptive variable selection using nonnegative garrote. *Control Eng. Pract.*21(9), 1157–1164. https://doi.org/10.1016/j.conengprac.2013.05.006

Wang, Y., Fang, Z., Liang, B.,Tsang, E. P., 2014. Remediation ofhexavalent chromium contaminated soil by stabilized nanoscale zerovalent iron prepared from steel pickling waste liquor. *J. Chem. Eng.*247, 283–290. https://doi.org/10.1016/j.cej.2014.03.011

Zambianchi, M., Durso, M., Liscio, A.et al., 2017. Graphene oxide doped polysulfone membrane adsorbers for the removal of organic contaminants from water. *J. Chem. Eng.*326, 130–140. https://doi.org/10.1016/j.cej.2017.05.143

Zhang, L., Zhao, B., Jiang, C., Yang, J., Zheng, G., 2015. Preparation and transport performances of high-density, aligned carbon nanotube membranes. *Nanoscale Res. Lett.*10(266), 1–8. DOI 10.1186/s11671-015-0970-8

Zhen, X., Liu, L., Wang, X., Zhong, G., Tang, J., 2019. Fates and ecological effects of current-use pesticides CUPs in a typical river-estuarine system of Laizhou Bay, North China. *Environ. Pollut.* 252, 573–579. https://doi.org/10.1016/j.envpol.2019.05.141

Zheng, M.et al., 2020. Simultaneous control of soil erosion and arsenic leaching at disturbed land using polyacrylamide modified magnetite nanoparticles. *Sci.Total Environ.* 702, Article ID134997. https://doi.org/10.1016/j.scitotenv.2019.134997

7 Nanotechnology for Sustainable Environment

Sankara Rao Miditana, Saivenkatesh Korlam, A. Ramesh Babu, Neha Agarwal, and Vijendra Singh Solanki

7.1 INTRODUCTION

Our planet faces a multitude of environmental challenges. A growing global population puts strain on natural resources like water, land, and energy. Industrial activity pollutes air, water, and land with toxins, greenhouse gases, and hazardous waste. Unsustainable consumption patterns, characterized by excessive use of resources and waste generation, further exacerbate these issues. These factors combine to create a crisis of climate change, biodiversity loss, and pollution. The Earth's temperature is rising due to greenhouse gas emissions, disrupting weather patterns, melting glaciers, and raising sea levels, with devastating consequences for ecosystems and human societies. Habitat destruction, pollution, and climate change are driving mass extinctions at an alarming rate, jeopardizing the vital services provided by ecosystems. Contamination of air, water, and soil with industrial and agricultural pollutants poses serious health risks to humans and wildlife.

In the face of these pressing environmental issues, innovative solutions are urgently needed. Nanotechnology has emerged as a promising path forward. Nanotechnology revolves around understanding and controlling matter at the nanoscale, a realm where dimensions range from roughly 1 to 100 nm (Miditana 2023). For perspective, a human hair is about 80,000 nm wide. At this miniscule scale, materials exhibit unique physical, chemical, and biological properties that differ significantly from their bulk counterparts. This distinct behavior unlocks a vast potential for applications across various disciplines, with particular promise for environmental solutions.

The core principle of nanotechnology lies in manipulating individual atoms and molecules to create novel structures and materials with tailored properties. This manipulation can be achieved through various techniques, including precise chemical reactions for building NPs with desired sizes and functionalities. Physical vapor deposition allows for the creation of thin films with nanoscale features by vaporizing materials and then condensing them onto a surface (Ichou et al., 2023). Nature itself serves as inspiration, with certain organisms naturally synthesizing NMs with specific properties. By studying these biological processes, scientists can develop biomimetic approaches to create environmentally friendly NMs. These techniques offer a versatile toolbox for designing NMs, with properties that are highly beneficial for environmental applications.

Nanotechnology presents a promising avenue for achieving environmental sustainability. This potential arises from two key strengths. First, it offers unparalleled control over material properties at the nanoscale (Pokrajac et al., 2021). For example, NPs with high surface areas can be designed to target and absorb pollutants, facilitating the remediation of contaminated water and soil (Roy et al., 2021). This control over material behavior paves the way for groundbreaking solutions to environmental challenges. Second, nanotechnology can address existing environmental burdens and prevent future ones. Industrial processes and waste products significantly contribute to environmental pollution. Fortunately, nanotechnology offers solutions for both cleaning up existing contamination and preventing future pollution. NPs can be employed to degrade pollutants, act as filters for water purification, and even function as environmental sensors for monitoring purposes (Beni and Jabbari, 2022).

DOI: 10.1201/9781003484189-7

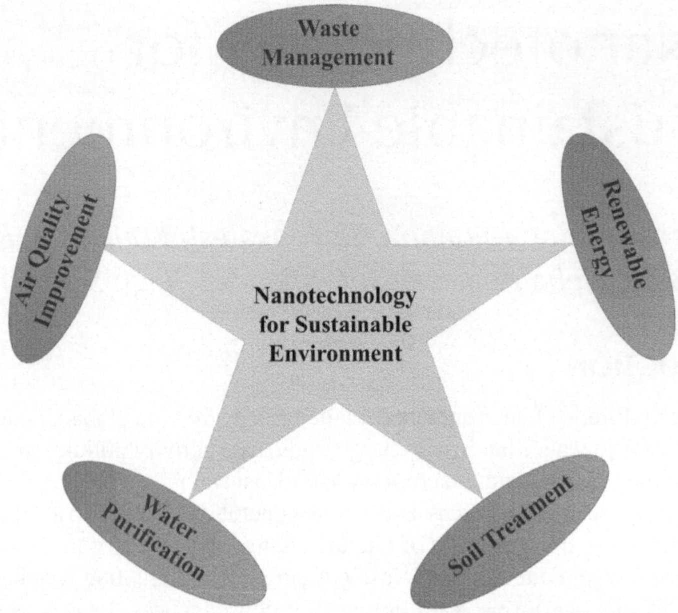

FIGURE 7.1 Sustainability via nanotechnology.

Beyond the realm of clean energy, nanotechnology offers significant advantages in resource efficiency and minimization. This technology allows for the development of highly functional materials with remarkably low material consumption. This has the potential to revolutionize manufacturing processes, leading to a more sustainable use of natural resources and a significant reduction in waste generation (Hegab et al., 2023). However, it is critical to acknowledge that venturing into nanotechnology for sustainability requires a prudent and responsible approach. Potential environmental and health risks associated with the use of NMs necessitate thorough investigation and the development of responsible management strategies before widespread adoption can be considered. This cautious approach will ensure that nanotechnology fulfils its promise for a sustainable future.

While nanotechnology has undoubtedly offered significant advancements, the world grapples with a critical issue: growing demand for essential resources like water, food, and energy. This extends to an increased need for various goods and services, including communication technology, transportation, and basic necessities like housing and healthcare. The current challenge lies in achieving improved living standards for all, while minimizing the environmental impact of human activity (Raghav et al., 2020). In this context, nanotechnology emerges as a potential game-changer. Its ability to manipulate materials at the nanoscale offers promising solutions to the pressing global issues. By enabling more efficient resource utilization, cleaner energy production, and improved environmental remediation strategies, nanotechnology holds the key to a more sustainable future (Pokrajac et al., 2021; Raghav et al., 2020).

This chapter delves into the current environmental applications of nanotechnology across various facets of human life, such as air, water, soil treatments. It further explores the potential advancements in these applications to achieve sustainable development goals. By examining the existing impact of nanotechnology and its future trajectory, this chapter aims to illuminate a path toward a more sustainable future. Figure 7.1 reveals the sustainability through nanotechnology.

7.2 NANOTECHNOLOGY IN WATER PURIFICATION

NMs are highly efficient in water treatment and purification by offering versatile solutions for a wide range of contaminants. Studies by Iravani et al. (2021) and Nishu et al. (2023) have shown the effectiveness of various NMs, including graphene oxide and carbon nanotubes in removing pollutants from water sources (Iravani, 2021; Nishu, 2023). These materials' large surface areas and unique properties enable them to act as adsorbents, catalysts, and filtration membranes, ultimately improving water quality. However, for sustainable and safe water treatment practices, challenges like scalability, cost-effectiveness, and potential environmental impacts of NMs need careful evaluation. Powerful techniques utilizing NMs include nanoadsorbents, metal and metal oxide NPs, and membrane filtration technologies. These techniques are discussed below.

7.2.1 Nanoadsorbents

Nanoadsorbents are powerful new materials for cleaning up water and wastewater. They are especially good at removing organic pollutants and heavy metals. This is because they have a very large surface area, which allows them to trap more contaminants. Nanoadsorbents can be made from a variety of materials, including carbon nanotubes (CNTs), metal based nano adsorbents, polymeric nanoadsorbents and zeolites. These are highly promising for developing more efficient and compact water treatment systems.

7.2.1.1 Carbon Nanotubes (CNTs)

CNTs are revolutionary material for developing next-generation water purification systems, particularly for small-scale applications. Their high surface area allows them to effectively capture various contaminants like salts, organic pollutants, and even bacteria (Arora and Attri, 2020). Remarkably, some CNTs can tackle multiple pollutants simultaneously. Furthermore, CNTs kill bacteria through a safe, non-chemical method and can be regenerated for extended use. These features, coupled with their suitability for simpler systems, make them ideal for point-of-use water purification devices. However, large-scale adoption remains limited, due to their high production costs.

7.2.1.2 Polymeric Nanoadsorbents

Dendrimers, star-shaped polymeric nanoadsorbents, are efficient in water purification. Their structure allows for targeted removal of pollutants: bulky internal cavities trap organic compounds, while tailored external branches capture heavy metals. Studies have demonstrated the removal of copper ions from water using dendrimer-ultrafiltration systems. These adsorbents are also regenerable through simple pH adjustments. However, their complex synthesis limits commercial availability, with only a few emerging companies in China currently producing them. Researchers have also explored combining dendrimers with biocompatible materials like chitosan to create effective bio-adsorbents for removing dyes from wastewater (Jain 2017). These biohybrid materials boast high efficiency (up to 99% removal for certain dyes) and environmental friendliness due to their biodegradability.

7.2.1.3 Zeolites

Combining zeolites and silver offers a powerful one-two punch for water purification. Zeolites, with their microscopic cages, act as a reservoir for silver NPs (Coutiño-Gonzalez et al., 2017). These silver ions are then gradually released into the water, where they attack and inhibit the growth of harmful microbes. This approach, exemplified by products like Agion®, has proven effective in water disinfection. Zeolites themselves can also be miniaturized into NPs for water treatment applications (Rahman et al., 2022). These nanozeolites can be used as adsorbents to remove contaminants or employed in wastewater treatment processes. The research suggests that zeolite-silver composites and nanozeolites hold promise for developing next-generation water purification solutions.

Compared to others, zeolites and nano-metals offer advantages in cost and compatibility with existing infrastructure (Munir et al., 2024). They can be used in existing filter systems without major modifications. However, CNTs, though highly effective for some pollutants, are currently too expensive for widespread use. Additionally, integrating CNTs requires extra filtration steps to ensure they aren't released into the environment. While all these materials are generally considered safe, their size, shape, and modifications can influence potential ecotoxicity. Each new modification needs to be evaluated for safety.

7.2.2 Nano Metal and Metal Oxides

Nanoscale metal oxides are revolutionizing water treatment with their superior capabilities. These NPs boast a large surface area, efficiently capturing a wide range of contaminants like heavy metals, radioactive materials, and arsenic (Naseem and Durrani, 2021). Additionally, certain metal oxides, like nano iron hydroxide, are highly durable and abrasion-resistant. Commercially available products like ArsenXnp, which combines iron oxide NPs with polymers, demonstrate the real-world application of this technology for arsenic removal (Alka et al., 2021).

Nanosilver offers another exciting avenue for water purification, particularly for point-of use systems. It exhibits strong antimicrobial activity against a broad spectrum of bacteria, with minimal human toxicity. Widely used in water filters and algaecides for decades, nanosilver is EPA-approved for various applications (Nagar and Thalappil, 2020). A significant advantage is that it doesn't require additional energy sources like ultraviolet light for disinfection. However, limitations exist. The antimicrobial effect weakens as silver ions are released over time, necessitating filter or device replacements. Additionally, nanosilver lacks a "depot effect," meaning it doesn't leave behind residual protection, potentially impacting long-term water quality.

Nano-titanium dioxide (TiO_2) presents a unique approach to water treatment through photocatalysis. When activated by ultraviolet light, TiO_2 triggers chemical reactions that degrade organic contaminants and microorganisms (Roy, 2022). Furthermore, TiO_2 is a low-cost, stable material with a long lifespan. However, limitations exist. Traditional TiO_2 requires ultraviolet light for activation, which can be energy-intensive. Additionally, current TiO_2 technology shows lower overall effectiveness compared to other methods like ozonation (Sari et al., 2024).

Magnetic NPs offer promise for cleaning contaminated groundwater. These NPs can be injected directly into the ground and removed using a magnetic field, allowing for targeted treatment (Al Aani, et al., 2020). This magnetic recovery offers a potentially faster and more cost-effective approach compared to traditional pump-and-treat methods. However, limitations exist for magnetic NPs like nano-zero valent iron (nZVI). While highly reactive for contaminant removal, nZVI has a short lifespan and requires further research on stabilization methods.

The widespread adoption of nanometals and nanometal oxides in water treatment hinges on overcoming some key challenges. Developing highly effective nano-TiO_2 activated by visible light would significantly improve photocatalysis. Enhancing the stability of nZVI NPs is crucial for their long-term effectiveness in groundwater remediation. Finally, while generally considered low for the materials discussed here, the potential environmental impact of any stabilizers used with these NPs needs careful evaluation.

7.2.3 Membrane Filtration Processes

Membrane separation processes are a mainstay in water and wastewater treatment, offering a reliable and automated method for filtering unwanted substances. These membranes act as physical barriers, selectively allowing or rejecting materials based on their size and the pore size of the membrane itself. However, exciting new research is utilizing nanotechnology to push the boundaries of membrane filtration, focusing on improving efficiency and reducing a major challenge - fouling.

Membrane technologies encompass a broader range of filtration techniques, including reverse osmosis (RO), ultrafiltration (UF), and microfiltration (MF). RO membranes, as reviewed by Lia et al. (2010) and Teow et al. (2021), excel in desalting brackish and seawater when high pressure is exerted to force water molecules through these semi-permeable membranes, leaving behind salts and contaminants (Lia and Wang 2010; Teow et al., 2021). UF and MF membranes, with larger pore sizes compared to NF and RO, are effective in removing colloids, bacteria, and macromolecules from water, contributing to enhancing water quality in industrial, municipal, and environmental applications.

Nanofiltration (NF) and membrane technologies have emerged as pivotal tools in water treatment and purification, offering selective filtration capabilities at the nanoscale level. Studies by Agboola et al. (2014) and Yadav et al. (2022) have delved into the principles and applications of NF, which relies on semi-permeable membranes to separate ions, molecules, and particles, based on their size and charge (Agboola et al., 2014; Yadav et al., 2022). NF membranes, with pore sizes typically ranging from 1 to 10 nm, are particularly effective in removing divalent ions, organic compounds, and fine particles from water sources, making them valuable in desalination, wastewater treatment, and drinking water production.

One area of advancement is nanofiltration, a pressure-driven technique where membranes reject particles and molecules smaller than 0.5-1 nm (Jiang et al., 2017). These membranes employ a unique charge-based repulsion mechanism to separate various ions. This makes them ideal for applications like softening water, removing color, odor, and heavy metals from groundwater, and even desalination, for converting seawater into potable water.

Another innovation is the development of nanocomposite membranes. These membranes combine existing matrix materials with embedded nanofillers, often inorganic NPs like metal oxides (Al_2O_3, TiO_2) or zeolites. These nanofillers offer a multitude of benefits: increased surface area for improved efficiency, enhanced mechanical and thermal stability for polymeric membranes, improved water permeability due to increased hydrophilicity, and even antibacterial properties (nanosilver, CNTs) or photocatalytic contaminant degradation (TiO_2) to reduce fouling.

Surface functionalization presents another approach. This technique modifies the membrane surface with chemicals that can oxidize organic contaminants, preventing the formation of fouling layers. Additionally, nanoscale functional layers containing silver or titanium dioxide can be deposited to achieve similar anti-fouling and decontamination effects. These layers are particularly advantageous, due to their minimal material requirements and ability to maintain the underlying membrane structure. Researchers are also exploring exciting new frontiers in membrane technology. Bio-nanocomposite membranes utilize highly selective proteins immobilized on their surfaces, offering enormous potential for recovering valuable metals like platinum from dilute solutions. Self-assembled block copolymer membranes involve the self-organization of block copolymers to create membranes with precisely controlled nanopore structures for superior selectivity and permeate efficiency. This method is still in its early stages, with production limited to small laboratory batches.

Finally, an emerging approach explores the controlled growth of biofilms on nanofiber surfaces for targeted degradation of pollutants like antibiotics. The biocompatible nature and size similarity of nanofibers to microorganisms make them ideal substrates for this application. Despite their effectiveness, challenges, such as membrane fouling, energy consumption, and maintenance costs, persist in membrane technologies. Ongoing research efforts, as evidenced by studies like Ahmed et al. (2023) and Yusuf et al. (2020), focus on developing advanced membrane materials, improving membrane durability and performance, and optimizing membrane processes to overcome these challenges (Ahmed et al., 2023; Yusuf et al., 2020). The continuous innovation and integration of NMs in membrane technologies hold promise for achieving sustainable and efficient water treatment solutions in the face of evolving environmental challenges.

7.3 NANOTECHNOLOGY IN AIR QUALITY IMPROVEMENT

The fight for clean air is getting a boost from nanotechnology. It is constantly developing new and innovative ways to utilize the unique properties of materials at the nanoscale for air pollution mitigation. NPs can be used in air filters and purification systems to trap and remove pollutants, such as particulate matter (PM), volatile organic compounds (VOCs), and gases (Attia et al., 2024). Nanocatalysts play a crucial role in catalytic converters used in vehicles and industrial processes to convert harmful gases like nitrogen oxides (NOx) and carbon monoxide (CO) into less harmful substances. NMs enable the development of highly sensitive and selective sensors for detecting air pollutants in real-time. Nanosensors can detect various gases, particulates, and VOCs at low concentrations, providing valuable data for air quality monitoring and management. Various NMs are employed in air quality remediation, including nanofilters, nanoadsorbents, nanosensors, and nanocatalysts. Emerging nanotechnologies show significant promise in mitigating air pollution by providing innovative solutions across various domains.

7.3.1 Nanoporous Materials for Adsorption

Air pollution is a major global concern, and nanoporous materials are emerging as powerful tools for remediation. These materials are like tiny sponges with an incredibly high surface area, which allows them to effectively capture and adsorb pollutants from the air. Nanoporous materials, such as metal-organic frameworks (MOFs), carbon-based NMs and magnetic NPs, are being explored for their exceptional adsorption capacities toward various existing contaminants in air.

7.3.1.1 Metal-Organic Frameworks (MOFs)

MOFs exhibit exceptional potential for air quality management (AQM) applications, provided effective synthesis strategies are employed to maximize their identified features. The synthesis of MOFs involves various critical variables, including the choice of metal ions, organic linkers, solvents, synthetic or post-synthetic methods, and experimental conditions. Modifications to MOF structures, such as functionalization, pre-functionalization of organic linkers, mixing of different metal nodes, and thin film coating, have been reported to enhance their performance in AQM significantly (Baumann et al., 2019). These structural modifications enable MOFs to retain specific targets strongly, improve adsorption selectivity through coordination bonding, acid-base/electrostatic interactions, and open metal sites on the pore surface. However, despite their efficacy, the high synthesis costs of MOFs relative to conventional AQM agents raise concerns about their practical applications on a large scale, particularly regarding energy and financial competitiveness (Sharanyakanth and Radhakrishnan, 2020).

In addition to adsorption applications, MOFs hold promise as heterogeneous catalysts for AQM, although challenges such as waste disposal and high costs remain. Specific synthetic strategies aimed at selecting and arranging basic building blocks within MOFs can enhance their catalytic activity and specificity. The host-guest mechanism in MOFs' heterogeneous catalysts, driven by their physicochemical properties and responsiveness to environmental stimuli, presents opportunities for tailored catalysis (Alhumaimess, 2020). Functional engineering of MOFs is essential to harness their excellent catalytic properties while addressing challenges associated with waste disposal and cost, paving the way for their practical utilization in future AQM endeavors (Kumar et al., 2016).

The utilization of sorbents stands out among various AQM control techniques as a well-developed and effective method for treating and eliminating toxic air pollutants. Commercial sorbent materials like activated carbon, zeolites, polymers, activated alumina, and silica gel have been extensively studied and employed across industries for pollutant storage, separation, purification, and removal. However, existing conventional sorbents may fall short of meeting the increasingly stringent environmental regulations and "greener" energy requirements of future technological advancements. Challenges, such as the disposal of spent sorbents and the need for higher quality standards in AQM, drive the exploration of alternative sorption materials, leading to a focus on MOFs as promising solutions in the literature (Kumar et al., 2016).

7.3.1.2 Carbon-Based NMs

The fight for clean air is taking a fascinating turn with the rise of carbon-based NMs. These tiny warriors offer unique strategies for air quality remediation. One significant approach involves their use as efficient adsorbents for pollutants like VOCs and PM. Materials such as activated carbon, carbon nanotubes, and graphene, known for their high surface areas and porosities, exhibit excellent adsorption capacities. Functionalization of these carbon NMs further enhances their adsorption selectivity and overall performance in capturing harmful gases from the air (Shen et al., 2024).

In addition to pollutant capture, carbon-based NMs also contribute to air quality improvement through catalytic conversion processes. Certain NMs like CNTs and graphene-based materials possess catalytic activity that enables them to convert pollutants such as VOCs and NOx into less harmful or inert compounds like CO_2 and N_2. This catalytic conversion strategy offers a promising avenue for reducing the environmental impact of air pollutants by transforming them into benign substances (Maduraiveeran and Wei, 2021).

Moreover, carbon-based NMs are integral to advanced sensing technologies used in air quality monitoring. These NMs exhibit high sensitivity and selectivity in detecting trace levels of pollutants, facilitating real-time assessment and management of air quality. By leveraging the unique properties of carbon-based NMs, such as their conductivity and surface reactivity, sensors can provide accurate and reliable data for informed decision-making in air quality control measures. The development and application of carbon-based NMs in sensing technologies signify a significant advancement in monitoring environmental pollution and ensuring healthier living environments (Hejazi et al., 2023).

7.3.1.3 Magnetic Nanoparticles (NPs)

Magnetic NPs are pivotal in the current strategy for air quality remediation due to their distinctive properties that facilitate efficient pollutant management. The battle against air pollution is gaining a magnetic edge with the rise of magnetic NPs. These microscopic warriors (billions of times smaller than a meter) offer unique properties for air purification (Jiang et al., 2018). One strategy utilizes their targeted attraction. Coating them with specific molecules creates "magnets" for pollutants like harmful VOCs found in paints. This targeted approach allows for the removal of specific air contaminants.

These NPs, often iron-based or ferrite NPs, serve as effective adsorbents for pollutants like heavy metals, VOCs and PM. Their high surface area and magnetic nature enable easy separation post adsorption, streamlining the purification process. Additionally, magnetic NPs find application as catalysts or catalyst supports in air purification systems, aiding catalytic reactions that convert harmful gases, such as NOx and SO_2, into less toxic compounds or capture VOCs through oxidation processes. This dual functionality enhances their effectiveness in mitigating air pollution (Chen and Zhu, 2020). Furthermore, magnetic NPs play a vital role in sensing and monitoring air quality parameters in real time. Integrated into sensor platforms, they boost sensitivity and enable selective detection of specific pollutants, contributing significantly to effective air quality management strategies. Their versatility in pollutant removal, catalytic applications, and sensing technologies underscores their importance in advancing air quality remediation efforts and ensuring healthier living environments.

7.3.2 Nanofibrous Membranes for Filtration

Nanofiber-based membranes are gaining traction for air filtration applications due to their high porosity, small pore sizes, and excellent mechanical properties. These membranes can effectively trap PM, bacteria, aerosols, and other contaminants, contributing to cleaner air environments (Yang et al., 2024). Recent developments in nanofibrous membrane technology have focused on enhancing filtration efficiency, durability, and sustainability.

7.3.2.1 Enhanced Filtration Efficiency

Recent advancements have led to the development of nanofibrous membranes with increased surface area and fine pore structures, resulting in higher filtration efficiencies for fine particles and pollutants. Functionalization of nanofibers with coatings or additives further improves their filtration performance, allowing for the removal of a wide range of airborne contaminants (Balamurugan et al., 2011).

7.3.2.2 Durability and Longevity

Research efforts have also focused on enhancing the durability and longevity of nanofibrous membranes to withstand prolonged use and maintain filtration efficiency over time. Strategies such as electrospinning techniques, polymer blending, and nanocomposite structures contribute to membrane durability and resistance to mechanical stress, ensuring sustained air quality improvement (Scaffaro and Citarrella, 2023).

7.3.2.3 Sustainable Filtration Solutions

The development of eco-friendly and sustainable nanofibrous membranes is a recent trend in air quality remediation. Biodegradable polymers and green manufacturing processes are being explored to reduce environmental impact while maintaining high filtration efficiency, aligning with the growing demand for environmentally friendly filtration solutions (Lyu et al., 2021).

7.3.3 Nanostructured Photocatalysts for Air Purification

Air pollution remains a significant public health concern. Nanostructured photocatalysts are emerging as a promising technology for air remediation, offering a way to harness the power of light to purify the air we breathe. These materials, with structures engineered on the nanoscale, possess unique properties that make them highly effective in breaking down pollutants. Nanostructured photocatalysts play a vital role in air purification by utilizing light energy to initiate chemical reactions that break down pollutants into harmless substances (Miditana et al., 2019). Advancements in nanostructured photocatalysts, such as metal oxides (e.g., zinc oxide, ZnO) and perovskite materials, enable efficient degradation of pollutants under light irradiation. These photocatalysts offer sustainable and energy-efficient solutions for removing organic pollutants and harmful gases from air streams (Karthikeyan et al., 2020). Recent developments in this field have focused on enhancing the efficiency, stability, and selectivity of nanostructured photocatalysts for effective air quality remediation.

7.3.3.1 Efficiency Enhancement

Advances in nanostructuring techniques, such as nanocomposites and heterojunctions, have significantly improved the efficiency of photocatalysts. Nanostructured materials with high surface areas and tailored band structures exhibit enhanced light absorption and charge separation, leading to improved pollutant degradation rates (Jaison et al., 2023).

7.3.3.2 Stability and Durability

Nanostructured photocatalysts focus on enhancing the stability and durability to ensure long-term performance in harsh environmental conditions. Strategies such as surface passivation, doping, and morphology control contribute to improved photocatalyst stability and resistance to photo corrosion (Wei et al., 2023).

7.3.3.3 Selective Photocatalysis

Recent developments include the design of nanostructured photocatalysts with selective photocatalytic activities, targeting specific pollutants while minimizing side reactions. This selectivity is achieved through precise control of catalyst composition, morphology, and surface properties (Fermoso et al., 2020).

7.3.4 NP-Based Sensors for Air Quality Monitoring

NP-based sensors, including metal oxide NPs and quantum dots are revolutionizing air quality monitoring by providing real-time, sensitive, and selective detection of various pollutants. These sensors offer portability, low cost, and high accuracy, making them valuable tools for pollution mitigation

TABLE 7.1
Different Nanosensors are Employed to Detect Toxic Gases (Kumar et al., 2016)

Gas Detected	Structure	Concentration(ppm)	Sensitivity
NO_2	Graphene-SnO_2 Nanocomposites	1	24.7
NO_2	SnO_2/graphene Nanocomposites	5	26,342
NO_2	RGO-polythiophene Thin film	10	26.36
NO_2	RGO/Au Nanocomposite	5	1.33
NO_2	ZnO Nanorods	100	622
SO_2	TiO_2/rGO Nanocomposite	1	10.08
H_2S	rGO/hexagonal WO_3 Nanosheets composite	10	45
H_2S	Cu NPs decorated SWCNTs Nanotube	5	11
H_2S	SnO_2-rGO Nanofibers	5	34

efforts (Saleem et al., 2022). Table 7.1 outlines the diverse nanosensors utilized for detecting harmful gases, such as NO_2, SO_2, and H_2S.

NP-based sensors have revolutionized air quality monitoring by offering high sensitivity, selectivity, and real-time detection capabilities for various pollutants. Recent developments in this field have focused on enhancing sensor performance, expanding detection capabilities, and improving sustainability in air quality monitoring systems.

7.3.4.1 Enhanced Sensitivity and Selectivity

Advances in NP-based sensors have led to increased sensitivity and selectivity for detecting trace levels of pollutants, such as VOCs, nitrogen oxides (NOx), PM, and gases like carbon monoxide (CO) and sulphur dioxide (SO_2). Functionalized NPs and nanocomposite materials enhance sensor response and specificity, allowing for accurate pollutant quantification in diverse environments (Saleem et al., 2022).

7.3.4.2 Multi-Parameter Monitoring

Recent developments include the integration of NP-based sensors into multi-parameter monitoring systems capable of measuring various air quality parameters simultaneously. These sensors offer real-time data on pollutant levels, humidity, temperature, and atmospheric pressure, providing comprehensive insights into environmental conditions (Singh et al., 2021).

7.3.4.3 Sustainable Sensor Platforms

Sustainable sensor platforms using eco-friendly NP materials and low-power consumption designs are emerging as a trend in air quality monitoring. These platforms reduce environmental impact while maintaining high-performance sensing capabilities, contributing to sustainable and efficient monitoring solutions (Chen et al., 2021).

7.4 NANOTECHNOLOGY IN SOIL TREATMENT

Nanotechnology plays a crucial role in soil remediation, helping to address soil contamination and enhance soil quality. In this discussion, let's delve into the ways NMs contribute to this field.

7.4.1 High Reactivity and Specific Area

Nanotechnology offers a powerful solution for cleaning up contaminated soil. The secret lies in the unique properties of NMs – their incredibly small size. This translates to a much larger surface area compared to their volume (high specific surface area), thereby, high reactivity to interact more readily with pollutants in the soil. This superior reactivity is what makes NMs more effective than traditional methods for soil remediation. Their small size also allows for easier penetration into the soil, particularly beneficial for in situ (on-site) remediation techniques (Lee et al., 2024).

7.4.2 Types of NMs Used in Soil Reclamation

Nanotechnology offers a diverse arsenal of tools for soil remediation, each with its own strengths against specific contaminants. Four prominent types of NMs used in soil reclamation are carbon nanotubes, zerovalent iron NPs, magnetic NPs, and zeolites.

7.4.2.1 Carbon Nanotubes (CNTs)

Carbon nanotubes are cylindrical carbon structures with exceptional mechanical strength and high surface area. In soil reclamation, CNTs are used for their adsorption capacity and ability to immobilize contaminants. They can adsorb heavy metals, organic pollutants, and nutrients, reducing their bioavailability and potential ecological impact (Ogunsola et al., 2024). Additionally, CNTs can improve soil structure and water retention, enhancing overall soil quality.

7.4.2.2 Zerovalent Iron NPs (nZVI)

Zerovalent iron NPs (nZVIs) are effective in soil reclamation due to their reactivity with various contaminants. They can degrade organic pollutants through processes like chemical reduction and oxidation, transforming them into less harmful forms (Chen et al., 2023). Offering a sustainable remediation approach, nZVIs are particularly useful for treating soil contaminated with chlorinated compounds, heavy metals, and pesticides.

7.4.2.3 Magnetic NPs

Magnetic NPs, such as iron oxide NPs, have magnetic properties that enable easy separation from soil after remediation processes. These NPs can be functionalized to target specific contaminants, enhancing their adsorption and removal efficiency (Mohammad et al., 2017). Magnetic NPs are used in soil reclamation for contaminant immobilization, remediation monitoring, and controlled-release applications.

7.4.2.4 Zeolites

Zeolites are microporous aluminosilicate minerals with high cation exchange capacity and adsorption capabilities. In soil reclamation, zeolites are employed for nutrient retention, soil conditioning, and contaminant immobilization (Gomes et al., 2016). They can adsorb heavy metals, ammonia, and organic compounds, reducing their mobility and potential leaching into groundwater.

7.4.2.5 Choosing the Right Weapon

The most effective NM for soil remediation depends on the specific type of contaminant present. CNTs offer exceptional adsorption, nZVI tackles heavy metals through reduction reactions, magnetic NPs enable targeted removal, and zeolites act as molecular sieves for specific pollutants. By understanding the strengths of each type, we can choose the right nanoweapon for the job, leading to more effective and targeted soil remediation.

7.4.3 Fate of Soil Contaminants

The presence and behavior of contaminants in soil are influenced by various factors, including their source and the soil's properties. Contaminants can exist in solid, liquid, or gaseous phases, with transport into the soil mediated by factors such as soil pH, clay content, organic matter content, and chemical compound properties (Chacón et al., 2022). Diffuse pollution, stemming from natural phenomena or wastewater discharge, and point source pollution, associated with industrial activities, are common origins of soil contaminants. The transportation of pollutants into the soil primarily occurs through mechanisms like volatilization, leaching, and suspension. Heavy metals, persistent organic pollutants, and inorganic pollutants are particularly problematic, as they resist degradation and persist in soil for extended periods due to anthropogenic activities and soil composition (Reijonen et al., 2016).

The fate of contaminants in soil is significantly influenced by aging processes, where pollutants transform into more stable solid phases and become sequestered (Ltifi et al., 2014). This aging process, a permanent transformation, affects soil properties and reduces its permeability, limiting the diffusion and adsorption of contaminants. The interaction between soil organic matter, black carbon, inorganic components, and soil structure plays a crucial role in this aging phenomenon (Ltifi et al., 2014). Furthermore, the pH value of soil is a key factor in governing the mobility and concentration of heavy metals. Higher pH levels decrease the availability and solubility of heavy metals but may increase their toxicity and mobility (Buss et al., 2018).

Clay minerals also contribute to heavy metal transportation in soil, with specific adsorption and ion exchange processes facilitated by clay mineral properties. Organic matter content influences heavy metal adsorption in soil, forming complexes that affect metal mobility and concentration (Bayouli et al., 2020). The functional groups in metal ions also impact their adsorption capacity in organic matter, with negatively charged groups being more readily adsorbed and stabilized at higher pH levels (Biswas et al., 2018).

7.4.4 Role of Nanotechnology in the Removal of Heavy Metals from Soil

Nanotechnology plays a pivotal role in the removal of heavy metals from soil, offering effective and efficient solutions to address environmental contamination. NMs such as NPs and nanotubes have high surface areas and unique adsorption properties, making them ideal nano-adsorbents for targeting heavy metal ions in soil (Ajith et al., 2021). These nano-adsorbents exhibit enhanced sorption capacities, due to their small size and high reactivity, allowing them to adsorb a larger quantity of heavy metals per unit mass compared to conventional adsorbents (Xu et al., 2016). Surface modification techniques can further enhance the adsorption selectivity and efficiency of NMs by functionalizing them with specific ligands or functional groups (Wadhawan et al., 2020).

In addition to nano-adsorbents, NPs play a crucial role in nano-enabled soil amendments, such as nanocomposites or nanofertilizers. These nano-enabled products not only improve soil quality and promote plant growth but also contribute to reducing the bioavailability of heavy metals in soil (Dimkpa and Bindraban, 2018). Nanostructured catalysts are also instrumental in catalytic degradation processes, transforming heavy metal compounds into less toxic forms through advanced oxidation processes (AOPs) (Xiao et al., 2023). These nanocatalysts enhance remediation efficiency by facilitating the breakdown of complex organic pollutants associated with heavy metals.

Furthermore, nanotechnology is integrated into various remediation techniques, such as soil washing, electrokinetic remediation, and phytoremediation. NPs improve the effectiveness of these methods by enhancing the mobilization, extraction, or immobilization of heavy metal contaminants in soil. Nanotechnology-based approaches often require lower amounts of remediation agents, energy, and resources compared to traditional methods, leading to reduced waste generation and environmental impact (Nilimaa, 2023). However, challenges such as NP mobility, long-term environmental impacts, and safety considerations need to be addressed to ensure responsible and effective use in soil remediation projects.

7.5 NANOTECHNOLOGY IN RENEWABLE ENERGY

Renewable energy sources like solar, wind, and hydrogen offer a sustainable path toward a cleaner future. However, efficiency and cost limitations have hampered their widespread adoption. Nanotechnology, the manipulation of matter at the atomic and molecular scale, is emerging as a game-changer in this domain. This section explores the exciting potential of nanotechnology in enhancing the efficiency and affordability of renewable energy technologies.

7.5.1 Solar Energy - Enhanced Solar Cell Efficiency

Traditional solar cells utilize silicon wafers, which are expensive and have limitations in light absorption. Nanotechnology offers solutions through various methods, as depicted in Figure 7.2.

- *Silicon Nanocrystals*: Embedding silicon NPs within the solar cell structure increases light absorption and improves efficiency (Mose, 2023).
- *Quantum Dots*: These semiconductor NPs can capture a wider range of light wavelengths, leading to more efficient conversion of sunlight into electricity (Serrano et al., 2009).
- *CNTs*: Their exceptional electrical conductivity can enhance electron transport within solar cells, further boosting efficiency (Serrano et al., 2009).
- *Self-Cleaning Surfaces*: NPs can be used to create superhydrophobic coatings for solar panels, repelling dust and grime, thereby minimizing maintenance and maximizing energy production (Zhang et al., 2008).

7.5.1.1 Mechanisms of Enhancing Solar Cell Efficiency by NPs

NPs play a crucial role in enhancing solar cell efficiency through various mechanisms such as light trapping, surface plasmon resonance, and charge carrier transport.

- *Light Trapping*: NPs, due to their small size, possess unique light-trapping properties. When strategically placed on the surface of solar cells, NPs can scatter and reflect incident sunlight, thereby increasing the optical path length within the cell (Khan et al., 2019). This

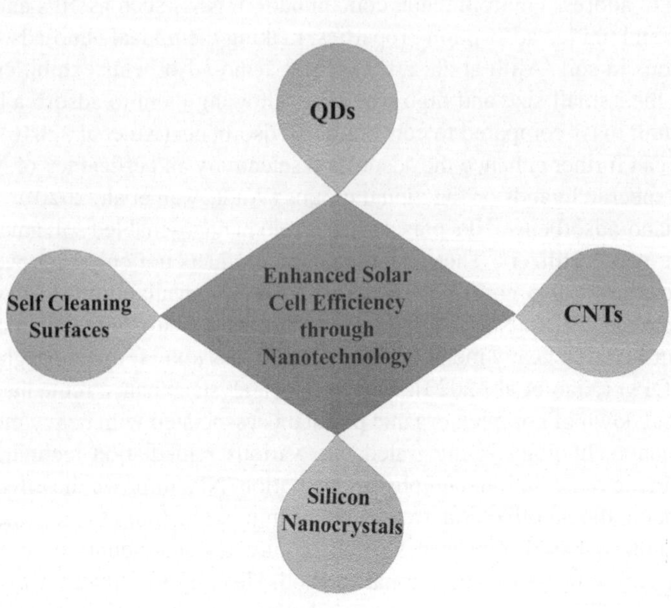

FIGURE 7.2 Solar cell efficiency via nanotechnology.

phenomenon allows for more efficient absorption of photons, particularly those with wavelengths that match the bandgap of the semiconductor material used in the solar cell.

- *Surface Plasmon Resonance (SPR)*: Metal NPs, such as gold or silver, exhibit localized surface plasmon resonance (SPR) when exposed to light. This SPR effect can enhance the absorption of light and improve the generation of electron-hole pairs within the semiconductor material of the solar cell. By controlling the size, shape, and composition of NPs, researchers can tune the SPR wavelength to coincide with the solar spectrum, thereby increasing light absorption and overall efficiency
- *Charge Carrier Transport*: NPs can also facilitate the efficient transport of charge carriers (electrons and holes) within the solar cell structure. Through processes like quantum confinement and surface passivation, NPs can minimize recombination losses and improve charge extraction efficiency, leading to higher overall power conversion efficiencies in solar cells (Eslami and Matloub, 2023).

7.5.2 Wind Energy - Improved Turbine Materials

Nanotech-enabled materials can be lighter and stronger, allowing for the construction of larger and more efficient wind turbines. Additionally, these materials can enhance durability and reduce maintenance costs

- *Superhydrophobic Coatings*: Similar to solar panels, wind turbine blades can benefit from nanotech coatings that repel ice accumulation, improving efficiency and reducing downtime during cold weather
- *NPs as composite materials for turbine blades*: One of the key areas where NPs are applied is in composite materials used for turbine blades. By incorporating NPs into the composite matrix, engineers can enhance the mechanical properties of the material, such as strength, stiffness, and fatigue resistance. This results in more durable and lightweight turbine blades that can withstand the harsh conditions of wind energy generation more effectively (Albusalih et al., 2023).
- *Surface Property Modification*: NPs can also be used to modify the surface properties of turbine components. For example, coating turbine blades with NMs can reduce drag, leading to improved aerodynamic performance and increased energy capture efficiency. Additionally, NPs can be employed in lubricants and coatings to reduce friction and wear in turbine bearings and gears, thereby extending their lifespan and reducing maintenance requirement (Ahmadi et al., 2019).

7.5.3 Hydrogen Energy - Artificial Photosynthesis

Nanocatalysts can be designed to mimic the natural process of photosynthesis, splitting water molecules into hydrogen using sunlight. This clean and efficient method of hydrogen production holds immense potential for future energy needs (Serrano et al., 2009).

NPs, particularly those made from semiconducting materials such as titanium dioxide (TiO_2), cadmium sulphide (CdS), and copper indium gallium selenide (CIGS), exhibit unique optical and electronic properties, due to their small size and high surface area-to-volume ratio. These properties enable efficient absorption of solar radiation across a broad spectrum of wavelengths, including visible and near-infrared light, which is crucial for driving the photochemical reactions involved in hydrogen production (Xu and Li, 2018).

Moreover, NPs facilitate the separation of photoexcited charge carriers (e- and h^+), preventing their recombination and thereby improving the overall efficiency of the photocatalytic process. Through surface modification and doping techniques, the charge transfer kinetics at the NP-electrolyte interface can be optimized, further enhancing the efficiency of hydrogen generation (Hisatomi et al., 2014).

Additionally, NPs serve as active sites for catalyzing the redox reactions involved in water splitting, where water molecules are dissociated into oxygen and hydrogen. Metal NPs, such as platinum (Pt), nickel (Ni), and cobalt (Co), are commonly employed as cocatalysts to promote the hydrogen evolution reaction (HER) and oxygen evolution reaction (OER) on the surface of semiconductor photocatalysts, thus accelerating the rate of hydrogen production (Miditana et al., 2024).

7.6 NANOTECHNOLOGY IN WASTE MANAGEMENT

Waste management is a rising global concern, with traditional methods often reaching capacity and struggling to address emerging waste streams like NMs. Nanotechnology, however, offers promising solutions for waste treatment, recycling, and disposal.

Nanotechnology offers a two-pronged approach to waste management: developing new, more sustainable materials and creating innovative treatment methods for existing waste. NMs can be engineered to replace conventional materials in packaging, reducing overall waste generation (Pushparaj et al., 2022). Furthermore, nanotechnologies like nanofiltration hold promise for efficiently treating industrial waste, heavy metals, and toxic compounds present in leachate (liquid that drains from landfills).

7.6.1 The Crucial Role of NPs in Waste Management

NPs, with their unique properties and vast applications across various industries, have revolutionized numerous fields. However, their miniscule size raises concerns about their potential environmental and health impacts. Effective NP waste management becomes critical to ensure the sustainable development of nanotechnology.

- *Environmental Concerns*: Due to their small size and high reactivity, NPs can easily enter the environment through various pathways, including wastewater discharge, industrial emissions, and product degradation. Their persistence and potential accumulation in ecosystems raise concerns about their impact on soil fertility, aquatic life, and even human health through the food chain (Part et al., 2018).
- *Health Risks*: NPs can potentially penetrate biological barriers and interact with cells, leading to unknown health consequences. Inhalation of airborne NPs is a significant concern, with potential for respiratory issues and other health problems Walser et al., 2012).

7.6.2 NP Waste Management Strategies

- *Prevention at Source*: The most effective approach is to minimize NP release during production and use. This can involve developing closed-loop manufacturing systems, optimizing product life cycles, and substituting NPs with safer alternatives when possible (Bakaraki Turan et al., 2019).
- *Treatment Technologies*: Various treatment technologies are being explored to capture and remove NPs from wastewater and industrial emissions. Techniques like filtration, membrane separation, and coagulation can be employed (Oberdörster et al., 2005).
- *Safe Disposal*: Safe disposal of NP waste requires specialized facilities and protocols. Encapsulation, solidification, and incineration under controlled conditions are some potential methods (Jain et al., 2021).

7.7 CONCLUSION

This chapter has explored the pivotal role of nanotechnology in fostering a sustainable environment. By delving into the applications of NMs across various domains, including water, air, and soil remediation, waste management, and renewable energy, we have highlighted how nanotechnology can address critical environmental challenges with unparalleled efficacy and innovation. Nanotechnology offers advanced purification techniques for water treatment, effective pollutant degradation, monitoring air quality, precise toxin neutralization in soil remediation, and revolutionary waste management through material breakdown and recycling. It also enhances renewable energy with more efficient solar cells, batteries, and energy storage systems. Despite its potential, nanotechnology applications must be approached cautiously, with thorough assessment of environmental and health impacts and strengthened regulatory frameworks for safe implementation. In conclusion, nanotechnology stands as a powerful tool in the quest for environmental sustainability. By continuing to innovate and applying nanotechnological solutions responsibly, we can address global environmental issues and pave the way for a greener and more sustainable future.

REFERENCES

Ahmadi, M.H., Ghazvini, M., Nazari, M.A., Ahmadi, M.A., Pourfayaz, F., Lorenzini, G., Ming, T., 2019. Renewable energy harvesting with the application of nanotechnology: A review. *Int. J. Energy Res.* 43(4), 1387–1410. https://doi.org/10.1002/er.4282

Ahmed, M. A., Amin, S., Mohamed, A. A., 2023. Fouling in reverse osmosis membranes: Monitoring, characterization, mitigation strategies and future directions. *Heliyon.* 9(4), e14908. https://doi.org/10.1016/j.heliyon.2023.e14908

Ajith, M.P., Aswathi, M., Priyadarshini, E., Rajamani, P., 2021. Recent innovations of nanotechnology in water treatment: A comprehensive review. *Bioresour. Technol.* 342, 126000. https://doi.org/10.1016/j.biortech.2021.126000

Al Aani, S., Mustafa, T. N., Hilal, N., 2020. Ultrafiltration membranes for wastewater and water process engineering: A comprehensive statistical review over the past decade. *J. Water Process Eng.* 35, 101241. https://doi.org/10.1016/j.jwpe.2020.101241

Albusalih, D., Al Dulaimi, Z., Al-Shamani, A., 2023. Review on design and manufacturing of wind turbine for renewable energy applications. *AIP Conf. Proc.* 2776, 050002. https://doi.org/10.1063/5.0136639

Alhumaimess, Mosaed S., 2020. Metal–organic frameworks and their catalytic applications. *J Saudi Chem Soc.* 24(6), 461–473. https://doi.org/10.1016/j.jscs.2020.04.002

Alka, S., Shahir, S., Ibrahim, N., Ndejiko, M. J., Vo, D.-V. N., Manan, F. A., 2021. Arsenic removal technologies and future trends: A mini review. *J. Clean. Prod.* 278, 123805. https://doi.org/10.1016/j.jclepro.2020.123805

Arora, B., Attri, P., 2020. Carbon nanotubes (CNTs): A potential NM for water purification. *J. Compos. Sci.* 4(3), 135. https://doi.org/10.3390/jcs4030135

Agboola, O., Maree, J., Mbaya, R., 2014. Characterization and performance of nanofiltration membranes. *Environ. Chem. Lett.* 12, 241–255. https://doi.org/10.1007/s10311-014-0457-3

Attia, Y. A., Ezet, A. E., Saeed, S., Galmed, A. H., 2024. Nano carbon-modified air purification filters for removal and detection of particulate matters from ambient air. *Sci. Rep.* 14, 621. https://doi.org/10.1038/s41598-023-50902-x

Bakaraki Turan, N., Erkan, H.S., Onkal Engin, G., Bilgili, M.S., 2019. NPs in the aquatic environment: Usage, properties, transformation and toxicity: A review. *Process Saf. Environ. Prot.* 130, 238–249. https://doi.org/10.1016/j.psep.2019.08.014

Balamurugan, R., Sundarrajan, S., Ramakrishna, S., 2011. Recent trends in nanofibrous membranes and their suitability for air and water filtrations. *Membranes.* 1(3), 232–248. https://doi.org/10.3390/membranes1030232

Bayouli, I.T., Gómez-Gómez, B., Bayouli, H.T., Pérez-Corona, T., Meers, E., Ammar, E., Ferchichi, A., Albarrán, Y.M., 2020. Heavy metal transport and fate in soil-plant system: Study case of industrial cement vicinity, *Tunisia. Arab. J. Geosci.* 13, 75. https://doi.org/10.1007/s12517-019-4898-7

Biswas, B., Qi, F., Biswas, J.K., Wijayawardena, A., Khan, M.A.I., Naidu, R., 2018. The fate of chemical pollutants with soil properties and processes in the climate change paradigm: A review. *Soil Syst.* 2(51), 1–20. https://doi.org/10.3390/soilsystems2030051

Beni, A. A., Jabbari, H., 2022. NMs for environmental applications. *Results Eng.* 15, 100467. https://doi.org/10.1016/j.rineng.2022.100467

Baumann, A.E., D.A. Burns, Liu, B., Thoi, V. S., 2019. Metal-organic framework functionalization and design strategies for advanced electrochemical energy storage devices. *Commun Chem* 2: 86. https://doi.org/10.1038/s42004-019-0184-6

Buss, W., Shepherd, J.G., Heal, K.V., Masek, O., 2018. Spatial and temporal microscale pH change at the soil-biochar interface. *Geoderma*. 331, 50–52. https://doi.org/10.1016/j.geoderma.2018.06.016

Chen, D., Zhu, Q., 2020. Magnetic NPs for environmental remediation: A review. *J. Hazard. Mater*. 390, 121974. https://doi.org/10.1016/j.jhazmat.2019.121974

Chen, Q., Zhou, M., Pan, Y., Zhang, Y., 2023. Ligand-enhanced zero-valent iron for organic contaminants degradation: A mini review. *Processes*. 11(2), 620. https://doi.org/10.3390/pr11020620

Chen, X., Leishman, M., Bagnall, D., Nasiri, N., 2021. Nanostructured gas sensors: From air quality and environmental monitoring to healthcare and medical applications. *NMs*. 11(8), 1927. https://doi.org/10.3390/nano11081927

Chacón, L., Reyes, L., Rivera-Montero, L., Barrantes, K., 2022. Chapter 5 - Transport, fate, and bioavailability of emerging pollutants in soil, sediment, and wastewater treatment plants: Potential environmental impacts. In: *Emerg. Contam. Environ.: Chall. Sustain. Pract*. pp. 111–136. https://doi.org/10.1016/B978-0-323-85160-2.00020-2

Coutiño-Gonzalez, E., Baekelant, W., Steele, J. A., Kim, C. W., Roeffaers, M. B. J., Hofkens, J., 2017. Silver clusters in zeolites: From self-assembly to ground-breaking luminescent properties. *Acc. Chem. Res*. 50(9), 2353–2361. https://doi.org/10.1021/acs.accounts.7b00295

Dimkpa, C.O., Bindraban, P.S., 2018. Nanofertilizers: New products for the industry? *J. Agric. Food Chem*. 66(26), 6462–6473. https://doi.org/10.1021/acs.jafc.7b02150

Eslami, F., Matloub, S., 2023. Advancing perovskite solar cell performance: Enhanced efficiency and stability through superimposed PbS QDs. *Sol. Energy*, 265, 112152. https://doi.org/10.1016/j.solener.2023.112152

Fermoso, J., Sánchez, B., Suarez, S., 2020. Air purification applications using photocatalysis. *Nanostruct. Photocatalysts: From Mater. to Appl. Solar Fuels Environ. Remediat. Micro Nano Technol*., pp. 99–128. https://doi.org/10.1016/B978-0-12-817836-2.00005-3

Gomes, H.I., Mayes, W.M., Rogerson, M., Stewart, D.I., Burke, I.T., 2016. Alkaline residues and the environment: A review of impacts, management practices and opportunities. *J. Clean. Prod*. 112(Part 4), 3571–3582. https://doi.org/10.1016/j.jclepro.2015.09.111

Hejazi, M. A., Osman Eksik, Ç. T., Caner, Ü., Levent, T., 2023. Carbon-based NMs in gas sensing applications. *Emergent Mater* 6, 45–77. https://doi.org/10.1007/s42247-023-00454-7

Hegab, H., Shaban, I., Jamil, M., Khanna, N., 2023. Toward a sustainable future: Strategies, indicators, and challenges for implementing sustainable production systems. *Sustain. Mater. Technol*. 36, e00617. https://doi.org/10.1016/j.susmat.2023.e00617

Hisatomi, T., Kubota, J., Domen, K., 2014. Recent advances in semiconductors for photocatalytic and photoelectrochemical water splitting. *Chem. Soc. Rev*. 43(22), 7520–7535. https://doi.org/10.1039/C3CS60378D

Ichou, H., Arrousse, N., Berdimurodov, E., Aliev, N., 2023. Exploring the advancements in physical vapor deposition coating: A review. *J. Bio-Tribo-Corros*. 10, n. pag. https://doi.org/10.1007/s40735-023-00806-0

Iravani, S., 2021. NMs and nanotechnology for water treatment: Recent advances. *Inorg. Nano-Met. Chem*. 51(12), 1615–1645. https://doi.org/10.1080/24701556.2020.1852253

Jain, K., Patel, A.S., Pardhi, V.P., Flora, S.J.S., 2021. Nanotechnology in wastewater management: A new paradigm towards wastewater treatment. *Molecules*. 26(6), 1797. https://doi.org/10.3390/molecules26061797

Jain, K., 2017. Dendrimers: Smart nanoengineered polymers for bioinspired applications in drug delivery. biopolym.-based compos. *Drug Deliv. Biomed. Appl*. 169–220. https://doi.org/10.1016/B978-0-08-101914-6.00007-7

Jaison, A., Mohan, A., Lee, Y.C., 2023. Recent developments in photocatalytic nanotechnology for purifying air polluted with volatile organic compounds: Effect of operating parameters and catalyst deactivation. *Catalysts*. 13(2), 407. https://doi.org/10.3390/catal13020407

Jiang, Bo., Luning, L., Yi, XI., Zhang, N., Chen Y., Pei, Lu., Dayi, Z., 2018. Advances of magnetic NPs in environmental application: Environmental remediation and (Bio)sensors as case studies. *Environ Sci Pollut Res*. 25, 30863–30879. https://doi.org/10.1007/s11356-018-3095-7

Jiang, S., Li, Y., Ladewig, B. P., 2017. A review of reverse osmosis membrane fouling and control strategies. *Sci. Total Environ*. 595, 567–583. https://doi.org/10.1016/j.scitotenv.2017.03.235

Karthikeyan, C., Arunachalam, P., Ramachandran, K., Al-Mayouf, A.M., Karuppuchamy, S., 2020. Recent advances in semiconductor metal oxides with enhanced methods for solar photocatalytic applications. *J. Alloys Compd*. 828, 154281. https://doi.org/10.1016/j.jallcom.2020.154281

Khan, I., Saeed, K., Khan, I., 2019. NPs: Properties, applications and toxicities. *Arab. J. Chem*. 12(7), 908–931. https://doi.org/10.1016/j.arabjc.2017.05.011

Kumar, P., Ki-Hyun, K., Kwon, E. E., Szulejko, J. E., 2016. Metal–Organic frameworks for the control and management of air quality: Advances and future direction. *J Mater Chem A*, 345–361. https://doi.org/10.1039/C5TA07068F

Lee, H., Sam, K., Coulon, F., De Gisi, S., Notarnicola, M., Labianca, C., 2024. Recent developments and prospects of sustainable remediation treatments for major contaminants in soil: A review. *Sci. Total Environ.* 912, 168769. https://doi.org/10.1016/j.scitotenv.2023.168769

Lia, D., Wang, H., 2010. Recent developments in reverse osmosis desalination membranes. *J. Mater. Chem.* 20, 4551–4566. https://doi.org/10.1039/B924553G

Ltifi, M., Abichou, T., Tisot, J.P., 2014. Effects of soil aging on mechanical and hydraulic properties of a silty soil. *Geotech. Geol. Eng.* 32, 1101–1108. https://doi.org/10.1007/s10706-014-9784-1

Lyu, C., Zhao, P., Xie, J., Dong, S., Liu, J., Rao, C., Fu, J., 2021. Electrospinning of nanofibrous membrane and its applications in air filtration: A review. *NMs.* 11(6), 1501. https://doi.org/10.3390/nano11061501

Maduraiveeran, G., Wei, J., 2021. Carbon NMs: Synthesis, properties and applications in electrochemical sensors and energy conversion systems. *Mater Sci Eng B* 272, 115341. https://doi.org/10.1016/j.mseb.2021.115341

Miditana, S.R. 2023. Synthesis, characterization, and antibacterial properties of surfactant encapsulated Mn Mg Co-doped titania NPs. *Iran J Catal*, 13 (4), 449–461. https://doi.org/10.30495/ijc.2023.1992163.2029

Miditana, S.R., Tirukkovalluri, S.R., Alim, A.S., Imandi, M.R., 2019. Photocatalytic degradation of Allura Red by Mn–Ni co-doped nanotitania under visible light irradiation. *Int. J. Innov. Technol. Explor. Eng.* 8(12), 650–657. https://doi.org/10.35940/ijitee.K2458.1081219

Miditana, S.R., Tirukkovalluri, S.R., Raju, I.M., Alim, S.A., 2024. Photocatalytic degradation of Orange-II by surfactant assisted Mn/Mg co-doped TiO2 NPs under visible light irradiation. *Curr. Chem. Lett.* 13, 265–276. https://doi.org/10.5267/j.ccl.2023.6.003

Mohammed, L., Gomaa, H.G., Ragab, D., Zhu, J., 2017. Magnetic NPs for environmental and biomedical applications: A review. *Particuology.* 30, 1–14. https://doi.org/10.1016/j.partic.2016.06.001

Mose, P., 2023. The applications of nanotechnology in renewable energy. *J. Phys. Sci.*, 5(1), 1–12. https://doi.org/10.47941/jps.1626

Munir, N., Javaid, A., Abideen, Z., Duarte B., Jarar H., El-Keblawy A., Sheteiwy M.S., 2024. The potential of zeolite nanocomposites in removing microplastics, ammonia, and trace metals from wastewater and their role in phytoremediation. *Environ. Sci. Pollut. Res.* 31, 1695–1718. https://doi.org/10.1007/s11356-023-31185-1

Nagar, A., Thalappil, P., 2020. Clean water through nanotechnology: Needs, gaps, and fulfillment. *ACS Nano.* 14(6), 6420–6435. https://doi.org/10.1021/acsnano.9b01730

Naseem, T., Durrani, T., 2021. The role of some important metal oxide NPs for wastewater and antibacterial applications: A review. *Environ. Chem. Ecotoxicol.* 3, 59–75. https://doi.org/10.1016/j.enceco.2020.12.001

Nilimaa, J., 2023. Smart materials and technologies for sustainable concrete construction. *Dev. Built Environ.*, 15, 100177. https://doi.org/10.1016/j.dibe.2023.100177

Nishu, S. K., 2023. Smart and innovative nanotechnology applications for water purification. *Hybrid Adv.* 3, 100044. https://doi.org/10.1016/j.hybadv.2023.100044

Oberdörster, G., Oberdörster, E., Oberdörster, J., 2005. Nanotoxicology: An emerging discipline evolving from studies of ultrafine particles. *Environ. Health Perspect.* 113(7), 823–839. https://doi.org/10.1289/ehp.7339

Ogunsola, S.S., Oladipo, M.E., Oladoye, P.O., Kadhom, M., 2024. Carbon nanotubes for sustainable environmental remediation: A critical and comprehensive review. *Nano-Struct. Nano-Objects.* 37, 101099. https://doi.org/10.1016/j.nanoso.2024.101099

Part, A.M., Olivito, K.J., Pius, D.R., 2018. NM waste management. *Environ. Sci. Technol. Lett.* 5(2), 86–90. https://doi.org/10.1021/acs.estlett.7b00593

Pokrajac, L., Abbas, A., Chrzanowski, W., Dias, G. M., Eggleton, B. J., Maguire, S., Maine, E., Malloy, T., Nathwani, J., Nazar, L., Sips, A., Sone, J., van den Berg, A., Weiss, P. S., Mitra, S., 2021. Nanotechnology for a sustainable future: Addressing global challenges with the International Network4Sustainable Nanotechnology. *ACS Nano.* 15, 18608–18623. https://doi.org/10.1021/acsnano.1c10919

Pushparaj, V., Manikandan, M., MubarakAli, N., Kasim, M.G., Jayavel, S., 2022. Polymer nanocomposite packaging systems: A review. *Compos. Commun.* 33, 100933. https://doi.org/10.1016/j.cocom.2022.100933

Raghav, S., Yadav, P. K., Kumar, D., 2020. Nanotechnology for a sustainable future. In *Handbook of NMs for Manufacturing Applications. Micro Nano Technol.* pp 465–492. Elsevier. https://doi.org/10.1016/B978-0-12-821381-0.00019-3

Rahman, R. O. A., El-Kamash, A. M., Hung, Y.-T., 2022. Applications of nano-zeolite in wastewater treatment: An overview. *Water.* 14(2), 137. https://doi.org/10.3390/w14020137

Reijonen, I., Metzler, M., Hartikainen, H., 2016. Impact of soil pH and organic matter on the chemical bioavailability of vanadium species: The underlying basis for risk assessment. *Environ. Pollut.* 210, 371–379. https://doi.org/10.1016/j.envpol.2015.12.046

Roy, A., Sharma, A., Yadav, S., Jule, L. T., Krishnaraj, R., 2021. NMs for remediation of environmental pollutants. *Bioinorg. Chem. Appl.* 2021, n. pag. https://doi.org/10.1155/2021/1764647

Roy, J., 2022. The synthesis and applications of TiO_2 NPs derived from phytochemical sources. *J. Ind. Eng. Chem.* 106, 1–19. https://doi.org/10.1016/j.jiec.2021.10.024

Sari, Y., Gareso, P. L., Armynah, B., Tahir, D., 2024. A review of TiO2 photocatalyst for organic degradation and sustainable hydrogen energy production. *Int. J. Hydrog. Energy.* 55, 984–996. https://doi.org/10.1016/j.ijhydene.2023.11.126

Saleem, H., Zaidi, S.J., Ismail, A.F., Goh, P.S., 2022. Advances of NMs for air pollution remediation and their impacts on the environment. *Chemosphere.* 287(Part 2), 132083. https://doi.org/10.1016/j.chemosphere.2021.132083

Scaffaro, R., Citarrella, M.C., 2023. Nanofibrous polymeric membranes for air filtration application: A review of progress after the COVID-19 pandemic. *Macromol. Mater. Eng.* 308, 2300072. https://doi.org/10.1002/mame.202300072

Serrano, E., Rus, G., García-Martínez, J., 2009. Nanotechnology for sustainable energy. *Renew. Sust. Energ. Rev.* 13(9), 2373–2384. https://doi.org/10.1016/j.rser.2009.06.003

Sharanyakanth, P.S., Radhakrishnan, M., 2020. Synthesis of Metal-Organic Frameworks (MOFs) and its application in food packaging: A critical review. *Trends Food Sci Technol* 104, 102–116. https://doi.org/10.1016/j.tifs.2020.08.004

Shen, Y., Jiawei Kang, L. G., Fan Qiu, Y. F., Zhang, S., 2024. Recent progress on the application of MOFs and their derivatives in adsorbing emerging contaminants. *Sep Purif Technol* 350, 127955. https://doi.org/10.1016/j.seppur.2024.127955

Singh, D., Dahiya, M., Kumar, R., Nanda, C., 2021. Sensors and systems for air quality assessment monitoring and management: A review. *J. Environ. Manage.* 289, 112510. https://doi.org/10.1016/j.jenvman.2021.112510

Teow, Y. H., Sum, J. Y., Ho, K. C., Mohammad, A. W., 2021. Chapter 3 - Principles of nanofiltration membrane processes. *Osmosis Eng.* 53–95. https://doi.org/10.1016/B978-0-12-821016-1.00014-0

Wadhawan, S., Jain, A., Nayyar, J., Mehta, S.K., 2020. Role of NMs as adsorbents in heavy metal ion removal from wastewater: A review. *J. Water Process Eng.* 33, 101038. https://doi.org/10.1016/j.jwpe.2019.101038

Walser, T.C., Stupp, N., Vaughan, W.M., 2012. Finding the appropriate use of NMs toward a more sustainable world. *ACS Nano.* 6(11), 991–1004. https://doi.org/10.1021/nn302933v

Wei, Y., Meng, H., Wu, Q., Bai, X., Zhang, Y., 2023. TiO2-based photocatalytic building material for air purification in sustainable and low-carbon cities: A review. *Catalysts.* 13(12), 1466. https://doi.org/10.3390/catal13121466

Xu, L., Qi, X., Li, X., Bai, Y., Liu, H., 2016. Recent advances in applications of NMs for sample preparation. *Talanta.* 146, 714–726. https://doi.org/10.1016/j.talanta.2015.06.036

Xiao, W., Li, B., Yan, J., Wang, L., Huang, X., Gao, J., 2023. Three-dimensional graphene composites: Preparation, morphology and their multi-functional applications. *Compos. Part A Appl. Sci. Manuf.* 165, 107335. https://doi.org/10.1016/j.compositesa.2022.107335

Xu, Z., Li, C., 2018. Photocatalytic water splitting on metal oxide-based semiconductor photocatalysts. In *Metal Oxides in Heterogeneous Catalysis*. Metal Oxides. pp. 355–399. https://doi.org/10.1016/B978-0-12-811631-9.00007-7

Yadav, D., Hazarika, S., Ingole, P. G., 2022. Recent development in nanofiltration (NF) membranes and their diversified applications. *Emergent Mater.* 5, 1311–1328. https://doi.org/10.1007/s42247-021-00302-6

Yang, L., Li, X., Yu, J., Zhang, S., Ding, B., 2024. Transparent and antibacterial air filters based on nanofiber/nets membrane. *Compos. Commun.* 48, 101916. https://doi.org/10.1016/j.coco.2024.101916

Yusuf, A., Sodiq, A., Giwa, A., Eke, J., Pikuda, O., De Luca, G., Di Salvo, J. L., Chakraborty, S., 2020. A review of emerging trends in membrane science and technology for sustainable water treatment. *J. Cleaner Prod.* 266, 121867. https://doi.org/10.1016/j.jclepro.2020.121867

Zhang, X., Shi, F.X., Niu, J., Jiang, Y., Wang, Z., 2008. Superhydrophobic surfaces: from structural control to functional application. *J. Mater. Chem.* 18, 621–633. https://doi.org/10.1039/B711226B

8 Functionalized Nanomaterials
Properties to Perspective Applications in Environmental Remediation

Amrit Krishna Mitra and Rama Saha

8.1 INTRODUCTION

To attain environmental cleanliness goals, we rely heavily on material science, especially in the realm of NMs, which have seen significant advancements in the past decade. Human activities across industries, agriculture, waste management, and urban settings have led to a surge in environmental toxins, posing threats to ecosystems and human well-being. To counter this, society needs to implement remedial actions and develop new systems capable of handling these pollutants effectively. Tackling the accumulation of toxins from various human activities over the past century necessitates concerted efforts to clean up contaminated areas and develop robust new systems (Ghosh et al., 2023). Nanotechnology provides great promise for environmental conservation by analyzing and regulating emissions, developing green technologies, and remediating waste sites and damaged water sources (Khan et al., 2021). It has the ability to constantly detect and mitigate environmental pollutants, cleansing the air and water by removing poisons. To eliminate harmful metal ions, various methods such as ion exchange, reverse osmosis, precipitation filtration, biosorption, coagulation, and extraction can be used, with adsorption emerging as the most effective due to its cost-effectiveness, efficiency, and simplicity in removing even minute amounts of heavy metal ions. Functionalized and engineered NMs have the potential to transform a variety of sectors, including environmental engineering, water purification, drinking water treatment, industrial wastewater treatment, and biomedical engineering (Palit and Hussain, 2020). These sophisticated materials are the next generation of intelligent materials. They cover a wide range of categories and are effective nanofillers. This includes: (i) Metal NPs (such as Pt, Pd, Ni, Ru, Al, Ag, Au, and Cu) having unique chemical, optical, and electrical properties; (ii) Metal oxide NMs (such as TiO_2, Fe_2O_3, Al_2O_3, ZnO, and SiO_2) created using sol-gel or hydrothermal techniques; (iii) Bimetallic NMs; (iv) Carbon-based NMs (such as fullerene, carbon nanotubes, and graphene sheets); (v) NMs based on zeolite and silica with mesoporous structures; VI) Ceramic NMs; (vii) Polymeric NMs; (viii) Bio-nano materials; (ix) Metallic-organic frameworks; and (x) Core-shell NMs (Singh et al., 2013; Gupta et al., 2021; Hua et al., 2012).

NPs are often manufactured using either a top-down or bottom-up strategy, with chemical or physical processes (Arole and Munde, 2014). However, these processes frequently involve the use of dangerous chemicals or large amounts of energy, which presents environmental and economic issues. Biological synthesis is a superior alternative since it is more environmentally friendly, cost-effective, and does not include dangerous ingredients or require excessive energy (Shanker et al., 2016). This method has potential for a variety of applications, particularly in biomedicine. Many microbes and plants have been effectively used in this method, with bacteria, fungi, yeast, and algae commonly used to create silver, gold, copper, palladium, and iron nanostructures. Graphene oxide (GO) and graphene nanosheets (GNs) have attracted worldwide attention in recent years. Graphene is a two-dimensional carbon NM made up of single layers of sp^2-hybridized carbon atoms organized

DOI: 10.1201/9781003484189-8

in six-membered rings. It has excellent mechanical, thermal, and electrical properties. In contrast, GO is graphene that has been functionalized with multiple oxygen-containing groups. NMs' high surface area to volume ratio considerably increases their sorption efficiency. Adding functional groups like amine (-NH_2), carboxylic (-COOH), thiol (-SH), and methyl to NM surfaces has increased performance (Darwish and Mohammadi, 2018; Amieva et al., 2016).

This research focuses on developing novel NMs for heavy metal ion removal (e.g., MOFs, nZVI, g-C3N4, and MXenes) and investigating how environmental conditions influence sorption behaviour. Recent study has shown that commercially available ACF, when surface-functionalized or impregnated with metals, can catalyze the oxidation of SO_2, NO, and VOCs. To improve performance, researchers are investigating surface modification of NMs, with functional groups such as amine (-NH_2), carboxylic (-COOH), thiol (-SH), and methyl. These diverse properties highlight the wide range of applications for manufactured NPs.

8.2 PREPARATION OF NPS

NPs are prepared using a variety of techniques, including inert gas condensation, ion sputtering, spray pyrolysis, laser pyrolysis, low-temperature reactive synthesis, flame synthesis, sol-gel process, mechanical alloying/milling, mechanochemical synthesis, and electrodeposition. Researchers choose the sol-gel technique because it produces high-purity products and allows for control over molecular structure. In contrast, biological synthesis uses a single-step bio-reduction method that is both environmentally benign and energy efficient. Traditional approaches are slower and less efficient at reducing metal ions to their elemental forms than biological methods. As a result, a combination of biological and non-biological processes may be used in manufacturing, taking into account operational selection, reaction conditions, prices, scalability, particle size, and dispersion (Zhang et al. 2016).

8.2.1 Physical Methods

Process components typically include raw materials for grinding, surfactants to coat the surface and avoid agglomeration and overheating, and fluid transporters that act as reaction solvents. Dry and wet processes are commonly employed in mechanical grinding to produce NPs (Saleh 2016). Laser ablation, aerosol-aided deposition, electron beam induced reduction, and flame synthesis have become attractive methods for resolving environmental concerns about organic solvent use, reagent risks, and stabilizer pollution (Fominski et al., 2018). However, these approaches have limitations, including uneven outcomes in nanomaterial synthesis, sample dilution, and difficulties in controlling the size, shape, and quantity of the NMs produced.

8.2.2 Chemical Methods

Hydrothermal, solvothermal, solid state, sol-gel, and molten salt processes are all widely used (Zhou et al., 2017). Furthermore, wet chemical reduction is widely used because of its capacity to manipulate size and shape, modify precursor concentration and reaction temperature, and include organic or inorganic ligands. However, chemical processes have limitations for large-scale production, due to the excessive use of hazardous chemicals that pose threats to the environment and biological health. To address this issue, biocompatible 'green reagents' such as ascorbic acid and sodium citrate can improve catalytic characteristics and yield by maintaining strict solvent and reagent purity (Qian et al., 2021). In the year 2020, Filipczak et al. reported that the combination of tannic acid and sodium citrate regulates the nucleation, development, and stability of silver NPs during synthesis. Physical-chemical techniques or combinations of chemical and physical approaches, such as heterojunctions like mechanical milling, are also used in nanomaterial manufacturing (Das et al., 2019).

8.2.3 Natural Production of NPs

The use of plant extracts, bacteria, and fungi constitutes biogenic approaches for producing silver NPs. These methods are environmentally benign and sustainable, since they prevent the use of harmful or poisonous substances. Researchers are particularly interested in silver NPs because of its unique qualities, such as size, structure, and antibacterial activities. Plant extracts, fungus, and bacteria are the three principal sources of silver NP manufacturing, with these biological approaches seeing increased use in recent years (Thangavelu et al., 2020).

8.2.3.1 Viruses

Viruses have unique properties that make them ideal models for biologically manufacturing metallic NMs, such as their size, numerous chemical groups that can be modified, and monodispersity. Thangavelu et al., in the year 2020, investigated the use of the squash leaf curl China virus, a plant pathogenic virus, to manufacture gold (Au) and silver (Ag) NPs with band gap values in the semiconductor region, based on electrical conductivity measurements. Love et al., in the year 2014, used the tobacco mosaic virus to produce NPs in Au and Ag solutions, which were supplemented with plant cell extracts from *Nicotiana benthamiana* and *Hordeum vulgare* (Love et al., 2014). The presence of the virus caused a reduction in the size of the biosynthesized NMs, as observed (Qidwai et al., 2018).

8.2.3.2 Bacteria

Bacteria are widely investigated in biosynthesis research because of their versatility in the synthesis of inorganic chemicals. They produce both extracellular and intracellular NPs in both conditions, making them ideal for large-scale development. Ameen et al. in the year 2020 used the bacterila *Cuprividus sp.* to produce silver (Ag) NPs for antimicrobial applications. Ag NPs' spherical form and amide functional groups help to reduce ionic species extracellularly (Kimber et al., 2018). The discovery that the efficiency of Ag NPs did not differ significantly among strains highlights the need to investigate alternative NPs for similar aims and to incorporate other pathogens in nano biomedical research. Show et al., in the year 2015, used thermophilic bacteria (BKH1) as a biological model to show that tetraethyl orthosilicate could be converted into silicon dioxide (SiO_2) NPs at ambient temperature (Show et al., 2015). These NPs exhibit two absorption bands and an amorphous nature, showing potential for use as optical probes.

Biosynthesis produces SiO_2 NPs, which reduces energy usage and increases cost-effectiveness for businesses. This technique also produces ethyl alcohol, a desirable byproduct. Kimber et al. in 2018 achieved the manufacture of Cu NPs (20–40 nm) with *Shewanella oneidensis* bacteria, demonstrating efficient and ecologically friendly nanomaterial synthesis (Ghomi et al., 2019).

8.2.3.3 Fungus and Yeast

Fungi are selected due to their resilience, ability to accumulate chemicals, low cost, efficient metabolism, and ease of cultivation. They also excel in secreting extracellular enzymes and proteins. The advantages of fungal-mediated biosynthesis include adaptive structures from both extracellular and intracellular reducing enzymes, as well as biomimetic mineralization. Raliya et al., in the year 2015, used *Aspergillus flavus TFR* to economically and environmentally generate TiO_2 NPs (Raliya et al., 2015). Similarly, Ghomi et al., in 2019, used Penicillium species for the production of ZrO_2 NPs (Chau et al., 2022). Penicillium species have been shown to produce spherical extracellular ZrO_2 NPs (Korbekandi et al., 2016). Clarance et al.in 2020 used *Fusarium solani ATLOY-8*, isolated from *Chonemorpha fragrans*, to create Au NPs (Al-Radadi, 2021). The increased surface area of fungal mycelium explains why fungi are preferred over bacteria and algae in biosynthesis. This knowledge is expected to accelerate the rate of ion transformation into metallic NPs (Makarov et al., 2014).

8.2.4 NP Production Based on Plants

Plant extracts derived from leaves, flowers, fruits, tubers, roots, latex, and other sources serve as suitable precursors for the biogenic synthesis of NPs because they contain a variety of naturally occurring plant biomolecules that aid in the reduction of metal ions to their nano-form (Carmona et al., 2017). Researchers have successfully created several metallic and metal oxide NPs from various plant components and extracts, including those from *Aloe vera*, *Carica papaya*, Tea, *Nelumbo nucifera*, *Allium sativum*, *Moringa oleifera*, *Garcinia mangostana*, *Vitex negundo*, *Acalypha indica*, *Actaea racemosa* (black cohosh), *Sansevieria trifasciata*, *Impatiens scapi-flora*, *Alternanthera dentata*, *Acorus calamus*, *Boerhavia diffusa*, *Ziziphora tenuior, and Gelidium amansii*. For instance, stable silver NPs were synthesized using aqueous extracts from the native medicinal plant *Buddleja globosa*. Silver nitrate salt was used as a precursor in the creation of NPs with constant size and shape. Plant extracts used in this synthesis included *Thevetia peruviana* latex, *Citrullus lanatus* fruit rind extract, *Gum karaya* extract, *Hibiscus rosa-sinensis* flower extract, potato extract, *Dioscorea alata tuber* extract, and *Solanum xanthocarpum* berry extract. This plant-based technique is a viable alternative to microbial synthesis because of its faster rate, stability, and adaptability for large-scale production (Li et al., 2011).

8.2.5 NP Production Based on Micro-Emulsion

To produce NPs from immiscible liquids, a microemulsion of surfactants, co-surfactants, solvents, and water is used. Surface derivatization of superparamagnetic iron oxide NPs allow for the synthesis of SPIONs with homogeneous size, shape, and good dispersibility, hence preventing agglomeration. Cubes, hollow spheres, spheres, and needle-shaped particles have all been produced. The pH of the reaction mixture and the oil/water ratio can both be modified to control the microemulsion. For example, to make SPIONs (Super Paramagnetic Iron Oxide Nanoparticles), the oil phase was cyclohexane, the aqueous precursors were $FeSO_4$/Fe $(NO_3)_3$, and the surfactants were polyoxyethylene isooctyl ether phosphate (NP5) and nonoxynol-9 phosphate (NP9), using an oil-in-water emulsion. The pH of the media influences the micelle structure, with reported formations comprising needle and rod geometries, as well as hollow spheres. Unlike superparamagnetic behavior, the hollow microspheres exhibited ferromagnetic properties. A key advantage of the microemulsion technique lies in its capacity to adjust the aqueous phase, thereby influencing NP size. Various morphologies can be achieved through precise environmental control. It‘s important to note that morphology changes may affect magnetic properties.

8.2.6 NP Production Based on Sonochemical Synthesis

Sonochemical synthesis is the process of activating a liquid with ultrasonic vibrations to generate high temperatures and partial pressures, hence promoting oxidation, reduction, breakdown, and hydrolysis. The size, shape, and magnetic characteristics of particles are significantly dependent on ultrasonic frequency, sonication duration, and reaction temperature. SPIONs with an average size of 10 nm were created utilizing iron (II) acetate in deoxygenated water and high ultrasonic horn radiation. The technique involves Fe (II) oxidation by water molecules, which is then transformed into hydrogen peroxide by ultrasonic absorption, resulting in SPION NP production. MFe_2O_4 nanocrystals with diameters ranging from 15–20 nm to 20–50 nm were synthesized in an aqueous $FeCl_2$/urea solution. Other ways for preparing SPIONs include aerosol techniques, electrochemical methods, sol-gel processes, flow injection syntheses, and microwave-assisted synthesis (Sadegh et al., 2017).

8.3 MECHANISM OF 'GREEN SYNTHESIS' FOR METALS AND THEIR OXIDE NPS

The synthesis of NPs using different microorganisms involves specific procedures (Table 8.1). Initially, enzymes capture metallic ions either on the surface or inside microbial cells. Subsequently, these captured metal ions undergo reduction to form metal NPs.

TABLE 8.1
Green Synthesis of Metallic NPs from Various Plants Extracts (Singh et al., 2018)

Order	Plant Origin	NPs	Size (nm)	Morphology	Applications
1	*Aloe barbadensis Miller* (Aloe vera)	Gold and silver	10–30	Spherical, triangular	Cancer hyperthermia, optical coatings
2	*Acalypha Indica*	Silver	20–30	Spherical	Antibacterial activity against water borne pathogens
3	Apiin extracted from henna leaves	Silver and gold	39	Spherical, triangular, and quasi-spherical	Hyperthermia of cancer cells and IR-absorbing optical coatings
4	*Azadirachta indica* (neem)	Gold, silver and silver-gold alloys	5–35 and 50–100	Spherical, triangular, hexagonal	Remediation of toxic metals
5	*Cinnamomum camphora* (camphor tree)	Gold and silver	55–80	Triangular, spherical (Au), and quasi-spherical	----
6	*Carica papaya* (papaya)	Silver	60–80	Spherical	----
7	*Citrus limon* (lemon)	Silver	<50	Spherical, spheroidal	------
8	*Coriandrum sativum* (coriander)	Gold	6.75–57.91	Spherical, triangular, truncated triangular, decahedral	Drug delivery, tissue/tumor imaging, photo thermal therapy
9	*Eucalyptus citriodora* (neelagiri)	Silver	20	Spherical	Antibacterial
10	*Garcinia mangostana* (mangosteen)	Silver	35	Spherical	Antimicrobial activity against E. coli and S. aureus
11	*Morus* (mulberry)	Silver	15–20	Spherical	Antimicrobial activity against E. coli, B. subtilis
12	*Nelumbo nucifera* (lotus)	Silver	25–80	Spherical, triangular, truncated triangular, decahedra	Larvicidal activity against malaria and flariasis vectors
13	*Ocimum sanctum* (tulsi; root extract)	Silver	10±2 and 5±1.5 nm	Spherical	Catalytic reduction
14	*Ocimum sanctum* (tulsi; leaf extract)	Gold and silver	30 and 10–20	Crystalline, hexagonal, triangular and spherical	Biolabeling, biosensor
15	*Psidium guajava* (guava)	Gold	25–30	Mostly spherical	-----
16	*Terminalia catappa* (almond)	Gold	10–35	Spherical	Biomedical field

The following theory discusses how *verticillium sp*. or algal biomass produces silver and gold NPs with the help of microorganisms (Ishak et al., 2019): (a) Initially, electrostatic interactions between ions and negatively charged cell wall enzymes caused silver or gold ions to be trapped on the surface of fungal cells, (b) ten ions of gold or silver were bioreduced to produce nuclei of either metal. The two fundamental components in NP synthesis are nicotinamide adenine dinucleotide (NADH) and NADH-dependent nitrate reductase. *B. licheniformis* produced bioreduced silver NPs through the action of nitrate reductase. However, the bioreduction mechanisms involved in the formation of metal salt ions and the subsequent metallic NPs produced by bacteria are unknown (Kanchi and Ahmed, 2018).

8.4 TYPES OF NMS IN ENVIRONMENTAL MONITORING

The most pertinent illustrations of cutting-edge nanostructures and NMs are on their properties and their environmental uses. In this section, specific examples are reviewed and discussed in the following subsections.

8.4.1 Carbon Nanotubes and Nanofibers

Carbon nanotubes (CNTs) are classified into two types: single-walled (SWNTs) and multiwalled (MWNTs), which are used as nanoscale building blocks in analytical devices. CNTs are ideal for chemical and biological sensing due to their high surface area, nanoscale size, and outstanding electrical properties. They facilitate direct electronic communication between redox proteins and electrode surfaces, enhance biomolecule binding, and increase electrocatalytic activity. As a result, interference effects are reduced, and analytes such as pesticides can be detected with low applied voltage without the use of electronic mediators. Enzymes such as acetylcholinesterase (AChE) and organophosphate hydrolyse (OPH) can be encapsulated within carbon nanotube/hybrid composites for electrochemical biosensor applications. In terms of sensitivity, applied potential, and surface fouling, CNTs outperformed macroscale materials. Pesticides, such as organophosphorus, phenolic compounds, and herbicides, are some of the environmental applications (Jawed et al., 2020). The nature and conditions of the experiment influence the adsorbent's ability to absorb organic contaminants. CNT-impregnated chitosan hydrogel beads (CSBs) have been developed to eliminate Congo red (CR) and have a higher maximum capacity (450.4 mg/g) than chitosan without impregnation. A new generation of CSBs with superior mechanical properties was developed using sodium dodecyl sulfate and MWCNTs. Single wall carbon nanotubes (SWCNTs) absorb organic contaminants more effectively due to their larger specific surface area. SWCNTs are more effective at removing benzene and toluene, with maximum adsorption capacities of 9.98 and 9.96 mg/g, respectively. When compared to commercial AC, carbons NPs have significantly higher sorption effectiveness. Oxidized carbon nanotubes (CNTs) were used to demonstrate high sorption capacity and efficiency for Pb (II), Cd(II), and Cr(VI) from water. Polyethylenimine-multiwalled carbon nanotube polymeric nanocomposite (PEI-MWCNTs) was made from functionalized MWCNTs.

These NMs, with an average diameter of 25 nm, pore volume of 2×10^{-3} cm^3/g, and surface area of 1.70 m^2/g, successfully removed 99% of Cr (VI) from a 10 ppm metallic solution in 60 minutes at a pH below 4. The adsorption capacity (Qm) may remove up to 40 mg/g of Cr (VI). After coating the adsorbent with polydopamine (PDA) in an alkaline solution, PEIs were conjugated on the CNTs-PDA surface using the Michael addition procedure to form CNTs-PDA-PEI. The presence of amide groups throughout the PAMAM structure was associated to enhance metal adsorption. Magnetic carbon NPs were also developed and tested for the removal of heavy metals. Magnetic Fe-MWCNTs were produced by doping MWCNTs with zero-valent iron particles (FeO) in a CVD reactor. In 60 minutes, at a dose of 0.2 g/L and 10 ppm of the different metallic solutions, these magnetic Fe-MWCNTs removed 74% and 77% of As (III) and As (V), respectively (Luo et al., 2016).

Although acidic pH diminishes the efficacy of metal ion removal, direct adsorption remains effective. Surface functionalization removes non-functionalization lag and alters surface charge, allowing for quicker adsorption. Because of its larger surface area and zeta potential, NH_2 functionalization is more effective at neutral pH in removing divalent metal ions. Although CNTs may adsorb a wide range of metals from water and wastewater, it can be challenging to develop cost-effective materials and recycle them once dangerous metals have been adsorbed. To address these issues, high adsorption capacity materials must be developed or changed to efficiently and easily remove hazardous metals during separation and reuse.

8.4.2 Metal NPs

Metal NPs (NPs) such as gold, silver, and iron are widely studied in nanotechnology, due to their electrical, optical, catalytic, and thermal properties. Ag NPs are recognized for their antibacterial properties, which vary depending on their size and form. Au NPs are utilized in signal amplification, bio-immobilization matrixes in biosensors, and colorimetric and fluorescence tests. A battery-powered nanomaterial surface energy transfer (NSET) sensor was developed for detecting and screening mercury in soil, water, and fish. The NSET system coupled AuNPs with an organic dye (Rhodamine) as a donor to improve metal ion detection sensitivity. In the presence of mercury, self-adsorbed dye molecules are released from the NPs' surface, causing discrete variations in the fluorescence signal proportional to the concentration of Hg(II). Other NPs, such as zero-valent iron and bimetallic iron NPs, were used to remove heavy metal ions such as As(III), Pb(II), Cu(II), Ni(II), and Cr, as well as chlorinated organics and polychlorinated biphenyls (PCBs) (Hrbac et al., 2007).

8.4.3 Magnetic NPs

Magnetic iron oxide NPs have several environmental applications, including cell labeling, biosensing, cleaning, and separations. They must be uniform in size, well-distributed, and chemically stable. Recent advances have allowed for greater control over shape, size, dispersion, and chemical composition. Natural products can be synthesized or obtained in a variety of methods from different companies. Using them as capture probes in immunomagnetic separations is an interesting application. Immuno-functionalized NPs bind to target analytes, concentrating them before separating them using a magnetic field. Low detection limits are made easier, and sample preparation is simplified. Environmental contaminants adsorbed on the surface of a particle can be easily separated using a magnet (Falcaro et al., 2016). For example, when arsenic is deposited on iron oxide NPs, low magnetic fields are a good way to remove it. Magnetic NP carriers allow biomolecules to be separated from the medium and targeted for delivery to a specific spot. In the field of biosensors, the electrocatalytic properties of different ferric oxides at a glassy carbon electrode (GCE) showed performance characteristics similar to modified electrodes with Prussian Blue (Chavali and Nikolova, 2019).

8.4.4 Metal Oxides

Since the discovery of titanium dioxide's photocatalytic properties, other semiconductor photocatalysts have been researched and used in environmental applications. These include CeO_2, MnO_2, WO_3, ZnO, and SnO_2. Nano-TiO_2 has been used to make hollow spheres, wires, and hole-filled spheres. TiO_2 is commonly used in dielectric gate materials, photovoltaic cells, photocatalysis, and photo-assisted organic molecule degradation. Commercially accessible TiO_2 photocatalysts, such as Degussa P-25, have two common applications: gas sensors and photodegradation of aqueous organic contaminants. Doped TiO_2 NPs are used as biocides and in soil and water remediation operations (He and Hwang, 2014). *Micrococus lylae* and *Escherichia coli* showed 99% and 95% efficiency after 30-minute and 1-hour exposures, respectively. The semiconductor CeO_2 was used in hydrocarbon oxidation clean-up, gas sensing, and heterogeneous catalysis applications. Its sensing capabilities were attributed to structural stability, high oxygen mobility, and ionic defects that functioned as oxygen traps. TiO_2-doped CeO_2 catalysts were employed to catalyze the moist oxidation of phenol. It has been demonstrated that ZnO can detect and degrade organic contaminants in water. Electrochemical sensors are the most common biosensing application for metal oxide NPs (NPs). Hydrogen peroxide sensing is an important aspect of environmental processes, and both MnO_2 and CeO_2 have catalytic active sites for this purpose (Lim et al., 2018).

8.4.5 Quantum Dots

Carbon-based quantum dots (QDs) are a fascinating category of advanced carbon NPs, typically around 10 nm in size. Recently, carbon dots (CDs) have garnered attention due to their simple synthesis, strong compatibility with biological systems, and numerous potential applications. They possess a range of advantageous physical and chemical properties, including customizable photoluminescence, low quantum yield, excellent biocompatibility, catalytic activity, affordability, low toxicity, plentiful functional groups (such as amino, hydroxyl, and carboxyl), high stability, and photoinduced electron transfer. With the global population on the rise and the consequent increase in food consumption leading to substantial waste generation, addressing this environmental challenge has become paramount. While boosting agricultural and technological outputs addressed the food production issue, scientists sought ways to transform waste into materials that uphold sustainability principles. This quest led to the discovery of CDs sourced from various organic materials (such as banana peels, orange peel waste, sugarcane bagasse, *Trapa bispinosa* peels, bread, and jaggery), which hold promise for diverse environmental sensing applications (Permatasari et al., 2021).

Colloidal inorganic nanocrystals, or QDs, are fluorophores with unique photophysical properties. In environmental science, they are used as sensing probes for nitroaromatic explosives, phenols, pesticides, and trace metal ions. Size, composition, and surface states all influence their optical and spectroscopic properties, and any physical or chemical disruption inhibits fluorescence. QDs covered with inorganic semiconductor materials with higher band gaps, such as (CdSe)ZnS, form core-shell structures that improve chemical and photo-stability, as well as photoluminescence quantum yield. QDs used for metal analysis include CdS for Cu(II), Ag(II), and I-; CdSe for Cu(II), Ag(I), CN-, triethylamine, and benzylamine; and CdS-ZnS and CdTe for Cu(II). In addition, QDs can be coupled with antibodies and enzymes to form QDs-enzyme conjugates that are sensitive to substrates and enzyme inhibitors (Zhao et al., 2011).

8.4.6 Graphene NMs

Graphene, generated from graphite, has excellent mechanical, electrical, and thermal properties, as well as a honeycomb network structure. Heavy metals are removed from water and wastewater in several forms, including graphene oxide (GO) and reduced graphene oxide (r-GO). The reduction product, r-GO, has more structural faults but is more easily modified by functional groups like -OH, -NH_2, and -COOH, whereas the oxidative product, GO, contains carbonyl and epoxy groups. The work investigates heavy metal removal with graphene-based NMs. Using the Langmuir isotherm, the Qm value for Zn (II) removal was determined to be 243.9 mg/g. Hg (II) removal was investigated utilizing a GO and iron-manganese (GO/Fe-Mn) composite. The -OH groups on the surface of graphene oxide/Fe-Mn adsorbed Hg (II), making mercury removal from metallic solutions easier due to ligand exchange with $Hg(OH)_2$. Adding different functional groups to GO can improve its adsorption behaviour. Due to high concentration of COOH groups in GO obtained by directly oxidizing graphite with $K_2Cr_2O_7$, it has the highest Qm value for Pb (II). Thiol-functionalized r-GO has been identified as a novel adsorbent for Hg (II) removal. The heavy metal removal process mediated by graphene-based NMs is built on surface metal hydroxide precipitation and electrostatic interactions, both of which rely on surface area and charge. Economic research is needed into the reusability and recyclability of graphene-based NMs, as well as industrial uses, including wastewater treatment (Fang et al., 2010).

8.4.7 Surface Modification of SPIONs

Surface modified SPIONs are used in environmental research, nanobiology, and nanomedicine due to their unique properties, which include solubility, biocompatibility, fluorescence, and adsorption capacity (Singh et al., 2018). These applications require specialized coatings due to coupling, grafting, and electrostatic interactions. Coatings are applied on a variety of materials, including organic nonpolymeric, polymeric, and inorganic molecules, via electrostatic interactions, van der Waals interactions, and covalent attachment achieved through grafting and silanization. The biocompatibility of Fe_3O_4 NPs has been improved by coating them with 3-aminopropyl triethoxysilane (APTES) and inorganic elements such as gold and silicon. The Layer-by-Layer methodologies as well as other static electro-charge based surface modification methods are more sensitive to ionic strength and pH fluctuations. In general, covalently linked NPs are more robust and resistant to these changes. Surface modification can alter the functional groups on the surface of NPs, increasing their resistance to aggregation. Because hydrophilic coating materials can react or bind to dangerous substances in wastewater, it is critical to select them carefully when treating water (Biao et al., 2018).

8.5 ENVIRONMENTAL APPLICATIONS OF ADVANCED NMS

Applications of nanotechnology have the potential to dramatically impact society. Nanotechnology finds wide-ranging applications across various industries such as information technology, communication, food production, energy, pharmaceuticals, and medicine. Moreover, NMs offer new avenues for mitigating environmental pollution, including the development of eco-friendly, high-density hydrogen fuel cells, durable construction materials, and advanced drug delivery systems. Due to their exceptional ability to alter physiochemical properties, NPs and nanodevices are utilized in diverse areas such as hydrogen fuel cells, cancer therapy, vaccinations, nanoscale electronics, and nanographene batteries. By working at the nanoscale, nanotechnology enables the modification of molecules and substances, enhancing their mechanical properties or accessing anatomical regions that are otherwise difficult to reach. Table 8.2 enumerates various applications of different NMs.

The rising utilization of NPs in both household and industrial products poses a potential risk of environmental contamination. These applications contribute to the direct release of harmful substances into the soil and groundwater, posing significant environmental hazards. Assessing the environmental risk involves evaluating the pathways of NP exposure, as well as their persistence, reactivity, ecotoxicity, and mobility. Advances in nanotechnology have led to innovative methods for enhancing environmental remediation. This chapter will exclusively focus on the environmental applications of nanotechnology. Figure 8.1 illustrates various approaches to environmental remediation, while Table 8.3 provides a list of different NMs and their applications in this field.

Nanotechnology has been applied in a variety of sectors due to the improved properties of materials at the nanoscale that differ from their bulk nature. Nanotechnology research has recently had an impact on practically every basic field, including but not limited to environmental remediation, food technology, electrical transformers, medicine, and agriculture. Green nanotechnology has emerged as the primary alternative for mitigating environmental problems since it removes harmful and toxic antecedents from its process and is affordable also (Dutta et al., 2022).

8.5.1 Nanotechnology for Heavy Metal Detection and Removal

It is widely understood that heavy metals (such as Ni, Cu, Fe, Cr, Zn, Co, Pb, Hg, and Mn) can contaminate air, soil, and water. Their presence originates from various sources, including mining waste, vehicle emissions, natural gas, paper, plastic, coal, and dye industries. Even at trace levels measured in parts per million (ppm), specific metals like lead, copper, cadmium, and mercury ions pose heightened toxicity risks. Consequently, it is crucial to detect these hazardous metals in aquatic and biological environments to implement appropriate remediation strategies. Traditional analytical methods utilizing instrumental systems generally offer good sensitivity in multi-element analysis. However, the experimental setups required for such analyses are typically expensive, labor-intensive, skill-reliant, and stationary (Peng et al., 2016). Metallic NPs, with their size variability and distance-dependent optical properties, have become the preferred choice for detecting heavy metal ions in

TABLE 8.2
Applications of NMs in Various Fields (Mandal and Ganguly. 2011)

Field of Application	Principles of Application	NMs
Solar cell	Convert photon energy into electricity	Ag NW, Cu NW
Biomedicine	Gene Delivery	CNTs
Medical	Selective reactivity with certain biomolecules and antiviral activity	Fullerenes
Agricultural	Fertilizer developer	SiO2, ZnO, CuO, Fe and Mg
Construction	Increase the strength of material	SiO^2, Fe^2O^3
Cutting Tools	For hard material	Tungsten Carbide
Environmental	Increase growing rate of plants	CNTs
Elimination of Pollutants	Used as catalysts to react with toxic gases	Titanium oxide, calcium carbonate mixed silicon-based polymer
Energy	Alternative energy storage media	CNTs
Textile	Coating textiles such as nylon, to provide antimicrobial characteristics	Ultra hydrophobic water-resistant and strain resistance fabrics
Cosmetics	Used in sunscreen and in cosmetic industry	Titanium oxide and Zinc oxide
Car tires	Mechanical reinforcement	Carbon block
Car Bumpers	To make car exterior lighter	Nanosized clay
Cancer Diagnosis	Biomedical Imaging used for cancer and tumour detection	Gold and iron-based NPs
Transistor	For amplifying or switching electrical power	Si and Ge

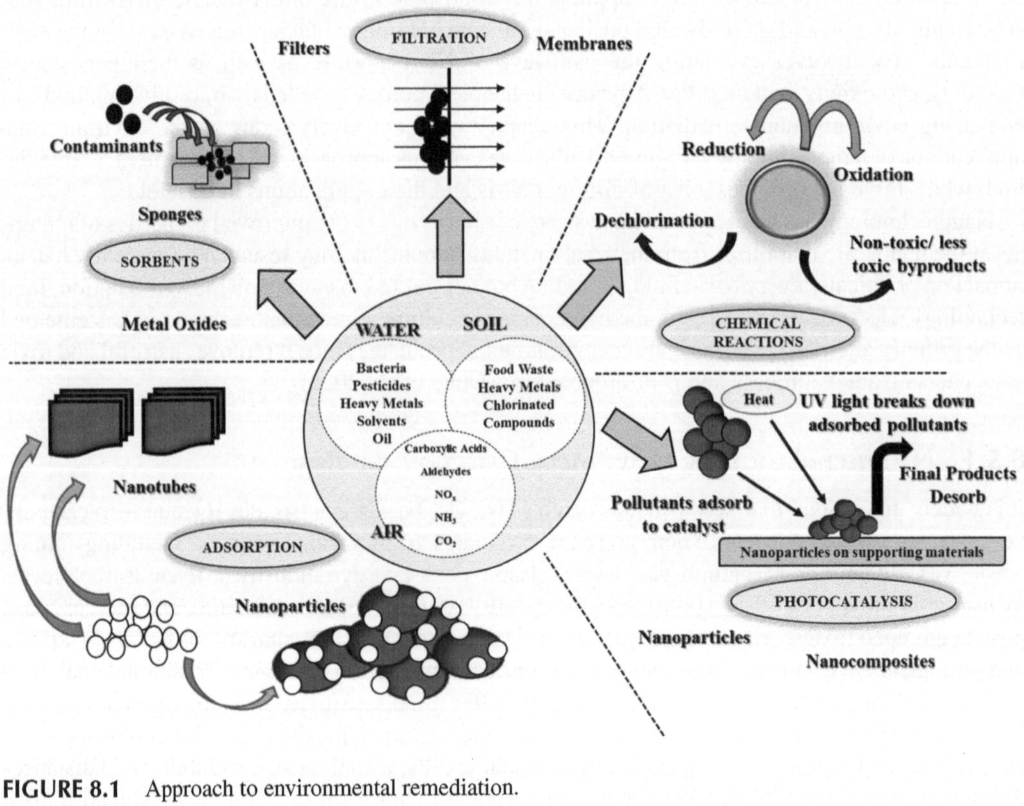

FIGURE 8.1 Approach to environmental remediation.

TABLE 8.3
Different NMs and Their Application in Environmental Remediation (Das et al., 2015)

Type of NMS		Materials	Applications
Inorganic NMs	Metal and Metal-based NMs	Ag NPs/Ag ions	Water disinfectant – *E. coli*
		TiO_2 NPs	Water disinfectant, soil- MS-2 phage, *E. coli*, hepatitis B virus, aromatic hydrocarbons, biological nitrogen, phenanthrene
		Metal-doped TiO_2	Water contaminants- 20chlorophenol, endotoxin, *E. coli*, Rhodamine B, *Staphylococcus aureus*
		Titanate nanotubes	Gaseous- Nitric Oxide
		Binary mixed oxides	Water- Methylene blue dye
		Iron-based	Water- Heavy metals, chlorinated organic solvents
		Bimetallic NPs	Water, soil- chlorinated and brominated contaminants
	Silica NMs	Amine-modified xerogels	Gaseous- CO_2, H_2S
		Amine-modified aluminosilicates and porous silica	Gaseous- CO_2, aldehydes, ketones
		Carboxylic acid-functionalized mesoporous silica	Wastewater- cationic dyes, heavy metals
		Amino-functionalized mesoporous silica	Wastewater- heavy metals
		Thiol-functionalized mesoporous silica	Wastewater- heavy metals
Carbon NMs	Graphene	Pristine Graphene	Water- Fluoride
		Graphene Oxide	Water/ Gaseous- SO_x, H_2, NH_3, heavy metals, pesticides, pharmaceuticals
		ZnO- graphene/CdS-graphene	Water- heavy metals
		TiO_2- graphene	Gaseous- Benzene
	CNTs		Adsorb Pb^{2+}, Cd^{2+}, Ni^{2+}, Cu^{2+}
			Adsorb cationic dyes and charged dyes
Polymer NMs		Amphiphilic polyurethane NPs	Soil- Polynuclear aromatic hydrocarbons
		PAMAM Dendrimers	Wastewater- heavy metals
		Amine-modified PDLLA-PEG	Gaseous- VOCs
		Polyamine-modified Cellulose	Gaseous- VOCs
		Polymer nanocomposites (PNCs)	Water- metal ions, dyes, microorganisms

water systems contaminated by such pollutants. These NPs offer simplicity, cost-effectiveness, and high sensitivity, particularly at sub-parts per million levels, when employed as colorimetric sensors for heavy metal ions in environmental samples (Khorramifar et al., 2023).

Heavy metal pollution in water sources is mostly caused by industrial effluents containing hazardous metal ions such as mercury, arsenic, copper, lead, and cadmium. Heavy metals are very hazardous to human health, due to their inherent toxicity and tendency to bioaccumulate in the bodies of living beings. Cd (II) ions were successfully eliminated using silver NPs derived from $AgNO_3$ and *Ficus benjamina* leaf extracts. Cadmium ion removal depended on adsorbent dosage, heavy metal concentration, pH, agitation speed, and contact length. Gold NPs functionalized with proanthocyanidin have demonstrated outstanding performance in removing methylene blue dye and heavy metal ions, such as Ni^{2+}, Cu^{2+}, Cd^{2+}, and Pb^{2+}. Fe_3O_4 magnetic nanorods anchored on DMSA were utilized to remove the dangerous heavy metal lead Pb^{2+}, which adsorbs best at pH 5 and 300 K. *Spirulina platensis* was utilized to produce cadmium sulphide (CdS) NPs that showed photocatalytic activity in the breakdown of malachite green coloring. This was accomplished without the need

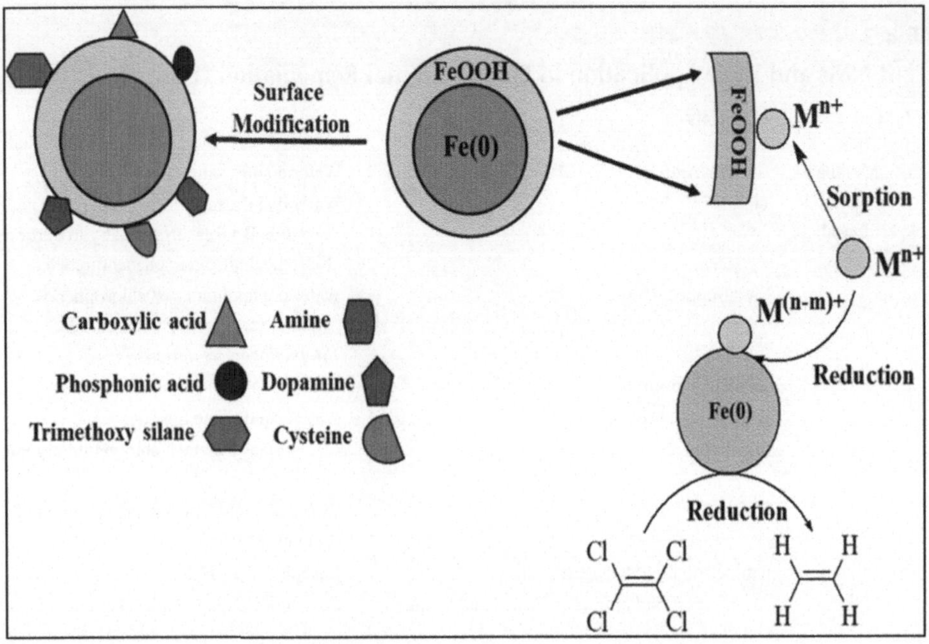

FIGURE 8.2 FeNPs-assisted degradation of heavy metals and chlorinated pollutants in aqueous solutions.

for a sulfur precursor. Nanotechnology can reduce in-situ pollutant concentrations while shortening the time and cost of cleaning up large-scale polluted areas. This opens up the prospect of employing nanotechnology to clean up hydrocarbon pollution.

Using various plant extracts, silver NPs (Ag NPs) were synthesized, which functioned as colorimetric sensors for heavy metal ions in water, including cadmium, chromium, mercury, calcium, and zinc (Cd^{2+}, Cr^{3+}, Hg^{2+}, Ca^{2+}, and Zn^{2+}). The presence of zinc and mercury ions (Zn^{2+} and Hg^{2+}) could be detected visually through the color changes induced by the as-synthesized Ag NPs. Similarly, Ag NPs derived from fresh mango leaves and dried leaves (MF-Ag NPs and MD-Ag NPs, respectively) exhibited selective sensing for lead and mercury ions (Pb^{2+} and Hg^{2+}). Furthermore, Ag NPs produced from green tea and pepper seed extract (GT-Ag NPs) demonstrated selective sensing capabilities for mercury, lead, and zinc ions (Hg^{2+}, Pb^{2+}, and Zn^{2+}). Figure 8.2 illustrates an instance where iron NPs (FeNPs) are utilized to remove heavy metals and chlorinated contaminants from water.

8.5.1.1 General Mechanism of the Removal of Heavy Metal Ions

***Removal of Chromium (VI)*:** Chromium is a widely used metal in a variety of industries, including electroplating, dyeing, tanning leather, cement, and photography. It is classified into two types: trivalent and hexavalent; the latter is more dangerous. Maghemite NPs were synthesized using the sol-gel process to extract Cr(VI) from aqueous solutions containing Ni(II) and Cu(II). The NPs' BET surface area measured 198 m^2/g. The Langmuir adsorption isotherm equation fit the adsorption data rather well. Table 8.4 demonstrates the optimal pH and maghemite NP adsorption capability for removing Cr and other metals from aqueous solutions. Adsorption occurs in all three scenarios due to electrostatic forces; nevertheless, maghemite has the highest adsorption capability for Cu(II) ions.

According to FTIR (Fourier Transform Infrared Spectroscopy) and XPS (X-ray photoelectron spectroscopy) studies, Cr(VI) adsorbed on NPs via outer sphere complexation, with the Langmuir model explaining the adsorption isotherm (Ahmadi et al., 2017).

***Removal of Cd(II)*:** Cadmium pollution of water due to human activities is a severe environmental issue. Industries that discharge cadmium include metal finishing, metallurgy, alloying, electroplating, pigments, batteries, fertilizers, textile operations, tanneries, zinc

TABLE 8.4
Adsorption Capacity of Maghemite (γ-Fe_2O_3) NPs for Different Metals (Hu et al., 2007)

Metal	pH	q_e(mg/g)	Adsorption Mechanism
Cr (VI)	2.5	17.0	Due to electrostatic attraction and ion exchange
Cu(II)	6.5	26.0	Due to electrostatic attraction and ion exchange
Ni(II)	8.5	23.6	Electrostatic attraction

smelting, and acid mine drainage. It not only causes nausea and vomiting, but it also harms the kidneys, liver, and bones. Thiolactic acid-modified titanium dioxide NPs were used to remove cadmium from water. $TiCl_4$ hydrolysis was utilized to produce nanocrystalline TiO_2, which was then treated with thiolactic acid. Chemisorption was employed to adsorb Cd(II) ions onto akaganeite nanocrystals, with an adsorption capacity of 17.1 mg/g (Permatasari et al., 2021). Akaganeite's sorbent capacity increased with temperature, efficiency, and solution pH. Cd(II) ions were recovered from simulated wastewater samples via flocculation techniques. Various functional groups, including hydroxyl, alkoxy, and oxy groups, are responsible for cationic exchange in NPs.

***Removal of Hg(II)*:** The principal sources of mercury contamination in the environment include metallurgical operations, paint manufacturing, wood pulping, fossil fuel combustion, and mercury and gold mining. Mercury and its compounds are extremely hazardous to human health, and mercury has been identified as one of the leading pollutants (Miretzky and Fernandez Cirelli, 2009). It has also been demonstrated to affect the gastrointestinal system and cause severe chronic conditions such as renal failure. Mercury's hazardous effects are determined by its molecular form. Mercury has an impact on crucial cell processes as well. The use of structured alumina NPs is to primarily remove inorganic mercury from wastewater. The sol-gel technique was employed to create alumina NPs, and the particle size ranged from 5 to 100 nm. It was discovered that increasing the amount of adsorbent boosts adsorption from 93.88% to nearly 100%, and that the flocculation technique using colloidal alumina can completely remove mercury.

***Removal of As(III) and As(V)*:** Because of its toxicity, arsenic pollution in the environment is a severe issue. Arsenic exists in two oxidation states in the environment: As(III) and As(V). Although As(V) is apparently more toxic than its trivalent sister, both forms are exceedingly dangerous to humans. Arsenic pollution in natural streams comes from geothermal activity and mineral weathering. Some enterprises that emit arsenic into the environment include fertilizer and pesticide manufacturing, gold mining, petroleum and refining, and ceramic color creation. It can cause liver disease, skin cancer, lung cancer, neurological abnormalities, and neurological impairments, in addition to lower limb paralysis (.

At 298 K, synthetic akaganeite (β-FeOOH) was found to have an adsorption capacity of 120 mg/g and to effectively remove As(V) ions. The adsorption data followed the Langmuir adsorption isotherm, with akaganeite nanocrystals outperforming other materials. Temperature and pH fluctuations affect the adsorption of As(V). The Freundlich isotherm model explained the adsorption of As(III) and As(V) on nanocrystalline TiO_2, which has a higher adsorption capacity due to its larger surface area. It was discovered that nanoscale zerovalent iron could remove As(V), converting it into magnetite or maghemite in around 90 days. The adsorption of arsenic revealed a pseudo-first-order response, with nanoscale zero valent iron outperforming micron-sized zero valent iron.

***Removal of Cu(II)*:** Cu(II) is an important chemical, but larger amounts have been shown to be hazardous to humans and may cause difficulties with the central nervous system. High

Cu(II) concentrations can induce irritation, dermatitis, itching, as well as hepatic and renal damage. The adsorption process followed the Langmuir model. An FTIR examination of gum Arabic NPs indicated that the surface hydroxyl groups of Fe_3O_4 were linked to the carboxylic groups of gum arabic. The amine group of gum arabic was shown to be surface complexed, allowing ions to adsorb on the NPs. In the case of magnetic NPs, this reaction is linked to the iron oxide surface hydroxyl groups.

8.5.2 Nanotechnology for Hydrocarbon Degradation and Removal

Because of its hazardous and carcinogenic components, hydrocarbon pollution poses a major threat to the ecosystem. Petroleum is a common source, as it comprises a complex mixture of hydrocarbon compounds. Nanotechnology has proven to be an effective method for remediating hydrocarbon pollutants, since it reduces pollutant concentrations while also reducing clean-up time and expenses. Silver NPs were used in photocatalysis to treat polycyclic aromatic hydrocarbons (PAHs) like pyrene, anthracene, and phenanthrene, as well as anthracene, a hazardous solid PAH (Hassan et al., 2015). The formation of silver NPs with garlic extract is more successful than with wet garlic extract, resulting in increased absorption of petroleum hydrocarbons (PAHs). The green approach was more effective at eliminating phenanthrene, pyrene, and anthracene, with an ideal effectiveness level of more than 85%. When zero-valent iron NPs were made with extracts from green mango peels, the total petroleum hydrocarbons were destroyed by more than 90% in about a week. These NPs proved to be more effective in terms of removal efficacy, cost-effectiveness, and environmental hazards (Desalegn et al., 2018).

8.5.3 Application of CNTs (Carbon Nanotubes) in Photocatalysis

Photocatalysis, which uses semiconductors, is one of the most advanced approaches for wastewater remediation. There are numerous semiconductor materials employed, including Fe_3O_4, ZnO, and TiO_2, but their quantum efficiency is low and their UV photoreaction is slow. CNTs show significant promise as new catalysis materials due to their large specific surface area, hollow tube shape, high chemical stability, increased quantum efficiency, and nano-size. They can also increase the light absorption zone (Zhang et al., 2022).

8.5.4 Graphene and Graphene Oxide-Based Adsorbents for the Purification of Heavy Metal-Contaminated Water

Graphene is the thinnest substance, having strength 200 times that of steel. It is a one-atom-thick multilayer hexagonal lattice composed of carbon atoms. Sir Andre Geim and Sir Konstantin Novoselov discovered graphene in 2004, and their contribution was recognized with a Nobel Prize in 2010. Graphene (2-D) has several applications in almost every field, including health and environmental sciences, touch screens, mobile devices, LCDs, semiconductors, computer chips, batteries, energy generation, water filters, supercapacitors, solar cells, and batteries. These 2-D graphene-based materials are gaining popularity in the water treatment sector, due to their unique physicochemical properties, which include electrical properties, a large surface area, thermal mobility, high mechanical strength, and customizable surface chemistry.

Diana et al. in the year 2016 employed graphene oxide, nickel, and platinum NPs to create a self-propelled microbot system (Lee et al., 2022). Each layer has a specific function, such as self-propelling, collecting Pb^{2+} ions in heavy metals, or manipulating microbots with an external magnetic field. The chemical has antimicrobial properties as well. When iron oxide magnetite NPs Fe_3O_4 were introduced to the nanocomposites, they removed 98% of the Pb^{2+} ions from water with a capacity of 126.6 mg/g (Nqombolo and Nomngongo, 2023).

8.5.5 Removal of Pollutant Dyes

Colors created in industrial wastewater are poisonous and carcinogenic, making textile effluents a severe health and environmental issue. Conventional treatment methods are costly, time-consuming, and generate a significant amount of sludge, which adds to secondary pollution. Sixty percent of the organic pollutants used in various industries are dyes, both anionic and cationic, which are mostly used in the process of producing fabric pigmentation. Approximately 15% of dyes are wasted and released into the hydrosphere, contaminating the environment. In environmental chemistry, regulating dye-containing effluents is a tough task. Metal and metal oxide semiconductor NPs are being used to oxidize dangerous pollutants in response to the increased need for clean and safe drinking water. Metal oxide semiconductor NPs with high surface area to mass ratios that promote organic pollutant adsorption, such as ZnO, TiO_2, SnO_2, WO_3, and CuO, have been employed to enhance photocatalytic activity of synthetic dyes.

Because of the numerous active sites on the surfaces of these NPs, their surface energy increases, leading to enhanced removal of contaminants even at low concentrations. This implies that a smaller amount of nanocatalyst is needed for purifying contaminated water compared to bulk materials. Furthermore, metal NPs, such as those derived from *Z. armatum* leaf extract, exhibit heightened photocatalytic breakdown of various pollutant dyes (González et al., 2015).

8.5.6 Activated Carbon in Wastewater Treatment

Initially, activated carbon (AC) was used as a sorbent; however, CNTs, fullerenes, and graphene were used as nanosorbents to overcome the challenges of heavy metal and dye removal at the ppb level (Wong et al., 2018). Typically, AC is produced using readily available carbonaceous precursors such as coal, wood, coconut shells, and agricultural waste. It also has a significant surface area and porosity. AC is frequently used in water treatment to remove organic and inorganic pollutants from effluent streams. Furthermore, it has extremely weak acidic ion exchange properties, allowing it to adsorb contaminants from wastewater and eliminate metal impurities. AC formed from coconut sawdust was used as an adsorbent to extract Cr(VI) from an aqueous solution. When gamma radiation was used to develop a novel sodium polyacrylate grafted AC, additional functional groups appeared on the surface, improving the AC's ability to absorb metal ions. AC materials are promising for heavy metal and dye removal due to their high sorption ability and low price content.

8.5.7 Application of (SPIONS) in Water Treatment

The primary drivers for integrating SPIONs into water treatment processes include their surface functional groups, size-related effects, and the convenience of extracting them from water using an external magnetic field. In the subsequent sections, we'll explore the applications of SPIONs in eliminating heavy metal ions, detecting and eradicating pesticide residues, purging microcystin, and managing oily water (Leonel et al., 2021).

8.5.7.1 Absorption of Toxic Water Soluble Molecules

Microcystin (MCs), a highly hazardous substance, is produced by cyanobacterial blooms, which are prevalent in many eutrophic environments. While traditional methods like coagulation and sand filtering effectively remove individual cyanobacterial cells, they fail to eliminate dissolved toxins. Activated carbon can be used to remove MCs, but due to insufficient accessible mesopores for adsorbing large MCs, it requires high dosages. Ozonation and chlorination can also eliminate MCs, but they necessitate large doses, potentially generating carcinogenic byproducts. Magnetic NPs offer promise in toxin removal due to their easy preparation, functionalization, and recovery. High-performance liquid chromatography (HPLC) is used to assess the efficiency of MCs removal.

8.5.7.2 Oily Wastewater Treatment

Oily wastewater originates from various sources, such as petroleum chemical industries, steel production, coking plants, gas stations, machining operations, oil extraction processes, and leaks from transportation vehicles. If this oily effluent is directly discharged into water bodies, it forms a thick layer of oil film on the surface. This not only hinders gas exchange in the water but also inhibits photosynthesis in nearby plants, leading to the depletion of aquatic life and environmental pollution. Oil in wastewater exists in four states: soluble, emulsifiable, dispersed, and as oil slick. Treatment strategies for oily wastewater are determined by its characteristics, as well as environmental and economic considerations. Magnetic particles have emerged as promising tools for oil removal due to their low energy consumption, high removal efficiency, and minimal secondary pollution. Additionally, magnetic NPs have been effective in removing oil from various surfaces, including rocks, feathers, and animal hair contaminated with crude oil (Abuhasel et al., 2021).

8.5.7.3 Novel Draw Solutions for Forward Osmosis

Osmosis, a natural phenomenon involving the movement of water across a selectively permeable membrane, has been known and utilized by humans since ancient times. Unlike conventional methods relying on hydraulic pressure differences, forward osmosis (FO) utilizes the osmotic pressure gradient ($\Delta\pi$) across the membrane. This process requires a selectively permeable membrane and a chemical potential difference between the two sides of the membrane, which is generated by draw solutes or agents. Currently, most forward osmosis solutes in use are saccharides and heat-sensitive salts, which pose challenges for energy- or labor-intensive recycling. An ideal draw solute should possess high solubility in water, yield a significant osmotic pressure upon dissolution, require minimal energy for recovery, and be reusable after recovery to be practical. One potential solution is magnetoferritin, a protein that can be efficiently extracted from aqueous streams using a magnetic field. However, its heat sensitivity and susceptibility to bacterial degradation limit its applicability in other contexts (Holloway et al., 2007).

8.5.8 NMs in Remediation of Air Pollution

Nanotechnology plays a significant role in air purification by utilizing nanocatalysts with increased surface areas for gaseous processes. These catalysts expedite chemical reactions, converting harmful emissions from vehicles and industrial facilities into safer gases. Filters, characterized by their compact size and porous structure, play a vital role in mitigating air pollution. They permit gases to pass through while trapping particles within their structure. Toxins are addressed by filters through mechanisms such as surface contact, inertial force, and electric charge generated by the particle and filter structure. However, filters can lead to pressure drops and consume considerable energy.

Microorganisms and contaminants can be efficiently removed using carbon nanotube-based membranes and nanofilters, featuring nanopores ranging from 1 to 10 nm in size. Carbon nanotubes offer the potential for large-scale application, due to their 100-fold faster gas capture rate compared to conventional methods. Nanomembrane technology enables the separation and purification of pollutants and gases across various industries, preventing their release into the environment. Nanosensors, intelligent devices capable of detecting and responding to nanoscale physical changes, play a crucial role in identifying toxins. NPs, tiny three-dimensional materials acting as sensors, can detect toxins in an area, facilitating early detection and mitigation measures (Gabal et al., 2020).

The commercial sector faces various risks, including hazardous spills, but advancements in sensor technology have led to faster and more accurate biosensors. Smart dust represents a new class of nanosensors capable of swiftly detecting hazardous gas molecules using lightweight nano computers. These ultrafine particles, constructed from silicon, transmit data wirelessly to a central server at a critical data transmission rate of at least one kilobit per second (kbps). Solar-powered nanosensors have an hourly flight time and can transmit data for up to 20 kilometers, including information on humidity, contaminants, and climatic variables. This data can be utilized to regulate air pollution levels in specific locations (Jung and Kim, 2020).

8.5.9 Antimicrobial Activity of NMs

The escalating resistance of germs to conventional antibiotics and antiseptics has spurred numerous investigations aimed at enhancing antimicrobial effectiveness. In vitro tests for antibacterial properties reveal that metallic NPs effectively inhibit various microbial species. The efficacy of these NPs in combating bacteria hinges on two key factors: the material used in their fabrication and the size of the particles. Over time, microbial resistance to antimicrobial drugs has steadily increased, posing a significant threat to public health. Examples of antimicrobial drug resistance include methicillin, sulphonamide, penicillin, and vancomycin (Rajak et al., 2020). Current antibiotics face numerous challenges, including resistance from biofilms and mutants resistant to multiple drugs. Bacterial capacity to develop resistance to drugs implies a rapid decline in the effectiveness of antibiotics. Hence, diseases in organisms persist despite the administration of high antibiotic doses aimed at combating bacteria. Another significant contributor to multidrug resistance against high antibiotic doses is biofilms. Drug resistance typically arises in infectious conditions such as gingivitis and lung infections. Utilizing NPs stands out as the most promising strategy for preventing or mitigating microbial drug resistance. As depicted in Figure 8.3, metallic NPs can impede or entirely hinder the formation of biofilms and multidrug resistance through various mechanisms (Shaikh et al., 2019).

Various types of NPs, including those containing metals, releasing nitric oxide (NO NPs), and containing chitosan (chitosan NPs), employ multiple mechanisms to combat microorganisms simultaneously. Due to their diverse modes of action, NPs are effective against drug resistance. Consequently, microorganisms would need to possess multiple gene mutations within their cells simultaneously to evade the mechanisms of NPs. However, it is rare for multiple biological gene mutations to occur simultaneously within the same cell. Table 8.5 outlines various mechanisms exhibited by NPs. Among the most highly regarded inorganic NPs are silver NPs, prized for their efficacy as antibacterial, antifungal, antiviral, and anti-inflammatory agents. The primary processes leading to bacterial cell death include: (a) denaturation of the outer membrane of bacteria; (b) formation of pits or gaps in the bacterial cell membrane, causing fragmentation; and (c) interactions between Ag NPs and enzyme groups containing disulfide and sulfhydryl. Additionally, antibacterial activity is influenced by NP shape. It is observed that truncated triangular NPs, with their

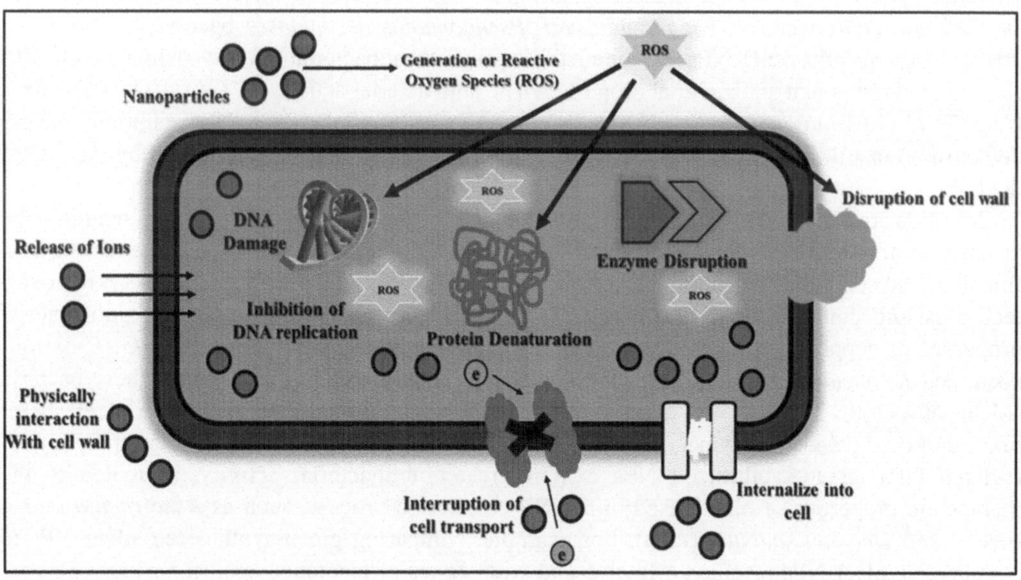

FIGURE 8.3 Different mechanisms of metal antibacterial action of NP.

TABLE 8.5
Multiple Antibacterial Activity Mechanisms for Different Types of Metallic NPs (Slavin et al., 2017)

S. No.	NPs	Multiple Mechanisms
1.	Chitosan-containing NPs	(a) Decreases the activities of metalloproteins (b) Due to its positive charge, chitosan binds with DNA in bacterial and fungal cells inhibiting transcription of mRNA, resulting in protein translation.
2.	Copper-containing NPs	(a) Higher concentration of Cu^{2+} ions can produce ROS (b) Copper interacts with amine and carboxyl groups that are present on microbes such as *B. subtilis*.
3.	Titanium dioxide-containing NPs (TiO_2 NPs)	(a) In the photocatalysis process, TiO_2 NPs generate ROS including hydrogen peroxide (H_2O_2) and hydroxyl radicals (OH-) upon exposure to near-UV and UVA radiation.
4.	Magnesium-containing NPs	(a) MgF_2 NPs can cause lipid peroxidation and a drop in cytoplasmic pH, which raises membrane potential. (b) MgX_2 NPs can cause lipid peroxidation of the microbial cell envelope by generating ROS.

high-atom-density surfaces, exhibit increased antibacterial activity and heightened reactivity in natural environments.

Due to their non-toxic nature, ability to be functionalized, polyvalent effects, and photothermal activity, the production of gold NPs (AuNPs) greatly aids in the development of efficient antibacterial drugs. Reactive oxygen species-related processes are not associated with the antibacterial activity of AuNPs. Researchers have attempted to bind AuNPs to bacterial membranes and alter membrane potential, resulting in decreased ATP levels, to investigate the antibacterial potential of these NPs. Moreover, this binding prevented tRNA from attaching to the ribosome.

The antibacterial properties of zinc oxide (ZnO), copper oxide (CuO), and iron oxide (Fe_2O_3) NPs were studied by Azam et al. in 2012 against gram-positive (*Staphylococcus aureus* and *Bacillus subtilis*) and gram-negative (*Escherichia coli*, *Pseudomonas aeruginosa*) bacteria (Azam et al., 2012). Consequently, ZnO NPs exhibited the strongest antibacterial activity, while Fe_2O_3 NPs showed the least potent antibacterial properties. The antibacterial activity of ZnO (19.89±1.43 nm), CuO (29.11±1.61 nm), and Fe_2O_3 (35.16±1.47 nm) NPs followed this order. These findings clearly demonstrate that the antibacterial activity of each sample is significantly influenced by the size of the NPs.

ZnO NPs are expected to exert antimicrobial action through several mechanisms, including the production of reactive oxygen species (ROS), release of zinc ions on the surface, membrane dysfunction, and cell entry. Furthermore, the antibacterial activity of ZnO NPs depends on their surface area and concentration. Mahapatra et al. in the year 2008 confirmed the antibacterial properties of copper oxide NPs against various bacterial species, including *Shigella*, *P. aeruginosa*, and *K. pneumoniae*, demonstrating appropriate antibacterial action against these bacteria (Mahapatra et al., 2008). It was hypothesized that NPs would penetrate the bacterial cell membrane and disrupts vital enzymes, leading to cell death. When compared to commercially manufactured NPs, green-synthesized NPs exhibit greater antibacterial activity, attributed to the therapeutic properties of plants used in the NP production process, such as *Azadirachta indica* (neem) and *Ocimum sanctum* (tulsi). For example, comparing green-synthesized silver NPs to commercial silver NPs revealed effective and wide zones of clearance against various bacterial strains.

Hospitals and other facilities that handle patients with infectious microorganisms may pose a public health risk, due to the high concentrations of these pathogens in their wastewater. Therefore, it is crucial for society to ensure proper wastewater treatment to remove harmful pollutants before discharging it into the sewage system. The prevalent methods for wastewater treatment can be broadly categorized as physical (such as heating, UV light, etc.), chemical (including hypochlorite/chlorination, ozonation, etc.), or biological. The primary drawbacks of the first two methods lie in their complex implementation and expensive maintenance, often due to issues like pipe system corrosion. Biological approaches have gained prominence as viable alternatives owing to their eco-friendliness. Many of these methods lack specific targets, employing live microorganisms in filtration processes through biofilm production or in fermentation operations to eliminate harmful substances.

Hoang et al. in the year 2016, reported on the utilization of lysing enzymes in combination with Fe_3O_4MNPs (magnetic NPs) as a biological treatment to target Mycobacterium tuberculosis (MTB) in contaminated wastewater (Hoang et al., 2016). This work represented advancement over previous studies by employing a combination of anti-TB antibody molecules and magnetic amino NPs. The compound NP-NH_2-anti-TB demonstrated the ability to selectively capture MTB, which could then be eliminated by the addition of lysing enzymes. The primary role of the NPs is to bring these molecules closer together, facilitating the coupling of antibody-captured bacteria and enhancing the effectiveness of lysing enzymes on the same surface area. This approach aims to overcome diffusion limitations and increase the reaction rate. Additionally, Hoang et al. conducted an initial assessment of the new approach in a wastewater model by introducing *Mycobacterium bovis* from the *Bacillus Calmette-Guérin* vaccine, into wastewater samples obtained from a hospital and a facility treating tuberculosis patients. The study's findings suggested potential applications for the NP-NH_2-anti-TB combination with enzymes to efficiently treat wastewater contaminated with MTB.

8.5.10 Applications of NMs in Agro-Based Sectors

Research has illustrated that NPs offer an innovative solution to various emerging global challenges, including those in agriculture. Operating at the nanoscale enhances properties such as cation exchange capacity, increased diffusion, ion adsorption, and complexation. Currently, nanotechnology development in the agricultural sector focuses on creating environmentally friendly nano-fertilizers that efficiently deliver nutrients and ions to plant cells, as well as on plant gene transformations to introduce desirable traits like drought resistance and accelerated growth cycles (Ainomugisha et al., 2024).

Addressing the increasing demand for food resulting from global population growth necessitates innovations in sustainable farming practices that yield higher outputs. However, these innovations must avoid long-term negative effects, such as poisonous runoff, bioaccumulative toxicity, or the degradation of arable land and water supplies. Research is underway to integrate nanotechnology into agriculture to meet these demands. Figure 8.4 depicts various applications of nanotechnology in agriculture.

Due to their remarkably high surface area to volume ratios and superior absorption efficiency into targeted plants, nano-fertilizers offer a valuable option. A study on phosphorus nano-fertilizers reported absorption efficiencies of up to 90%, making them highly sought-after in fertilizer formulations. Another advantage of nano-fertilizers is the delayed release of nutrients into plants over a 40–50 day period, compared to the 4–10 day timeframe of conventional fertilizers. This not only proves economically advantageous by reducing resources needed for fertilizer transportation, but also requires less fertilizer overall.

Crops treated with nano-fertilizers, both conventional and otherwise, have shown improved health due to their enhanced nutrient uptake capacity. For instance, in one study comparing a control group using traditional equivalents with a potato-specific nano-fertilizer containing elements like

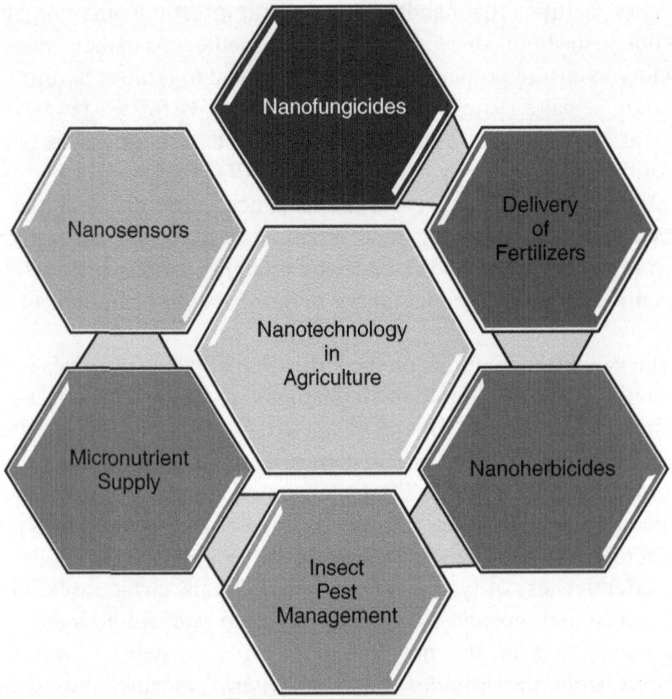

FIGURE 8.4 Various applications of nanotechnology in agriculture.

potassium, phosphorus, nitrogen, and magnesium, the potato crop treated with nano-fertilizer exhibited higher yield and better water use and agronomic efficiency. This efficiency is measured in terms of yield increases per unit of administered nutrient. Additionally, the study revealed that potatoes treated with nano-fertilizers contained higher levels of nutrients, including ascorbic acid and starch.

Another investigation focused on the application of iron-based nano-fertilizers to black-eyed peas and found significant improvements in root stability and leaf chlorophyll content, thereby enhancing photosynthesis. Similarly, the use of a different nano-fertilizer increased the rate of photosynthesis in maize crops, as indicated by the concentration of soluble carbohydrates. Nanotechnology has emerged as a popular choice for optimizing and modifying domestic plants, largely due to its significant contributions to genetic engineering and plant transformations. While past plant genetic modifications have primarily relied on methods like Agrobacterium or gene guns (biolistics), these approaches face limitations in versatility and compatibility, particularly with chloroplast/mitochondrial gene transformations. Additionally, they may risk cell or organelle damage, particularly with biolistics (Pandey et al., 2024).

Although biolistics and Agrobacterium are effective for certain plant species, more advanced techniques are under investigation, leveraging NMs for a more precise and less intrusive delivery method. By circumventing legal constraints associated with GMOs and utilizing carbon nanotubes (CNTs) and various porous NP (NP) delivery mechanisms, these techniques offer the potential for higher-throughput plant transformation. Conventional methods of plant transformation, such as Agrobacterium and biolistics, carry the risk of DNA incorporation into the plant genome, resulting in transgenic plants classified as genetically modified organisms (GMOs). Consequently, research into non-incorporative or DNA-free genetic modifications has become a critical area of study. Numerous other applications of NMs in environmental remediation exist. **Table 8.6** provides a comprehensive list of various NP models and their effectiveness in removing specific pollutants from the environment.

TABLE 8.6
Various Other Applications of Nanotechnology (Taran et al., 2021)

Type of NPS	Removal Target
Ag-doped TiO_2	2,4,6-Trichlorophenol
Ag-doped TiO_2 nanofibers	Methylene blue dye
Cu/Fe/Ag-doped TiO_2	Nitrate (NO_3^-)
Silica NPs prepared by mixing salicylic acid and hyper-branched poly (propylene imine)	Removal of polycyclic aromatic hydrocarbons (PAHH) such as pyrene and phenanthrene and Pb^{2+}, Hg^{2+}, Cd^{2+}, $Cr_2O_7^{2-}$ from contaminated aqueous solutions
PAMAM dendrimer composite membrane consisting of chitosan and a dendrimer	Separation of CO_2 from a feed gas mixture of CO_2 and N_2 on porous substrates.
Fe^0 coated with carboxymethyl cellulose polymer matrix	Hexavalent chromium (Cr^{6+}) from aqueous solutions
Gold coated with chitosan polymer	Zn^{2+}, Cu^{2+} form aqueous solutions
Poly (methacrylic acid)-grafted chitosan/bentonite	Th^{4+}
Carbon nanotubes/ Al_2O_3nanocomposites	Fluorides
Multiwall carbon nanotube (MWCTs)	Zn^{2+}

8.6 CHALLENGES AND FUTURE PERSPECTIVES

The advancement of human development and industrialization has led to the exploitation of natural resources, prompting the utilization of NPs in environmental litigation via biotechnology. NPs, capable of functioning on both small and large scales, offer sustainable and holistic solutions to this challenge. When properly produced, various NPs exhibit minimal harmful effects and can effectively remove contaminants such as landmines and sewage spills. There is potential to explore their suitability for achieving lower manufacturing costs (Bolisetty et al., 2019).

Although NPs possess exceptional catalytic properties that have shown promise in reducing environmental toxins, further research is needed to identify the optimal conditions for their biogenesis and environmental application to achieve consistent success rates. More investigation is required to determine the ideal circumstances for NP biosynthesis and utilization. In addressing the removal of organic dyes and heavy metal ions, researchers should focus on developing new NMs and nanocomposites with high surface area, functional groups, and sorption capacity. The risks posed by microbes to human health and safety are significant, thus materials with enhanced stability and the ability to remove multiple pollutants simultaneously should be developed. Affordable, efficient, and recyclable adsorbents need to be engineered for everyday use. A wide array of water purification treatment methods should be developed to meet the demands of environmental contamination.

8.7 CONCLUSION

The presence of heavy metal ions and dyes in wastewater poses significant concerns for environmental conservation and public health. Despite ongoing efforts, achieving optimal conditions for the removal of these ions remains elusive. NMs have garnered considerable attention in this regard, due to their unique properties, including large surface areas and small particle sizes, which result in a high number of adsorption active sites. Adsorption processes utilizing NPs are highly efficient, easy to implement, and effective for removing both organic and inorganic contaminants from wastewater. It is likely that these types of adsorbents will soon be widely adopted for commercial use in wastewater treatment. Carbon-based materials, in particular, exhibit high biocompatibility with the environment and living organisms, making them promising candidates for such applications.

The process of metal ion adsorption is influenced by various factors, such as pH, contact time, and the type of adsorbent used. Magnetic NPs, owing to their surface functional groups that can interact and target specific substances, have emerged as promising candidates for numerous applications, including water purification. Their magnetic responsiveness enables energy-efficient separation. Smaller particles, due to their greater surface impact and structural changes, offer enhanced benefits in water treatment. However, challenges remain in designing NP architectures tailored to specific applications. Surface-modified superparamagnetic iron oxide NPs (SPIONs) have shown significant potential in water treatment, particularly as stimuli-responsive agents that could facilitate the use of forward osmosis in pressure-driven osmosis and desalination processes. Further research is needed to scale up production and optimize structural designs for practical implementation.

ACKNOWLEDGEMENT

The corresponding author (Amrit Krishna Mitra) is thankful for the financial assistance provided by the Department of Science & Technology and Biotechnology, Government of West Bengal, India (Memo No. 860(Sanc.)/STBT-11012(25)/5/2019-ST SEC dated 03/11/2023).

REFERENCES

Abuhasel, Khaled, Mohamed Kchaou, Mohammed Alquraish, Yamuna Munusamy, and Yong Tzyy Jeng. "Oily wastewater treatment: Overview of conventional and modern methods, challenges, and future opportunities." *Water* 13, no. 7 (2021): 980.

Ahmadi, Mehdi, Maryam Hazrati Niari, and Babak Kakavandi. "Development of maghemite NPs supported on cross-linked chitosan (γ-Fe2O3@ CS) as a recoverable mesoporous magnetic composite for effective heavy metals removal." *Journal of Molecular Liquids* 248 (2017): 184–196.

Ainomugisha, Safiki, Moses Matovu, and Musa Manga. "Application of green agro-based NPs in cement-based construction materials: A systematic review." *Journal of Building Engineering* (2024): 108955. https://doi.org/10.1016/j.jobe.2024.108955

Al-Radadi, Najlaa S. "Green biosynthesis of flaxseed gold NPs (Au-NPs) as potent anti-cancer agent against breast cancer cells." *Journal of Saudi Chemical Society* 25, no. 6 (2021): 101243. https://doi.org/10.1016/j.jscs.2021.101243

Amieva, Edgar Jimenez-Cervantes, Juventino López-Barroso, Ana Laura Martínez-Hernández, and Carlos Velasco-Santos. "Graphene-based materials functionalization with natural polymeric biomolecules." *Recent Advances in Graphene Research* 1 (2016): 257–298.

Arole, Vinita Munde, and Surekha Vishnupant Munde. "Fabrication of NMs by top-down and bottom-up approaches-an overview." *Journal of Materials Science* 1 (2014): 89–93.

Azam, Ameer, Arham S. Ahmed, Mohammad Oves, Mohammad S. Khan, Sami S. Habib, and Adnan Memic. "Antimicrobial activity of metal oxide NPs against Gram-positive and Gram-negative bacteria: a comparative study." *International Journal of Nanomedicine* (2012): 6003–6009. https://doi.org/10.2147/IJN.S35347

Biao, Linhai, Shengnan Tan, Qinghuan Meng, Jing Gao, Xuewei Zhang, Zhiguo Liu, and Yujie Fu. "Green synthesis, characterization and application of proanthocyanidins-functionalized gold NPs." *NMs* 8, no. 1 (2018): 53. https://doi.org/10.3390/nano8010053

Bolisetty, Sreenath, Mohammad Peydayesh, and Raffaele Mezzenga. "Sustainable technologies for water purification from heavy metals: review and analysis." *Chemical Society Reviews* 48, no. 2 (2019): 463–487. 10.1039/c8cs00493e

Carmona, Erico R., Noelia Benito, Tanya Plaza, and Gonzalo Recio-Sánchez. "Green synthesis of silver NPs by using leaf extracts from the endemic Buddleja globosa hope." *Green Chemistry Letters and Reviews* 10, no. 4 (2017): 250–256. https://doi.org/10.1080/17518253.2017.1360400

Chau, Tan Phat, Geetha Royapuram Veeraragavan, Mathiyazhagan Narayanan, Arunachalam Chinnathambi, Sulaiman Ali Alharbi, Baskaran Subramani, Kathirvel Brindhadevi, Tipsukon Pimpimon, and Surachai Pikulkaew. "Green synthesis of Zirconium NPs using Punica granatum (pomegranate) peel extract and their antimicrobial and antioxidant potency." *Environmental Research* 209 (2022): 112771. https://doi.org/10.1016/j.envres.2022.112771

Chavali, Murthy S., and Maria P. Nikolova. "Metal oxide NPs and their applications in nanotechnology." *SN applied sciences* 1, no. 6 (2019): 607. https://doi.org/10.1007/s42452-019-0592-3

Darwish, Maher, and Ali Mohammadi. "Functionalized nanomaterial for environmental techniques." In Chaudhery Mustansar Hussain and Ajay Kumar Mishra (Eds.), *Nanotechnology in environmental science* (2018): 315–350. https://doi.org/10.1002/9783527808854.ch10

Das, Anupritee, Madhu Kamle, Ajay Bharti, and Pradeep Kumar. "Nanotechnology and it's applications in environmental remediation: an overview." *Vegetos* 32 (2019): 227–237. https://doi.org/10.1007/s42535-019-00040-5

Das, Sumistha, Biswarup Sen, and Nitai Debnath. "Recent trends in NMs applications in environmental monitoring and remediation." *Environmental Science and Pollution Research* 22 (2015): 18333–18344.

Desalegn, Biruck, Mallavarapu Megharaj, Zuliang Chen, and Ravi Naidu. "Green mango peel-nanozerovalent iron activated persulfate oxidation of petroleum hydrocarbons in oil sludge contaminated soil." *Environmental technology & innovation* 11 (2018): 142–152. https://doi.org/10.1016/j.eti.2018.05.007

Dutta, Vishal, Ritesh Verma, C. Gopalkrishnan, Min-Hao Yuan, Khalid Mujasam Batoo, R. Jayavel, Ankush Chauhan, Kun-Yi Andrew Lin, Ravindran Balasubramani, and Suresh Ghotekar. "Bio-inspired synthesis of carbon-based NMs and their potential environmental applications: A state-of-the-art review." *Inorganics* 10, no. 10 (2022): 169.

Falcaro, Paolo, Raffaele Ricco, Amirali Yazdi, Inhar Imaz, Shuhei Furukawa, Daniel Maspoch, Rob Ameloot, Jack D. Evans, and Christian J. Doonan. "Application of metal and metal oxide NPs@ MOFs." *Coordination Chemistry Reviews* 307 (2016): 237–254. : https://doi.org/10.1016/j.ccr.2015.08.002

Fang, Chen, Omid Veiseh, Forrest Kievit, Narayan Bhattarai, Freddy Wang, Zach Stephen, Chun Li, Donghoon Lee, Richard G. Ellenbogen, and Miqin Zhang. "Functionalization of iron oxide magnetic NPs with targeting ligands: their physicochemical properties and in vivo behavior." *Nanomedicine* 5, no. 9 (2010): 1357–1369. https://doi.org/10.2217/nnm.10.55

Fominski, V. Yu, R. I. Romanov, D. V. Fominski, D. V., & Shelyakov, A. V. "Preparation of MoSex> 3/Mo-NPs catalytic films for enhanced hydrogen evolution by pulsed laser ablation of MoSe2 target." *Nuclear Instruments and Methods in Physics Research Section B: Beam Interactions with Materials and Atoms* 416 (2018): 30–40. http://doi.org/10.1016/j.nimb.2017.12.002

Gabal, Esraa, Somenath Chatterjee, Farah K. Ahmed, and Kamel A. Abd-Elsalam. "Carbon nanomaterial applications in air pollution remediation." In *Carbon NMs for Agri-Food and Environmental Applications*, pp. 133–153. Elsevier, 2020.

Ghomi, Ahmad Reza Golnaraghi, Mohammad Mohammadi-Khanaposhti, Hossein Vahidi, Farzad Kobarfard, Mahdieh Ameri Shah Reza, and Hamed Barabadi. "Fungus-mediated extracellular biosynthesis and characterization of zirconium NPs using standard penicillium species and their preliminary bactericidal potential: A novel biological approach to NP synthesis." *Iranian journal of pharmaceutical research: IJPR* 18, no. 4 (2019): 2101. https://doi.org/10.22037/ijpr.2019.112382.13722

Ghosh, Shampa, Jitendra Kumar Sinha, Soumya Ghosh, Kshitij Vashisth, Sungsoo Han, and Rakesh Bhaskar. "Microplastics as an emerging threat to the global environment and human health." *Sustainability* 15, no. 14 (2023): 10821.

González, Joaquín A., María E. Villanueva, Lidia L. Piehl, and Guillermo J. Copello. "Development of a chitin/graphene oxide hybrid composite for the removal of pollutant dyes: adsorption and desorption study." *Chemical Engineering Journal* 280 (2015): 41–48. https://doi.org/10.1016/j.cej.2015.05.112

Gupta, Kanika, Pratiksha Joshi, Rashi Gusain, and Om P. Khatri. "Recent advances in adsorptive removal of heavy metal and metalloid ions by metal oxide-based NMs." *Coordination Chemistry Reviews* 445 (2021): 214100.

Hassan, Saad SM, Waleed IM El Azab, Hager R. Ali, and Mona SM Mansour. "Green synthesis and characterization of ZnO NPs for photocatalytic degradation of anthracene." *Advances in Natural Sciences: Nanoscience and Nanotechnology* 6, no. 4 (2015): 045012. https://doi.org/10.1088/2043-62

He, Xiaojia, and Huey-Min Hwang. "Engineered TiO 2 NPs: Their fate and effects in natural aquatic environments." *The Journal of Chemical Physics* 7 (2014): 1–20.

Hoang, Dong Quy, Tuong Vi Tran, Ngoc Quyen Tran, Cuu Khoa Nguyen, Thi Hiep Nguyen, Minh Dung Truong, Dai Lam Tran, Le Van Thu, and Dai Hai Nguyen. "Functionalization of Fe3O4 NPs with biodegradable chitosan-grafted-mPEG for paclitaxel delivery." *Green Processing and Synthesis* 5, no. 5 (2016): 459–466. https://doi.org/10.1515/gps-2016-0093

Holloway, Ryan W., Amy E. Childress, Keith E. Dennett, and Tzahi Y. Cath. "Forward osmosis for concentration of anaerobic digester centrate." *Water Research* 41, no. 17 (2007): 4005–4014. https://doi.org/10.1016/j.watres.2007.05.054

Hrbac, Jan, Vladimir Halouzka, Radek Zboril, Kyriakos Papadopoulos, and Theodor Triantis. "Carbon electrodes modified by nanoscopic iron (III) oxides to assemble chemical sensors for the hydrogen peroxide amperometric detection." *Electroanalysis: An International Journal Devoted to Fundamental and*

Practical Aspects of Electroanalysis 19, no. 17 (2007): 1850–1854. https://doi.org/10.1002/elan.200703938

Hu, Jing, Irene MC Lo, and Guohua Chen. "Performance and mechanism of chromate (VI) adsorption by δ-FeOOH-coated maghemite (γ-Fe2O3) NPs." *Separation and Purification Technology* 58, no. 1 (2007): 76–82. https://doi.org/10.1016/j.seppur.2007.07.023

Hua, Ming, Shujuan Zhang, Bingcai Pan, Weiming Zhang, Lu Lv, and Quanxing Zhang. "Heavy metal removal from water/wastewater by nanosized metal oxides: a review." *Journal of Hazardous Materials* 211 (2012): 317–331.

Ishak, NAI Md, S. K. Kamarudin, and S. N. Timmiati. "Green synthesis of metal and metal oxide NPs via plant extracts: An overview." *Materials Research Express* 6, no. 11 (2019): 112004. https://doi.org/10.1088/2053-1591/ab4458

Jawed, Aquib, Varun Saxena, and Lalit M. Pandey. "Engineered NMs and their surface functionalization for the removal of heavy metals: A review." *Journal of Water Process Engineering* 33 (2020): 101009. https://doi.org/10.1016/j.jwpe.2019.101009

Jung, Seojin, and Jooyoun Kim. "Advanced design of fiber-based particulate filters: Materials, morphology, and construction of fibrous assembly." In *Polymers* 12, no. 8 (2020): 1714. https://doi.org/10.3390/polym12081714

Kanchi, Suvardhan, and Shakeel Ahmed, eds. *Green Metal NPs: Synthesis, Characterization and Their Applications*. John Wiley & Sons, 2018.

Khan, Shamshad, Mu Naushad, Adel Al-Gheethi, and Jibran Iqbal. "Engineered NPs for removal of pollutants from wastewater: Current status and future prospects of nanotechnology for remediation strategies." *Journal of Environmental Chemical Engineering* 9, no. 5 (2021): 106160.

Khorramifar, Ali, Hamed Karami, Larisa Lvova, Alireza Kolouri, Ewa Łazuka, Magdalena Piłat-Rożek, Grzegorz Łagód et al. "Environmental engineering applications of electronic nose systems based on MOX gas sensors." *Sensors* 23, no. 12 (2023): 5716.

Kimber, Richard L., Edward A. Lewis, Fabio Parmeggiani, Kurt Smith, Heath Bagshaw, Toby Starborg, Nimisha Joshi et al. "Biosynthesis and characterization of copper NPs using Shewanella oneidensis: Application for click chemistry." *Small* 14, no. 10 (2018): 1703145. https://doi.org/10.1002/smll.201703145

Korbekandi, Hassan, Soudabeh Mohseni, Rasoul Mardani Jouneghani, Meraj Pourhossein, and Siavash Iravani. "Biosynthesis of silver NPs using Saccharomyces cerevisiae." *Artificial Cells, Nanomedicine, and Biotechnology* 44, no. 1 (2016): 235–239. http://doi.org/10.3109/21691401.2014.937870

Lee, Suhyun, Lakshmi Prasanna Lingamdinne, Jae-Kyu Yang, Janardhan Reddy Koduru, Yoon-Young Chang, and Mu Naushad. "Biopolymer mixture-entrapped modified graphene oxide for sustainable treatment of heavy metal contaminated real surface water." *Journal of Water Process Engineering* 46 (2022): 102631.

Leonel, Alice G., Alexandra AP Mansur, and Herman S. Mansur. "Advanced functional nanostructures based on magnetic iron oxide NMs for water remediation: A review." *Water Research* 190 (2021): 116693.

Li, Xue-Mei, Gaojie Xu, Yue Liu, and Tao He. "Magnetic Fe3O4 NPs: Synthesis and application in water treatment." *Nanoscience & Nanotechnology-Asia* 1, no. 1 (2011): 14–24.

Lim, Jihn Yih, N. M. Mubarak, E. C. Abdullah, Sabzoi Nizamuddin, and Mohammad Khalid. "Recent trends in the synthesis of graphene and graphene oxide based NMs for removal of heavy metals: A review." *Journal of Industrial and Engineering Chemistry* 66 (2018): 29–44. https://doi.org/10.1016/j.jiec.2018.05.028

Love, Andrew J., Valentine Makarov, Igor Yaminsky, Natalia O. Kalinina, and Michael E. Taliansky. "The use of tobacco mosaic virus and cowpea mosaic virus for the production of novel metal NMs." *Virology* 449 (2014): 133–139. https://doi.org/10.1016/j.virol.2013.11.002

Luo, Xiaogang, Xiaojuan Lei, Ning Cai, Xiuping Xie, Yanan Xue, and Faquan Yu. "Removal of heavy metal ions from water by magnetic cellulose-based beads with embedded chemically modified magnetite NPs and activated carbon." *ACS Sustainable Chemistry & Engineering* 4, no. 7 (2016): 3960–3969. https://doi.org/10.1021/acssuschemeng.6b00790

Mahapatra, Ojas, Megha Bhagat, C., and Arunachalam, Kantha D. "Ultrafine dispersed CuO NPs and their antibacterial activity." *Journal of Experimental Nanoscience* 3, no. 3 (2008): 185–193. https://doi.org/10.1080/17458080802395460

Makarov, Vladimir V., Andrew. J. Love, Olga V. Sinitsyna, Svetlana S. Makarova, Igor V. Yaminsky, Michael E. Taliansky, and Natalia O. Kalinina. ""Green" nanotechnologies: synthesis of metal NPs using plants." *Acta Naturae (англоязычная версия)* 6, no. 1 (20) (2014): 35–44.

Mandal, Gopa, and Tapan Ganguly. "Applications of NMs in the different fields of photosciences." *Indian Journal of Physics* 85 (2011): 1229–1245. https://doi.org/10.1007/s12648-011-0149-9

Miretzky, Patricia, and A. Fernandez Cirelli. "Hg (II) removal from water by chitosan and chitosan derivatives: A review." *Journal of Hazardous Materials* 167, no. 1–3 (2009): 10–23. https://doi.org/10.1016/j.jhazmat.2009.01.060

Nqombolo, Azile, and Philiswa Nosizo Nomngongo. "Graphene and graphene oxide–based nanoadsorbents in wastewater treatment." In *Adsorption Through Advanced Nanoscale Materials*, pp. 143–161. Elsevier, 2023. https://doi.org/10.1016/B978-0-443-18456-7.00007-9

Palit, Sukanchan, and Chaudhery Mustansar Hussain. "Functionalization of NMs for industrial applications: Recent and future perspectives." In Chaudhery Mustansar Hussain (Ed.), *Handbook of Functionalized NMs for Industrial Applications*, pp. 3–14. Elsevier, 2020.

Pandey, Garima, Navin Ahlawat, Sangeeta Bajpai, and Monika Kamboj. "Recent trends in synthesis and application of NMs for agri-food industries." In *Nanotechnology and NMs in the Agri-Food Industries*, pp. 253–282. Elsevier, 2024. https://doi.org/10.1016/B978-0-323-99682-2.00005-0

Peng, Jiyu, Fei Liu, Fei Zhou, Kunlin Song, Chu Zhang, Lanhan Ye, and Yong He. "Challenging applications for multi-element analysis by laser-induced breakdown spectroscopy in agriculture: A review." *TrAC Trends in Analytical Chemistry* 85 (2016): 260–272. https://doi.org/10.1016/j.trac.2016.08.015

Permatasari, Fitri Aulia, Muhammad Alief Irham, Satria Zulkarnaen Bisri, and Ferry Iskandar. "Carbon-based quantum dots for supercapacitors: Recent advances and future challenges." *NMs* 11, no. 1 (2021): 91. https://doi.org/10.3390/nano11010091

Qian, Lanting, Sharmila Durairaj, Scott Prins, and Aicheng Chen. "Nanomaterial-based electrochemical sensors and biosensors for the detection of pharmaceutical compounds." *Biosensors and Bioelectronics* 175 (2021): 112836. https://doi.org/10.1016/j.bios.2020.112836

Qidwai, Afifa, A. Pandey, R. Kumar, S. K. Shukla, and A. Dikshit. "Advances in biogenic NPs and the mechanisms of antimicrobial effects." *Indian Journal of Pharmaceutical Sciences* 80, no. 4 (2018). https://doi.org/10.4172/pharmaceutical-sciences.1000398

Rajak, Bablu Lal, Rahul Kumar, Manashjit Gogoi, and Sanjukta Patra. "Antimicrobial activity of NMs." *Nanoscience in Medicine Vol. 1* (2020): 147–185. https://doi.org/10.1007/978-3-030-29207-2_5

Raliya, Ramesh, Pratim Biswas, and J. C. Tarafdar. "TiO2 NP biosynthesis and its physiological effect on mung bean (Vigna radiata L.)." *Biotechnology Reports* 5 (2015): 22–26. https://doi.org/10.1016/j.btre.2014.10.009

Sadegh, Hamidreza, Gomaa AM Ali, Vinod Kumar Gupta, Abdel Salam Hamdy Makhlouf, Ramin Shahryari-Ghoshekandi, Mallikarjuna N. Nadagouda, Mika Sillanpää, and Elżbieta Megiel. "The role of NMs as effective adsorbents and their applications in wastewater treatment." *Journal of Nanostructure in Chemistry* 7 (2017): 1–14.

Saleh, Tawfik A. "Nanomaterials for pharmaceuticals determination." *Bioenergetics* 5, no. 1 (2016): 1000226. https://doi.org/10.4172/2167-7662.1000226

Shaikh, Sibhghatulla, Nazia Nazam, Syed Mohd Danish Rizvi, Khurshid Ahmad, Mohammad Hassan Baig, Eun Ju Lee, and Inho Choi. "Mechanistic insights into the antimicrobial actions of metallic NPs and their implications for multidrug resistance." *International Journal of Molecular Sciences* 20, no. 10 (2019): 2468. https://doi.org/10.3390/ijms20102468

Shanker, Uma, Vidhisha Jassal, Manviri Rani, and Balbir Singh Kaith. "Towards green synthesis of NPs: From bio-assisted sources to benign solvents: A review." *International Journal of Environmental Analytical Chemistry* 96, no. 9 (2016): 801–835.

Show, Shilpi, Abiral Tamang, Trinath Chowdhury, Dipankar Mandal, and Brajadulal Chattopadhyay. "Bacterial (BKH1) assisted silica NPs from silica rich substrates: A facile and green approach for biotechnological applications." *Colloids and Surfaces B: Biointerfaces* 126 (2015): 245–250. https://doi.org/10.1016/j.colsurfb.2014.12.039

Singh, Jagpreet, Tanushree Dutta, Ki-Hyun Kim, Mohit Rawat, Pallabi Samddar, and Pawan Kumar. "'Green'synthesis of metals and their oxide NPs: Applications for environmental remediation." *Journal of Nanobiotechnology* 16 (2018): 1–24. http://dx.doi.org/10.4172/2167-7662.1000226

Singh, Sarika, K. C. Barick, and D. Bahadur. "Functional oxide NMs and nanocomposites for the removal of heavy metals and dyes." *NMs and Nanotechnology* 3 (2013): 20.

Slavin, Yael N., Jason Asnis, Urs O. Hńfeli, and Horacio Bach. "Metal NPs: Understanding the mechanisms behind antibacterial activity." *Journal of Nanobiotechnology* 15 (2017): 1–20. https://doi.org/10.1186/s12951-017-0308-z

Taran, Mojtaba, Mohsen Safaei, Naser Karimi, and Ali Almasi. "Benefits and application of nanotechnology in environmental science: An overview." *Biointerface Research in Applied Chemistry* 11, no. 1 (2021): 7860–7870. https://doi.org/10.33263/BRIAC111.78607870

Thangavelu, Raja Muthuramalingam, Rajendran Ganapathy, Pandian Ramasamy, and Kathiravan Krishnan. "Fabrication of virus metal hybrid NMs: An ideal reference for bio semiconductor." *Arabian Journal of Chemistry* 13, no. 1 (2020): 2750–2765. https://doi.org/10.1016/j.arabjc.2018.07.006

Wong, Syieluing, Norzita Ngadi, Ibrahim M. Inuwa, and Onn Hassan. "Recent advances in applications of activated carbon from biowaste for wastewater treatment: A short review." *Journal of Cleaner Production* 175 (2018): 361–375.

Zhang, Jing, Mingchong Dai, Shijie Zhang, Meng Dai, Peng Zhang, Shuguang Wang, and Zuoli He. "Recent progress on carbon-nanotube-based materials for photocatalytic applications: A review." *Solar RRL* 6, no. 9 (2022): 2200243.

Zhang, Xi-Feng, Zhi-Guo Liu, Wei Shen, and Sangiliyandi Gurunathan. "Silver NPs: Synthesis, characterization, properties, applications, and therapeutic approaches." *International Journal of Molecular Sciences* 17, no. 9 (2016): 1534. https://doi.org/10.3390/ijms17091534

Zhao, Guixia, Jiaxing Li, Xuemei Ren, Changlun Chen, and Xiangke Wang. "Few-layered graphene oxide nanosheets as superior sorbents for heavy metal ion pollution management." *Environmental Science & Technology* 45, no. 24 (2011): 10454–10462. https://doi.org.1021/es203439v

Zhou, Nan, Qingda An, Zuoyi Xiao, Shangru Zhai, and Zhan Shi. "Solvothermal synthesis of three-dimensional, Fe 2 O 3 NPs-embedded CNT/N-doped graphene composites with excellent microwave absorption performance." *RSC Advances* 7, no. 71 (2017): 45156–45169. https://doi.org/10.1039/C7RA06751H

9 Carbon Nanomaterials for Remediation of Water Pollutants

Saumya Agrahari, Anamika Srivastava, Divya Singh, Nirmala Kumari Jangid, and Manish Srivastava

9.1 INTRODUCTION

Even though water is a valuable and plentiful resource on the planet, there is a water shortage as a result of severe industrialization, unfavorable climate change and fast population growth (Kotia et al., 2020). As a result, it becomes increasingly difficult and necessary to provide clean, pure water, primarily drinking water, in underdeveloped nations. The World Health Organization conducted research that estimates that 1.1 billion individuals are without immediate access to sources of potable water (Thines et al., 2017). The weather conditions has been steadily changing due to the application of nanotechnology in the treatment of environmental toxins, particularly those found in water. This has a noticeable effect on both the wellness of people and other organisms. Nano adsorbent application is expected to spur the creation of new technologies and give rise to more potent instruments for the detection, prevention, and resolution of issues associated with water pollution, including wastewater (Hsu et al., 2023).

Microorganisms, organic chemicals, and inorganic harmful elements are the three main categories considered for defining water pollutants. Many inorganic metallic elements, including cadmium, copper, lead, chromium, and mercury, are included in the list of inorganic, hazardous elements. Pesticides, endocrine disruptors, organic dyes, detergents, medications, personal hygiene items, and prevalent industrial organic wastes (i.e., halogens, aromatics, and phenolics) are among the common organic pollutants found in water. In Stockholm, Sweden, in 2001, the US, the European Community, and ninety other nations signed the landmark Stockholm Convention, an international environmental agreement. Twelve major persistent organic pollutants (POPs), including chlordane, aldrin, endrin, dieldrin, hexachlorobenzene, mirex, polychlorinated dibenzo-p-dioxins, polychlorinated biphenyls (PCBs), hexachlor, polychlorinated dibenzofurans, and toxaphene, have been banned, and their production, use, and/or release will be restricted (Lu and Astruc, 2020).

Numerous negative effects, including cancer, renal damage, liver disease, miscarriages, anemia, encephalopathy, and nephritic syndrome, can be brought on by exposure to heavy metal ions (Baby et al., 2019). Lead acid cells, paper, polishing, and glass industries are the main sources of Pb^{2+} emitted into the environment. Water released from solar cells, battery electroplating, metallurgical processes, and fabric factories is typically contaminated with cadmium (Wang et al., 2012). Diseases like nephritis, stomach discomforts, liver damage, and nasal mucous ulcers are all brought on by chromium metal ions (VI) (Moreno-Castilla et al., 2004).

Membrane filtration, distillation, and absorption are the methods used in conventional water treatments. One of the exciting elements is carbon, which may form a variety of configurations with a broad range of regular features. The remarkable physical, chemical, thermal, and electrical characteristics of CNMs, such as graphene, oxides of graphene, carbon nanotubes, fullerenes, and activated carbons, have made them useful in a variety of applications, including storage of energy,

DOI: 10.1201/9781003484189-9

electronics, sensors, drug delivery, healthcare treatment, and purification of water (Baby et al., 2019). Because of their high conductivity, high strength, and low weight, CNMs have drawn significant attention in studies since the development of NMs and nanotechnology (Peng et al., 2020). Due to its properties, CNMs are inexpensive, environmentally benign, and possess strong sorption, quick kinetics, thermal stability, and conductivity; they hold great promise for technological innovation aimed at optimizing water treatment (Kotia et al., 2020).

CNMs include fullerenes, carbon nanotubes (CNTs), nanofibers, graphene, and nanodiamonds, among other forms with zero, one, two, and three dimensions (Ravi and Vadukumpully, 2016). Sp^2-bonded graphite and sp^3-bonded diamond are two distinct solid-state allotropes of carbon with unique structures and properties (Fatima and Mushtaq, 2023). The freely accessible activated carbon provided by agricultural waste streams and biomass sources is one advantage of carbon-derived nano adsorbents. These NPs are less expensive and useful for adsorption at the nanoscale. With a high adsorption rate, these adsorbents offer an improved adsorption capacity (Hussain et al., 2022). Environment-induced degradation of CNMs may impact their initial toxicity through physicochemical alterations. Currently, non-biodegradation and biodegradation are the two main categories under which CNM degradation occurs. The majority of non-biodegradation is chemical in nature. Enzymatic, microbial, and breakdown of cells are all forms of biodegradation. Enzyme degradation is mostly used in studies on the biodegradation of CNMs (Peng et al., 2020). This chapter highlights various kinds of CNMs and their synthesis via biological routes and application in the removal of aqueous pollutants from wastewater. It also covers the application of advanced sustainable NMs, particularly carbon nanotubes, and activated carbon, to remove contaminants and aqueous pollutants from wastewater.

9.2 FUNDAMENTAL CHARACTERISTICS OF CNMs

NMs of Zero-dimensional (0-D) are defined as those with all three dimensions < 100 nm; fullerene and quantum dots are two examples of such materials (Baby et al., 2019). The three main categories of CNMs are: advanced two dimensions graphene-based materials (graphene, graphene oxide, and associated modification compounds), C60 fullerene, and SWCNTs/MWCNTs (single-wall/multi-wall carbon nanotubes) and derivatives (Lu et al., 2020). CNMs have been extensively applied as efficient adsorption materials for wastewater clean-up because of their huge specific surface area, high free energy, porosity, and activity. The typical use of powdered activated carbon (PAC), which is widely utilized, is to purify drinking water (Figure 9.1). The functional groups of natural materials are modified on nanoscale surfaces to create new, promising materials at affordable costs (Lu et al., 2020). CNMs, including carbon nanotubes (CNTs), graphene oxide, carbon dots, fullerene, and activated carbon, are used in water purification technologies primarily because of their structural characteristics (Kotia et al., 2020).

9.2.1 Carbon Nanotubes (CNTs)

Many potential applications of CNTs include unique and strong porous structures that exist in the fields of hydrogen storage, adsorption, catalysis, and medicine. The CNTs exhibit robust adsorption characteristics for heavy metals, antibiotics, and petroleum in aquatic environments due to their expansive particular surface area, as well as extensive porosity structure. CNTs are grapheme (GRA) sheets bound into cylinders and coated at both ends with fullerene. (Peng et al., 2020). MWCNTs & SWCNTs are two types of CNTs. One of the most popular adsorbents for treating metal ions in water is CNTs, which have shown remarkable stability and removal effectiveness, even in the most adverse circumstances (Yu et al., 2018). For the primary time, Tian et al. created magnetic nanocomposites of poly(m-phenylenediamine)-coated iron oxide/acid-oxidized MWCNTs (PmPD/Fe_3O_4/o-MWCNTs) using in situ oxidative polymerization to remove hexavalent Cr (Tian et al., 2015).

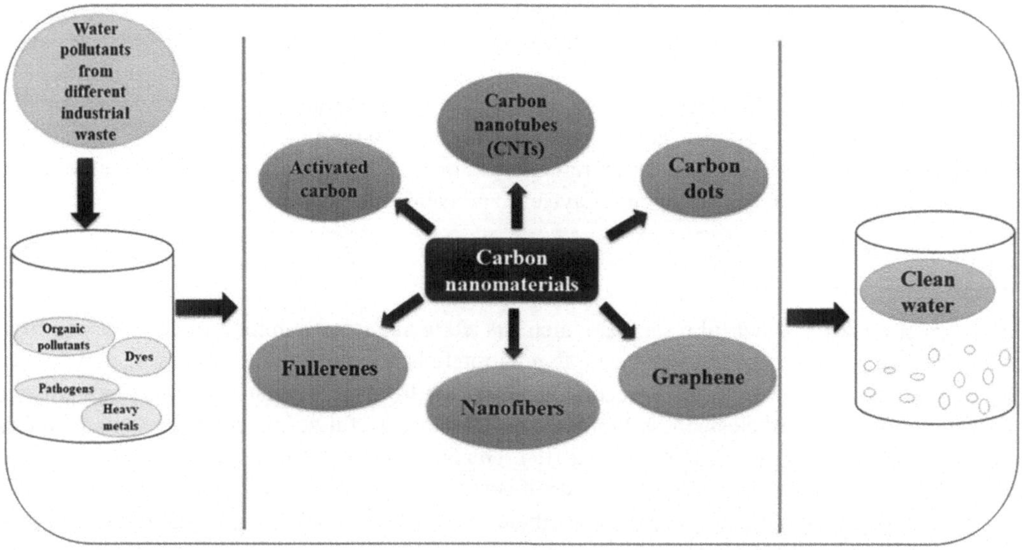

FIGURE 9.1 Schematic diagram of CNMs and their applications in the purification of wastewater.

9.2.2 Graphene Oxide

The Hummers method states that oxidants (i.e., concentrated sulfuric acid, potassium permanganate, and nitric acid) are commonly used to oxidize graphite to form graphite oxide (GO). GO has hydroxyl and epoxy groups on sp^3 hybridized carbon on the plane of basal, and carbonyl and carboxyl groups on sp^2 hybridized carbon near the sheet borders. It is also noticeably more oxygenated than pure graphite (Fatima and Mushtaq, 2023). Because of their extensive surface area, pi-pi arranging, bonding of hydrogen, and electrostatic attraction, as well as their nanoscale size, graphene-based nano-adsorbents are promising advanced materials for the remediation of organic pollutants from water (Baby et al., 2019). Lead (II), Cadmium (II), Copper (II), and Zinc (II) ions were removed from wastewater using GO nanocomposites (GO-Zr-P) functionalized with zirconium and phosphate (Abbo et al., 2021).

9.2.3 Carbon Dots

A recent addition to the carbon family, carbon dots (CDs) has garnered significant interest because of their exceptional and adaptable photoluminescence (PL), significant quantum yield (QY), low harmful effects, small size, notable biocompatibility, and plentiful inexpensive sources. These attributes have important applications in biomedicine, catalysis, optoelectronic devices, and anticounterfeiting (Liu et al., 2020). Amorphous and belonging to the zero-dimensional carbon family, dots have a standard size of less than 10 nm. The most commonly used methods for the simple production of dots include hydrothermal, chemical oxidation, laser ablation, and microwave methods. The quickest, least expensive, and most consistent way to achieve uniformly sized particles for dot preparation is through hydrothermal carbonization. For the most part, dots are characterized by exceptional solubility of water, non-toxicity, and edge effects that enable the detection of heavy metal ions. These are intriguing and distinctive qualities (Sari et al., 2023).

9.2.4 Fullerene

American chemists developed fullerene, a carbon NM with a zero-dimensional structure, in the mid-1980s. It is one of the novel and unique forms of carbon NMs used as an adsorbent. The most widely recognized member of the fullerene family is C60. These nanostructures are known to be the smallest stable carbon ones and are made up of sp^2-hybridized carbon atoms structured in closed hollow cages as pentagons and hexagons. They are found on the interface between molecules and NMs (Elessawy et al., 2020). The carbon atoms found in fullerenes are typically covalently bonded and exist in the sp^2-hybrid state. The fullerene that has been studied the most is the fullerene C60. The 60 carbon atoms that make up the extremely symmetric spherical molecule are arranged at the edges of 20 hexagons and 12 pentagons (Zaytseva and Neumann, 2016).

9.2.5 Activated Carbon

Agricultural waste is a plentiful resource, which has led to an increase in the manufacture of adsorbents based on biomaterials in recent years. In addition to being readily available, these agricultural products are inexpensive to acquire; creating an adsorbent from them doesn't cost much in terms of production costs. Nevertheless, these biomaterials are quite useful at low concentrations of metal and are recyclable with ease (Thines et al., 2017). The different physical characteristics of activated carbon, like surface area, size of pores, volume of pores, etc., and chemical characteristics, including functional groups and surface charge, are attributed to both chemical and physical activation processes (Srivastava et al., 2021).

9.3 PROCESS OF CNMS

A topic of interest is the fabrication of new-generation CNMs, including innovative substances with high efficiency and multiple functions. The growth of carbon-based compounds like graphene and carbon nanotubes depends critically on precursors, which are crucial in determining a technology's feasibility. CNTs have been synthesized using a variety of carbon precursors, including CH_4, C_6H_6, C_2H_2, xylene (C_8H_{10}), toluene ($C_6H_5CH_3$), etc. The synthesis of carbon-based compounds commonly used arc discharge, lateral polarization, CVD, and other techniques. Simply put, heterogeneous CVD processes entail passing a given proportion of a hydrocarbon, primarily C_6H_6, C_8H_{10}, $C_6H_5CH_3$, hexane (C_6H_{14}), ethanol, methanol, CH_4, C_2H_2, ethylene (C_2H_4), and CO, over small metal particles (Fe, Ni, Cu, Co,) in a furnace in a gaseous or liquid flow. This is typically done as a mixture with either H_2 or an inert gas like Ar (Kumar et al., 2016). Carbon NM synthesis has been achieved by a variety of physical, chemical, and electrochemical processes (Figure 9.2).

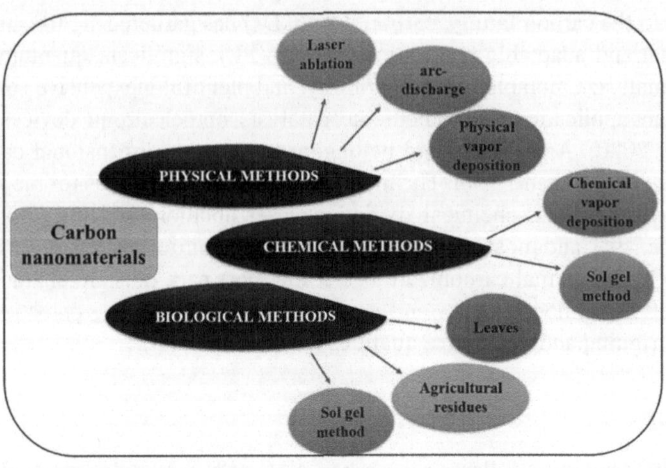

FIGURE 9.2 Various synthesis methods of CNMs.

9.3.1 Physical and Chemical Methods

There are five common ways to get CNTs. High current is transferred between two electrodes maintained in an inert atmosphere during the arc-discharge process. This produces heat that is high enough to induce carbon in the anode to evaporate and deposit on the cathode. Therefore, MWCNTs are produced. Another option is to create SWCNTs, but this requires altering the anode, which is typically comprised of graphite. A suitable catalyst made up of transition metal, like Fe or Co, is inserted into a hollow drilled in the core to facilitate the synthesis of SWCNTs in the gas phase (Janas, 2020).

Since carbon dioxide is a clean gas with a modest reaction rate at activation temperatures and is easy to handle, it is a desirable agent. In comparison to steam, carbon dioxide also results in increased carbon yields and microporosity levels. Higher temperatures may cause water molecules to break down materials more quickly, which could explain the lower AC output by steam activation (Zaini et al., 2021; Rashid and Ralph, 2017).

Conventionally, ablation, arc vaporization of graphite, combustion procedures, and CVD technologies are used to create fullerenes. Nevertheless, further development and advancements in this field are required because existing techniques are not particularly effective. In contrast to laser vaporization and arc discharge, the CVD method of producing fullerenes has the benefit of requiring a lower temperature. Both hot-filament CVD and microwave-enhanced CVD procedures were presented by Kleckley et al. as means of synthesizing fullerene (Manawi et al., 2018).

A variety of CVD methods, including liquid injection (LICVD) (Szabó et al., 2010), oxygen-supported (Byon et al., 2007), aerosol-supported (ACVD) (Meysami et al., 2015), hot wire (HWCVD) (Lehman et al., 2006), hot filaments (HFCVD) (Makris et al., 2004), and microwave plasma boosted (MWCVD) (Watanabe et al., 2010), are used to produce CNTs. Among the benefits of CVD are (Sheoran et al., 2022):

a. It is the main approach to mass-produce carbon nanotubes economically.
b. It employs a wide range of substrates and hydrocarbons across each of their different states, including solid, liquid, and gaseous.
c. It can be used to create entangled, straight, and coiled CNTs, among other shapes.

9.3.2 Biological Methods

Plants can be applied on an extensive basis by enabling the application of various plant parts, including leaves, roots, stems, latex, seeds, fruits, fruit peels, and agricultural residue. Plants have demonstrated good performance as reducing agents, due to enhanced stability as well as quick kinetics (Diel et al., 2021). Using a straightforward ball mill methodology and without any toxic substances or oxides, a unique and environmentally friendly method has been reported for the fabrication of extremely porous N-doped graphene/CNTs composite (Fouda et al., 2021). Alkallas et al. demonstrated a laser ablation approach to synthesize carbon nanotubes that were coated with nickel oxide NPs. The synthetic materials were applied to remove methyl orange coloring from wastewater after being examined by highly developed equipment (Gacem et al., 2022). Ghosh et al. synthesized synthetic wood nanotubes (SWNTs) and a Fe/Co zeolite catalyst, using eucalyptus oil as a natural precursor by pyrolysis process (Ghosh et al., 2007). Li et al. reported biosynthesized CNTs/$Ni(OH)_2$ nanostructured via a low-temperature hydrothermal method, which is used for electrochemical supercapacitors (Li et al., 2020).

Bhushan et al. (2021) synthesized a novel activated carbon by impregnating jack fruit peel charcoal with KOH via irradiation of microwave. A simple, eco-friendly technology uses banana peels in a one-step hydrothermal procedure to produce activated carbon and high-fluorescence carbon dots. The carbon dots, with an average size of 3–6 nm, develop strong photoluminescence in bright blue color. Also, they have applications in electronics, catalysis, and biosensing. Activated carbon used as an electrode in supercapacitors has an extensive specific surface area (294.6 m^2g^{-1}) and an especially porous structure (Nguyen et al., 2022).

Recently, there has been widespread exploitation of biomass, such as plant shrubberies and wheat stubble, as a source of carbon for carbon quantum dot (CQD) synthesis (Nguyen et al., 2021). Nguyen et al. 2021 reported the production of three biochars by slow pyrolysis process from coffee husks, mimosa bark, and wattle bark of Acacia auriculiformis, respectively, that were compared and evaluated for their ability to remove MB at pH 6-12. SEM, BET analysis, and FTIR spectroscopy were used to analyse the properties of the materials.

The thermal decomposition of sesame oil, turpentine oil, neem oil, jatropha oil, eucalyptus oil, palm oil, camphor, tea-tree extract, agricultural waste, insects, food products, waste food, etc., has allowed researchers to create high-quality graphene, SWCNTs, MWCNTs, and carbon dots. Typically, the formation of graphene and CNTs results from the catalytic action of fossil gaseous species that are formed when hydrocarbons thermally decompose. The production of graphene and CNTs is a crucial area of contemporary nanotechnology, utilizing naturally occurring precursors found in the surrounding environment. It is advantageous to convert natural precursors and waste materials into enhanced carbon nanotubes and graphene structures, as these potentially hold promise for a range of innovative electrical and energy applications (Nguyen et al., 2021). Granular activated carbons (GACs) and powdered activated carbons (PACs) are the two types of AC that are sold commercially. PACs have a higher capacity for adsorption, due to their large specific surface area and micro-porosity. Currently, a variety of pollutants, such as organic micro-pollutants, pharmaceuticals, arsenic, carcinogenic compounds, microplastics, heavy metals, color, and odor, are all effectively removed by using GAC as a water treatment material (Jjagwe et al., 2021).

Mandarin and orange peels, with qmax values of 196.08 and 378.89 mg/g, respectively, were chemically activated to produce Activated (AC) for MB dye adsorption. The surface area of the AC made from the kiwi peel was higher (306.18 m^2/g) than that of the papaya peel (15.26 m^2/g). It is evident that the types of precursor materials have an impact on the ability of adsorption of the AC because the kiwi peel had a lower adsorption capacity of Pb(II) (158.82 μg/g) than the papaya peel (38.31 mg/g) (Srivastava et al., 2021). Magnetic carbon adsorbents, polyurethane, nitro-phenol, yeast alcohol dehydrogenase, polyethylene with a low density, acetate, acetylene, and oleates, are among the various compounds that can be prepared using a range of carbon-based supports (or holders) (Reynel-Ávila et al., 2021).

9.4 APPLICATION OF CNMS IN THE REMEDIATION OF WATER POLLUTANTS

The adsorptive potential of f-CNTs for water pollution remediation was studied by K. M. Lee et al. (Lee et al., 2018). The use of activated carbons (ACs) results in their large surface area, compound complexity, and microporous structure. Water treatment facilities use ACs as adsorbents. There are several practical groups for phenol, carbonyl, lactone, carboxyl, quinone, etc., on the outside layer. The performance of adsorption of ACs depends on properties such as surface beneficial categories and pore size distribution. However, carbon may react with oxygen at average temperatures (≥ 300 °C) (Sheoran et al., 2022). Georgin et al. reported the removal of Basic Fuchsin dye by using powdered biosorbent obtained from the leaves of mandacaru (*Cereus jamacaru*) (Georgin et al., 2020).

Graphene NPs have a substantial surface area, high tensile strength, outstanding optical and electrical qualities, and ease of modification, which make them useful for treating wastewater. Toxic and dangerous combinations of various contaminants can be eliminated with the employment of Gr NMs as nano sorbents (Sheoran et al., 2022). Chemicals or microorganisms that are either naturally occurring or man-made, that are not regularly inspected but have the potential to be highly hazardous when found in water bodies, are known as CECs (contaminants of emerging concern). Most CECs are polar or semi-polar pollutants with high solubility in water, their removal by physical processes like sedimentation and flocculation is not practical (Jjagwe et al., 2021).

TABLE 9.1
Different Types of Carbon NMs and Their Application in Wastewater

S. No.	Carbon NMs	Synthesis Material	Water Pollutants	References
1.	Magnetic nanocomposites/SiO_2/ CNTs		Direct Blue 71, Reactive Blue 19	Abbasi (2017)
2.	Magnetized adsorbent	Orange Peel	Crystal violet (CV)	Ahmed et al. (2020)
3.	CNT-EB activated column	Eucalyptus	Methylene Blue, Eosin yellow	Yadav (2022)
4.	Multiwalled CNTs	(Several plant extracts) Green synthesis	Pharmaceutical and endocrine disruptors	Spaolonzi et al., (2022)
5.	MNP-GAC MNP-PAC	Green tea extracts	Copper	Shahrashoub and Bakhtiari, (2021)
6.	Activated carbon pellets (ACP)	Textiles sludges	Malachite green	Tang and Zaini, (2020)
7.	Activated carbon	waste pulp of *Citrus limetta*	Cr(VI)	Agrahari et al. (2023)
8.	MWCNT-PDA/GO hybrid aerogel	Green synthesis	Copper (II), lead (II)	Zhan et al. (2019)
9.	MWCNT decorated with Au/ Fe_3O_4 nanocomposites	Green synthesis	Pb(II)	Zondo et al. (2022)
10.	Magnetic fullerenes	PET waste bottles	Methylene blue, Acid Blue 25 dye	Elessawy et al. (2020)

GAC has been demonstrated to be a successful approach for removing pesticides and medicines, although membrane bioreactors, activated sludge, and aeration have demonstrated good removal efficiency towards endocrine disrupting chemicals (EDCs), personal care products (PCPs), and surfactants (Ahmed et al., 2017). Different types of carbon NMs and their application in wastewater treatment are described in **Table 9.1**.

9.4.1 Heavy Metals Removal

Chufa et al., 2021 reported the removal of fluoride from aqueous solutions using SWCNTs and MWCNTs. Fluoride removal effectiveness with functionalized carbon nanotubes has been seen to exceed 80% (Chufa et al., 2021). The adsorption of heavy metal ions from water is thought to be facilitated by fullerenes, which are considered useful NMs, due to their large surface area, lower aggregation tendency, and ability to penetrate spaces and defects between carbon nanoclusters through the penetration of adsorbates (Baby et al., 2019). Unfortunately, because of their high cost, their direct use is not profitable. This is why research has been done on modifying traditional sorbents (zeolites, activated carbons, etc.) to increase their ability to absorb by adding tiny quantities of fullerenes. The development of the sorbent surface's porous structure, an increase in hydrophobicity, and an improvement in the material's ability to remove heavy metals from aquatic environments were among the important benefits of adding fullerene. For example, the sorption capacity of activated carbons for Lead (II) and Copper (II) increased by 1.5–2.5 times after 0.001–0.004% of the fullerenes were added (Burakov et al., 2018). In their comparison research, Alekseeva et al. discovered that fullerenes were more effective in removing Cu^{2+} ions than nanocomposite-polystyrene films (Alekseeva et al., 2016). Anitha et al. performed a molecular dynamic simulation to investigate the capabilities of adsorption of heavy metal ions, such as Cadmium (II), Copper (II), Lead (II), and Mercury (II), from aqueous media. The simulation included both bare SWCNTs and their modified alternatives, such as SWCNTs-OH, SWCNTs-NH_2, and SWCNTs-COOH (Anitha et al., 2015). Cr(VI), Co, and Se ions found in wastewater can be effectively removed by combining activated

CNTs with NZVIs in a nanocomposite (Chadha et al., 2022). In the environment. chromium (Cr) is typically found as Cr(III). Trivalent chromium is hypo-toxic, when it transforms into Cr(VI), it becomes highly toxic. The removal of hexavalent Chromium is easier to accomplish in an acidic aqueous solution, and researchers have utilized carboxylated MWCNTs coated with chitosan to create a quick and efficient approach for Cr(VI) removal. One of them can adsorb up to 140 mg/g or more at its maximum (Maitlo et al., 2019; Huang et al., 2018).

Antibiotic residues are growing in the environment, due to their extensive use. Antibiotics have a hard time breaking down naturally (Peng et al., 2020). These days, antibiotic poisoning of the water supply has grown to be a significant issue. The low cost and straightforward adsorption process of antibiotics is commonly used because of their poor biodegradability in aquatic environments; π-π EDA (π-π electron donor-receptor), π-π electron coupling stacking, hydrophobic interaction, and other mechanisms are the key mechanisms by which CNTs absorb organic contaminants. CTC (Chlortetracycline hydrochloride) and TCN (tetracycline hydrochloride) adsorption are primarily caused by the interaction of π-π between the adsorbents (Xiong et al., 2018).

Lead (II), Cadmium (II), Copper (II), and zinc (II) ions were removed from wastewater using GO nanocomposites (GO-Zr-P) modified with zirconium and phosphate. The addition of phosphate and zirconium changed the fraction of GO that was layered and flat in these nanocomposites. The produced GO-Zr-P nanocomposites sheets got significantly wrinkled upon the incorporation of zirconium and phosphate, increasing their ability to be adsorptive for Lead (II), Cadmium (II), Copper (II), and Zinc (II) ions in contrast to pure GO. For 20 minutes, the metal ion removal capacity of GO-Zr-P nanocomposites was approximately 99% (Abbo et al., 2021).

Kadirvelu et al. (2001) discovered that the heavy metals, including Cd^{+2}, Hg^{+2}, Cu^{+2}, Pb^{+2}, and Ni^{+2}, are removed from industrial wastewater by adsorbing them onto AC made from solid agricultural waste. Kamboj et al. studied the removal of heavy metals, including chromium, copper, and nickel, from industrial wastewater by using AC made from a naturally occurring aquatic plant, water hyacinth (Kamboj and Tiwari, 2024).

9.4.2 Dyes Removal

The adsorption of dyes on carbon NMs has shown promising adsorption properties in recent decades. Many adsorbents, including zeolites, clays, polymers, and activated carbon, have been employed to remediate organic dye pollution (Jun et al., 2018). Park et al. effectively used titania to embellish the SWCNT aerogel and used it to remove MB from water (Park et al., 2016). MWCNTs-TiO_2 were synthesized by Duan et al., then used to photodegrade methyl orange (MO) (Duan et al., 2016). Using methylene blue and malachite green as typical organic dyes, researchers conducted comparative adsorption studies between graphite and GO and discovered that GO demonstrated significantly superior adsorption to graphite (Baby et al., 2019). Rhodamine B, crystal violet, and methylene blue are examples of cationic dyes that have been removed from water using GO. The adsorption capacities of MB, CV, and RhB were determined to be 199.2, 195.4, and 154.8 mg g−1, respectively, indicating that the higher the starting [dye], the higher the adsorption (Jin ct al., 2012).

Wang et al. fabricated magnetic carbon NMs by the hydrothermal method in which biologically regenerated glucose and ferric ammonium citrate were used as carbon and iron sources for the removal of organic dye rhodamine B (Wang et al., 2018). Ahlawat et al. prepared the three various materials, including AC, MWCNTs, and their composites (MWCNTs/AC), for the removal of cationic dyes and anionic dyes, MB and MO, respectively (Ahlawat et al., 2020). Furthermore, pollutants were adsorbed onto the surface-by-surface exchange reactions until the functional sites on the surface were completely occupied. At that point, the pollutants may diffuse within the adsorbent and engage in additional interactions with the functional groups. Surface modification is a crucial step toward enhancing the surface properties of raw CNMs by changing their surface, which helps to overcome their drawbacks (Jun et al., 2018).

Ma et al. studied the adsorption of MO and MB from aqueous solutions onto the surface of alkali-AC-CNTs (activated carbon nanotubes) with high surface area and an increase in number of mesopores. ACNTs have an adsorption capability of 399 mg/g, and 149 mg/g for MB and MO, respectively. The testing also demonstrated how strongly the external transfer of mass and diffusion between particles influences the adsorption process (Ma et al., 2012).

Yu et al. reported that the two carbon NMs CNTs and GnPs (graphene nanoplates) with CNFs (cellulose nanofibrils) were synthesized to remove cationic and anionic dye MB and Congo red, respectively. CNF-GnP has a maximum adsorption capability of 1178.5 mg. g^{-1} and 585.3 mg. g^{-1}, respectively (Yu et al., 2018).

9.4.3 Pathogens Removal

Although the exact mechanism of action is still being studied, carbon-based NMs also have antibacterial qualities. According to certain research, CNTs may be used in antimicrobial surface coatings or for disinfection. In particular, hybrid NPs of carbon nanotubes coated with silver (Ag) have demonstrated antibacterial properties (Zaytseva and Neumann, 2016). Brunet et al., introduced that the pathogenic bacteria in water can be killed through the use of hydrophilic modified fullerenes (C60) and a photocatalytic method. Due to the lower bond strength between carbon and hydrogen, fullerene molecules may be readily transformed into C–C bonds, making them perfectly clean, green materials for storing hydrogen (Brunet et al., 2009). MWCNTs have been demonstrated to have potent antibacterial characteristics.

Multiple parameters, including functional molecules on the surface, CNT purity, diameter, length, and density, influence the antibacterial capabilities of MWCNTs (Hirschfeld et al., 2017). The specific mechanism underlying MWCNTs' antibacterial ability is yet unknown, though. MWCNTs have diverse mechanisms of action on the microbial surface. First, by strongly affiliating with the membrane surfaces, the integrity of the cell membrane is affected. Secondly, causing the production of ROS (reactive oxygen species) in turn causes DNA damage and cell death. Third, the MWCNTs' impurity may play a significant role in their antibacterial activity. Gram-negative i.e. *E. coli* and gram-positive i.e. *E. faecalis* and *S. epidermidis* bacteria are not immune to the in vitro antibacterial activity of nanocomposites, i.e., chitosan, its derivatives, and MWCNT derivative 4. The agar diffusion technique was used to evaluate the bacteria, namely *E. Coli*, as gram-positive bacteria (Anzar et al., 2020).

Engel et al. introduced results that single-walled carbon nanotubes combined with iron oxides have been shown to have activity of antibacterial effects against *E. coli* bacteria (Engel et al., 2018). The antibacterial effectiveness against *E. Coli*, salmonella species, *Shigella flexineri*, and Klebsiella was examined using membrane filters coated with carbon nanotubes (Anzar et al., 2020). In particular, numerous investigations have demonstrated that MWNTs demonstrate a lower level of toxicity to human and animal cells than SWNTs. The death of pathogenic bacteria (*E. coli* K12) cells is suggested to be facilitated by the diameter of CNTs, as demonstrated by the significantly higher antibacterial activity of SWNTs compared to MWNTs (Kang et al., 2008). Yang et al. examined the antibacterial efficacy of three distinct SWCNT lengths (less than 1 μm, between 1 and 5 μm, and approximately 5 μm) against bacterial cells in suspension. From these experiments, the result clearly showed the higher antibacterial action of longer SWCNTs at the same weight concentration and treatment time dependence (Yang et al., 2010). It was shown by a thorough investigation into the mechanisms of Pb^{2+} adsorption on modified MWNT that Pb^{2+} interacts with chemical groups on the surfaces of the carbon nanotubes that carry oxygen (Smith et al., 2016).

9.5 FUTURE PROSPECTS

In the recent past, there has been a significant increase in the requirement for clean water sources, which has presented major obstacles to the measures implemented to provide access to clean water.

CNTs, GO, fullerenes, and AC are examples of CNMs with large surface areas that are capable of efficiently adsorbing heavy metals, dyes, organic molecules, and even microplastics from water. Future studies must address regulatory concerns about the negative environmental effects and potential health dangers of carbon NMs, as they are being employed more and more in water purification. It will be crucial for a secure and environmentally friendly deployment to comprehend the long-term actions and fate of these elements in aquatic ecosystems. Introducing nano-absorbents in water and wastewater treatment has the advantage of increasing system efficiency, using less area for purification, and providing stable operations under varying conditions (Hsu et al., 2023). The extensive use of carbon NPs in water filtration technologies will depend on future developments in their fabrication and economical manufacture. To address the requirement for massive applications, efforts might concentrate on affordable production methods. Thus, continued study in the fields of material science, nanotechnologies, and sustainability engineering points to a bright future for carbon NMs in the prevention of water pollution. Regulatory agencies, businesses, and researchers working together will be crucial to properly and successfully realizing this promise.

9.6 CONCLUSION

In conclusion, the removal of water contaminants from wastewater has great promise for the fabrication of carbon NMs employing sustainable and green technologies. These creative solutions improve the efficacy and efficiency of water filtration technology, while also addressing environmental issues related to conventional synthesis processes. Carbon NMs that are fabricated sustainably, including those made from waste materials or biomass, have several benefits. Reduction of the environmental impact and dependence on conventional raw materials are possible because they can be made from natural sources. Removal of heavy metals, dyes, organic contaminants, and microbes is among the uses of green-produced carbon NMs in water remediation. Addressing these issues and realizing the full potential of naturally synthesized carbon NMs for environmentally friendly water management will require ongoing research and cooperation between researchers. Fundamentally, green synthesized carbon NPs offer novel and sustainable approaches to protect water resources and enhance public health globally, marking a potential prospect in the area of remediation of water contamination.

REFERENCES

Abbasi, M., 2017. Synthesis and characterization of magnetic nanocomposite of chitosan/SiO_2/carbon nanotubes and its application for dye removal. *J. Clean. Prod.* 145, 105–113. https://doi.org/10.1016/j.jclepro.2017.01.046

Abbo, H. S., Gupta, K. C., Khaligh, N. G., Titinchi, S. J., 2021. Carbon NMs for wastewater treatment. *Chem. Bio. Eng. Reviews*, 8(5), 463–489. https://doi.org/10.1002/cben.202100003

Agrahari, S., Singh, A. K., Gupta, M. K., Afroz, M., Pandey, V., Singh, H., Tandon, P. K., 2023. Valorization of the waste pulp of Citrus limetta for decontamination of chromium from polluted water. *Mater. Today Proc.* https://doi.org/10.1016/j.matpr.2023.10.121

Ahlawat, W., Kataria, N., Dilbaghi, N., Hassan, A. A., Kumar, S., Kim, K. H. 2020. Carbonaceous nanomaterials as effective and efficient platforms for removal of dyes from aqueous systems. *Environ. Res. 181*, 108904. https://doi.org/10.1016/j.envres.2019.108904

Ahmed, M., Mashkoor, F., Nasar, A., 2020. Development, characterization, and utilization of magnetized orange peel waste as a novel adsorbent for the confiscation of crystal violet dye from aqueous solution. *Groundw. Sustain. Dev.* 10, 100322. https://doi.org/10.1016/j.gsd.2019.100322

Ahmed, M. B., Zhou, J. L., Ngo, H. H., Guo, W., Thomaidis, N. S., Xu, J., 2017. Progress in the biological and chemical treatment technologies for emerging contaminant removal from wastewater: A critical review. *J. Hazard. Mater.* 323, 274–298. https://doi.org/10.1016/j.jhazmat.2016.04.045

Alekseeva, O. V., Bagrovskaya, N. A., Noskov, A. V., 2016. Sorption of heavy metal ions by fullerene and polystyrene/fullerene film compositions. *Prot. Met. Phys. Chem. Surf.* 52, 443–447. https://doi.org/10.1134/S2070205116030035

Anitha, K., Namsani, S., Singh, J. K., 2015. Removal of heavy metal ions using a functionalized single-walled carbon nanotube: A molecular dynamics study. *J. Phys. Chem. A*. 119(30), 8349–8358. https://doi.org/10.1021/acs.jpca.5b03352

Anzar, N., Hasan, R., Tyagi, M., Yadav, N., Narang, J., 2020. Carbon nanotube-A review on Synthesis, Properties and plethora of applications in the field of biomedical science. *Sens. Int.* 1, 100003. https://doi.org/10.1016/j.sintl.2020.100003

Baby, R., Saifullah, B., Hussein, M. Z., 2019. Carbon NMs for the treatment of heavy metal-contaminated water and environmental remediation. *Nanoscale res. Lett.* 14(1), 341. https://doi.org/10.1186/s11671-019-3167-8

Bhushan, B., Nayak, A., Kotnala, S., 2021. Green synthesis of highly porous activated carbon from jackfruit peel: Effect of operating factors on its physico-chemical characteristics. *Mater. Today Proc.* 44, 187–191. https://doi.org/10.1016/j.matpr.2020.08.554

Brunet, L., Lyon, D. Y., Hotze, E. M., Alvarez, P. J., Wiesner, M. R., 2009. Comparative photoactivity and antibacterial properties of C60 fullerenes and titanium dioxide NPs. *Environ. Sci. Technol.* 43(12), 4355–4360. https://doi.org/10.1021/es803093t

Burakov, A. E., Galunin, E. V., Burakova, I. V., Kucherova, A. E., Agarwal, S., Tkachev, A. G., Gupta, V. K., 2018. Adsorption of heavy metals on conventional and nanostructured materials for wastewater treatment purposes: A review. *Ecotox. Environ. Safe.* 148, 702–712. https://doi.org/10.1016/j.ecoenv.2017.11.034

Byon, H. R., Lim, H., Song, H. J., Choi, H. C., 2007. A synthesis of high-purity single-walled carbon nanotubes from small diameters of cobalt NPs by using an oxygen-assisted chemical vapor deposition process. *Bull. Korean Chem. Soc.* 28(11), 2056.

Chadha, U., Selvaraj, S. K., Thanu, S. V., Cholapadath, V., Abraham, A. M., Manoharan, M., Paramsivam, V., 2022. A review of the function of using carbon NMs in membrane filtration for contaminant removal from wastewater. *Mater. Res. Exp.* 9(1), 012003. https://doi.org/10.1088/2053-1591/ac48b8

Chufa, B. M., Murthy, H. A., Gonfa, B. A., Anshebo, T. Y., 2021. Carbon nanotubes: A review on green synthesis, growth mechanism and application as a membrane filter for fluoride remediation. *Green Chem. Lett. Rev.* 14(4), 647–664. https://doi.org/10.1080/17518253.2021.1991484

Diel, J. C., Franco, D. S., Igansi, A. V., Cadaval Jr, T. R., Pereira, H. A., Nunes, I. D. S., Basso, C. W., Alves M. D. C. M., Morais, J., Pinto, D., Dotto, G. L., 2021. Green synthesis of carbon nanotubes impregnated with metallic NPs: Characterization and application in glyphosate adsorption. *Chemosphere*, 283, 131193. https://doi.org/10.1016/j.chemosphere.2021.131193

Duan, Q., Lee, J., Liu, Y., Qi, H., 2016. Preparation and photocatalytic performance of MWCNTs/TiO2 nanocomposites for degradation of aqueous substrate. *J. Chem.* 2016(1), 1262017. https://doi.org/10.1155/2016/1262017

Elessawy, N. A., El-Sayed, E. M., Ali, S., Elkady, M. F., Elnouby, M., Hamad, H. A., 2020. One-pot green synthesis of magnetic fullerene nanocomposite for adsorption characteristics. *J. Water Proc. Eng.* 34, 101047. https://doi.org/10.1016/j.jwpe.2019.101047

Engel, M., Hadar, Y., Belkin, S., Lu, X., Elimelech, M., Chefetz, B., 2018. Bacterial inactivation by a carbon nanotube–iron oxide nanocomposite: A mechanistic study using E. coli mutants. *Environ. Sci. Nano.* 5(2), 372–380. https://doi.org/10.1039/C7EN00865A

Fatima, T., Mushtaq, A., 2023. Efficacy and challenges of carbon-based NMs in water treatment: A review. *Int. J. Chem. Biochem. Sci*, 23, 232–248.

Fouda, A. N., El Shazly, M. D., Almaqwashi, A. A., 2021. Facile and scalable green synthesis of N-doped graphene/CNTs nanocomposites via ball milling. *Ain Shams Eng. J.* 12(1), 1017–1024. https://doi.org/10.1016/j.asej.2020.04.011

Gacem, A., Modi, S., Yadav, V. K., Islam, S., Patel, A., Dawane, V., Jameel, M., Inwati, G. K., Piplode, S., Solanki, V. S., Basnet, A., 2022. Recent advances in methods for synthesis of carbon nanotubes and carbon nanocomposite and their emerging applications: A descriptive review. *J. Nanomater.* 2022(1), 7238602. https://doi.org/10.1155/2022/7238602

Georgin, J., Franco, D., Drumm, F. C., Grassi, P., Netto, M. S., Allasia, D., Dotto, G. L., 2020. Powdered biosorbent from the mandacaru cactus (Cereus jamacaru) for discontinuous and continuous removal of Basic Fuchsin from aqueous solutions. *Powder Technol.* 364, 584–592. https://doi.org/10.1016/j.powtec.2020.01.064

Ghosh, P., Afre, R. A., Soga, T., Jimbo, T., 2007. A simple method of producing single-walled carbon nanotubes from a natural precursor: Eucalyptus oil. *Mater. Lett.* 61(17), 3768–3770. https://doi.org/10.1016/j.matlet.2006.12.030

Hirschfeld, J., Akinoglu, E. M., Wirtz, D. C., Hoerauf, A., Bekeredjian-Ding, I., Jepsen, S., Haddouti, EL., Limmer, A., Giersig, M., 2017. Long-term release of antibiotics by carbon nanotube-coated titanium alloy surfaces diminishes biofilm formation by Staphylococcus epidermidis. Nanomed: Nanotechnol. *Biol. Med.* 13(4), 1587–1593. https://doi.org/10.1016/j.nano.2017.01.002

Hsu, C. Y., Rheima, A. M., Mohammed, M. S., Kadhim, M. M., Mohammed, S. H., Abbas, F. H., Zainab, T. A., Mahdi, Z. M., Abbas, Z. S., Hachim, S. K., Ali, F. K., Mahmoud, Z. H., Kianfar, E., 2023. Application of carbon nanotubes and graphene-based nano adsorbents in water treatment. *BioNanoScience*. 13(4), 1418–1436. https://doi.org/10.1007/s12668-023-01175-1

Huang, Y. M., Lee, X., Macazo, F. C., Grattieri, M., Cai, R., Minteer, S. D., 2018. Fast and efficient removal of chromium (VI) anionic species by a reusable chitosan-modified multi-walled carbon nanotube composite Chem. *Eng. J.* 345, 708–708. https://doi.org/10.1016/j.cej.2018.01.133

Hussain, N., Bilal, M., Iqbal, H. M., 2022. Carbon-based NMs with multipurpose attributes for water treatment: Greening the 21st-century nanostructure materials deployment. *Biomater. Polym. Horiz.* 1(1), 48–58. https://doi.org/10.37819/bph.001.01.0131

Janas, D., 2020. From bio to nano: A review of sustainable methods of synthesis of carbon nanotubes. *Sustainability*. 12(10), 4115. https://doi.org/10.3390/su12104115

Jin, Q. Q., Zhu, X. H., Xing, X. Y., Ren, T. Z., 2012. Adsorptive removal of cationic dyes from aqueous solutions using graphite oxide. *Adsorpt. Sci. Technol.* 30(5), 437–447. https://doi.org/10.1260/0263-6174.30.5.437

Jjagwe, J., Olupot, P. W., Menya, E., Kalibbala, H. M., 2021. Synthesis and application of granular activated carbon from biomass waste materials for water treatment: A review. *J. Biores. Bioprod.* 6(4), 292–322. https://doi.org/10.1016/j.jobab.2021.03.003

Jun, L. Y., Mubarak, N. M., Yee, M. J., Yon, L. S., Bing, C. H., Khalid, M., Abdullah, E. C., 2018. An overview of functionalized carbon NM for organic pollutant removal. *J. Ind. Eng. Chem.* 67, 175–186. https://doi.org/10.1016/j.jiec.2018.06.028

Kadirvelu, K., Thamaraiselvi, K., Namasivayam, C., 2001. Removal of heavy metals from industrial wastewater by adsorption onto activated carbon prepared from agricultural solid waste. *Bioresour. Technol.* 76(1), 63–65. https://doi.org/10.1016/S0960-8524(00)00072-9

Kamboj, V., Tiwari, D. P., 2024. Removal of heavy metal (Cu, Cr, and Ni) ions from aqueous solution using derived activated carbon from water hyacinth. *Biomass Conver Bioref.* 14(4), 5075–5083. https://doi.org/10.1007/s13399-022-02702-y

Kang, S., Herzberg, M., Rodrigues, D. F., Elimelech, M., 2008. Antibacterial effects of carbon nanotubes: Size does matter! *Langmuir*. 24(13), 6409–6413. https://doi.org/10.1021/la800951v

Kotia, A., Yadav, A., Rohit Raj, T., Gertrud Keischgens, M., Rathore, H., Sarris, I. E., 2020. Carbon NPs as sources for a cost-effective water purification method: A comprehensive review. *Fluids*, 5(4), 230. https://doi.org/10.3390/fluids5040230

Kumar, R., Singh, R. K., Singh, D. P., 2016. Natural and waste hydrocarbon precursors for the synthesis of carbon-based NMs: Graphene and CNTs. *Renew. Sustain. Energy Rev.* 58, 976–1006. https://doi.org/10.1016/j.rser.2015.12.120

Lee, K. M., Wong, C. P. P., Tan, T. L., Lai, C. W., 2018. Functionalized carbon nanotubes for adsorptive removal of water pollutants. *Mater. Sci. Eng. B.* 236, 61–69. https://doi.org/10.1016/j.mseb.2018.12.004

Lehman, J. H., Deshpande, R., Rice, P., To, B., Dillon, A. C., 2006. Carbon multi-walled nanotubes grown by HWCVD on a pyroelectric detector. *Infrared Phys. Techn.* 47(3), 246250. https://doi.org/10.1016/j.infrared.2004.12.004

Li, W. S., Shih, Y. C., Cheng, H. C., 2020. Green synthesis of CNTs/Ni $(OH)_2$ nanostructures for electrochemical supercapacitors. *Chem. Phys. Lett.* 750, 137499. https://doi.org/10.1016/j.cplett.2020.137499

Liu, J., Li, R., & Yang, B., 2020. Carbon dots: A new type of carbon-based NM with wide applications. *ACS Cent. Sci.* 6: 2179–2195. https://doi.org/10.1021/acscentsci.0c01306

Lu, F., Astruc, D., 2020., Nanocatalysts and other NMs for water remediation from organic pollutants. *Coord. Chem. Rev.* 408, 213180. https://doi.org/10.1016/j.ccr.2020.213180

Ma, J., Yu, F., Zhou, L., Jin, L., Yang, M., Luan, J., Tang, Y., Fan, H., Yuan, Z., Chen, J., 2012. Enhanced adsorptive removal of methyl orange and methylene blue from aqueous solution by alkali-activated multi-walled carbon nanotubes. *ACS appl. Mater. Interfaces*. 4(11), 5749–5760. https://doi.org/10.1021/am301053m

Maitlo, H. A., Kim, K. H., Kumar, V., Kim, S., Park, J. W., 2019. NMs-based treatment options for chromium in aqueous environments. *Environ. Int.* 130, 104748. https://doi.org/10.1016/j.envint.2019.04.020

Makris, T. D., Giorgi, R., Lisi, N., Pilloni, L., Salernitano, E., Sarto, F., Alvisi, M., 2004. Carbon nanotube growth by HFCVD: Effect of the process parameters and catalyst preparation. *Diam. Relat. Mater.* 13(2), 305–310. https://doi.org/10.1016/j.diamond.2003.10.013

Manawi, Y. M., Ihsanullah, Samara A., Al-Ansari, T., Atieh, M. A., 2018. A review of carbon NMs' synthesis via the chemical vapor deposition (CVD) method. *Materials*, 11(5), 822. https://doi.org/10.3390/ma11050822

Meysami, S. S., Koós, A. A., Dillon, F., Dutta, M., Grobert, N., 2015. Aerosol-assisted chemical vapor deposition synthesis of multi-wall carbon nanotubes: III. Towards upscaling. *Carbon*. 88, 148–156. https://doi.org/10.1016/j.carbon.2015.02.045

Moreno-Castilla, C., Alvarez-Merino, M. A., López-Ramón, M. V., Rivera-Utrilla, J., 2004. Cadmium ion adsorption on different carbon adsorbents from aqueous solutions: Effect of surface chemistry, pore texture, ionic strength, and dissolved natural organic matter. *Langmuir*, 20(19), 8142–8148. https://doi.org/10.1021/la049253m

Nguyen, T. N., Le, P. A., Phung, V. B. T., 2022. Facile green synthesis of carbon quantum dots and biomass-derived activated carbon from banana peels: Synthesis and investigation. *Biomass Convers. Biorefin*. 12(7), 2407–2416. https://doi.org/10.1007/s13399-020-00839-2

Nguyen, T. T. H., Nguyen, X. C., Nguyen, D. L. T., Nguyen, D. D., Vo, T. Y. B., Vo, Q. N., Nguyen, T. D., Ly, Q. V., Ngo, H. H., Vo, D. N., Nguyen, T. P.., Kim, T., Van Le, Q., 2021. Converting biomass of agro wastes and invasive plants into alternative materials for water remediation. *Biomass Convers. Biorefin*. 1–16. https://doi.org/10.1007/s13399-021-01526-6

Park, H. A., Liu, S., Salvador, P. A., Rohrer, G. S., Islam, M. F., 2016. High visible-light photochemical activity of titania decorated on single-wall carbon nanotube aerogels. *RSC adv*. 6(27), 22285–22294. https://doi.org/10.1039/C6RA03801H

Peng, Z., Liu, X., Zhang, W., Zeng, Z., Liu, Z., Zhang, C., Liu, Y., Shao, B., Liang, Q., Tang, W., Yuan, X., 2020. Advances in the application, toxicity, and degradation of carbon NMs in the environment: A review. Environ Int, 134, 105298. https://doi.org/10.1016/j.envint.2019.105298

Rashid, M. H. O., Ralph, S. F., 2017. Carbon nanotube membranes: Synthesis, properties, and future filtration applications. *Nanomater*. 7(5), 99. https://doi.org/10.3390/nano7050099

Ravi, S., Vadukumpully, S., 2016. Sustainable carbon NMs: Recent advances and their applications in energy and environmental remediation. *J. Environ Chem Eng*, 4(1), 835–856. https://doi.org/10.1016/j.jece.2015.11.026

Reynel-Ávila, H. E., Camacho-Aguilar, K. I., Bonilla-Petriciolet, A., Mendoza-Castillo, D. I., González-Ponce, H. A., Trejo-Valencia, R., 2021. Engineered magnetic carbon-based adsorbents for the removal of water priority pollutants: An overview. *Adsorpt. Sci. Technol*. 2021, 9917444. https://doi.org/10.1155/2021/9917444

Sari, E. K., Tumbelaka, R. M., Ardiyanti, H., Istiqomah, N. I., Suharyadi, E., 2023. Green synthesis of magnetically separable and reusable Fe_3O_4/Cdots nanocomposites photocatalyst utilizing Moringa oleifera extract and watermelon peel for rapid dye degradation. *Carbon Resour. Convers*. 6(4), 274–286. https://doi.org/10.1016/j.crcon.2023.04.003

Shahrashoub, M., Bakhtiari, S., 2021. The efficiency of activated carbon/magnetite NPs composites in copper removal: Industrial waste recovery, green synthesis, characterization, and adsorption-desorption studies. *Micropor. Mesopor. Mater*. 311, 110692. https://doi.org/10.1016/j.micromeso.2020.110692

Sheoran, K., Kaur, H., Siwal, S. S., Saini, A. K., Vo, D. V. N., Thakur, V. K., 2022. Recent advances of carbon-based NMs (CBNMs) for wastewater treatment: Synthesis and application. *Chemosphere*. 299, 134364. https://doi.org/10.1016/j.chemosphere.2022.134364

Smith, M. K., Singh, V., Kalaitzidou, K., & Cola, B. A. 2016. High thermal and electrical conductivity of template fabricated P3HT/MWCNT composite nanofibers. *ACS Appl. Mater. Interfaces*, *8*(23), 14788–14794. https://doi.org/10.1021/acsami.6b01845

Spaolonzi, M. P., Duarte, E. D., Oliveira, M. G., Costa, H. P., Ribeiro, M. C., Silva, T. L., Silva, M. G. C., Vieira, M. G., 2022. Green-functionalized carbon nanotubes as adsorbents for the removal of emerging contaminants from aqueous media. *J. Clean. Prod*. 373, 133961. https://doi.org/10.1016/j.jclepro.2022.133961

Srivastava, A., Gupta, B., Majumder, A., Gupta, A. K., Nimbhorkar, S. K., 2021. A comprehensive review of the synthesis, performance, modifications, and regeneration of activated carbon for the adsorptive removal of various water pollutants. *J. Environ. Chem. Eng*., 9(5), 106177. https://doi.org/10.1016/j.jece.2021.106177

Szabó, A., Perri, C., Csató, A., Giordano, G., Vuono, D., Nagy, J. B., 2010. Synthesis methods of carbon nanotubes and related materials. *Materials*. 3(5), 3092–3140. https://doi.org/10.3390/ma3053092

Tang, S. H., Zaini, M. A. A., 2020. Development of activated carbon pellets using a facile low-cost binder for effective malachite green dye removal. *J. Clean. Prod*. 253, 119970. https://doi.org/10.1016/j.jclepro.2020.119970

Thines, R. K., Mubarak, N. M., Nizamuddin, S., Sahu, J. N., Abdullah, E. C., Ganesan, P., 2017. Application potential of carbon NMs in water and wastewater treatment: A review. *J. Taiwan Inst. Chem. Eng*. 72, 116–133. https://doi.org/10.1016/j.jtice.2017.01.018

Tian, Z., Yang, B., Cui, G., Zhang, L., Guo, Y., Yan, S., 2015. Synthesis of poly (m-phenylenediamine)/iron oxide/acid oxidized multi-wall carbon nanotubes for removal of hexavalent chromium. *RSC Adv*., 5(3), 2266–2275. https://doi.org/10.1039/C4RA10282G

Wang, J., Zhang, Q., Shao, X., Ma, J., Tian, G., 2018. Properties of magnetic carbon NMs and application in removal organic dyes. *Chemosphere*. 207, 377–384. https://doi.org/10.1016/j.chemosphere.2018.05.109

Wang, X., Guo, Y., Yang, L., Han, M., Zhao, J., Cheng, X., 2012. NMs as sorbents to remove heavy metal ions in wastewater treatment. *J. Environ. Anal. Toxicol*, 2(7), 154–158. 10.4172/2161-0525.1000154

Watanabe, T., Tsuda, S., Yamaguchi, T., Takano, Y., 2010. Microwave plasma chemical vapor deposition synthesis of boron-doped carbon nanotube. *Phys. C: Supercond. Appl*. 470, S608–S609. https://doi.org/10.1016/j.physc.2010.01.037

Xiong, W., Zeng, Z., Li, X., Zeng, G., Xiao, R., Yang, Z., Zhou, Y., Zhang, C., Cheng, M., Hu, L., Zhou, C., Qin, L., Xu, R., Zhang, Y., 2018. Multi-walled carbon nanotube/amino-functionalized MIL-53 (Fe) composites: remarkable adsorptive removal of antibiotics from aqueous solutions. *Chemosphere*. 210, 1061–1069. https://doi.org/10.1016/j.chemosphere.2018.07.084

Yadav, S. K., Dhakate, S. R., Singh, B. P., 2022. Carbon nanotube incorporated eucalyptus-derived activated carbon-based novel adsorbent for efficient removal of methylene blue and eosin yellow dyes. Biores. Technol. 344, 126231. https://doi.org/10.1016/j.biortech.2021.126231

Yang, C., Mamouni, J., Tang, Y., Yang, L., 2010. Antimicrobial activity of single-walled carbon nanotubes: length effect. Langmuir. 26(20), 16013–16019. 2010. https://doi.org/10.1021/la103110g

Yu, G., Lu, Y., Guo, J., Patel, M., Bafana, A., Wang, X., Qiu, B., Jeffryes, C., Wei Suying., Guo, Zhanhu., Wujcik, E. K., 2018. Carbon nanotubes, graphene, and their derivatives for heavy metal removal. *Adv. Compos. Hybrid Mater*., 1, 56–78. https://doi.org/10.1007/s42114-017-0004-3

Zaini, M. A. A., Zhi, L. L., Hui, T. S., Amano, Y., Machida, M., 2021. Effects of physical activation on pore textures and heavy metals removal of fiber-based activated carbons. *Mater. Today Proc*. 39, 917–921. https://doi.org/10.1016/j.matpr.2020.03.815

Zaytseva, O., Neumann, G., 2016. Carbon NMs: Production, impact on plant development, agricultural and environmental applications. *Chem. Biol. Technol*. 3, 1–26. https://doi.org/10.1186/s40538-016-0070-8

Zhan, W., Gao, L., Fu, X., Siyal, S. H., Sui, G., Yang, X., 2019. Green synthesis of amino-functionalized carbon nanotube-graphene hybrid aerogels for high-performance heavy metal ions removal. *Appl. Surf. Sci*. 467, 1122–1133. https://doi.org/10.1016/j.apsusc.2018.10.248

Zondo, B. Z., Sadare, O. O., Simate, G. S., Moothi, K., 2022. Removal of Pb^{2+} ions from synthetic wastewater using functionalized multi-walled carbon nanotubes decorated with green synthesized iron oxide-gold nanocomposite. *Water SA*. 48(3), 304–316. https://doi.org/10.17159/wsa/2022.v48.i3.3959

10 Nanosheet Composites for Wastewater Treatment
A Natural Biopolymer Approach

Orkun Pinar and Susana Rodríguez-Couto

10.1 INTRODUCTION

Water influences ecosystems, sustainable growth, the production of food and energy, and all aspects of life (Sahu et al., 2023). Most of the water is present as salty seawater, freshwater making up only about 2.5%–3.0% of the total water sources and being mostly found in the form of groundwater, ice caps, and glaciers. In addition, only ~1.0%, of this freshwater is available for human consumption. Water pollution, which can be caused by both naturally occurring contaminants and human activity, has resulted in a more severe freshwater shortage (Zeng et al., 2021). Water shortage is a major global challenge since it was estimated that six billion people could be suffering from water scarcity by 2050, because the global demand for water is predicted to rise by 55.0% (Meena et al., 2019; Al Harby et al., 2022). Global freshwater demand and consumption have increased in recent decades as a result of climate change, population growth, urbanization, and increased irrigated agriculture. Freshwater scarcity is a major issue that has a significant impact on human activity and economic growth (Hube et al., 2020). Additionally, according to the World Health Organization (WHO), more than 2 billion individuals live in water-stressed regions of the world and most of them have to drink water contaminated with feces. Moreover, the biggest threat to the safety of drinking water is the microbial pollution sourced from this kind of contamination.

Approximately 380 trillion liters of wastewater are generated annually worldwide. Asian countries lead the generation of wastewater with an approximate annual generation of 159 trillion liters, followed by European countries and North American countries with 68 and 67 trillion liters, respectively. In addition, in terms of urban wastewater, Asian counties are also in the first place of wastewater generation, with approximately 42.0% of the total global production (Qadir et al., 2020).

Industrial and agricultural wastes, which could primarily contain inorganic and organic pollutants such as heavy metals, phenolic compounds, dyes, soaps, oils, drugs, fertilizers, and pesticides, are significant causes of aquatic pollution (Salama et al., 2021). Dyes (Naim et al., 2022), heavy metal ions (Choi and Lee, 2022), chemicals dissolved in irrigation water (Runkle et al., 2018), and pharmaceuticals, such as antimicrobials, antidepressants, and active drugs ingredients (Angeles et al., 2020), which can cause toxic, lethal, and carcinogenic effects on humans and animals, even at low concentrations.

According to the 2020 and 2021 World Water Development Reports from the United Nations, the treatment of the organic matter in wastewater can lower GHG (greenhouse gas) emissions because, on a global scale, more than 80.0% of all wastewaters is released into the environment without being treated. Reusing wastewater after complete or partial treatment can also lower the energy demand for having access to the water. Nowadays, around 10.0%–20.0% of the total wastewater generated by human activity undergoes treatment and this treated water can be reused (Salama et al., 2021).

To overcome water pollution and ensure the quality of available water supplies, researchers have used different approaches for wastewater treatment, including adsorption, membrane filtration, ion exchange, electrostatic interactions, coagulation-flocculation, chemical oxidation, and biodegradation. Due to its low operation cost, flexible and simple design, and high efficiency, adsorption has

DOI: 10.1201/9781003484189-10

been generally regarded as the most effective method for treating wastewater (Wang W. et al., 2020c; Stephen and Palanisamy, 2021).

In general, there are four main stages of wastewater treatment: preliminary, primary, secondary, and tertiary (advanced) treatments. The main aim of the first step of wastewater treatment, during preliminary treatment (i.e., pre-treatment), is to remove large materials and coarse solids. As a result of this step, large, suspended, or floating solid particles are removed or reduced in size. These pieces could be plastics, paper, cloth, wood, sand, glass, or other kinds of large solids. In the primary treatment step, which has been identified as a critical stage to improve the sustainability of wastewater treatment processes, organic and inorganic suspended solids and phosphorous are removed via physical processes. Secondary wastewater treatment is the biological process in which the biodegradable compounds are removed. In this stage, microorganisms use dissolved organic compounds as nutrients for growth producing CO_2 and different by-products in a controlled environment. After the secondary wastewater treatment, residual carbon, nitrogen, and phosphorous types and concentrations could directly affect the next step of the treatment, which is called tertiary treatment. The secondary effluent characteristics are improved during the tertiary treatment to enable reuse. Chlorination, membrane filtration, ozonation, ultraviolet irradiation, microalgal cultivation, and photo-Fenton processes are among the tertiary treatment techniques considered on most research studies because of their high efficiency and reduced operational costs. The tertiary wastewater treatment removes inorganic substances, bacteria, viruses, and parasites, rendering the treated water secure for reusing, recycling, or environmental release (Zagklis and Bampos, 2022; Zhou et al., 2022).

Selecting the appropriate wastewater treatment method is complicated since several factors, including technical, social, economic, regulatory, governmental, and environmental considerations, must be taken into account. Traditionally, technological, and financial aspects were preponderant, but the other aforementioned factors need also to be considered for a more efficient wastewater treatment. On the other hand, this selection is an important task that needs extensive knowledge, expertise, and feasibility studies. The conventional techniques for treating wastewater primarily focus on enhancing biochemical, physical, and natural procedures. These methods aim to improve the effluent characteristics, such as chemical oxygen demand (COD), biological oxygen demand (BOD), total suspended solids (TSS), and turbidity. Wastewater can be treated using a variety of unit operations and unit processes for each treatment stage. Physical forces are used in unit operations, whereas chemical and/or physical forces are used to increase the effluent quality in unit processes (Wang et al., 2018; Ullah et al., 2020). However, traditional techniques could be challenging, either becoming more expensive or less effective to meet the current strict regulatory effluent limits, for instance removing heavy metals (Rajasulochana and Preethy, 2016) or complex organic dyes (Peramune et al., 2022). Due to the limitations of the traditional technologies, alternative, effective, and less expensive advanced technologies are in demand, considering the type and concentration of pollutant, heterogeneity of effluent, and required degree of treatment (Rajasulochana and Preethy, 2016).

While proposing new technologies, not only macro pollutants but also emerging contaminants (micropollutants), which are generally organic in nature and typically occur at trace concentrations and are not removed via conventional wastewater treatment approaches, should be considered (Rout et al., 2021). Overall, since water resources are essential for both human and animal survival, providing new methods for increasing wastewater treatment quality has become an urgent issue. In this sense, the implementation of new technologies is required to save energy and increase wastewater recovery yields.

10.2 APPLICATIONS OF BIOPOLYMER-BASED NANOTECHNOLOGIES FOR WASTEWATER TREATMENT

Polymers are high molecular weight molecules, called macromolecules, which are built up by connecting numerous small molecules, called monomers. The formation of polymers is known as the polymerization process. There are three classes of polymers, according to their origin: semi-synthetic polymers, synthetic polymers, and biopolymers. Semi-synthetic polymers are created through the modification of natural polymers, using chemical processes that modify their physical characteristics. Carboxymethyl starch, carboxymethyl cellulose, hydroxypropyl guar gum, and silicones are known examples of semi-synthetic polymers. As for synthetic polymers, they are obtained by the polymerization of chemical molecules in laboratories. Nylon®, Teflon™ (polytetrafluoroethylene), polyvinyl chloride, polyethylene, polystyrene, and synthetic rubber are important examples of these synthetic materials frequently used in different products (Rajeswari et al., 2021). Biopolymers are naturally occurring polymers composed of covalently attached monomeric units, which can be efficiently obtained from natural sources, and hold great potential for wastewater treatment as a promising replacement for petroleum-based materials. These eco-friendly biopolymers and their composites can be divided into different groups based on the sources, preparation techniques, and possible applications. They are advantageous in many aspects over their synthetic counterparts since such materials are naturally abundant, degraded by living organisms, obtained from renewable sources, and have a non-toxic character. Moreover, multiple functional groups present within these biomaterials make them ideal candidates for modification into various shapes and forms suitable for water treatment purposes such as hydrogels, aerogels, beads, and adsorbents. Among these biopolymers, the polysaccharide-based cellulose and chitosan and the protein-based keratin and soy are the most frequently used for wastewater treatment (Udayakumar et al., 2021; Zubair and Ullah, 2021).

10.2.1 Polysaccharide-Based Biomaterials for Wastewater Treatment

High molecular weight polymeric carbohydrates known as polysaccharides are made up of numerous monosaccharide units that are joined by glycosidic links. Additionally, the degree of polymerization (DP) refers to the number of monosaccharide units present in a polysaccharide that gives biopolymers different DPs and molecular weights (Albuquerque et al., 2020). Solubility and rheological characteristics of polysaccharides are regulated by different properties, such as the branching degree, polydispersity, the composition of monosaccharides, and the pattern of the glycosidic bond. The biodegradability and mechanical properties of polysaccharides can vary according to their formation as linear, which is made up of a straight chain of monosaccharides joined by a straight chain, or branched, which has a chain with arms and turns (Kabir et al., 2022; Seidi et al., 2022). Since they are easily available, sustainable, non-toxic, biocompatible, biodegradable, non-antigenic, non-carcinogenic, chiral, multifunctional, and highly reactive together with their strong chelating and adsorbing properties, they have become excellent candidates for wastewater treatment (Zubair and Ullah, 2021). Cellulose, starch, chitin, dextran, chitosan, alginate, and carrageenan are examples of polysaccharide-based biopolymers (Russo et al., 2021). Moreover, there are many kinds of carbon-based materials, some of which have unique structures or special physical, chemical, adsorption, and catalytic properties (e.g., graphene, carbon nanotubes, nanodiamonds, and activated carbon) and are of potential interest for use in different water treatment processes, including adsorption, catalysis, membrane separation, and microbial fuel cells (Dong et al., 2021).

Cellulose is a very adaptable biopolymer for composite applications due to its biocompatibility, high thermal stability, high tensile strength, high Young's modulus, and eco-friendliness. As a linear-chain polymer, cellulose is made up of β-D-glucopyranose ring units that are chemically connected by β-(1,4) glycosidic bonds. Each anhydroglucose unit contains three hydroxyl groups, allowing a hydrogen-bonded network to form, giving the polymer superior mechanical properties and high tensile strength (Krivdin, 2021). To increase the potential applications of cellulose and its composites, researchers have chemically modified it to enhance its properties, such as its adsorption capacity. The functional characteristics of cellulose have been improved by increasing its porosity, altering its surface with various functional groups, and using other forms of cellulose, including nanofibers (Sohouli et al., 2022). Moreover, it is crucial to build sustainable composites by using cellulose as the matrix and combining it with carbon materials for enhanced performance properties to overcome complications based on the separation of these materials from the water, surface inertness, agglomeration, inactivation, recovery, and the generation of secondary pollution (Dong et al., 2021). Despite cellulose and its ester derivatives, including cellulose acetate (CA), cellulose acetate phthalate (CAP), cellulose acetate butyrate (CAB), cellulose acetate trimellitate (CAT) and hydroxypropyl methylcellulose phthalate (HPMCP) were identified as hard to process, they have been commonly used in commercial products. Among them, CA membranes are frequently chosen because of their excellent water affinity, superior transport characteristics, low protein adsorption, suitable mechanical strength, high hydrophilicity with desalting nature, and excellent film-forming properties (Khalil et al., 2017; Rajeswari et al., 2021). Cellulose-based polymers can be used as adsorbents, flocculants, and oil/water separation membranes for water treatment in the oil and gas sectors (Peng et al., 2020). Among them, nanocelluloses, which have nanocrystal, nanofibril, and bacterial-based forms according to preparation techniques, processing conditions, size, and morphology, present a high surface area that can be functionalized, making them promising biomaterials as adsorbents (Hamidon et al., 2022).

Based on the applied form of cellulose in water treatment, cellulose-based materials can be approximately divided into three major groups: one-dimensional fibers, two-dimensional membranes, and three-dimensional gels in terms of macroscopic dimension. Among them, the one-dimensional fibers can be prepared by electrospinning, wet spinning, and dry spray-wet spinning (Dong et al., 2021). Tamaddon et al. (2020) proposed $ZnFe_2O_4$@methyl cellulose as a magnetic photocatalyst to degrade metronidazole from wastewater. They found degradation percentages of 92.65% and 71.12% for synthetic and real samples, respectively, operating under optimal conditions (i.e., initial [MNZ]: 5 mg/L, photocatalyst loading: 0.3 g, pH: 7.0). Additionally, the magnetic photocatalyst was reused 6 times with no significant loss of its catalytic efficiency. This makes the developed catalyst very interesting for its potential use on a large scale.

Two-dimensional (2D) materials offer new opportunities for the next generation of high-performance membrane technology applications in industrial separation processes, where polymeric and ceramic membranes characterized by enhanced permeability, selectivity, and stability have been used the most. Several assembly techniques, including vacuum filtration, spin coating, and hot dropping, have been utilized to manufacture 2D membranes (Cheng et al., 2021). Thus, Chen Y. et al. (2022a) fabricated nanofibrous membranes by the vacuum filtration of graphene oxide (GO)/peptide nanofibers (PNFs)/cellulose nanofibers (CNF)/ZrO_2 nanohybrids. It was shown that the developed hybrid nanofibrous membranes provided a promising performance for adsorbing and removing the anion fluoride (< 200 mg/L) from water samples with a removal efficiency of up to 99.0%. This makes them promising materials for application in fluoride-containing wastewater.

Gao et al. (2021) prepared GO-cellulose nanocrystal (CNC) hybrid membranes (GO/CNC) through a simple vacuum filtration process and used them for the removal of antibiotics (e.g., sulfamethoxazole, levofloxacin, and norfloxacin) from artificial wastewater containing 100 mg/L of the salts NaCl, NH_4Cl, $MgCl_2$, and $CaCl_2$ and 10 mg/L of each antibiotic. The membrane rejection found was 74.8%, 90.9%, and 97.2% for sulfamethoxazole, levofloxacin, and norfloxacin respectively, with the concomitant passing of some nutrients, such as NO_3^- and $H_2PO_4^-$ through the hybrid membrane. Hence, the produced membrane could be a useful approach for resource recovery from wastewater. In addition, it presented a good antifouling behavior which indicated its potential use for long-term wastewater treatment applications.

Three-dimensional (3D) cellulose gels, typically existing as hydrogels or aerogels, have been widely utilized in diverse fields such as separation, adsorption, catalysis, photoelectrical sensing, and biomedicine (Sudhaik et al., 2023). Ching et al. (2018) produced 3D cellulose hydrogels by fibrillating microcrystalline cellulose (MCC), with an ultrasound probe, for Methylene Blue (MB) removal from aqueous solutions. The developed hydrogel showed high adsorption capacity (98 mg/g in 10 min) for MB (10 mg/L), which, according to the authors, was similar to that achieved using the commercial activated carbon DARCO 100 mesh. Nonetheless, an economic comparison of the developed hydrogel with the commercial one has not been done, which is critical to assess its marketability. Likewise, the advantages of the developed material over the existing commercial one should have been pointed out.

Among polysaccharide-based biopolymers, chitin and chitosan are also very attractive for the removal of pollutants from aqueous solutions, due to their availability, affordability, environmentally safe characteristics, biodegradability, non-toxicity, and biocompatibility, together with their excellent physical and mechanical features. Chitin is similar to cellulose, except its second carbon binds to an acetyl amino group as opposed to a hydroxyl group in cellulose, where the chitin monomer is N-acetyl-glucosamine and the cellulose monomer is D-glucose. Additionally, after cellulose, chitin is one of the most abundant biopolymers on Earth, as it can be obtained from skeletons of crustaceans and cartilage-like fish tissues. Chitosan is also a natural polysaccharide-based biopolymer, which is the partially N-deacetylated derivative of chitin, composed of glucosamine copolymers and N-acetyl glucosamine with one primary amino and two free hydroxyl groups for each C-3 and C-6 building unit. Natural resources available for recovery of chitosan are crustacean shells, insects, and fungi (Sarode et al., 2019; Hasan et al., 2022). Due to their high concentrations of amino and hydroxyl functional groups and their low cost, chitin and chitosan derivatives have drawn a lot of attention as efficient bio adsorbents for the removal of a variety of aquatic pollutants. They have been mostly used for heavy metal ion adsorption from water and wastewater (Shaikh and Shaikh, 2022). The capability of chitin to bind uranium and a variety of metal ions, including cadmium, chromium, copper, lead, nickel, and zinc, was indicated in the literature (Na et al., 2020). Among chitin-chitosan based works, Zhang et al. (2019) prepared a novel poly (acrylic acid) grafted chitosan and biochar composite (PAA/CTS/BC) for selective adsorption of heavy metals. This material showed high adsorption capability with maximum capacities of 312.50 mg Cr (III)/g, 476.19 mg Pb (II)/g, 370.37 mg Cd (II)/g, 111.11 mg Cu (II)/g, 114.94 mg Zn (II)/g, 99.01 mg Ni (II)/g, 138.89 mg Mn (II)/g, and 135.14 mg Co (II)/g. They also tested the developed composite PAA/CTS/BC to treat wastewater from a metal finishing factory containing different heavy metals and found that both wastewater purification and metal recycling was attained. Thus, PAA/CTS/BC has potential to treat heavy metal-containing wastewater.

Starch is another critical natural polysaccharide-based biopolymer found in plants and algae, particularly in stalks, roots, seeds, and leaves. Since it is low cost, abundant, reusable, and biologically degradable, starch appears as a promising candidate in different areas, including wastewater treatment (Jiang et al., 2020). The natural form of native starch is semicrystalline granules, composed of amylose and amylopectin. Amylose is a linear polymer consisting of α-1,4-linked glucosyl units, while amylopectin comprises branched chains linked via α-(1→6) to the linear parts of starch. Amylopectin is the major polysaccharide type, which determines starch granule characteristics. The amylopectin and amylose ratio can vary greatly, ranging from trace amounts to 70.0% for the amylose (Apriyanto et al., 2022). Growing interest in sustainable water treatment has spotlighted natural polymer-based flocculants like starch, valued for its availability, cost-effectiveness, and environmental benefits. Starch is particularly advantageous because it can be easily modified through esterification, etherification, and graft copolymerization to enhance its water treatment capabilities. Up to date, extensive research has been conducted concerning the impact of various external environmental factors, including the pH level, temperature, and dose, on the effectiveness of starch-based flocculants in achieving desirable flocculation results (Du et al., 2017). Starch can also serve as an adsorbent when modified, since native starch cannot be utilized directly for this purpose due to its limited specific surface area, inadequate water solubility, low molecular weight, and absence of reactive functional groups. Among the different modification approaches, physical methods, including mechanical activation treatment, microwave radiation treatment, sonication, and extrusion, and chemical methods, including grafting, cross-linking oxidation, and esterification, as well as enzyme treatments, are commonly employed to enhance the adsorption capacity of native starch (Fang et al., 2022).

There are different works in the literature focused on starch-based polymers for wastewater treatment applications. Among them, Zhou et al. (2019) produced PAFC-Starch-g-p(AM-DMDAAC) (polyaluminum ferric chloride-starch graft copolymer with acrylamide and dimethyl diallyl ammonium chloride) to be applied it as a coagulant for synthetic textile wastewater treatment. In their study, they successfully removed over 80% of the dyes Brilliant Blue KN-R, Yellow M-3RE, Dark Blue M-2GE, and Green KE-4B. That starch-based material outperformed other coagulants like PAFC-PAM, PAFC-PDMDAAC, and PAFC-P(AM-co-DMAAC), as well as PAFC alone, and required less coagulant compared to conventional coagulation-flocculation processes. The authors highlighted the environmental benefits of the biodegradable PAFC-Starch-g-p(AM-DMDAAC), suggesting it as a promising and eco-friendly alternative to traditional coagulants.

Yeamin et al. (2021) compared the adsorption capacity of the synthetic polymeric adsorbents polyaniline (PAni) and PAni/starch composites and the natural low-cost adsorbents wheat flour (WF), rice flour (RF), graham flour (GF), puffed rice (PR), turmeric powder (TP), pure starch, and starch nanoparticles (SN) on MB and Orange Green (OG). Although the starch-based adsorbents, including WF, RF, GF, PR, TP, pure starch, and SN, were cost-effective and biodegradable alternatives, PAni and its composites were more effective, considering the adsorption degrees of MB and OG dyes. Only TP showed an exceptionally comparable adsorption efficiency to PAni on MB. Therefore, the developed material lacks versatility, and other low-cost alternatives need to be explored.

Kongseng et al. (2023) prepared starched-based hydrogel composites using non-biodegradable polyacrylamide and biodegradable cassava starch (CS) and PA, obtaining a maximum water absorption of 74 g/g at 30 wt% CS. The developed hydrogel adsorbed both cationic and anionic dyes, including MB, Crystal Violet (CV), CR, and Reactive Orange (RO) in four hours. Particularly on MB, the maximum adsorption capacity was found to be 993 mg/g, with an initial concentration of 5500 mg/L. In addition, after coating the hydrogel with ZnO photocatalyst particles, the dyes adsorbed on the hydrogel composite were degraded by 90.0% in 12 hours under UV light and in only 1.5 hours under sunlight. This photodegradation was kept for three cycles under UV light and four cycles under sunlight, while adsorption remained higher than 80.0% for eight cycles. Additionally, the hydrogel composite presented antibacterial activities.

Alginate is a common polysaccharide, with nontoxic and biodegradable characteristics, found in the outer cell wall of brown macroalgae, which comprises guluronic acid and mannuronic acid units, connected by β-1,4-glycosidic bonds. As a naturally occurring biopolymer, alginate has proven to be effective in removing various heavy metals, dyes, and various organic/inorganic pollutants from aqueous solutions. Alginate-based natural biopolymers represent a class of biological materials that offer significant benefits in terms of efficient adsorption, ion exchange, precipitation, reverse osmosis, evaporation, oxidation, and biosorption processes for pollutants (Thakur, 2021). Thus, Wang Z. et al. (2020d) reported the effective removal (>80.0%) of the heavy metal ions Pb (II), Cu (II), Cd (II), Co (II), and Cr (III) in single experiments and high selectivity (>90.0%) for Pb (II), Cu (II), and Cd (II) in competitive experiments with sodium alginate-meso-2,3-dimercaptosuccinic acid hybrid aerogel (SA-DMSA). Huang et al. (2018) also prepared a novel composite adsorbent of Ca-DTCS/ALG beads by immobilization of DTPA-modified chitosan micro-gels (DTCS) into the alginate matrix to be assessed for copper adsorption in comparison with DTCS powder and plain alginate (ALG) beads. The maximum adsorption capacities of Ca-DTCS/ALG and DTCS were determined as 107.5 and 106.0 mg/L, respectively.

Carrageenan is an anionic polysaccharide that originates from seaweed with 15.0%–40.0% ester sulfate. Since there are three types of carrageenan, the sulfate content is an important parameter to determine the carrageenan type, as κ-carrageenan with one sulfate group per disaccharide, ι-carrageenan with two sulfate groups per disaccharide, and λ-carrageenan with three sulfate groups per disaccharide (Sohouli et al., 2022). Biopolymers composed of carrageenan have notable features, such as gelation capabilities, effective stabilization, thickening properties, and the ability to strongly bind to protein structures. Consequently, these biopolymers are highly valued for a diverse range of applications in various industries (Jiang et al., 2021). Thus, Lapwanit et al. (2018) investigated the ability of a hydrogel composed of κ-carrageenan and poly (glycidyl methacrylate) (CG/PG), which had multiple functional groups, to remove the cationic dye MB. The results revealed that the CG/PG ratio, initial dye concentration, pH value of the solution, treatment time, temperature, and amount of adsorbent used were critical factors on the efficiency of MB adsorption. Moreover, the CG/PG hydrogel exhibited improved adsorption ability and durability in contrast to pure CG, with a maximum MB adsorption of 166.62 mg/L. Also, it was found that, after five cycles of repeated use, CG/PG hydrogel adsorption capacity was kept at 83.3% related to the first cycle. Duman et al. (2020) tested an agar/κ-carrageenan hydrogel for the removal of MB from aqueous solutions. A maximum dye adsorption capacity of 242.3 mg/g was obtained when operated at 35°C and pH 7.0. The authors claimed that agar/κ-carrageenan hydrogel can be regarded as a sustainable and efficient adsorbent material for the elimination of cationic dyes from aqueous solutions.

10.2.2 Protein-Based Biopolymers for Wastewater Treatment

Keratin, collagen, zein, gelatin, fibrinogen, microfibrillar proteins, and wheat gluten are among the protein-based biopolymers (Udayakumar et al., 2021). Over the past twenty years, there has been substantial growth in the attention of biomaterials derived from proteins. This was largely due to their exceptional characteristics, such as being naturally abundant, biodegradable, and non-toxic, together with their easy modification owing to the presence of multiple functional groups (Saha et al., 2019). These remarkable characteristics render them important biological materials for water treatment. In this sense, various types of proteins, including keratin, silk proteins, soy proteins, albumin, and gelatin have been investigated for their industrial applications in terms of wastewater treatment (Zubair and Ullah, 2021). Among the aforementioned protein-based biopolymers, keratin is a fibrous protein found abundantly in natural sources, such as wool, hair, feathers, horns, and nails. The keratin polypeptide structure is constituted by multiple amino acids, each one bearing a distinct side chain with an exclusive chemical structure, bonding property, reactivity, and charge. The primary function of these side chains lies in adsorption processes and not in the formation of polypeptides. Moreover, the molecular composition of keratin sets it apart from other proteins like collagen, due to its significant concentration of cysteine that contains sulfur and constitutes around 7%–20% of the total amino acid residues present. The oxidized state of cysteine residues is bonded through disulfide covalent bonds. These bonds provide protein strength with mechanical, thermal, and chemical stability (Wang et al., 2016). With various chemical treatments, the side chains of keratin can exhibit diverse functional properties that enable the elimination of pollutants from wastewater (Donner et al., 2019). In this context, Donner et al. (2019) built functional biopolymers using chemically modified chicken feathers to eliminate any harmful trace elements from synthetic wastewater and wastewater from an oil sand process. In that work, unmodified poultry feathers (SM-03) added to synthetic wastewater spiked with Ni (II), Co (II), Zn (II), Cu (II), Pb (II), V (V), Cr (VI), As (III), and Se (IV) (30–50 μg/L each) removed over 82.0% of Pb (II), Ni (II), Co (II), and Zn (II), whereas modified poultry feathers removed 68%–100.0% of Se (IV), V (V), and Cr (VI). The authors stated that the produced biopolymers could be a suitable alternative for the treatment of wastewater on a large scale, due to the easy chemical modification and availability of the aforementioned raw material. But more experiments are needed to know their real industrial applicability. Similarly, Zahara et al. (2021) used keratin from chicken feathers to produce one keratin-based and seven keratin-based chemically modified biopolymers to treat synthetic wastewater containing toxic trace elements. Further, the developed biopolymers were applied to treat synthetic wastewater spiked with the metals Ni (II), Co (II), Zn (II), Cu (II), Cd (II), V (V), Cr (VI), As (III), and Se (IV) (100 μg/L each). All the developed biopolymers were able to remove all the trace metals tested with removal percentages ranging from 60.0% to 95.0%. They indicated that those developed keratin-based biopolymers showed an important potential for the effective treatment of wastewater containing multi metal pollutants, as well as reutilizing feather waste generated by the poultry industry.

Another important protein-based biopolymer for various applications including wastewater treatment is soy protein, since soybean contains high protein and oil levels (Choudhary et al., 2022). Thus, various forms of soy protein, such as soy flour, soy protein concentrate (SPC), and soy protein isolate (SPI), are prominent examples, with 50.0%, 70.0%, and 90.0% of protein content, respectively. Soy proteins consist of twenty amino acids that allow for the integration of various functional groups, such as carboxyl, hydroxyl, and amide at either the side chains or backbone (Mayilswamy and Kandasubramanian, 2022). Soy proteins have received significant interest due to their sustainable nature, affordability, wide accessibility, biodegradable properties, and ease of modification as a by-product resulting from the processing of soybean oil. Soy protein research has primarily focused on enhancing the strength of protein materials and expanding their water resistance capacity and flexibility via chemical or enzymatic modification. Moreover, soy protein-derived resins possess polar functional groups, which demonstrate satisfactory bonding capabilities and enable the creation of composite materials featuring enhanced mechanical performance (Mayilswamy and Kandasubramanian, 2022; Tian et al., 2022). There are several works based on this valuable protein in the wastewater treatment literature. Among those papers, Liu et al. (2017) prepared SPI (soy protein isolate)/PEI (polyethyleneimine) composite hydrogels using a chemical crosslinking method. They showed that SPI/PEI composite hydrogels exhibited efficient adsorption of Cu (II) ions from aqueous solutions, especially when the PEI percentage was 50.0%. Moreover, the SPI/PEI composite hydrogel with 50.0% PEI exhibited exceptional selectivity in eliminating Cu (II) ions even in the presence of other competitive heavy metal ions like Zn (II), Cd (II), and Pb (II). In addition, they showed the reusability of the recovered Cu (II) to fabricate copper nanoparticles with catalytic activity to reduce 4-nitrophenol to 4-aminophenol. Hence, the developed composite material would not only remove Cu (II) from wastewater but also recycle them.

Silk is an attractive biomaterial that possesses remarkable features, including not only high strength, elasticity, and light weight but being biocompatible, biodegradable, non-toxic, and easily available. Silk fibers are mostly produced by specific abdominal glands of silkworms and spiders via ducts, spinnerets, or spigots and represent an important biopolymer filament in nature. Silkworm silk fibers contain two important proteins, namely sericin and fibroin. Sericin is a globular and water-soluble protein that constitutes around 20%–30.0% of the total silk protein, while fibroin makes up approximately 70.0%–80.0% of the overall silk fiber (Gore et al., 2019). Silk fibers are frequently utilized in the silk industry, due to their high capacity for adsorbing dyes. Consequently, functionalized silk fibroin (SF) composites can be effectively employed to remove dye pollutants from wastewater generated by different industries and textile production processes. In the literature, it has been determined that SF exhibits a high degree of efficacy when absorbing certain types of dyes, such as CR, MB, Methyl Orange (MO), and Reactive Green (RG) (Gupta et al., 2023), while sericin could be a promising biopolymer for the removal of heavy metals (Koley et al., 2016).

Kwak et al. (2016) prepared silk sericin and kraft lignin (SS/KL) blend beads by means of coagulation methods for the removal of Cr (VI) from an aqueous solution. The found that the SS/KL beads presented higher adsorption capacity for Cr (VI) than that attained by sericin beads. A 50:50 SS/KL ratio was found as the optimum ratio in that work with an adsorption capacity of 65.72 mg/g for Cr (VI). Additionally, over 90.0% of Cr (VI) ions were recovered from the Cr (VI) adsorbed SS/KL beads. Also, the SS/KL beads remained stable for more than seven adsorption-desorption cycles and with a recycling efficiency of 82.0%. Therefore, it seems that the developed SS/KL beads can be very suitable for the removal of copper-containing wastewater.

Bruder et al. (2021) built multifunctional aerogels from SF extracted from *Bombyx mori* silkworm cocoon. Then, cryo-assembly assisted electro-spraying was applied to produce micro-aerogels with a wrinkled surface. Those aerogels showed effective adsorption capacities for Cu^{2+} (186.7 mg/g in 240 min), the anionic dye Methylene Orange (811.3 mg/g), and the organic solvent chloroform (1138 g/g). The authors claimed that the developed aerogels could be an ecological and affordable alternative for multipollutant adsorption.

10.3 WASTEWATER TREATMENT APPLICATIONS OF NATURAL BIOPOLYMER ASSEMBLED NANOSHEET NANOCOMPOSITES

Currently, various types of nanomaterials have been employed in wastewater treatment of industrial effluents, surface water, groundwater, and drinking water, considering the nature of the pollutants and the desired efficacy of treatment level with high efficiency, environmental friendliness, and cost-effectiveness (Anjum et al., 2019; Bhat et al., 2022). In the field of wastewater treatment, at present, a variety of nanomaterials, such as bioactive nanoparticles, biomimetic membranes, nano-adsorbents, nanocatalysts, nanostructured catalytic membranes, and molecularly imprinted polymers, are actively utilized. These innovative nanomaterials serve as effective tools for eliminating both inorganic and organic compounds from polluted water sources. In addition to removing toxic metal ions and disease-causing microorganisms present in wastewater, they also play a crucial role in eliminating other types of contaminants that pose significant risks to human health and the environment (Saraswat, 2022).

In general, there are four main classes of nanomaterials for wastewater treatment processes. The first class consists of nano-adsorbents that are commonly utilized to eliminate heavy metals from wastewater. Graphene, activated carbon, carbon nanotubes, ferric oxides, manganese oxide, zinc oxide, and magnesium oxide are examples of such adsorbents. The second class comprises nano-catalysts, including photocatalysts, electrocatalysts, Fenton-based catalysts, and chemical oxidants, which exhibit a high degree of efficacy in eliminating organic as well as inorganic pollutants. As for the third class, nanomembranes, such as carbon nanotube membranes, electrospun nanofibers, and hybrid nano-membranes, have been employed to efficiently eliminate dyes, heavy metals, and foulants. The use of nanotechnology in combination with biological technologies, including algal membrane bioreactors, anaerobic digestion, and microbial fuel cells, is categorized in the fourth class (Anjum et al., 2019).

Nanomaterials exhibit extensive surface area, enhanced adsorption capacities, magnetic capabilities, chemical durability, and advanced antibacterial features (Ahmed et al., 2022; El-Aswar et al., 2022). Although nanotechnology has emerged as a pioneering technological solution to overcome the challenges of water scarcity and wastewater management, a broad range of research focus, comprehensive understanding, and new technologies are required. Carbon nanomaterials and metal/metal oxide nanoparticles are widely used in wastewater treatment. However, the chemical synthesis of these nanomaterials often involves harsh reagents, solvents, and reducing agents, raising ecological and toxicological concerns. Considering these, green synthesis of nanomaterials offers a sustainable, renewable, biodegradable, easily available, eco-friendly, and cost-effective alternative approach for producing nanoparticles using biological materials, such as biopolymers, plant extracts, and certain bacterial and fungal strains. Moreover, green synthesis of nanomaterials also offers enhanced biocompatibility and stability to composite materials (Rana et al., 2022). Therefore, in addition to the general characteristics of nanomaterials (i.e., large surface area, high adsorption capacity, chemical stability, and antimicrobial properties), green synthesized nanomaterials also have the asset value of being ecological, sustainable, and cost-effective (Bhat et al., 2022). In this sense, nanosheets, which are widely recognized as a form of nanomaterials, have advantageous features, including a high surface area with active oxygen-containing groups that make them highly suitable for producing functional and strong polymeric composites with exceptional mechanical/thermal/electrical conductivity, and good photo-/thermo-catalytic activities (Agboola et al., 2021).

Nanosheets, which consist of 2D nanomaterials, are considered a promising material to build functional membranes for molecular separation and can be created through the application of vacuum suction, polymer induction, or external-force-driven methods. To enhance both permeation rate and separation efficiency, several approaches have been suggested for modifying the aperture size and surface properties of them (Cheng et al., 2021; Nakagawa et al., 2021). They can be used for the treatment of water containing common pollutants, such as heavy metals (Azamat et al., 2015), or many organic hazardous dyes that possess high levels of toxicity, carcinogenicity, and mutagenicity (Chen Z. et al., 2022b).

Recently, increased attention has been paid to natural biopolymer-derived nanomaterials obtained from sustainable resources as potential substitutes for fossil-derived polymers. These sustainable alternatives offer the advantages of being both easily available and adjustable to meet specific requirements (Russo et al., 2021). At this point, natural biopolymer-assembled nanosheets are gaining attention for their eco-friendly, and sustainable characteristics. To fabricate these bio-modified nanosheets nanocomposites as hybrid nanomaterials, several natural biopolymers are assembled with nanosheets (Figure 10.1). Among biopolymers, the most abundant one, cellulose, with its different forms including nanocellulose, cellulose nanocrystals, cellulose nanofibrils, and carboxymethylcellulose (CMC), etc., is the most utilized biopolymer for the fabrication of hybrid nanocomposites Thus, Hernández-Francisco et al. (2020) fabricated an entangled cellulose nanofibril and cellulose nanosheet (CNF/CNS) from blue agave (*Agave tequilana Weber var*) for its potential use as a biosorbent. The developed nanomaterial was applied to remove Pb (II) from aqueous solutions with a maximum concentration of 1000 mg/L at pH 5.0–5.5, 380 rpm, and 25°C for 24 h. Under these parameters, the maximum adsorption capacity was found as 43.55 mg/g. This study underscored the potential of lignocellulosic byproducts in nanobiomaterials for heavy metal extraction from contaminated water, although more studies are needed to explore broader applications.

Wang H. et al. (2020a) reported the fabrication of an eco-friendly polymer nanocomposite hydrogel strengthen by cellulose nanocrystals (CNCs) and graphitic-like carbon nitride ($g\text{-}C_3N_4$) nanosheet ($g\text{-}C_3N_4$/CNCs-H) for the adsorption of MB from an aqueous solution. They also compared the adsorption capacities of $g\text{-}C_3N_4$/CNCs-H (sodium CMC (CMC-Na)/polyvinyl alcohol (PVA)/C_3N_4/

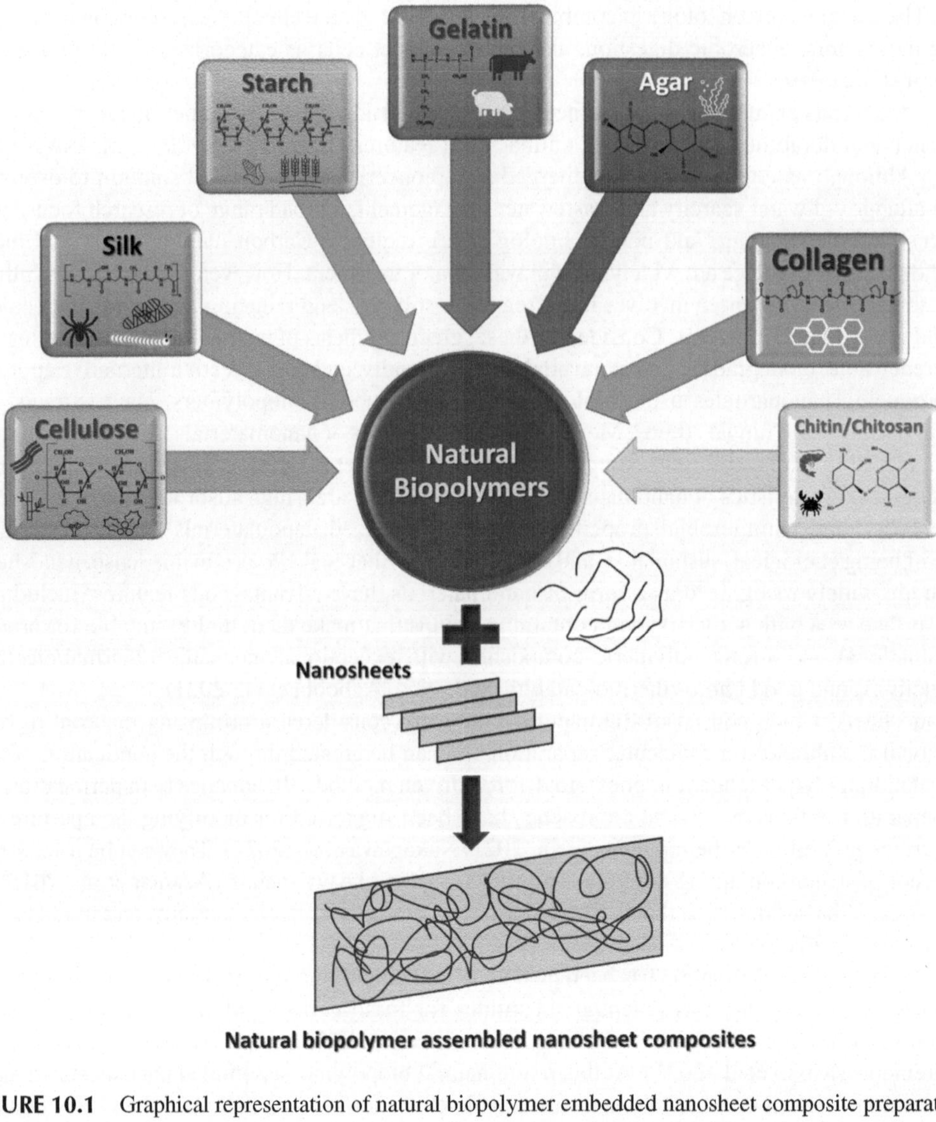

FIGURE 10.1 Graphical representation of natural biopolymer embedded nanosheet composite preparation.

CNCs) hydrogels with CNCs-H hydrogel and CMC-Na)/PVA hydrogel on MB. The g-C_3N_4/CNCs-H hydrogel demonstrated favorable adsorption capacity towards MB within pHs 5.0–10.0 and poor adsorption at lower pH. It was found that the MB adsorption rate of the g-C_3N_4/CNCs-H hydrogel was faster than that of the CMC-Na/PVA hydrogel. The introduction of cellulose nanocrystals and g-C_3N_4 to the CMC-Na/PVA hydrogel greatly improved the adsorption capacity of MB (10 mg/L) higher than 232.5 mg/g using 1.0% g-C_3N_4/CNCs-H hydrogel at room temperature. This improvement was due to CNCs and g-C_3N_4 in the adsorbent contributed to the presence of active functional groups, which significantly enhanced the efficiency of MB adsorption. This study illustrated the potential of hydrogels to remove contaminants from water, emphasizing the need for field validation in real environments.

Ma Q. et al. (2021a) reported the use of a CMC-based hydrogel, manufactured by polymerization of 2D MMTs, AA, MBA, and CM, for the removal of MB from water. It was shown that the adsorption rate and the adsorption capacity of the CMC-based hydrogel increased significantly for higher concentrations of 2D MMTs in the hydrogel and, thus, the MB adsorption capacity of the synthesized hydrogel containing 30% of 2D MMT was 410 mg/g. In addition, after five cycles, 90% of MB removal was still obtained. The good stability and reusability of the produced hydrogel suggested its suitability for commercial applications.

Ma Z. et al. (2021b) prepared a 3D recyclable g-C_3N_4@ cellulose aerogel for the photocatalytic degradation of MB and MO. The g-C_3N_{44}@cellulose aerogel showed excellent photocatalytic activity, achieving a removal ≥ 99.0% for MO, RhB, and MO, for an initial concentration of 20 mg/L at room temperature in 180 and 240 min, respectively. Furthermore, the g-C_3N_4@cellulose aerogel showed remarkable recyclability and structural robustness with almost no decrease in photocatalytic efficacy after 10 cycles for MB removal. This indicated that it could be appropriate for industrial applications.

Chen Z. et al. (2022b) prepared an eco-friendly sugarcane cellulose (SBC)/(CMC-Na) adsorbent hydrogel strengthened with carbon nitride (g-C_3N_4) (g-C_3N_4@SBC/CMC), using a sol gel method. The composite hydrogel displayed a highly selective adsorption of the cationic dye MB in MB/MO or MB/RhB mixtures. Thus, MB removal in the mixture MB/RO (10 mg/L each) was 99.07% whereas MO was only removed 6.91% (within 20 min). Likewise, MB removal in the mixture MB/RhB (10 mg/L each) was 97.06%, while RhB was only 11.15%. The adsorbent demonstrated high stability and reusability, as it maintained its high adsorption capacity even after seven cycles. Therefore, it could have great potential in wastewater treatment.

Zhang et al. (2023) prepared an adsorbent gel using bamboo nanocellulose (BCNF) derived from waste bamboo paper and montmorillonite nanosheet (MMTNS) cross-linked by polyethyleneimine (PEI) to evaluate the adsorption capacity of MB and Cu (II) from aqueous solutions. Adsorption experiments were performed at 25°C for 24 h at pH 5.0 for Cu (II) and pH 10.0 for MB for an initial concentration of 100 mg/L each one. Based on those experiments, the maximum adsorption capacities were 361.9 and 254.6 mg/g for MB and Cu (II), respectively. After five cycles, the removal capability for MB was 49.3% and for Cu (II) 47.1%, which was ascribed to their incomplete desorption. The developed adsorbent could have potential for the treatment of wastewater containing dyes and heavy metals.

Shi et al. (2023) prepared a nanofiltration membrane based on polyethylenimine (PEI)-mica nanosheets/cellulose nanofibers (CNF) via facile vacuum filtration for dye/salt separation. The nanofiltration performance was evaluated with different dyes (0.02 g/L) and salt solutions (1 g/L). The PEI-mica/CNFs membrane showed high water permeability (62.18 L/m^2/h/bar) and exceptional dye/salt separation efficiency. Thus, it showed dye retentions for Coomassie Brilliant Blue R250 (BB), CR, Evans Blue (EB), and Indigo Carmine (IC) as 99.65%, 98.89%, 98.64%, and 97.95%, respectively, while low salt rejections were obtained as 3.05% for $MgCl_2$, 3.42% for NaCl, 5.89% for $CaCl_2$, 7.05% for KCl, and 15.63% for Na_2SO_4. Additionally, the composite membrane demonstrated strong antifouling behavior and stability in terms of operation in dye/salt mixtures, and, thus, it could have great potential for effective textile wastewater treatment.

There is an urgent need for the advancement and implementation of water treatment technologies and materials that are both efficient and environmentally friendly, since there is an uncontrollable increase in industrial contamination and pathogens in water. In this sense, chitosan with its cost-effectiveness, low toxicity, and biocompatibility, could be a good biopolymer alternative to produce sustainable nanobiocomposites (Xing et al., 2022). In this sense, Vigneshwaran et al. (2020) produced a novel (2D/2D) CS@g–C_3N_4/ MXene (MX) hybrid composite, using g-C_3N_4/MX nanosheets together with CS as a catalyst for the degradation of the recalcitrant dyes MB and RhB from aqueous solutions under visible light irradiation. This hybrid composite (0.1 g/L) exhibited high photocatalytic activity, leading to MB and RhB degradation around 99.0% in 60 min and 98.5% in 40 min at 24.5°C, respectively, using an initial concentration of 50 mg/L of dye and a visible light irradiation of 0.15 mW/cm^2 intensity. Besides, CS@g-C_3N_4/MX displayed an outstanding photo-stability over five consecutive cycles of dye degradation, with undetectable loss of photocatalytic activity. Moreover, they showed that the organic dyes were oxidized to non-toxic products. Therefore, this novel hybrid composite could be a promising photocatalyst composite for sustainable water treatment.

Wang S. et al. (2020b) prepared biosurfactant-functionalized Ti_2CTX MXene nanosheets using three inexpensive and renewable biosurfactants, a cationic surfactant (chitosan, CS), an anionic surfactant (lignosulfonate, LS), and a non-ionic surfactant (enzymatic hydrolysis lignin, EHL) to enhance the adsorption of Pb (II) ions from aqueous solutions on Ti_2CTX MXene. Based on the adsorption optimization studies, the optimum adsorbent concentration, temperature, pH, and contact time were determined as 1.6 g/L, 30°C, pH 5.0, and 24 h, respectively. A maximum adsorption capacity of 232.9 mg/g was obtained for EHL-functionalized Ti_2CTX MXene nanosheets for a 1000 mg/L initial concentration of Pb (II) under optimized conditions. These functionalized nanosheets showed promise as an effective adsorbent for environmental applications.

Wang W. et al. (2020c) proposed a porous hydrogel adsorbent, in which exfoliated montmorillonite nanosheets (MMTNS) were incorporated to a carboxymethyl cellulose-chitosan system for the removal of MB. The features of MMTNS, such as providing negative charges on their surface and positive charges on the edges, being small in size, and a possibility of strong dispersion (i.e., in cellulose-based systems), were beneficial for the preparation of a homogeneous composite. High removal (97.0%) of MB removal was attained when using 0.2 g/L hydrogel within 360 min at pH values from 8.0 to 10.0 and up to 60°C even after five cycles of adsorption-regeneration. The use of this hydrogel has been found to enhance the effectiveness of polysaccharide polymers in terms of possible material design and their potential applications in wastewater treatment.

Mittal et al. (2021) prepared GO nanosheet-based nanocomposite hydrogels (NCH) of CS and CMC for the adsorption of MB (cationic dye) and MO (anionic dye) from wastewater. It was observed that 99.0% of MB was adsorbed on CS/CMC-NCH (0.4 g/L) hydrogel at pH 7.0, while about 82.0% of MO was adsorbed on CS/CMC-NCH (0.6 g/L) at pH 3.0 for initial dye concentrations of 50 mg/L. Also, the maximum adsorption capacities obtained were 655.98 mg/g for MB and 404.52 mg/g for MO at 25°C. Additionally, CS/CMC-NCH were used for 20 continuous adsorption-desorption cycles, and adsorption efficiency only decreased in the last three cycles, showing 93.0% adsorption for MB and 90% for MO. Therefore, CS/CMC-NCH is a very promising adsorbent for the removal of both cationic and anionic dyes.

Xing et al. (2022) synthesized a multifunctional GO nanosheet/CS/copper-based polyoxometalate (Cu-POM) nanocomposite that showed a promising potential for the removal of different dyes, mostly within the first 15 min, including MB, Gentian Violet (GV), Basic Fuchsin (BF) with 96.86%, 97.77%, and 96.47% at an initial concentration of 20 mg/L and ambient temperature. That nanocomposite also degraded up to 78.92% of tetracycline (TC) and 76.26% of norfloxacin (NC) at an initial concentration of 50 mg/L within 60 min. This developed adsorbent also possessed antibacterial activity against sensitive bacterial strains such as *Escherichia coli*, *Staphylococcus aureus*, *Bacillus subtilis*, and *Pseudomonas aeruginosa* with an approximately 100% growth inhibition after exposure to a concentration range of 25–100 μg/mL for one h. Moreover, this bactericidal effect was also observed on two drug-resistant *E. coli* strains, 99.93% for kanamycin-resistant strain and 97.94% for ampicillin-resistant strain after six h. This approach offered a successful method for formulating POM-based composites with antibacterial and adsorption properties that can be used in water treatment and the removal of diverse pollutants. On the other hand, this should be tested in a real industrial environment to know their actual potential.

Wani et al. (2023) synthesized superparamagnetic Fe_3O_4-MoS_2 nanosheets with CS (Fe_3O_4-MoS_2@CS) for the removal of Cr (VI) and ciprofloxacin (CIP) in simulated and real wastewater (from a local lake in Aligarh, India). In that paper, the Fe_3O_4-MoS_2@CS nanomaterial not only presented a maximum adsorption capacity for Cr (VI) of 75.4 mg/g at pH 4.0 and 190.7 mg/g for CIP at pH 7.0, at room temperature, but it was also successful for the removal of Cr (VI) (97.0% ± 3%) and CIP (62.0% ± 3.0%) from simulated wastewater. The Fe_3O_4-MoS_2@CS hybrid nanocomposite exhibited enhanced adsorption capacity in comparison to Fe_3O_4-MoS_2, while maintaining optimum reaction conditions. The Fe_3O_4-MoS_2@CS nanocomposite not only improved adsorption efficiency but also exhibited exceptional magnetic recyclability, allowing for easy recovery from wastewater using an external magnet. These properties make it a sustainable and promising candidate for wastewater treatment and clean production processes, although scale-up studies are required to assess its viability on a larger scale.

Othman et al. (2023) developed a new adsorbent called MCL "Ti_3C_2Tx MXene (M)/chitosan(C)/lignosulfonate adsorbent (L)" which consisted of Ti_3C_2Tx MXene nanosheets decorated with C/L nanospheres. This composite aims to enhance the biocompatibility and stability of MXenes in water and promote sustainability. In that work, the preferential adsorption affinity of MCL toward different heavy metal ions was observed during the competitive adsorption process in multi-metal systems. During Cu (II) ions removal, when the MCL dosage was used at 0.05 g/L, the adsorption capacity reached up to 102.0 mg/g, with an adsorption efficiency of 11.0%. On the other hand, if 1 g/L MCL was used, the adsorption efficiency was found as 93.0%, while the adsorption capacity was reduced to 46.9 mg/g. In addition, for the simultaneous removal of multiple heavy metals, a solution containing Pb (II), Cu (II), Cr (VI), Ni (II), and Co (II) at 10 mg/L each was tested. Operating at pH 7.0, 293 K (~19.85°C) for 2 h and 0.1 g/L of MCL dosage with very limited adsorption sites caused the highest removal efficiencies of 60.2% and 53.0% for the Pb (II) and Cr (VI), respectively. Finally, the metal ion adsorption trend observed was Pb (II) > Cr (VI) ≈ Cu (II) > Ni (II) ≈ Co (II). They also further tested MCL adsorbent in real conditions involved spiking secondary treated sewage from a Doha City treatment plant with 50 mg/L of Cu (II) and 1 g/L MCL removed 100% of Cu (II), showing increased removal efficiencies with higher dosages. This study suggested MCL, with its low-cost bioadsorbent components, as a promising replacement for current market products in heavy metal removal from wastewater.

There has also been substantial attention directed toward the development of ultralight aerogels in water pollution treatment, due to their desirable characteristics, such as low bulk density, high porosity, and excellent functional performance. For this purpose, Si et al. (2023) utilized a large surface-area metal framework (ZIF-8) onto cellulose nanofibers (CNF) and CS for wastewater treatment. Based on the 3D porous structure, ZIF-8@CNF/CS aerogel (Si-CCZA1) (with 1 wt% ZIF-8 content) showed a high saturation adsorption capacity between 35.99 g/g and 74.55 g/g for oils (over 88.0% of the adsorption capacity after 20 cycles) with an oil/water separation efficiency of 99.8%. They proposed that the approach in their work could be a good option for producing affordable, scalable, and eco-friendly biomass-based materials for the treatment of oily water contamination.

Starch is another low-cost biopolymer for biopolymer-based nanocomposite fabrication. In this sense, Hosseinzadeh and Ramin (2018) also prepared a novel polysaccharide-based nanocomposite hydrogel adsorbent (NHA) through a free radical graft copolymerization of acrylamide (AM) monomer onto starch backbones in the presence of GO nanosheets and nano-hydroxyapatite (n-HAp). This n-HAp incorporated starch-graft-poly AM/GO was used for the removal of Malachite Green (MG) (a cationic dye) from an aqueous solution. The maximum dye adsorption was 297 mg/g within 60 min with 3 wt% of n-Hap content and an initial dye concentration of 100 mg/L at pH 10.0. Due to their high regeneration capability, biocompatibility, biodegradability, porosity, water content, and water uptake capacity as well as low production costs, the synthesized nanocomposites can be employed as very effective adsorbents for the removal of cationic dyes from industrial effluents.

Moreover, starch nanocrystals, prepared by the hydrolysis of starch, are in use due to their various advantageous features, such as biocompatibility, biosecurity, biodegradability, abundance, renewability, and affordability. On this basis, dialdehyde starch nanocrystals (DASNCs) and GO (reduced form, RGO) nanosheets were combined to create compressible, mechanically strong, and extremely porous aerogels that were employed as supercapacitor electrodes and effective adsorbents (Chen et al., 2019). The adsorbent DASNCs-RGO hybrid aerogel (SRGO-1) (500 mg/L concentration) was used at pH 5.0 and 25°C for 48 h to remove RB and CV with maximum adsorption capacities at 539 mg/g and 318 mg/g for RB and CV, respectively. Considering their properties, the hybrid aerogels created in that work could be good candidates for flexible electrode materials for supercapacitors, biomedical, and environmental cleanup materials.

In an enzyme-nanosheet/alginate-based work, Wang et al. (2018) fabricated poly (vinyl alcohol)/sodium alginate/chitosan/montmorillonite nanosheets hydrogel beads for MB removal. In that work, the maximum adsorption capacity of MB on hydrogel beads was 137.15 mg/g and the MB removal percentage was 90.0%, after 250 min, for an MMTNS amount of 29.7%, compared to 68.16% and 42.46% for MMTNS amounts of 17.44% and 9.55%, indicating that MMTNS may improve the MB removal percentage on hydrogel beads. The optimum conditions for the MB removal were determined as 1.0 g of adsorbent, 30°C, initial pH of 8.0, and initial concentration of 25 mg/L for MB. The fabricated hydrogel also displayed good stability and reusability after five cycles.

Shahzad et al. (2019) employed a novel approach to produce 2-D titanium carbide MXene core (Ti_3C_2Tx) shell aerogel for the elimination of mercuric ions. These spheres were developed using a simple method that involved utilizing varying concentrations of Ti_3C_2Tx MXene nanosheets and sodium alginate. The MX-SA (4:20) spheres exhibited a notable maximum adsorption capacity of 932.84 mg/g for Hg (II) (25 mg/L) at 180 rpm, pH 6 and 298°K in 24 h. Moreover, the adsorbent demonstrated exceptional efficiency in removing heavy metals using 50 mg of adsorbent in a 30 mL aqueous solution containing 3 mg/L of each metal ion for 24 h at pH 4.67 and room temperature. Thus, 100% removal efficiency was achieved for Hg (II), Cd (II), and Cr (III), over 90.0% for Pb (II) and Cu (II), over 80% for Zn (II), Co (II), and Ni (II), and 37.0% for As (III). The authors also indicated that the process of spheres fabrication could be expanded and implemented on a large scale to address the remediation of various types of pollutants.

Miao et al. (2022) utilized montmorillonite nanosheets, gelatin, and sodium alginate as starting materials to fabricate a 3D porous hydrogel, using 3D printing technology to remove Pb (II) ions. The maximum adsorption capacity for Pb (II) was at 134 mg/g using 1000 mg/L initial concentration of ions Pb (II) and 4 g/L hydrogel concentration in 600 min at 25°C and pH 5.0. Moreover, the hydrogel also showed a good adsorption capacity of 161.47 mg/g toward Pb (II) from aqueous solutions (100 mg/L initial concentration) using a fix-bed column with a flow rate of 2 mL/min and 2.7943 g of the 3D-hydrogel, demonstrating its potential to be used in real-industrial scenarios.

Since modern industry faces a huge problem with environmental pollutants, particularly water pollutants, there is a critical requirement for novel water purification technologies through the development of eco-friendly functional materials, with affordable production techniques. At this point, silk could be an alternative as a highly promising biopolymer derived from silkworm cocoons with its known two proteins: fibroin and sericin. In one of the silk-based studies in the literature, Koley et al. (2016) reported a green and simple fabrication method of silk protein sericin-mediated hierarchical hydroxyapatite-based bio-hybrid flowers with nanosheet petals providing a large surface area and porosity to enhance the adsorption capability of toxic heavy metals and CR from wastewater. During the preparation, the hybrid flowers created using sericin at a concentration of 1 mg/mL were all smaller, more evenly developed, and spherically shaped (SP-hybrid). The morphologies of manufactured flowers significantly changed when sericin concentration was decreased to 0.1 and 0.01 mg/mL, simulating the bloom of natural flowers. Moreover, their surface shape was also changed when there was a change in sericin concentration. The hybrid flowers showed over 99.5% efficiency in adsorbing Pb (II) using 5 mg of flowers and 15 mg/L of heavy metal solutions for two h, with 99.0% adsorption occurring within 15 min. They also effectively removed Pb (II), Cd (II), and Hg (II) with 99.8%, 75.3%, and 64.3% efficiency, respectively, and adsorbed 50% of CR from wastewater within four h. This approach highlighted the potential of using industrial waste sericin to create value-added materials for environmental and economic benefits.

Zhao F. et al. (2020) reported the fabrication of multilayered inorganic-organic hybrid membranes, using metallic molybdenum disulfide (MoS2) and silk nanofibrils for water purification. In that work, the adsorption capacities for the hybrid membrane MoS_2–SNF (5:4) for different heavy metal ions were found as 759.1 mg/g, 425.0 mg/g, 249.4 mg/g, 246.2 mg/g, 153.5 mg/g, and 22.2 mg/g for Au (III), Pd (II), Hg (II), Ag (I), Pb (II), and Cu (II), respectively. The concentrations of heavy metal ions in deionized water were 2 mM for Pb (II), Hg (II), and Cu (II) and 5 mM for Au (III), Pd (II), and Ag (I). The rejection rates of MoS_2–SNF (5:4) for Au (III) and Pd (II) reached 99.21% and 99.85%, respectively, while the other four ions had rejection percentages as 63.11% for Hg (II), 53.92% for Ag (I), 44.41% for Pb (II), and 43.99% for Cu (II). In terms of dye separation, for cationic dyes, including Rhodamine 6G (R6G), Alcian Blue 8GX (AB), MB and CV, the hybrid membrane showed a very good efficiency with rejection rates over 95.0%, while they were over 84.0% for Brilliant Blue G (BB) and CR anionic dyes.

As an economically friendly and sustainable biopolymer, agar is another option for hydrogel formulation in terms of water decontamination. In this sense, Zhang et al. (2017) produced a metal-organic framework (MOF)-containing agar aerogel with photocatalytic activity by regulating the growth of zeolitic imidazolate framework (ZIF-8) on carbon nitride (C_3N_4) nanosheets. In that work, the inclusion of C_3N_4 nanosheets in the hybrid aerogel not only controlled the growth of ZIF-8 crystals but also presented excellent reusability. The hybrid hydrogel, Hgel-50 (aerogel loaded with 50.0% ZIF-8/C_3N_4) showed a maximum adsorption of 287.35 mg/g for CR at an initial concentration of 1000 mg/L, while the maximum adsorption was 154.87 mg/g for MB, with an initial concentration of 500 mg/L at 25°C.

Gelatin is a water-soluble natural polymer, obtained by the hydrolysis of the insoluble fibrous protein, collagen. Considering this, Nafti Mateur et al. (2020) described the fabrication of porous gelatin membranes for filtration. In their work, hexagonal boron nitride nanosheets (h-BNNS) were utilized for the stabilization of castor oil micro-droplets in a gelatin solution. Based on the complexation–ultrafiltration experiments for Fe (II) (50 mg/L, initial concentration) removal from aqueous solutions, a high removal rate of 97.0% was obtained in the presence of polyacrylic acid (PAA) (200 mg/L) at pH 5.0 and 25°C with 3.0 bars of transmembrane pressure. Gelatin and castor oil, the two primary ingredients, are inexpensive and enable the preparation of biodegradable disposal filters, offering a sustainable solution for the removal of industrial contaminants from water, aligning with the principles of eco-friendly and low cost.

10.4 CONCLUDING REMARKS

In summary, natural biopolymer-assembled nanosheet composites offer a promising platform for advanced wastewater treatment technologies. Studies in the literature based on natural biopolymer-assembled nanosheet composites for wastewater treatment have yielded promising results and significant potential for addressing the global challenge of water pollution (Table 10.1). Preparation techniques and surface modifications further enhanced the adsorption capacities and selectivity of the natural biopolymer-assembled nanosheet composites, enabling effective treatment of heavy metals, organic contaminants, and dyes. By integrating nanotechnology with natural biopolymers, such as cellulose, chitosan, starch, silk, agar, gelatin, and their derivatives, the fabricated composites exhibited enhanced performance, reusability, and stability, making them economically viable and environmentally sustainable options for possible large-scale applications. The cost-effectiveness and ease of regeneration of these materials could provide practicality in modern wastewater

TABLE 10.1
Natural Biopolymer-assembled Nanosheet Nanocomposites Designed for Wastewater Treatment Applications

Natural Biopolymer(s)	Nanosheet	Composite	Application	References
Sugarcane cellulose (SBC) Sodium carboxymethylcellulose (CMC-Na)	Carbon nitride ($g-C_3N_4$) nanosheets	$g-C_3N_4$@SBC/CMC hydrogel	**MB removal**: 99.07% **MO removal**: 6.91% **MB/RhB removal**: 97.06%/11.15%	Chen Z.et al. (2022b)
Cellulose nanofibril (CNF)	Cellulose nanosheets (CNS)	CNF/CNS composite	**Pb (II) adsorption capacity**: 43.55 mg/g	Hernández-Francisco et al. (2020)
Cellulose nanocrystals (CNCs) Sodium carboxymethylcellulose (CMC-Na)	Carbon nitride ($g-C_3N_4$) nanosheets	$g-C_3N_4$/CNCs-H hydrogel	**MB adsorption capacity**: 232.5 mg/g	Wang H. et al. (2020a)
Carboxymethyl cellulose (CMC)	2D Montmorillonite nanosheets	CMC-MMTs hydrogel	**MB adsorption capacity**: 410 mg/g **MB removal**: 100%	Ma Q. et al. (2021a)
Cellulose nanofibers	Carbon nitride ($g-C_3N_4$) nanosheets	3D $g-C_3N_4$@ cellulose aerogel	**Photocatalytic degradation**: **MB/MO/RhB**: > 99.0%	Ma Z. et al. (2021b)
Nanocellulose from bamboo	Montmorillonite nanosheet cross-linked by polyethyleneimine (PEI)	BMP Adsorbent Gel	**MB adsorption capacity**: 361.9 mg/g **Cu (II) adsorption capacity**: 254.6 mg/g	Zhang et al. (2023)
Cellulose nanofibers	Polyethylenimine (PEI)-mica nanosheets	The PEI-mica/CNFs membrane	**Dye retentions**: BB (Coomassie Brilliant Blue R250): 99.65 % CR: 98.89 % EB: 98.64 % IC: 97.95% **Salt rejections**: $MgCl_2$: 3.05% NaCl: 3.42% $CaCl_2$: 5.89% KCl: 7.05% Na_2SO_4: 15.63%	Shi et al. (2023)
Chitosan (CS)	Carbon nitride ($g-C_3N_4$)/ MXene (MX) nanosheets	(2D/2D) CS@$g–C_3N_4$/MX hybrid composite	**Photocatalytic degradation**: MB: 99.0% RhB: 98.5%	Vigneshwaran et al. (2020)
Chitosan (CS) Lignosulfonate (LS) Enzymatic hydrolysis lignin (EHL)	Ti_2CTX MXene nanosheets	EHL-functionalized Ti_2CTX nanosheets	**Pb (II) adsorption capacity**: 232.9 mg/g	Wang S. et al. (2020b)
Carboxymethyl cellulose (CMC) Chitosan (CS)	Montmorillonite nanosheets (MMTNS)	TiO_2@MMTNS/CMC/CS hydrogel	**MB adsorption capacity**: 283.97 mg/g **MB removal**: 97.0%	Wang W. et al. (2020c)
Carboxymethyl cellulose (CMC) Chitosan (CS)	Graphene oxide (GO) nanosheet	CS/CMC-NCH hydrogel	**MB adsorption capacity**: 99.0%/655.98 mg/g **MO adsorption capacity**: 82%/404.52 mg/g	Mittal et al. (2021)

(Continued)

TABLE 10.1 (Continued)

Natural Biopolymer(s)	Nanosheet	Composite	Application	References
Chitosan (CS)	Graphene oxide (GO) nanosheets	Graphene oxide (GO) nanosheet/ chitosan (CS)/ copper-based polyoxometalate (Cu-POM) nanocomposite	**MB removal**: 96.86% **GV removal**: 97.77% **BF removal**: 96.47% **TC removal**: 78.92% **NC removal**: 76.26% **Growth inhibition**: *E. coli, S. aureus, B. subtilis, P. aeruginosa:* ~100% Kanamycin-resistant *E. coli*: 99.93% Ampicillin-resistant *E. coli*: 97.94%	Xing et al. (2022)
Chitosan (CS)	Fe_3O_4-MoS_2 nanosheets	Fe_3O_4-MoS_2@CS	**Cr (VI) adsorption capacity-removal**: 175.4 mg/g-97.0±3% **CIP adsorption capacity-removal**: 190.7 mg/g-62.0±3.0%	Wani et al. (2023)
Chitosan(C) Lignosulfonate (L)	Ti_2CTX MXene nanosheets	Ti_3C_2Tx MXene/ chitosan/ lignosulfonate adsorbent (MCL) nanospheres	**Pb (II) removal**: 60.2% **Cr (VI) removal**: 53.0% **Cu (II) removal from real wastewater**: 42.0% to 81.0%	Othman et al. (2023)
Cellulose nanofibers (CNF) Chitosan (CS)	MOF (Metal-organic framework) nanosheets	ZIF-8@CNF/CS aerogel	**Oil adsorption capacity**: 35.99 g/g to 74.55 g/g **Oil/water separation efficiency**: 99.8%	Si et al. (2023)
Starch-graft-poly(acrylamide)	Graphene oxide (GO) nanosheets	NHA hydrogel adsorbent	**MG adsorption capacity**: 297 mg/g	Hosseinzadeh and Ramin (2018)
Starch nanocrystals (DASNCs)	Graphene oxide (GO) (reduced form, RGO) nanosheets	DASNCs-RGO hybrid aerogel	**RB adsorption capacity**: 539 mg/g **CV adsorption capacity**: 318 mg/g	Chen et al. (2019)
Sodium alginate (SA) Chitosan (CS)	Montmorillonite nanosheets (MMTNS)	Poly (vinyl alcohol)/sodium alginate/chitosan/ montmorillonite nanosheets hydrogel beads	**MB adsorption capacity**: 137.15 mg/g **MB removal**: 90.0%	Wang et al. (2018)
Sodium alginate (SA)	Ti_3C_2Tx MXene nanosheets	MX-$SA_{4:20}$	**Hg (II) adsorption capacity**: 932.84 mg/g **Hg (II), Cd (II), and Cr (III) removal**: 100% **Pb (II) and Cu (II) removal**: > 90.0% **Zn (II), Co (II), and Ni (II) removal**: >80.0% **As (III) removal**: 37.0%	Shahzad et al. (2019)

(Continued)

TABLE 10.1 (Continued)

Natural Biopolymer(s)	Nanosheet	Composite	Application	References
Sodium alginate (SA) Gelatin	Montmorillonite nanosheets (MMTNS)	3D porous hydrogel	**Pb (II) adsorption capacity**: 134 mg/g **Pb (II) adsorption capacity (in a fix bed column)**: 161.47 mg/g	Miao et al. (2022)
Silk protein-sericin	Hydroxyapatite (HAP) based	Hydroxyapatite (HAP)-based bio-hybrid flowers with nanosheet petals	**Pb (II) adsorption capacity**: 99.5% **CR adsorption capacity**: Pale color in 12h (50.0% in 4h) **Pb (II) removal**: 99.8% **Cd (II) removal**: 75.3% **Hg (II) removal**: 64.3%	Koley et al. (2016)
Silk nanofibrils (SNF)	Metallic molybdenum disulfide (MoS_2) nanosheets	MoS_2–SNF	**Au (III) adsorption capacity - rejection** 759.1 mg/g - 99.21% **Pd (II) adsorption capacity - rejection**: 425.0 mg/g – 99.85% **Hg (II) adsorption capacity - rejection**: 249.4 mg/g - 63.11% **Ag (I) adsorption capacity - rejection**: 246.2 mg/g - 53.92% **Pb (II) adsorption capacity - rejection**: 153.5 mg/g -44.41% **Cu (II) adsorption capacity - rejection**: 22.2 mg/g - 43.99% **R6G, AB, MB, CV rejection rate**: > 95.0% **Brilliant blue G (BB) and CR rejection rate**: > 84%	Zhao F. et al. (2020)
Agar	Zeolitic imidazolate framework (ZIF-8) on carbon nitride (C_3N_4) nanosheets	Hgel-50 (aerogel loaded with 50% ZIF-8/C_3N_4)	**CR adsorption capacity**: 287.35 mg/g **MB adsorption capacity**: 154.87 mg/g	Zhang et al. (2017)
Collagen	Hexagonal boron nitride nanosheets (h-BNNS)	h-BNNS/gelatin porous membranes	**Fe (II) removal**: 97.0% (in the presence of PAA)	Nafti Mateur et al. (2020)

Abbreviations: MB: Methylene Blue; **MO**: Methyl Orange; **RhB or RB**: Rhodamine B; **CR**: Congo Red, **EB**: Evans Blue; **IC**: Indigo Carmine; **GV**: Gentian Violet; **BF**: Basic Fuchsin; **TC**: Tetracycline; **NC**: Norfloxacin; **CIP**: Ciprofloxacin; **MG**: Malachite Green; **CV**: Crystal Violet; **R6G**: Rhodamine 6G, **AB**: Alcian Blue 8GX; **PAA**: poly (acrylic acid).

treatment processes. Continued R&D efforts in this area are crucial to further optimize their performance and ensure their long-term sustainability. Therefore, collaborative efforts among researchers, stakeholders, and policymakers are essential in deciphering the full potential of these composites. Using the power of nature and sustainable materials, these composites present an exciting frontier in wastewater treatment, with the goal of uncontaminated water resources, offering a greener and healthier future by saving both human health and the environment.

ACKNOWLEDGEMENT

This work was funded by the Mikkeli University Consortium (MUC), with the project number 23B350E6YT10.

REFERENCES

Agboola, O., Fayomi, O.S.I., Ayodeji, A., Ayeni, A.O., Alagbe, E.E., Sanni, S.E., Okoro, E.E., Moropeng, L., Sadiku, R., Kupolati, K.W., Oni, B.A., 2021. A review on polymer nanocomposites and their effective applications in membranes and adsorbents for water treatment and gas separation. *Membranes*. 11(2), 139. https://doi.org/10.3390/membranes11020139

Ahmed, S.F., Mofijur, M., Ahmed, B., Mehnaz, T., Mehejabin, F., Maliat, D., Hoang, A.T., Shafiullah, G.M., 2022. Nanomaterials as a sustainable choice for treating wastewater. *Environ. Res*. 214, 113807. https://doi.org/10.1016/j.envres.2022.113807

Al Harby, N.F., El-Batouti, M., Elewa, M.M., 2022. Prospects of polymeric nanocomposite membranes for water purification and scalability and their health and environmental impacts: A review. *Nanomaterials*. 12(20), 3637. https://doi.org/10.3390/nano12203637

Albuquerque, P.B.S., de Oliveira, W.F., dos Santos Silva, P.M., dos Santos Correia, M.T., Kennedy, J.F., Coelho, L.C.B.B., 2020. Epiphanies of well-known and newly discovered macromolecular carbohydrates: A review. *Int. J. Biol. Macromol*. 156, 51–66. https://doi.org/10.1016/j.ijbiomac.2020.04.046

Angeles, L.F., Mullen, R.A., Huang, I.J., Wilson, C., Khunjar, W., Sirotkin, H.I., McElroy, A.E., Aga, D.S., 2020. Assessing pharmaceutical removal and reduction in toxicity provided by advanced wastewater treatment systems. *Environ. Sci.: Water Res. Technol*. 6(1), 62–77. https://doi.org/10.1039/C9EW00559E

Anjum, M., Miandad, R., Waqas, M., Gehany, F., Barakat, M.A., 2019. Remediation of wastewater using various nano-materials. *Arabian J. Chem*. 12(8), 4897–4919. https://doi.org/10.1016/j.arabjc.2016.10.004

Apriyanto, A., Compart, J., Fettke, J., 2022. A review of starch, a unique biopolymer–Structure, metabolism and in planta modifications. *Plant Sci*. 318, 111223. https://doi.org/10.1016/j.plantsci.2022.111223

Azamat, J., Sattary, B.S., Khataee, A., Joo, S.W., 2015. Removal of a hazardous heavy metal from aqueous solution using functionalized graphene and boron nitride nanosheets: Insights from simulations. *J. Mol. Graphics Modell*. 61, 13–20. https://doi.org/10.1016/j.jmgm.2015.06.012

Bhat, S.A., Sher, F., Hameed, M., Bashir, O., Kumar, R., Vo, D.V.N., Ahmad, P., Lima, E.C., 2022. Sustainable nanotechnology based wastewater treatment strategies: Achievements, challenges and future perspectives. *Chemosphere*. 288, 132606. https://doi.org/10.1016/j.chemosphere.2021.132606

Bruder, V., Ludwig, T., Opitz, S., Christoffels, R., Fischer, T., Maleki, H., 2021. Hierarchical assembly of surface modified silk fibroin biomass into micro-, and milli-metric hybrid aerogels with core-shell, janus, and composite configurations for rapid removal of water pollutants. *Adv. Mater. Interfaces*. 8(5), 2001892. https://doi.org/10.1002/admi.202001892

Chen, Y., Dai, G., Gao, Q., 2019. Starch nanoparticles–graphene aerogels with high supercapacitor performance and efficient adsorption. *ACS Sustainable Chem. Eng*. 7(16), 14064–14073. https://doi.org/10.1021/acssuschemeng.9b02594

Chen, Y., Yang, G., Liu, B., Kong, H., Xiong, Z., Guo, L., Wei, G., 2022a. Biomineralization of ZrO2 nanoparticles on graphene oxide-supported peptide/cellulose binary nanofibrous membranes for high-performance removal of fluoride ions. *Chem. Eng. J*. 430, 132721. https://doi.org/10.1016/j.cej.2021.132721

Chen, Z., Pan, Y., Cai, P., 2022b. Sugarcane cellulose-based composite hydrogel enhanced by g-C_3N_4 nanosheet for selective removal of organic dyes from water. *Int. J. Biol. Macromol*. 205, 37–48. https://doi.org/10.1016/j.ijbiomac.2022.02.035

Cheng, L., Liu, G., Zhao, J., Jin, W., 2021. Two-dimensional-material membranes: manipulating the transport pathway for molecular separation. *Acc. Mater. Res*. 2(2), 114–128. https://doi.org/10.1021/accountsmr.0c00092

Ching, T.W., Haritos, V., Tanksale, A., 2018. Ultrasound-assisted conversion of cellulose into hydrogel and functional carbon material. *Cellulose*. 25, 2629–2645. https://doi.org/10.1007/s10570-018-1746-y

Choi, W.S., Lee, H.J., 2022. Nanostructured materials for water purification: Adsorption of heavy metal ions and organic dyes. *Polymers*. 14(11), 2183. https://doi.org/10.3390/polym14112183

Choudhary, R., Pandey, O.P., Brar, L.K., 2022. Soy flour–based biochars for enhanced removal of anionic and cationic dyes. *Biomass Convers. Biorefin*., 1–17. https://doi.org/10.1007/s13399-022-03371-7

Dong, Y.D., Zhang, H., Zhong, G.J., Yao, G., Lai, B., 2021. Cellulose/carbon composites and their applications in water treatment: A review. *Chem. Eng. J.* 405, 126980. https://doi.org/10.1016/j.cej.2020.126980

Donner, M.W., Arshad, M., Ullah, A., Siddique, T., 2019. Unravelled keratin-derived biopolymers as novel biosorbents for the simultaneous removal of multiple trace metals from industrial wastewater. *Sci. Total Environ*. 647, 1539–1546. https://doi.org/10.1016/j.scitotenv.2018.08.085

Du, Q., Wei, H., Li, A., Yang, H., 2017. Evaluation of the starch-based flocculants on flocculation of hairwork wastewater. *Sci. Total Environ*. 601–602, 1628–1637. https://doi.org/10.1016/j.scitotenv.2017.06.029

Duman, O., Polat, T.G., Diker, C.Ö., Tunç, S., 2020. Agar/κ-carrageenan composite hydrogel adsorbent for the removal of Methylene Blue from water. *Int. J. Biol. Macromol*. 160, 823–835. https://doi.org/10.1016/j.ijbiomac.2020.05.191

El-Aswar, E.I., Ramadan, H., Elkik, H., Taha, A.G., 2022. A comprehensive review on preparation, functionalization and recent applications of nanofiber membranes in wastewater treatment. *J. Environ. Manage*. 301, 113908. https://doi.org/10.1016/j.jenvman.2021.113908

Fang, K., Deng, L., Yin, J., Yang, T., Li, J., He, W., 2022. Recent advances in starch-based magnetic adsorbents for the removal of contaminants from wastewater: A review. *Int. J. Biol. Macromol*. 218, 909–929. https://doi.org/10.1016/j.ijbiomac.2022.07.175

Gao, H., Wang, Y., Afolabi, M.A., Xiao, D., Chen, Y., 2021. Incorporation of cellulose nanocrystals into graphene oxide membranes for efficient antibiotic removal at high nutrient recovery. *ACS Appl. Mater. Interfaces*. 13(12), 14102–14111. https://doi.org/10.1021/acsami.0c20652

Gore, P.M., Naebe, M., Wang, X., Kandasubramanian, B., 2019. Progress in silk materials for integrated water treatments: Fabrication, modification and applications. *Chem. Eng. J.* 374, 437–470. https://doi.org/10.1016/j.cej.2019.05.163

Gupta, S., Nighojkar, A., Mayilswamy, N., Kandasubramanian, B., 2023. Recent trends in the application of silk-based composites for remediation of toxic contaminants from wastewater. *J. Polym. Environ*. 31, 2243–2272. https://doi.org/10.1007/s10924-022-02741-6

Hamidon, T.S., Adnan, R., Haafiz, M.M., Hussin, M.H., 2022. Cellulose-based beads for the adsorptive removal of wastewater effluents: a review. *Environ. Chem. Lett*. 20(3), 1965–2017. https://doi.org/10.1007/s10311-022-01401-4

Hasan, S., Boddu, V.M., Viswanath, D.S., Ghosh, T.K. 2022. *Chitin and Chitosan: Science and Engineering*, Springer International Publishing AG, Switzerland.

Hernández-Francisco, E., Bonilla-Cruz, J., Márquez-Lamas, U., Suárez-Jacobo, Á., Longoria-Rodríguez, F., Rivera-Haro, J., Russell, P., Ali, Z., Yin, C.Y., Lara-Ceniceros, T.E., 2020. Entangled cellulose nanofibrils/nanosheets derived from native mexican agave for lead (II) ion removal. *Cellulose*. 27, 8785–8798. https://doi.org/10.1007/s10570-020-03373-6

Hosseinzadeh, H., Ramin, S., 2018. Fabrication of starch-graft-poly (acrylamide)/graphene oxide/hydroxyapatite nanocomposite hydrogel adsorbent for removal of malachite green dye from aqueous solution. *Int. J. Biol. Macromol*. 106, 101–115. https://doi.org/10.1016/j.ijbiomac.2017.07.182

Huang, Y., Wu, H., Shao, T., Zhao, X., Peng, H., Gong, Y., Wan, H., 2018. Enhanced copper adsorption by DTPA-chitosan/alginate composite beads: Mechanism and application in simulated electroplating wastewater. *Chem. Eng. J.* 339, 322–333. https://doi.org/10.1016/j.cej.2018.01.071

Hube, S., Eskafi, M., Hrafnkelsdóttir, K.F., Bjarnadóttir, B., Bjarnadóttir, M.Á., Axelsdóttir, S., Wu, B., 2020. Direct membrane filtration for wastewater treatment and resource recovery: A review. *Sci. Total Environ*. 710, 136375. https://doi.org/10.1016/j.scitotenv.2019.136375

Jiang, J.L., Zhang, W.Z., Ni, W.X., Shao, J.W., 2021. Insight on structure-property relationships of carrageenan from marine red algal: A review. *Carbohydr. Polym*. 257, 117642. https://doi.org/10.1016/j.carbpol.2021.117642

Jiang, T., Duan, Q., Zhu, J., Liu, H., Yu, L., 2020. Starch-based biodegradable materials: Challenges and opportunities. *Adv. Ind. Eng. Polym. Res*. 3(1), 8–18. https://doi.org/10.1016/j.aiepr.2019.11.003

Kabir, S.F., Rahman, A., Yeasmin, F., Sultana, S., Masud, R.A., Kanak, N.A., Haque, P. 2022. Occurrence, distribution, and structure of natural polysaccharides, in: M. Naeem, T. Aftab, M. M. A. Khan. (Eds.), *Radiation-Processed Polysaccharides*. Academic Press, pp. 1–27.

Khalil, H.P.S., Tye, Y.Y., Saurabh, C.K., Leh, C.P., Lai, T.K., Chong, E.W.N., Fazita, M.R., Hafiidz, J.M., Banerjee, A., Syakir, M.I., 2017. Biodegradable polymer films from seaweed polysaccharides: A review on cellulose as a reinforcement material. *eXPRESS Polym. Lett.* 11(4), 244–265. https://doi.org/10.3144/expresspolymlett.2017.26

Koley, P., Sakurai, M., Takei, T., Aono, M., 2016. Facile fabrication of silk protein sericin-mediated hierarchical hydroxyapatite-based bio-hybrid architectures: excellent adsorption of toxic heavy metals and hazardous dye from wastewater. *RSC Adv.* 6(89), 86607–86616. https://doi.org/10.1039/C6RA12818A

Kongseng, P., Amornpitoksuk, P., Chantarak, S., 2023. ZnO/Cassava starch-based hydrogel composite for effective treatment of dye-contaminated wastewater. *Macromol. Mater. Eng.* 308(2), 2200481. https://doi.org/10.1002/mame.202200481

Krivdin, L.B., 2021. Computational NMR of carbohydrates: Theoretical background, applications, and perspectives. *Molecules.* 26(9), 2450. https://doi.org/10.3390/molecules26092450

Kwak, H.W., Shin, M., Yun, H., Lee, K.H., 2016. Preparation of silk sericin/lignin blend beads for the removal of hexavalent chromium ions. *Int. J. Mol. Sci.* 17(9), 1466. https://doi.org/10.3390/ijms17091466

Lapwanit, S., Sooksimuang, T., Trakulsujaritchok, T., 2018. Adsorptive removal of cationic methylene blue dye by kappa-carrageenan/poly (glycidyl methacrylate) hydrogel beads: preparation and characterization. *J. Environ. Chem. Eng.* 6(5), 6221–6230. https://doi.org/10.1016/j.jece.2018.09.050

Liu, J., Su, D., Yao, J., Huang, Y., Shao, Z., Chen, X., 2017. Soy protein-based polyethylenimine hydrogel and its high selectivity for copper ion removal in wastewater treatment. *J. Mater. Chem. A.* 5(8), 4163–4171. https://doi.org/10.1039/C6TA10814H

Ma, Q., Wang, W., Ge, W., Xia, L., Li, H., Song, S., 2021a. Preparation of carboxymethyl cellulose-based hydrogel supported by two-dimensional montmorillonite nanosheets for methylene blue removal. *J. Polym. Environ.* 29(12), 3918–3931. https://doi.org/10.1007/s10924-021-02166-7

Ma, Z., Zhou, P., Zhang, L., Zhong, Y., Sui, X., Wang, B., Ma, Y., Feng, X., Xu, H., Mao, Z., 2021b. A recyclable 3D gC_3N_4 based nanocellulose aerogel composite for photodegradation of organic pollutants. *Cellulose.* 28, 3531–3547. https://doi.org/10.1007/s10570-021-03748-3

Mayilswamy, N., Kandasubramanian, B., 2022. Green composites prepared from soy protein, polylactic acid (PLA), starch, cellulose, chitin: A review. *Emergent Mater.* 5(3), 727–753. https://doi.org/10.1007/s42247-022-00354-2

Meena, R.A.A., Kannah, R.Y., Sindhu, J., Ragavi, J., Kumar, G., Gunasekaran, M., Banu, J.R., 2019. Trends and resource recovery in biological wastewater treatment system. *Bioresour. Technol. Rep.* 7, 100235. https://doi.org/10.1016/j.biteb.2019.100235

Miao, Y., Peng, W., Wang, W., Cao, Y., Li, H., Chang, L., Huang, Y., Fan, G., Yi, H., Zhao, Y., Zhang, T., 2022. 3D-printed montmorillonite nanosheets based hydrogel with biocompatible polymers as excellent adsorbent for Pb (II) removal. *Sep. Purif. Technol.* 283, 120176. https://doi.org/10.1016/j.seppur.2021.120176

Mittal, H., Al Alili, A., Morajkar, P.P., Alhassan, S.M., 2021. GO crosslinked hydrogel nanocomposites of chitosan/carboxymethyl cellulose: A versatile adsorbent for the treatment of dyes contaminated wastewater. *Int. J. Biol. Macromol.* 167, 1248–1261. https://doi.org/10.1016/j.ijbiomac.2020.11.079

Na, Y., Lee, J., Lee, S.H., Kumar, P., Kim, J.H., Patel, R., 2020. Removal of heavy metals by polysaccharide: A review. *Polym.-Plast. Technol. Mater.* 59(16), 1770–1790. https://doi.org/10.1080/25740881.2020.1768545

Nafti Mateur, M., Gonzalez Ortiz, D., Jellouli Ennigrou, D., Horchani-Naifer, K., Bechelany, M., Miele, P., Pochat-Bohatier, C., 2020. Porous gelatin membranes obtained from pickering emulsions stabilized with h-BNNS: Application for polyelectrolyte-enhanced ultrafiltration. *Membranes.* 10(7), 144. https://doi.org/10.3390/membranes10070144

Naim, M.M., Al-Harby, N.F., El Batouti, M., Elewa, M.M., 2022. Macro-reticular ion exchange resins for recovery of direct dyes from spent dyeing and soaping liquors. *Molecules.* 27(5), 1593. https://doi.org/10.3390/molecules27051593

Nakagawa, K., Kunimatsu, M., Yasui, K., Yoshioka, T., Shintani, T., Yasui, T., Kamio, E., Hung, W.S., Lee, K.R., Edman Tsang, S.C., Matsuyama, H., 2021. HNb3O8 nanosheet–graphene oxide composite membranes for molecular separation. *ACS Appl. Nano Mater.* 4(4), 3455–3466. https://doi.org/10.1021/acsanm.0c03348

Othman, Z., Mackey, H.R., Mahmoud, K.A., 2023. MXene/chitosan/lignosulfonate (MCL) nanocomposite for simultaneous removal of Co (II), Cr (VI), Cu (II), Ni (II) and Pb (II) heavy metals from wastewater. *2D Mater.* 10(2), 024004. https://doi.org/10.1088/2053-1583/acbd62

Peng, B., Yao, Z., Wang, X., Crombeen, M., Sweeney, D.G., Tam, K.C., 2020. Cellulose-based materials in wastewater treatment of petroleum industry. *Green Energy Environ.* 5, 37–49. https://doi.org/10.1016/j.gee.2019.09.003

Peramune, D., Manatunga, D.C., Dassanayake, R.S., Premalal, V., Liyanage, R.N., Gunathilake, C., Abidi, N., 2022. Recent advances in biopolymer-based advanced oxidation processes for dye removal applications: A review. *Environ. Res.* 215, 114242. https://doi.org/10.1016/j.envres.2022.114242

Qadir, M., Drechsel, P., Jiménez Cisneros, B., Kim, Y., Pramanik, A., Mehta, P., Olaniyan, O., 2020, February. Global and regional potential of wastewater as a water, nutrient and energy source. *Nat. Resour. Forum.* 44, 40. https://doi.org/10.1111/1477-8947.12187

Rajasulochana, P., Preethy, V., 2016. Comparison on efficiency of various techniques in treatment of waste and sewage water–A comprehensive review. *Resour.-Effic. Technol.* 2(4), 175–184. https://doi.org/10.1016/j.reffit.2016.09.004

Rajeswari, A., Christy, E. J. S., Pius, A. 2021. Biopolymer blends and composites: Processing technologies and their properties for industrial applications, in: Thomas, S., Gopi, S., Amalraj, A. (Eds.), *Biopolymers and Their Industrial Applications*. Elsevier, pp. 105–147.

Rana, A., Hasan, I., Koo, B.H., Khan, R.A., 2022. Green synthesized CeO2 nanowires immobilized with alginate-ascorbic acid biopolymer for advance oxidative degradation of crystal violet. *Colloids Surf. A: Physicochem. Eng.* 637, 128225. https://doi.org/10.1016/j.colsurfa.2021.128225

Rout, P.R., Zhang, T.C., Bhunia, P., Surampalli, R.Y., 2021. Treatment technologies for emerging contaminants in wastewater treatment plants: A review. *Sci. Total Environ.* 753, 141990. https://doi.org/10.1016/j.scitotenv.2020.141990

Runkle, B.R., Suvočarev, K., Reba, M.L., Reavis, C.W., Smith, S.F., Chiu, Y.L., Fong, B., 2018. Methane emission reductions from the alternate wetting and drying of rice fields detected using the eddy covariance method. *Environ. Sci. Technol.* 53(2), 671. https://doi.org/10.1021/acs.est.8b05535

Russo, T., Fucile, P., Giacometti, R., Sannino, F., 2021. Sustainable removal of contaminants by biopolymers: A novel approach for wastewater treatment. *Processes.* 9(4), 719. https://doi.org/10.3390/pr9040719

Saha, S., Zubair, M., Khosa, M.A., Song, S., Ullah, A., 2019. Keratin and chitosan biosorbents for wastewater treatment: A review. *J. Polym. Environ.* 27, 1389–1403. https://doi.org/10.1007/s10924-019-01439-6

Sahu, A., Dosi, R., Kwiatkowski, C., Schmal, S., Poler, J.C., 2023. Advanced polymeric nanocomposite membranes for water and wastewater treatment: A comprehensive review. *Polymers.* 15(3), 540. https://doi.org/10.3390/polym15030540

Salama, A., Abouzeid, R., Leong, W.S., Jeevanandam, J., Samyn, P., Dufresne, A., Bechelany, M., Barhoum, A., 2021. Nanocellulose-based materials for water treatment: Adsorption, photocatalytic degradation, disinfection, antifouling, and nanofiltration. *Nanomaterials.* 11(11), 3008. https://doi.org/10.3390/nano11113008

Saraswat, S. 2022. Application of nano-biotechnology in wastewater treatment: An overview, in: Rai, J. P. N., Saraswat, S. (Eds.), *Nano-Biotechnology for Waste Water Treatment: Theory and Practices*, Water Science and Technology Library, Springer, Cham., pp. 41–75.

Sarode, S., Upadhyay, P., Khosa, M.A., Mak, T., Shakir, A., Song, S., Ullah, A., 2019. Overview of wastewater treatment methods with special focus on biopolymer chitin-chitosan. *Int. J. Biol. Macromol.* 121, 1086–1100. https://doi.org/10.1016/j.ijbiomac.2018.10.089

Seidi, F., Yazdi, M.K., Jouyandeh, M., Habibzadeh, S., Munir, M.T., Vahabi, H., Bagheri, B., Rabiee, N., Zarrintaj, P., Saeb, M.R., 2022. Crystalline polysaccharides: A review. *Carbohydr. Polym.* 275, 118624. https://doi.org/10.1016/j.carbpol.2021.118624

Shahzad, A., Nawaz, M., Moztahida, M., Jang, J., Tahir, K., Kim, J., Lim, Y., Vassiliadis, V.S., Woo, S.H., Lee, D.S., 2019. Ti_3C_2Tx MXene core-shell spheres for ultrahigh removal of mercuric ions. *Chem. Eng. J.* 368, 400–408. https://doi.org/10.1016/j.cej.2019.02.160

Shaikh, V.A., Shaikh, I.V. 2022. Polysaccharides: Sustainable green-composite biomaterials, in: Baskar, C., Ramakrishna, S., La Rosa, A. D. (Eds.), *Encyclopedia of Green Materials*. Springer Nature, Singapore, pp. 1–11.

Shi, Q., Zhang, N., Wang, D., Zhang, J., Li, Y., Wang, Z., 2023. Polyethylenimine-mica nanosheets/cellulose nanofibers stacked loose nanofiltration membrane for effective dye/salt separation. *Desalination.* 551, 116410. https://doi.org/10.1016/j.desal.2023.116410

Si, R., Luo, H., Zhang, T., Pu, J., 2023. High hydrophobic ZIF-8@ cellulose nanofibers/chitosan double network aerogel for oil adsorbent and oil/water separation. *Int. J. Biol. Macromol.* 238, 124008. https://doi.org/10.1016/j.ijbiomac.2023.124008

Sohouli, E., Irannejad, N., Ziarati, A., Ehrlich, H., Rahimi-Nasrabadi, M., Ahmadi, F., Luque, R., 2022. Application of polysaccharide-based biopolymers as supports in photocatalytic treatment of water and wastewater: a review. *Environ. Chem. Lett.* 20(6), 3789–3809. https://doi.org/10.1007/s10311-022-01456-3

Stephen, D.P., Palanisamy, S.B., 2021. Advances in biopolymer composites and biomaterials for the removal of emerging contaminants. *Phys. Sci. Rev.* 8(8), 1789–1809. https://doi.org/10.1515/psr-2021-0056

Sudhaik, A., Raizada, P., Ahamad, T., Alshehri, S.M., Nguyen, V.H., Van Le, Q., Thakur, S., Thakur, V.K., Selvasembian, R., Singh, P., 2023. Recent advances in cellulose supported photocatalysis for pollutant mitigation: A review. *Int. J. Biol. Macromol.* 226, 1284–1308. https://doi.org/10.1016/j.ijbiomac.2022.11.241

Tamaddon, F., Mosslemin, M.H., Asadipour, A., Gharaghani, M.A., Nasiri, A., 2020. Microwave-assisted preparation of $ZnFe_2O_4$@ methyl cellulose as a new nano-biomagnetic photocatalyst for photodegradation of metronidazole. *Int. J. Biol. Macromol.* 154, 1036–1049. https://doi.org/10.1016/j.ijbiomac.2020.03.069

Thakur, S., 2021. An overview on alginate based bio-composite materials for wastewater remedial. *Mater. Today: Proc.* 37, 3305–3309. https://doi.org/10.1016/j.matpr.2020.09.120

Tian, H., Li, X., Yan, J., Yu, X., Liu, H., Li, J., Guo, G., 2022. Natural protein-based biocomposites with superior paramagnetism and microwave absorbing properties by incorporating nanosized Fe_2O_3. *Macromol. Mater. Eng.* 307(3), 2100748. https://doi.org/10.1002/mame.202100748

Udayakumar, G.P., Muthusamy, S., Selvaganesh, B., Sivarajasekar, N., Rambabu, K., Sivamani, S., Sivakumar, N., Maran, J.P., Hosseini-Bandegharaei, A., 2021. Ecofriendly biopolymers and composites: Preparation and their applications in water-treatment. *Biotechnol. Adv.* 52, 107815. https://doi.org/10.1016/j.biotechadv.2021.107815

Ullah, A., Hussain, S., Wasim, A., Jahanzaib, M., 2020. Development of a decision support system for the selection of wastewater treatment technologies. *Sci. Total Environ.* 731, 139158. https://doi.org/10.1016/j.scitotenv.2020.139158

Vigneshwaran, S., Karthikeyan, P., Park, C.M., Meenakshi, S., 2020. Boosted insights of novel accordion-like (2D/2D) hybrid photocatalyst for the removal of cationic dyes: Mechanistic and degradation pathways. *J. Environ. Manage.* 273, 111125. https://doi.org/10.1016/j.jenvman.2020.111125

Wang, B., Yang, W., McKittrick, J., Meyers, M.A., 2016. Keratin: Structure, mechanical properties, occurrence in biological organisms, and efforts at bioinspiration. *Prog. Mater. Sci.* 76, 229–318. https://doi.org/10.1016/j.pmatsci.2015.06.001

Wang, H., Li, J., Ding, N., Zeng, X., Tang, X., Sun, Y., Lei, T., Lin, L., 2020a. Eco-friendly polymer nanocomposite hydrogel enhanced by cellulose nanocrystal and graphitic-like carbon nitride nanosheet. *Chem. Eng. J.* 386, 124021. https://doi.org/10.1016/j.cej.2020.124021

Wang, S., Liu, Y., Lü, Q.F., Zhuang, H., 2020b. Facile preparation of biosurfactant-functionalized Ti_2CTX MXene nanosheets with an enhanced adsorption performance for Pb (II) ions. *J. Mol. Liq.* 297, 111810. https://doi.org/10.1016/j.molliq.2019.111810

Wang, W., Ni, J., Chen, L., Ai, Z., Zhao, Y., Song, S., 2020c. Synthesis of carboxymethyl cellulose-chitosan-montmorillonite nanosheets composite hydrogel for dye effluent remediation. *Int. J. Biol. Macromol.* 165, 1–10. https://doi.org/10.1016/j.ijbiomac.2020.09.154

Wang, W., Zhao, Y., Bai, H., Zhang, T., Ibarra-Galvan, V., Song, S., 2018. Methylene blue removal from water using the hydrogel beads of poly (vinyl alcohol)-sodium alginate-chitosan-montmorillonite. *Carbohydr. Polym.* 198, 518–528. https://doi.org/10.1016/j.carbpol.2018.06.124

Wang, Z., Wu, S., Zhang, Y., Miao, L., Zhang, Y., Wu, A., 2020d. Preparation of modified sodium alginate aerogel and its application in removing lead and cadmium ions in wastewater. *Int. J. Biol. Macromol.* 157, 687–694. https://doi.org/10.1016/j.ijbiomac.2019.11.228

Wani, A.A., Khan, A.M., Manea, Y.K., Singh, M., 2023. Facile synthesis of layered superparamagnetic Fe_3O_4-MoS_2 nanosheets on chitosan for efficient removal of chromium and ciprofloxacin from aqueous solutions. *J. Water Process Eng.* 51, 103340. https://doi.org/10.1016/j.jwpe.2022.103340

Xing, C., Chang, J., Ma, M., Ma, P., Sun, L., Li, M., 2022. Ultrahigh-efficiency antibacterial and adsorption performance induced by copper-substituted polyoxomolybdate-decorated graphene oxide nanocomposites. *J. Colloid Interface Sci.* 612, 664–678. https://doi.org/10.1016/j.jcis.2021.12.175

Yeamin, M.B., Islam, M.M., Chowdhury, A.N., Awual, M.R., 2021. Efficient encapsulation of toxic dyes from wastewater using several biodegradable natural polymers and their composites. *J. Cleaner Prod.* 291, 125920. https://doi.org/10.1016/j.jclepro.2021.125920

Zagklis, D.P., Bampos, G., 2022. Tertiary wastewater treatment technologies: A review of technical, economic, and life cycle aspects. *Processes.* 10(11), 2304. https://doi.org/10.3390/pr10112304

Zahara, I., Arshad, M., Naeth, M.A., Siddique, T., Ullah, A., 2021. Feather keratin derived sorbents for the treatment of wastewater produced during energy generation processes. *Chemosphere.* 273, 128545. https://doi.org/10.1016/j.chemosphere.2020.128545

Zeng, M., Chen, M., Huang, D., Lei, S., Zhang, X., Wang, L., Cheng, Z., 2021. Engineered two-dimensional nanomaterials: An emerging paradigm for water purification and monitoring. *Mater. Horiz.* 8(3), 758. https://doi.org/10.1039/D0MH01358G

Zhang, L., Tang, S., He, F., Liu, Y., Mao, W., Guan, Y., 2019. Highly efficient and selective capture of heavy metals by poly (acrylic acid) grafted chitosan and biochar composite for wastewater treatment. *Chem. Eng. J.* 378, 122215. https://doi.org/10.1016/j.cej.2019.122215

Zhang, W., Shi, S., Zhu, W., Huang, L., Yang, C., Li, S., Liu, X., Wang, R., Hu, N., Suo, Y., Li, Z., 2017. Agar aerogel containing small-sized zeolitic imidazolate framework loaded carbon nitride: A solar-triggered regenerable decontaminant for convenient and enhanced water purification. *ACS Sustainable Chem. Eng.* 5(10), 9347–9354. https://doi.org/10.1021/acssuschemeng.7b02376

Zhang, X., Li, F., Zhao, X., Cao, J., Liu, S., Zhang, Y., Yuan, Z., Huang, X., De Hoop, C.F., Peng, X., Huang, X., 2023. Bamboo nanocellulose/montmorillonite nanosheets/polyethyleneimine gel adsorbent for methylene blue and Cu (II) removal from aqueous solutions. *Gels*. 9(1), 40. https://doi.org/10.3390/gels9010040

Zhao, F., Peydayesh, M., Ying, Y., Mezzenga, R., Ping, J., 2020. Transition metal dichalcogenide–silk nanofibril membrane for one-step water purification and precious metal recovery. *ACS Appl. Mater. Interfaces*. 12(21), 24521–24530. https://doi.org/10.1021/acsami.0c07846

Zhou, L., Zhou, H., Yang, X., 2019. Preparation and performance of a novel starch-based inorganic/organic composite coagulant for textile wastewater treatment. *Purif. Technol.* 210, 93–99. https://doi.org/10.1016/j.seppur.2018.07.089

Zhou, Q., Sun, H., Jia, L., Wu, W., Wang, J., 2022. Simultaneous biological removal of nitrogen and phosphorus from secondary effluent of wastewater treatment plants by advanced treatment: A review. *Chemosphere*. 296, 134054. https://doi.org/10.1016/j.chemosphere.2022.134054

Zubair, M., Ullah, A., 2021. Biopolymers in environmental applications: Industrial wastewater treatment, in: Thomas, S., Gopi, S., Amalraj, A. (Eds.), *Biopolymers and Their Industrial Applications*. Elsevier, pp. 331–349.

11 Biocomposite Nanomaterials for Sustainable Environment

Rajesh Singh, Rahul Maurya, Ashwini Kumar Mishra, Bhumika Kumar, Huma Shafi, Vijay Kumar, and Chandrashekhar Y Jagtap

11.1 INTRODUCTION

Our Common Future, a report by the United Nations World Commission on Environment and Development (2015), coined the term 'sustainable development'. According to the report, sustainable development was defined as "A development that meets the needs of the present without compromising the ability of future generations to meet their own needs." It is also known as the Brundtland Report. Harmony between the demographic developments and the productive potential of the ecosystem is the key to achieving sustainable development. Overexploitation of resources by the present population will compromise the essential needs of the future. Thus, technological advancements may solve some problems (Vilaplana, Strömberg, and Karlsson 2010; Borowy 2013). Various countries across the globe have implemented several policies and goals to promote sustainable development at the international level. A report by the Economic and Social Council of the United Nations' Commission on Sustainable Development highlighted the policy options and measures that could be taken at the international level to support sustainable development, which include: encouraging energy efficiency; improving access to energy; promoting the development, use and transfer of advanced and cleaner energy technologies-- especially renewable energy and advanced energy technologies, which are crucial for improving access to modern energy services; improving energy efficiency to help with industrial growth initiatives and reduce air pollution and its effects on the local, regional, and global atmosphere; remove institutional and market-related obstacles to improve energy efficiency and ways to improve and transfer energy-efficient technologies; improving the efficiency with which energy is produced and used and to significantly reduce atmospheric emissions; increasing industrial efficiency, renewable energy and cleaner fossil fuel technologies; and other topics such as health and access to public services, including education for children and women (Kaasa 2007). The United Nations has also set up 17 Sustainable Development Goals with 169 associated targets for the 2030 Agenda for Sustainable Development, which include safe drinking water and sanitation, improved hygiene; sufficient, secure, affordable, and nutritious food for all; resilient, safe, and sustainable human habitats; reliable and sustainable energy, universal access to affordable; and the utilization of all-natural resources and patterns of production and consumption. The agenda is implemented for today and future generations (Cf 2015; United Nation 2015)

In recent decades, there has been a growing interest among academics and businesspeople in adopting greener and more sustainable practices. Exploiting natural resources has garnered significant attention to mitigate the enormous use of fossil fuels linked to reserve exhaustion and environmental issues. Utilizing sustainable raw materials has emerged as one of the core objectives of Green Chemistry (Silva et al. 2021; Zhang et al. 2016).

The manipulation of matter at the atomic or molecular level and using materials and structures with nanoscale dimensions, typically between 1 and 100 nm, are called nanotechnology. Owing to their minuscule size, NPs exhibit distinct physical and chemical characteristics, including high interfacial reactivity or large surface area-to-volume ratios. These characteristics offer hope for

DOI: 10.1201/9781003484189-11

innovative and enhanced environmental technology (Zhang and Fang 2010). The biological synthesis via nanobiotechnology processes can significantly boost NP production without using toxic, harsh, and expensive chemicals commonly used in conventional physical and chemical processes (Shah et al. 2015). High-performing nano-range bio-composites offer a space for creatively and sustainably producing innovative functional materials. By integrating nanotechnology and biotechnology, researchers have been able to harness the unique properties of BNMs for various environmental applications. The use of biocomposite in synthesizing and assembling NPs has emerged as a clean, nontoxic, and environmentally friendly approach. Recently, there has been an increasing focus on developing sustainable solutions for various environmental challenges. One area of research that has gained significant attention is the development of BNMs (Bashir and Liu 2015). BNMs refer to materials typically composed of a combination of at least one organic (biological) and one inorganic (NM) component at the nanoscale. These materials offer a promising avenue for addressing environmental issues, such as pollution, energy efficiency, and waste management. BNMs have several advantages in terms of sustainability and environmental protection. They are derived from renewable and recycled resources, reducing the reliance on non-renewable raw materials and minimizing waste. Furthermore, the synthesis processes for BNMs are benign and energy-effective, minimizing the carbon footprint and reducing overall environmental impact. Compared to traditional synthetic composites, "green biocomposites" or "Eco composites" have less adverse ecological effects and may eventually replace them. The two primary benefits of green biocomposites over synthetic ones are their biodegradability and biocompatibility; they have low density and hardness and are less irritating to the skin and respiratory system. Additionally, biocomposites benefit the global economy by creating goods that use less energy and offer a higher quality of life (Satyanarayana, Arizaga, and Wypych 2009; Rawat and Singh 2022). The future integration of biorefineries and green chemistry will play a crucial role in ensuring the availability of a wide range of raw materials to prepare sustainable BNMs. Overall, this chapter of the book will provide insights into the potential of BNMs for sustainable environmental solutions.

11.2 CONSIDERATIONS FOR THE DESIGN OF BNMS

The biocomposite materials are typically classified into polymer matrix biocomposites, ceramic matrix biocomposites, and metal matrix biocomposites. Polymer matrix biocomposites, where natural or synthetic polymers serve as the matrix material, are often reinforced with natural fibers, NPs, or other additives. Ceramic matrix biocomposites, consisting of a ceramic matrix reinforced with biodegradable fibers or particles, are commonly used in bone tissue engineering and dental applications. Metal matrix biocomposites are reinforced with biodegradable materials to enhance mechanical properties and biocompatibility (Ali et al. 2021).

The design of BNMs encompasses diverse strategies aimed at tailoring their properties and functionalities for biomedical applications. Bottom-up assembly methods leverage molecular self-assembly or synthesis techniques to precisely engineer NMs from their constituent components, allowing for fine control over structure and composition. Conversely, top-down fabrication techniques manipulate bulk materials or larger structures to create nanoscale features, enabling the customization of BNMs with specific morphologies and dimensions. Smith et al. provide insight into using standard and innovative top-down techniques in conjunction with peptide or protein self-assembly, paving the way for developing a new generation of functional materials with tailored properties and applications in biomedicine (Smith et al. 2011). Hybridization combines multiple material types to create synergistic effects, enhancing mechanical strength, conductivity, or biological activity (Mohammadi et al. 2017). Surface functionalization techniques modify the surface properties of NMs to improve biocompatibility, enable targeted interactions, or impart stimuli-responsive behavior. Template-assisted synthesis utilizes templates or

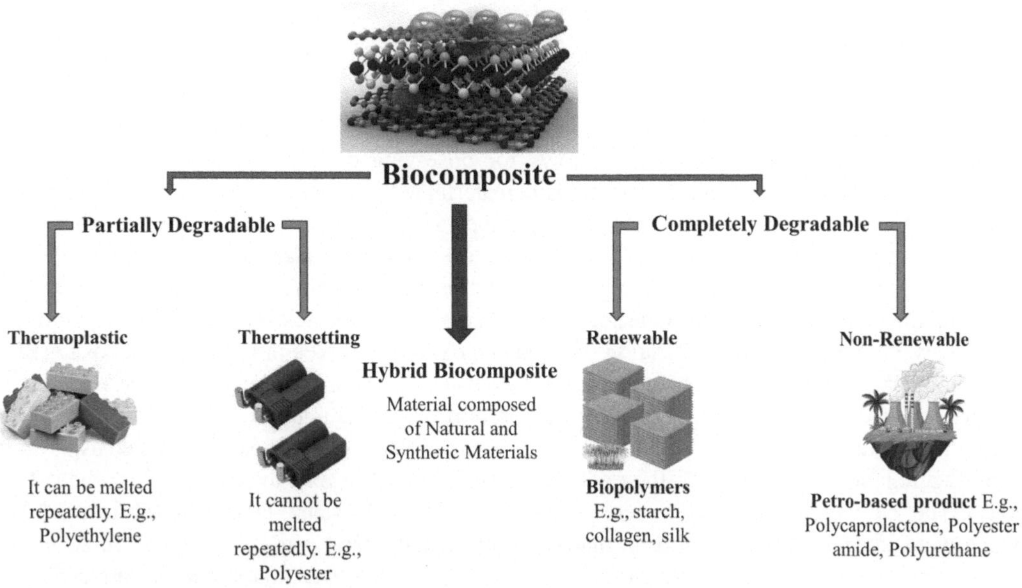

FIGURE 11.1 Classification of the bio-composite materials.

scaffolds to guide NM growth, offering precise control over nanostructure formation. Green synthesis methods emphasize sustainability by employing eco-friendly approaches, such as biological agents or environmentally benign chemicals (Wu et al. 2015; Mehta, Mahajan, and Hivrale 2021).

Both organic and inorganic water pollutants can be successfully removed using nanocomposites made from carbon, metal oxides, or polymeric materials, which are created using green synthesis techniques. Their superior catalytic and adsorption capacities aid in mineralizing hazardous organic compounds. Researchers have thus far employed several bio-composites that were created using diverse techniques in various water treatment procedures. This chapter discusses the use of green bio-composites in conjunction with various remediation strategies to eliminate water contaminants (Rawat and Singh 2022). Computational design and modeling techniques aid in predicting material properties and optimizing design parameters, facilitating the rational design of BNMs tailored to specific biomedical needs (Dai and Chen 2023). Collectively, these approaches empower researchers to develop versatile and effective NMs for applications ranging from drug delivery and tissue engineering to diagnostics and biosensing. Bio-composite materials can be classified based on their degradability (see Figure 11.1). Completely biodegradable biocomposites are made with both biodegradable fibers and a biodegradable matrix, allowing the entire material to decompose naturally. These biocomposites often use bio-based polymers like soy or starch plastics. Partially biodegradable options use a combination of materials. The biofibers might degrade naturally, but the matrix, often a synthetic resin, might not. This issue creates a situation where the remaining matrix needs to be separated from the degraded fibers for proper disposal.

11.3 CHARACTERIZATION OF BNMs

Characterizing BNMs is crucial in understanding their structure, composition, and performance. BNMs can be described using conventional and sophisticated methods viz Nuclear Magnetic Resonance (NMR), Atomic Force Microscopy (AFM), Fourier Transform Infrared Spectroscopy (FTIR), etc., for the characterization of properties like structural, adsorptive, thermal, gelling, mechanical, rheological, etc. Here is an overview of how BNMs can be characterized:

11.3.1 Structural Characterization

BNMs exist in numerous forms, based on the production techniques. BNMs with various morphologies, including NPs, nanotubes, nanowires, nanoflowers, nano cubes, etc., have undergone substantial research and demonstrated significant technological significance in numerous applications. SEM (Scanning Electron Microscopy) allows for observing detailed surface features and the arrangement of different elements in a composite material at the nanoscale level. The SEM analysis approach employs electron selective detection methods to achieve nanoscale resolution and chemical characterization of nano-sized biological entities (NBCs) (Jaksch 2011; Bhattacharya, Germinario, and Winter 2008). SEM micrographs are useful for examining the porosity of BNMs, including the homogeneity, interconnectedness, and size of pores. One example is the utilization of SEM analysis to examine the infiltration of graphene oxide (GO) NPs into the biocomposite (BC) polymer matrix, as well as their uniformity and even dispersion within the structure of BC/GO nanocomposites and fiber alignment. When a higher proportion of GO is used in the BC matrix, it can cause the NPs to agglomerate. The NPs' specific contact surface is significantly reduced as a result of this aggregation, which impairs their adherence and interaction with the matrix (Rashidian et al. 2021). The ZnO and Ag:ZnO NPs were prepared by Buşilă, Mariana, et al. The fabricated NPs, which were integrated into a chitosan matrix, were characterized using field emission scanning electron microscopy (FE-SEM). Regarding the BNMs samples, it was noted that the chitosan peak had a lower intensity and a blue-shift. This suggests that the Ag and ZnO NPs within the chitosan matrix are in the nanoscale size range, which aligns with the findings from the SEM pictures. The BNMs under investigation exhibited strong antibacterial properties and had great potential for application in the medical field (Buşilă et al. 2015). A composite material consisting of coconut shell NPs and epoxy resin was created utilizing a compression molding approach. The fractured surface of prepared BNMs was analyzed for its mechanical, physical, and microstructure characteristics using SEM-EDS (Scanning Electron Microscopy--Energy Dispersive Spectroscopy). The coconut shell BNMs were composed primarily of carbon and oxygen, as revealed by EDS. The carbon content varies as milling time changes, which is intriguing (Ismail et al. 2020).

Transmission Electron Microscopy (TEM) surpasses SEM in terms of resolution, allowing for a more complete examination of the internal structure and morphology of BNMs at the nanoscale. TEM images offer high-resolution imaging at the nanometre scale, often even below a nanometer. This enables efficient screening of a huge number of particles, hence preventing significant sampling problems. The technique entails utilizing a high-energy electron beam to subject the sample to bombardment. Nevertheless, as TEM images represent the objects as projected along the direction of the input beam, properly determining the thickness of the particles can provide a challenge. The variation in the intensity of the beam that strikes the viewing screen, and the subsequent creation of an image depends on the amount of energy absorbed by the sample (Kaushik et al. 2015; Venkateshaiah et al. 2020). The *in-situ* nano reinforced Green BNMs were prepared by (Kaur, M. et al.) from natural keratin and montmorillonite (MMT)/cellulose nanocrystals (CNC) that were characterized for their morphology, one of the examples of the application of TEM in the characterization of BNMs. In this study, the prepared bio NM specimens were immersed in a polymer resin and then cut into thin slices with a thickness of 80 nm, using an ultra-microtome. A sample slice was placed on a copper support grid with a fine mesh for examination. The TEM examinations were crucial in confirming the degree of dispersion (exfoliation/intercalation) of prepared BNMs (Kaur, Arshad, and Ullah 2018).

Atomic Force Microscopy (AFM) enables the analysis of surface topography and roughness at the nanoscale, offering valuable insights into the surface features of biocomposite materials. AFM offers valuable insights into the structure, surface characteristics (such as roughness and transparency), mechanical behaviour, and adhesion qualities of NBCs. Additionally, it offers precise assessment of the dimensions and distribution of NBCs. AFM, which can measure at scales less than a nanometer and detect forces as little as a Pico newton, provides numerous distinct benefits. In addition, it can do high-resolution mechanical testing in conjunction with imaging. Additionally, it works well under special in-situ circumstances, such photographing wet or fluid-layered samples. Nevertheless, the scanning process of AFM might potentially introduce artifacts caused by contamination layers, surface shear, temperature fluctuations, or intricate surface topography. Exceptionally small scanning increments are required for high resolution AFM scanning. Beyond the non-contact or tapping mode's distance restriction, scanning large, uneven regions or intricate overall shapes may provide challenges (Venkateshaiah et al. 2020). AFM is a valuable technique for determining the dimensions of cellulose particles at the nanoscale. Oksman and colleagues discovered that using AFM analysis will be a good substitute for TEM or SEM for the characterization of cellulose nanowhiskers (a type of BNMs) because the AFM does not have any restrictions in terms of contrast and resolution. The morphology of the these BNMs appeared distinct as compared to TEM images (Visakh and Thomas 2010).

X-ray Diffraction (XRD) is employed for the examination of the crystalline arrangement of NMs, yielding insights into their crystal phases, crystal dimensions, and crystallographic alignment, such as silver NPs. Polymers made from natural materials are known to have varying amounts of unstructured amorphous parts and organized crystalline areas. These structural features have a direct impact on the properties and uses of their NBCs. The polymer plasticity and flexibility are significantly influenced by the existence of amorphous regions, but the elasticity and rigidity of these materials are affected by the crystalline regions (Abhilash, Rajender, and Suresh 2016). The amorphous/crystalline ratio of naturally occurring polymers and their related non-bonded contacts (solvent evaporation, lyophilization, etc.) are significantly influenced by purification and drying processes. XRD is an analytical method employed to ascertain the properties of materials, such as their solid state, crystal size and form, and phase identification, using quantitative phase analysis. The mathematical basis of XRD is Bragg's Equation (Burapapadh, Takeuchi, and Sriamornsak 2016): $n\lambda = 2d \sin \theta$. The variables in this context are defined as follows: n represents the order of reflection, which can take on values of 1, 2, 3, and so on. λ represents the wavelength of the incident beam. d represents the distance between parallel lattice planes. Lastly, ϴ represents the angle between the incident beam and a lattice plane, which is commonly referred to as the Bragg angle (Ahmad et al. 2020). The arrangement of atoms in the crystal lattice dictates the precise locations of the peaks observed in an XRD pattern. As the material achieved greater symmetry, the number of peaks in its diffraction pattern decreased. The peak intensities connected to the diffraction intensity are determined by the configuration of atoms inside the crystal lattice (Son et al. 2020). Calcium phosphate NMs were prepared by the microwave method and evaluated for effects of microwaves on their crystallinity, morphology, and lattice structure by XRD analysis. The XRD pattern showed a change in composition of calcium phosphate BNMs, which resembles another mineral, crystalline monetite (anhydrous form of dicalcium phosphate) (Reardon, Huang, and Tang 2015).

The presence of functional groups in BNMs is identified and chemical bonds are examined using Fourier Transform Infrared Spectroscopy (FTIR), facilitating the comprehension of the chemical composition and interactions between different components. NBC molecules must absorb light in the infrared portion of the electromagnetic radiation spectrum and convert it to molecular vibration to use infrared spectroscopy (IR) as a characterization tool. The measurement of this absorption is done by observing its dependence on wavelength, usually expressed in wave numbers ranging from 4000 to 600 cm^{-1}. This absorption is indicative of the chemical bonds present in a sample, such as stretching or binding. By employing a method, the wave number raw data is converted into an infrared spectrum, which acts as a distinctive "molecular fingerprint" that aids in the identification of the structure of organic samples. To prepare a solid sample for analysis, it can either be thinly sliced or put onto a KBr window, or it can be coarsely powdered with potassium bromide (KBr) and squeezed into a transparent disk. Simultaneously, liquid samples are either directly monitored or diluted using a solvent that is transparent to infrared (IR) radiation (Kumar et al. 2019; Jaleh and Fakhri 2016; Isra 2022). Bertolacci et al. investigated the morphology of a nanocomposite using Electron Microscopy techniques (Figure 11.2). The presence and distribution of platinum NPs (PtNPs) within the matrix

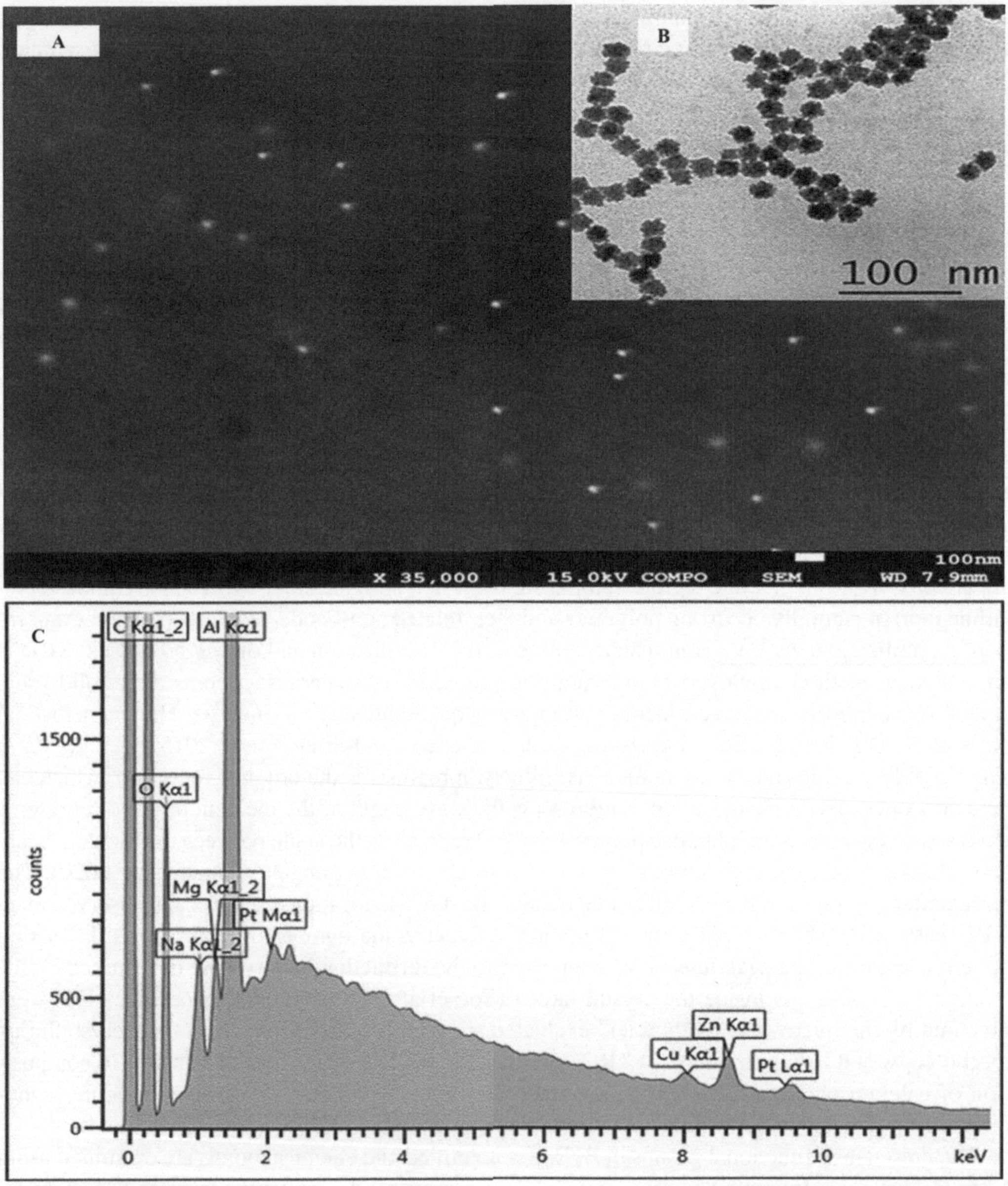

FIGURE 11.2 (a): SEM analysis using backscattering imaging mode Image. (b): TEM imaging. (c): EDS spectrum reprinted from Bertolacci et al. (© 2020).

were confirmed through SEM analysis using backscattering imaging mode (see Figure 11.2a). TEM imaging (see Figure 11.2b) provided additional confirmation of the size and shape of the NPs. PtNPs were found to be uniformly dispersed within the matrix, as indicated by arrows, and their presence was further validated by detecting Pt peaks in the EDS spectrum (see Figure 11.2c), confirming successful NP incorporation. Additionally, the EDS spectrum revealed peaks corresponding to Mg and Na, which are characteristic of the hydrogel matrix, while Al, Zn, and Cu peaks were attributed to procedural contamination. Overall, the combined microscopy and spectroscopic analyses provided comprehensive insights into the morphology and composition of the nanocomposite material (Bertolacci, Valentini, and Pompa 2020). Muralishwara et al. performed the AFM and XRD study to

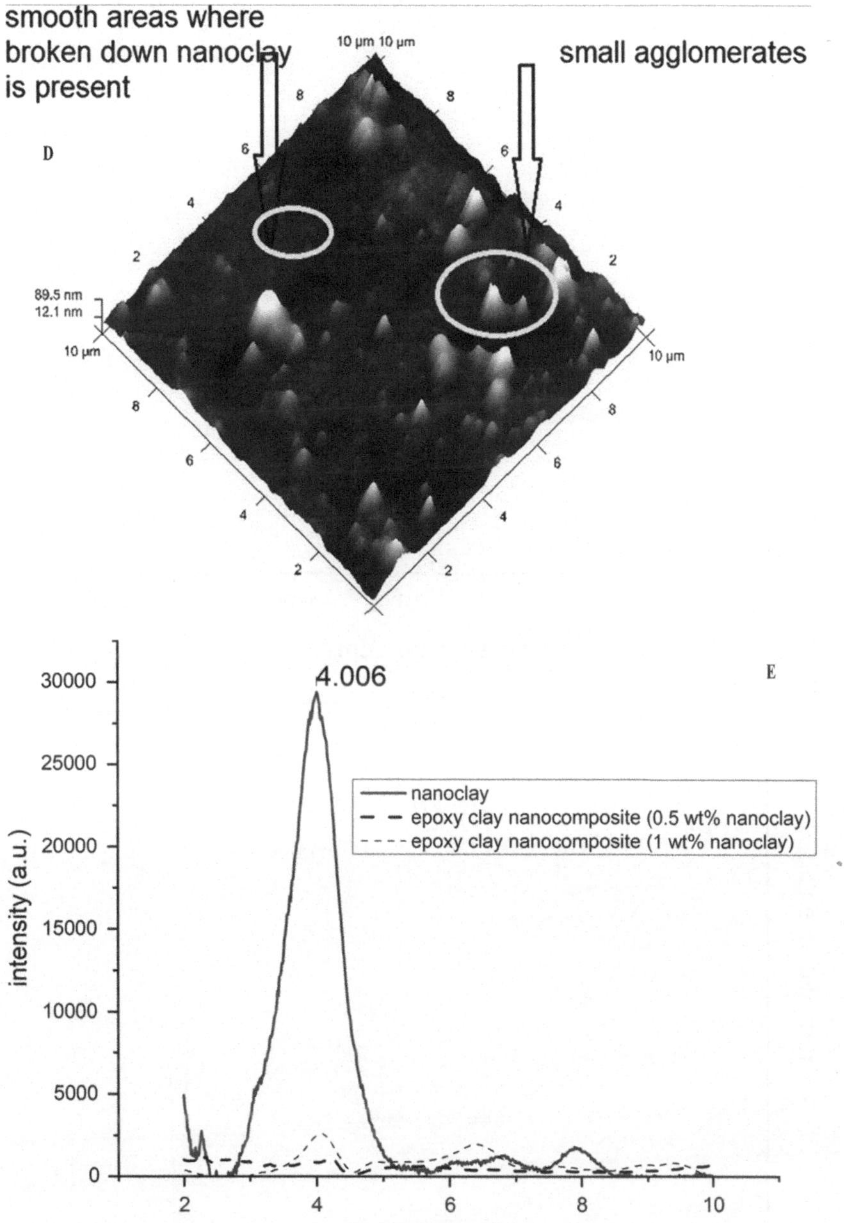

FIGURE 11.2 (Continued) (d): AFM pictures of a surface coated with 0.5 weight percent nanoclay in epoxy clay nanocomposite. (e): XRD spectra of nanoclay and epoxy clay nanocomposites.

investigate the topography image of epoxy clay nanocomposites (see Figure 11.2d). Figure 11.2e displays XRD spectra of two materials: An epoxy clay nanocomposite containing 0.5 and 1 weight percent of nanoclay and organo-modified montmorillonite nanoclay (Nanomer 1.31PS). At 2θ = 4.006°, a clear diffraction peak is seen, which is equivalent to the (001) lattice spacing of nano clay, or a d-spacing of 2.204 nm. In the nanocomposite with 0.5 wt.% nano clay, no discernible peaks are observed, indicating an exfoliated structure of nano clay within the epoxy resin. However, a low-intensity peak is detected in the nanocomposite containing 1 wt.% nanoclay, suggesting an

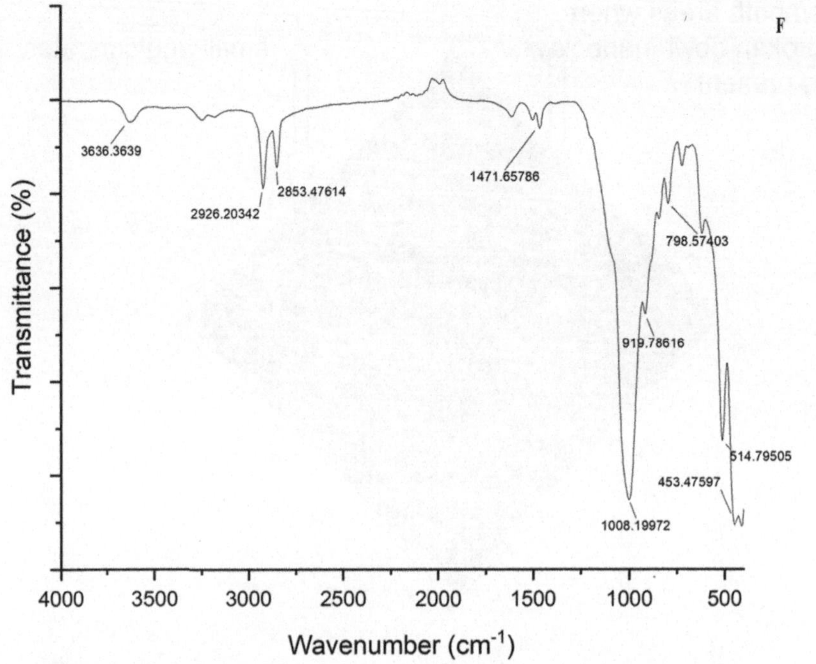

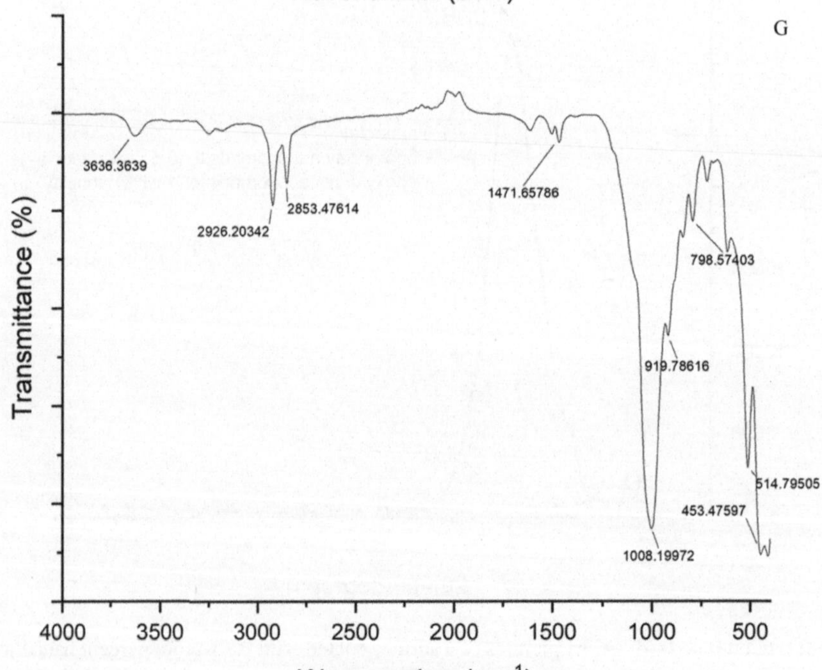

FIGURE 11.2 (Continued) (f): FTIR spectra of nanoclay (g): epoxy clay nanocomposites (1 wt.% nano clay).

intercalated structure. Given the very low intensity relative to the baseline, it is inferred that the nanoclay is transitioning toward a mixture of intercalated and exfoliated structures. This analysis provides valuable insights into the nanocomposite's microstructure and the distribution of nanoclay within the epoxy matrix. The FTIR spectra of nanoclay and epoxy clay nanocomposites shown in Figure 11.2f–h exhibit two prominent peaks observed at 454 and 515 cm^{-1}, attributed to the Si–O

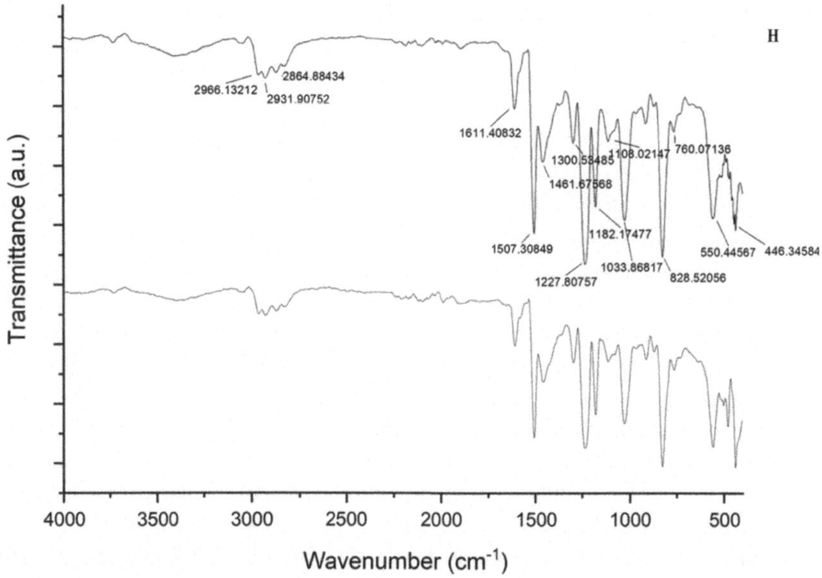

FIGURE 11.2 (Continued) (h): epoxy clay nanocomposites (0.5 wt.% nano clay). Image reprint from Muralishwara et al. (© 2022) open access article distributed under the Creative Commons Attribution 4.0 International License.

bending and Al–O stretching vibrations, respectively. These peaks provide key information about the chemical structure and molecular structure of the nanoclay material, aiding in the characterization and understanding of its properties and potential applications (Muralishwara et al. 2022).

11.3.2 Chemical Composition Analysis

Ultraviolet-Visible (UV-VIS) spectroscopy examination exposes the bio-composite materials sample to electromagnetic waves in the UV-visible region (usually 190–900 μm) in a straightforward, inexpensive, and quick manner. The visible range is roughly 400–800 nm, while the UV range typically spans 100–400 nm. Depending on the chemical makeup of the substance being exposed to light, molecules either absorb or transmit light when exposed to it. A unique wavelength that corresponds to the distinctive functional group of the scanned biocomposite is used to generate a specific spectrum. The chemical makeup of the sample and its microstructure-related light scattering qualities are factors that influence its spectral features. Functional components that absorb light in the UV–VIS spectrum should be present in the sample. Additionally, this technique allows for quick observation of modifications to the biocomposite when it is subjected to temperature and pH variations (Venkatachalam 2016; Cazón et al. 2022).

The elemental content and distribution of the elements inside BNMs can be ascertained by using Energy-Dispersive X-ray Spectroscopy (EDX) in conjunction with SEM or TEM (Orasugh, Ghosh, and Chattopadhyay 2020). EDX is a technique that examines the properties of X-rays produced by primary electrons. Examining the amount and energy of X-rays released, which reveal details about the atomic structures of the elements contained in the sample, establishes the chemical composition of the material. EDX is a significant technique used in SEM, ranking third in importance. It provides more precise quantitative findings for composite surfaces. Although EDX can offer information about the elemental composition of a limited region, it is primarily used to analyze the elemental composition of NBCs. An experiment was conducted by Manju et al. where they created NBCs by combining poly (lactic acid) (PLA) with sepiolite at various concentrations (ranging from 1 wt.% to 7 wt.%). They used SEM-EDX analysis to confirm the presence of sepiolite in the PLA. The analysis of the bare sepiolite spectrum revealed the absence of any further contaminants in the sepiolite sample. The EDX spectra of NBCs also verified the existence of sepiolite, based on the detection of peaks associated with carbon (C), silicon (Si), and magnesium (Mg) (Kurukavak 2022).

Determining surface chemistry and interactions is aided by the information that X-ray Photoelectron Spectroscopy (XPS) offers regarding the elemental makeup and chemical states of the outermost layer of biocomposite materials. One of the most popular techniques for studying solid-state materials is XPS. Following Einstein's photoelectric effect, it calculates the kinetic energy of electrons released by photoelectric effects brought on by X-ray illumination. One of the distinctive properties of the elements to determine the quantity of elements and chemical interactions in the system is the energy of discharged core electrons. Because of this, XPS is also known as electron spectroscopy for chemical analysis. Because the photoelectron generated inside a solid sample may be distributed by the interaction with other electrons in the system, the photoelectron's mean free path is incredibly short at a few nanometers. As a result, it is typically used for both quantitative and qualitative investigation of the surface composition and chemical structure of the sample. Its spectroscopic principle allows for non-destructive analysis without causing damage to the materials, which is an advantage (Ratner and Castner 2009; Lee et al. 2008). A study was performed by Bensalem, S., et al. where they have characterized the structural and surface characteristics of chitosan/montmorillonite NBCs, using XPS analysis. Upon conducting this analysis, it was determined that the NBCs in question exhibited surfaces that were highly enriched with montmorillonite, a type of clay mineral. Furthermore, it was found that chitosan was effectively intercalated within the clay galleries of the host material. This suggests that the NBCs possess a well-defined and stable structure, which may have potential applications in various fields, including biomedicine and nanotechnology (Bensalem et al. 2017).

Inductively Coupled Plasma Mass Spectrometry (ICP-MS) quantifies trace elements or ions released from biocomposite materials, which is particularly important for assessing their biocompatibility and potential toxicity (Hebert et al. 2023). Nelson et al. analyzed NPs (NPs) in various environments using single-particle inductively coupled plasma mass spectrometry (spICP-MS). This technique offers a wealth of information, including the particles' number, concentration, and size, along with the concentration of dissolved elements. The researchers have developed custom gold NP reference materials (RMs) dispersed in toluene. Analysis of these RMs using spICP-MS yielded a particle size data compared to transmission electron microscopy (TEM) measurements. The excellent agreement between the two techniques highlights the potential of spICP-MS for characterizing NPs in hydrocarbon samples (Nelson et al. 2021).

11.3.3 Mechanical and Thermal Properties

Techniques such as compression testing, Nano-indentation and tensile testing can be used to estimate the mechanical properties of BNMs, including tensile strength, and hardness.

Tensile strength and breaking lengthening: Tensile strength is a useful tool for measuring the consistency of flexibility (lengthening before breakage), polymer chains, and resistance to extension.

Thermal Analysis: The thermal characteristics of biocomposite materials, such as their decomposition temperature, thermal stability, and heat capacity, are investigated using thermogravimetric analysis (TGA) and differential scanning calorimetry (DSC). These are a set of techniques that look at how a solid sample changes with temperature and time. Heat of reaction (enthalpy (H)), amorphous state (Glass Transition (Tg)), crystallinity (melting point), thermal stability/degradation, etc., are among the data that can be obtained (Corcione and Frigione 2012). The solid transitions are determined by DSC as a function of time and temperature which is used in the identification of melting, decomposition, and solid-solid transitions (polymorphism, crystallization, etc.). Solvent evaporation is typically the first transition seen in DSC, while polymer breakdown may be the cause of the final thermal peaks (Kumar, Singh, and Nanda 2020). Performing the thermal examination after the sample has cooled down allows for the detection of solid-state transitions, assuming that the sample is stable and shows no indications of degradation (gas evolution, color change, etc.). In a DSC thermogram, the kinetic process of crystallization is indicated by an exothermic peak. The composites' thermal phase change is shown by the exothermic (heat emitted) and endothermic (heat absorption) peaks and magnitudes. The Tg, crystallization temperature (Tc), degree of crystallization (Xc), and fusion temperature (Tm) are the main thermal data that were retrieved from this investigation. It is also possible to calculate the composite's heat capacity and enthalpy fluctuation (Neto et al. 2021). Habibi, Youssef, et al. grafted NBCs of poly(3-caprolactone) cellulose nanocrystals with the evaluation of long-term stability using DSC analysis. The formation of a crystalline structure at the surface of the nanocrystals were linked to the grafted poly(3-caprolactone) chains; they adopt a crystalline brush-like configuration extending from the surface of the NPs displayed by a low magnitude endothermic peak (Habibi et al. 2008). Numerous naturally occurring polymers have varying levels of crystallinity. The Tg, this is the temperature at which an amorphous state typically softens, can be measured by DSC to distinguish between these degrees. The solid sample's molecular mobility is blamed for this. Numerous physical properties, including solubility, drug-polymer interaction, drug release, stability during storage, and many more, are significantly impacted by this transition.

11.3.4 Biological Characterization

11.3.4.1 Cell Viability and Proliferation Assays

Cell viability assays such as MTT, Alamar Blue, or Live/Dead staining are used to assess the cytotoxicity and biocompatibility of BNMs (Behzadi et al. 2017; Tarrahi, Mahjouri, and Khataee 2021).

11.3.4.2 Cellular Uptake Studies

It is possible to use fluorescence microscopy and flow cytometry to investigate how cells absorb and internalize BNMs (Behzadi et al. 2017, Perrigue et al. 2023). A nano biomaterial known as Sepiolite occurs naturally as silicate with nanofibrous properties. It can serve as a nanocarrier, since it could be spontaneously absorbed by mammalian cells, because of its small size at the nanometer scale, as shone by Castro-Smirnov, F.A., Ayache, J., Bertrand, JR. et al. This team of researchers also confirmed that endocytic and non-endocytic pathways can be used to internalize the sepiolite into mammalian cells, with macro pinocytosis being a prominent pathway. Remarkably, sepiolite-mediated gene transfer became twice as successful when the cells were subjected to endocytosis inhibitors, such as drugs like chloroquine (Castro-Smirnov et al. 2017).

11.3.4.3 In Vivo Studies

Animal studies are conducted to evaluate the biocompatibility, biodistribution, and therapeutic efficacy of biocomposite materials in living organisms (Tarrahi, Mahjouri, and Khataee 2021; Kumar, Sharma, and Maitra 2017a). Overall, a comprehensive characterization approach combining multiple techniques is essential to understand the structure, composition, properties, and biological behaviour of BNMs, facilitating their development for various applications. BNMs offer a promising solution for a sustainable environment. Combining the strength and versatility of bio-based fibers with the unique properties of NMs allows these materials to be tailored for a extensive of applications (shown in Table 11.1). Research in this field is ongoing, with scientists constantly striving to improve production methods, explore new material combinations, and unlock the full potential of BNMs for a greener future.

11.4 BNMS APPLICATIONS IN FARMING, FOOD, AND PHARMACEUTICAL INDUSTRY

Researchers are developing innovative ways to deliver nutrients directly to crops to address global food demand. Biodegradable nanocarriers made from natural materials are showing promise in this area. These carriers slowly release nutrients to plant roots, minimizing waste and maximizing the availability of essential nutrients for increased crop productivity (Dutta et al. 2022). Biocomposites, a new technology in agriculture, offer a promising solution for increasing crop yields and improving

TABLE 11.1
Application of BNMs

Name of Biocomposite Material	Origin Type	NM	Application	References
Cellulose	Plant (Wood pulp, agricultural waste)	Nanofibrils	Lightweight automotive parts, packaging films	(Lozano Fernandez and Miskolczi 2022)
Chitin	Animal (Crustacean shells)	Nanofibers	Biomedical implants, drug delivery systems	(Azuma et al. 2014)
Polycaprolactone/Starch	Plant	Nanofibrils	Food packaging, bio-medical	(Ali Akbari Ghavimi et al. 2015)
Bacterial Nanocellulose-Polylactic Acid	Bacterial	Nanofibrils	Wound dressings, tissue engineering scaffolds, 3D printing	(Wang et al. 2020)
Fungal Mycelium-Biodegradable Polymer	Fungal (Mushroom mycelium)	Mycelium-based foam and sandwich composites	construction, manufacturing, agriculture, and biomedical	(Yang, Park, and Qin 2021)
Bio-polyester	Mineral	Reinforced with carbon nanotubes	biodegradable sensors, EMI materials,	(Qazi et al. 2020)
Lignin Nanofiber-Bio-epoxy	Plant (Wood waste)	Nanocomposites	High-strength composites for construction applications	(Ebrahimnezhad-Khaljiri and Ghadi 2024)
Silk Fibroin	Animal (Silkworm cocoons)	Hydroxyapatite/silk fibroin nanocomposite	Drug delivery systems, wound healing patches	(Mobika et al. 2020)
Nanocellulose	Recycled Material	Nanofibrils	Biodegradable furniture, lightweight composites	(Wei et al. 2014)
Nanocrystalline Cellulose-Polyhydroxybutyrate	Plant (Agricultural waste)	Nanocrystalline cellulose (NCC)	Medical implants with improved biocompatibility	(Malmir et al. 2019)

soil health. These materials combine natural polymers with NPs, creating tools for seed coating, pest control, controlled fertilizer release, and nanosensors to monitor soil conditions (Moreira et al. 2022). Biodegradable films made with nano-biocomposites can act as an antimicrobial barrier, extending shelf life and reducing food spoilage. They can even house tiny sensors that detect spoilage markers, providing real-time information on food quality. Additionally, nano-biocomposite coatings can be applied directly to food products, creating a shield against moisture, oxygen, and UV light, further enhancing freshness and quality. This technology reduces food waste and promotes safer consumption (Kumar, Kaur, and Bhatia 2017b).

11.5 VARIOUS CATEGORIES OF POLYMER-BASED BNM IN ENVIRONMENTAL REMEDIATION

Polymer-based nano-biocomposite materials have unique properties, such as high surface area, mechanical strength, tunable surface chemistry, and biocompatibility, which have gained significant attention in environmental remediation. They offer promising solutions for various environmental challenges, including water purification, soil remediation, and air pollution control (Singh 2018). Examples of common BNMs are listed in Table 11.2. Applications of polymer-based BNMs in ecological remediation:

TABLE 11.2
Polymer-Based BNM in Environmental Remediation

S.N.	Biomaterial	System	Characterization	Result	Finding	Ref.
1	Cellulose and Chitosan	microbeads	Adsorption isotherm XRD SEM zeta potential FTIR	Size: 2 μm SA: 10.4 m^2/g ZP: −63 mV (NP), +66 mV (CCMB-0.25:1) RE: 91.5% and 88.4%, using 14 and 9 g/L of CCMB-0.25:1 MT: 120 min Ads Model: Langmuir isotherm Removal of direct blue 78 dye pH range for MAC of DB78 dye:3–4	CCMB, chitosan, and nanocellulose have the potential to be very effective sorbents for eliminating anionic pollutants and promising dye adsorbents for wastewater treatment.	(Bassyouni et al. 2022)
2	Green synthesis by *Phoenix dactylifera* waste	zinc oxide NPs (ZnO-NPs)	ED X XPS UV FTIR XRD DSC/TG analysis Raman	Size:30 nm UV abs: 381 nm XRD 2θ values with plane: 31.7(1,0,0), 34.3 (0,0,2), 36.2(1,0,1), 47.5 (1,0,2), 56.5 (1,1,0), 62.8 (1,0,3) and 67.9(1,1,2) corresponded to hexagonal ZnO crystal. EDX analysis: 63.3% (Zn) and 36.7% (O) Endothermic peaks:139.6, 211.5 and 368.2°C	NPs demonstrated significant antibacterial effects	(Bassyouni et al. 2022)
3	Ethylene diamine tetra acetate dianhydride cross-linked chitosan- and agarose (7:3 weight ratio)-based aerogel membrane decorated with α-FeOOH- and γ-AlOOH-based material	biopolymers and task-specific Fe–Al-based nanocomposites	SEM, DLS FTIR PXRD BET	Size: ~420 nm SA:384.14 $m^2 \cdot g^{-1}$ of 2.49 nm Density: 0.021 $g \cdot m^{-2}$ Pore volume: 0.239 $cm^{-3} \cdot g^{-1}$ pore size: 0.4^{-2} nm For cationic dyes such as MG and MB, the C-AG membrane exhibited removal efficiencies: 65 ± 3 and 74% ± 4% with fluxes 1026 ± 27 and 1085 ± 21 $L \cdot m^{-2} \cdot h^{-1}$, respectively. swelling ratio of 1974% pure water flux of 19,228 $L \cdot m^{-2} \cdot h^{-1}$	Purification of pollutants such as fluoride, arsenate, and dyes.	(Mruthunjayappa et al. 2020)
4	Silver and Sod. Alginate	Silver NPs (AgNPs)-alginate composite beads	UV IR TEM	Size: 2.4 ± 0.4 mm (AR) 2.8 ± 0.2 mm (SGR) DE: over 5-log with a 1 min HRT Antimicrobial effect: *Escherichia coli*	Excellent disinfection performance	(Lin et al. 2013)

Continued

TABLE 11.2 (Continued)

S.N.	Biomaterial	System	Characterization	Result	Finding	Ref.
5	Alginic acid	Oxide-based alginate composites	FTIR, SEM, BET, XRD and TGA	Adsorption conc. at pH: 11.8 mg/g at pH 5.9 (6% GO-CA) Isoelectronic point and pH (3.7) pH < 3.7 (anion exchanger) pH >3.7 (cation exchanger)	Eliminated the water-soluble pollutants, such as water-soluble antibiotics, organic dyes like methylene blue, Eriochrome Black-T, Eriochrome Blue Black R, methyl orange, and phenolic compounds like bisphenol A and 2,4-dichlorophenol, as well as heavy metal ions like cadmium (Cd), arsenic (As), cobalt (Co), and lead (Pb2+).	(Thakur 2021)
6	Cellulose, Chitosan	chitosan and bacterial cellulose, Nanocomposites loaded with silver sulfadiazine	FTIR, RAMAN spectroscopy, SEM, TEM, and DLS	Size: 50–150 nm (SEM) 310–478 nm (DLS) Zeta: −5.82 mV DR: 85%–95% in 30 min swelling degree 360%	wound dressing	(Stanescu et al. 2021)
7	sodium alginate, Cellulose	Cellulose Nanocrystal/Silver/ Alginate Bio nanocomposite film (CNC/Ag/Alg)	FTIR, FESEM, TEM, OM XRD, TGA, UV Visibility, FWS, Moisture Absorption, WVP	Thickness: 12 μm Thermal degradation: 178°C–249°C and 530°C–677°C, with Tmax: 232°C and 622°C. FWS: 56.36, MA: 13.24%, WVP: 4.60, TS: 39.40, EB 14.70%	Films could block UV light and water, and can be utilized as food packaging materials.	(Yadav, Liu, and Chiu 2019)
8	*Sargassum vulgare*	NP	EDX, SEM, FTIR, size particles, and Zeta potential	Size: 85–50 nm DI with efficacy: 91.08% and 88.89%	Effective against sugar beet root diseases	(Abdelwahab, Rabie, and Mohamed 2023)

Abbreviations of Table 2: Surface area: SA Removal efficiencies: RE Mixing Time: MT Brunauer–Emmett–Teller: BET Scanning Electron Microscope: SEM Nanocellulose Particle: NP Maximum Adsorption Capacity: MAC Simultaneous-Gelation-Reduction: SGR Adsorption–Reduction: AR Disinfection Efficiency: DE Hydraulic Retention Time: HRT Drug Release: DR. film water solubility (FWS) Tensile Strength (TS) water vapour permeation (WVP) moisture absorption (MA) and Elongation at break (EB) (%) for Alg Ag/Alg CNC/Alg and CNC/Ag/Alg films. Dynamic light scattering: DLS Disease incidence: DI. Field Emission Scanning Electron Microscopy: FESEM Mean Pore Diameter: MPD

11.5.1 Nanocomposite Membranes for Water Purification

Fabrication of Polymer-based nanocomposite materials, such as graphene, carbon nanotubes, or metal oxides, are used for water purification through processes like ultrafiltration, nanofiltration, and reverse osmosis. These membranes exhibit enhanced filtration efficiency, antimicrobial properties, and fouling resistance (Pandey, Shukla, and Singh 2017).

11.5.2 Adsorbent Nanocomposites for Wastewater Treatment

Scientists are developing advanced materials called BNMs to clean wastewater. These materials combine natural polymers, with NPs like zeolites, activated carbon, or metal-organic frameworks. Thanks to their creative design, they can remove dyes, chemical compounds, heavy metals, and other contaminants from the water. They are particularly attractive because they can hold many contaminants (high adsorption capacity), target specific pollutants (selectivity), and be reused multiple times (Adesina Adegoke et al. 2020).

11.5.3 Polymer Nanocomposite Hydrogels for Soil Remediation

Hydrogel-based nanocomposites incorporating polymers with NPs such as clay minerals or metal NPs are employed for soil remediation purposes, including soil stabilization, nutrient delivery, and contaminant immobilization (Kim et al. 2022). These hydrogels can absorb and retain water, improve soil structure, and facilitate the controlled release of nutrients or remediation agents (Sethi et al. 2023).

11.5.4 Nanocomposite Catalysts for Air Pollution Control

Polymer-based nanocomposite catalysts containing NPs like titanium dioxide, zeolites, or metal NPs are used for catalytic degradation of air pollutants such as nitrogen oxides (NOx), volatile organic compounds (VOCs), and particulate matter (PM). These catalysts exhibit high catalytic activity, selectivity, and durability for air purification applications (Li et al. 2021).

11.5.5 Nanocomposite Sorbents for Oil Spill Remediation

Polymer-based nanocomposite sorbents incorporating graphene, carbon nanotubes, or silica NPs are developed for oil spill cleanup and remediation (Elhenawy et al. 2021). These sorbents can selectively adsorb oil from water surfaces, forming oil-absorbing meshes or aerogels for efficient oil recovery and environmental remediation (Damavandi and Soares 2022).

11.5.6 Polymer Nanocomposite Filtration Media for Air and Water Purification

Nanocomposite materials composed of polymers and NPs are utilized in filtration media for air and water purification applications. These filtration systems employ nanofibers, membranes, or filters with enhanced mechanical strength, chemical stability, and filtration efficiency for removing particulate matter, pathogens, and pollutants from air and water streams (Pandey, Shukla, and Singh 2017).

11.5.7 Biodegradable Nanocomposites for Sustainable Remediation

BNMs incorporating biodegradable polymers and environmentally friendly NPs are developed for sustainable environmental remediation applications. These materials offer eco-friendly alternatives for pollution control, waste management, and remediation processes, with reduced environmental impact (Rawat and Singh 2022). Overall, polymer-based BNMs are promising to address various environmental challenges by providing innovative water purification, soil remediation, air pollution control, and sustainable environmental remediation solutions.

11.5.8 Green BNMs in Drinking and Wastewater Treatment

Green BNMs have emerged as a highly promising drinking and wastewater treatment solution. Their environmentally friendly nature and remarkable remediation capabilities have been extensively researched and demonstrated in various settings, making them an excellent alternative to traditional treatment methods. BNMs are composed of natural substances and can be synthesized by microorganisms, making them a sustainable and eco-friendly option for effective water treatment (Gautam et al. 2019). These NMs typically consist of biodegradable polymers, such as chitosan, cellulose, or alginate, combined with natural or bio-derived NPs like graphene, nanocellulose, or clay minerals (Supriya et al. 2023; Norizan et al. 2022). One significant advantage of these materials lies in their biocompatibility and low toxicity, ensuring minimal environmental impact during production, usage, and disposal. Green BNMs are utilized in drinking water treatment to remove contaminants, including heavy metals, organic pollutants, and microbial pathogens (Gautam et al. 2019). Their high surface area and functional groups facilitate efficient adsorption, filtration, and disinfection processes, resulting in purified water that meets safety standards for human consumption. Additionally, these materials often exhibit antimicrobial properties, further enhancing their effectiveness in ensuring water quality (Namasivayam et al. 2023).

In wastewater treatment, green BNMs are crucial in removing pollutants and improving water quality before discharge into the environment (Zhang et al. 2023). They can be employed in various treatment processes, such as adsorption, coagulation-flocculation, membrane filtration, and photocatalysis. These materials effectively adsorb or degrade contaminants, reduce turbidity, and enhance the biodegradability of wastewater, leading to cleaner effluents and reduced environmental pollution (Kolya and Kang 2023). Moreover, green BNMs offer advantages, such as renewability, biodegradability, and sustainability, aligning with the principles of green chemistry and environmental stewardship. These materials contribute to developing more sustainable water treatment technologies by utilizing renewable resources and eco-friendly synthesis methods. Furthermore, their potential for functionalization and customization allows for tailored solutions to specific water treatment challenges, enhancing their versatility and applicability in diverse environmental contexts (Dar et al. 2020).

11.5.9 Graphene-Based BNMs: Current Challenges and Prospects

The distinct chemical and physical characteristics of graphene, a two-dimensional carbon atom material, make it a desirable choice for creating BNMs. Due to its high mechanical strength, surface area, and electrical conductivity, graphene has the potential to significantly improve the performance and properties of biopolymers or natural materials. Incorporating graphene into these materials can enhance their mechanical strength, increase their biocompatibility with living organisms, and improve their environmental sustainability. In summary, the unique properties of graphene make it a promising additive in developing innovative and sustainable biocomposite materials (Azizi-Lalabadi and Jafari 2021). The use of graphene-based BNMs holds great promise in various fields. However, the application of such materials is hindered by several challenges. One of the most significant challenges is achieving scalability, cost-effectiveness, and environmental sustainability in graphene production. Another crucial challenge is ensuring the proper dispersion and compatibility of graphene within biocomposite matrices without compromising its unique properties. Despite these challenges, ongoing research efforts aim to develop scalable, cost-effective, and environmentally friendly synthesis methods. Furthermore, advancements in nanocomposite fabrication techniques and a better understanding of graphene-biocomposite interactions offer opportunities to tailor the properties of graphene-based BNMs for specific applications (Orsu and Koyyada 2020).

11.6 CONCLUSION AND FUTURE PROSPECTIVE

BNMs are emerging as a promising option to create a more sustainable environment. These materials offer a unique combination of biodegradability, high performance, and diverse functionalities that can be utilized to address a variety of environmental challenges. From water purification and pollution remediation to energy production and lightweight construction materials, the potential applications of BNMs are extensive and constantly evolving. As research in this field progresses, we can expect even more innovative and effective solutions to emerge, which will pave the way for a cleaner and healthier planet.

BNMs are emerging as a game-changer in the fight for a sustainable environment. These innovative materials combine natural, biodegradable polymers with powerful NPs. This unique blend offers a sustainable solution: the biopolymers decompose naturally, reducing waste, while the NPs enhance the material's ability to tackle environmental challenges. BNMs can revolutionize environmental issues by addressing everything from cleaning polluted water to developing sustainable energy sources. Moreover, BNMs hold promise for sustainable resource management by enabling the development of biodegradable packaging materials, eco-friendly construction materials, and bio-based composites. These materials can reduce reliance on non-renewable resources, minimize waste generation, and promote circular economy principles by facilitating the recycling and reuse of materials. Additionally, BNMs can potentially drive advancements in renewable energy technologies. By incorporating NMs into energy storage devices, such as batteries and supercapacitors, or enhancing the efficiency of photovoltaic cells through light-absorbing nanocomposites, these materials can contribute to the transition toward clean and sustainable energy sources. Furthermore, integrating BNMs into agricultural practices holds promise for enhancing soil health, increasing crop yields, and reducing environmental impact. NM-based soil amendments and nutrient delivery systems can improve soil fertility, water retention, and plant nutrient uptake, promoting sustainable agriculture practices and food security. In conclusion, the prospects of BNMs for creating a sustainable environment are characterized by innovation, versatility, and potential impact across various sectors. Continued research and development efforts to optimize material properties, scalability, and environmental performance will be essential for realizing the full potential of these materials in addressing global environmental challenges and building a more sustainable future.

REFERENCES

Vilaplana, F., Strömberg, E., Karlsson, S., Environmental and resource aspects of sustainable biocomposites, *Polymer Degradation and Stability* 95(11) (2010) 2147–2161.

Borowy, I., *Defining Sustainable Development for Our Common Future: A History of the World Commission on Environment and Development (Brundtland Commission)*. Routledge, London, England 2013. https://doi.org/10.4324/9780203383797

Kaasa, S.M., The UN Commission on Sustainable Development: Which mechanisms explain its accomplishments?, *Global Environmental Politics* 7(3) (2007) 107–129.

Cf, *Transforming Our World: The 2030 Agenda for Sustainable Development, United Nations*: New York, NY, USA (2015). https://sustainabledevelopment.un.org/post2015/transformingourworld/publication

United Nations. Transforming our world: the 2030 Agenda for Sustainable Development, 2015. https://sdgs.un.org/2030agenda

Silva, A.C.Q., Silvestre, A.J.D., Vilela, C., Freire, C.S.R., Natural polymers-based materials: A contribution to a greener future, *Molecules* 27(1) (2021).

Zhang, M., Xie, L., Yin, Z., Khanal, S.K., Zhou, Q., Biorefinery approach for cassava-based industrial wastes: Current status and opportunities, *Bioresource Technology* 215 (2016) 50–62.

Zhang, L., Fang, M., Nanomaterials in pollution trace detection and environmental improvement, *Nano Today* 5(2) (2010) 128–142.

Shah, M., Fawcett, D., Sharma, S., Tripathy, S.K., Poinern, G.E.J., Green synthesis of metallic nanoparticles via biological entities, *Materials* 8(11) (2015) 7278–7308.

Bashir, S., Liu, J., Chapter 1 - Nanomaterials and their application, in: J.L. Liu, S. Bashir (Eds.), *Advanced Nanomaterials and Their Applications in Renewable Energy*, Elsevier, Amsterdam, 2015, pp. 1–50.

Satyanarayana, K.G., Arizaga, G.G.C., Wypych, F., Biodegradable composites based on lignocellulosic fibers: An overview, *Progress in Polymer Science* 34(9) (2009) 982–1021.

Rawat, S., Singh, J., Chapter 11 - Green biocomposite materials for sustainable remediation application, in: J.R. Koduru, R.R. Karri, N.M. Mubarak, E.R. Bandala (Eds.), *Sustainable Nanotechnology for Environmental Remediation*, Elsevier 2022, pp. 273–297, https://doi.org/10.1016/C2020-0-01522-7

Ali, M.S., Al-Shukri, A., Maghami, M., Gomes, C., Nano and bio-composites and their applications: A review, *IOP Conference Series: Materials Science and Engineering*, IOP Publishing, 2021, p. 012093.

Smith, K.H., Tejeda-Montes, E., Poch, M., Mata, A., Integrating top-down and self-assembly in the fabrication of peptide and protein-based biomedical materials, *Chemical Society Reviews* 40(9) (2011) 4563–4577.

Mohammadi, M.R., Nojoomi, A., Mozafari, M., Dubnika, A., Inayathullah, M., Rajadas, J., Nanomaterials engineering for drug delivery: A hybridization approach, *Journal of Materials Chemistry B* 5(22) (2017) 3995–4018.

Wu, G., Li, P., Feng, H., Zhang, X., Chu, P.K., Engineering and functionalization of biomaterials via surface modification, *Journal of Materials Chemistry B* 3(10) (2015) 2024–2042.

Mehta, M.R., Mahajan, H.P., Hivrale, A.U., Green synthesis of chitosan capped-copper nano biocomposites: Synthesis, characterization, and biological activity against plant pathogens, *BioNanoScience* 11(2) (2021) 417–427.

Dai, X., Chen, Y., Computational biomaterials: Computational simulations for biomedicine, *Advanced Materials* 35(7) (2023) 2204798.

Jaksch, H., The contrast mechanisms of LL-BSE electrons in FE-SEM characterization of polymer, single proteins, and oxidization states of elements, *Microscopy and Microanalysis* 17(S2) (2011) 902–903.

Bhattacharya, D., Germinario, L.T., Winter, W.T., Isolation, preparation and characterization of cellulose microfibers obtained from bagasse, *Carbohydrate Polymers* 73(3) (2008) 371–377.

Rashidian, E., Babaeipour, V., Chegeni, A., Khodamoradi, N., Omidi, M., Synthesis and characterization of bacterial cellulose/graphene oxide nano-biocomposites, *Polymer Composites* 42(9) (2021) 4698–4706.

Buşilă, M., Muşat, V., Textor, T., Mahltig, B., Synthesis and characterization of antimicrobial textile finishing based on Ag:ZnO nanoparticles/chitosan biocomposites, *RSC Advances* 5(28) (2015) 21562–21571.

Ismail, I., Arliyani, Z.J., Mursal, N.G.O., Abdullah, C.K., Fazita, M.R.N., Khalil, H.P.S., Properties and characterization of new approach organic nanoparticle-based biocomposite board, *Polymers*, 2020;12(10):2236. https://doi.org/10.3390/polym12102236. PMID: 32998404; PMCID: PMC7600339.

Kaushik, M., Fraschini, C., Chauve, G., Putaux, J.-L., Moores, A., Transmission electron microscopy for the characterization of cellulose nanocrystals, *The Transmission Electron Microscope-Theory and Applications* (2015) 30, 130–163.

Venkateshaiah, A., Padil, V.V.T., Nagalakshmaiah, M., Waclawek, S., Černík, M., Varma, R.S., Microscopic techniques for the analysis of micro and nanostructures of biopolymers and their serivatives, *Polymers*, 2020, *12*(3), 512. https://doi.org/10.3390/polym12030512

Kaur, M., Arshad, M., Ullah, A., In-situ nanoreinforced green bionanomaterials from natural keratin and montmorillonite (MMT)/cellulose nanocrystals (CNC), *ACS Sustainable Chemistry & Engineering* 6(2) (2018) 1977–1987.

Visakh, P.M., Thomas, S., Preparation of bionanomaterials and their polymer nanocomposites from waste and biomass, *Waste and Biomass Valorization* 1(1) (2010) 121–134.

Abhilash, V., Rajender, N., Suresh, K., Chapter 14 - X-ray diffraction spectroscopy of polymer nanocomposites, in: S. Thomas, D. Rouxel, D. Ponnamma (Eds.), *Spectroscopy of Polymer Nanocomposites*, William Andrew Publishing 2016, pp. 410–451. https://doi.org/10.1016/B978-0-323-40183-8.00014-8

Burapapadh, K., Takeuchi, H., Sriamornsak, P., Development of pectin nanoparticles through mechanical homogenization for dissolution enhancement of itraconazole, *Asian Journal of Pharmaceutical Sciences* 11(3) (2016) 365–375.

Ahmad, M., Gani, A., Hassan, I., Huang, Q., Shabbir, H., Production and characterization of starch nanoparticles by mild alkali hydrolysis and ultra-sonication process, *Scientific Reports* 10(1) (2020) 3533.

Son, D., Cho, S., Nam, J., Lee, H., Kim, M., X-ray-based spectroscopic techniques for characterization of polymer nanocomposite materials at a molecular level, *Polymers*, 2020, 12 (5), 1053. https://doi.org/10.3390/polym12051053

Reardon, P.J.T., Huang, J., Tang, J., Mesoporous calcium phosphate bionanomaterials with controlled morphology by an energy-efficient microwave method, *Journal of Biomedical Materials Research Part A* 103(12) (2015) 3781–3789.

Kumar, S., Boro, J.C., Ray, D., Mukherjee, A., Dutta, J., Bionanocomposite films of agar incorporated with ZnO nanoparticles as an active packaging material for shelf life extension of green grape, *Heliyon* 5(6) (2019) e01867.

Jaleh, B., Fakhri, P., Infrared and Fourier transform infrared spectroscopy for nanofillers and their nanocomposites, 2016, In Sabu Thomas, Didier Rouxel, Deepalekshmi Ponnamma, (Eds.), *Spectroscopy of Polymer Nanocomposites*, William Andrew Publishing, 112–129. https://doi.org/10.1016/B978-0-323-40183-8.00005-7

Isra, D., Physicochemical characterization of nanobiocomposites, in: M.M.E. Magdy, K. Selcan (Eds.), *Biocomposites, IntechOpen*, Rijeka, 2022, p. Ch. 2, http://doi.org/10.5772/intechopen.104003

Bertolacci, L., Valentini, P., Pompa, P.P., A nanocomposite hydrogel with catalytic properties for trace-element detection in real-world samples, *Scientific Reports* 10(1) (2020) 18340.

Muralishwara, K., Sudhakar, Y.N., Kini, U.A., Sharma, S., Gurumurthy, B.M., Moisture absorption and spectroscopic studies of epoxy clay nanocomposite, *Polymer Bulletin* 79(7) (2022) 5587–5611.

Venkatachalam, S., Chapter 6 - Ultraviolet and visible spectroscopy studies of nanofillers and their polymer nanocomposites, in: S. Thomas, D. Rouxel, D. Ponnamma (Eds.), *Spectroscopy of Polymer Nanocomposites*, William Andrew Publishing 2016, pp. 130–157. http://doi.org/10.1016/B978-0-323-40183-8.00006-9

Cazón, P., Cazón, D., Vázquez, M., Guerra-Rodriguez, E., Rapid authentication and composition determination of cellulose films by UV-VIS-NIR spectroscopy, *Food Packaging and Shelf Life* 31 (2022) 100791.

Orasugh, J.T., Ghosh, S.K., Chattopadhyay, D., Chapter 10 - Nanofiber-reinforced biocomposites, in: B. Han, S. Sharma, T.A. Nguyen, L. Longbiao, K.S. Bhat (Eds.), *Fiber-Reinforced Nanocomposites: Fundamentals and Applications*, Elsevier 2020, pp. 199–233. eBook ISBN: 9780128199107

Kurukavak, Ç.K., Surface and mechanical characterization of bionanocomposites for advanced applications: A review, *Proceedings of the Institution of Mechanical Engineers, Part L: Journal of Materials: Design and Applications* 236(10) (2022) 1949–1966.

Ratner, B., Castner, D., Electron Spectroscopy for Chemical Analysis, 2009, in *Surface Analysis: The Principal Techniques*, 2nd Edition, Editor(s): John C. Vickerman, Ian S. Gilmore, 47–112. John Wiley & Sons, Ltd. https://doi.org/10.1002/9780470721582.ch3

Lee, J., Sohn, S., Yun, H.J., Shin, H.J., Degradation studies on high-voltage-driven organic light-emitting device using in situ on-operation method with scanning photoelectron microscopy, *Applied Physics Letters* 93(13) (2008), 3310.

Bensalem, S., Hamdi, B., Del Confetto, S., Iguer-Ouada, M., Chamayou, A., Balard, H., Calvet, R., Characterization of chitosan/montmorillonite bionanocomposites by inverse gas chromatography, *Colloids and Surfaces A: Physicochemical and Engineering Aspects* 516 (2017) 336–344.

Hebert, D., Nelson, J., Diehl, B.N., Zito, P., Single-Particle ICP-MS/MS application for routine screening of nanoparticles present in powder-based facial cosmetics, *Nanomaterials*, 2023, 13 (19), 2681. https://doi.org/10.3390/nano13192681

Nelson, J., Saunders, A., Poirier, L., Lopez-Linares, F., Analysis of gold nanoparticles in a hydrocarbon solvent by single particle-inductively coupled plasma mass spectrometry (spICP-MS) and TEM, *SN Applied Sciences* 3(2) (2021) 161.

Corcione, C.E., Frigione, M., Characterization of nanocomposites by thermal analysis, *Materials*, 2012, 5 (12), 2960–2980. https://doi.org/10.3390/ma5122960

Kumar, A., Singh, P., Nanda, A., Hot stage microscopy and its applications in pharmaceutical characterization, *Applied Microscopy* 50(1) (2020) 12.

Neto, J.S.S., de Queiroz, H.F.M., Aguiar, R.A.A., Banea, M.D., A review on the thermal characterisation of natural and hybrid fiber composites, *Polymers*, 2021, 13 (24), 4425. https://doi.org/10.3390/polym13244425

Habibi, Y., Goffin, A.-L., Schiltz, N., Duquesne, E., Dubois, P., Dufresne, A., Bionanocomposites based on poly (ε-caprolactone)-grafted cellulose nanocrystals by ring-opening polymerization, *Journal of Materials Chemistry* 18(41) (2008) 5002–5010.

Behzadi, S., Serpooshan, V., Tao, W., Hamaly, M.A., Alkawareek, M.Y., Dreaden, E.C., Brown, D., Alkilany, A.M., Farokhzad, O.C., Mahmoudi, M., Cellular uptake of nanoparticles: Journey inside the cell, *Chem Soc Rev* 46(14) (2017) 4218–4244.

Tarrahi, R., Mahjouri, S., Khataee, A., A review on in vivo and in vitro nanotoxicological studies in plants: A headlight for future targets, *Ecotoxicol Environ Saf* 208 (2021) 111697.

Perrigue, P.M., Henschke, A., Grześkowiak, B.F., Przysiecka, Ł., Jaskot, K., Mielcarek, A., Coy, E., Moya, S.E., Cellular uptake and retention studies of silica nanoparticles utilizing senescent fibroblasts, *Sci Rep* 13(1) (2023) 475.

Castro-Smirnov, F.A., Ayache, J., Bertrand, J.-R., Dardillac, E., Le Cam, E., Piétrement, O., Aranda, P., Ruiz-Hitzky, E., Lopez, B.S., Cellular uptake pathways of sepiolite nanofibers and DNA transfection improvement, *Scientific Reports* 7(1) (2017) 5586.

Kumar, V., Sharma, N., Maitra, S.S., In vitro and in vivo toxicity assessment of nanoparticles, *International Nano Letters* 7(4) (2017a) 243–256.

Lozano Fernandez, M.E., Miskolczi, N., Production of cellulose nano-fibers and its application in poly-lactic-acid: Property improvement by new types of coupling agents, *Polymers* 14(9) (2022) 1887.

Azuma, K., Ifuku, S., Osaki, T., Okamoto, Y., Minami, S., Preparation and biomedical applications of chitin and chitosan nanofibers, *Journal of Biomedical Nanotechnology* 10(10) (2014) 2891–2920.

Ali Akbari Ghavimi S., Ebrahimzadeh, M.H., Solati-Hashjin, M., Osman, N.A., Polycaprolactone/starch composite: Fabrication, structure, properties, and applications, *Journal of Biomedical Materials Research Part A* 103(7) (2015) 2482–2498.

Wang, Q., Ji, C., Sun, L., Sun, J., Liu, J., Cellulose nanofibrils filled poly (lactic acid) biocomposite filament for FDM 3D printing, *Molecules* 25(10) (2020) 2319.

Yang, L., Park, D., Qin, Z., Material function of mycelium-based bio-composite: A review, *Frontiers in Materials* 8 (2021) 737377.

Qazi, R.A., Khan, M.S., Shah, L.A., Ullah, R., Kausar, A., Khattak, R., Eco-friendly electronics, based on nanocomposites of biopolyester reinforced with carbon nanotubes: A review, *Polymer-Plastics Technology and Materials* 59(9) (2020) 928–951.

Ebrahimnezhad-Khaljiri, H., Ghadi, A., Recent advancement in synthesizing bio-epoxy nanocomposites using lignin, plant oils, saccharides, polyphenols, and natural rubbers: A review, *International Journal of Biological Macromolecules* (2024), 256, 128041. https://doi.org/10.1016/j.ijbiomac.2023.128041

Mobika, J., Rajkumar, M., Priya, V.N., Sibi, S.L., Substantial effect of silk fibroin reinforcement on properties of hydroxyapatite/silk fibroin nanocomposite for bone tissue engineering application, *Journal of Molecular Structure* 1206 (2020) 127739.

Wei, H., Rodriguez, K., Renneckar, S., Vikesland, P.J., Environmental science and engineering applications of nanocellulose-based nanocomposites, *Environmental Science: Nano* 1(4) (2014) 302–316.

Malmir, S., Barral, L., Bouza, R., Esperanza, M., Seoane, M., Feijoo-Bandín, S., Lago, F., Poly (3-hydroxybutyrate-co-3-hydroxyvalerate)/cellulose nanocrystal films: Artificial weathering, humidity absorption, water vapor transmission rate, antimicrobial activity and biocompatibility, *Cellulose* 26 (2019) 2333–2348.

Dutta, S., Pal, S., Panwar, P., Sharma, R.K., Bhutia, P.L., Biopolymeric nanocarriers for nutrient delivery and crop biofortification, *ACS Omega* 7(30) (2022) 25909–25920.

Moreira, J., Kuntzler, S., Vargas, B., Comitre, A., Costa, J.A., Morais, M., Nanofiber-reinforced bionanocomposites in agriculture applications, In: Muthukumar, C., Thiagamani, S.M.K., Krishnasamy, S., Nagarajan, R., Siengchin, S. (eds) *Polymer Based Bio-nanocomposites. Composites Science and Technology*. Springer, Singapore. (2022), 311–332. https://doi.org/10.1007/978-981-16-8578-1_17

Kumar, N., Kaur, P., Bhatia, S., Advances in bio-nanocomposite materials for food packaging: A review, *Nutrition & Food Science* 47(4) (2017b) 591–606.

Singh, P.P., *Environmental Remediation by Nanoadsorbents-Based Polymer Nanocomposite, New Polymer Nanocomposites for Environmental Remediation*, Elsevier 2018, pp. 223–241. https://doi.org/10.1016/B978-0-12-811033-1.00010-X

Bassyouni, M., Zoromba, M.S., Abdel-Aziz, M.H., Mosly, I., Extraction of nanocellulose for eco-friendly biocomposite adsorbent for wastewater treatment, *Polymers* 14 (9).(2022). https://doi.org/10.3390/polym14091852

Mruthunjayappa, M.H., Sharma, V.T., Dharmalingam, K., Sanna Kotrappanavar, N., Mondal, D., Engineering a biopolymer-based ultrafast permeable aerogel membrane decorated with task-specific Fe–Al nanocomposites for robust water purification, *ACS Applied Bio Materials* 3(8) (2020) 5233–5243.

Lin, S., Huang, R., Cheng, Y., Liu, J., Lau, B.L.T., Wiesner, M.R., Silver nanoparticle-alginate composite beads for point-of-use drinking water disinfection, *Water Research* 47(12) (2013) 3959–3965.

Thakur, S., An overview on alginate based bio-composite materials for wastewater remedial, *Materials Today: Proceedings* 37 (2021) 3305–3309.

Stanescu, P.O., Radu, I.C., Leu Alexa, R., Hudita, A., Tanasa, E., Ghitman, J., Stoian, O., Tsatsakis, A., Ginghina, O., Zaharia, C., Shtilman, M., Mezhuev, Y., Galateanu, B., Novel chitosan and bacterial cellulose biocomposites tailored with polymeric nanoparticles for modern wound dressing development, *Drug Deliv* 28(1) (2021) 1932–1950.

Yadav, M., Liu, Y.-K., Chiu, F.-C., fabrication of cellulose nanocrystal/silver/alginate bionanocomposite films with enhanced mechanical and barrier properties for food packaging application, *Nanomaterials*, 2019, 9 (11), 1523. https://doi.org/10.3390/nano9111523

Abdelwahab, N., Rabie, W., Mohamed, F., Fabrication and characterization of novel biocomposite based on Sargassum vulgare for controlling sugar beet root diseases, *Chemical and Biological Technologies in Agriculture* 10(1) (2023) 52.

Adesina Adegoke, K., Samuel Agboola, O., Ogunmodede, J., Oluyomi Araoye, A., Solomon Bello, O., Metal-organic frameworks as adsorbents for sequestering organic pollutants from wastewater, *Materials Chemistry and Physics* 253 (2020) 123246.

Kim, S.M., Rhie, Y.H., Kong, S.M., Kim, Y.S., Na, Y.H., Synthesis of nanocomposite hydrogels for improved water retention in horticultural soil, *ACS Agricultural Science & Technology* 2(6) (2022) 1206–1217.

Sethi, S., Singh, A., Medha, B.S. Kaith, S. T., Khullar, S., Natural polymer-based nanocomposite hydrogels as environmental remediation devices, in: U. Shanker, C.M. Hussain, M. Rani (Eds.), *Handbook of Green and Sustainable Nanotechnology: Fundamentals*, Developments and Applications, Springer International Publishing, Cham, 2023, pp. 407–441.

Li, Q., Yan, X., Shi, M., Wang, Q., Liu, H., Lin, Z., Huang, X., Recent advances in metal/ceria catalysts for air pollution control: mechanism insight and application, *Environmental Science: Nano* 8(10) (2021) 2760–2779.

Elhenawy, S., Khraisheh, M., AlMomani, F., Hassan, M.K., Al-Ghouti, M.A., Selvaraj, R., Recent developments and advancements in graphene-based technologies for oil spill cleanup and oil-water separation processes, *Nanomaterials (Basel)* 12(1) (2021), https://doi.org/10.3390/nano12010087

Damavandi, F., Soares, J.B.P., Polystyrene magnetic nanocomposite blend: An effective, facile, and economical alternative in oil spill removal applications, *Chemosphere* 286(Pt 2) (2022) 131611. https://doi.org/10.1016/j.chemosphere.2021.131611

Pandey, N., Shukla, S.K., Singh, N.B., Water purification by polymer nanocomposites: An overview, *Nanocomposites* 3(2) (2017) 47–66.

Gautam, P.K., Singh, A., Misra, K., Sahoo, A.K., Samanta, S.K., Synthesis and applications of biogenic nanomaterials in drinking and wastewater treatment, *Journal of Environmental Management* 231 (2019) 734–748.

Supriya, N.V. Rane, A. Chaturvedi, S.K. Vanka, A. Kumari, biobased graphene for synthesis of nanophotocatalysts in the treatment of wastewater: A review and future perspective, in: P.D. Pathak, S.A. Mandavgane (Eds.), *Biorefinery: A Sustainable Approach for the Production of Biomaterials, Biochemicals and Biofuels*, Springer Nature Singapore, Singapore, 2023, pp. 203–232.

Norizan, M.N., Shazleen, S.S., Alias, A.H., Sabaruddin, F.A., Asyraf, M.R., Zainudin, E.S., Abdullah, N., Samsudin, M.S., Kamarudin, S.H., Norrrahim, M.N., Nanocellulose-based nanocomposites for sustainable applications: A review, *Nanomaterials*, 2022 12 (19), 3483. https://doi.org/10.3390/nano12193483

Namasivayam, S.K.R., Prakash, P., Babu, V., Paul, E.J., Bharani, R.S.A., Kumar, J.A., Kavisri, M., Moovendhan, M., Aquatic biomass cellulose fabrication into cellulose nanocomposite and its application in water purification, *Journal of Cleaner Production* 396 (2023) 136386.

Zhang, Z., Ahmed, A.I.S., Malik, M.Z., Ali, N., Khan, A., Ali, F., Hassan, M.O., Mohamed, B.A., Zdarta, J., Bilal, M., Cellulose/inorganic nanoparticles-based nano-biocomposite for abatement of water and wastewater pollutants, *Chemosphere* 313 (2023) 137483.

Kolya, H., Kang, C.-W., Next-generation water treatment: Exploring the potential of biopolymer-based nanocomposites in adsorption and membrane filtration, *Polymers*, 2023, 15 (16), 3421. https://doi.org/10.3390/polym15163421

Dar, O.A., Malik, M.A., Talukdar, M.I.A., Hashmi, A.A., Chapter 19 - Bionanocomposites in water treatment, in: K. Mahmood Zia, F. Jabeen, M.N. Anjum, S. Ikram (Eds.), *Bionanocomposites*, Elsevier 2020, pp. 505–518, https://doi.org/10.1016/B978-0-12-816751-9.00019-2

Azizi-Lalabadi, M., Jafari, S.M., Bio-nanocomposites of graphene with biopolymers; fabrication, properties, and applications, *Advances in Colloid and Interface Science* 292 (2021) 102416.

Orsu, P., Koyyada, A., Recent progresses and challenges in graphene based nano materials for advanced therapeutical applications: A comprehensive review, *Materials Today Communications* 22 (2020) 100823

12 Bio-Nanomaterials in Environmental Remediation

Nisha Saini, Anjali Yadav, Nirmala Kumari Jangid, and Jaya Dwivedi

12.1 INTRODUCTION

In the twenty-first century, environmental pollution presents a substantial threat to both human populations and other living species on Earth. Due to increasing anthropogenic activities, every environmental domain like air, water, or soil is confronting significant challenges, due to continual and inappropriate alterations in their natural state (Kumar and Bharadvaja, 2019). Industrial, residential, and agricultural activities that are necessary to sustain human populations indiscriminately introduce pollutants into the environment. In 2016, water pollution killed an estimated 1.8 million people around the world. In the same year, there were more than 13.7 million fatalities worldwide caused by an unhealthy environment (Pandey et al., 2021). Research has demonstrated a direct correlation between contaminants and environmental health, encompassing human and other biological entities.

In this regard, nanotechnology presents a viable and appropriate option for the remediation of pollutants. Scientific research has been more interested in nanotechnology due to its effectiveness and adaptability.

NMs are currently a focus of research, with possible uses in biomedicine, increased agricultural output, and improved water treatment and clean-up. When discharged into water bodies, wastewater from industrial sources contains a concentrated mixture of heavy metals and dyes that could be harmful to aquatic life, as well as human health. Compounds with high toxicity and low degradability that are mutagenic, carcinogenic, and genotoxic are frequently found in toxic wastewater (Nguyen et al., 2022; Yadav et al., 2024). In recent decades, NMs have garnered attention as a promising absorbent for environmental remediation efforts. NMs have garnered interest owing to their distinctive attributes, including size, shape, and distribution. Furthermore, they manifest mechanical, magnetic, thermal, and optical characteristics. Typically, NPs are fabricated employing either a top-down or bottom-up methodology. The top-down process involves breaking up bulk material using tools like milling, grinding, and machining, whereas the bottom-up process involves putting tiny pieces together to create larger structures, such as the formation of NPs. Reducing agents are commonly employed in physical and chemical procedures to synthesize NPs by reacting with precursor material (Mathur et al., 2022).

However, because of their high toxicity, these processes produced by products that could be hazardous to humans and the environment. Thus, it is essential to create NMs that produce the least amount of waste or hazardous byproducts. Therefore, there is a discernible shift in favor of using eco-friendly reducing agents for the elimination process, due to their non-toxic nature and lower risk to human health, such as genotoxicity, lung toxicity, and carcinogenicity.

The great stability and low toxicity of biological substances are attributed to the presence of bioactive components (Bolade et al., 2020). Bio NMs are especially well-suited for treating industrial wastewater because of their environmentally beneficial qualities. These days, bio-NM-based technologies are thought to offer a possible replacement for traditional therapeutic approaches. Bio NMs are leading the advancement of the swiftly growing field of nanotechnology. Owing to their small size, high surface area to volume ratio, quantum characteristics, and low toxicity, bio NMs have special physicochemical properties that make them ideal for a variety of uses in the fields of health,

DOI: 10.1201/9781003484189-12

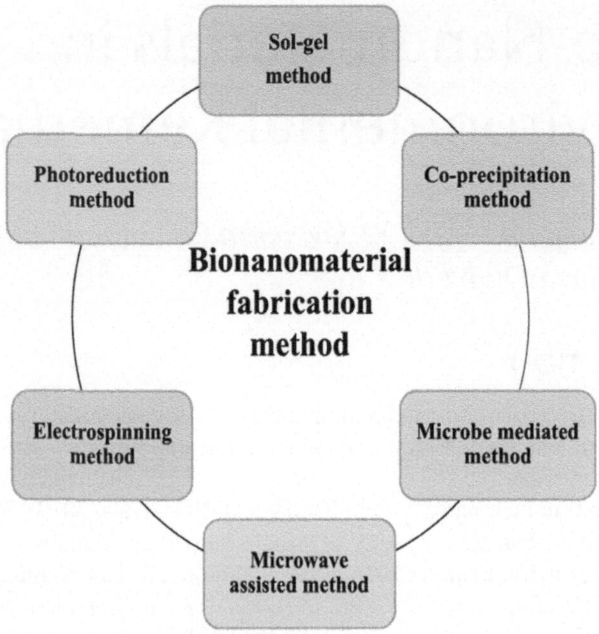

FIGURE 12.1 Methods for the preparation of bio NMs.

biology, chemistry, energy, and environmental sciences. The unique attributes of bio NMs have positioned them as environmentally sustainable options in NMs research. Several methods exist for fabricating intelligent NMs (Figure 12.1). The current chapter concentrates on the development and environmental remediation prospects of bio NMs.

12.2 BIO-NMs

The use of NMs, which are usually between 1 and 100 nanometers in size, has been made easier by nanotechnology. Because of their small size, large surface area, and significant reactivity, NMs are effective adsorbents in a variety of applications.

There are several methods for creating NMs, such as chemical, biological, and physical processes (Bolade et al., 2020) (Figure 12.2). Due to their large surface area, small size, and significant reactivity, NMs are effective adsorbents in a variety of applications. Any material's surface area per unit mass increases as it is reduced to a nanoscale dimension (Saraswat et al., 2023; Srivastava et al., 2023). As a result, the greater exposure of materials to surrounding chemicals can change their reactivity. NMs are utilized in environmental remediation because of their diverse architectures and morphologies, which enable the removal of various harmful elements. Figure 12.3 represents various advantages of bio-NMs. Moreover, NMs possess the capability to access contaminated regions inaccessible to microparticles.

Because of their unique thermal and electrical properties, NMs produced by both physical and chemical techniques were initially exploited in the electronics industry. Nonetheless, their large surface area, ease of modification, and superior microbial disinfection capacity make them effective at removing organic and inorganic pollutants. This shows that carbon-based NMs and other changes have the potential for novel applications in water treatment. NMs generated from chemicals, such as graphene and titanium dioxide, have a wide range of applications. However, some metal-based NPs are persistent and do not degrade, which raises several environmental concerns (Khan et al., 2019). There are drawbacks to the chemical synthesis methods used to create NMs, including the possibility of contamination from precursor molecules and the release of toxic byproducts. Therefore, there

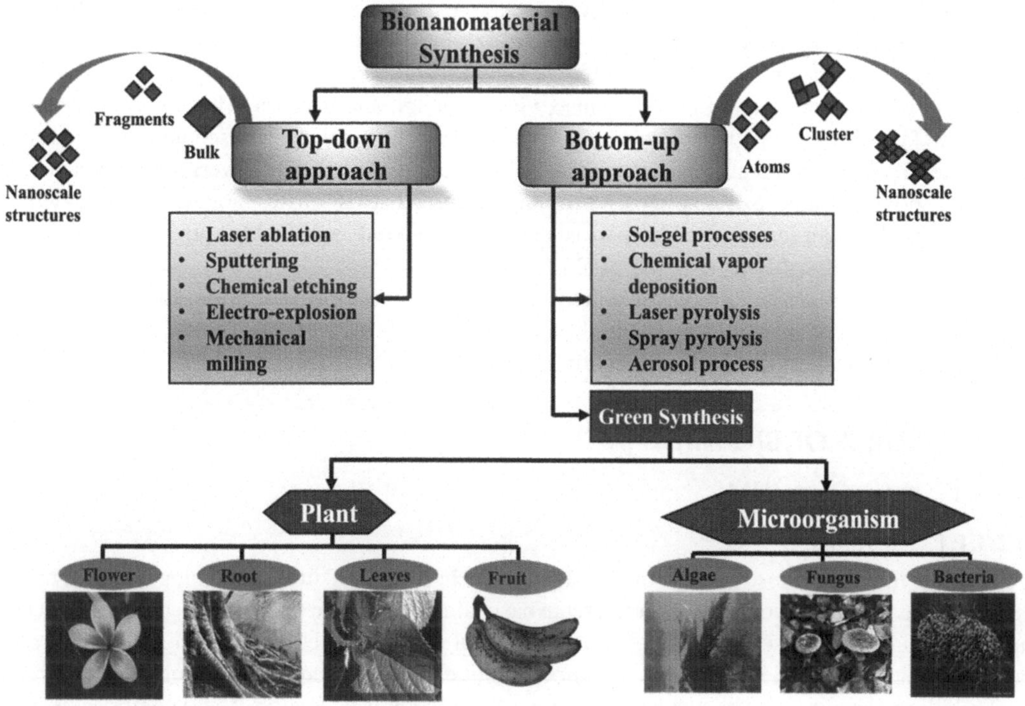

FIGURE 12.2 Synthesis of Bio NMs.

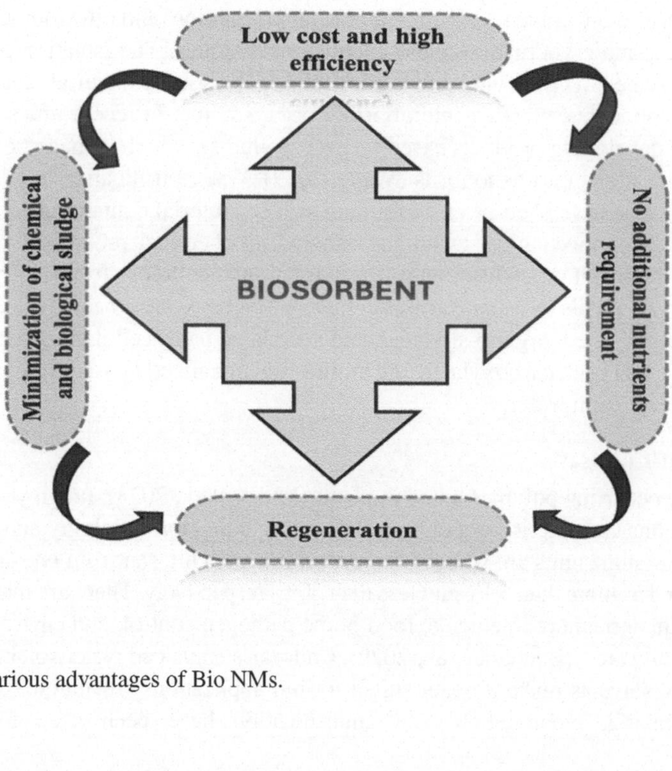

FIGURE 12.3 Various advantages of Bio NMs.

is a pressing need for extremely effective, economical, and ecologically friendly methods of producing materials at the nanoscale.

Biologically derived reducing agents (Bio-based NMs) are required to effectively eliminate toxins. These NMs can be found in plants, algae, viruses, fungi, and bacteria. They are referred to as "bio-nano factories," distinguished by their great efficiency due to their macroscopic nature, particularly excelling in heavy metal clean-up applications (Saraswat et al., 2023). Diverse microorganisms, including flora and their extracts, agricultural and agro-industrial waste, biomolecules, enzymes, and algae, including micro-seaweeds, microalgae, phytoplankton, zooplankton, and diatoms, can be used to create bio-based NMs (Roopan et al., 2013; Zhang et al., 2022).

Materials obtained from biological sources are referred to as bio-based NMs. Both organic and inorganic NPs can be produced by living things. NMs are created when biological species interact with inorganic materials through a variety of intracellular and extracellular processes.

12.3 SOURCE OF BIO NMs

12.3.1 Plant-Based NMs

12.3.1.1 Cellulose

Among polymers, cellulose stands out as a significant chemical constituent and biopolymer. Thirty-six cellulose molecules come together to create basic units, which are then crammed into 20–50 nm micro-fibrillated cellulose structures. The structural stability of the cellulose chain is ensured by its arrangement, which is similar to that of nanocrystalline rods. Consequently, cellulose microfibrils and cellulose nanocrystals, the two primary types of nanocellulose, emerge. These comprise the following nine categories of cellulose-based particles: wood fiber, microcrystalline cellulose, plant fibers, micro-fibrillated cellulose, algal cellulose nanocrystals, and tunicate cellulose NPs (Zhang et al., 2013).

Nanocellulose exhibits cellulose characteristics, like a crystalline structure, hydrophilicity, and extensive surface area. Extraction methods for nanocellulose include top-down approaches and various physical (such as microwave treatment, gamma ray irradiation, and ultrasonication), mechanical (including crushing and cryocrushing), and chemical techniques. The isolation process comprises two fundamental stages: (i) purifying and homogenizing the source material, and (ii) isolating the pure cellulosic components, whether microfibrillar or crystalline. In recent times, researchers have been interested in developing products based on nano-cellulose, which is made possible by the conversion of cellulose fibers into nanofibrils (Wang, 2019). Water purification applications involving NPs based on cellulose have been extensively studied. The material containing cellulose is first pre-treated by a variety of mechanical techniques, including chemical processes, cryocrushing, and refining. The pre-treatment procedure entails the extraction of cellulose from other constituents. For the sorption of heavy metals in water, various cellulose NM types have been used, such as carboxylated, thiolated, succinylated, organo-silylated, and acetylated nano-cellulose. Furthermore, cationic nano-cellulose that has been carboxylated and amino-functionalized is used to separate or eliminate organic contaminants from water.

12.3.1.2 Chitin/Chitosan

Chitin, a naturally occurring polymer found in plants (Yadav et al., 2023), fungi, yeast, and arthropod exoskeletons, exhibits characteristics akin to cellulose, including low solubility and chemical reactivity. *N*-acetyl-glucosamine units are what distinguish chitin and chitosan from one another. Chitin and chitosan are defined as more than 50% and less than 50%, respectively. There are many uses for chitin/chitosan, such as in agriculture, medicine, food-borne pathogen control, and environmental remediation (purification of water) (Bandara et al., 2020). Chitosan's enhanced water solubility and compatibility with organic solvents make it better suited for bio applications. Techniques like hydrothermal methods, cross-linking, precipitation, and emulsification have been used to prepare chitin/

chitosan-based nanomedicines (NMs). Chitin/chitosan-based NMs have garnered significant attention in environmental remediation research, offering promising prospects for diverse applications. Also, bio-based NMs derived from alginates, casein, starch, and soy protein isolates have also been investigated for their wide-ranging bio-applications. The goal of the industrial packaging system of today is to offer methods for more dependable and safe procedures. Chitin-derived NMs have been widely employed and investigated for this purpose. An additional use for nanocarriers is the efficient delivery of medications to their intended locations, such as curcumin, a significant polyphenol. Drug delivery often involves the utilization of specific protein-based biopolymers and polysaccharide-based bioNMs (MoballeghNasery et al., 2020).

12.3.2 Biogenic Synthesis of NMs

Microorganisms, including fungi, algae, plants, bacteria, and others, can create NMs both extracellularly and intracellularly. These bioinspired nanomedicines, or NMs, are becoming more powerful instruments, with more potential to solve environmental problems in an eco-friendly and cost-effective way.

12.3.2.1 Bacteria

Currently, many techniques are used to remediate water, including exposure to UV radiation and chlorination. Membrane-based procedures, however, are significantly simpler than other techniques. They are also believed to be scalable, less polluting, and environmentally beneficial. They are widely used in the treatment of water and wastewater because they require less energy (Alves et al., 2020). As a biological reservoir, bacterial cellulose (BC) facilitates the synthesis of NMs, particularly bio-cellulose membranes. Similar to plant cellulose, bacterial fermentation or metabolic activities can easily produce BC. Remarkably, BC is easier to make than plant cellulose as it doesn't contain hemicelluloses, lignin, or pectin.

12.3.2.2 Fungi Mediated Synthesis

A recently developed area known as "myconanotechnology" investigates the use of fungi in the cost-effective synthesis of NPs (NPs) with notable stability and precise size control. With their ability to absorb metal ions, remarkable metal tolerance, and diverse nature, fungi are attracting a lot of interest in mycosynthesis of NPs. Fungal species, akin to bacteria; employ metal biomineralization and biotransformation mechanisms in the production of mycogenic NPs. There are numerous studies documenting the effective mycogenic synthesis of metallic NPs for the pre-concentration of common environmental pollutants. Yadav et al. (2015) observed that fungi secrete a variety of proteins, enzymes, polypeptides, and metabolic intermediates, which aid in the synthesis, stabilization, and immobilization of metallic NPs. This enables the NPs to maintain their capabilities as nano adsorbents and nanocatalysts. As an illustration, Bhargava et al. used A. japonicus AJP01 to quickly create a "nanogold-fungal composite," which enhanced the catalytic reduction of hexacyanoferrate (III) and 4-nitrophenol with $NaBH_4$ (Bhargava et al., 2015). In this instance, the fungal strain performed two tasks to avoid aggregation drawbacks: it reduced Au(III) to Au(0) and anchored the resulting Au NPs onto the fungal hybrid. In an additional study, the soil fungus C. oxysporum AJP03 was utilized to generate quasi-spherical gold NPs with an average size of 72.32 ± 21.80 nm (Bhargava et al., 2016).

12.3.2.3 Algae

Since protecting the environment comes first, using bio-based or green synthesized non-metallic NPs is thought to be the best choice. Algae are employed in the synthesis of NMs, in addition to other techniques. Employing algae either as living cell factories or utilizing algal extracts as natural reducing agents is proposed as an environmentally friendly and effective method for synthesizing various metallic NPs. Currently, marine polysaccharides find applications in diverse nanobiotechnological domains, encompassing tissue engineering, gene and drug delivery, biosensors, and water treatment. Polysaccharides like fucoidan, carrageenan, ulvan, agarose, and alginate are abundant in a variety of marine organisms, particularly algae. Their natural abundance, safety, non-toxicity, and biodegradability make these polysaccharides essential in the synthesis of NMs. Marine polysaccharides can be skillfully transformed into a range of nanostructures, including NPs, nanofibers, beads, and sponges, to address a variety of nanobiotechnology needs. While marine algae have historically been valued for their nutritional, additive, and medicinal properties, they are now gaining increased recognition in NP synthesis research (AlNadhari et al., 2021).

Fluid concentrations from two algae species—*Turbinariaconoides* and *Sargassumtenerrimum*—were used as a reducing agent in the production of Au-NPs in a work by Ramakrishna et al. (Ramakrishna et al., 2016). Additionally, it was found that these biosynthesized Au-NPs exhibited efficiency in degrading p-nitroaniline, 4-nitrophenol, sulforhodamine, and rhodamine B. The synthesis of NMs based on algae has, however, received little research and has only a few uses in the elimination of environmental contaminants (Li et al., 2021). As a result, research on algae and the synthesis of NMs is still on going.

12.3.3 Agricultural Waste as Biosorbents

The advantages of using agricultural waste materials for wastewater treatment are several, including cost-effectiveness, environmental friendliness, renewability, chemical stability, and simplicity of access. These agricultural materials are effective in the adsorption process because of their structural properties, which include a large surface area, great porosity, and a variety of functional groups (Saravanan et al., 2021). The bioactive components, which include lignin, pectin, cellulose, and hemicellulose, interact with metal ions to facilitate their removal. Agricultural waste materials that are ideal for this use include rice husk, sugar beet pulp, wheat bran, bark, dry leaves, and other fruit and vegetable peels. The tannin-rich bark enhances the adsorption process. The use of agricultural waste materials for pollutant removal has been extensively studied.

12.3.4 Industrial Waste Materials as Biosorbents

Products made from industrial waste are powerful sorbents that take out toxins from the air. These substances can have their surfaces chemically altered or activated to increase the effectiveness of adsorption. Byproducts are typically obtained as industrial waste materials. These materials can be acquired at a reasonable price because of their wide availability. Bark, rich in tannins, facilitates the adsorption process. The use of agricultural waste materials to remove pollutants from wastewater streams has been extensively studied. Sawdust binds to reactive colors with the aid of organic components like cellulose, hemicellulose, and lignin. Sawdust's charge can be either positive or negative based on the pH of the mixture. pH is important in the sorption process. Adsorbents can also be made from other industrial products, including fly ash, sugarcane bagasse, sludge, furnace slag, etc. (Regkouzas and Diamadopoulos, 2019).

12.4 APPLICATIONS

12.4.1 Removal of Water Pollutants

12.4.1.1 Dye

The chromogenic compounds with an aromatic ring structure that give color to any material are called dyes. Chromogenic sites, which encompass functional groups like azo, nitro, carboxyl, aromatic amine, or methylene groups, define the properties of dyes. These compounds find application in various industries, including textile, food, pulp and paper, pharmaceuticals, and dye cinematography. They are also used in scientific laboratories as substances for intercalating, tracking, and

TABLE 12.1
Dye Detection in Effluents and Removal Using Bio-fabricated NMs

Bio-fabricated Nanocatalyst	Reductant/Capping Agent/Stabilizer	Dye	Efficiency/ Parameter	References
CuO NPs	Psidiumguajava	Nile blue	93%, 100min	Singh et al.(2019)
CuO NPs	Psidiumguajava	RY160	80%,120min	
CuNPs	Escherichia sp. SINT7	Congo red malachite green Direct blue-1 Reactive black-5	97.7% 90.55% 88.42% 83.61%	Noman et al.(2020)
SnO_2 NPs	Piper betle	Reactive yellow 186	92.17%,160min	Singh et al.(2018)
CuO-NPs	Solanumnigrum	Methylene blue	97%, 50min	Muthuvel et al.(2020)
AgNPs	Bromothymol blue	Anthocyanin	20min	Al-Thabaiti and Khan (2020)
	Methylene Blue Dye	Activated Carbon Prepared from Citrulluslanatus Rind	98%	Üner et al. (2016)
	Congo red Dye	TiO_2NPs loaded onto AC prepared from bio-waste watermelon rind	101.45%	Masoudian et al.(2019)
	Basic Brown 16 Dye	Durian (Duriozibethinusmurray) Shell	74.26%	Gopalakrishnan et al. (2020)

detection. One significant source of environmental heavy metal contamination is the dye industry. Before dye-contaminated water transfers chemical contaminants to receiving lands, it is usually preferable to clean it up (Jangid et al., 2020;Liu et al.,2020). The main causes of water pollution are the industrial effluents released into the environment. To clarify the underlying mechanisms, this part focuses on investigating the degradation of highly hazardous colors utilizing bio-fabricated NMs (Table 12.1).

Green synthesis produces NPs by reducing a metal salt to NPs and using phytochemicals for capping and reducing purposes. The hydroxyl and carboxylate groups found in phenolic compounds, flavonoids, and other biomolecules have a strong affinity for metals and function as a capping and reducing agent for NPs. In a synergistic process, an array of extracellular secretory enzymes, proteins, peptides, small sugars, secondary metabolites, and low molecular weight compounds from microbial and plant sources perform vital functions in reducing metal salts and alleviating pollutant toxicity (Zaib et al., 2021). Reactive oxygen species (ROS) are produced by green-synthesized NPs by the photo-catalytic technique, which is one of the processes involved in the breakdown process of dyes. Superoxide and hydroxyl radicals are two examples of these ROS that interact with dye molecules and cause their destruction (Anchan et al., 2019).

12.4.1.2 Phenolic Compounds Removal

Several prevalent biosorbents, including rice husk, chitosan, chitin, and Azollafiliculoides, have been employed to eliminate phenolic compounds from wastewater. Additionally, in a related investigation, tendu leaf waste demonstrated notable efficacy as a biosorbent in phenol removal, with maximum sorption capacities of 7.69 and 31.35 mg/g for raw and modified materials, respectively. Adebayo and Areo (2021) documented the synthesis of an adsorbent designed for eliminating 4-nitrophenol and phenol from aqueous solutions using clay and coconut shells. The adsorption isotherm was found to be represented by the Liu isotherm, but the Avrami fractional kinetic model governed the adsorption kinetics. Maximum adsorption capacities for 4-nitrophenol and phenol were found to be 476.9 and 1665 mg/g, respectively, in the investigation.

A bio-derived silica/calcium carbonate ($CSeSiO_2/CaCO_3$) nanocomposite made from residual eggshells (ES) and rice husks (RH) was used to eliminate phenol, as reported by Birma et al. (Bwatanglang et al., 2021). According to the study's findings, the $CSeSiO_2/CaCO_3$ nanocomposite outperformed other adsorbents in terms of phenol removal efficiency. According to the pseudo-second-order kinetic model, the adsorption process occurs. Phenol was shown to have a sorption capacity of 14.06, 10.64, and 10.33 mg/g on the $CSeSiO_2/CaCO_3$ nanocomposite, $RHSiO_2$, and $ESCaCO_3$, respectively. Furthermore, the current adsorbent is under the Freundlich and Langmuir isotherm models.

12.4.1.3 Pesticides

Using biosorbents to eliminate pesticides, such oxamyl, ethroprophos, DDT, and its derivatives, Aldrin, malathion, lindane, quinalphos, dieldrin, endrin, and several others, has been investigated in the research. It was successfully possible to extract the 4,4-DDT-containing aqueous solution using wood sawdust and cork waste. These materials are significantly impacted by the biosorbent's size. Ethroprophos was bound using chitosan derived from prawn shells (Chen et al., 2011). The removal of odor and color from prawn shells was achieved by treatment with $KMNO_4$ and oxalic acid, followed by deactivation with HCl and deproteinization with NaOH. The material showed a maximum sorption capacity of 85.47 mg/g when wood sawdust and cork waste were used to make an aqueous solution.

Haq et al. found that neem seed powder (NSP) biosporptively removed the pesticide halosulfuron methyl from aqueous media. 54% removal efficiency was found at a pH of 3 (Haq et al., 2021). The pseudo-second-order kinetic model was consistent with the adsorption kinetics, and the Langmuir model was consistent with the isotherm. It was discovered that halosulfuron methyl has a maximum adsorption capability of 200 mg/g. This confirms that it is one of the most affordable, environmentally benign, and efficient adsorbents available for the elimination of halosulfuron methyl. Boudesocque et al. utilized lignocellulosic substrates sourced from local agroindustry to achieve the removal of pesticides such as dimetomorph (4-[3-(4-chlorophenyl)-3-(3,4-dimethoxyphenyl)acryloyl]morpholine), terbumeton (N-(1,1-dimethyl)-N-ethyl-6-methoxy-1,3,5-triazine-2,4-diamine), isoproturon (3-(4-isopropylphenyl)-1,1-dimethylurea) and diethylterbumeton (N-(1,1-dimethylethyl)-6-methoxy-1,3,5-triazine-2,4-diamine) (Boudesocque et al., 2008). Adsorption kinetics shows that pesticides are absorbed at rates of 1 to 8 grams per kilogram of lignocellulosic substrate. **Table 12.2** gives a summary of the chemicals that have been removed with the help of Bio NMs.

TABLE 12.2
Bio NMs in the Removal of Pesticides

Adsorbents	Pollutants	Dosage	pH	Time	References
Fe_3O_4/graphene nanocomposite	Triazine	0.5 g	5.0	70 min	Boruah et al.(2017)
Fe/Pd bimetallic NPs	DDT	0.55 g	10.0	30 min	Ulucan-Altuntas and Debik(2020)
Chitosan-ZnO NPs	Permethrin	0.5 g	7.0	90 min	Dehaghi et al.(2014)
Copper chitosan nanocomposite	Malathion	1.0 g	2.0	16 h	Jaiswal et al.(2012)
Pd NP loaded TiO_2	Paraoxon	0.02 g	–	120 min	Keihan et al.(2017)
LFCOs NPs	Vitavax	3 g	7.0	40 min	Tavakkoli and Yazdanbakhsh(2013)

12.4.1.4 Organic Herbicides

Numerous research endeavours have explored the efficacy of biosorbents in alleviating the impact of organic herbicides, including paraquat, metribuzin, and moline. A scientist used rice husks coated with methacrylic acid in a study to create a biosorbent that would remove paraquat.

Huang et al. synthesized a composite of chitosan and a metal-organic framework modified by ionic liquid for the adsorption of 2,4-dichlorophenoxyacetic acid (Huang et al., 2020a,b). It was found that the adsorption isotherm and kinetics agreed with the pseudo-second-order kinetic model and the Temkin isotherm, respectively. Lago et al. suggested that atrazine removal could be achieved by employing recycled waste and bio waste in permeable barriers (Lago et al., 2021). In contrast, 20% acetonitrile was required to fully recover atrazine adsorbed on sepiolite. Atrazine was completely removed from cork (98% and 94%) and pine bark (94%). Sepiolite demonstrated the highest adsorption capacity at 23.3 mg/g, while cork and pine bark followed with capacities of 13.1 and 14.8 mg/g, respectively. The recycling potential of sorbents was evaluated by conducting adsorption-desorption cycles. After two regeneration cycles, sepiolite showed the highest efficiency, eliminating 81% of atrazine; pine came in second with 78% and cork with 54%. Mehmandost et al. (2020)used biosorbents made from HeracleumPersicum to remove paraquat and diquat from water sources. It was discovered that even after five reuses, both biosorbents maintained efficiency in the removal of herbicides that was greater than 82%. The findings demonstrate that low ionic strength reduces sorption capacity due to biosorbent aggregation.

12.4.1.5 Pharmaceutical Pollutants

Pharmaceutical compounds have lately appeared as a novel type of organic micropollutant in water. If these substances enter the water distribution system, they pose a threat to living organisms. Such micropollutants encompass antibiotics, steroids, hormones, and other endocrine-disrupting chemicals, which can pose significant risks to both human health and wildlife. The nano-based method has presented a promising answer to this challenge (Malik et al., 2017).

Recently, D. vulgaris was employed to synthesize bio-Pt and bio-Pd-nanocatalysts for the removal of three pharmaceutical pollutants: ciprofloxacin, sulfamethoxazole, and 17β-estradiol. These synthesized NPs exhibited high effectiveness, achieving clearance rates of approximately 94.85% for 17β-estradiol and sulfamethoxazole, and 70% for ciprofloxacin. Diatrizoate is classified as a persistent micropollutant because of its high chemical stability and resistance to biological breakdown, emphasizing the significance of removing it from effluent streams. To overcome this, an effort was made to produce palladium NPs (Pd-NPs) using *S. oneidensis*(De Corte et al., 2011).

12.4.1.6 Polycyclic Aromatic Hydrocarbons (PAHs)

Many different types of biosorbents have been used to remove different PAHs from aqueous solutions, including naphthalene, phenanthrene, anthracene, pyrene, fluorene, chrysene, and fluoranthene. The efficiency of *E. intestinalis*, green macroalgal biomass, as a biosorbent for cleaning freshwater and seawater was evaluated by Boleydei and co-workers (2018). According to their research, seawater has a greater capacity to sorb crude and used oil than freshwater does. They also observed that the isotherms fit the Langmuir isotherm model, whereas the adsorption kinetics adheres to the pseudo-second-order kinetic model. Three PAH compounds—benzopyrene, naphthalene, and phenanthrene—have the capacity to be absorbed by green algae (Zhang et al., 2019). Nevertheless, the removal of components of algal lipids led to a 25% decrease in adsorption capacity. Cai et al. investigated the adsorption kinetics of biochar made from crab shells with mesoporous structures (Cai et al., 2019). Their research showed that 93.9 mg of diesel oil could be adsorbed by the biochar. A high specific surface area (2441 m^2/g), a large pore volume (1682 m^2/g), and different surface properties defined by hydroxyl and carboxyl functional groups all contribute to this capability.

12.4.1.7 Removal of Heavy Metals

Much research has been conducted to eliminate metal ions from contaminated water utilizing bio fabricated NMs. In addition to harmful organic pollutants, toxic inorganic pollutants include heavy metals such as $Cr^{6}+$, Cd^{2+}, Cu^{2+}, Zn^{2+}, Hg^{2+}, Pb^{2+}, Cr^{3+}, and Mn^{2+}. In recent times, NMs/NPs (NMs/NPs) have been employed for the removal of heavy metals (Table 12.3).

Lignin extracted from rice husk was employed in the production of a double network hydrogel adsorbent (RH-CTS/PAM gel) aimed at removing Pb(II), Cd(II), and Cu(II), achieving maximum sorption capacities of 374.32, 268.98, and 196.68 mg/g, respectively. The exceptional adsorption performance of RH-CTS/PAM gel was attributed to macro-porous structure and high buried water content (Ma et al., 2019). Wastes from peanuts were also considered to be efficient adsorbents for the elimination of metal ions from the contaminated water. Activated carbon prepared from peanut shells was found to show effective elimination of Cd(II) from the wastewater, attaining the utmost sorption capacity of 55.43 ± 6.82 mg/g (da Gama et al., 2018). $FeCl_3$-modified pomelo peel -(FPP) displayed efficient Cr(VI) removal, with a maximum sorption capacity of 21.55 mg/gram. The hydrolysis of $FeCl_3$ increases the specific surface area and adsorption sites of FPP, hence increasing its sorption capacity (Wang et al., 2020). Chang et al. investigated the removal of Cu(II) and Ni(II) ions with bentonite and bentonite/GO composites (Chang et al., 2020). Using bentonite, they attained a maximum sorption capacity of 248.9 mg/g for Cu(II) and 215.8 mg/g for Ni(II), whereas the sorption capacity of the bentonite/GO composite was 558.4 mg/g for Cu(II) and 402.5 mg/g for

TABLE 12.3
BiofabricatedNanocomposites for Heavy Metal Removal

Adsorbent	Heavy Metal	Surface area (m^2/g)	C_o (mg/L)	q_{max} (mg/g)	References
Activated carbon prepared by chestnut shell	Chromium (VI)	989.4	50	33	Niazi et al. (2018)
Modified mangosteen shell	Nickel (II)	209	10	57.14	Anitha et al. (2021)
Modified Jatoba fruit shell	Lead (II)	N.R	50–800	48.75	Souza et al. (2017)
	Cadmium(II)	N.R	50–800	30.27	
Corncob activated carbon	Chromium (VI)	826.6041	0.5–10	9.985	Li et al. (2018)
Modified teakwood sawdust	Chromium (VI)	999.65	35–290	72.46	Ramirez et al. (2020)
Activated carbon prepared from *Ulva lactuca*	Lead (II)	345.40	60	83.3	Ibrahim et al. (2016)
	Chromium (III)	345.40	60	82	
	Copper (II)	345.40	60	84.7	
	Cadmium (II)	345.40	60	84.6	
Acid-activated *SterculiavillosaRoxb.* Shells	Chromium (VI)	971.97	50–250	163.51	Patra et al. (2019)
Modified spent coffee powder	Arsenic (V)	N.R.	1–100	83.3	Hao et al. (2017)
	Copper (II)	N.R.	1–100	200.1	
ZVI NPs supported on modified waste rock wool	Chromium (VI)	102	10–200	197.69	Zhou et al. (2018)
Coconut husk based activated carbon	Mercury (II)	N.R	40–450	44.9	Guo et al. d(2017)

Ni(II) ions. Aquatic biomass has been studied and demonstrated to remove metal ions from many types of water. Ibrahim et al. examined the removal of Pb(II), Cr(III), Cu(II), and Cd(II) utilizing marine algae Ulva lactuca powder (AP) and Ulva lactuca-derived activated carbon (AAC) (Ibrahim et al., 2016). Using AP and AAC, the maximum adsorption capacity was 68.9 and 83.3 mg/g for Pb(II), 60.9 and 82 mg/g for Cr(III), 64.5 and 84.7 mg/g for Cu(II), and 62.5 and 84.6 mg/g for Cd(II). Magnetized peanut husks were used to remove Cd(II) and Pb(II) from contaminated water, with maximal sorption capacities of 7.68 and 28.3 mg/g, respectively (Rozumová et al., 2016).

12.4.2 Removal of Air Pollutants

Air pollution poses a significant threat to global public health, with detrimental effects on the human population. As per the World Health Organization (WHO), about 600,000 children below the age of five succumb annually due to air pollution. Moreover, there is an increasing risk of airborne infections like asthma, acute respiratory diseases, lung cancer, and chronic respiratory diseases, given that 92% of the global population lives in regions where air pollution surpasses acceptable thresholds. Particulate matter (PM), which is accountable for over 4.2 million fatalities each year, is one of the primary perpetrators of air pollution, along with ozone, sulphur dioxide, nitrogen dioxide, polycyclic aromatic hydrocarbons (PAHs), and volatile organic compounds (VOCs)(García-Mayagoitia et al., 2020). Hybrid and nanocomposite techniques are being used to create bioactive and biodegradable biomaterials, which are becoming more and more necessary as global pollution levels rise. Several approaches are being developed to meet this demand, one of which is Hydrothermal Carbonization (HTC), which takes place in the presence of water under moderate pressure and temperature.

Many industries including medication delivery, CO_2 sequestration, sensor technology, and catalysis, use hydrothermal carbonization (HTC). Carbon nanotube synthesis has also been accomplished through the use of Chemical Vapor Deposition (CVD), another noteworthy method. A variety of contaminants are included in the category of air pollution, such as sulphur dioxide, hydrocarbons, heavy metals, nitrogen oxides, sulphur monoxide (CO), and chlorofluorocarbons (CFCs).

Materials that are pollutant-specific, biodegradable, non-toxic, inexpensive, easily synthesized, recyclable, and stable are necessary for efficient air decontamination (Guerra et al., 2018). Furthermore, air biofiltration poses three main challenges:

i) Development of filter materials,
ii) Process optimization, and
iii) Studies focusing on microbiological enhancement.

The use of bio fabricated NMs for air pollution treatment, including bio-nanocomposites and bio-layered nanofibrous structures, is explained in more detail in the section that follows. Air pollution is a major environmental and societal issue that requires practical solutions. Particulate matter (PM), which can enter human lungs and bronchi, stands out among these problems as a primary cause of air pollution. Cars and factories that release smoke are the main sources of PM pollution. Various materials have been created to reduce particulate matter in the atmosphere, frequently utilizing environmentally benign and sustainable resources. Petroleum-based polymers used as air filters, such as polypropylene (PP), polyethylene (PE), polystyrene (PS), and polyamide (PA), are among the examples [Khalid et al., 2017]. Materials used as membranes in air filtration systems include bacterial cellulose (BC) and soy protein isolate (SPI).

The remarkable ability of the membrane to capture PM is attributed to the dipole interaction and electrostatic attraction between the functional groups of soy protein and PM. Moreover, the mechanical and physical properties of Acetobacter xylinum-produced bacterial cellulose are unique. By adding 4 weights percent acrylic acid to SPI, the particulate removal efficiency of 99.94% for PM2.5 and 99.95% for PM10 are achieved in the SPI/BC composite film, which is comparable to the air penetration rate of disposable masks (Liu et al., 2017). Artificially generated nanofibers are used in the production of innovative air filtration materials by electrospinning technology. These materials are classified as porous three-dimensional nanofibrous aerogels and two-dimensional constructions. Notable advantages of these membrane filters include high porosity, large surface area, small pore size, and a network of interconnected holes that traps fine particles(Duan et al., 2018). Membranes made of Ag/PAN (Silver NPs on Polyacrylonitrile) are an example of such air filters. Owing to its capacity to remove NPs from the air and successfully fight bacteria, these membranes can be used in masks, biomedical applications, electronics, and indoor air purification (Bortolassi et al., 2019).

12.4.3 Removal of Soil Pollutants

Bioremediation stands out as an environmentally friendly soil treatment method, offering benefits such as safety, absence of secondary pollution, and cost-effectiveness. The three main methods used to treat petroleum-contaminated soil are phytoremediation, plant-microbial remediation, and microbial remediation. Bacterial cellulose (BC) added to petroleum-contaminated soils did not speed up the bioremediation process (Saum et al., 2018). However, the utilization of plants (mesquite tree seedlings), compost, and combinations thereof enhanced the phytoremediation of petroleum-contaminated soil. Barati et al. examined and compared the efficacy of poultry manure and its bacterial cellulose (BC) derivative for degrading petroleum hydrocarbons in contaminated soils (Barati et al., 2018). They used barley plants in a greenhouse setting for five months to evaluate deterioration at different total petroleum hydrocarbon (TPH) levels. The PMB-treated soil showed the highest dry weight, microbial respiration, and potential for decomposing total petroleum hydrocarbons (TPHs) when the two additions were compared. They also explored the impact of plant growth (oat and barley) and bacterial cellulose (BC) derived from poultry manure on total petroleum hydrocarbon (TPH) degradation and microbial respiration rate in contaminated soil (Barati et al., 2018). This discovery indicates a notable enhancement in both the average microbial respiration rate and the percentage of total petroleum hydrocarbons (TPHs) reduction in soil treated with bacterial cellulose (BC). In comparison to treatments without BC, there was an improvement of 37.73% in the microbial respiration rate and 21.76% in the reduction of TPHs for barley 45.18% in the microbiological respiration rate and 20.36% in the reduction of TPHs for oat plants. Heavy metal pollution of soil is a serious problem. Soil mercury concentrations grow noticeably as a result of fossil fuel combustion and air dispersion, usually by a factor of 3–10.

Mercury-contaminated soil has been remedied using a variety of techniques, such as biological methods, thermal treatment, stabilization or solidification, and soil washing. Recently, Wang et al. investigated the use of biogenically produced selenium NPs generated from *Citrobacter freundii Y9* for the bioremediation of Hg-contaminated soil (Wang et al., 2017a; Wang et al., 2017b).

12.5 DISPOSAL OF BIOSORBENTS

Only a few studies on biosorbent regeneration and reutilization exist, and none of them address the eventual disposal of biosorbents. In the context of the widespread industrial application of biosorbents, careful consideration must be given to the proper disposal methods to minimize environmental impact (Huang et al., 2020a,b). Various biosorbent disposal options include landfill deposition, incineration, fertilizer use, regeneration, and reuse, among others. Because of the high lignin and cellulose content of biosorbents, incineration is often chosen as the first stage. This successfully reduces volume and mass, while facilitating metal recovery (Fernández-González et al., 2019). The biomass was revealed to have a higher carbon content and lower oxygen content, which is beneficial for increasing thermal energy output. Furthermore, it had low sulphur and nitrogen content, indicating that corrosive and poisonous gasses were emitted to a minimum during combustion. As a result, pyrolysis emerges as a promising method for extracting thermal energy from biomass. Furthermore, landfill deposition is a low-cost and simple method for disposing of excess biosorbents. Toxic metal ions are removed from biosorbents before disposing of in landfills. Alternative disposal strategies, such as burial in soil or dispersion on the ground for natural degradation, have been proposed as potential final disposal options. While this strategy is less expensive, spontaneous adsorbent degradation can take a long period. Another possibility is to use waste adsorbents as biofertilizers. Biosorbents formed from bacteria, forestry, and agricultural waste commonly contain organic chemicals that can improve soil quality. Thus, using waste adsorbents as biofertilizers not only solves the problem of final disposal but also improves the soil structure.

Biosorbents are most commonly disposed of by regenerating and reusing them. Desorption allows biosorbents to restore their adsorption capability, extending their useful life and reducing waste formation. However, with subsequent cycles, biosorbents may lose mass, modifying their physicochemical properties and potentially decreasing adsorption efficacy over time (Bădescu et al., 2018). An alternative disposal option is to encapsulate and enclose the waste adsorbent in an inert substance. Choosing appropriate source materials is critical for biosorbent preparation, processing, and disposal. Priority should be given to enhance the effectiveness and capacities of adsorbent materials in removing environmental pollutants. Ensuring proper maintenance of the adsorption process, including parameters impacting adsorption efficacy, is critical for increasing adsorption rates.

12.6 ECONOMY ANALYSIS

Biosorbentsare chosen based on their economic feasibility. Biosorption techniques' cost-effectiveness is influenced by factors such as biomass cultivation expenses and the immobilization process. Commercial biosorbent materials were shown to be more inexpensive than activated carbon and ion-exchange resins. Biomass immobilization is a simple and cost-effective method, with biosorbent costs mostly determined by the source of raw biomass. The economic feasibility of a biosorption process is determined by its efficacy, regeneration capability, and kinetics. However, the fermentation process is inconsistent in delivering a sustainable and cost-effective source of biomass for such applications.

The increased complexity of biomass regeneration, reuse, and recycling creates difficulties. Nonetheless, regeneration improves the efficiency of the biosorption process. Biosorbents are used to produce valuable materials; nevertheless, extended usage of regenerated biomass across five cycles led to diminishing effectiveness, resulting in a lower biosorption rate (de Andrade et al., 2018). To address this difficulty, a new eluent for recovering adsorbed pollutants such as metal ions and dyes is required. Initially, biosorption was viewed as a way to repurpose low-cost, underutilized resources originating from industrial or agricultural waste.

Furthermore, a complete understanding of the biosorption mechanism and the variables influencing it should be the top priority of scientific research. When evaluating the financial effects of the biosorption process, it is crucial to take into account variables such as energy consumption, regeneration, process optimization, treatment methods, disposal, transportation, and desorption. The size and features of the treatment approach will determine the initial capital investment and treatment processing costs. Waste materials have been employed as feedstock for the manufacturing of biosorbents to reduce processing costs. Potential sources of biosorbent production include agricultural leftovers, industrial byproducts, home waste, bacterial waste from fermentation businesses, sludge from processing industries, and fungal waste from the food processing sector (Zhang and Leiviska, 2020).

Managing trash is a significant concern in many sectors. Utilizing waste as biosorbent materials is a viable method for managing waste disposal difficulties while also reducing biosorbent production costs. Another costly element is feedstock preparation, which requires pre-treatment for specific materials before processing. The pre-treatment method used is determined by the qualities of the feedstock and can result in significant costs. As a result, establishing a cost-effective, simple, and efficient pre-treatment approach is critical. The pre-treatment procedure and primary treatment facility must be cost-effective. Additionally, disposal costs should be considered in the total cost assessment of the biosorption process. As the biosorbent material goes through continuous cycles and reaches full usage, replenishment is required, and the spent or fully utilized biosorbent can be reused for other goods. The cost of biosorbent synthesis varies according to the material composition, with more energy consumption during biosorption, resulting in higher costs. Biosorption appears as a more economically viable alternative to known technologies, such as electrodialysis, reverse osmosis, and precipitation.

12.7 CONCLUDING REMARKS

Climate change has shifted the research community's focus to global well-being and food security. Bio NMs, valued for their unique properties, such as low toxicity, high stability, and extensive availability in natural reservoirs, have emerged as critical materials for tackling environmental pollution issues. Scientists have shown that bio NMs, which have special properties, are just as efficient as or even more so than physical and chemical NMs. They have the potential to reduce pollution by removing heavy metals and other impurities from the environment. The biosorption process still faces many obstacles today, including as commercialization, scaling up for large-scale applications, and improving performance in real-world settings. Furthermore, biosorption offers advantages for large-scale applications, due to its low cost and enhanced removal effectiveness. Overcoming obstacles in bio-NMs' commercialization will be addressed by achieving successful large-scale synthesis. Moreover, thorough analyses of how bio-NMs interact with various environmental matrices are essential. Future research will emphasize the importance of toxicological assessments of processed bio NMs.

REFERENCES

Adebayo, M.A. and Areo, F.I., 2021. Removal of phenol and 4-nitrophenol from wastewater using a composite prepared from clay and Cocos nucifera shell: Kinetic, equilibrium and thermodynamic studies. *J. Environ. Sustain.*3, 100020. https://doi.org/10.1016/j.resenv.2021.100020

Al-Thabaiti, S.A. and Khan, Z., 2020. Biogenic synthesis of silver NPs, sensing and photo catalytic activities for bromothymol blue. *J. Photochem. Photobiol.*3, 100010. https://doi.org/10.1016/j.jpap.2020.100010

AlNadhari, S., Al-Enazi, N.M., Alshehrei, F. and Ameen, F., 2021. A review on biogenic synthesis of metal NPs using marine algae and its applications. *Environ. Res.*194, 110672. https://doi.org/10.1016/j.envres.2020.110672

Alves, A.A., Silva, W.E., Belian, M.F., Lins, L.S.G. and Galembeck, A., 2020. Bacterial cellulose membranes for environmental water remediation and industrial wastewater treatment. *Int. J. Environ. Sci. Technol.*17, 3997–4008.

Anchan, S., Pai, S., Sridevi, H., Varadavenkatesan, T., Vinayagam, R. and Selvaraj, R., 2019. Biogenic synthesis of ferric oxide NPs using the leaf extract of Peltophorumpterocarpum and their catalytic dye degradation potential. *Biocatal. Agric. Biotechnol.*20, 101251. https://doi.org/10.1016/j.bcab.2019.101251

Anitha, D., Ramadevi, A. and Seetharaman, R., 2021. Biosorptive removal of Nickel (II) from aqueous solution by Mangosteen shell activated carbon. *Mater. Today Proc.*45, 718–722. https://doi.org/10.1016/j.matpr.2020.02.748

Bădescu, I.S., Bulgariu, D., Ahmad, I. and Bulgariu, L., 2018. Valorisation possibilities of exhausted biosorbents loaded with metal ions–a review. *J. Environ. Manage.*224, 288–297. https://doi.org/10.1016/j.jenvman.2018.07.066

Bandara, S., Du, H., Carson, L., Bradford, D. and Kommalapati, R., 2020. Agricultural and biomedical applications of chitosan-based NMs. *NMs.*10(10), 1903. https://doi.org/10.3390/nano10101903

Barati, M., Bakhtiari, F., Mowla, D. and Safarzadeh, S., 2018. Comparison of the effects of poultry manure and its biochar on barley growth in petroleum-contaminated soils. *Int. J. Phytorem.*20(2), 98–103. https://doi.org/10.1080/15226514.2017.1337069

Bhargava, A., Jain, N., Gangopadhyay, S. and Panwar, J., 2015. Development of gold NP-fungal hybrid based heterogeneous interface for catalytic applications. *Process Biochem.*50(8), 1293–1300. https://doi.org/10.1016/j.procbio.2015.04.012

Bhargava, A., Jain, N., Khan, M.A., Pareek, V., Dilip, R.V. and Panwar, J., 2016. Utilizing metal tolerance potential of soil fungus for efficient synthesis of gold NPs with superior catalytic activity for degradation of rhodamine B. *J Environ Manage.*183, 22–32. https://doi.org/10.1016/j.jenvman.2016.08.021

Bolade, O.P., Williams, A.B. and Benson, N.U., 2020. Green synthesis of iron-based NMs for environmental remediation: A review. *Environ. Nanotechnol. Monit. Manag.*13, 100279. https://doi.org/10.1016/j.enmm.2019.100279

Boleydei, H., Mirghaffari, N. and Farhadian, O., 2018. Comparative study on adsorption of crude oil and spent engine oil from seawater and freshwater using algal biomass. *Environ. Sci. Pollut. Res.*25, 21024–21035. https://doi.org/10.1007/s11356-018-2281-y

Bortolassi, A.C.C., Nagarajan, S., de Araújo Lima, B., Guerra, V.G., Aguiar, M.L., Huon, V., Soussan, L., Cornu, D., Miele, P. and Bechelany, M., 2019. Efficient NPs removal and bactericidal action of electrospun nanofibers membranes for air filtration. *Mater. Sci. Eng. C.*102, 718–729. https://doi.org/10.1016/j.msec.2019.04.094

Boruah, P.K., Sharma, B., Hussain, N., and Das, M.R., 2017. Magnetically recoverable Fe_3O_4/graphene nanocomposite towards efficient removal of triazine pesticides from aqueous solution: Investigation of the adsorption phenomenon and specific ion effect. *Chemosphere.*168, 1058–1067. https://doi.org/10.1016/j.chemosphere.2016.10.103

Boudesocque, S., Guillon, E., Aplincourt, M., Martel, F. and Noël, S., 2008. Use of a low-cost biosorbent to remove pesticides from wastewater. *J. Environ. Qual.*37(2), 631–638. https://doi.org/10.2134/jeq2007.0332

Bwatanglang, I.B., Magili, S.T. and Kaigamma, I., 2021. Adsorption of phenol over bio-based silica/calcium carbonate (CS-SiO_2/$CaCO_3$) nanocomposite synthesized from waste eggshells and rice husks. *PeerJ Physical Chemistry.*3, 17. https://doi.org/10.1016/j.jhazmat.2011.01.114

Cai, L., Zhang, Y., Zhou, Y., Zhang, X., Ji, L., Song, W., Zhang, H. and Liu, J., 2019. Effective adsorption of diesel oil by crab-shell-derived biochar NMs. *Materials.*12(2), 236. https://doi.org/10.3390/ma12020236

Chang, Y.S., Au, P.I., Mubarak, N.M., Khalid, M., Jagadish, P., Walvekar, R. and Abdullah, E.C., 2020. Adsorption of Cu (II) and Ni (II) ions from wastewater onto bentonite and bentonite/GO composite. *Environ. Sci. Pollut. Res.*27(26), 33270–33296. https://doi.org/10.1007/s11356-020-09423-7

Chen, B., Yuan, M. and Liu, H., 2011. Removal of polycyclic aromatic hydrocarbons from aqueous solution using plant residue materials as a biosorbent. *J. Hazard. Mater.*188(1-3), 436–442. https://doi.org/10.1016/j.jhazmat.2011.01.114

Da Gama, B.M.V., do Nascimento, G.E., Sales, D.C.S., Rodríguez-Díaz, J.M., de Menezes Barbosa, C.M.B. and Duarte, M.M.M.B., 2018. Mono and binary component adsorption of phenol and cadmium using adsorbent derived from peanut shells. *J. Clean. Prod.*201, 219–228. https://doi.org/10.1016/j.jclepro.2018.07.291

De Andrade, J.R., Oliveira, M.F., da Silva, M.G. and Vieira, M.G., 2018. Adsorption of pharmaceuticals from water and wastewater using nonconventional low-cost materials: a review. *Ind. amp; Eng. Chem. Res.*57(9), 3103–3127. https://doi.org/10.1021/acs.iecr.7b05137

De Corte, S., Hennebel, T., Fitts, J.P., Sabbe, T., Bliznuk, V., Verschuere, S., van der Lelie, D., Verstraete, W. and Boon, N., 2011. Biosupported bimetallic Pd–Au nanocatalysts for dechlorination of environmental contaminants. *Environ. Sci. Technol.*45(19), 8506–8513. https://doi.org/10.1021/es2019324

Dehaghi, S.M., Rahmanifar, B., Moradi, A.M. and Azar, P.A., 2014. Removal of permethrin pesticide from water by chitosan–zinc oxide NPs composite as an adsorbent. *J. Saudi Chem. Soc.*18(4), 348–355. https://doi.org/10.1016/j.jscs.2014.01.004

Duan, G., Fang, H., Huang, C., Jiang, S. and Hou, H., 2018. Microstructures and mechanical properties of aligned electrospun carbon nanofibers from binary composites of polyacrylonitrile and polyamic acid. *J. Mater. Sci. Lett.*53(21), 15096–15106. https://doi.org/10.1007/s10853-018-2700-y

Fernández-González, R., Martín-Lara, M.A., Moreno, J.A., Blázquez, G. and Calero, M., 2019. Effective removal of zinc from industrial plating wastewater using hydrolyzed olive cake: scale-up and preparation of zinc-based biochar. *J. Clean. Prod.*227, 634–644. https://doi.org/10.1016/j.jclepro.2019.04.195

García-Mayagoitia, S., Pérez-Hernández, H., Medina-Pérez, G., Campos-Montiel, R.G. and Fernández-Luqueño, F., 2020. Bio-NMs in the air pollution treatment. In *NMs for Air Remediation.*227–248. https://doi.org/10.1016/B978-0-12-818821-7.00012-9

Gopalakrishnan, Y., Al-Gheethi, A., Abdul Malek, M., Marisa Azlan, M., Al-Sahari, M., Radin Mohamed, R.M.S., Alkhadher, S. and Noman, E., 2020. Removal of basic brown 16 from aqueous solution using durian shell adsorbent, optimisation and techno-economic analysis. *Sustainability.*12(21), 8928. https://doi.org/10.3390/su12218928

Guerra, F.D., Attia, M.F., Whitehead, D.C. and Alexis, F., 2018. Nanotechnology for environmental remediation: materials and applications. *Molecules.*23(7), 1760. https://doi.org/10.3390/molecules23071760

Guo, Y., Wang, Z., Zhou, X. and Bai, R., 2017. Removal of mercury (II) from aqueous solution with three commercial raw activated carbons. *Rev. Chem. Intermed.*43, 2273–2297. DOI 10.1007/s11164-016-2761-y

Hao, L., Wang, P. and Valiyaveettil, S., 2017. Successive extraction of As (V), Cu (II) and P (V) ions from water using spent coffee powder as renewable bioadsorbents. *Sci. Rep.*7(1), 42881. https://doi.org/10.1038/srep42881

Haq, A.U., Saeed, M., Usman, M., Zahoor, A.F., Anjum, M.N., Maqbool, T., Naheed, S. and Kashif, M., 2021. Mechanisms of halosulfuron methyl pesticide biosorption onto neem seeds powder. *Sci. Rep.*11(1), 9960. https://doi.org/10.1038/s41598-021-88929-7

Huang, H., Jia, Q., Jing, W., Dahms, H.U. and Wang, L., 2020b. Screening strains for microbial biosorption technology of cadmium. *Chemosphere.*251, 126428. https://doi.org/10.1016/j.chemosphere.2020.126428

Huang, X., Feng, S., Zhu, G., Zheng, W., Shao, C., Zhou, N. and Meng, Q., 2020a. Removal of organic herbicides from aqueous solution by ionic liquid modified chitosan/metal-organic framework composite. *Int. J. Biol. Macromol.*149, 882–892. https://doi.org/10.1016/j.*ijbiomac*.2020.01.165

Ibrahim, W.M., Hassan, A.F. and Azab, Y.A., 2016. Biosorption of toxic heavy metals from aqueous solution by Ulva lactuca activated carbon. *Egypt. J. Basic Appl. Sci.*3(3), 241–249. https://doi.org/10.1016/j.ejbas.2016.07.005

Jaiswal, M., Chauhan, D. and Sankararamakrishnan, N., 2012. Copper chitosan nanocomposite: synthesis, characterization, and application in removal of organophosphorous pesticide from agricultural runoff. *Environ. Sci. Pollut. Res.*19, 2055–2062. DOI 10.1007/s11356-011-0699-6

Jangid, N.K., Yadav, A., Jadoun, S., Srivastava, A. and Srivastava, M., 2020. Outdoor pollution management by nanotechnology. In *Impact of Textile Dyes on Public Health and the Environment.*258–277. https://doi.org/10.4018/978-1-7998-0311-9.ch012

Keihan, A.H., Rasoulnezhad, H., Mohammadgholi, A., Sajjadi, S., Hosseinzadeh, R., Farhadian, M. and Hosseinzadeh, G., 2017. Pd NP loaded TiO_2 semiconductor for photocatalytic degradation of Paraoxon pesticide under visible-light irradiation. *J. Mater. Sci.: Mater. Electron.*28, 16718–16727. https://doi.org/10.1007/s10854-017-7585-z

Khalid, B., Bai, X., Wei, H., Huang, Y., Wu, H. and Cui, Y., 2017. Direct blow-spinning of nanofibers on a window screen for highly efficient PM2.5 removal. *Nano Lett.*17(2), 1140–1148. https://doi.org/10.1021/acs.nanolett.6b04771

Khan, I., Saeed, K. and Khan, I., 2019. NPs: Properties, applications and toxicities. *Arab. J. Chem.*12(7), 908–931. https://doi.org/10.1016/j.arabjc.2017.05.011

Kumar, L., BharadvajaN., 2019. Enzymatic bioremediation: A smart tool to fight environmental pollutants. In *Smart Bioremediation Technologies.*99–118. https://doi.org/10.1016/B978-0-12-818307-6.00006-8

Lago, A., Silva, B. and Tavares, T., 2021. Cleaner approach for atrazine removal using recycling biowaste/waste in permeable barriers. *Recycling.*6(2), 41. https://doi.org/10.3390/recycling6020041

Li, H., Gao, P., Cui, J., Zhang, F., Wang, F. and Cheng, J., 2018. Preparation and Cr (VI) removal performance of corncob activated carbon. *Environ. Sci. Pollut. Res.*25, 20743–20755. https://doi.org/10.1007/s11356-018-2026-y

Li, S.N., Wang, R., and Ho, S.H., 2021. Algae-mediated biosystems for metallic NP production: From synthetic mechanisms to aquatic environmental applications. *J. Hazard. Mater.*420, 126625. https://doi.org/10.1016/j.jhazmat.2021.126625

Liu, X., Souzandeh, H., Zheng, Y., Xie, Y., Zhong, W.H. and Wang, C., 2017. Soy protein isolate/bacterial cellulose composite membranes for high efficiency particulate air filtration. *Compos. Sci. Technol.*138, 124–133. https://doi.org/10.1016/j.compscitech.2016.11.022

Liu, Y., Huang, L., Mahmud, S. and Liu, H., 2020. Gold NPs biosynthesized using Ginkgo biloba leaf aqueous extract for the decolorization of azo-dyes and fluorescent detection of Cr (VI). *J. Clust. Sci.*31(2), 549–560. https://doi.org/10.1007/s10876-019-01673-x

Ma, J., Li, T., Liu, Y., Cai, T., Wei, Y., Dong, W. and Chen, H., 2019. Rice husk derived double network hydrogel as efficient adsorbent for Pb (II), Cu (II) and Cd (II) removal in individual and multicomponent systems. *Bioresour. Techno.*290, 121793. https://doi.org/10.1016/j.biortech.2019.121793

Malik, B., Pirzadah, T.B., Kumar, M. and Rehman, R.U., 2017. Biosynthesis of NPs and their application in pharmaceutical industry. *Nanotechnology: Food and Environmental Paradigm.*235–252. https://doi.org/10.1007/978-981-10-4678-0_13

Masoudian, N., Rajabi, M. and Ghaedi, M., 2019. Titanium oxide NPs loaded onto activated carbon prepared from bio-waste watermelon rind for the efficient ultrasonic-assisted adsorption of congo red and phenol red dyes from wastewaters. *Polyhedron.*173, 114105. https://doi.org/10.1016/j.poly.2019.114105

Mathur, S., Singh, D. and Ranjan, R., 2022. Remediation of heavy metal (loid) contaminated soil through green nanotechnology. *Front. Sustain. Food Syst.*6, 932424. https://doi.org/10.3389/fsufs.2022.932424

Mehmandost, N., García-Valverde, M.T., Soriano, M.L., Goudarzi, N., Lucena, R., Chamjangali, M.A. and Cardenas, S., 2020. Heracleumpersicum based biosorbent for the removal of paraquat and diquat from waters. *J. Environ. Chem. Eng.*8(6), 104481. https://doi.org/10.1016/j.jece.2020.104481

MoballeghNasery, M., Abadi, B., Poormoghadam, D., Zarrabi, A., Keyhanvar, P., Khanbabaei, H., Ashrafizadeh, M., Mohammadinejad, R., Tavakol, S. and Sethi, G., 2020. Curcumin delivery mediated by bio-based NPs: A review. *Molecules*.25(3), 689. https://doi.org/10.3390/molecules25030689

Muthuvel, A., Jothibas, M. and Manoharan, C., 2020. Synthesis of copper oxide NPs by chemical and biogenic methods: Photocatalytic degradation and in vitro antioxidant activity. *Nanotechnol. Environ. Eng*.5(2), 14. https://doi.org/10.1007/s41204-020-00078-w

Nguyen, N.T.T., Nguyen, L.M., Nguyen, T.T.T., Liew, R.K., Nguyen, D.T.C. and Van Tran, T., 2022. Recent advances on botanical biosynthesis of NPs for catalytic, water treatment and agricultural applications: A review. *Sci. Total Environ*.827, 154160. https://doi.org/10.1016/j.scitotenv.2022.154160

Niazi, L., Lashanizadegan, A. and Sharififard, H., 2018. Chestnut oak shells activated carbon: Preparation, characterization and application for Cr (VI) removal from dilute aqueous solutions. *J. Clean. Prod*.185, 554–561. https://doi.org/10.1016/j.jclepro.2018.03.026

Noman, M., Shahid, M., Ahmed, T., Niazi, M.B.K., Hussain, S., Song, F. and Manzoor, I., 2020. Use of biogenic copper NPs synthesized from a native Escherichia sp. as photocatalysts for azo dye degradation and treatment of textile effluents. *Environ Pollut*.257, 113514. https://doi.org/10.1016/j.envpol.2019.113514

Pandey, A., Brauer, M., Cropper, M.L., Balakrishnan, K., Mathur, P., Dey, S., Turkgulu, B., Kumar, G.A., Khare, M., Beig, G. and Gupta, T., 2021. Health and economic impact of air pollution in the states of India: The Global Burden of Disease Study 2019. *The Lancet. Planetary health*.5(1), 25–38. https://doi.org/10.1016/S2542-5196(20)30298-9

Patra, C., Medisetti, R.M.N., Pakshirajan, K. and Narayanasamy, S., 2019. Assessment of raw, acid-modified and chelated biomass for sequestration of hexavalent chromium from aqueous solution using SterculiavillosaRoxb. shells. *Environ. Sci. Pollut. Res*.26, 23625–23637. https://doi.org/10.1007/s11356-019-05582-4

Ramakrishna, M., Rajesh Babu, D., Gengan, R.M., Chandra, S. and Nageswara Rao, G., 2016. Green synthesis of gold NPs using marine algae and evaluation of their catalytic activity. *J. nanostructure chem*.6, 1-13.3 DOI 10.1007/s40097-015-0173-y

Ramirez, A., Ocampo, R., Giraldo, S., Padilla, E., Flórez, E. and Acelas, N., 2020. Removal of Cr (VI) from an aqueous solution using an activated carbon obtained from teakwood sawdust: Kinetics, equilibrium, and density functional theory calculations. *J. Environ. Chem. Eng*.8(2), 103702. https://doi.org/10.1016/j.jece.2020.103702

Regkouzas, P. and Diamadopoulos, E., 2019. Adsorption of selected organic micro-pollutants on sewage sludge biochar. *Chemosphere*, 224, 840–851. https://doi.org/10.1016/j.chemosphere.2019.02.165

Roopan, S.M., Madhumitha, G., Rahuman, A.A., Kamaraj, C., Bharathi, A. and Surendra, T.V., 2013. Low-cost and eco-friendly phyto-synthesis of silver NPs using Cocos nucifera coir extract and its larvicidal activity. *Ind Crops Prod*.43, 631–635. https://doi.org/10.1016/j.indcrop.2012.08.013

Rozumová, L., Životský, O., Seidlerová, J., Motyka, O., Šafařík, I. and Šafaříková, M., 2016. Magnetically modified peanut husks as an effective sorbent of heavy metals. *J. Environ. Chem. Eng*.4(1), 549–555. https://doi.org/10.1016/j.jece.2015.10.039

Saraswat, P., Singh, S., Prasad, M., Misra, R., Rajput, V.D. and Ranjan, R., 2023. Applications of bio-based NMs in environment and agriculture: A review on recent progresses. *Hybrid Adv*.100097. https://doi.org/10.1016/j.hybadv.2023.100097

Saravanan, A., Kumar, P.S., Yaashikaa, P.R., Karishma, S., Jeevanantham, S. and Swetha, S., 2021. Mixed biosorbent of agro waste and bacterial biomass for the separation of Pb (II) ions from water system. *Chemosphere*.277, 130236. https://doi.org/10.1016/j.chemosphere.2021.130236

Saum, L., Jiménez, M.B. and Crowley, D., 2018. Influence of biochar and compost on phytoremediation of oil-contaminated soil. *Int. J. Phytorem*.20(1), 54–60. https://doi.org/10.1080/15226514.2017.1337063

Singh, J., Kaur, N., Kaur, P., Kaur, S., Kaur, J., Kukkar, P., Kumar, V., Kukkar, D. and Rawat, M., 2018. Piper betle leaves mediated synthesis of biogenic SnO_2 NPs for photocatalytic degradation of reactive yellow 186 dye under direct sunlight. Environ. Nanotechnol. *Monit. Manag*.10, 331–338. https://doi.org/10.1016/j.enmm.2018.07.001

Singh, J., Kumar, V., Kim, K.H. and Rawat, M., 2019. Biogenic synthesis of copper oxide NPs using plant extract and its prodigious potential for photocatalytic degradation of dyes. Environ. Res.177, 108569. https://doi.org/10.1016/j.envres.2019.108569

Souza, I.P., Cazetta, A.L., Pezoti, O. and Almeida, V.C., 2017. Preparation of biosorbents from the Jatoba (Hymenaeacourbaril) fruit shell for removal of Pb (II) and Cd (II) from aqueous solution. *Environ. Monit. Assess*.189, 1–16. https://doi.org/10.1007/s10661-017-6330-7

Srivastava, M., Srivastava, A., Jangid, N.K., Yadav, A. and Sunidhi, 2023. Classification and application of advanced composite materials. *Adv. Compos*.1–36. https://doi.org/10.1007/978-3-031-42731-2_1

Tavakkoli, H. and Yazdanbakhsh, M., 2013. Fabrication of two perovskite-type oxide NPs as the new adsorbents in efficient removal of a pesticide from aqueous solutions: Kinetic, thermodynamic, and adsorption studies. *Microporous Mesoporous Mater*.176, 86–94. https://doi.org/10.1016/j.micromeso.2013.03.043

Ulucan-Altuntas, K. and Debik, E., 2020. Dechlorination of dichlorodiphenyltrichloroethane (DDT) by Fe/Pd bimetallic NPs: Comparison with nZVI, degradation mechanism, and pathways. *Front. Environ. Sci. Eng*.14, 1–13. https://doi.org/10.1007/s11783-019-1196-2

Üner, O., Geçgel, Ü. and Bayrak, Y., 2016. Adsorption of methylene blue by an efficient activated carbon prepared from Citrulluslanatus rind: Kinetic, isotherm, thermodynamic, and mechanism analysis. *Water, Air, Soil Pollut*.227,1–15. DOI 10.1007/s11270-016-2949-1

Wang, D., 2019. A critical review of cellulose-based NMs for water purification in industrial processes. *Cellulose*.26, 687–701. https://doi.org/10.1007/s10570-018-2143-2

Wang, Q., Zhou, C., Kuang, Y.J., Jiang, Z.H. and Yang, M., 2020. Removal of hexavalent chromium in aquatic solutions by pomelo peel. *Water Sci. Eng*.13(1), 65–73. https://doi.org/10.1016/j.wse.2019.12.011

Wang, R., Wang, S., Tai, Y., Tao, R., Dai, Y., Guo, J., Yang, Y. and Duan, S., 2017a. Biogenic manganese oxides generated by green algae Desmodesmus sp. WR1 to improve bisphenolA removal. *J. Hazard. Mater*.339, 310–319. https://doi.org/10.1016/j.jhazmat.2017.06.026

Wang, X., Zhang, D., Pan, X., Lee, D.J., Al-Misned, F.A., Mortuza, M.G. and Gadd, G.M., 2017b. Aerobic and anaerobic biosynthesis of nano-selenium for remediation of mercury contaminated soil. *Chemosphere*.170, 266–273. https://doi.org/10.1016/j.chemosphere.2016.12.020

Yadav, A., Kon, K., Kratosova, G., Duran, N., Ingle, A.P. and Rai, M., 2015. Fungi as an efficient mycosystem for the synthesis of metal NPs: progress and key aspects of research. *Biotechnol.Lett*.37, 2099–2120. https://repo.knmu.edu.ua/handle/123456789/9878

Yadav, A., Raghav, S., Jangid, N.K., Srivastava, A., Jadoun, S., Srivastava, M. and Dwivedi, J., 2023. Myricaesculentaleaf extract: Assisted green synthesis of porous magnetic chitosan composites for fast removal of Cd (II) from water: Kinetics and thermodynamics of adsorption. *Polymers*.15(21), 4339. https://doi.org/10.3390/polym15214339

Yadav, A., Raghav, S., Srivastava, A., Jangid, N.K. and Srivastava, M., 2024. Comparative study of nickel-iron composite with Fe_3O_4 NPs for the adsorption of chromium from aqueous medium. *Water, Air, Soil Pollut*.235(4), 1–22. https://doi.org/10.1007/s11270-024-07017-4

Zaib, M., Akhtar, A., Maqsood, F. and Shahzadi, T., 2021. Green synthesis of carbon dots and their application as photocatalyst in dye degradation studies. *Arab. J. Sci. Eng*.46(1), 437–446. https://doi.org/10.1007/s13369-020-04904-w

Zhang, C., Lu, J. and Wu, J., 2019. Adsorptive removal of polycyclic aromatic hydrocarbons by detritus of green tide algae deposited in coastal sediment. *Sci. Total Environ*.670, 320–327. https://doi.org/10.1016/j.scitotenv.2019.03.296

Zhang, J., Xia, A., Chen, H., Nizami, A.S., Huang, Y., Zhu, X., Zhu, X. and Liao, Q., 2022. Biobased carbon dots production via hydrothermal conversion of microalgae Chlorella pyrenoidosa. Sci. Total Environ.839, 156144. https://doi.org/10.1016/j.scitotenv.2022.156144

Zhang, R. and Leiviska, T., 2020. Surface modification of pine bark with quaternary ammonium groups and its use for vanadium removal. *Chem. Eng. J*.385, 123967. https://doi.org/10.1016/j.cej.2019.123967

Zhang, Y., Nypelö, T. and Salas, C., 2013. Cellulose nanofibrils. J. Renew. Mater. 195–211. https://doi.org/10.7569/JRM.2013.634115

Zhou, L., Li, R., Zhang, G., Wang, D., Cai, D. and Wu, Z., 2018. Zero-valent iron NPs supported by functionalized waste rock wool for efficient removal of hexavalent chromium. *J. Chem. Eng*.339, 85–96. https://doi.org/10.1016/j.cej.2018.01.132

13 Nanocomposites and Environmental Pollution Control

Nachiappan Senthilnathan, Rakesh Namdeti, and Uma Reddy Meka

13.1 INTRODUCTION

The NCs are substances that are basically in the solid phase, made with one of their dimensions in nanometer size. These NCs were made up of nanoparticles (NPs), nanoclays, and/or nanofibers. In general, when the size of a certain constituent of substances is less than 'critical size', these substances show an extraordinary capability in certain properties (Schmidt, Shah, and Giannelis 2002). The NCs are made of metallic, polymeric, or ceramic substances. Use of polymeric substances can overcome certain defects and show some synergistic effect by keeping the NCs primary features intact. Generally, glass fiber, organic fiber, etc., were chosen in the dispersed phase in reinforcing a substance. Thus, the matrix material is made of a continuous phase made of metallic, polymeric, and/ or ceramic matrices (Wang, Zheng, and Zheng 2011). These NCs are considered to be emerging candidates toward a better future as they overcome the limitations shown by conventional micro-composites, as well as mono-lithics. The NCs possess several advantages over other conventional composites, such as: (1) excellent surface and volume ratio, (2) high mechanical strength, (3) enhanced optical properties, etc. Along with these positive attributes, NCs show certain disadvantages, such as: (1) toughness, (2) impact performance due to incorporation of NPs inside the bulk matric composite substance (Omanović-Mikličanin et al. 2020). These NCs are classified as (Figure 13.1):

1. Metallic NCs
2. Polymeric NCs
3. Ceramic NCs.

13.2 FABRICATION METHODS OF METALLIC NCS

Metallic NCs are created by combining metallic NPs or nanostructures with a matrix material, which is usually a polymer, ceramic, or another metallic substrate. Several methods are used to create metallic NCs, including:

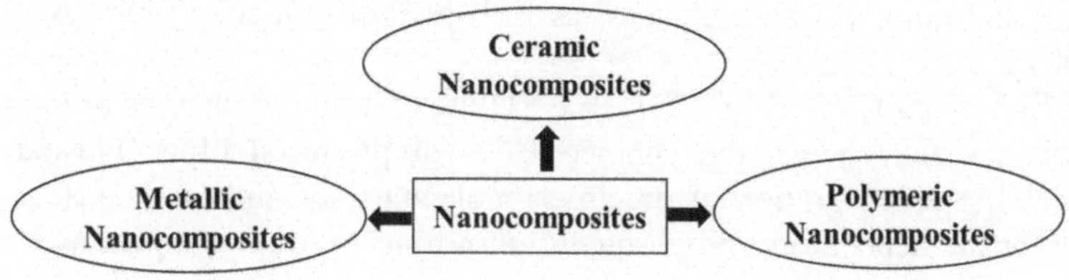

FIGURE 13.1 Major classification of NCs.

DOI: 10.1201/9781003484189-13

13.2.1 Chemical Vapor Deposition (CVD)

CVD is a versatile and commonly used process in materials science and engineering for depositing thin films and coatings on surfaces. This technique involves the chemical reaction of gaseous precursor molecules at high temperatures to produce a solid substance on the surface of a substrate. CVD has applications in a variety of industries, including semiconductor production, optics, aerospace, and coatings technology, due to its ability to manufacture high-quality, homogenous films with perfect control over composition, thickness, and morphology (Sivakumar et al. 2004).

13.2.1.1 Principles of CVD

The fundamental principles underlying CVD involve the following key steps:

1. **Precursor Delivery**: Gaseous precursor molecules containing the appropriate elements or compounds are supplied into a reaction chamber, usually through a carrier gas like hydrogen or nitrogen.
2. **Transport and Reaction**: The precursor molecules are carried to the substrate's surface, where they conduct chemical reactions catalyzed by thermal energy or plasma activation. These reactions cause the deposition of solid-phase material on the substrate surface.
3. **Film Growth and Nucleation**: As precursor molecules react and deposit on the substrate surface, thin film growth happens via nucleation and subsequent layer-by-layer growth mechanisms. The deposition process is affected by temperature, pressure, precursor concentration, and substrate morphology.
4. **Film Morphology and Properties**: Various process parameters, such as temperature gradients, gas flow rates, and substrate characteristics, influence the deposited film's morphology, microstructure, and properties. Control over these factors is critical for achieving optimal film characteristics and homogeneity.

13.2.1.2 Types of CVD

CVD encompasses several variations tailored to specific applications and materials requirements:

1. **Low-Pressure CVD (LPCVD)**: LPCVD uses lower pressures, usually below atmospheric pressure, to improve uniform gas flow and precise control over deposition parameters. This process is widely used in semiconductor device manufacture to deposit thin layers of silicon dioxide, silicon nitride, and polysilicon (Zhang and Chen 2009).
2. **Atmospheric Pressure CVD (APCVD)**: APCVD runs at or near atmospheric pressure, making it easy and inexpensive to produce thin films and coatings on a large scale. This approach is appropriate for depositing oxides, nitrides, and metals onto glass substrates for display and optical applications (Baiju et al. 2005).
3. **Metal Organic CVD (MOCVD)**: Metal alkoxides and β-diketonates are used to build thin films of metallic and compound semiconductor materials with exact stoichiometry and composition. This process is frequently used to create epitaxial layers for semiconductor and optoelectronic devices, as well as thin-film solar cells (Ananthakumar et al. 2004).
4. **Plasma-Enhanced CVD (PECVD)**: PECVD uses plasma discharges to improve deposition kinetics and change the characteristics of produced films. Plasma activation accelerates the disintegration of precursor molecules and increases surface reactions, allowing for the deposition of high-quality thin films at lower temperatures and higher growth rates.

13.2.1.3 Applications for CVD

Chemical Vapor Deposition has a wide range of applications across industries and technologies, including:

1. **Semiconductor Fabrication**: In the semiconductor sector, CVD is widely employed to deposit thin films of silicon dioxide, silicon nitride, polysilicon, and other metal layers for integrated circuit fabrication, interconnects, and device encapsulation.
2. **Optical Coatings and Thin Films**: CVD is used to apply optical coatings, anti-reflection coatings, and transparent conductive films to glass and plastic substrates for optical components, lenses, mirrors, and solar panels.
3. **Surface Modification and Functionalization**: CVD allows the deposition of functional coatings and surface modifications to improve the mechanical, electrical, and chemical properties of substrates, such as corrosion resistance, wear protection, and biocompatibility in biomedical implants and devices (Warrier & Anil Kumar, 2001).
4. **Thin-Film Solar Cells and Energy Devices**: CVD is used to manufacture thin-film solar cells, photovoltaic devices, and energy storage materials, providing cost-effective and scalable manufacturing techniques for sustainable energy technology.

CVD is a fundamental process in materials synthesis and thin-film deposition, providing fine control over layer characteristics, composition, and shape. Its adaptability and vast uses across several industries highlight its importance in the advancement of materials science and engineering. Continued research and innovation in CVD technology are expected to generate significant improvements in materials development, device manufacturing, and technological innovation across a variety of industries.

13.2.2 Electrochemical Deposition

Electrochemical deposition, also known as electrodeposition or electroplating, is a versatile and frequently used process for controlling the deposition of metals and alloys onto conductive substrates. This electrochemical method uses electrolysis principles to selectively deposit metal ions from an electrolyte solution onto a cathodic substrate, resulting in a uniform and adherent metal coating. Electrochemical deposition has a wide range of applications in a variety of industries, including electronics, automotive, aerospace, and ornamental coatings, due to its ability to manufacture thin films, coatings, and functional surfaces with exact composition, thickness, and morphology.

13.2.2.1 Principles of Electrochemical Deposition

The electrochemical deposition process consists of many main steps:

1. **Electrolyte Preparation**: An electrolyte solution comprising dissolved metal ions, usually in the form of metal salts or complex ions, is prepared to aid in the electroplating process. The electrolyte content and pH are optimized to support the necessary electrochemical processes while controlling the deposition parameters (Gao, Chen, and Li Puma 2009).
2. **Electrode Setup**: The electrochemical cell is set up with the substrate to be plated as the cathode (negative electrode) and a suitable inert or sacrificial anode (positive electrode). The cathode and anode are immersed in electrolyte solution and connected to an external power supply.
3. **Electrochemical Reaction**: When a direct current (DC) voltage is applied across the electrodes, the electric field causes metal cations in the electrolyte solution to migrate toward the cathode. At the cathode surface, metal cations undergo reduction processes, acquiring electrons and forming solid metal atoms or clusters (Dou et al. 2011).

4. **Metal Deposition**: As deposited metal atoms or clusters collect on the cathode surface, they progressively create a continuous metal coating or film. The deposition rate and thickness of the plated layer are affected by current density, electrolyte concentration, temperature, and agitation (Belessi et al. 2007).
5. **Post-Deposition Treatment**: After electrochemical deposition, the plated substrate may go through post-treatment processes, including rinsing, drying, annealing, or surface finishing to improve adhesion, mechanical characteristics, and surface aesthetics.

13.2.2.2 Types of Electrochemical Deposition

Electrochemical deposition includes a variety of processes customized to specific applications and material requirements:

Electroplating: Electroplating is the process of depositing a metal coating onto a conductive substrate usually for decorative, corrosion protection, or functional purposes. Chrome, nickel, copper, gold, and silver are among the most often plated metals, each with its own set of qualities and aesthetic appeal (Guglielmi, Kickelbick, and Martucci 2014).

Electroless Plating: Electroless plating, also known as autocatalytic plating, deposits metal coatings using chemical reduction reactions that do not require an external power source. This approach is especially useful for plating nonconductive materials or intricate geometries where electrodeposition is problematic.

Pulse and Pulse Reverse Plating: Pulse and pulse reverse plating techniques modulate current density and polarity throughout the deposition process to achieve fine control over coating parameters, such as grain size, strain, and composition. These methods provide advantages in deposit homogeneity, adhesion, and microstructural refinement (Jeon, Yi, and Oh 2003).

Selective Deposition and Patterned Electroplating: Selective deposition techniques allow for the localized deposition of metal coatings on specific regions of the substrate surface, making it easier to fabricate patterned features, microstructures, and interconnects for microelectronics, MEMS (Micro-Electro-Mechanical Systems), and advanced packaging applications.

13.2.2.3 Applications of Electrochemical Deposition

Electrochemical deposition has numerous uses in diverse industries and technology, including:

Electronics and Semiconductor Devices: Electroplating is widely utilized in the manufacture of printed circuit boards (PCBs), interconnects, solder bumps, and semiconductor packaging, where it imparts corrosion resistance, conductivity, and solderability to metal components.

Automotive and Aerospace Components: Electroplated coatings are used on automotive and aerospace components to improve corrosion protection, wear resistance, and aesthetics. Chromium, nickel, and zinc coatings are widely utilized on car trim, wheels, and aerospace fasteners (Lee and Park 2003).

Microelectronics and MEMS Devices: Electrochemical deposition is used to fabricate microelectronic devices, MEMS sensors, and actuators, allowing for the precise thickness and homogeneity of conductive, insulating, and barrier layers.

Decorative and Functional Coatings: Electroplated coatings are used in decorative finishes, jewelry, architectural hardware, and consumer products to give aesthetic appeal, tarnishing resistance, and durability.

Electrochemical deposition is a versatile and necessary process for the controlled deposition of metal coatings on surfaces, providing precise control over composition, shape, and functionality. Its vast uses in a variety of industries highlight its importance in the advancement of materials science, electronics, and industrial technology. Continued research and innovation in electrochemical deposition techniques are expected to accelerate advances in surface engineering, functional coatings, and device integration, paving the way for improved performance, sustainability, and technological innovation (Chen and He 2001).

13.2.3 Sol-Gel Method

The sol-gel method is a versatile and frequently used approach for synthesizing inorganic and hybrid organic-inorganic materials like sols, gels, and thin films. This solution-based technique converts a precursor solution, known as a sol, into a gel network by controlled hydrolysis and condensation processes. The sol-gel approach has several advantages, including the capacity to create materials with specific compositions, structures, and properties, as well as scalability and compatibility with a variety of substrates and processing conditions (Ennas et al. 1998).

13.2.3.1 Principles of the Sol-Gel Method

The sol-gel process involves several key steps:

Precursor Selection: The sol-gel process starts with the selection of appropriate precursor molecules, which are usually metal alkoxides, metal chlorides, or organosilanes that undergo hydrolysis and condensation processes to generate the desired oxide or hybrid materials. Precursor selection effects the ultimate product's composition, structure, and qualities.

Hydrolysis: In the presence of water, the precursor molecules hydrolyze, resulting in the creation of metal hydroxides or organosilane species. Hydrolysis reactions include the breaking of metal-oxygen or silicon-oxygen bonds in precursor molecules, which produces reactive hydroxyl groups.

Condensation: After hydrolysis, the hydroxyl groups of nearby precursor molecules undergo condensation processes, which establish chemical bonds and connect the precursor species into a three-dimensional network. Condensation processes produce metal oxide or hybrid organic-inorganic networks, which create the gel phase.

Gel Formation: As the condensation events continue the precursor solution transitions from a liquid sol to a semi-solid gel. The gel network captures solvent molecules and precursor species, resulting in a porous structure with a large surface area and a consistent composition.

Drying and Aging: The gel is dried and aged to eliminate solvent molecules and stimulate further condensation reactions, resulting in densification and strengthening of the gel network. Drying methods include air drying, freeze drying, and supercritical drying, depending on the finished material's desired properties (Chiang and Ma 2002).

Post-Treatment and Functionalization: The dried gel's characteristics and structure can be tailored for specific purposes using post-treatment techniques such as calcination, sintering, or surface functionalization. Heat treatment may be used after processing to eliminate residual organics, crystallize the material, or introduce dopants and functional groups.

13.2.3.2 Types of Sol-Gel Materials

The sol-gel method enables the synthesis of a wide range of materials, including:

Metal Oxides: Metal oxide materials created using the sol-gel process include silica (SiO_2), alumina (Al_2O_3), titania (TiO_2), zirconia (ZrO_2), and a variety of mixed metal oxides. These materials are used in catalysis, sensors, coatings, and optoelectronics because of their large surface area, thermal stability, and chemical inertness.

Hybrid Organic-Inorganic Materials: These materials mix inorganic metal oxide networks with organic polymer chains or ligands, resulting in synergistic characteristics and functions.

Examples include silica-based hybrids, organosilanes, and metal-organic frameworks (MOFs), which have improved mechanical flexibility, optical transparency, and customized surface chemistry (Nazir et al. 2008).

NCs and Thin Films: The sol-gel process allows for the insertion of nanoscale fillers, NPs, and reinforcements into the gel matrix, resulting in NCs with enhanced mechanical, electrical, and thermal properties. Sol-gel thin films are deposited onto substrates utilizing processes such as spin coating, dip coating, and chemical vapor deposition, which provide fine control over layer thickness and shape.

13.2.3.3 Applications of Sol-Gel Materials

Sol-gel-derived materials have several uses in a variety of sectors and technology, including:

Catalysis and Adsorption: Sol-gel-derived metal oxide materials are effective catalysts, adsorbents, and supports for chemical processes, environmental remediation, and gas sensing.

Optics and Photonics: Because of their optical transparency, low refractive index, and tunable optical properties, sol-gel thin films and hybrid materials are used in optics, photonics, and photonic devices, such as anti-reflective coatings, waveguides, optical fibers, and solar cells.

Biosensors and Biomedical Devices: Sol-gel-derived materials are used in biosensors, bioimaging, drug delivery systems, and tissue engineering scaffolds because of their biocompatibility, surface functionalization, and controlled release features.

Coatings and Surface Modifications: Sol-gel coatings and surface treatments offer corrosion protection, wear resistance, and functionalization for metals, ceramics, glass, and polymers used in automotive, aerospace, and architectural applications.

The sol-gel process is a versatile and powerful methodology for creating a wide range of inorganic and hybrid materials with specific compositions, structures, and properties. Its solution-based nature, scalability, and diversity make it an appealing choice for materials synthesis and processing in a variety of sectors and research fields. Continued research and innovation in sol-gel chemistry, processing techniques, and applications are expected to propel future advances in materials science, nanotechnology, and functional materials to solve societal and technical concerns.

13.2.4 Ball Milling

Ball milling is a mechanical technique used to synthesize, modify, and treat materials by applying mechanical energy in the form of high-energy collisions between grinding balls and the substance to be milled. This versatile approach is widely used in materials science, nanotechnology, chemistry, and engineering because of its capacity to produce finely separated powders, enhance chemical reactions, and induce structural alterations in a controlled manner.

13.2.4.1 Principles of Ball Milling

The essential idea of ball milling is the mechanical comminution of materials via the repetitive impact and frictional forces generated by the grinding balls as they collide with the material particles in a revolving container, usually a cylindrical jar or vial. The grinding balls, which are commonly composed of steel, ceramic, or other hard materials, are put within the milling container alongside the material to be processed and an appropriate milling media, such as liquid or gas.

During ball milling, the rotation of the milling container transmits kinetic energy to the grinding balls, causing material particles to distort, shatter, and comminute. High-energy impacts between the grinding balls and the material reduce particle size, produce fine powders, and activate chemical reactions, such as solid-state reactions, alloying, and mechanochemical transformations.

13.2.4.2 Types of Ball Milling

Ball milling encompasses various techniques and configurations tailored to specific applications and materials requirements:

Planetary Ball Milling: Planetary ball mills use many grinding jars mounted eccentrically on a solar wheel that rotates in the opposite direction of the jar. This arrangement allows for high-energy milling and homogenous mixing of powders, resulting in the formation of homogeneous alloys, composites, and nanomaterials.

High-Energy Ball Milling: High-energy ball milling uses specialized ball mills with high-speed spinning vials and balls to provide intense mechanical milling under regulated atmosphere or vacuum conditions. High-energy ball milling allows for rapid kinetics, solid-state processes, and material amorphization, culminating in the formation of metastable phases, nanocrystalline materials, and complex microstructures.

Attritor Mills: Attritor mills use grinding media like beads or balls in a stirred milling chamber to achieve high-energy milling and particle dispersion in a liquid medium. Attritor mills are ideal for wet grinding, dispersion, and particle size reduction in suspensions and slurries, with benefits in terms of scalability, energy efficiency, and process control.

13.2.4.3 Applications of Ball Milling

Ball milling finds diverse applications across various fields and industries, including:

Nanomaterials Synthesis: Ball milling is commonly used to synthesize nanomaterials such as NPs, NCs, and nanostructured materials via mechanical alloying, solid-state processes, and bottom-up techniques. Ball milled nanomaterials have distinct features and functions that make them suitable for applications in electronics, catalysis, energy storage, and biomedical engineering.

Powder Metallurgy and Alloy Development: Ball milling is important in powder metallurgy and alloy development because it facilitates the homogenization, mixing, and refinement of metal powders, allowing for the synthesis of advanced alloys, intermetallic compounds, and metal matrix composites with tailored properties and microstructures.

Catalysis and Chemical Reactions: Ball milling stimulates chemical reactions and catalytic processes via mechanochemical activation, solid-state diffusion, and surface interactions, resulting in the synthesis of catalysts, photocatalysts, and functional materials used in environmental remediation, energy conversion, and chemical synthesis.

Material Processing and Mechanical Properties: Ball milling is used to refine materials and adjust their mechanical properties, such as hardness, toughness, and ductility, using grain refinement, phase transitions, and microstructural control. Ball-milled materials have superior mechanical performance and wear resistance for use in automotive, aerospace, and structural engineering.

Ball milling is a versatile and powerful technology for material production, processing, and modification, with fine control over particle size, structure, and characteristics. Its extensive application across fields emphasizes its importance in improving materials research, nanotechnology, and engineering. Continued research and innovation in ball milling techniques, equipment design, and process optimization are expected to promote additional advances in materials synthesis, functional materials, and technological innovation to solve societal and technical concerns.

13.3 METALLIC NCS: PROPERTIES AND FUTURE OUTLOOK

Metallic NCs have a variety of unique features due to the synergistic actions of the metallic NPs and matrix material. These properties include the following:

13.3.1 Enhanced Mechanical Strength

Enhanced mechanical strength is a material's or structure's ability to endure external pressures and stresses without deformation or failure. This feature is critical in many industries, including aerospace, automotive, construction, and manufacturing, where materials are subjected to extreme working conditions. Materials with higher mechanical strength are more resistant to tension, compression, bending, and shear pressures, allowing them to withstand hostile conditions and structural loads. Alloying, heat treatment, and nanomaterial reinforcement are used to improve mechanical strength by adjusting material composition, microstructure, and defect density. Furthermore, advances in materials science and engineering allow for the creation of composite materials, laminates, and engineered structures with customized mechanical properties to satisfy specific performance needs. Enhanced mechanical strength improves the longevity, dependability, and safety of products and infrastructure, promoting innovation and growth across a wide range of technical applications.

13.3.2 Improved Electrical Conductivity

Improved electrical conductivity refers to a material's increased ability to conduct electricity with low resistance. This feature is critical in many fields, including electronics, power transmission, and telecommunications, where effective electrical conduction is required. Materials with improved electrical conductivity have low electrical resistance, allowing electric current to flow with minimal losses and heat creation. Doping, alloying, and nanostructuring techniques are used to increase charge carrier density and reduce scattering processes inside the material lattice, hence improving electrical conductivity. In addition, the creation of conductive polymers, carbon-based materials, and metal alloys with optimal microstructures and dopant concentrations helps to improve electrical conductivity. Enhanced electrical conductivity allows for the design and implementation of high-performance electronic devices, energy storage systems, and conductive coatings, which drives technological and innovative improvements in a variety of industries.

13.3.3 Enhanced Thermal Stability

Enhanced thermal stability refers to a material's capacity to tolerate high temperatures and thermal stress without experiencing major deterioration or structural changes. This feature is critical in applications that expose materials to high temperatures, such as aerospace, automotive, electronics, and industrial processes. Materials with high thermal stability are resistant to thermal expansion, oxidation, breakdown, and phase transitions, ensuring long-term reliability and performance in hostile settings. Thermal stability is improved through techniques like as alloying, surface coatings, and thermal treatment, which increase material microstructure, chemical composition, and heat dissipation capabilities. Furthermore, the development of improved ceramics, refractory materials, and high-temperature polymers with customized thermal properties helps to improve thermal stability. Improved thermal stability enables the design and manufacture of components, coatings, and structural materials that can endure severe temperatures, accelerating innovation and progress in a variety of technological applications.

13.3.4 Tailorable Surface Properties

Tailorable surface qualities are the capacity to change and tailor the characteristics of a material's surface to match specific needs and applications. Surface qualities like as adhesion, wetting behavior, chemical reactivity, and biocompatibility all have a significant impact on interactions with the surrounding environment. Surface attributes are tailored using a variety of techniques and surface modification methods, including chemical treatments, physical coatings, and surface patterning, to meet desired functionality and performance requirements. Controlling surface roughness, chemistry, and topography allows materials scientists and engineers to improve surface wettability, increase adhesion, prevent corrosion, and impart antibacterial or antifouling qualities. Tailorable surface properties are used in a variety of industries, including automotive, aerospace, biomedical, and electronics, where surface engineering techniques enable the development of advanced coatings, functional surfaces, and tailored interfaces that are optimized for specific applications and environments.

13.3.5 Future Outlook of Metallic NCs

Finally, metallic NCs are a rapidly developing research area with enormous potential in a variety of industrial areas. Continued developments in fabrication techniques, together with a better understanding of their properties and uses, are expected to accelerate the general acceptance of metallic NCs in the coming years. However, constraints, such as scalability, cost-effectiveness, and environmental sustainability, necessitate further investigation and innovation in order to fully exploit the capabilities of metallic NCs for tackling critical societal and environmental issues.

13.4 POLYMERIC NCs: ADVANTAGES, APPLICATIONS AND FUTURE OUTLOOK

Attracting significant interest in the scientific community in the past years, polymeric NCs are a category of materials unique for their excellent attributes and their broad array of applications. As fascinating as they are versatile, these materials are a composite of a polymer matrix infused with NPs, typically on a nanometer scale, creating a singularly potent synergy(Shahabaz et al. 2022).

This fusion of NPs within the polymer matrix significantly amplifies the properties of the emerging NC. This enhancement ushers in the era of stronger mechanical robustness, superior thermal resistance, and improved electrical conductivity. This revolution is highly significant in industries like aerospace, automotive, electronics, and healthcare, where the demand for state-of-the-art materials with superior performance is continuously on the rise.

Polymeric NCs are composite materials crafted from a polymer matrix incorporated with nanoscale fillers or reinforcements. The nanoscale fillers, which could be inorganic substances like clay minerals, carbon nanotubes, or metal NPs, are dispersed within the polymer matrix to conceive a versatile, robust material with exceptional characteristics. These nanoscale fillers have the potential to significantly augment the physical and chemical properties of the final composite material, including mechanical robustness, thermal resistance, and barrier properties. By modulating the size and concentration of the nanofillers, researchers can adjust and enhance the overall performance of the polymeric NC, tailoring it to meet the specific requirements of diverse applications (Sharma et al. 2023).

Understanding the concept and significance of polymeric NCs is critical in the creation and design of advanced materials with precisely engineered features. The integration of nanoscale fillers into the polymer matrix sparks new possibilities for creating materials with increased functionality and performance. For instance, carbon nanotubes' addition can substantially increase the electrical conductivity of the composite, opening it up for applications in fields like electronics and energy storage. The introduction of metal NPs can boost the material's catalytic activity, enabling its use in varied chemical processes (M. Darwish et al. 2022a). By deciphering the behavior and inserts of the nanoscale fillers in the polymer matrix, researchers can unlock the full potential of polymeric NCs and open new avenues in pollution control engineering.

Moreover, polymeric NCs offer a host of advantages over traditional composite materials. Owing to the nanoscale size of the fillers, they distribute more uniformly throughout the polymer matrix, leading to improved mechanical features like enhanced stiffness, toughness, and wear resistance.

Additionally, the introduction of nanofillers can enhance composite thermal stability, enabling it to resist higher temperatures without significant degradation. Besides, polymeric NCs manifest superior barrier properties, qualifying as ideal candidates for applications requiring resistance to gas permeation, moisture absorption, and UV degradation.

In conclusion, the field of polymeric NCs shows immense promise in materializing advanced materials with customized properties for specific applications. The incorporation of nanoscale fillers into the polymer matrix facilitates the creation of materials with exceptional performance and functionality. By understanding the principles and features of polymeric NCs, scientists and engineers can unlock transformative advancements across industries, including air, water and soil pollution control. With continued research and innovation, polymeric NCs have the potential to revolutionize the field of materials science and pave the way for futuristic and sustainable solutions for several environmental issues.

13.4.1 Importance of Polymeric NCs in Various Industries

Polymeric NCs have emerged as extraordinarily vital and irreplaceable materials across diverse industries, attributable to their unparalleled properties and performance. In the aerospace sector, these phenomenal composites find indispensable use in manufacturing lightweight yet incredibly high-strength components, leading to a considerable reduction in fuel utilization and emissions. Moreover, they actively contribute toward enhancing fuel efficiency, durability, and safety within the automotive industry, making them an integral part of its operations. In addition to these remarkable applications, the excellent electrical conductivity and thermal management attributes can be used in several industries (Rajput et al. 2022).

13.4.2 Advantages, Potential Applications and Future Perspectives of Polymeric NCs

Polymeric NCs boast numerous advantages over conventional polymer materials, positioning them highly in demand for various applications. Incorporation of nanoscale fillers augments the mechanical attributes of the composite, leading to increased tensile strength, flexural strength, and impact resistance. They also showcase superior thermal stability with higher heat deflection temperatures and lower thermal expansion coefficients compared to unadulterated polymers (Cazan, Enesca, and Andronic 2021). Additionally, polymeric NCs can offer enhanced barrier properties against gas seepage and water vapor transmission (Hemath et al. 2020). Moreover, their tunable electrical conductivity and dielectric properties make them ideal for applications in electronics and sensors. The potential applications of polymeric NCs are vastly diverse, encompassing automotive components, packaging materials, construction components, electronics, aerospace components, medical devices, and more, offering a wide scope for innovation, development, and exploration across various fields. With ongoing research and improvements in nanotechnology, the opportunities for employing polymeric NCs are virtually limitless, promising constant growth, enhancement, and adaptation to meet evolving requirements and challenges across multiple industries. The expanding domain of polymeric NCs opens the doors to new possibilities, enabling the creation of groundbreaking materials that surpass current possibilities. By harnessing the unique attributes of nanoscale fillers, scientists and engineers can develop advanced composites with unparalleled strength, durability, and utility. The versatility of polymeric NCs permits custom-designed and customized formulations that meet specific application needs (Ogbonna et al. 2022). Whether creating lightweight, fuel-efficient automotive parts or developing high-performance packaging materials, polymeric NCs applications continue to expand and evolve with each new breakthrough. As research in nanotechnology makes progress, new fillers and manufacturing techniques are discovered, leading to even more innovative possibilities for polymeric NCs. This field's future is bright, with the potential to create materials stronger, lighter, and more environmentally friendly than ever before. As industries recognize the benefits of polymeric NCs, investment in research and development continues to rise, opening the doors to exciting possibilities for collaborations and technological advancements (Yadav et al. 2023). The journey towards unlocking the full potential of polymeric NCs is an ongoing one, as scientists and engineer's cross hurdles and discover new ways to employ these extraordinary materials (Nagaraju et al. 2023). Together, they are shaping a future in which polymeric NCs play a transformative role in shaping the world around us. Regardless of whether it's through enhancing the performance and sustainability of everyday products or transforming entire industries, the impact of polymeric NCs is undeniable. By embracing the possibilities offered by these advanced materials, we can pave the way for a more efficient, sustainable, and technologically advanced future (Erol et al. 2023).

13.4.2.1 Applications in Water Treatment

Polymeric NCs are a class of polymer-based materials that have dispersed in situ or ex situ inorganic nanofillers. These inorganic nanofillers range from commonly employed dispersed inorganic sheets or platelets (e.g., clay, graphene oxide, etc.) to dispersed inorganic particles like silica, metal oxides, etc. Polymeric NCs have been used for a multitude of applications, like gas storage, reinforcement, packaging, tires, coatings, and so forth. Recently, these NCs are also being heavily researched for water treatment applications like heavy metal ion separation, dye removal, and pollutant neutralization. Polymeric NCs have shown great potential for improving the efficiency and selectivity of water treatment processes. The prime driving force for this interest in research, development, and utilization of the polymeric NCs-involved materials for water and wastewater treatment is the fact that these NCs can not only have high mechanical and thermal strength achievable by the reinforcing filler materials, but can also be tunable, functional, and possess tailored hydrophilicity or hydrophobicity, and the nanoscale dimensions (Filipe, Alberto, and Ferreira 2021).

One of the hallmark applications of synthetic crosslinked dense polymer networks (CDPNs) is in the field of membrane-based molecular separation. They are exploited for water purification, desalination, and membrane gas separation. The feature of CDPNs that makes them suitable for these applications is their selective diffusion behavior toward transported molecules, as this selective diffusion can enable high separation performance (Mei et al., 2022). Separation in dense polymers like CDPNs is often defined as a transport phenomenon that follows Fickian diffusion with an aspect ratio (L/d) of <3, where L is the mesh size of the polymer network and d is the size of the transported molecule. When the network of a dense polymer is non-deformable, the mesh size L of the polymer network is fixed independent of the temperature, molecular weight of the polymer, or its polymerization conditions. This leads to L being very small on the order of 1-10 nm.

13.4.2.2 Applications in Air Pollution Control

Photocatalyst polymeric NCs are being adopted for the fabrication of a wide range of products, such as coatings, packaging materials, layer on textile and glass, automobiles, furniture, and so on. When polymeric NCs are incorporated with NPs (TiO2, ZnO, Bi2O3, ZnFe2O4, etc.) their degradation efficiency is enhanced. In those cases, the degradation efficiency is dependent on several factors like structural design, in particular the properties of the target pollutants. Products like copolymer supports with an immobilized photocatalyst can easily be removed from the water treatment plant after termination of photocatalytic degradation by conventional methods. Such supports can degrade consequent pollutants and eliminate secondary pollution source issues.

NCs can ideally be employed in the construction of various filters for the complete removal of several pollutants from the air because of their associated advantages, such as mechanical toughness, good electrical conductivity, and superior thermal properties, etc. Kim and co-workers incorporated carboxylate-functionalized multiwalled carbon nanotubes (MWCNTs) bifunctional oxygen- and nitrogen-doped reduced graphene oxide-based composite. The surface modification of CNTs, not only improved the affinity between the polymer matrix and nanofiller, but also resulted in hydrophilic nature, and enhanced the oxidation and adsorption of SO2 and NO.

The air pollution problems have stirred widespread concern, as air pollution is considered one of the main threats to all living organisms and has a significant impact on the environment, human health, and quality of life throughout the world. So, preventing or removing air impurities has become an urgent issue (Filipe, Alberto, and Ferreira 2021). Nanotechnology has gained attention recently and received significant interest in air pollution control because of having improved properties over traditional materials or methods with respect to remediation of several pollutants.

13.4.2.3 Applications in Waste Management

(In the last few years, the synthesis and study of NCs have steadily gained academic and commercial interest. Nano scaled fillers confer benefits such as high flame-retardant properties to the material and a reinforcement effect arising from their extraordinary high surface area. The choice of the used polymeric matrix for NC synthesis highly influences the obtained material behavior. Moreover, few works have been addressed to all-polystyrene NC study. Swelling tests via solvent sorption, X-ray scattering and microscopy of solutions, have given consistent results. The experimental observations have been partially ascribed to poor nanofiller/matrix interfacial adhesion and to the stiffness of the polyhedral oligomeric silsesquioxane. The enhanced sorption has been related to the fact that the characteristic length of the filler, l(≃5nm), is nearly equal to the radius of gyration of the polymer chains. The whole set of experimental observations shows the potential of an all-styrene NC. However, real NCs usually encapsulate agglomerates of NPs and bare NPs as well. The rarer all-polystyrene nanofiller provides only a theoretical nanofiller configuration that does not correspond to realistic systems.

To address these environmental problems, one way is to make the biodegradable polymer by using natural molecules, which can act as renewable resources. The second way is to recycle the existing polymeric waste by using appropriate technologies. In both ways, polymeric technology is certainly an answer for the future. The synthetic polymers are made from of petroleum. With the increase in the microbial population throughout the world, the use of petroleum resources is depleting day by day. For that reason, it is important to find an alternative source of polymers. In livestock industries, which are mostly depend on agriculture, the fat accumulation of some animals would be one source to make polymeric materials. The bacterial biomass bio-synthesis of polymeric materials like polyhydroxyalkanoates is proposed from renewable resources.

Circulatory Management of Polymer Waste: Recycling into fine fibers and their applications were discussed by Šišková et al. (2021). Due to the non-biodegradability of synthetic polymers, management of polymeric waste has become a thrust area in the current environmental scenario. There are new attempts to recycle and reuse the existing polymeric waste. At the same time, new polymeric technologies are being developed to recycle the natural polymer waste instead of dumping it into land, which can spoil the fertility of the soil. The landfill system and the incineration system are causing environmental problems. If the plastic waste goes to the landfill, then there is a possibility of leaching harmful chemical compounds and heavy metals in soil, which is polluting ground water and soil. The incineration of synthetic polymers causes environmental problems like air pollution, odor generation, acid rain formation, and generation of poisonous gas like dioxins and furans.

13.4.2.4 Applications in Renewable Energy

The future advancement trends and strategies of polymeric NCs are also articulated. NCs represent a relevant pathway toward the advanced material science emerging in recent decades, successfully combining organic polymers with NP reinforcement for large-area film applications under realistic conditions. Correct tuning of the interaction between the NP and polymer phases can strongly improve the performances of the film active layers. To date, the complete transition from lab-scale synthesis to scalable NC formulation for OPVs and OLEDs devices is still hampered by uppermost requirements regarding the dimension of the filler particles and surface functionalization, such as massive production and high NP concentrations in the films (Vogiatzis and Theodorou, 2014). Integrating the NP design to enable multifunctional filler molecules to play different roles is one key point to improve the performances of the cells. It has been demonstrated that utilizing the elaborated NP topology is an efficient means to produce printable film with high electrical conductivities under realistic conditions. These proofs of concept provide the versatility of polymer-compatible octopus-shaped NP, able to maintain a low structure factor in polymer matrices without strictly spoiling the geometry-dependent charge transports.

Polymeric NCs have emerged as a notable solution for renewable energy applications. For instance, NC materials are beneficial in dielectric energy storage applications, due to enhanced energy storage capabilities, charge-discharge efficiency, and good discharging stability (Fan et al. 2022). Surface modification and extensive interfacial contact between the polymer matric and the nanofillers result in a remarkable increase of NC dielectric permittivity, breakdown strength, and energy density. Modifying topological structure of NCs with proper toughening agents to enhance strain/energy conversion coefficient and elastic deformation window as well as eliminate hysteresis losses is found to be an efficient way to pursuit high strain sensitivity and comprehensive-based performances. The high-frequency electromagnetic response of the electromagnetic wave absorbing materials is based on the individual intrinsic characteristics of the filler materials, the hybridization of the filled materials and charge transport thereof among the hybrid components. The coupling between the light (electromagnetic wave) and the guest (filler) due to the dipole caused by the intrinsic and/or extrinsic electrons is a major contributory mechanism.

13.4.2.5 Applications in Green Packaging

To date, almost all these anticipated bio-NC (i.e., bio-based NC systems) have presented good functional properties, such as a solvent-dependent process capability, increased thermal capability, antimicrobial capability, antioxidant capability, mechanical strength, moisture sorption capability, and high biodegradability. But biopolymer-based bio-NCs do not have adequate surface properties, limiting their production capacity or impacting the storage/durability of receptive products. Thus, further research is needed to develop the necessary functional groups. Such elements should include the eco-functional groups that can still be supplied by environmentally friendly processes during biopolymer-based NC formation. For consumer compatibility in regard to both environmental and anti-microbial action, polymer based biodegradable materials and biopoly-nanomaterial phrases need to be considered too.

NCs have emerged as an important advancement/progress in polymer science, and a large number of them have been developed over the last decade, as mentioned above (Filipe, Alberto, and Ferreira 2021). Green NCs, which include biodegradable polymers and renewable/abundant natural fillers, are of great interest for researchers to produce biodegradable and eco-friendly materials (Kumar and Gautam 2019). For green packaging, biodegradable biopolymer-based composite materials processed by nano technology or multifunctional hybrid materials based on bio-composites or multi-phase composite materials have been produced. One way to improve the properties of green biopolymers in food packaging materials may be to produce NCs and hybrid NCs (Armentano et al. 2024). Antimicrobial packaging is used to extend the shelf life of food by preventing bacterial adhesion, preventing bacterial growth, deactivating existing bacteria, and preventing spore metabolism. In the forming of each type of microbial packaging, antimicrobial materials such as metals, metal oxides/ions, organic antimicrobial elements, quorum sensing inhibitors / RNA-cleaving enzymes, and natural and synthetic polymers, antimicrobial bio-NPs, and antimicrobial NPs can be caste as a packaging. These include, but are not limited to, drug-based antimicrobial components (e.g., essential oils, pyranocoumarins, aldehydes, phenolic components, polyphenols, glycosides, glucosides, enzymes) in plant and/or animal extracted phenols (tannins, flavonoids etc.).

13.4.2.6 Applications in Environmental Monitoring

Due to their unique properties, such as high surface area, enhanced mechanical strength, and tailored chemical reactivity, Polymeric NCs have emerged as powerful tools in **environmental monitoring**. Polymeric NCs are considered to be a charismatic class of inorganic nanofillers and organic polymers that are highly efficient, cost effective and accurate for monitoring different environmental pollutants, however, its potential has not been fully tapped in environmental protection applications (Emam et al. 2022). They are the leading materials for the contaminant treatment and sensing (M. S. A. Darwish, Mostafa, and Al-Harbi 2022b). Both polymeric NCs and polymer fillers have combined synergic features that offer scope for combating the massive problematic environmental pollutants. (Agboola et al. 2021). Every segment of the polymeric NCs, in contrast to the pristine polymers, performs a distinguishable use in environmental protection sectors and each of these polymeric NCs and their composite materials with the fillers perform major effect and have been expanded as a possible formless catalyst, anion exchanger/absorbent, and sequestrant in neutralizing radical-loaded surroundings, as evanescent supports in pollutant ("cations and anions") detection. Many recent studies have well documented the use of NMs and NCs for designing sensors for many toxic pollutants in environmental monitoring (Shakeel et al., 2022a & Shakeel et al. 2022b).

13.4.2.7 Future Outlook of Polymeric NCs

NCs have been widely used for the removal of heavy oils, which are usually used for insulation purposes in high voltage power transmission lines. The loss of these oils, with more than three hundred million metric tons lost annually, is a great challenge due to their possible environmental and technology-related problems. They can be trapped from water surfaces by a sophisticated underwater super oleophilic, polyurethane sponge prepared by reduction action. The method is applied through a one-step concurrent polymerization method. Adsorption methods involve adsorption of heavy oil onto nanostructures. NCs remove metals via various mechanisms, including chelation, ion exchange, adsorption, and electrostatic interaction. The affinity and capacity of the NCs depends on their chemical composition, polymers structure, chelating agents, and the technique used for assembling the hydrogel (Agboola et al. 2021). The hydrogel usually impresses the largest contribution because of its appearance in large amounts acting as a binding agent. Long term research focuses on developing wider, more environmentally friendly applications, such as sugarcane, rice, or wheat for the development of sustainable biodegradable hydrogels. Further improvement of the selective recovery of metals by the hydrogels can be achieved by adding organic ligands and metal complexes, which introduce functional groups used by the hydrogels for metal recovery.

13.5 CERAMIC NCs (CNCS)

CNCs are materials in which the ceramics are combined with nanoscale particles or fibers to improve their properties. These nanoscale reinforcements can considerably enhance the mechanical, thermal, electrical, and optical properties of the ceramics. The inclusion of NPs to a ceramic matrix can lead to an enhancement of the mechanical strength, toughness, wear resistance, and thermal stability of the ceramic. One of the major advantages of CNCs is their enhanced surface area to volume ratio, which leads to better solubility, multifunctional capabilities, enhanced electrical as well as thermal conductivity, and surface catalytic activity, can be intensified (El-Sadik and El-Sadik 2012). These positive attributes make ceramic NCs as an emerging candidate in wide range of applications in environment, electronics, aerospace industries, cutting tools, wear-resistant coatings, etc. The CNCs produce synergistic effect when the ceramic matrix is coupled with nanoscale reinforcements, to produce better performance characteristics than conventional micro-composites or monolithic ceramics. The factors such as the distribution, alignment, and interfacial bonding between the ceramic matrix and the NPs affect the effectiveness of these reinforcements used in CNCs.

13.5.1 Applications of CNCs in Pollution Control

Ceramic NCs have emerged as a promising material in wastewater treatment, due to their enhanced characteristics from the combination of ceramic properties with the benefits of nanotechnology. These are some ways in which ceramic NCs contribute to wastewater treatment:

13.5.1.1 CNCs in Membrane Technologies

CNCs were recently used in membrane technologies for microfiltration, as well as ultrafiltration of oily wastewaters and other suspensions. The high mechanical strength and chemical stability of CNCs make them operate efficiently under extreme and harsh conditions. Enhanced filtration utilizing CNCs in wastewater treatment is primarily due to the unique material properties that these composites offer.

CNC membranes can achieve improved filtration efficiencies, due to their small pore size and high surface area. This leads to effective removal of fine particulates, pathogens, and other contaminants from water. On the other hand, these CNCs are chemically inert and can withstand extreme corrosive wastewater environments and high temperatures during cleaning and sterilization.

The CNCs can be engineered to resist fouling by modifying the surface properties, in which accumulation of substances that clog filter pores can be prevented, thus maintaining improved filtration efficiency over time. These CNCs have good mechanical strength and are durable during their use in environmental applications. CNCs have high mechanical robustness, which makes them less susceptible to breakage, as well as less deformation under high-pressure filtration conditions, which lead to longer lifespan and lower maintenance of CNCs.

CNCs are particularly effective in treating oily wastewaters; therefore, an efficient separation of oil and water phases can be achieved. Thus, they are highly useful for industrial applications where oily effluent is a matter of concern. By structure modifications in CNCs, the membranes can be engineered to be selectively permeable, allowing certain molecules to pass through while retaining other specific molecules in virtue of size, charge, or chemical affinity. Recently, Baig et al. (2024) showed a simple and effective method of using MIL-125-NH_2 metal organic framework on alumina NC for oily wastewater treatment. These researchers found that a CNC membrane displayed more than 99% oil rejection.

CNCs are highly environmentally sustainable, as these substances can often be washed and reused several times, CNC filters contribute to more sustainable wastewater treatment systems by cutting down operating cost as well as waste generation. OGUNDIPE (2023)developed a novel kaolin clay, silver oxide and ZnO nanocomposite filter for treating domestic wastewater.

13.5.1.2 CNCs in Effective Adsorption

CNCs are emerging as promising candidates in wastewater treatment, due to their unique adsorption capabilities. CNCs, being made up of nanoscale particles, provide an enhanced surface area compared to bulk materials. Thus, presence of enormous active sites leads to better adsorption of contaminants present in wastewater. CNCs can be designed with certain properties tailored for adsorption, such as pore size, surface chemistry, porosity, etc. Thus, when these parameters are precisely controlled, then adsorption efficiency for targeted contaminants in wastewaters can be significantly improved.

Ceramics are well known for their chemical stability, and they are highly resistant to extreme environments. When NPs are incorporated into ceramic matrices, their stability can be further improved, thus allowing them to withstand harsh conditions often encountered during wastewater treatment processes. By using surface functionalization and modification, CNCs can demonstrate selective adsorption toward specific pollutants without disturbing other compounds present in the stream. This selective approach enhances overall efficiency of wastewater treatment; therefore, the generation of secondary pollutants can be substantially minimized. Recently, Pirozzi et al. (2023) employed novel magnetic metal based CNCs as an adsorbent for effective adsorption of Gallic acid. The composite was prepared using synthetic zeolite 4A, that is $Na_{12}Al_{12}Si_{12}O_{48}\cdot 27H_2O$, framework type - Linde Type A, and the adsorption efficiency depends mainly on pH.

CNCs can often be regenerated effectively and can be reused several times without scarifying the adsorption capacity. This unique property of CNCs reduce the operational costs and makes them economically effective for large-scale wastewater treatment applications. In comparison with conventional adsorbents such as activated carbon, etc., these CNCs are often eco-friendly and more versatile. They can be produced from abundant, as well as low-cost substances, and their use reduces dependency on non-renewable resources. CNCs can be effectively used in diverse pollutant control, such as heavy metals, organic pollutants, dyes, pharmaceutical wastes, etc.

13.5.1.3 CNCs in Catalytic Degradation

Certain CNCs have unique catalytic properties that can break down carcinogenic organic recalcitrant into less toxic substances. They can be applied in advanced oxidation processes (AOPs) such as photocatalysis, ozonation, etc., to treat wastewaters. In photocatalysis, semiconductor-based NPs like TiO_2, ZnO can be incorporated into the ceramic matrices. Recently, (Fernandes et al. 2023) developed a mesoporous ceramic membrane of diatomite/TiO_2/Nb_2O_5 NCs and treated the contaminated water. Here, diatomite was employed as ceramic matrix, and TiO_2 and Nb_2O_5 were incorporated as NPs. Simultaneous adsorption and catalytic degradation of recalcitrant dye occurred during their use. Recently, Singh et al. (2024) developed nano TeO_2-ZnO NCs and employed it in the presence of sunlight for the removal of toxic pollutants present in wastewater. 91% Cr removal was achieved with this NC, and around 90% bacterial pathogenicity removal was achieved in this work. When these NPs were exposed to sunlight/UV light, they generate reactive oxygen species that are highly efficient in degrading recalcitrant present in wastewaters. Similarly, researchers employed CNCs for Fenton and Fenton-like reactions by using iron oxide NPs for enhanced degradation of organic pollutants. Recently, Seroglazova et al. (2023) synthesized novel $PrFeO_3$/CeO_2 NCs and employed it for the removal of methyl violet dye. They found high cyclic stability of synthesized NCs and proposed the possible mechanism of photo-Fenton like decomposition of dye used in the presence of novel composite.

13.5.2 Challenges in Using CNCs

CNCs are emerging as promising candidate in diverse environmental applications, and there are certain challenges to using CNCs that need to be addressed by the research community for successful implementation. First of the primary challenges is production of CNCs with controlled composition, structure, and morphology in large scale for commercialization. Scaling up the production without comprising the quality and unique properties of CNCs in a cost effective manner is highly required to get good results. Similarly, CNCs must provide sufficient stability, as well as durability, during exposure toward extreme treatment conditions. Prevention of leaching and degradation of NPs from composite matrix is crucial for practical applications of CNCs. Recently, Omar et al. (2024) elaborated the challenges in implementation of CNCs on water purification and provided technical solutions for several challenges encountered. They also shown the future direction of CNCs for a better environment. (Solaiman et al. 2024) recently reviewed the current applications, future perspectives of CNC membranes for remediation of textile dye wastewater. They elaborated the production methodology and mechanism involved in membrane production.

The selectivity and catalytic activity of CNCs toward targeted pollutants is a more challenging one. Selection of suitable nanomaterials and matrix is required to maximize the catalytic efficiency and minimize the degradation and deactivation of the catalyst during reactions. When CNCs are used in membrane-based technologies, these are more susceptible to fouling by organic matters, suspended solids, microbial growth, etc. This issue can be resolved by developing effective antifouling strategies and cleaning protocols, which leads to the enhanced life span of membranes. The cost consideration of CNCs is another major challenge for adopting them for wastewater treatment. High cost involved in synthesis and fabrication techniques must be controlled effectively, using low cost raw materials. The potential environmental, as well as health risks associated with NPs and the compliance with regulatory standards for water quality and treatment, must be ensured for the approval of public and regulatory bodies. Sameera Khathoon et al. (2024) elaborated the environmental, legal, health, and safety issues involved in nanoceramics, and discussed that the industrial applications of nanoceramics starts from lab to real time applications.

Lastly, integrating CNCs with existing infrastructure facility is another major concern that needs to be addressed properly. Careful planning, engineering, and designing the conventional treatment methodologies, retrofitting existing facilities, and optimization in integrating CNCs in existing system can substantially enhance the overall performance and efficiency of the process.

Thus, these challenges can be effectively addressed by interdisciplinary research and collaboration among a diverse community that include engineers, policymakers, industrial stakeholders, public, etc. When these challenges are properly handled, CNCs can be emerged as one of the best candidates for handling environmental issues in effective manner.

13.5.3 Future Outlook of CNCs

CNCs will continue to play a predominant role in waste water treatment processes. They offer advanced materials for removing diverse pollutants, like heavy metals, organic compounds, and microorganisms, from contaminated water sources. These materials can be employed in several treatment processes, such as adsorption, catalytic degradation, membrane filtration, and advanced oxidation processes, to improve the quality of water and ensure safe potable water supplies for the general public across the world.

CNCs are a promising candidate for mitigating air pollution by capturing harmful gases, particulate matter, and volatile organic compounds (VOCs) from industrial emissions, vehicle exhausts, and indoor air environments. CNC-based adsorbents, catalytic converters, and air filtration systems can significantly reduce the concentration of diverse air pollutants and mitigate the health risks associated with airborne contaminants, which are going to be a big threat for the human community in the near future.

In soil remediation applications, CNCs can be emerged as a reliable alternative toward immobilization, degradation, or extraction of contaminants such as heavy metals, organic pollutants, and pesticides from contaminated soils and sediments, in near future. CNC-based adsorbents, catalysts, and nanostructured membranes can offer promising solutions for restoration of soil quality, protecting ecosystems, and safeguarding agricultural productivity.

The future improvements in CNCs will emphasize green and sustainable production methodologies, and eco-friendly and low-cost raw materials' effective use. Implementation of energy-efficient processes using CNCs can minimize the environmental impacts; thus, sustainable pollution control technologies can be developed effectively. With effective integration of renewable resources, recyclable materials, and low-energy fabrication techniques, CNCs can contribute to environmental sustainability and climate resilience.

REFERENCES

Agboola, Oluranti, Ojo Sunday Isaac Fayomi, Ayoola Ayodeji, Augustine Omoniyi Ayeni, Edith E Alagbe, Samuel E Sanni, Emmanuel E Okoro, et al. 2021. "A Review on Polymer NCs and Their Effective Applications in Membranes and Adsorbents for Water Treatment and Gas Separation." *Membranes* 11: 139. doi:10.3390/membranes11020139

Ananthakumar, S., K. Prabhakaran, U. S. Hareesh, P. Manohar, and K. G.K. Warrier. 2004. "Gel Casting Process for Al2O3–SiC NCs and Its Creep Characteristicis." *Materials Chemistry and Physics* 85 (1). Elsevier: 151–57. doi:10.1016/J.MATCHEMPHYS.2003.12.022

Armentano, Ilaria, Debora Puglia, Francesca Luzi, Carla Renata Arciola, Francesco Morena, Sabata Martino, and Luigi Torre. 2024. "NCs Based on Biodegradable Polymers." *Mdpi.ComI Armentano, D Puglia, F Luzi, CR Arciola, F Morena, S Martino, L TorreMaterials, 2018•mdpi.Com.* Accessed June 24. doi:10.3390/ma11050795

Baig, Umair, Abdul Waheed, Lukka Thuyavan Yogarathinam, and Isam H. Aljundi. 2024. "Tailored Alumina NC Membranes Featuring MIL-125-NH2 Metal-Organic Frameworks for Oily Wastewater Treatment." *Npj Clean Water* 2024 7:1 7(1). Nature Publishing Group: 1–12. doi:10.1038/s41545-024-00321-w

Baiju, K. V., C. P. Sibu, K. Rajesh, P. K. Pillai, P. Mukundan, K. G. K. Warrier, & W. Wunderlich. 2005. "An aqueous sol–gel route to synthesize nanosized lanthana-doped titania having an increased anatase phase stability for photocatalytic application." *Materials chemistry and physics*, *90*(1), 123–127.

Belessi, V., D.A. Lambropoulou, I. Konstantinou, A. P. Katsoulidis, P. Pomonis, D. Petridis, and T.A. Albanis. 2007. "Structure and Photocatalytic Performance of TiO2/Clay NCs for the Degradation of Dimethachlor." *Applied Catalysis B: Environmental* 73 (3–4). Elsevier: 292–99. doi:10.1016/J.APCATB.2006.12.011

Cazan, Cristina, Alexandru Enesca, and Luminita Andronic. 2021. "Synergic Effect of TiO2 Filler on the Mechanical Properties of Polymer NCs." *Polymers 2021*, 13 (12). Multidisciplinary Digital Publishing Institute: 2017. doi:10.3390/POLYM13122017

Chen, Dong Hwang, and Xin Rong He. 2001. "Synthesis of Nickel Ferrite NPs by Sol-Gel Method." *Materials Research Bulletin* 36 (7–8). Pergamon: 1369–77. doi:10.1016/S0025-5408(01)00620-1

Chiang, Chin Lung, and Chen Chi M. Ma. 2002. "Synthesis, Characterization and Thermal Properties of Novel Epoxy Containing Silicon and Phosphorus NCs by Sol–Gel Method." *European Polymer Journal* 38 (11). Pergamon: 2219–24. doi:10.1016/S0014-3057(02)00123-4

Darwish, Marjan, Camila Mafla-Gonzalez, Belmin Kolenovic, Adrianna Deremer, Daniel Centeno, Tianchi Liu, Deok Yang Kim, et al. 2022a. "Rapid Synthesis of Metal NPs Using Low-Temperature, Low-Pressure Argon Plasma Chemistry and Self-Assembly." *Green Chemistry* 24 (20). The Royal Society of Chemistry: 8142–54. doi:10.1039/D2GC02592B

Darwish, Mohamed S.A., Mohamed H. Mostafa, and Laila M. Al-Harbi. 2022b. "Polymeric NCs for Environmental and Industrial Applications." *International Journal of Molecular Sciences 2022*, 23 (3). Multidisciplinary Digital Publishing Institute: 1023. doi:10.3390/IJMS23031023

Dou, Binlin, Valerie Dupont, Weiguo Pan, and Bingbing Chen. 2011. "Removal of Aqueous Toxic Hg(II) by Synthesized TiO2 NPs and TiO2/Montmorillonite." *Chemical Engineering Journal* 166 (2). Elsevier: 631–38. doi:10.1016/J.CEJ.2010.11.035

El-Sadik, Abir, and Abir El-Sadik. 2012. "Novel Promises of Nanotechnology for Tissue Regeneration." *Tissue Regeneration - From Basic Biology to Clinical Application*, March. IntechOpen. doi:10.5772/25865

Emam, Hossam E, Ahmed K Badawi, Adedapo Oluwasanu Adeola, and Philiswa Nosizo Nomngongo. 2022. "Advanced Polymeric NCs for Water Treatment Applications: A Holistic Perspective." *Mdpi.ComAO Adeola, PN NomngongoPolymers, 2022•mdpi.Com.* doi:10.3390/polym14122462

Ennas, G., A. Musinu, G. Piccaluga, D. Zedda, D. Gatteschi, C. Sangregorio, J. L. Stanger, G. Concas, and G. Spano. 1998. "Characterization of Iron Oxide NPs in an Fe2O3–SiO2 Composite Prepared by a Sol–Gel Method." *Chemistry of Materials* 10 (2). American Chemical Society: 495–502. doi:10.1021/CM970400U

Erol, Mustafa, Sibel Demiroğlu Mustafov, Salih Alper Akalın, Begüm Uzunbayır, Erdem Tevfik Özdemir, Berkay Coşkun, and Zeynep Ertekin. 2023. "Carbon Nanostructures for Automotive and Aerospace Applications." *Handbook of Functionalized Carbon Nanostructures*. Springer, Cham, 1–29. doi:10.1007/978-3-031-14955-9_67-1

Fan, Youjun, Zhonghui Shen, Xinchen Zhou, Zhenkang Dan, Le Zhou, Weibin Ren, Tongxiang Tang, Shanyong Bao, Cewen Nan, and Yang Shen. 2022. "Highly Sensitive Strain Sensor from Topological-Structure Modulated Dielectric Elastic NCs." *Advanced Materials Technologies* 7 (1). John Wiley and Sons Inc. doi:10.1002/ADMT.202101190

Fernandes, Yuri L.R.L., Mariana C.L. Silva, Rafael A. Raimundo, Joyce Cavalcante, Jayara B.Q. Tomaz, Meysam Mashhadikarimi, Marco A. Morales, Fabiana V. da Motta, Wilson Acchar, and Uílame U. Gomes. 2023. "Development of Mesoporous Ceramic Membranes of Diatomite/TiO2/Nb2O5 NCs for the Treatment of Contaminated Water." *Journal of Environmental Chemical Engineering* 11 (6). Elsevier: 111161. doi:10.1016/J.JECE.2023.111161

Filipe, José António, Manuel Alberto, and M Ferreira. 2021. "Analysis of Nanosciences and Nanotechnology and Their Applications." https://arxiv.org/abs/2111.03425v1

Gao, Bin, George Z. Chen, and Gianluca Li Puma. 2009. "Carbon Nanotubes/Titanium Dioxide (CNTs/TiO2) NCs Prepared by Conventional and Novel Surfactant Wrapping Sol–Gel Methods Exhibiting Enhanced Photocatalytic Activity." *Applied Catalysis B: Environmental* 89 (3–4). Elsevier: 503–9. doi:10.1016/J.APCATB.2009.01.009

Guglielmi, Massimo, Guido Kickelbick, and Alessandro Martucci. 2014. "Sol-Gel NCs." *Sol-Gel NCs*, January. Springer New York, 1–227. doi:10.1007/978-1-4939-1209-4/COVER

Hemath, Mohit, Sanjay Mavinkere Rangappa, Vinod Kushvaha, Hom Nath Dhakal, and Suchart Siengchin. 2020. "A Comprehensive Review on Mechanical, Electromagnetic Radiation Shielding, and Thermal Conductivity of Fibers/Inorganic Fillers Reinforced Hybrid Polymer Composites." *Polymer Composites* 41 (10). John Wiley & Sons, Ltd: 3940–65. doi:10.1002/PC.25703

Jeon, Hyung Jun, Sung Chul Yi, and Seong Geun Oh. 2003. "Preparation and Antibacterial Effects of Ag–SiO2 Thin Films by Sol–Gel Method." *Biomaterials* 24 (27). Elsevier: 4921–28. doi:10.1016/S0142-9612(03)00415-0

Kumar, Prakash, and Sanjeev Gautam. 2019. "Developing ZnO NP Embedded Antimicrobial Starch."

Lee, Jin Hong, and Byung Ok Park. 2003. "Transparent Conducting ZnO:Al, In and Sn Thin Films Deposited by the Sol–Gel Method." *Thin Solid Films* 426 (1–2). Elsevier: 94–99. doi:10.1016/S0040-6090(03)00014-2

Mei, X.; Li, C.; Peng, P.; Wang, J.; He, E.; Qiu, Z.; Xia, D.; Zhao, Q.; Shen, D. 2022, "*Bombyx mori* C-Type Lectin (BmIML-2) Inhibits the Proliferation of *B. mori* Nucleopolyhedrovirus (BmNPV) through Involvement in Apoptosis." *Int. J. Mol. Sci.* 24, 2224. https://doi.org/10.3390/ijms24032224

Nagaraju, Sharath Ballupete, H. C. Priya, Yashas Gowda Thyavihalli Girijappa, and Madhu Puttegowda. 2023. "Lightweight and Sustainable Materials for Aerospace Applications." *Lightweight and Sustainable Composite Materials: Preparation, Properties and Applications*, January. Woodhead Publishing, 157–78. doi:10.1016/B978-0-323-95189-0.00007-X

Nazir, Rabia, Muhammad Mazhar, M. Javed Akhtar, M. Raza Shah, Nawazish A. Khan, M. Nadeem, Muhammad Siddique, Mazhar Mehmood, and N. M. Butt. 2008. "Superparamagnetic Bimetallic Iron-Palladium Nanoalloy: Synthesis and Characterization." *Nanotechnology* 19 (18). Nanotechnology. doi:10.1088/0957-4484/19/18/185608

Ogbonna, V. E., A. P.I. Popoola, O. M. Popoola, and S. O. Adeosun. 2022. "A Review on the Recent Advances on Improving the Properties of Epoxy NCs for Thermal, Mechanical, and Tribological Applications: Challenges and Recommendations." *Polymer-Plastics Technology and Materials* 61 (2). Taylor & Francis: 176–95. doi:10.1080/25740881.2021.1967391

Ogundipe, F.O., Saidu M., Abdulkareem A. S., and Busari A.O. 2023. "Development Of Beneficiated Kaolin Clay, Silver Oxide And Zinc Oxide Nanocomposite Adsorbents To Remove Chemical And Biochemical Oxygen Demands Concentration In Domestic Wastewater". *Bima Journal of Science And Technology (2536-6041)* 7(2): 86–112. https://doi.org/10.56892/bima.v7i2.430

Omanović-Mikličanin, Enisa, Almir Badnjević, Anera Kazlagić, and Muhamed Hajlovac. 2020. "NCs: A Brief Review." *Health and Technology* 10 (1). Springer: 51–59. doi:10.1007/S12553-019-00380-X/METRICS

Omar, Najib Meftah Almukhtar, Mohd Hafiz Dzarfan Othman, Zhong Sheng Tai, Mohamad Fairus Rabuni, Ahmed Omar Aswaye Amhamed, Mohd Hafiz Puteh, Juhana Jaafar, Mukhlis A. Rahman, and Tonni Agustiono Kurniawan. 2024. "Overcoming Challenges in Water Purification by NC Ceramic Membranes: A Review of Limitations and Technical Solutions." *Journal of Water Process Engineering* 57 (January). Elsevier: 104613. doi:10.1016/J.JWPE.2023.104613

Pirozzi, Domenico, Michele Pansini, Antonello Marocco, Serena Esposito, Gabriele Barrera, Paola Tiberto, Paolo Allia, and Filomena Sannino. 2023. "Adsorption of Gallic Acid by Tailor-Made Magnetic Metal-Ceramic NCs." *Journal of Molecular Liquids* 371 (February). Elsevier: 121083. doi:10.1016/J.MOLLIQ.2022.121083

Rajput, Anupama, Upma, Sudheesh K. Shukla, Nitika Thakur, Anamika Debnath, and Bindu Mangla. 2022. "Advanced Polymeric Materials for Aerospace Applications." *Aerospace Polymeric Materials*, August. John Wiley & Sons, Ltd, 117–36. doi:10.1002/9781119905264.CH5

Sameera Khathoon, A., Darul Raiyaan Darul, Mohamed Khalith Mohamed, and Kantha D. Arunachalam. 2024. "Industrial Applications of Nanoceramics: From Lab to Real-Time Utilization—the Environmental, Legal, Health, and Safety Issues of Nanoceramics." *Industrial Applications of Nanoceramics*, January. Elsevier, 411–23. doi:10.1016/B978-0-323-88654-3.00027-5

Schmidt, Daniel, Deepak Shah, and Emmanuel P. Giannelis. 2002. "New Advances in Polymer/Layered Silicate NCs." *Current Opinion in Solid State and Materials Science* 6 (3). Elsevier: 205–12. doi:10.1016/S1359-0286(02)00049-9

Seroglazova, Anna S., Maria I. Chebanenko, Vladimir N. Nevedomskyi, and Vadim I. Popkov. 2023. "Solution Combustion Synthesis of Novel PrFeO3/CeO2 NC with Enhanced Photo-Fenton Activity under Visible Light." *Ceramics International* 49 (10). Elsevier: 15468–79. doi:10.1016/J.CERAMINT.2023.01.132

Shahabaz, S. M., Pradeep Kumar Shetty, Nagaraja Shetty, Sathyashankara Sharma, S. Divakara Shetty, and Nithesh Naik. 2022. "Effect of Alumina and Silicon Carbide NP-Infused Polymer Matrix on Mechanical Properties of Unidirectional Carbon Fiber-Reinforced Polymer." *Journal of Composites Science* 6 (12). Multidisciplinary Digital Publishing Institute: 381. doi:10.3390/JCS6120381

Ahmad Shakeel, Komal Rizwan, Ujala Farooq, Shahid Iqbal, Ataf Ali Altaf (2022a),Advanced polymeric/inorganic nanohybrids: An integrated platform for gas sensing applications, *Chemosphere*, 294, 133772, https://doi.org/10.1016/j.chemosphere.2022.133772

Ahmad Shakeel, Komal Rizwan, Ujala Farooq, Shahid Iqbal, Tanveer Iqbal, Nasser S. Awwad, Hala A. Ibrahium. (2022b). Polymer based nanocomposites: A strategic tool for detection of toxic pollutants in environmental matrices, *Chemosphere*, 303, Part 1, 2022, 134923, https://doi.org/10.1016/j.chemosphere134923

Sharma, Shubham, Akarsh Verma, Sanjay Mavinkere Rangappa, Suchart Siengchin, and Shigenobu Ogata. 2023. "Recent Progressive Developments in Conductive-Fillers Based Polymer NCs (CFPNC's) and Conducting Polymeric NCs (CPNC's) for Multifaceted Sensing Applications." *Journal of Materials Research and Technology* 26 (September). Elsevier: 5921–74. doi:10.1016/J.JMRT.2023.08.300

Singh, Aishwarya, Suraj Kumar Modi, Preeti Joshi, Bhavani Prasad Nenavathu, Manu Smriti Singh, Swati Verma, and Mohammad Rafe Hatshan. 2024. "Sunlight Mediated Removal of Toxic Pollutants from Yamuna Wastewater Using Efficient Nano TeO2–ZnO NCs." *Chemosphere* 348 (January). Pergamon: 140658. doi:10.1016/J.CHEMOSPHERE.2023.140658

Šišková, Alena Opálková, Petra Peer, Anita Eckstein Andicsová, Igor Jordanov, Piotr Rychter, Patricia Krawczak, Sary Awad, Florentin Berthet, and José-Marie Lopez-Cuesta. 2021. "Circulatory Management of Polymer Waste: Recycling into Fine Fibers and Their Applications." *Materials* 14 (16). Multidisciplinary Digital Publishing Institute: 4694. doi:10.3390/MA14164694

Sivakumar, Sugananth, Krishna Gopakumar Warrier, S Sivakumar, C P Sibu, P Mukundan, P Krishna Pillai, and K G K Warrier. 2004. "Nanoporous Titania–Alumina Mixed Oxides—an Alkoxide Free Sol–Gel Synthesis." *ElsevierS Sivakumar, CP Sibu, P Mukundan, PK Pillai, KGK WarrierMaterials Letters*, 2004•Elsevier. doi:10.1016/j.matlet.2004.03.050

Solaiman, Joynab Mohammed, Natarajan Rajamohan, Mohammad Yusuf, and Hesam Kamyab. 2024. "NC Ceramic Membranes as Novel Tools for Remediation of Textile Dye Waste Water – A Review of Current Applications, Machine Learning Based Modeling and Future Perspectives." *Journal of Environmental Chemical Engineering* 12 (2). Elsevier: 112353. doi:10.1016/J.JECE.2024.112353

Vogiatzis, Georgios G. and Doros N. Theodorou. 2014. Local Segmental Dynamics and Stresses in Polystyrene C60 Mixtures." *Macromolecules*, 47 (1), 387–404. doi:10.1021/ma402214

Wang, Ru-Min, Shui-RRong Zheng, and Yujun George Zheng. 2011. *Polymer Matrix Composites and Technology*. https://books.google.com/books?hl=en&lr=&id=uHhwAgAAQBAJ&oi=fnd&pg=PP1&dq=Polymer+matrix+composites+and+technology.+Woodhead+Publishing+Limited&ots=abBBnd6zg0&sig=IGKJkBnEbnZKm87GlBUDyXY-iLw

Warrier, K. G.K., and G. M. Anilkumar. 2001. "Densification of Mullite–SiC NC Sol–Gel Precursors by Pressureless Sintering." *Materials Chemistry and Physics* 67 (1–3). Elsevier: 263–66. doi:10.1016/S0254-0584(00)00447-8

Yadav, Ramkumar, Mayank Singh, Deepika Shekhawat, Seul Yi Lee, and Soo Jin Park. 2023. "The Role of Fillers to Enhance the Mechanical, Thermal, and Wear Characteristics of Polymer Composite Materials: A Review." *Composites Part A: Applied Science and Manufacturing* 175 (December). Elsevier: 107775. doi:10.1016/J.COMPOSITESA.2023.107775

Zhang, Huanjun, and Guohua Chen. 2009. "Potent Antibacterial Activities of Ag/TiO2 NC Powders Synthesized by a One-Pot Sol-Gel Method." *Environmental Science and Technology* 43 (8): 2905–10. doi:10.1021/ES803450F

14 Photocatalytic Activity of Nanomaterials in Environmental Remediation

Prasann Kumar and Joginder Singh

14.1 INTRODUCTION

Increasing environmental pollution is a significant and urgent issue causing various adverse effects, such as declining air quality and polluted water sources. Given the detrimental impacts of anthropogenic activities, there is an urgent need for paradigm-shifting solutions that can save the environment without any further negative impacts. Nanomaterials (NMs) have been growing more promising for environmental remediation due to their uniqueness and flexibility of applications in addressing this challenge. These materials have a relatively high surface area to volume ratio owing to their small size, and their unique properties make them highly effective for the sorption and degradation of pollutants (Puri and Gupta, 2023; González-González et al., 2023). NMs not only serve as a tool for environmental remediation but also offer a range of unique benefits, including greater operational efficiency, reduced energy consumption, and the ability to target specific pollutants in a more focused manner. NMs supported latest technologies serve as important instruments for a variety of applications, such as biomedical applications, wastewater remediation, and environmental remediation. These technologies provide a variety of distinct advantages, such as improved operational effectiveness, decreased energy usage, and the capability to address particular pollutants selectively. These NMs can be altered according to the targeted contaminants, whether heavy metals, organic compounds, or both, offering a tailored approach to remedial efforts. In addition, NM-based remediation techniques are often associated with lower environmental risks than traditional methods, such as incineration, physical and chemical methods, bioremediation etc., by reduced emission of the toxic by products that are detrimental to the environment.

A promising approach to environmental remediation is photocatalytic activity of NMs that effectively degrade pollutants in water, air, and even soil, while also enabling energy production and storage. This discussion regarding NMs in environmental remediation has been mainly focused on their photocatalytic activities. In this mechanism, chemical reactions that break down pollutants are triggered by light energy. Utilizing sunlight as a renewable energy source to facilitate the remediation process, this pollutant metabolism method is sustainable, renewable, and cost-effective. Researchers utilize photocatalytic NMs to propose viable and scalable approaches to tackle pollution in various environmental matrices. Since our core focus of this chapter is to examine the fundamentals of photocatalytic activity and its potential application in real-time ecological remediation, we cover some fundamental principles and plausible designs to enhance charge carrier separation. This article delves into the mechanisms governing the photocatalytic process, emphasizing the role of NMs as facilitators of the degradation reactions of pollutants (Rajput et al. 2024; Singhwal and Rana 2024). Understanding the fundamental principles of this approach will pave the way for developing creative remediation strategies for promoting environmental protection and human health. At the center of NM-based ecological remediation is photocatalytic activity, which drives the degradation of

DOI: 10.1201/9781003484189-14

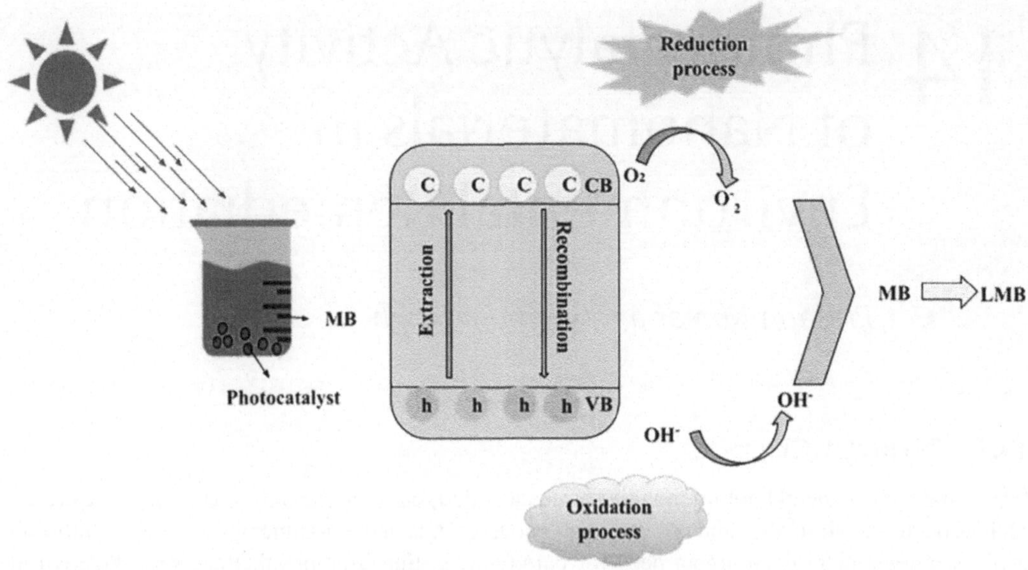

FIGURE 14.1 Photocatalytic activity of NMs.

(Source: created by authors)

pollutants through chemical reactions activated by light. So, the success of this process depends on the fundamental properties of some specific NMs known as photocatalysts, which degrade organic and inorganic pollutants with illumination.

The basic principle of photocatalysis is to generate electron-hole pairs upon the absorption of photons. The electron-hole pairs perform the redox reactions to convert the target pollutants into harmless products. Photocatalysis effectiveness is based on the NM's ability to absorb light over an extensive range of wavelengths, including ultraviolet and visible range. This allows for the effective use of energy sources for the degradation of pollutants. Furthermore, the surface properties and composition of NMs play a crucial role in determining their photocatalytic performance, owing to their profound effects on factors such as reaction kinetics, selectivity, and stability. With researchers continuing to explore and improve the photocatalytic properties of NMs, the opportunities for scalable and sustainable solutions to pollution challenges become more and more favourable, thus advancing towards a cleaner and healthier environment (Figure 14.1).

14.2 NMS IN PHOTOCATALYSIS

Due to their distinctive properties and varied conventional applications, NMs have become valuable components in the war against pollution-based issues. Unlike their larger counterparts, nanoscale materials have exhibited many novel physical, chemical, and biological properties. This opened up new avenues for environmental remediation through enhanced capture, breakdown, and detection of dangerous chemicals. Compared with classical remediation techniques, NMs provide more advantages (Krishna Veni et al., 2023; Belhameid et al., 2024; Sohn et al., 2024). NMs have a small volume unit with large surface areas and tenable surface chemistry and functionalities, which make them more effective against various pollutants in complex environmental conditions. Adsorption of contaminants is crucial for pollution remediation; thus, NMs are critical for reducing pollution. NMs with high specific surface areas (e.g., porous metal oxides, carbon nanotubes (CNTs), and graphene-based materials) can adsorb air, water, and soil pollutants.

These materials' small sizes and large surface area-to-volume ratios have many active sites available for pollutant molecules to lengthen, thus benefitting their removal from the environment. Moreover, the surface chemistry of NMs can be changed to increase their interaction with specific pollutants, making them ideal for targeted adsorption and removal of specific pollutants. Pollutant degradation and adsorption by NMs are being increasingly used in practice (Chen, Zhang, and Huang, 2023; Meng et al., 2024, Meng et al., 2024). The process relies on photocatalysis, using light energy to catalyze chemical reactions that degrade the contaminants into harmless byproducts. Titanium dioxide (TiO2), zinc oxide (ZnO) and cadmium sulphide (CdS) are highly overused as they can produce reactive oxygen species (ROS) under light exposure as light-responsive photocatalysts alternatives. ROS subsequently react with organic and inorganic pollutants to initiate degradation pathways, which are complex and harmful pollutants that biodegrade into more minor, less toxic compounds. NM has emerged as a green and sustainable approach to remediating pollutants because it uses readily available resources (sunlight) to accelerate degradation. Considering the above, NMs also fulfill an essential task by promptly providing effective sensing and monitoring for detection and quantification of environmental pollutants. The excellent characteristics of NMs, such as high sensitivity, selectivity and low response time, make them potential candidates for detecting air, water and soil pollutants (Vinoth and Pandikumar 2024). The sensing elements in NM-based sensors are often CNTs, quantum dots, or metal NPs. These sensors will generate valuable data for environmental monitoring, enabling the detection of pollution events, which can be followed by rapid remediation. NMs play an essential role in innovative pollutant capture, degradation, and detection strategies, thus relevant in addressing pollution-related challenges. Such tools' innovative characteristics and potential multifunctional applications are essential in generating sustainable and effective remediation solutions, contributing towards a greener and healthier ecosystem for current and future generations. As environmental remediation is the most critical aspect, sustainable and transformative photocatalytic activity must be emphasized (Kumari, 2023; Sharma et al., 2023; Fathima Khyrun et al., 2024; Tehseen et al., 2024). A novel methodology shows that the power of light can be harnessed to promote chemical reactions to decompose pollutants, which can provide an environmentally friendly, renewable, and efficient resolution to pollution issues.

Sustainable photocatalytic activity is how we can move away from traditional remediation techniques, which require highly energy-consuming processes for recovery or for producing toxic byproducts. This approach, however, uses renewable energy sources, such as solar energy, to induce the degradation reactions of pollutants, leading to less environmental impact and obtaining more sustainable characteristics. As the dominant and widespread fountain of energy globally, solar energy will provide sustainable photocatalytic activity. These photocatalytic processes also use sunlight rather than requiring external energy inputs and thus contribute to reducing reliance on fossil fuels and greenhouse gas emissions that can occur with traditional remediation methods. Promoting clean and renewable energy sources mitigates the environmental footprint of pollution removal activities and supports global climate change mitigation strategies (Aswini et al., 2023; Thind et al., 2023; Corsi et al., 2023; Aissa et al., 2023). In addition, sustainable photocatalytic activity is critical to realize a fundamental solution to pollution, as it is consistent with treating the root source of the environmental issues rather than their superficial manifestations. Conventional remedial approaches tend to contain or eliminate toxins, which can mitigate pollution for the moment but not resolve its systemic roots. This photocatalysis degrades pollutants into harmless byproducts using chemical reactions that neutralize them in the environment. This approach clears impurities from the atmosphere (like taking out the trash) while preventing their build-up by restoring ecosystem homeostasis and preserving environmental system performance. In addition, sustainable photocatalytic activity is excellent for efficiently eliminating complex and recalcitrant pollutants that are difficult to remediate by traditional approaches. Specific contaminants, such as organic compounds or emerging contaminants, are resistant to degradation and can remain in the environment for extended periods. This behavior is a serious risk to human and ecological health.

Photodegradation is a widely applicable practical strategy employed to decompose these pollutants using the reactivity of photocatalysts that can be activated by light to start degradation processes and metabolize complex structures into more straightforward and harmless species. Therefore, this skill is crucial for dealing with new environmental challenges and helping the ecosystems to remain viable in the long time scales (Chauhan et al. 2024; Ahmaruzzaman, Yadav, and Babita Devi 2024; Dutta et al. 2022). The critical part of sustainable photocatalytic activity is to provide sustainable, cost-effective solutions to remediate decentralized pollution. Photocatalytic reactors can be applied in different settings, in situ or ex-situ, such as in water treatment facilities, industrial flyovers, or contaminated sites. Additionally, coupling photocatalytic processes with current infrastructure or engineered systems reduces the need for costly adjustments or investments in novel technology. These features constitute the sustainability and feasibility of photocatalysis in solving pollution issues at local, regional, and global levels, as well as the scale of applications related to photocatalysis. Sustainable and transformational photocatalytic activity is required, since it offers the potential to tackle pollution in an eco-friendly, renewable, and effective manner. By using solar energy for the chemical transformation of by-products, photocatalysis provides a viable and renewable alternative for pollution remediation compared to traditional remediation methods. It is also a practice that carries a much lower environmental footprint, maintaining the health of the ecological systems. This is also a unique and revolutionary solution because it addresses the real reason for pollution, eliminating the pollutant from its home and stopping pollution from collecting. Due to its adaptability, efficiency, and cost-effectiveness, photocatalysis is anticipated to represent powerful, distinctive, sustainable, and environmentally beneficial solutions to current and future environmental challenges (Figure 14.2).

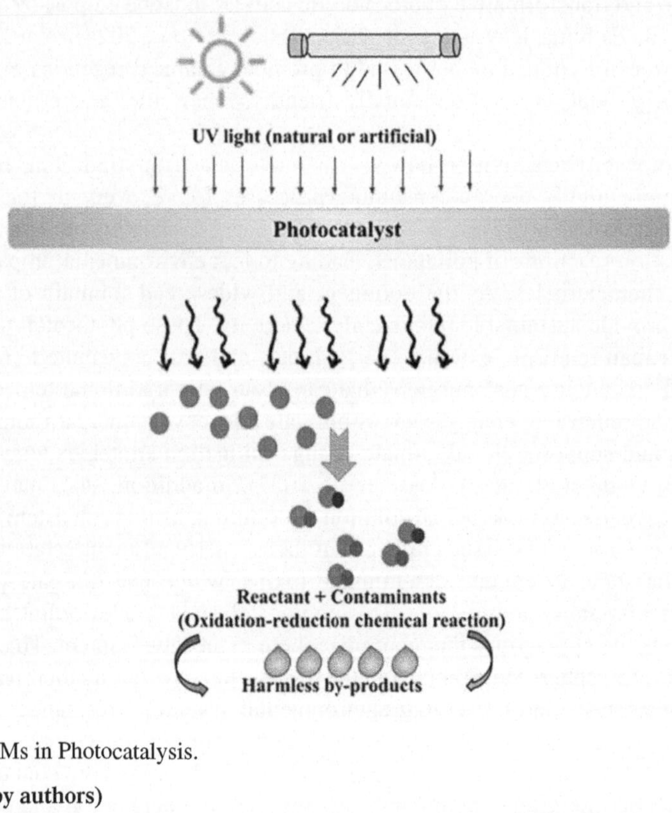

FIGURE 14.2 NMs in Photocatalysis.

(Source: created by authors)

14.3 MECHANISMS OF PHOTOCATALYSIS

The principles governing the interaction between light and NMs and between NMs and target pollutants are the basis of photocatalytic activity of NMs. The main principle behind photocatalysis is the ability of certain NMs, called photocatalysts, to absorb light energy at the appropriate time to activate chemical reactions, giving rise to by-products coming from pollutants' degradation into innocuous metabolites. It will be necessary for the effective design and optimization of photocatalytic systems for environmental remediation and other applications to deeply explore the principles underlying the photocatalytic process (Yu et al. 2022; Zelekew et al. 2023; Budimir and Prekodravac 2024). Only those photons with energy greater than or equal to the material's bandgap are absorbed. This phenomenon is the basic mechanism for the photocatalytic activity of NMs. A photon absorbed by a semiconductor NM creates an electron-hole pair (i.e., excited electrons transiting from the valence to the conduction band. Excitation generates free charge carriers (electrons and holes), which are then delivered and coupled to chemical reactions (e.g., contaminant breakdown). Excitation generates pairs of photogenerated electrons and holes, which are spatially separated through the internal electric field in the NM. They migrate to the surface, where they can participate in reduction reactions. Simultaneously, cavities form in the bulk material migrate to the surface, enabling their participation in oxidation reactions. These charge carriers are primarily responsible for the generation of ROS, including hydroxyl radical (•OH), superoxide radical ($•O^{2-}$), and photogenerated holes (h^+). These ROS play crucial roles in the degradation of the pollutant (Saleem et al. 2022; R. Devi et al. 2024; Arsenov et al. 2023; Rathinam Thiruppathi Venkadajapathy and Sivaperumal 2024). The redox reactions of light-generated electrons and holes with adsorbed pollutants or other surface species can fragment pollutants into less complex, non-toxic compounds.

Electrons can cleave O_2 molecules or other elements to form ROS, which operate and oxidize natural contaminants. Alternatively, holes can oxidize water or hydroxide ions to generate hydroxyl radicals, which oxidize organic pollutants. These redox reactions herein occur on the photocatalyst surface, which is stimulated and promoted by active sites and suitable reaction conditions. The photocatalytic activity of NMs and selectivity toward certain pollutants heavily rely on their surface chemistry. Surface reactions, the adsorption of contaminants and the migration of charge carriers are influenced by surface defects, functional groups and surface states (Djellabi et al. 2021; Rasool et al. 2024; Teixeira et al. 2024). Surface modification, doping, or formation of heterostructures can modulate the surface chemistry of the NMs, thus producing NMs with favorable photocatalytic efficiencies with customized reactivities toward specific pollutants. In photocatalytic configurations, maximizing performance critically depends on the efficient mass transfer of reactants and products. High specific surface areas and availabilities of surface sites of NMs can promote the adsorption of pollutants and diffusion of reactants into the activities of photocatalytic reactions. Furthermore, the configuration of photocatalytic reactors (e.g., reactor geometry and flow rate, illumination conditions) can play an essential role in influencing light distribution and reactant transfer, which could directly control the efficiency of the process. For such long-term photocatalytic applications, the photostability and durability of NMs need to be considered. However, the unstable light irradiation property of the NMs can also let the photocatalytic activity slow down with the time for photooxidation, photo corrosion, or surface reconstruction. Various strategies can be followed to improve the photostability and stability of photocatalytic systems. These are related to surface passivation, heterostructure formation, and encapsulation.

This work protects NMs from photo corrosion and extends their working lives. It is well known that a combination of a diverse set of fundamental factors, such as bandgap excitation, reactive species generation, redox reactions, surface chemistry, mass transfer, and reactor design, governs the photocatalytic activity of NMs. By understanding and optimizing these principles, researchers can design more efficient and selective photocatalytic systems for environmental remediation, water purification, air treatment, and beyond. Using such tools will help fight pollution challenges and continue to shift the needle for sustainability. Solar energy is significant in catalytic reactions or processes with clean and renewable resources, like photocatalysis. Solar energy is vast, inexhaustible, and the main driving force for many catalytic reactions, which can provide a green and clean alternative to traditional energy. Photocatalysis is a surface reaction driven by sunlight to achieve chemical transformation in the system; most photocatalysts are semiconductors. In illumination conditions using sunlight or artificial sources, these NMs are photoexcited when absorbing photons with energy equal to or higher than the bandgap. This process allows the transferring of energy from absorbed photons, leading to the liberation of electrons from the valence band into the conduction band and, hence, the formation of electron-hole pairs (Balasurya et al. 2023; Y.-Z. Wang et al. 2024; Mane et al. 2024; Güell et al. 2023). The absorbed solar energy generates reactive species that may initiate redox processes with oxidizing/reducing adsorbed reactants or pollutants. Electrons can lead to the reduction of oxygen molecules or other atoms, producing ROS including hydroxyl (•OH) and superoxide radicals ($•O^{2-}$). Moreover, holes may produce additional reactive ROS via the oxidation of water or hydroxide ions.

ROS are the key mediators of chemical transformations, from the degradation of organic pollutants and conversion of harmful emissions to benign products to the synthesis of high-value chemicals and fuels. With many benefits over classical energy sources, including sustainability and accessibility, the reaction process is not only a low-cost renewable star source; this strategy is suitable for a wide range of solar energy applications in catalytic reactions. Unlike finite fossil fuels, solar energy is abundant and dispersed, including in developing nations, and can free us from burning fossil fuels to meet the planet's energy needs without putting undue strain on natural resources or accelerating environmental degradation. In addition, solar energy is cleaned by nature because no greenhouse gas or any other hazardous emission is generated in its production and usage, indicating solar energy can be a sustainable use of energy to power the catalytic processes. Moreover, solar energy is inherently decentralized. It can be harvested at or near the point of consumption. This characteristic contributes to decreasing dependence on centralized energy systems and distribution grids. However, the decentralization process also paves the way for energy independence and resilience, especially in geographically isolated or off-grid regions, where energy sources may be traditional or even scarce and unreliable. Solar photovoltaic and thermal systems have greatly improved, causing the cost of energy from solar to be remarkably affordable, comparable to fossil fuels and other conventional energy sources (Nie et al. 2024; Suvathi et al. 2024; Arsenov et al. 2023; Rathinam; Djellabi et al. 2021).

Solar energy supports sustainable and energy-efficient processes for groups, including environmental remediation, chemical synthesis, and renewable energy production in catalytic reactions. To solve the above challenges of environmental pollution at a low treatment cost, photocatalysis has been proposed as an innovative tool to drive pollutant degradation reaction by a high-solar-energy input without the need for further energy input. In turn, solar-powered catalytic conversion of biomass to fuels and chemicals would assist in mitigating climate change and lower reliance on nonrenewable energy sources. Catalysts that convert sunlight as an energy source are leading the way for the development of greener and environmentally friendly processes toward addressing global challenges such as pollution, climate change, and energy security. By harnessing solar energy to facilitate catalytic transformations, researchers and engineers can devise novel strategies that promote sustainability, improve resource efficiency, and lay the groundwork for a cleaner and more resilient future for future generations. The generation of ROS plays a vital role in pollutant degradation, especially in photocatalytic processes containing NMs as catalysts.

ROS are oxygen-containing ionic or molecular species with free electrons that render them highly reactive. This gives them the property of a potent oxidizing agent, which can ignite and keep a process going. The progress of ROS is necessary to destroy organic and inorganic substances, oxidising the contaminants to more simplified and less poisonous compounds of the redox process in the pollutant degradation step. A comprehensive insight into the production and function of ROS is essential for the advancement of photocatalytic systems and for devising practical approaches for environmental remediation. In photocatalytic processes, ROS are primarily produced by photogenerated charge carriers, electrons, and holes on the surface of semiconductor NMs. As electrons in the valence band are excited by the conduction band, electron-hole pairs are generated. Photogenerated electrons and holes are active and undergo redox reactions with adsorbed pollutants or surface species. This process, therefore, generates the so-called ROS. The electrons at the surface of the photocatalyst can, in turn, reduce oxygen molecules or other species to produce ROS, e.g., superoxide radicals ($\bullet O^{2-}$) or hydroperoxyl radicals ($\bullet HO^2$). ROS are potent oxidants that can also connect with organic pollutants, destroying complicated compounds into more straightforward and less harmful substances, such as carbon dioxide (CO_2) and water (H_2O). In addition, the holes generated can also oxidize water molecules (H2O) or hydroxide ions (OH−), resulting in highly reactive hydroxyl radicals (•OH) that can oxidize a variety of organic pollutants (Sathish et al. 2024; Y.-Z. Wang et al. 2024; Djellabi et al. 2021).

ROS play an essential role in the decomposition of contaminants by generating and transferring the oxidative crack, leading to harmless by-products from contaminants. However, most conventional remediation methods drive pollutants by a succession of physical phases or transform them into less toxic but persistent species. Photocatalytic methods, with the help of ROS, are a very efficient approach to breaking down pollutants into complementary products. Microorganisms can integrate pollutants into their structure (i.e., a process known as mineralization), which helps log them out of the food chain, thus potentially reducing the impact of these pollutants on human beings and ecosystems. Furthermore, in the existence of various organic and inorganic pollutants (e.g., persistent organic pollutants (POPs), pharmaceuticals, pesticides, and industrial chemicals), the reactivity of ROS has a wide range, generating a large number of oxidated products. The ROS-based photocatalytic methodologies have high versatility and can be applied in different environmental matrices (e.g., air, water, and soil) to treat various environmental pollution types.

The ROS-based photocatalytic system displays various advantages like selection, efficiency, and environmental feasibility, all illustrating great potential for environmental remediation and water treatment. Nevertheless, generating ROS in photocatalytic processes can also generate secondary contaminants or byproducts based on reactant properties and the reaction conditions. Fine-tuning reaction pathways for minimizing the formation of undesired byproducts while guaranteeing the overall sustainability of the environmentally benign photocatalytic procedure still needs to be considered. Pollutant degradation efficiency may be improved, and secondary pollutant generation efficiency may be reduced by strategies such as catalyst design, reaction optimization and post-treatment methods. The generation of ROS is a particularly crucial process in photocatalytic reactions that remove contaminants (Rajput et al. 2024; Chen, Zhang, and Huang 2023; Devi, et al. 2024). Photocatalysts harness the oxidizing potential of ROS to transform organic and inorganic pollutants into less noxious radicals efficiently. It offers an environmentally friendly and solvent-free solution to pollution challenges. Insights into how ROS work to degrade pollutants are vital to optimizing photocatalytic systems and for a breakthrough in environmental remediation technology (Figure 14.3).

14.4 DIVERSE NMS IN PHOTOCATALYSIS

Semiconductor NPs represent a class of NMs with unique electronic, optical, and plasmonic properties with potential applications in photocatalysis, sensing, electronics, and optoelectronics. The tenable bandgap is the core feature of their operation, defining their ability to absorb and emit light and their conductivity. Semiconductor NPs comprise elements from one of the following groups: III-V, II-VI, or IV-VI of the periodic table. These can be TiO_2, ZnO, cadmium sulphide (CdS), lead sulphide (PbS), etc. Such materials exhibit unique electronic arrangements, which give rise to unique light-emitting and conduction properties, making them suitable building blocks for nanomachines and systems. TiO_2 NPs have a wide energy bandgap (3.2 eV for the anatase phase), allowing them to effectively capture ultraviolet (UV) radiation and generate electron-hole pairs. These pairs of electrons and holes then participate in redox, corresponding to adsorbed pollutants (R. Devi et al. 2024; Zelekew et al. 2023; Aissa et al. 2023; Gloria et al. 2023; Yu et al. 2022).

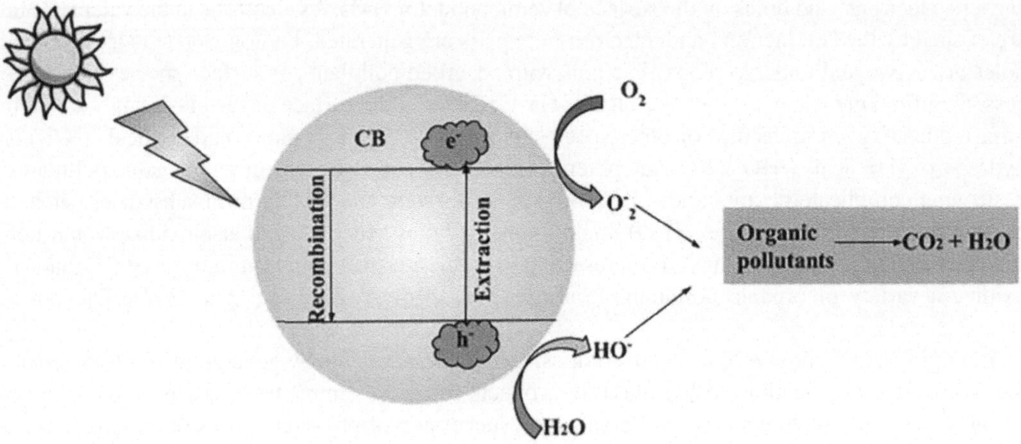

FIGURE 14.3 Mechanisms of Photocatalysis.

(Source: created by authors)

The broad applicability of this photocatalytic activity includes environmental remediation, water purification, self-cleaning surfaces, and solar energy conversion. As biocompatible NPs with minimal toxicity, TiO_2 NPs are also considered to have considerable biomedical potential, with applications such as drug delivery, imaging, and cancer therapy. An important example of a semiconductor NP is ZnO, a direct bandgap semiconductor ~3.4 eV with unique piezoelectric behavior. These and other applications, sensors, actuators, energy harvesting devices, and transparent conductive coatings led to renewed interest in ZnO NPs. The unique piezoelectricity and semiconducting properties make it a promising candidate for energy-efficient electronics, microelectromechanical systems (MEMS), and nanogenerators. Due to its natural antibacterial and antifungal activity, ZnO NPs are great for applications like healthcare, antimicrobial coatings, and water treatment. CdS, a semiconductor NP, has been actively studied because of its unique optoelectronic properties, including a direct bandgap of ~2.4 eV and a high photoluminescence quantum yield. CdS NPs show prominent absorbance in the solar spectral range of electromagnetic radiation and are used in multiple applications, including solar cells, light-emitting diodes (LEDs), photodetectors, and photocatalysis (Meng et al. 2024; Ahmad et al. 2023; Chen, Zhang, and Huang 2023; Sohn et al. 2024). The toxicity of cadmium has led to the investigation of other semiconductor materials with similar optical properties and lower environmental impact. PbS NPs are semiconductor materials with unique electronic and optical properties. They also possess a tenable bandgap that covers the visible to near-infrared region, from 0.4 to 1.6 eV. PbS NPs exhibit pronounced quantum confinement effects, leading to precise bandgap and optical property tunability by size (dimension) and composition. The NPs have several properties that make them extremely promising for applications, including infrared photodetectors, photovoltaics, thermal imaging, and sensing. The work that is being done on several semiconductor NPs like silicon quantum dots, copper indium gallium selenide (CIGS) NPs, and perovskite NPs does bode well for possible new ways in which to harness the beneficial properties of semiconductor materials across many fields (Jie et al. 2024; Mane et al. 2024; Y.-Z. Wang et al. 2024; Balasurya et al. 2023). With the potential to transform many of the world's sectors, such as renewable energy, environmental remediation, health care, and electronics, NMs offer hope for a promising future. However, addressing challenges relating to the toxicity, stability, scalability, and environmental impact of semiconductor NPs is crucial for effectively using their potential in practical applications. Owing to their unique physicochemical characteristics and broad applications, metal oxides are highly effective photocatalysts for environmental cleanup, water purification, and renewable energy generation.

These materials, comprising metal cation–oxygen anion bonds, offer a variety of electronic, optical, and surface properties that make them suitable for mediating photocatalytic reactions under light. Metal oxides are efficient photocatalysts: fundamental properties, mechanisms of photocatalytic activity, performance enhancement strategies, and applications. Many materials with a wide range of crystal structures, compositions, and properties belong to the family of metal oxides. Some examples of these are TiO_2, ZnO, Ferric Oxide (Fe_2O_3), Tungsten Oxide (WO_3), etc. The metal oxides we are dealing with show unique electronic band structures that can be accessed using valence and conduction bands separated by a bandgap. The band gap plays a vital role in how the material can absorb light and generate photoexcited electron-hole pairs, which is key to their photocatalytic activity. In addition, the metal oxides' surface chemical structure and morphology play a crucial role in determining their photocatalytic performance, influencing several thermodynamic/factorial aspects such as adsorption, charge transfer, and reaction kinetics. Metal oxides are photocatalytic, producing electron-hole pairs upon excitation with photons, whose energy is equivalent to or greater than the metal oxide bandgap. The electrons and holes generated by the light then migrate to the surface of the metal oxide. Here, they participate in redox reactions with the reactants or pollutants that rise to the surface. Electrons then reduce oxygen molecules or other molecules, generating ROS, such as hydroxyl radicals (•OH) or superoxide radicals (•O2-). Conversely, holes oxidize water or hydroxide ions, producing more ROS. ROS are essential for triggering and assisting chemical reactions, transforming pollutants into harmless derivatives. Metal oxide photocatalysis has been improved through many methods, including doping, heterostructure formation, surface modification, and cocatalyst deposition. Incorporating foreign elements allows the alteration of the lattice's electronic structure, bandgap, and surface properties, which is the definition of doping in the metal oxide system. These properties can enhance the light absorption, charge separation, and catalytic activity (Singhwal and Rana 2024; Fathima Khyrun et al. 2024; Aswini et al. 2023). Heterostructure formation is a necessary process involving the combination of metal oxides with different semiconductor materials to form interfaces with specific electronic properties and enhanced charge transfer kinetics. Surface modification techniques like coating with noble metals or organic molecules, can improve metal oxides' stability, adsorption properties, and photocatalytic efficiency. In addition, cocatalysts like metals or metal oxides can also be applied onto the surface of metal oxides, which form an interface with the metal and can facilitate charge transfer and promote selectivity of reactions. The metal oxides of steel have been widely used in environmental remediation to decompose organic contaminants, water and air purification, and remove toxic pollutants.

Because of its superior photocatalytic performance and stability, TiO_2 is extensively utilized for the photocatalytic degradation of organic pollutants such as dyes, pesticides, and pharmaceuticals. ZnO is a very attractive photocatalyst for environmental applications because of its high absorption of UV light and efficient ROS production. Metal oxides also decompose harmful gases; purify contaminated water sources, and their applications in the photocatalytic oxidation of Volatile Organic Compounds (VOCs). Metal oxides serve many vital functions as catalysts for renewable energy generation, including solar energy conversion, hydrogen production via photocatalytic water splitting, and environmental remediation. Metal oxides, such as Fe_2O_3 and WO_3, demonstrate great potential. They can trap solar energy and act as photocatalysts in water splitting to form H_2 gas, an environmentally friendly, green, sustainable energy source. Furthermore, metal oxide-based photocatalysts have been used in DSSC (dye-sensitized solar cells) and perovskite solar cells. These act as photoanodes or electron transport materials, allowing for efficient energy capture of energy from light and charge separation. Metal oxides possess outstanding photocatalytic qualities; however, they face many limitations that hinder their practical applications (Rasool et al. 2024; Rathinam Thiruppathi Venkadajapathy and Sivaperumal 2024). These problems include limited absorption of visible light, rapid recombination of charges, and susceptibility to photo corrosion under prolonged irradiation. These challenges call for wresting innovative approaches like bandgap engineering, cocatalyst design, and nanostructuring to improve the efficiency of light harvesting, charge separation, and the durability of the photocatalyst. Moreover, it is also necessary to limit the development of synthesis methods, optimize reactor design, and develop cost-effective photocatalytic systems to promote the transition of metal oxide-based photocatalysis from the laboratory to the real world.

Metal oxides are a versatile and efficient photocatalyst in various domains like environmental remediation, water purification, renewable energy generation, etc. Since they have unique characteristics and many researchers are working on new ways to increase efficiency and overcome current challenges, they are potential candidates for solving global pollution, energy and sustainability problems. Delving deeper into using metal oxides as photocatalysts can revolutionize various industries and foster a more sustainable and durable future. Understanding their behavior in photocatalytic reactions is crucial for the full utilization of the properties of carbon-based NMs for environmental remediation, energy conversion, and other applications. Carbon-based NMs, including graphene, CNTs, carbon dots, and carbon quantum dots (CQDs), have unique structural, electronic, and optical properties, making them candidates for photocatalysis (Kundu and Karak 2022; Mane et al. 2024; Y.-Z. Wang et al. 2024; Rasool et al. 2024). Carbon NMs have carbon atoms that are primers with sp2 hybridization and exist in various nanostructures like 0-dimensional CQDs, 2-dimensional graphene sheets, and 1-dimensional CNTs.

Notable features of NMs, including high surface areas, good electrical conductivity, and several specific electronic structures, make them ideal candidates for subsequent photocatalytic applications. Graphene, a single layer of carbon atoms arranged in a lattice, has high mechanical strength, good thermal conductivity, and high electron mobility, making it an excellent potential substrate for photocatalytic reactions. They are cylindrical nanostructures formed by rolling graphene sheets. This class of nanostructures features high aspect ratios, large surface areas, and remarkable electrical and mechanical properties. As a result, they are suitable for various photocatalytic applications. In contrast, carbon NPs with quasi-zero dimensions can be classified as carbon dots and CQDs. These NPs display tenable photoluminescence, possess good biocompatibility, and are readily synthesized. Hence, they are attractive to applications such as photocatalytic water splitting, pollutant degradation, sensing, etc. However, due to their capability to absorb light energy and convert it to drive chemical processes, carbon-based NMs are also photocatalysts (Boruah, Yadav, and Das 2020; Krishna Veni et al. 2023; Sharma et al. 2023).

Carbon-based NMs can absorb photons and generate electron-hole pairs upon illumination. The electron-hole pairs can then participate in the redox reactions of adsorbed reactants or pollutants. Photogenerated electrons can lead to the formation of ROS such as hydroxyl radicals (•OH) or superoxide radicals (•O^{2-}) upon reduction of oxygen molecules or other species. The photogenerated holes can oxidize water or hydroxide ions, producing additional reactive oxygen species (ROS). ROS play a crucial role in starting and propagating photocatalytic reactions, leading to the degradation of pollutants or the production of valuable chemicals and fuels. Such carbon-based NM synthesis methods vary according to structural, functional, and application necessities. Graphene can be synthesized by different methods, such as chemical vapour deposition CVD), liquid-phase exfoliation, or reduction of graphene oxide. Several techniques, like arc discharge, laser ablation, and chemical vapour deposition, can produce CNTs. Carbon dots and CQDs can be synthesized using simple methods such as hydrothermal or solvothermal reactions, advanced techniques like laser ablation, or chemical oxidation of carbon precursors. These synthesis methods enable the size, shape, surface composition, and electronic properties of the carbon-based NMs to be manipulated; properties can be tailored for specific photocatalytic applications. Photocatalysis is a broad field suitable for various carbon-based NM applications, such as environmental remediation, water splitting, carbon dioxide reduction, and organic synthesis. Carbon-based NMs are used for environmental remediation by degrading organic pollutants, removing heavy metals, and disinfecting water and air. Graphene-based photocatalysts have shown excellent photocatalytic efficiency in degrading organic dyes, pesticides, and pharmaceuticals. This is due to their large specific surface area, good electrical conductivity, and intense interaction with pollutants. Other ways CNTs were used include

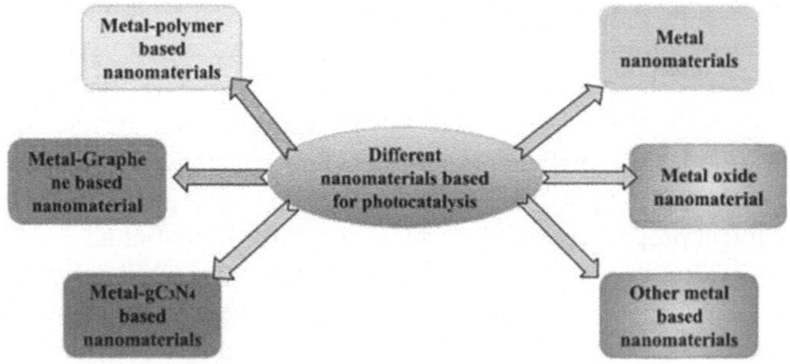

FIGURE 14.4 Diverse NMs in Photocatalysis.

(Source: created by authors)

the degradation of a range of contaminants, heavy metal removal and oil-water separation, as they have been reported to exhibit high adsorption capacities, mechanical robustness and ease of functionalization. Photocatalytic water splitting is possible with carbon-based NMs.

This method uses solar energy to convert water into gaseous compounds of hydrogen and oxygen and their usage in environmental purification. Their remarkable properties, including good electron conductivity, high surface area, and adequate charge carrier transportation kinetic have made graphene-based photocatalysts popular candidates for HER and OER in photocatalytic water-splitting systems (Rathinam Thiruppathi Venkadajapathy and Sivaperumal 2024). CNTs aroused significant interest in photocatalytic water splitting because of the high aspect ratios, good electrical conductivities, and promising developments in their integration into photoelectrochemical devices. Moreover, carbon-based NMs have been employed to generate value-added chemicals and fuels using photocatalytic carbon dioxide reduction reactions (CO_2RR). Light-irradiated, in common with graphene-based photocatalysts, demonstrates reducing carbon dioxide to methane, methanol, and other hydrocarbons. Domestication is deemed to be an attractive and sustainable way of capturing carbon. The adjustable photoluminescence, high specific surface area, and catalytic properties of carbon dots and CQDs make them promising for carbon dioxide reduction (CO_2RR). Based on their unique structural, electronic, and optical properties, carbon-based NMs play a key role in photocatalytic processes by enabling the efficient conversion of light and charge into initiators for chemical reactions. Multi-faceted synthesis pathways, tenable properties of carbonaceous NMs, and diverse applications in confronting the global challenge of environmental pollution, energy transformation, and sustainable chemistry have led to considerable potential for carbonaceous NMs. Continuous studies focusing on understanding the origin, improving synthesis methods, and exploring new potential applications will broaden the applicability of carbon-based NMs in photocatalysis, leading to progress toward a greener and more sustainable future (Figure 14.4).

14.5 PRACTICAL USES OF NM-BASED PHOTOCATALYSIS

Water purification is crucial in providing access to clean and potable drinking water, especially in areas suffering from water deficiency or contamination problems. Because of their unique physicochemical properties, large surface area-to-volume ratios, and tenable surface chemistry, NMs have the potential to revolutionize water purification technologies. These substances have significant applications, as both their degrading potency for pollutants and their capacity for microorganism mortality qualify them as common agents during water purification processes; therein lies the challenges and perspectives for the future that are being addressed in the text. Potable water is necessary for the well-being and health of humans. However, freshwater pollutants, including heavy metals, organic contaminants, pathogens, and microorganisms, are one of the significant risks to public health and the environment.

Traditional water treatment approaches, although efficient, can be limited by cost and environmental impact. Nanotechnology offers new approaches to overcome these hurdles through the unique characteristics of NMs in water treatment systems. Silver, TiO_2, and Fe_2O_3 NPs are among the best options for removing these pollutants and disinfection in water purification systems because of their excellent adsorption and catalytic properties. Composite materials combine the traits of NPs with the unique properties of the polymer or other matrix, resulting in combined mechanical stability, enhanced mechanical properties, stability, and selectivity (Rajput et al. 2024; Belhameid et al. 2024; Ahmad et al. 2023; Dutta et al. 2022). Such examples are CNT-polymer composites, graphene-based materials, and nanocomposite membranes. Nanomembranes (thin-filed deposited from NMs) such as graphene oxide (GO), CNT, and nanoporous ceramics have unique characteristics of high permeability, selectivity, and fouling-resistance. These characteristics make them ideal for water filtration and separation processes. Because NMs have high surface area-to-volume ratios and are tunable for surface chemistry, they are suitable for adsorbing water contaminants (heavy metals, organic pollutants, and dyes) in water by physical or chemical interactions. For example, noble metals can have catalytic properties that allow the degradation of organic pollutants or further conversion of harmful species into less toxic byproducts by light irradiation or another stimulus. For example, silver and copper metal NPs are widely known for their great anti-microbiological phenomena of attacking bacteria, viruses, and germs. These properties make them suitable as disinfectants and microbial during the water for potable application use NMs in point-of-use point-of-entry water treatment systems.

NM based technologies are broadly employed in several industrial processes, including wastewater treatment, desalination, and resource recovery. They were created based on pollutant minimization, the use of water, and resource recovery. The use of NMs in environmental remediation assists in removing contaminants from polluted water bodies, restoring ecological balance, and minimising the impact of human activity on the environment. However, mass applications of NM-based water treatment technologies face several challenges, including issues of scalability, cost-effectiveness, and environmental impact, as well as regulatory barriers (Jie et al. 2024; Rathinam Thiruppathi Venkadajapathy and Sivaperumal 2024). To overcome these challenges, a better understanding of synthesis techniques that can easily be scaled, properties of NMs and their resulting performance, potential environmental risks from treatment methods, and process limitations are necessary so that the source of freshwater, as well as cost-effective and practical techniques can be utilized. NMs show excellent potential in water treatment technologies and solving problems related to water quality and availability worldwide. When scholars and practitioners use our unique characteristics, we can create incredible and sustainable solutions that offer the world's societal community clean and safe drinking water. Further research and joint efforts to harness the power of NMs for purifying water for all water users worldwide need to ramp up. Nanocatalysts are a promising tool for air detoxification that provides new merits for tackling air pollution issues and effective green methods for eliminating malicious atmospheric pollutants. Nanocatalysts also share similar very desired properties; the high ratio of surface area to volume and favorable catalytic activity enables their catalytic transformation of toxic gases (like NOx and CO) and VOCs into chemically non-toxic or less harmful products using catalytic oxidation, reduction or other chemical conversions. This section discusses the application of nanocatalysts in air detoxification. Nanocatalysts are essential for detoxifying air since they facilitate catalytic reactions to convert hazardous pollutants into less toxic or inert compounds. Mechanisms of action are conditioned by the particular catalytic characteristics of the NMs and the nature of the targeted pollutants. Nanocatalysts speed up oxidation, widely defined as the process of gaseous pollutants (carbon monoxide (CO), nitrogen oxides (NOx), sulphur dioxide (SO_2) and VOCs) being oxidized into CO_2, N_2, H_2SO_4 and H_2O in the presence of O_2 and other oxidants [8]. Nanocatalysts enter the picture, as they play a vital role in the reduction of nitrogen oxides (NOx) and SO_2 into N_2 and H_2SO_4, respectively. This is achieved via reduction reactions using reducing agents such as CO, H_2 or hydrocarbons. The surfaces of the nanocatalysts adsorb gaseous pollutants onto themselves and then catalytically transform the adsorbed species into less toxic compounds (Rajput et al. 2024; Belhameid et al. 2024; Ahmad et al. 2023; Dutta et al. 2022). For eliminating persistent airborne contaminants like CO, NOx, SO_2, O_3, VOCs, hydrogen sulphide (H_2S), and ammonia (NH_3) in the atmosphere, nanocatalysts have taken a dominant position over conventional catalysts. Nanocatalysts can accelerate the oxidation or reduction of particulate matter (PM) pollutants such as soot, diesel particles, and airborne NPs into relatively harmless substances. Bottom-up and top-down techniques for synthesizing nanocatalysts have been developed to allow the composition, morphology, and catalytic properties to be tuned.

Nanocatalysts are usually formed by dissolving precursor compounds and stimulating chemical reactions that lead to nanoscale precipitation. Nanocatalysts can be prepared in a sol by hydrolyzing metal alkoxides or other precursors. Drying of this leads to gelation in the presence of a tellurite, forming nanoscale powders/films. This is how nanocatalysts grow on such substrates: the gaseous precursor compounds are decomposed in the high-temperature reactor; and catalytic layers of coatings in the nanometre range are deposited on the substrates. They are usually synthesized using templated approaches, which help control nanocatalysts' sizes, structures, and orientations. Nanoporous membranes, CNTs and polymer matrices are some nanocatalysts. Methods around air purification utilize different micro catalysts, with the investigation travelling through active adoption and removing rubbish from indoor and outside air pools. Nanocatalysts are placed inside automotive catalytic converters that prevent CO, NOx and VOCs from being released into outside air via vehicle exhausts. Nanocatalysts could also be incorporated into air filtration systems to facilitate oxidation and/or reduction of specific gaseous pollutants and enhance PM and gas removal efficiency. Portable air cleaners for homes and commercial buildings are being upgraded with nanocatalysts that can kill indoor air pollutants, including VOCs, formaldehyde and odors. Nanocatalysts are integrated into industrial processes to manage the emission of SO_2, NOx, and VOCs, among other pollutants from factories, power plants, and chemical plants. The nanocatalysts can undergo deactivation or degradation with time owing to fouling, poisoning or sintering and a resultant loss of catalytic activity and catalytic efficiency (J. Wang et al. 2020; Kumar Sharma, Sharma, and Shibani 2023). Most nanocatalysts have environmental footprints from resource depletion, energy use, and waste profiles that may complicate their manufacture, use, and disposal. The balance between the cost and integration into air purification systems may hinder the widespread usage of nanocatalysts, particularly in developing countries or low-resource environments. In air purification, such nanocatalysts need extensive testing, standardization, and certification for their efficacy, following regulatory norms directed towards health and the environment. Research efforts in the future need to be concentrated on overcoming these challenges in the form of stable, high-performance, cost-effective nanocatalysts and the establishment of sustainable, scalable manufacturing processes.

Practically, continuous advancements in NMs' synthesis, characterization, and integration technologies will help develop and integrate state-of-the-art air purification systems. These systems will use the unique properties of nanocatalysts to improve human health and air quality. This kind of application is found in the real world. A case study can serve as solid proof of the ability of nanocatalysts to address an urgent environmental and industrial challenge. They showcase actual implementations of nanocatalysts in various situations, reinforcing their practical implications and benefits. This review combines these case studies to showcase nanocatalysts' efficacy across various application fields. Example: Nanocatalysts in catalytic converters of automobiles paved a long way through the new era; we know how much emissions have reduced these days compared to the earlier eras. The reaction, catalyzed by platinum group metal supported on high surface area substrates, in which CO, NOx and VOCs transform to gaseous products that are less noxious (e.g., CO_2, N_2, and H_2O vapour). High efficiency three-way catalytic converters have resulted in a significant decline in tailpipe emissions of CO), NOx, and VOCs from the automobile exhaust system significantly improving air quality and public health. The nanocatalyst-based catalytic converter is one of the latest technologies in the automotive industry, and it is usually used to minimize automobile pollution. In the catalytic oxidation of VOCs, VOCs are emitted into the air from industrial processes such as painting, printing, and chemical manufacture. This off-gassing can be highly toxic to both the environment and general health. Reduce emissions and comply with legislation by catalytically combusting VOCs into a less poisonous constituent. The system reached a high VOC removal efficiency of over 95%; as such, it effectively reduced energy consumption and operating costs and demonstrated the feasibility and efficacy of nanocatalyst-based emission control technologies. NPs with photocatalytic properties, such as TiO_2 NPs, are commonly used in water purification via nanocatalysis. These nanocatalysts are used to degrade organic pollutants, disinfection of water, and abatement of contaminants via photocatalytic oxidation reactions when irradiated with light. Information about organic pollutants in effluent streams treated by TiO_2 NPs in photocatalytic oxidation at a municipal wastewater treatment plant saw the development of a nanocatalyst-based system to break down organic pollutants, pharmaceutical residues, and industrial chemicals, including decreasing effluent contaminants and improving water quality to meet regulatory requirements. Besides, nanocatalysts play a crucial role in photocatalytic water-splitting reactions that produce hydrogen. This includes light-triggered water splitting and the consequent production of molecular hydrogen and oxygen species [43]. Semiconductor NPs, CdS, and WO_3 were used to facilitate the photoelectrochemical reactions associated with the water-splitting process. Because of a lack of infrastructure and know-how, H_2 use for fuel applications remains limited; therefore, a photocatalytic water-splitting system has been devised in a nanocatalyst-based research lab to produce H_2 using sunlight on-site. The hydrogen production of the system based on CdS NPs supported on a conductive substrate was efficient. This understanding illustrates the capabilities of this system for deliverable renewable, sustainable energy nanocatalysts coated filters in indoor air purification. Nanocatalysts applied to filters for air purification were also used for outdoor air purification. They could remove particulate matter (PM), VOCs, biological contaminants, and nanocatalyst reactions to deactivate or transform hazardous pollutants into harmless species. In a commercial office block, an upgrade was made in the HVAC systems with air filters coated with a nanocatalyst to provide a healthy interior environment to their occupants. With these filters, PM, VOCs, and microbial contaminants were captured, which lowered the risks of diseases, respiratory illnesses, and even discomfort and unproductivity from varied air pollutants in the workplace. These empirical examples and case studies also validate the success of nanocatalysts for environmental, health, and sustainability challenges across industries and applications. Using the distinct properties and catalytic power of the NMs, new ideas are being generated to reduce pollution and save resources, leading to a cleaner and healthier environment for the coming generation (Figure 14.5).

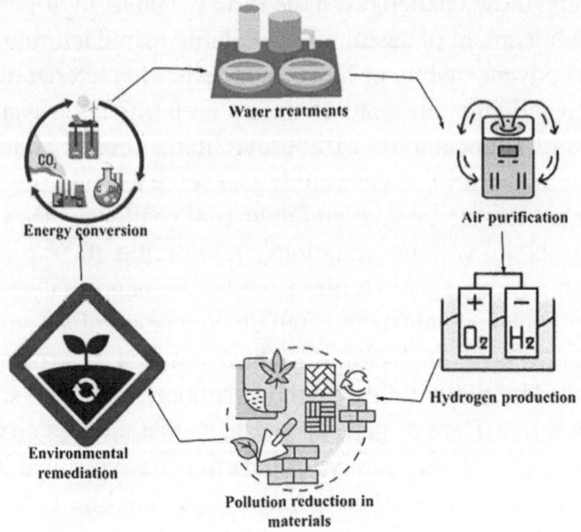

FIGURE 14.5 Practical Uses of NM-Based Photocatalysis.

(Source: created by authors)

14.6 ENVIRONMENTAL IMPACTS AND LONG-TERM EFFECTS

For data science to foster the widespread usage of NM-based photocatalysis, it becomes essential to understand and convey the most significant advantages of the technology in facilitating its environmental impact. Even more generally, it is necessary to evaluate the ecological footprint of NM-based photocatalysis over its entire life cycle; this encompasses stages ranging from synthesis to disposal (including its use in environmental remediation, water purification, and renewable energy generation applications). Such assessment considers the toxicity of NMs, ecological effects, energy consumption, pollution caused, and behavior of the NM in the environment. Round-ups will facilitate evidence-based decision-making by influencing policymakers, politicians, and stakeholders, ensuring NM-based photocatalysis's protected and sustainable incorporation. NMs deemed harmful to humans or environmental health are referred to as "toxic." The physicochemical nature of NMs (e.g., dimensions, morphology, surface characteristics or chemical composition) and their interaction with biological systems in the context of photocatalysis must be evaluated to obtain current knowledge regarding the toxicity behavior of the NMs. Toxicological studies of NMs focus on their cytotoxicity, genotoxicity, and immunotoxicity, and oxidative stress in organisms. Drawing upon this, these studies seek to assess the NMs' potential to bioaccumulate and biomagnify through the various levels of ecological food webs. Such tests can create thorough toxicity assessments that can help developers design risk management strategies to minimize exposure and manage potential hazards of the apparent use of NMs in catalysis applications. This necessitates a holistic understanding of the environmental factors of NMs to predict their behavior and mobility across the different environmental compartments (air, water, soil, and sediments). The fate of nano aluminium studies (e.g., how NPs aggregate, settle, dissolve, and transform) influences their dispersion and life span in the subsurface and the biosphere. Future monitoring and modeling efforts should: 1) track NMs through their lifecycle from emission sources to sites of final deposition, 2) quantify their capacity for long-range transport and 3) bioaccumulation and biomagnification and 4) identify hotspots for environmentally relevant exposure. Population with a moderate interest in NMs might need more quantitative information on their fate and significant factors affecting them to tailor emission control measures and avoid environmental pollution risks. Energy consumption related to photocatalysis using NMs involves quantifying energy inputs associated with producing NMs used for photocatalysis, the intensity of photocatalytic reactions, and the system's operation. Life Cycle Assessment (LCA) methodologies are employed to assess the energy intensity of photocatalytic systems throughout their production, deployment, and disposal processes. It consists of several stages, including extraction of raw materials, manufacturing or production processes, transportation, and product treatment at the end of its life cycle.

Assessment of energy efficiency metrics (e.g., energy return on investment (EROI) and net energy balance) provides insight into NM-mediated photocatalysis's broader energy performance and environmental sustainability. One way to mitigate the environmental impact of large-scale photocatalytic technologies is to implement miniaturized and modular photocatalytic processes, which promote a more efficient use of energy (labor) throughout the catalysis and offer the possibility of integration with renewable energy systems. The implications of resource-to-waste processes are investigated in NM-based photocatalysis through the analysis of waste stream composition, management, and generation over the entire operation of photocatalytic systems. Waste characterization studies establish the composition and volume of the waste generated throughout NM production life cycles, including production, manufacturing, usage, and end-of-life (decommissioning). Green chemistry concepts, recycling systems, and material reclamation programs are examples of approaches to waste minimization (Rathinam Thiruppathi Venkadajapathy and Sivaperumal 2024). Based on these properties, proper waste management protocols, such as waste segregation, treatment, and disposal, should be developed to ensure the safe handling and removal of waste streams containing NMs. One example of this in practice is a corrosion-resistant material placed as an anode to prevent the leaching of copper. Evaluation of possible ecological consequences when NM was first discovered, typical toxicity assays for using NM in a wide range of ecotoxicological research, and photocatalysis-related ecological interactions in NM are essential for evaluating ecotoxicological impacts. Environmental toxicity studies assess the effects of NMs on the health, behavior, reproduction or survival of various organisms spanning the food web, including bacteria, algae or aquatic invertebrates, fish or terrestrial plants. Ecological risk assessments evaluate the likelihood and magnitude of harmful effects on the structure and function of an ecosystem resulting from NM exposure. This assessment incorporates aspects such as contact route(s), NM exposure levels, and the susceptibilities of the organisms in question. Such NM-based photocatalysis requires mitigation measures that deter potential environmental harm through environmental monitoring, risk management strategies, and regulatory frameworks.

Assessment of the environmental impact of photocatalysis based on NMs requires applying multiple tools to assess the toxicity of NMs, their environmental impact, energy consumption, waste production, and ecosystem interactions. Scientific knowledge, risk assessment methods, and stakeholder engagement precede comprehensive assessments. Quantitative health risk assessment models play a key role in informing decisions and guiding future development of safe and responsible uses of nanotechnology. Fostering ongoing research, surveillance, and regulatory oversight to guarantee derisked and sustainable deployment of NM-based photocatalysis for global environmental remediation and preventing relevant intolerable risks to human health and the environment should be a top priority. Understanding the persistent effects of NM-based photocatalysis on environmental complexes is crucial for evaluating the long-term sustainability and ecological impact of such technologies in the long run. The photocatalysis approach in the NM is a novel technique for the cleaning precession that demands investigation for its ability to sanitize pollutants. Nevertheless, it is essential to carefully assess potential risks to ecological systems to reduce adverse effects on biodiversity, ecological services, and ecological functions. This paves the way for long-term studies and monitoring to understand the cumulative impact, ecological responses, and feedback mechanisms associated with the exposures. Integrating ecological principles, studies in the field, and modeling approaches can give researchers a broad understanding of the complex interactions between NMs and ecosystems. This information can drive decisions based on empirical environmental health and biodiversity protection data. Ecological risk assessment (ERA) provides an integrated framework for analyzing possible hazards, exposure pathways, and ecological impacts from NMs to ecological systems. ERA integrates information regarding NM characteristics, ecological interactions, exposure scenarios, and species sensitivity to estimate the likelihood and magnitude of adverse impacts on ecological receptors, including individual organisms, populations, communities, and ecosystems. Tiered approaches, such as probabilistic risk assessment (PRA) and species sensitivity distribution (SSD) analysis, objectively assess the risks and uncertainties posed by NM exposure.

These can help guide risk management strategies and mitigate environmental harm. Diversity, abundance, distribution, and structures of species can be directly and indirectly affected by exposure to NMs, which collectively influence the environmental ecological system. So far, the legacy of NM on biodiversity metrics (including species richness, evenness, and composition) of terrestrial, aquatic, and marine systems has been studied through long-term monitoring studies that record changes in exposure gradients. Community-level studies focus on the effects of NMs on demographic variables, reproductive success, and mortality rates of key indicator species, providing long-term demographic information about population dynamics and community resilience. To better understand the long-term effects of NM-based photocatalysis on the operation of ecosystem functions, it is essential to evaluate changes in ecosystem processes, biogeochemical cycles, and the deliverance of ecosystem services (Sharma et al. 2023; Aswini et al. 2023; Hassan et al. 2023; Dutta et al. 2022). Functional trait-based approaches assess ecosystems' resistance and recovery capacity against NM-induced perturbations. This identifies key functional groups and ecological thresholds that regulate ecosystem functioning and resilience. NM Bioaccumulation and Biomagnification within food webs affect trophic interactions and energy flow between ecosystems discussion of ecological research on the trophic transfer of NM. Primary producers are the initial consumers of

NMs in the food chain, consuming NMs through both consumption and trophic transfer processes, following which NMs are transferred to higher trophic levels in the ecosystem, including herbivores, carnivores, and secondary and tertiary consumers. The implications of indirect impacts of NM exposure on ecological stability and resilience are assessed as trophic cascade effects and incorporate feedbacks, including changes to food webs, trophic interactions, and predator-prey dynamics. NMs in photocatalysis may change habitats' structural, compositional, and qualitative composition, affecting the habitat suitability, distribution, and resilience of the species' habitats.

Field surveys and remote sensing techniques assess changes involving key habitat characteristics such as vegetation cover, soil properties, water quality, and geomorphological features following exposure. Habitat modeling approaches predict habitats' potential distribution and suitability in different NM exposure scenarios. This information informs conservation, restoration, and efforts to reduce habitat loss and fragmentation. Recognizing feedback and ecological responses following NM exposure is pivotal for understanding the enduring interactions between NMs and the environment. Integrated modeling methods (including dynamic systems modeling, agent-based modeling, and ecosystem service modeling) are available to simulate the effects of exposure to NPs, environmental stressors, feedback loops, nonlinear interactions, and emergent properties of ecosystems (Rajput et al. 2024; Rathinam Thiruppathi Venkadajapathy and Sivaperumal 2024). The ability of ecosystems to adapt and respond to disturbances induced by NMs is being evaluated before organisms, which not only consists of feedback mechanisms like nutrient cycling or species interactions but also serves as indicators of ecosystem resilience by quantifying feedback control mechanisms such as nutrient cycling, species interactions, and ecosystem resilience. Adaptive management strategies are iteratively used to ensure appropriate monitoring, assessment, and management cycles that address uncertainties and dynamic changes during NM-based photocatalysis. Intervention, adaptive management, and assessment are part of an integrated approach that includes stakeholder participation, decision analysis, and ecosystem-based management to identify management goals, priority actions, and adaptive action and build a resilient, sustainable ecology. Collaborated governance structures, interdisciplinary research partnerships, and knowledge co-production processes enable the co-design and co-management of NM-based photocatalysis to tackle emerging environmental issues and promote ecosystem health. A multifaceted approach deploying ecological principles, field studies, modeling methods and adaptive management strategies is crucial for fully understanding the long-term impact of NM-based photocatalysis on ecological systems. By assessing ecological risk, biodiversity loss, ecological function, trophic dynamics, habitat modification, feedback loops and adaptive management strategies, researchers can establish evidence-based frameworks for protecting ecosystems in the context of new emerging nanotechnology applications. Ongoing investigation, surveillance, and collaboration are crucial to filling knowledge gaps, reducing uncertainties, promoting sustainable improvements in NM-facilitated photocatalysis and protecting biodiversity and ecosystem services for future generations. The sustainability assessment of environmental remediation should provide a comprehensive evaluation of the potential benefits of remediation technologies and their environmental, social, and economic risks. Sidewalks in front of buildings become heavily polluted and must be remediated to reduce long-term pollution, restore ecological systems, and protect human health. However, they can also come with unintended consequences, trade-offs, and uncertainties that must be addressed to ensure sustainability. Such an extensive assessment requires integrating scientific knowledge, stakeholder perspectives, risk management strategies, and adaptive governance approaches to enhance benefits and reduce risks. A preventive approach can become the basis for sustainable environmental remediation practices that better address contamination challenges and reduce adverse impacts within ecosystems and communities (Aissa et al. 2023; Dutta et al. 2022). Remediation technologies destroy, decompose, or confine harmful pollutants in the soil, water, air, and sediments to reduce pollution (in all its forms) in the environment and prevent direct exposure. Remediation can help restore degraded habitats, increase species diversity, and enhance ecological resilience by loosening ecological imbalances and supporting native species. Remediation treatments improve air and water quality, reduce exposure to toxic pollutants, and reduce the human health impact of disease and illness from pollution. Remedial actions aim to fulfil the nation's adherence to environmental and legal requirements, which help minimize the legal and reputational risk to those responsible for the action. Cleanup projects can create job opportunities, drive economic growth, and increase property values by restoring contaminated land and promoting sustainable development. Remediation activities, such as habitat destruction, soil erosion and chemical disturbances, can cause ecological disruptions, alterations in habitat conditions and effects on unintended species. Workers, communities, and bystanders at remediation sites can be at risk from hazardous substances, airborne pollutants, and contaminated water. Remediation technologies may also be limited in effectiveness, efficiency, and scalability, leading to incomplete contaminant removal, treatment failures, or residual contamination. This may lead to remediation projects burdened by regulatory requirements, permitting processes, and stakeholder engagement, which may increase implementation time and cost. Remediation activities may lead to displacement, social conflicts, and inequitable distribution of benefits and burdens, disrupting local economies, social structures, and cultural practices (Rathinam Thiruppathi Venkadajapathy and Sivaperumal 2024; Khan et al. 2023). For example, to reduce the introduction of contaminants into the environment and control the spread of pollution, source reduction, prevention, and containment should be given priority. Risk assessment methodologies, exposure modeling, and toxicity assessments should thus be used to prioritize remedial actions and target human health and ecological risk.

It is essential to ensure transparency, inclusion, and accountability and to bring in the different stakeholders in the decision-making process. These stakeholders include local communities, indigenous groups, regulatory agencies, and industry partners. Adaptive management strategies featuring monitoring, feedback systems, and iterative improvement processes are suggested to respond to uncertainties and change conditions and maximize remediation benefits. Conduct LCAs to evaluate remediation technologies' environmental, economic, and social impact over their entire life cycle, including raw-material extraction to disposal. Equity, social justice, and community empowerment must be embraced in the remediation planning and implementation process to ensure disproportionate environmental burdens and the pursuit of ecological equity. Engineered barriers, liners, and caps are essential to contain contaminated areas and mitigate the migration of pollutants into surrounding areas. Here, you can deploy practical remediation technologies, such as bioremediation, phytoremediation, chemical oxidation, and thermal treatment, to render contaminated media acceptable (e.g., reduce pollutant levels to acceptable limits). Relatedly, comprehensive environmental monitoring, sampling, and surveillance initiatives should be implemented to track the level of contaminants, actively assess the success of remediation efforts, and identify any emerging risks or unintended consequences.

Deliver risks, uncertainties, and updates on each issue to stakeholders in clear, timely, and readily available formats to encourage trust, collaboration, and well-informed decision-making. It is necessary to establish long-term stewardship, institutional controls, and monitoring programs to maintain the effectiveness and integrity of remediation and avoid later re-contamination. The case studies and the collaborative project highlight the importance of multi-stakeholder partnerships, cooperative networks, and knowledge-sharing platforms to foster joint problem-solving, innovation, and capacity-building in environmental management. Engaging local communities, indigenous knowledge holders, and marginalized groups also lies at the center of participatory planning processes that integrate diverse perspectives to empower stakeholders and address social inequality. Train co-management frameworks, co-design processes, and shared decision-making structures to enable collaborative governance, equitable material ownership, and flexible management of environmental risks. Reforms in the adaptations of regulatory frameworks, permitting processes, and enforcement procedures are needed to enable adaptive management alternatives, stimulate experimentation, and allow adaptive learning in environmental governance (Boruah, Yadav, and Das 2020; Krishna Veni et al. 2023). However, a sustainable environmental remediation balance cannot be achieved without a holistic and flexible approach that embraces environmental, social, economic, and governance dimensions. Implementing sustainable remediation principles, risk management strategies, and adaptive governance approaches enables stakeholders to address ecological contamination challenges effectively

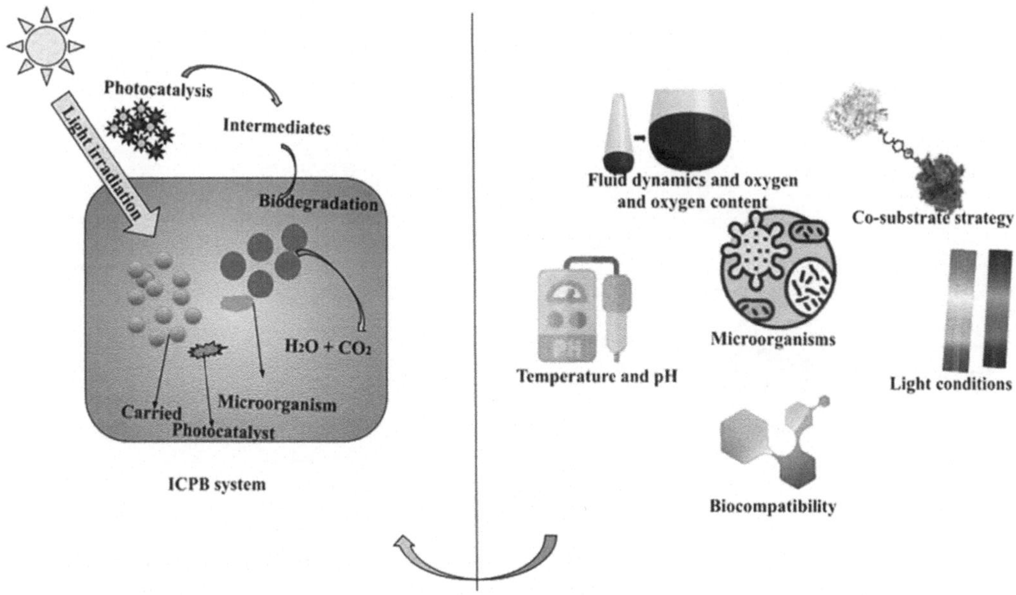

FIGURE 14.6 Environmental Impacts and Long-Term Effects.

(Source: created by authors)

while safeguarding the environment's health, human well-being, and social equity. Progressing and expanding sustainable remediation practices and supporting resilient, inclusive, and eco-efficient models for environmental management will require continued research, monitoring, and interactions (Figure 14.6).

14.7 CASE STUDIES AND EMPIRICAL EVIDENCE

Much empirical evidence supports the efficacy of NMs in pollutant degradation. Such studies prove that nanotechnology solutions are effective in remediation, pollution control, and environmental protection. The empirical evidence in the current study represents strong support for the ability of NMs to sufficiently degrade, remove, or sequester a large variety of pollutants and types of pollution. These include organic pollutants, heavy metals, pathogens, emerging pollutants, etc., in the soil, water, air, and sediment matrices. By analyzing empirical data from laboratory studies, pilot-scale experiments, and real-world applications, researchers can verify the NMs' effectiveness, mechanisms of action, and performance in breaking down pollutants and provide evidence-based guidance for environmental management and remediation decision-making. Empirical studies have shown NMs to be effective in degrading organic contaminants, such as petroleum hydrocarbons, polycyclic aromatic hydrocarbons (PAHs), pesticides, pharmaceuticals, and personal care products. Figure 14.7 illustrates the photocatalytic properties of semiconductor NPs, including TiO_2 and ZnO; upon UV or visible light irradiation, these NPs activate oxidative reactions, which decompose organic pollutants. Reductive dichlorination and hydrogenation reactions can reduce chlorinated solvents, nitroaromatic compounds, or other persistent pollutants, using metal NPs, including palladium (Pd), platinum (Pt), etc. Carbonaceous NMs such as GO and CNTs enhance the microbial degradation process of organic pollutants by providing sorption sites, electron transfer, and redox mediation capabilities. Through the processes of sequestration, immobilization, or transformation, NMs represent promising strategies to remediate contaminated environments from toxic metals such as lead (Pb), mercury (Hg), cadmium (Cd), arsenic (As), and chromium (Cr). The adsorption of

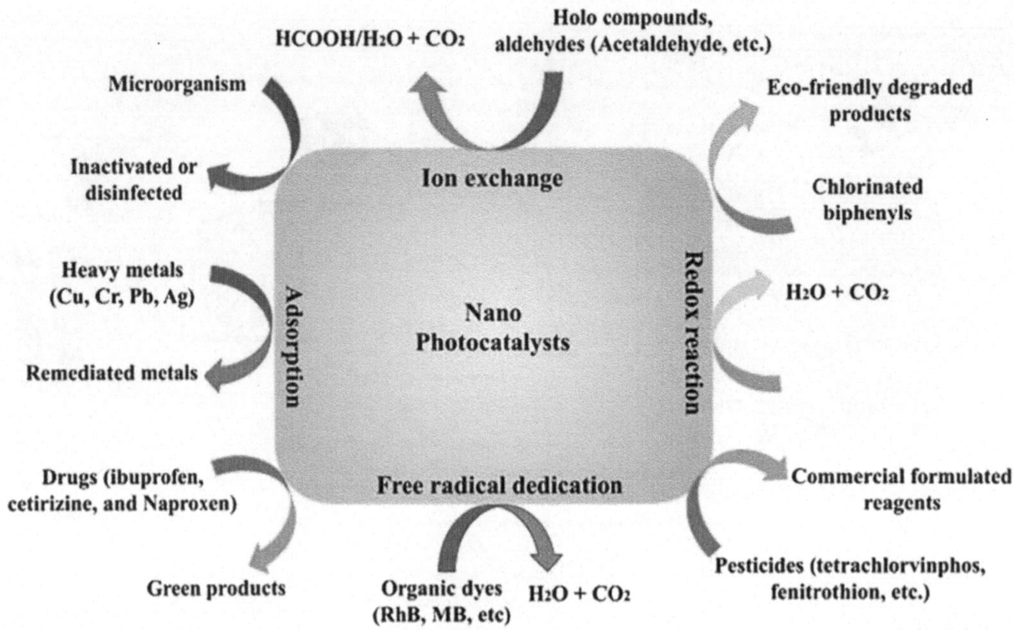

FIGURE 14.7 Photocatalytic degradation of pollutant.

(Source: created by authors)

heavy metals on to the surfaces of metal oxides such as Fe_3O_4 and MnO_2 results from electrostatic interactions, ion exchange, and surface complexation. NPs such as zero-valent iron (ZVI) and calcium carbonate ($CaCO_3$) can induce precipitation and co-precipitation reactions with insoluble metal hydroxides, carbonates, and sulphides. Reducing NMs like nano zero-valent iron (nZVI) enhance the conversion of toxic metal ions into less mobile and less hazardous forms, e.g., metals or metal sulphides, through reduction and immobilization processes (Kundu and Karak 2022; Rathinam Thiruppathi Venkadajapathy and Sivaperumal 2024.

Because of these properties, NMs can deactivate, disinfect, and eliminate pathogenic microorganisms in water, air, and surfaces, such as bacteria, viruses, fungi, and protozoa. The antimicrobial activity of silver NPs (AgNPs) exerts several mechanisms, such as cell membrane disruption, oxidative stress, and enzyme inhibition, culminating in the death of microbial cells. Metal oxide NPs, ZnO, and CuO, have different antimicrobial activity. They generate ROS and degrade the microbial cell membranes. Nanocomposite materials integrating NMs with antimicrobial agents or functional coatings induce enhanced pathogen inactivation and reduced microbial biofilm formation on surfaces and filtration membranes. If you want to get into detail, NMs efficiently eliminate emerging contaminants from water, wastewater, or aquatic habitats, such as pharmaceuticals, personal care items, endocrine-disrupting compounds, and microplastics. Among Advanced oxidation processes (AOPs) based on NMs, namely photocatalysis, electrocatalysis, and sonocatalysis, NMs have shown important efficiency in the degradation and mineralization of emerging pollutants. This is accomplished by producing very reactive radicals and oxidants in these processes. An illustrative example is NM-modified membranes, including GO membranes and CNT membranes that enhance the removal efficiency of emerging pollutants via a combination of adsorption, sieving, and electrostatic repulsion mechanisms. Functionalized NPs dispersed in sorbent materials or immobilized on surfaces can improve emerging pollutants' adsorption capacity and selectivity in water treatment and remediation processes. Evidence supporting the efficacy of filed NMs in environmental remediation comes from real-world case studies and field-scale applications (Singhwal and Rana 2024; Belhameid et al. 2024; Meng et al. 2024). Pilot-scale experiments and field trials have shown that

NMs (especially nZVI and nano-scale iron sulphide (FeS)) effectively intercept contaminants and remedy the groundwater at sites contaminated with chlorinated solvents, heavy metals, and petroleum hydrocarbons. NMs such as nanostructured sorbents, nano-enabled amendments and nano-emulsions formulations demonstrate the potential for improving soil remediation efficiency while limiting contamination risks due to leaching, volatilization, and secondary pollution in contaminated soil environments. Large-scale treatment facilities, distributed systems, and individual devices are equipped with water treatment NM technologies (e.g., membrane filtration, adsorption, photocatalysis, etc.) for improving water quality, removing harmful contaminants, and ensuring the safety of drinking water sources. Their studies are significant as they will help further understand the processes in the interactions between NMs and pollutants and the degradation pathways.

These findings are essential, as they provide new insights into the physicochemical, biochemical, and environmental aspects that underlie the efficiency of NMs. In addition, reaction kinetics, degradation rates, and reaction mechanisms are investigated to quantify degradation efficiency, reaction kinetics, and reaction pathways involved in pollutant degradation processes mediated by NMs. Advanced spectroscopic, microscopic, and surface analytic techniques could enhance our understanding of NM surface characteristics, oxidative/reactive sites, and interfacial interaction processes in pollutant adsorption, catalysis, and transformation. The role of NM's fate, transport, and transformation in various environmental matrices, such as soil, water, air and sediment, has been studied to assess their behavior and long-term presence in complex ecological systems. The performance of NMs in pollutant degradation and environmental remediation in different contamination conditions and environmental matrices is substantiated by experimental data obtained from laboratory studies, pilot-scale tests, and field applications. Researchers, through the synthesizing and analyzing empirical data, can validate whether or not NMs are effective, help elucidate the mechanisms by which they decompose, and refine remediation strategies that could be employed to challenge ecological issues. However, such approaches based on NM require extensive complementary research, multidisciplinary collaboration, and new technology developments. These parameters ensure that approaches like these will be safe, sustainable, and practical for solving global pollution problems. Actual world implementations of environmental remediation projects provide insight into the challenges, successes, and best practices of deploying remediation technologies in real-world settings. By reviewing case studies, pilot projects, and field trials, stakeholders learn about project outcomes, evaluate the effectiveness of mitigation measures, and make sound decisions on future projects. Each lesson addresses different dimensions of remediation: technical, regulatory, financial, social, environmental, etc. They highlight the importance of adaptive management, stakeholder involvement, and interdisciplinary teamwork to achieve sustainable remediation goals. The importance of site characterization and assessment cannot be undermined, considering the contaminants, hydrogeological conditions, and challenges at the site. A remediation project at a former industrial facility revealed unexpected subsurface variability, prompting additional study and modification of the remediation approach to locate high concentrations in the subsurface matrix and address migration pathways. Applying these technologies and devising unique solutions that consider the site can sensitively align with the site's specific conditions and are crucial to achieving timely and cost-effective remediation objectives. A groundwater remediation project assessed various technologies (e.g., in situ chemical oxidation (ISCO), enhanced bioremediation, monitored natural attenuation) to target complex contaminant petro penetrating plumes and improve treatment benefits. Proper implementation, operation, and maintenance of remediation systems are essential to ensure treatment effectiveness, system reliability, and regulatory compliance over time.

The operational issues faced by a soil vapour extraction (SVE) system deployed at a contaminated site, were related to equipment malfunctioning and the lack of maintenance, which highlighted the importance of monitoring and resolving problems promptly. Continuous long-term monitoring, performance evaluation, and adaptive management are essential for adequately evaluating remediation progress, assessing treatment effectiveness, and identifying emerging challenges (Khan et al. 2023; Rathinam Thiruppathi Venkadajapathy and Sivaperumal 2024). Rebounding contaminant concentrations in treated areas were detected by a long-term groundwater monitoring program, which led to a reassessment of remediation objectives, system design, and monitoring protocols. Compliance with regulatory requirements, permitting processes, and stakeholder engagement obligations is crucial for project authorizations, financial closure, and successful project outcomes. Permitted activities related to the remediation project met many delays and regulatory challenges. Including permission violations, resistance from the local stakeholders and failures of proper documentation. These challenges demonstrate the impact of proactive support of regulatory obligations and transparent and honest communication. By employing sound risk communication, stakeholder engagement, and community involvement processes, you promote transparency, trust and collaboration around the projects with its stakeholders and affected communities. An outreach campaign that offered communication channels answered questions, and responded to community concerns, seeking feedback from stakeholders and increasing acceptance of remediation at a contaminated site. Now-launched remediation projects may lead to unforeseen environmental and social impacts, requiring further effort to identify and mitigate negative consequences for natural systems, ecosystems and regional populations. The unintentional effect of wetland restoration projects on local wildlife habitats and ecological processes highlights the importance of environmental monitoring, adaptive management, and habitat restoration measures. Encouraging technology transfer, knowledge sharing, and capacity-building initiatives can help facilitate the adoption of innovative remediation technologies, foster collaboration, and accelerate the advancement of the industry. A knowledge-sharing platform was created for professionals in the remediation industry to share best practices, lessons learned and technical know-how (Khan et al. 2023; Rathinam Thiruppathi Venkadajapathy and Sivaperumal 2024). This platform was designed to facilitate innovation and foster continuous improvement of remediation practices. Long-term monitoring and management programs can help ensure remediation activities' effectiveness, integrity, and sustainability.

These programs help mitigate the risks of contaminant rebound, recontamination, and system failure. A post-remediation monitoring program detected residual contamination, delays in ecological recovery, and emergence of pollutants. Adaptive management actions were implemented in response to these results, including habitat enhancement and source control. Insights from implementing environmental remediation projects worldwide underscore the importance of site characterization, technology selection, stakeholder engagement, regulatory compliance, risk communication, cost management, and long-term monitoring in achieving sustainable remediation outcomes. By integrating empirical evidence, sharing best practices, and facilitating collaboration, stakeholders can have more tailored remediation strategies that are better at minimizing environmental impacts and ensuring the longevity and resilience of remediation work. Dealing with a changing set of environmental problems and achieving sustainable remediation technology globally will require continuing education, flexible organization, and knowledge sharing (Nie et al. 2024; Mane et al. 2024; Kumar Sharma).

14.8 SIGNIFICANCE AND FUTURE PROSPECTS

NM-based photocatalysis is promising for addressing complex pollution problems, improving remediation efficiency, and promoting environmental sustainability. NMs are unique in their physicochemical properties and tenable catalytic capabilities and are applicable for the decontamination of organic pollutants, removal of toxic metals, water disinfection, and air pollution. With the catalytic implementation of photocatalysis, it uses light as its energy source to activate redox reactions, enhancing the degradation of pollutants by producing high-energy radicals and oxidants on the surface of NMs (Rajput et al. 2024; Belhameid et al. 2024; Tehseen et al. 2024). This approach has a promising remediation applicability in multiple environmental compartments, such as soil, water, air, and sediments. Compared to conventional remediation methods, it offers many advantages, such as higher selectivity, rapid reaction kinetics, and lower energy consumption. Applications of photocatalysis with NM can be utilized by researchers, engineers, policymakers, etc., to improve environmental remediation, facilitate sustainable development, conserve the environment's health, and preserve humankind's health. Applying NMs in photocatalysis provides a practical and environmentally friendly approach for degrading diverse organic contaminants, including petroleum hydrocarbons, polycyclic aromatic hydrocarbons (PAHs), pesticides, pharmaceuticals, and industrial chemicals.

Photocatalytic activity of semiconductor NPs such as TiO_2, ZnO, and Fe_2O_3 is induced by ultraviolet (UV) or visible light. Such characteristics enable selective degradation of organic compounds through oxidation, mineralization and decomposition mechanisms. Photogenerated ROS, for instance, hydroxyl radicals (•OH) and superoxide radicals (•O2), generate cleavage of chemical bonds, converting organic pollutants into innocuous final products like CO_2, H_2O and inorganic ions (Sathish et al. 2024; Balasurya et al. 2023; Teixeira et al. 2024). The oxidative degradation process is selective and effective at translating food waste into non-threatening end products. This technique reduces environmental contamination and limits the danger of biological exposure. Photocatalysis involving NMs holds promise as an efficient technology for removing toxic metals and metalloids from contaminated environments, including groundwater, soil, sediment, and industrial waste. Metal oxide NPs, such as TiO_2, ZnO, and cerium oxide (CeO_2), have high rates of retention and redox activity, allowing them to bind, isolate, and reduce metal ions to immobile, less toxic forms. Light-induced photocatalytic reactions on the NM surface generate hot carriers, which are pairs of electrons and holes. This process allows metal ions to be reduced into elemental metals or metal oxides and thus to precipitate or co-precipitate as insoluble phases. In addition, using NMs in photocatalysis increases the decomposition rate of metal-containing chelators and high-energy organic complexing agents, thereby facilitating the release and removal of bound metal species from contaminated matrices. Additionally, the combination of adsorption and photocatalytic reduction process enhances the efficiency and selectivity of metal ion recovery, consequently mitigating the risk of groundwater contamination, soil instability, and deleterious ecological impact. This work outlines the potential for implementing NM-based photocatalysis as an affordable and effective solution for water and wastewater treatment, reducing the load of microbial pathogens and safeguarding public health. Metal and metal oxide NPs, such as silver Ag, ZnO, and TiO_2, can also be phototoxic. NPs successfully disrupt the growth, replication, and survival of bacteria, viruses, fungi, and protozoa. Photogenerated ROS, such as singlet oxygen and hydroxyl radicals (•OH), can damage membranes of microbial cells, generating oxidative damage to cellular components and deactivating pathogenic microorganisms via photooxidation and photodissociation. NMs based photokilling is an enjoyable alternative to conventional disinfection methods, which include chlorination and ultraviolet (UV) irradiation (Singhwal and Rana 2024; Belhameid et al. 2024; Tehseen et al. 2024). These benefits include enhanced microbial inactivation rates, broader antimicrobial activity, and reduced disinfectant byproduct (DBP) formation. This cross-cutting methodology is relevant to various water treatment processes, including disinfection of drinking water, wastewater, and recreational water.

Its adoption results in better water quality, reduced cases of waterborne diseases, and more positive public health outcomes. Applying photocatalysis with NMs can contribute to air pollution control strategies by mitigating the flow of VOCs, NOx, SO_2 and other air pollutants from industrial, transportation, and indoor sources. When subjected to UV or visible light, semiconductor NPs (TiO_2, ZnO, WO_3) exhibit photocatalytic activity by oxidizing and decomposing gaseous pollutants into harmless gaseous residues and mineral salts. Photocatalytic reactions that generate hydroxyl radicals (•OH) and oxygen radicals (•O2) occur on the surfaces of the NM. These ROS then react with organic compounds, nitrogen oxides, and sulphur dioxide, resulting in the rupture of chemical bonds and the transformation of contaminants into less toxic or inert species. Combining NM-mediated photocatalysis with various air purification technologies (e.g., photocatalytic coatings, air filters, and photocatalytic reactors) can improve pollutant removal efficiencies, reduce energy demand and improve indoor and outdoor air's thermal and chemical conditions. This strategy complements traditional air pollution control measures such as emission controls, fuel switching and source reduction approaches. It encourages sustainable urban development, preserves the environment, and improves public health (M. S. Devi, Srinivasan, and Muthuvel 2023; Corsi et al. 2023;

Thind et al. 2023). Utilizing NMs in photocatalysis will follow the principles of green chemistry and environmental sustainability as it promotes resource efficiency, pollution prevention, and renewable energy use in a remediation process. Examples of semiconducting NPs include TiO_2 or ZnO; these are relatively inexpensive, non-toxic, and environmentally friendly materials that can be produced by cheap, high-scale, and eco-friendly method. As solar energy is an eco-friendly and renewable resource for photocatalytic reactions, less reliance on fossil fuels will be provided, and the carbon emission associated with remediation treatment can be effectively reduced. NM-mediated photocatalysis operates under mild reaction conditions, such as room temperature and atmospheric pressure, in contrast to thermal or chemical oxidation processes. Such a feature reduces the energy needed, reducing operational costs and environmental impact.

Additionally, using NMs in photocatalysis results in minimal waste and byproducts, thus promoting the formation of harmless end products and reducing the potential environmental risks associated with secondary pollution. Using a sustainable and eco-friendly method can be a rewarding step for enhancing the technical growth of ecological remediation, stimulating environmental innovation, and realizing the developmental goals of sustainability. Therefore, efforts to advance the technology and commercialization of the photocatalytic remediation of the environment using new materials represent a step toward sustainable photocatalytic approaches to environmental remediation. This, in turn, fosters research and development activities, creates investment opportunities, and helps the nanotechnology sector grow. NM synthesis, surface engineering, and optimization of photocatalytic performance also aid in generating next-generation photocatalysts with enhanced activity, stability, and selectivity for pollutant degradation. NMs have also been applied to several remediation technologies, such as water treatment controls, air purification, and self-cleaning surfaces. This has resulted in growth in market opportunities and commercial adoption for those applications. Collaborative research partnerships, public-private collaborations, and technology transfer initiatives further facilitate knowledge exchange and enhance technology transfer and market diffusion of NM-based photocatalysis.

Such initiatives accelerate the transformation of research discoveries into practical solutions and applications in the real world. This path of innovation-driven growth provides for economic development, job creation, and sustainable growth that addresses these pressing environmental challenges and promotes environmental stewardship (Y.-Z. Wang et al. 2024; Khan et al. 2023; Kumar Sharma, Sharma, and Shibani 2023). Using their unique properties, NMs can make new advances in different areas, such as the degradation of organic pollutants, elimination of toxic metals, water disinfection, and air pollution abatement by focusing on solar energy-driven photocatalytic reactions that have high capacities the researchers, engineers, and policymakers are under the great potential to find new solutions. This interdisciplinary approach opens possibilities to influence technological innovation for progress toward achieving sustainable development goals and safeguarding ecological health and human well-being by integrating principles of green chemistry and environmental sustainability in both protection measures and the development of new environmental technologies. Continued research, collaboration, and investment are essential as we are required to realize the full potential of NM-based photocatalysis and to address global environmental challenges in the 21st century. The efforts to contribute to sustainable and practical environmental remediation methods involve developing, implementing, and improving innovative technologies, policies, and strategies for solving pollution problems while ensuring environmental sustainability, social justice, and economic efficiency. By incorporating these elements, contributions to sustainable remediation enable communities to protect vital ecosystems and enhance their resilience to environmental threats. These efforts include approaches ranging from green chemistry to nature-based solutions to community-based and adaptive management approaches to achieve lasting environmental gains and improve the quality of life for present and future generations. Sustainable remediation programs focus on utilizing green chemistry and clean technologies to reduce environmental impact, conserve resources, and promote the use of safer alternatives to conventional remediation strategies (Singhwal and Rana 2024; Kumari 2023; M. S. Devi, Srinivasan, and Muthuvel 2023). It highlights the design of chemical products and processes used in a preventive way that reduces the generation of hazardous substances and waste and their impact on the environment. Advancements in green remediation include the development of biodegradable surfactants, natural sorbents, and bio-based catalysts to mitigate contaminants efficiently.

Along with this has been the attentiveness toward making more efficient low-energy and low-waste treatment technologies such as phytoremediation, bioremediation, and electrokinetic remediation. Sustainable remediation emphasises the use of nature-based solutions and ecosystem-based approaches to address pollution, as well as the rehabilitation of degraded habitats and the enhancement of the environment's resilience. Nature-based solutions leverage natural processes, ecological function, and green infrastructure to mitigate environmental impacts, improve water quality, and enhance biodiversity. Examples include wetland restoration, riparian buffers and green stormwater management (Saleem et al. 2022; Budimir and Prekodravac 2024; Tyagi et al. 2024). These strategies enhance the ability of soils to absorb water, reduce runoff and allow pollutants to be filtered out and naturally degraded. Ecosystem restoration is chiefly concerned with the rehabilitation of degraded ecological systems, the restoration of environmental services and the prediction of the environmental connectivity required to enhance the survival of indigenous species, restore habitats, and increase environmental productivity. By investing in nature-based solutions and ecological recovery initiatives, stakeholders can achieve multiple environmental and socio-economic benefits, including improved water quality, flood abatement, carbon sequestration, recreation, and community involvement and stewardship opportunities.

Effective remediation programs are not only designed for environmental protection but also consider equity, social inclusion, and participatory planning approach, meaning that the people living in the affected areas would have to be involved in the decision-making process, and their concerns are considered in the plan design. For greater transparency, inclusivity and accountability, community-led approaches involve the participation of diverse stakeholders, including residents, indigenous peoples, and local organizations, in the planning, implementation, and monitoring. Community-centered remediation includes a range of efforts related to knowledge transfer, support for marginalized communities, and integrating local knowledge and values into remediation plans. Such initiatives include community advisory groups, citizen science programs and participatory mapping exercises. Through innovative engagement and collaboration, sustainable remediation initiatives can also help to build trust, strengthen social cohesion, and successfully address environmental justice.

This approach will help equalize the remediating actions through all community members, thus increasing social cohesion. This approach will employ adaptive management principles within a risk-based decision-making framework to deal with uncertainties and realize optimal remediation and resilient systems performance. Adaptive management means learning, monitoring, and adjusting corrective measures based on feedback from tracking data, empirical studies, and stakeholder input. Much progress has been made in adaptive remediation, including developing decision support tools, modeling approaches, and risk assessment methodologies. These tools are used to evaluate various remedies, establish the sequencing of interventions, and optimize resource allocation. Stakeholders use risk-based strategies to identify, assess, and manage environmental threats related to contaminated sites (Boruah, Yadav, and Das 2020; Fathima Khyrun et al. 2024). It allows them to describe interventions to human health and environmental metrics and reconcile investments in remediation strategies to leverage maximum benefits, given resource constraints. Shifting to adaptive management and risk-based decision-making allows sustainable remediation initiatives to effectively chart a course through such complexity, uncertainty, and change. So, this is a step toward strengthening resilience, creating more flexibility, and encouraging innovation in dealing with these environmental challenges. Undoubtedly, sustainable remediation activities will depend on policies, regulations, and governance frameworks that should support and help maintain ecological integrity, ensure meeting ecological standards, and provide incentives to adopt best practices in the planning and executing remediation activities. Contributions to policy support, including establishing regulatory guidance, technical standards, and performance metrics to evaluate the effectiveness of remediation, protect public health, and minimize environmental risks. Institutional capacity-building programs aim to strengthen regulatory agencies, train ecological specialists, and contribute to cross-sectoral synergies. These efforts improve governance, enforcement, and coordination of remediation activities. Policy support and institutional capacity can also be enhanced to promote sustainable remediation practices, stimulate innovation, and overcome the barriers to implementation. This also will encourage transparency, accountability, and public involvement in environmental decision-making. Sustainable remediation approaches highlight knowledge transfer, education, and outreach activities to promote awareness, capacity building, and support optimal environmental stewardship and remediation strategies (Sathish et al. 2024; Kumar Sharma, Arsenov et al. 2023).

In education and learning, well-known contributions include developing training courses, workshops, and teaching materials. These programs are intended to ensure the proper dissemination of technical knowledge, promote skills development, and empower affected parties to participate in remediation. Making connections among practitioners, researchers, and policymakers, you can create opportunities for knowledge-sharing through conferences, seminars, and online forums to share experiences, lessons learned, and innovative solutions. This encourages collaboration and the sharing of thoughts. Investments in educational and knowledge-sharing programs that can help stakeholders create a skilled workforce, promote continued learning and drive innovation in environmental remediation (Krishna Veni et al. 2023; Corsi et al. 2023; Sohn et al. 2024). It will also help us with our sustainable development and responsibilities to the environment for generations to come. The evolution of sustainable and practical interventions in environmental clean-up includes initiating, deploying, and improving innovative technologies, policies, and processes designed to address pollution problems while sustaining ecological sustainability, social equity, and economic viability. By implementing green chemistry, nature-based solutions, and approaches that enhance community engagement, adaptive management, policy support, and education, stakeholders can achieve long-term environmental improvements, build resilience, and promote well-being for current and future generations. Through collaborative efforts and integrated action, sustainable remediation initiatives can contribute to a healthier and more resilient environment and a legacy of environmental stewardship and responsible resource use for future generations.

Future directions and environmental remediation advancements are expected to meet the dynamic challenges of new pollution, improve remediation efficiency, and advance sustainable development goals. The future integration of environmental remediation relies on diverse emerging technologies and interdisciplinary collaboration. As researchers, engineers, policymakers, and stakeholders work together in this exciting space, several key areas emerge where opportunity and progress are key to shaping the future of environmental remediation, with the potential to drive technological innovation, policy reform, and transformative change. Potential approaches include meaningful and impactful interdisciplinary research, emerging technologies, policy interventions, and societal trends to support resilience, sustainability, and equity in preventing environmental contamination and promoting ecosystem health and human well-being. Emerging trends in other areas include advanced materials and nanotechnology, which can allow for new remediation technologies with improved performance, selectivity, and efficiency (Sathish et al. 2024; Kumar Sharma, Arsenov et al. 2023). NMs, including graphene, CNTs, and metal oxides, have unique physicochemical properties and electrocatalytic behaviors that facilitate selective removal, immobilization, and transformation of pollutants in contaminated environments. The advances in synthesis, functionalization, and characterization techniques for NMs provide the tools to design tailored nanocomposites, hybrid materials, and engineered NPs with the required stability, reactivity, and environmental compatibility. Moreover, the combination of nanotechnology with other creative disciplines, including biotechnology, materials engineering, and computer science, offers the potential for designing next-generation removal strategies, such as bio-inspired NMs, self-assembling nanocomposites, and intelligent sensors for real-time monitoring and control of cleanup strategies. Prospects for promoting green remediation and sustainable practice emphasize implementing cost-effective, socially acceptable, and environmentally friendly practices in stewardship, planning, implementation, and management of remediation. The green remediation principles, wherein pollution prevention, resource conservation, and ecosystem restoration are preferred for remediation, minimize environmental footprint and energy consumption and enhance ecosystem services. Green technologies (e.g., phytoremediation, biochar amendment, and natural attenuation) provide low-impact, nature-based solutions for removing, sequestration, and degrading contaminants in soil, water, and air matrices. Moreover, embedding circular economy principles, LCA, and environmental footprint analysis into remediation decision-making processes promotes holistic, cradle-to-grave approaches that optimize resource (re) use, reduce waste generation, and maximize environmental and economic benefits.

Chances for advancing new remediation strategies and techniques include creating integrated remediation systems and integrating physical, biological, and chemical remediation processes to maximise removal efficiency. These methods include electrochemical remediation techniques (for example, electrocoagulation, electrokinetic, and electrochemical oxidation capable of degrading, extracting, or immobilizing contaminants in soil and groundwater in situ (Krishna Veni et al. 2023; Corsi et al. 2023; Sohn et al. 2024). This process allows microorganisms to couple their metabolic activity with the extracellular electron transfer process to generate electricity or provide energy for contaminants bioremediation (degradation of organic pollutants, metal ions recovery, nutrient removal, etc.) in an anaerobic field. Innovations such as CRISPR-Cas9 enable targeted modification of the environment. They can enhance the biodegradation capabilities of microbial communities, mitigate contaminant impacts, or restore ecosystem functions for pollutants that are difficult to degrade or emerging contaminants of concern. This includes the integration of sensor networks and remote sensing technologies with big data analytics platforms to leverage digitalization and data analytics to improve the efficiency of remediation performance, ecological monitoring, and real-time prediction of contaminant fate and transport in space and time. Digital twin technologies, which generate virtual representations of physical remediation systems and environmental processes, allow for simulation, visualization, and optimization of the remediation operation, enhancing decision-making, risk assessment, and resource allocation. Moreover, ML algorithms, AI, and data-driven modeling approaches provide predictive insights, anomalies and insight detection, and optimization recommendations for remediation strategies, providing adaptive management, early warning systems and automated control of remediation processes. Leveraging digitalization and data analytics as remediation tools can increase efficiency, lower costs, and drive better environmental solutions while enacting transparency, accountability, and stakeholder involvement in remediation projects. These may include adopting adaptive, science-based regulations, incentives, and standards fostering innovation, technology transfer, and sustainable practices as prospects for policy innovation and regulatory reform in environmental remediation. Risk-based decision-making, performance-based standards, and valuation of ecosystem services can be implemented through programmatic regulatory frameworks that facilitate flexible approaches to remediation planning and implementation, providing an opportunity for innovation and cost-effectiveness. Moreover, financial intercessions, including tax credits, grants, and subsidies, promote private sector involvement in the research and advancement of remediation technologies, catalyzing innovation and encouraging market competitiveness (Nordin et al. 2022; Rathinam Thiruppathi Venkadajapathy and Sivaperumal 2024; R. Devi et al. 2024).

Furthermore, international cooperation, standard harmonization, and knowledge-sharing initiatives facilitate transboundary collaboration, capacity-building, and technology transfer to address pollution challenges and achieve environmental sustainability targets globally. At the same time, governments, regulatory agencies, and international organizations can facilitate an environment that enables sustainable remediation practices, enabling innovation ecosystems, and developing public-private partnerships for practical solutions to environmental contamination through policy innovation and regulatory reform. Prospects involve public education, outreach, and communication efforts that educate and empower communities and stakeholders about remediation processes and environmental literacy and promote participation in citizen science and environmental remediation projects. Public engagement initiatives provide opportunities for dialogue, build trust, and encourage collaboration among different stakeholders, such as residents, indigenous communities, and civil society organizations (e.g., community forums, public consultations, and environmental education programs). Inclusive and participatory approaches to mediation decision-making and project planning ensure that local knowledge, cultural values, and community preferences are integrated into remediation strategies. This leads to increased project buy-in, social equity, and long-term sustainability. In addition, citizen science programs, crowdsourcing platforms, and mobile applications allow the public to help collect, monitor, and report environmental quality indicators, allowing communities to support and even lead ecological monitoring, enforcement, and advocacy initiatives. Through raising public awareness and engagement, stakeholders can leverage social capital, mobilise resources, and promote collective action to remediate environmental contamination and pursue environmental justice. The future and opportunities for environmental remediation provide a glimpse into the potential for tackling new and emerging pollution concerns, increasing the efficiency and effectiveness of remediation efforts and advancing broad goals for sustainability. This multidisciplinary lens encourages the exploration of interdisciplinary research, leveraging emerging technologies, innovative policy interventions, and evolving societal trends to inspire resilience, sustainability, and equity to combat environmental contamination and nurture ecosystem health and human well-being. The sustainable remediation prospects can be bright and prosperous through teamwork and mutual action to build a healthier and urgent world for future generations.

14.9 CONCLUSION

The employment of NMs for photocatalytic applications is a robust and adaptable approach to environmental cleanup, offering innovative remedies to complex pollution problems across various natural environments. Like semiconductor NPs, NMs use light to generate reactive radicals and oxidants. This process is necessary for degrading organic pollutants, removing toxic metals, disinfecting water, and reducing pollution. Thus, the NMs utilization in such photocatalytic processes represents a promising and versatile means to ameliorate the contaminants in various environments, such as soil, water, air, and sediments. This approach exhibits several advantages over conventional remediation processes, including higher selectivity, faster reaction kinetics, and reduced energy input (the requirement of saline or acidic conditions, etc.) compared to classical remediation methods. In addition, the NMs-based photocatalysis is driven by elementary principles of sustainability and green chemistry and thus provides eco-friendly, waste-producing, resource-hungry and eco-harmful processes. Using renewable energy sources using non-toxic materials, abundantly available for photocatalytic remediation of contaminants, helps achieve sustainable development goals and protects human health, environmental integrity, and well-being. Moreover, nanotechnology and material science innovations continue to track the remediation field. This advancement has led to the next generations of novel nanocomposites, hybrid materials, and designed NPs with tailored properties and enhanced performance. Addressing the hurdles such as scalability, stability, and/or risks raised from NM-based photocatalysis, such as their NP release, fate, and environmental toxicity, is quite significant. However, this would involve more rigorous investigation, further collaboration between fields and the introduction of regulation, to progress the field ethically, as well as to facilitate the safe and effective use of NMs to cleanse the world. In summary, the photocatalytic activity of NMs is promising as a sustainable paradigm shift strategy to resolve pollution and enhance environmental protection. Nanotechnology and light-driven catalysis can bring us closer to our planet's cleaner, healthier, and more sustainable technosphere.

REFERENCES

Ahmad, Muhammad Arslan, Muhammad Adeel, Noman Shakoor, Rabia Javed, Muhammad Ishfaq, Yutao Peng, Muhammad Zain, et al. 2023. "Modifying Engineered NMs to Produce next Generation Agents for Environmental Remediation." *Science of The Total Environment* 894: 164861. https://doi.org/10.1016/j.scitotenv.2023.164861

Ahmaruzzaman, Md, Gaurav Yadav, and Th Babita Devi. 2024. "Chapter 23 - NM as an Emerging Green Catalyst in Environmental Remediation." In *Micro and Nano Technologies*, edited by Shadpour Mallakpour and Chaudhery Mustansar – Industrial Applications of Nanoceramics Hussain, 425–42. Elsevier. https://doi.org/10.1016/B978-0-323-88654-3.00028-7

Aissa, Mohamed Ali Ben, A Modwi, Kamal Taha, Nuha Elamin, Rasha A AbuMousa, and Mohamed Bououdina. 2023. "Environmental Remediation Applications of MxOy-GC3N4 Nanocomposites (M = Mg, Ti, and Zn): Photocatalytic Activity for Indigo Carmine Dye Degradation." *Diamond and Related Materials* 136: 109988. https://doi.org/10.1016/j.diamond.2023.109988

Arsenov, Danijela, Jelena Beljin, Danica Jović, Snežana Maletić, Milan Borišev, and Ivana Borišev. 2023. "NMs as Endorsed Environmental Remediation Tools for the Next Generation: Eco-Safety and Sustainability." *Journal of Geochemical Exploration* 253: 107283. https://doi.org/10.1016/j.gexplo.2023.107283

Aswini, Ravi, Annamalai Padmanaban, Sankar Vigneshwaran, Hector Valdes, and SaravanaVadivu Arunachalam. 2023. "A Review on Versatile Nano-Photocatalysts for Environmental Remediation: Carbon-Decorated Bismuth-Based NMs." *Nano-Structures & Nano-Objects* 35: 100991. https://doi.org/10.1016/j.nanoso.2023.100991

Balasurya, S, Abdullah A AL-ghamdi, Mohammad K Okla, Saud A Al-Amri, Mostafa A Abdel-Maksoud, Mohammed Aufy, and S Sudheer Khan. 2023. "Efficient Photocatalytic Activity of Au@Mg Nanospheres on Mineralization of Polystyrene: A Sustainable Remediation Strategy on Sunlight-Induced Photodegradation, Environmental Toxicity, and Sensing of Cefixime." *Journal of Water Process Engineering* 51: 103350. https://doi.org/10.1016/j.jwpe.2022.103350

Belhameid, Ahmed, Ángela Inmaculada López-Lorente, Adel Megriche, Rafael Lucena, and Soledad Cárdenas. 2024. "Photocatalytic/Sorption-Self-Cleaning Activity of Cellulose Decorated with TiO2 and M@TiO2 (M=Au, Ag) Polymeric Nanocomposites." *Journal of Cleaner Production*, 142190. https://doi.org/10.1016/j.jclepro.2024.142190

Boruah, Purna K, Archana Yadav, and Manash R Das. 2020. "Magnetic Mixed Metal Oxide NMs Derived from Industrial Waste and Its Photocatalytic Applications in Environmental Remediation." *Journal of Environmental Chemical Engineering* 8 (5): 104297. https://doi.org/10.1016/j.jece.2020.104297

Budimir, Milica D, and Jovana R Prekodravac. 2024. "10 - Photocatalytic Properties of Zero-Dimensional Carbon–Based NMs: Application as Catalysts/Adsorbents in Water Treatment." In *Woodhead Publishing Series in Electronic and Optical Materials*, edited by Kuruvilla Joseph, Runcy Wilson, Gejo George, and Saritha – Zero-Dimensional Carbon NMs Appukuttan, 291–355. Woodhead Publishing. https://doi.org/10.1016/B978-0-323-99535-1.00011-1

Chauhan, Ankush, Garima Rana, Vishal Dutta, Amita Kumari, Subha Krishna Rao, Rajasekaran Subbarayan, Karthikeyan Ravi, Satheesh Selvaraj, and Suresh Ghotekar. 2024. "Recent Trends in Phyto-Mediated Iron-Based NMs for Environmental Remediation and Biomedical Applications." *Inorganic Chemistry Communications* 160: 111976. https://doi.org/10.1016/j.inoche.2023.111976

Chen, Fang, Yihe Zhang, and Hongwei Huang. 2023. "Layered Photocatalytic NMs for Environmental Applications." *Chinese Chemical Letters* 34 (3): 107523. https://doi.org/10.1016/j.cclet.2022.05.037

Corsi, Ilaria, Iole Venditti, Francesco Trotta, and Carlo Punta. 2023. "Environmental Safety of Nanotechnologies: The Eco-Design of Manufactured NMs for Environmental Remediation." *Science of The Total Environment* 864: 161181. https://doi.org/10.1016/j.scitotenv.2022.161181

Devi, M Sindhu, S Srinivasan, and A Muthuvel. 2023. "Selenium NM Is a Promising Nanotechnology for Biomedical and Environmental Remediation: A Detailed Review." *Biocatalysis and Agricultural Biotechnology* 51: 102766. https://doi.org/10.1016/j.bcab.2023.102766

Devi, Raman, Mamta Bulla, Sunil Kumar, Ajay Kumar Mishra, and Vinay Kumar. 2024. "Chapter 13 - Zinc Oxide-Based NMs for Photocatalytic Applications." In *Advanced Functional Materials and Methods for Photodegradation of Toxic Pollutants* edited by Ajay Kumar Mishra, Pardeep Singh, Pankaj Raizada, Sethumathavan Vadivel, and Rangabhashiyam – Advanced Functional Materials and Methods for Photodegradation of Toxic Pollutants Selvasembian, 327–59. Elsevier. https://doi.org/10.1016/B978-0-323-95953-7.00015-1

Djellabi, Ridha, Roberto Giannantonio, Ermelinda Falletta, and Claudia Letizia Bianchi. 2021. "SWOT Analysis of Photocatalytic Materials towards Large Scale Environmental Remediation." *Current Opinion in Chemical Engineering* 33: 100696. https://doi.org/10.1016/j.coche.2021.100696

Dutta, Vishal, Jyothis Devasia, Ankush Chauhan, Jayalakshmi M, Vasantha V L, Ayush Jha, Aatika Nizam, Kun-Yi Andrew Lin, and Suresh Ghotekar. 2022. "Photocatalytic NMs: Applications for Remediation of Toxic Polycyclic Aromatic Hydrocarbons and Green Management." *Chemical Engineering Journal Advances* 11: 100353. https://doi.org/10.1016/j.ceja.2022.100353

Fathima Khyrun, S M, A Jegatha Christy, Jeyanthinath Mayandi, and Suresh Sagadevan. 2024. "Synergistic Effect of Nano-Floret CeO2/ZnO Nanocomposite as an Efficient Photocatalyst for Environmental Remediation." *Ceramics International* 50 (7, Part B): 11817–32. https://doi.org/10.1016/j.ceramint.2024.01.086

Gloria, Diego Coelho Sanches, Carlos Henrique Vieira Brito, Thais Aline Prado Mendonça, Tayra Rodrigues Brazil, Raquel Aparecida Domingues, Nirton Cristi Silva Vieira, Elias Barros Santos, and Maraísa Gonçalves. 2023. "Preparation of TiO2/Activated Carbon NMs with Enhanced Photocatalytic Activity in Paracetamol Degradation." *Materials Chemistry and Physics* 305: 127947. https://doi.org/10.1016/j.matchemphys.2023.127947

González-González, Reyna Berenice, Lidia Yaritza Martínez-Zamudio, Jesús Alfredo Rodríguez Hernández, Georgia María González-Meza, Roberto Parra-Saldívar, and Hafiz M N Iqbal. 2023. "Pharmaceutical Pollution Fingerprinting and Waterbodies Remediation Using Waste-Derived Carbon Dots as Sustainable Advanced NMs." *Environmental Research* 238: 117180. https://doi.org/10.1016/j.envres.2023.117180

Güell, Frank, Andrés Galdámez-Martínez, Paulina R Martínez-Alanis, Ariadne C Catto, Luís F da Silva, Valmor R Mastelaro, Guillermo Santana, and Ateet Dutt. 2023. "ZnO-Based NMs Approach for Photocatalytic and Sensing Applications: Recent Progress and Trends." *Materials Advances* 4(17): 3685–707. https://doi.org/10.1039/d3ma00227f

Hassan, N S, A A Jalil, N F Khusnun, M B Bahari, I Hussain, M L Firmansyah, R E Nugraha, and S Rajendran. 2023. "Extra-Modification of Zirconium Dioxide for Potential Photocatalytic Applications towards Environmental Remediation: A Critical Review." *Journal of Environmental Management* 327: 116869. https://doi.org/10.1016/j.jenvman.2022.116869

Jie, Longfei, Xue Gao, Xiaoqing Cao, Shan Wu, Xiaoxing Long, Qiongyan Ma, and Jixin Su. 2024. "A Review of CdS Photocatalytic NMs: Morphology, Synthesis Methods, and Applications." *Materials Science in Semiconductor Processing* 176: 108288. https://doi.org/10.1016/j.mssp.2024.108288

Khan, Iltaf, Chunjuan Wang, Shoaib Khan, Jinyin Chen, Aftab Khan, Sayyar Ali Shah, Aihua Yuan, Sohail Khan, Mehwish K Butt, and Humaira Asghar. 2023. "Bio-Capped and Green Synthesis of ZnO/g-C3N4 Nanocomposites and Its Improved Antibiotic and Photocatalytic Activities: An Exceptional Approach towards Environmental Remediation." *Chinese Journal of Chemical Engineering* 56: 215–24. https://doi.org/10.1016/j.cjche.2022.07.031

Krishna Veni, K, S Agalya, R Kavitha, I Fatimah, S Sagadevan, and L C Nehru. 2023. "Efficient Photocatalytic Activity of Chitosan/Tin Oxide Nanocomposite for Environmental Remediation." *Inorganic Chemistry Communications* 154: 110865. https://doi.org/10.1016/j.inoche.2023.110865

Kumar Sharma, Nitish, Swedika Sharma, and N Shibani. 2023. "A Review on Energy Conversion Applications of Graphene-Based NMs and Environmental Remediation." *Materials Today: Proceedings*. https://doi.org/10.1016/j.matpr.2023.03.583

Kumari, Minashree. 2023. "Chapter 5 - Role of NMs in Environmental Remediation." In *Micro and Nano Technologies*, edited by Chandrabhan Verma, Jeenat Aslam, and Mohammad Ehtisham – Adsorption through Advanced Nanoscale Materials Khan, 83–99. Elsevier. https://doi.org/10.1016/B978-0-443-18456-7.00005-5

Kundu, Sukanya, and Niranjan Karak. 2022. "Polymeric Photocatalytic Membrane: An Emerging Solution for Environmental Remediation." *Chemical Engineering Journal* 438: 135575. https://doi.org/10.1016/j.cej.2022.135575

Mane, Vijay, Dnyaneshwar Dake, Nita Raskar, Ramprasad Sonpir, Elias Stathatos, and Babasaheb Dole. 2024. "A Review on Bi2O3 NM for Photocatalytic and Antibacterial Applications." *Chemical Physics Impact* 8: 100517. https://doi.org/10.1016/j.chphi.2024.100517

Meng, Fanpeng, Tingting Hao, Wenjie Tian, Jinsheng Zhao, Shaobin Wang, and Huayang Zhang. 2024. "Photocatalytic Aqueous Environmental Remediation via Two-Dimensional Carbon Nitride Nanosheets." *Surfaces and Interfaces* 44: 103717. https://doi.org/10.1016/j.surfin.2023.103717

Nie, Changliang, Xiaohan Wang, Ping Lu, Yukun Zhu, Xin Li, and Hua Tang. 2024. "Advancements in S-Scheme Heterojunction Materials for Photocatalytic Environmental Remediation." *Journal of Materials Science & Technology* 169: 182–98. https://doi.org/10.1016/j.jmst.2023.06.011

Nordin, Nurul Atikah, Mohamad Azuwa Mohamed, Mohd Nur Ikhmal Salehmin, and Siti Fairus Mohd Yusoff. 2022. "Photocatalytic Active Metal–Organic Framework and Its Derivatives for Solar-Driven Environmental Remediation and Renewable Energy." *Coordination Chemistry Reviews* 468: 214639. https://doi.org/10.1016/j.ccr.2022.214639

Puri, Nidhi, and Anjali Gupta. 2023. "Water Remediation Using Titanium and Zinc Oxide NMs through Disinfection and Photo Catalysis Process: A Review." *Environmental Research* 227: 115786. https://doi.org/10.1016/j.envres.2023.115786

Rajput, Priyadarshani, Pradeep Kumar, A K Priya, Smita Kumari, Seyede Roghie Ghadirnezhad Shiade, Vishnu D Rajput, Amin Fathi, et al. 2024. "NMs and Biochar Mediated Remediation of Emerging Contaminants." *Science of The Total Environment* 916: 170064. https://doi.org/10.1016/j.scitotenv.2024.170064

Rasool, Mohsin, Mirza Mahmood Baig, Muhammad Kaleem, Mehboob Hassan, Luqman Riaz, Muhammad Farooq Warsi, and Seung Goo Lee. 2024. "Photocatalytic and Adsorption Behavior of a New Efficient LaxNiyMn3–(X+y)O4/SiO2 Composite for Environmental Remediation Application." *Ceramics International* 50 (3, Part A): 4790–809. https://doi.org/10.1016/j.ceramint.2023.11.223

Venkadajapathy, Vimala Rathinam Thiruppathi, and Sivaperumal, Sivarmakrishnan. 2024. "Tailoring Functional Two-Dimensional Nanohybrids: A Comprehensive Approach for Enhancing Photocatalytic Remediation." *Ecotoxicology and Environmental Safety* 275: 116221. https://doi.org/10.1016/j.ecoenv.2024.116221

Vinoth, S, and A Pandikumar. 2024. "Recent Advances in Bismuth Oxyfluoride-Based Photocatalysts for Energy and Environmental Remediation." *Materials Today Chemistry* 36: 101924. https://doi.org/10.1016/j.mtchem.2024.101924

Saleem, Haleema, Syed Javaid Zaidi, Ahmad Fauzi Ismail, and Pei Sean Goh. 2022. "Advances of NMs for Air Pollution Remediation and Their Impacts on the Environment." *Chemosphere* 287: 132083. https://doi.org/10.1016/j.chemosphere.2021.132083

Sathish, T, N Ahalya, M Thirunavukkarasu, T S Senthil, Zakir Hussain, Md Irfanul Haque Siddiqui, Hitesh Panchal, and Kishor Kumar Sadasivuni. 2024. "A Comprehensive Review on the Novel Approaches Using NMs for the Remediation of Soil and Water Pollution." *Alexandria Engineering Journal* 86: 373–85. https://doi.org/10.1016/j.aej.2023.10.038

Sharma, Sunil Kumar, Amit Kumar, Gaurav Sharma, Tongtong Wang, Ana Iglesias-Juez, and Pooja Dhiman. 2023. "Recent Advances in G-C3N4/Metal Organic Frameworks Heterojunctions for High-Performance Photocatalytic Environmental Remediation and Energy Production." *Journal of Molecular Liquids* 382: 121890. https://doi.org/10.1016/j.molliq.2023.121890

Singhwal, Deepak, and Pawan S Rana. 2024. "Cadmium Oxide NPs Doped with Nickel and Silver: Photocatalytic Activity for Environmental Remediation." *Materials Chemistry and Physics* 314: 128895. https://doi.org/10.1016/j.matchemphys.2024.128895

Sohn, Erica Jungmin, Byung-Moon Jun, Seong-Nam Nam, Chang Min Park, Min Jang, Ahjeong Son, and Yeomin Yoon. 2024. "Photocatalytic Boron Nitride–Based NMs for the Removal of Selected Organic and Inorganic Contaminants in Aqueous Solution: A Review." *Chemosphere* 349: 140800. https://doi.org/10.1016/j.chemosphere.2023.140800

Suvathi, S, K Ravichandran, M Karunakaran, P K Praseetha, M Ayyanar, and S Gobalakrishnan. 2024. "Biowastes-Derived Enzyme-Powered Zinc Oxide and Titanium Oxide NMs Synthesis for Anticancer and Eco-Friendly Photocatalytic Activity." *Applied Materials Today* 36: 102024. https://doi.org/10.1016/j.apmt.2023.102024

Tehseen, Aqsa, Maira Liaqat, Tahir Iqbal, Iqra Maryam, Sumera Afsheen, Muhammad Shehzad Sultan, Asad Syed, Abdallah M Elgorban, and Ling Shing Wong. 2024. "Design, Fabrication, and Analysis of Bi2O3/MnO2 Hybrid NMs for Advanced Photocatalytic Applications." *Journal of Physics and Chemistry of Solids* 190: 112014. https://doi.org/10.1016/j.jpcs.2024.112014

Teixeira, J, P M Martins, R Fernández de Luis, E Falletta, M F Ordoñez, C L Bianchi, and S Lanceros-Méndez. 2024. "Nitrogen Oxide Remediation through Metal–Organic Frameworks with Bi-Functional Absorption and Photocatalytic Characteristics." *Chemical Engineering Journal* 486: 150400. https://doi.org/10.1016/j.cej.2024.150400

Thind, Sapanbir S, Mathias Paul, John B Hayden, Anuj Joshi, David Goodlett, and J Scott McIndoe. 2023. "A Highly Efficient Photocatalytic System for Environmental Applications Based on TiO2 NMs." *Industrial Chemistry & Materials* 1 (3): 431–42. https://doi.org/10.1039/d3im00053b

Tyagi, Shivanshi, Riti Thapar Kapoor, Swati Solanki, Aarushi Goyal, and Rachana Singh. 2024. "Chapter 2 - NM Mediated Wastewater Treatment: A New Frontier in Environmental Remediation." In *Plant and Soil Microbiome*, edited by Ajay Kumar, Joginder Singh, and Lucas Carvalho Basilio De – Microbiome-Based Decontamination of Environmental Pollutants Azevedo, 31–49. Academic Press. https://doi.org/10.1016/B978-0-443-21781-4.00009-8

Wang, Junhong, Xianzhao Shao, Junhai Liu, Qiang Zhang, Xiaohui Ji, and Guanghui Tian. 2020. "Mesoporous Magnetic G-C3N4 Nanocomposites for Photocatalytic Environmental Remediation under Visible Light." *Ecotoxicology and Environmental Safety* 205: 111147. https://doi.org/10.1016/j.ecoenv.2020.111147

Wang, Yan-Zhai, Syed Bilal Shah, Jun-Ying Liu, Hao Hu, and Yang-Chun Yong. 2024. "Biomineralization Induced Synthesis and Self-Assembly of Photocatalyst-Enzyme Hybrid System for Highly Efficient Environmental Remediation." *Applied Catalysis B: Environment and Energy* 351: 124015. https://doi.org/10.1016/j.apcatb.2024.124015

Yu, Shujun, Hao Tang, Di Zhang, Shuqin Wang, Muqing Qiu, Gang Song, Dong Fu, Baowei Hu, and Xiangke Wang. 2022. "MXenes as Emerging NMs in Water Purification and Environmental Remediation." *Science of The Total Environment* 811: 152280. https://doi.org/10.1016/j.scitotenv.2021.152280

Zelekew, Osman Ahmed, Haileyesus Hatano Haitosa, Xiaoyun Chen, and Yi-Nan Wu. 2023. "Recent Progress on Plant Extract-Mediated Biosynthesis of ZnO-Based Nanocatalysts for Environmental Remediation: Challenges and Future Outlooks." *Advances in Colloid and Interface Science* 317: 102931. https://doi.org/10.1016/j.cis.2023.102931

15 3D Printed Materials in Environmental Remediation

Prasann Kumar and Joginder Singh

15.1 INTRODUCTION

Human activities, industrialization, population growth and unsustainable use of resources, among other factors, have accelerated the pace of environmental change. Over the last 200 years, the release of Greenhouse Gases (GHG) like carbon dioxide, methane, and nitrous oxide into the atmosphere has not only significantly contributed to the depletion of the atmosphere and ozone layer but also led to the degradation of the environment. Burning fossil fuels for energy generation, transportation, and manufacturing are the main contributors to this emission (Nawaz et al. 2023; Devi, Srinivasan, and Muthuvel 2023; Zou et al. 2024). The questioned emissions, in turn, contribute to global warming and climate change that gives rise to urgent environmental challenges, such as a growing number of more severe weather events, rising sea levels, disruption of ecosystems and threats to biodiversity. There are ecological problems other than climate change. Pollution affects the quality of air, water, and soil. Industrial and agricultural practices, combustion of fossil fuels, and even the process of urbanization contribute to the air by emitting pollutant substances such as particulate matter, nitrogen oxides, sulphur dioxide, and volatile organic compounds (VOCs). Such harmful practices, including untreated sewage disposal, industrial effluent discharge, agricultural runoff and plastic waste, contaminate oceans, rivers, lakes, and groundwater and contribute to water pollution, which poses risks to aquaculture and aquatic ecosystems, human health, and drinking water supply. This may lead to degradation of land, which in turn harms the soil's fertility and hampers land productivity, causing desertification, soil erosion, and depletion of arable land. Deforestation is driven primarily by logging, agriculture, and urbanization, resulting in habitat loss, soil erosion, and biodiversity loss (Gonçalves et al. 2024; Ranjan et al. 2024; Brown et al. 2022). These are grave threats to the survival of many plant and animal species. The dangers, such as overfishing by fisheries and illegal harvesting of ocean biodiversity, continue to jeopardize marine ecosystems and the food sovereignty of millions who consume seafood as their primary source of protein. The extraction and exploitation of finite resources, minerals, metals, and fossil fuels drive the significant and widespread habitat destruction, pollution, and carbon emissions that all contribute to environmental degradation. The feedback loop phenomenon includes a positive feedback mechanism as the population grows with rapid urbanization, ecological issues are exacerbated and demands for resources, infrastructure, and ecology increase.

More energy consumption, waste generation, and pollution coincide with this growth. The prevalent unsustainable behavior in consumption seen in the societies, marked by discourses of disposability, the misuse of single-use plastics, and excessive consumption of goods and services, adds to the battering of the environment and the accumulation of waste in landfills, the oceans, and ecologies. Addressing these environmental challenges will take local, national, and international collaborations. Such actions also extend to adopting policies and regulations aimed at limiting GHG emissions, promoting the uptake of renewable energy and better waste management practices, conserving natural habitats and resources, and encouraging their responsible use (Siddiqui et al. 2023; da Silva et al. 2023; Siqueira et al. 2023). It is essential for creating awareness, education and sustainable habits. They hope to inspire people and organizations to do their bit to protect the natural environment from the harmful damage caused by humans. Most often, traditional methods are

DOI: 10.1201/9781003484189-15

ineffective in addressing the environmental crisis, resulting in a requirement for innovative tactics to resolve multifaceted issues successfully. First and foremost, we need a wild way to deal with the interdependent issues of the planet. Many environmental problems, including climate change, pollution, and habitat loss, are related and holistic solutions must consider the systemic drivers and effects. Creativity can aid in finding solutions for the immediate symptoms, deep causes, and synergies among environmental challenges. We must employ new strategies to balance better conservation, economic development, and social equity challenges. The potential of innovative solutions, such as 3DP (3-Dimensional Printing) applications, in addressing these challenges is a source of hope and inspiration.

Traditional methods present sustainability logic as an economic trade-off, disconnecting conservation initiatives and economic development agenda. On the contrary, innovative solutions can close this impasse by finding common motives to bolster sustainable progress and preserve environmental quality. Projects supporting renewable energy, green infrastructure, and sustainable agriculture can create jobs, boost the economy, and reduce environmental impacts. Moreover, it is crucial to use different strategies to build resilience and to facilitate adaptation to everyday ecological pressures (Prakash et al. 2023; Peng et al. 2022). Environmental shocks have been shown to occur with increasing frequency and intensity, and habitat degradation with the need to design new strategies to maximize resilience and adaptability at individual, community, and institutional levels. Nature-based solutions, such as green infrastructure, restoration of ecosystems, and sustainable land use practices, can enhance ecological and community resilience to the impacts of climate change. Moreover, developing such unique solutions is significant in engaging and inspiring hundreds of stakeholders to participate in conservation-related actions (Wu et al. 2021). Creative solutions, such as harnessing technology, social innovation, and participatory approaches, can achieve this leverage across individuals, communities, businesses, and governments to take collective action. The need for collective action is clear, and it is through our combined efforts that we can make a significant impact on the environment. For instance, digital platforms, citizen science initiatives, and gamification have the potential to raise awareness, facilitate behavioral change, and enable collaboration among stakeholders. Old solutions are no longer working or are unhealthy, and we need new ones to solve environmental problems. Embedding these enablers, such as creativity, innovation, and collaboration, into their working processes can allow fresh solutions to be developed in response to environmental degradation, the integration of sustainable development, the enhancement of resilience, and the promotion of social equity. Critical ideas, thoughts, and solutions will help determine the sustainability and resiliency of our children and those generations ahead of us as we navigate the ecological challenges of the 21st century.

The introduction of 3DP in environmental cleansing is an intersection where technological advancements are incredibly promising toward sustainability. 3DP, or additive manufacturing (AM), is the process of joining materials, layer by layer, to make a three-dimensional object from a digital model. Although initially valued for applications such as manufacturing, prototyping, and customization, 3DP technologies are now being investigated for their potential in environmental remediation. This technology opens fresh prospects for purging oceans of pollution, restoring lost habitats, and conserving resources. Customized filtration systems for air and water purification represent a crucial application of 3DP in ecological remediation. These traditional filtration systems are built using the same components, which sometimes may not be tailored to target specific contaminants or environmental conditions. 3DP technology can be used to develop complex and tailor-made filter designs that can trap multiple pollutants, including heavy metals, microplastics, and organic pollutants (Gonçalves et al. 2024, Wu et al. 2021, Chang et al. 2024, Mallakpour et al. 2023). Solutions such as using sustainable materials and manufacturing processes to 3D print filters can help mitigate the negative environmental impact of remediation technologies that are applied in traditional methods. This presents exciting opportunities, such as using 3DP to support marine environments in creating artificial reefs and habitat structures. Unique pressures from climate change, pollution, and loss of their natural settings are affecting coastal ecosystems such as coral reefs and mangrove forests. 3DP offers the potential for a low-cost manufacturing approach to synthetic reef structures with intricate textures and functionalities mirroring their naturally occurring habitats.

These are critical habitats for marine species and contribute to increasing ecological resilience. Moreover, 3D-printed reefs can be designed for favorable environmental conditions that promote coral health, such as surface texture, substrate composition, and structural complexity. This approach might even encourage the restoration of degraded reef ecosystems sooner, while 3DP also has the potential for soil remediation and land reclamation. Soil pollution, caused by industrial use or disposal of hazardous waste, mining, and implementation of improper waste disposal, poses a significant potential threat to human health, biodiversity, and the functioning of the ecosystem. 3DP technology opens new avenues for environmental remediation specialists, enabling them to create unique amendments, such as biochar aggregates, microbial inoculants, and nutrient-rich substrates applied with precision to tailor specific amendments for various contaminants of concern, thereby optimizing the remediation process. 3D-printed soil amendments may increase oil quality, support microbial activity, and enhance pollutant degradation and immobilization. Deep perennial roots fixate soil structure and quality for many decades, providing sequential, multi-decade benefits and facilitating sustainable land use practices, unlike artificial fertilizing agents that offer negligible time-frame benefits. Moreover, 3DP is critical for developing a circular economy and sustainable resource management, reducing dependence on primary resources and lesser waste generation and minimizing the environmental impacts associated with conventional manufacturing processes using waste materials as raw materials for 3DP techniques like recycled plastics, metals, and construction debris (Sreenivasan and Kandasubramanian 2023; Wan et al. 2022; Rivas-Sanchez et al. 2022; A. Kumar et al. 2023). In addition, 3DP facilitates just-in-time production and distributed manufacturing, thereby minimizing transportation emissions and curtailing the consumption of resources in the global supply chain. In recent years, the use of 3DP technology in environmental remediation has gained significant attention due to its potential to create novel, sustainable technologies for key environmental problems. Leveraging their know-how in 3DP flexibility, customization and ecological sustainability, they can develop solutions to pollution, regenerate natural territories and preserve ecosystems. The ongoing development of 3DP by technology developers, its many possible applications for sustainable practices, and its growing interest by environmental scientists and policy-makers represent an opportunity for linking prosecutors to promote introducing this technology to benefit the environment and future generations (Figure 15.1).

15.2 FUNDAMENTALS OF 3D PRINTED MATERIALS

3DP is a modern technique that creates 3D objects from 3D models or designs by depositing materials layer by layer. In contrast to traditional subtractive manufacturing processes, where the material is cut or drilled from a solid block, 3DP constructs objects layer-by-layer to offer unparalleled design flexibility, customization, and efficiency. The most exciting notion submerged in this investment has been its sheer development in the past many decades, resulting in a more significant approach, affordability, and adaptability at the exact moment. Thermoelectric technology finds utility in multiple sectors: manufacturing, healthcare, aerospace, automotive, and consumer durables. The core of the 3DP process involves having a digital model made via Computer-Aided Design (CAD) software or 3D scanning technology.

The slicing software chops the 3D digital model into thin, horizontal, cross-sectional layers. This creates instructions known as G-code, which the printer subsequently executes. After this, the material printer sequentially lays out material slices in the object's desired configuration. Depending on the 3DP technology and its application, various materials can be printed: plastics, metals, ceramics, resins, and composite materials (Wang et al. 2021; Ahmed and Jhung 2021; Kim et al. 2024; Vo et al. 2022). Various 3DP methods come with other benefits, restrictions, and use cases. The Fused

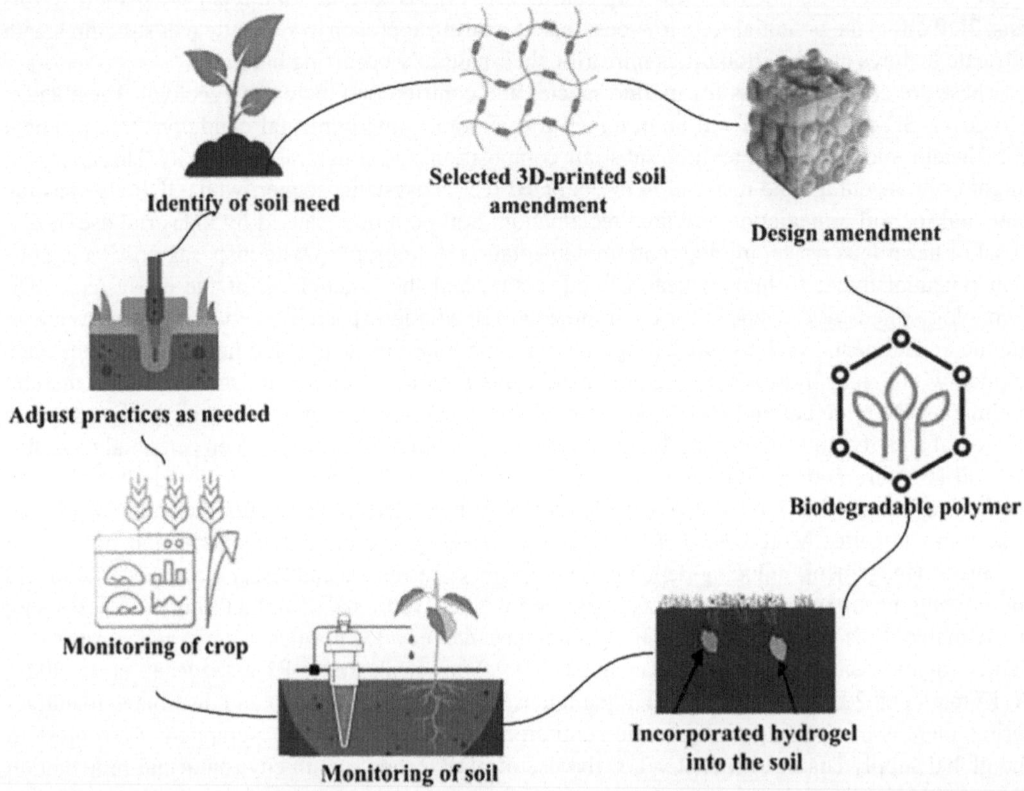

FIGURE 15.1 Utilization of 3D-printed soil amendments.
(Source: created by authors)

Deposition Modeling (FDM) 3DP technique uses a heated nozzle to deposit thermoplastic filaments on the build platform. The FDM is popular in rapid prototyping, hobbyist projects, and low-volume functional mechanical components and prototypes. Hence, Stereolithography (SLA) is the most common additive manufacturing process that uses a vat of photopolymer resin and employs a laser or light source to selectively reclaim the resin layer-by-layer to create detailed and precise items. Shape Memory Alloy (SLA) makes fine prototypes, dental appliances, jewelery, and small and complex components with complex shapes. The method behind selective laser sintering (SLS), is a 3DP technology that uses a literal high-powered laser to bind nylon powders, thermoplastics, metals, and ceramics into solid forms, layer by layer. SLS offers part versatility, material options, functional prototypes, and end and fit-in-feel parts for complex geometries. Moreover, it is essential to mention that there are different 3DP methods, including Digital Light Processing (DLP), Electron Beam Melting (EBM), and binder jetting. These technologies have separate expertise and might be helpful in entirely different industries and situations.

3DP technology is, in general, continuously revolutionizing how we design, produce, and deliver products with unrivaled opportunities for personalization, creativity, and environmental stewardship. As technology becomes more advanced and affordable, its applications will increasingly disrupt traditional manufacturing processes, accelerate product development timelines, and create new possibilities in many industries, from health care to aerospace, automotive, architecture, and consumer goods. 3DP combines various techniques to construct 3D objects in layers from a digital blueprint. Each has unique characteristics, strengths, weaknesses, and use cases across industries and scenarios. FDM is one of the simple methods for 3DP, hence, it is widely used and also commercially accessible (Rashid et al. 2024; S. Kumar et al. 2022; McDonagh et al. 2023; Jiang et al. 2024). This involves the deposition of thermoplastic filament through a nozzle at an elevated temperature onto the build platform. FDM technology is mainly used for fast prototyping, hobby projects, and inexpensive pieces and prototypes, selective layer curing, a photopolymer resin, and a UV light source.

The 3D object is detailed and accurate since the platform is raised step by step after each layer is cured. SLA is most commonly used to produce high-resolution prototypes, dental work, jewelery, and specific complex geometries, which makes it ideal for challenging components. SLS is a 3DP process that relies on a powerful laser that melts a stream of powder, such as nylon, thermoplastics, metals, and ceramics, into a solid object, layer by layer. The method has many applications, from functional prototyping to end-use parts and customized products with complex geometries. DLP is a digital light projector process that entirely cures the photopolymer resin matrix in layers, similar to SLA. Consequentially, you also get the prints much faster than with SLA. DLP finds its widest application in manufacturing dental models, jewelery and other intricate parts, where speed and precision are required. In binder jetting, a binding agent is selectively deposited into a powdered material such as sand, metal, or ceramic, which often bonds the powder particles. Once the object is printed in a layer, a new layer of powder is distributed evenly, and the process is repeated until the whole object gets printed. The binder jetting process is ideal for complex geometries, large-scale elements, molds and prototypes. EBM locally melts the metal powder in a high vacuum environment with an electron beam. This process primarily targets achieving metallic parts with exceptional mechanical properties, thus targeting high strength and density in general. Applications of experimentation-based manufacturing) is utilized in aerospace, automotive, and medical industries to create complex parts such as turbine blades, orthopaedic implants, and sophisticated structures. Material jetting involves spraying photopolymer or wax onto a building platform and curing or solidifying the material with UV light or heat. This approach facilitates the fabrication of multi-material components, showing different colors, textures, and properties, all in a single print step.

Material jetting technology can produce prototypes, appearance models, and end-use products with high-detail and smooth surface finish (Mohd Yusoff et al. 2023; Montaner and Hilton 2024; Vaishampayan et al. 2023; Jatoi et al. 2023). As mentioned earlier, just a few techniques are described in the vast world of these 3DPs. Every available process offers insights and uses unique design, manufacturing efficiencies, and product offerings. Presented here, 3DP technology is constantly evolving and predicted to disrupt industries, promote innovation, and generate new design, production and distribution opportunities. A wide range of materials is used in 3DP, such as thermoplastics, metals, ceramics, resins, and composites. It can vary depending on the application's needs (mechanical properties, softness, surface finish, temperature resistance, and chemical compatibility).

15.2.1 Thermoplastics

Due to their versatility, cost-effectiveness, and the broad spectrum of material properties they offer, thermoplastics presently represent the most frequently employed material family in 3DP processes. Here are some popular thermoplastics that include:

- *ABS (Acrylonitrile butadiene styrene)*: This polymer has a high degree of strength, toughness and impact resistance; ABS is one of the most common types of materials used in FDM 3DP for functional prototyping, consumer products, and end-use parts.
- *PLA (Polylactic Acid)*: PLA is produced from renewable resources like corn starch or sugar cane and is a biodegradable thermoplastic. By nature, such designs are easy to print, green, and applicable to prototyping, education, packaging, etc.

- PETG (Polyethylene Terephthalate Glycol): PETG will offer you the strength and durability of ABS; however, it is as easy to print with as PLA and is commonly used for making functional prototypes, mechanical components, and shopper products.
- *Nylon*: Nylon is a strong, stiff thermoplastic with excellent mechanical and chemical properties and flexibility78. The SLS and FDM 3DP segment produces parts with high impact resistance, low friction, and excellent wear properties.

15.2.2 Metals

The additive metal manufacturing term is 3DP metal; this process can produce high-strength; complex metal parts with good mechanical properties. Some frequently-used metals that can be printed in 3D are:

- *Stainless Steel*: The most widely used material for metal 3DP; stainless steel is strong, durable, and corrosion-resistant. It is also used extensively in aerospace, automotive, and medical industries to produce structural components, tooling, and implants.
- *Titanium*: Titanium has a high strength-to-weight ratio, is biocompatible, and offers excellent corrosion resistance. It produces highly durable, lightweight, high-performance components like turbine blades, implants, and structural components for aerospace, medical, and automotive applications.
- *Aluminium*: Balances strength, corrosion resistance, recovery from heat and electricity, and lightweight. It is used in the aerospace, automotive, and consumer goods industries to create structural parts, heat sinks, and electronic enclosures.
- *Cobalt Chrome*: Cobalt chrome alloys are biocompatible, powerful, and wear-resistant. They are used in medical and dental applications to manufacture implants, prosthetics, and restorations.

15.2.3 Ceramics

The ceramic 3DP process creates completely temperature-resistant, chemically inert, and biocompatible workpieces with excellent thermal and electrical characteristics. The following are the most commonly used ceramics in the 3DP process:

- *Alumina (Aluminum Oxide)*: Alumina is a standard ceramic used for its high strength, hardness, and chemical resistance. It makes aerospace, electronics, and medical applications insulating components, electrical substrates, and biomedical implants.
- *Zirconia (Zirconium Oxide)*: Zirconia has high strength, toughness, and biocompatibility. The dental, medical, and aerospace industries employ this to build crowns, dental implants, and structural components.

15.2.4 Resins

3DP processes using SLA and DLP using photopolymer resins cure or solidify when applied to UV light or heat. Many resin formulas are made for other properties, such as:

- *Standard Resins*: Standard resins have high resolution, a smooth surface finish, and sound detail reproduction. They are used to make visual models, prototypes, and jewelery.
- *Engineering Resins*: These are engineered to have specific desired mechanical properties, such as strength, flexibility, and temperature resistance. They are used for functional prototypes, end-use parts, and tooling for manufacturing.

- *Dental Resins*: Made especially for dental use, these materials are biocompatible (FDA-approved and certified) and are used in dental procedures, including crowns, bridges, and orthodontic devices.
- *High-Temperature Resins*: High-temperature resins exhibit excellent thermal stability and heat resistance, making them suitable for 3DP components that carry high temperatures (automotive components), especially tooling inserts.

15.2.5 Composites

Some of the composite materials used in 3DP are:

- *Carbon Fiber Reinforced Polymers (CFRP)*: CFRP has combined the properties of carbon fiber with the flexibility and ease of printing of thermoplastics. It produces rigid, light, high-strength, and stiff impact-resistant components in the aerospace, automotive, and sports industries.
- *Glass Fiber Reinforced Polymers (GFRP)*: Similar to CFRP, GFRP has a unique property, transparency, and electrical insulation. It is used for structural components, enclosures, and housing in the electronics, automotive, and consumer goods industries.

These are just a handful of materials used in 3DP processes. As 3DP technology matures, several materials and recipes are being developed and applied to meet the growing needs of specific industries and applications (Figure 15.2).

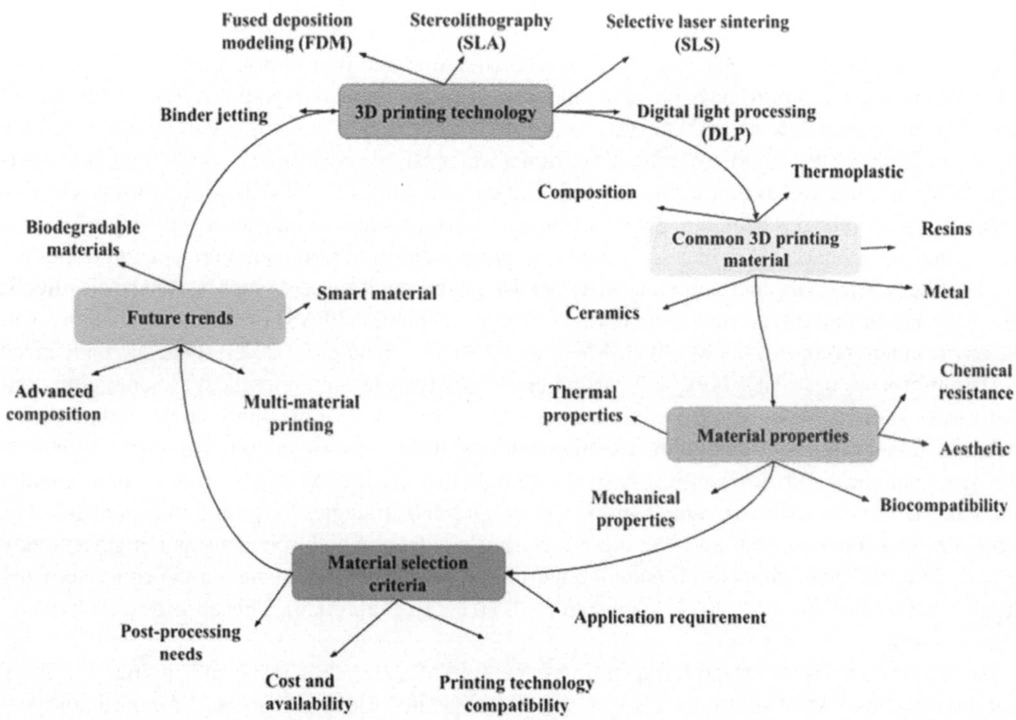

FIGURE 15.2 Fundamentals of 3D Printed Materials.

(Source: created by authors)

15.3 DESIGNS AND SYNTHESIS OF CUSTOMIZED MATERIALS

Materials that could potentially remediate the pollutants must be tailored to those contaminants for them to work, which is paramount in identifying effective remediation strategies for pollution issues. Each pollutant's specific chemical and physical nature calls for specialized materials removal or remediation. The knowledge of the unique properties of particular pollutants will enable scientists and engineers to create materials with superior properties that can specifically target and remediate such contaminants. Some examples of modification of materials toward specific environmental pollution are as follows: heavy metals, including lead, mercury, cadmium, and arsenic, damage ecological systems and human health, particularly with chronic exposure. The resultant materials can aquacade to achieve selective capture of heavy metal cations from wastewaters by designing for desired surface chemistries and functional groups of materials, e.g., activated carbon, zeolites, ion exchange resins, etc. It is thus possible to immobilize chelating agents (EDTA, DTPA, etc.) on solid supports to enhance soil and sediment complexation-extraction of heavy metals. Organic contaminants, such as pesticides, herbicides, industrial chemicals, and petroleum hydrocarbons, are toxic compounds with high environmental and human health risks. Using modified granular activated carbon (GAC) and biochar, specific surface functionalities enhance the adsorption and degradation of organic pollutants in aquatic and terrestrial systems.

Nanomaterials (NMs), such as graphene-based materials and zero-valent iron nanoparticles (ZVI NPs), exhibit potential for the remediation of organic pollutants via mechanisms such as catalytic degradation and oxidative reactions. Influences such as agricultural runoff and wastewater discharge of excess nitrogen and phosphorus have been significant components of nutrient pollution, eutrophication, algal blooms, and water quality degradation. Such materials include ion exchange resins, zeolites, and modified clays, which can selectively remove NO_3^- and PO_4^{-3} from agricultural runoff, stormwater, and wastewater effluents. Living processes for removing nutrients, such as constructed wetlands and biofiltration systems, use natural materials and biofilms to reduce excessive nutrients from water supplies via mechanisms such as denitrification and phosphorus uptake. Microplastics (MP; plastic fibres < 5 mm) have become ubiquitous contaminants of aquatic habitats, with significant risks to marine species and ecosystems. Here, we define a concept that can enable the facile microplastic characterization and extraction from a water sample based on the development of magnetic NPs and microbeads that can be coated with specific polymers. It will allow for the selective removal of microplastics from wastewater effluents and stormwater runoff via new filtration systems and membrane technology, achieved through the incorporation of porous materials and orientation of nanofibers. Emerging contaminants that persist in the environment, such as pharmaceuticals, personal care products, and per- and polyfluoroalkyl substances (PFAS) present new challenges to the environment (Baigorria et al. 2023; Finny et al. 2022; Zhang et al. 2023). One such advanced oxidation process uses TiO_2 NPs, and carbon-based catalysts to help photodegrade emerging contaminants under visible light and oxidative reactions. Water-soluble molecularly imprinted polymers (MIPs) can be custom-designed with selective binding sites to absorb and remove emerging pollutants from aqueous sources. We must design materials that selectively target certain environmental pollutants to develop effective remediation technologies and strategies for pollution challenges. The properties and functions that underlie these materials use from which scientists and engineers may provide innovative solutions to mitigate the harmful effects of contaminants on the ecosystem and human health (Farhana et al. 2024; Ogunsola et al. 2024; Li et al. 2024). This ensures a cleaner and healthier area.

Advanced composites research is a cutting-edge field of materials science and engineering study that provides innovative solutions to various applications in various industries. Advanced composites are made from two or more constituent materials, which, when combined, derive their characteristics from each other. These materials typically possess higher strength, stiffness, durability, and lightweight properties than conventional materials systems, making them suitable for demanding aerospace, automotive, marine, construction, and renewable energy applications. Due to their high strength-to-weight ratio, stiffness, and fatigue resistance, high-performance composites, including CFRP and Ceramic matrix composite (CMC), are commonly used in the aerospace industry. CFRP composites are used in different airframes, such as fuselages, wings, and empennages. The goal is to shave mass for more excellent fuel economy without sacrificing the aeroplane's strength and safety. It is produced via the compression of ceramic fibers into a ceramic matrix to form a CMC, which displays excellent thermal and chemical resistance; such properties make them suitable for heavyweight applications in extreme conditions. As a result, they can be used in high-temperature applications, such as in jet engines, rocket propulsion systems, and thermal protection sections (Shin et al. 2023; Kim et al. 2024; Rashid et al. 2024; Vo et al. 2022).

The adoption of advanced composites in the automotive sector is slow but gaining momentum. By using advanced composites, the vehicle's weight will be lightened by achieving high fuel efficiency and safety, and crashworthiness will also be increased. Automotive manufacturers choose carbon fiber composites for parts like chassis, body panels, and interior trim to fulfill lightweight targets and improve performance. Hydrogen composites, which use a combination of carbon fibers, glass or natural fibers, are a good compromise between cost, performance, and sustainability for automotive applications. Due to their high strength, corrosion resistance, and durability under extreme marine conditions, advanced composites are highly avoided in the composite industry (Zou et al. 2024; Gonçalves et al. 2023; Prakash et al. 2023; Siqueira et al. 2023). The structure of boat hulls, decks, masts, and other components typically contain FRP and carbon fiber composites that help reduce weight, provide buoyancy, and increase performance. High-performance composite materials can also create lightweight and high-speed ships, making these racing sailboats, luxury yachts, and military ships much easier and more responsive to run.

The future of building construction advanced composites transforms construction by enabling lightweight, durable, and energy-efficient building materials and structures. Used widely for improving the structural performance, durability, and aesthetics of building exteriors, cladding systems, bridges, and other infrastructure components, FRP and composite panels have come a long way since their inception. This flexibility means that contractors can use prefabricated methods for materials on-site to ensure fast assembly, less waste, and improved sustainability on site. Renewable energy technology includes modern composites that can help improve the efficiency, strength, and reliability of wind turbines harnessing energy from the breeze, for example, solar panels extracting energy from sunlight or hydroelectric infrastructure. Wind turbine blades use composite materials like fiberglass, carbon fiber, and epoxy resin to enhance energy capture, decrease weight, and resist challenging operating conditions. Composite materials are widely used to manufacture components for solar panels, tidal turbines, wave energy converters, and other applications because they are light and corrosion-resistant. In the meantime, this trend has been a major enabler of the explosion of renewable energy in Christian lands. Advanced composites are a newly found study area with the potential to contribute to technological development, innovation and sustainability in multiple and varying sectors. This process minimizes environmental impact and increases efficiency by using advanced composite materials' unique properties and capabilities to design and engineer new products and systems that better meet the requirements of a fast-paced global environment. This will continue to evolve from advances to R&D of advanced composites and offer further opportunities for innovation in this revolutionary material class (Figure 15.3).

15.4 APPLICATIONS IN ENVIRONMENTAL REMEDIATION

Environmental remediation includes, but is not limited to, removing contaminants such as pollutants from soil, water, and atmosphere. The primary mission is to make ecosystems normal again while protecting human health. The technology of 3DP provides innovative and creative solutions for several situations-oriented approaches to environmental remediation. For instance, micro-porous organic matter-supported biochar material, adsorbents, and bioremediation agents for various soil amendments can cater to specific pollutants and soil conditions by 3D printed materials. These

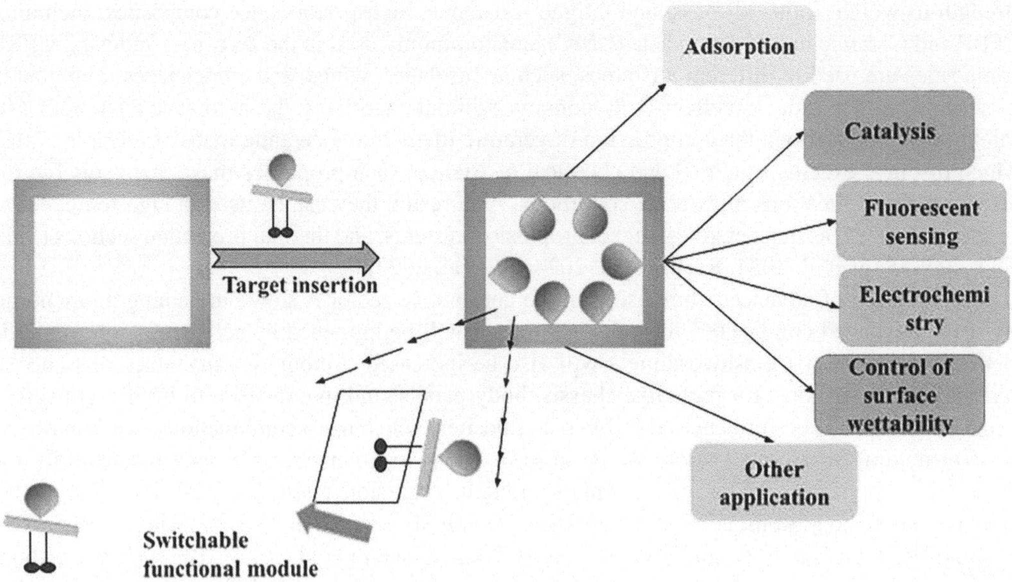

FIGURE 15.3 Design and Synthesis of Customized Materials.
(Source: created by authors)

materials can increase soil fertility, retain pollutants, promote microbial mechanisms for the decomposition of contaminants, and then replenish the soil quality and the ecological system functions. Contaminants present in polluted water bodies, including heavy metals, organic pollutants, and microplastics, have been successfully removed via the use of 3D-printed filtration systems and adsorbent materials (Zou et al. 2024; Dong et al. 2022; Iqbal, Bilal, and Zdarta 2022; da Silva et al. 2023; Brown et al. 2022). They are highly efficient, scalable, and can be customized according to the user requirements, enabling their practicality for large-scale water treatment in rural and urban areas.

Through 3DP technology, advanced air filtration systems and catalytic converters are produced to remove air pollutants, including particulate matter, VOCs, and NOx, from industrial and indoor air sources. Technology transfer on these systems can enhance the air quality, lower emissions, and reduce the health risks due to air pollution. Moreover, artificial reefs, structures, and erosion control barriers manufactured using 3DP can potentially rehabilitate damaged aquatic and terrestrial environments and improve biodiversity while counteracting the consequences of coastal erosion, habitat loss, and climate change. Features of habitats in these structures, such as those in coastal environments susceptible to degradation, serve to shelter aquatic organisms, provide support for shorelines, and enhance resilience for the whole ecosystem. 3DP technology is applied to use various rubbish, such as plastic waste, agricultural waste, and construction surplus wastage, to generate value-added products. This filament is used for multiple purposes, including 3DP, construction materials, and consumer goods. Hence, 3DP has emerged as a sustainable solution for landfill waste minimization and environmental pollution by reusing waste data. One significant advantage of 3DP technology in ecological remediation is its adaptability, which enables the targeted application of various pollutants. Selective removal of heavy metal ions from contaminated water sources and soil matrices (including lead, cadmium, mercury, and arsenic) is possible with 3D-printed adsorbent materials such as ion exchange resins, activated carbon, and functionalized NPs. Using physical adsorption, biodegradation, and oxidation, advanced 3D-printed filtration technology and catalytic reactors can serve for the complete removal of organic pollutants (pesticides, herbicides, industrial chemicals, petroleum hydrocarbons) in both liquid and gaseous phases.

This can facilitate the effective collection and enrichment of microplastics from water samples, wastewater effluents, and aquatic environments by developing specific pore dimensions and surface characteristics of filtration membranes and microplastic traps through 3D printing. These allow the material to be checked and microplastic contamination discarded. In the context of nutrient pollution and eutrophication in receiving waters (which are often a consequence of nutrient runoff from land), nutrient-specific 3D-printed ion exchange resins, zeolites, and absorbent materials are reasonably easily customizable to remove nitrate and/or phosphate from agricultural runoff effectively, stormwater and wastewater effluents (Phang et al. 2023; Ahmed and Jhung 2021; S. Kumar et al. 2022; Sahragard et al. 2024). Utilizing 3DP technology introduces several key advantages over conventional methods for processing polluted debris, such as chemical treatment, physical separation, and biological remediation. 3DP technology allows for the exact shaping and production of on-the-fly remediation solutions and then tailoring those to the contaminants present, the environment, and the defined remediation models. Compared to traditional solutions, this approach significantly improves efficiency and effectiveness. In turn, this minimizes negative consequences for the environment and resource consumption. However, we can practically use size customization by applying 3D-printed remediation systems. Such a quality allows flexibility and adaptability when effectively treating environmental issues of all types. The 3DP technology allows rapid prototyping and iterative design improvement, enabling timely development and evaluation of corrective actions before large-scale application. Thus, it speeds up the time and costs of remediation projects. Integrating 3D printing with digital technologies such as remote sensing, geographic information systems (GIS), and artificial intelligence (AI) can potentially improve environmental remediation monitoring, modeling, and decision-making processes. This allows for using data-driven methodologies and the immediate enhancement of response strategies (Figure 15.4).

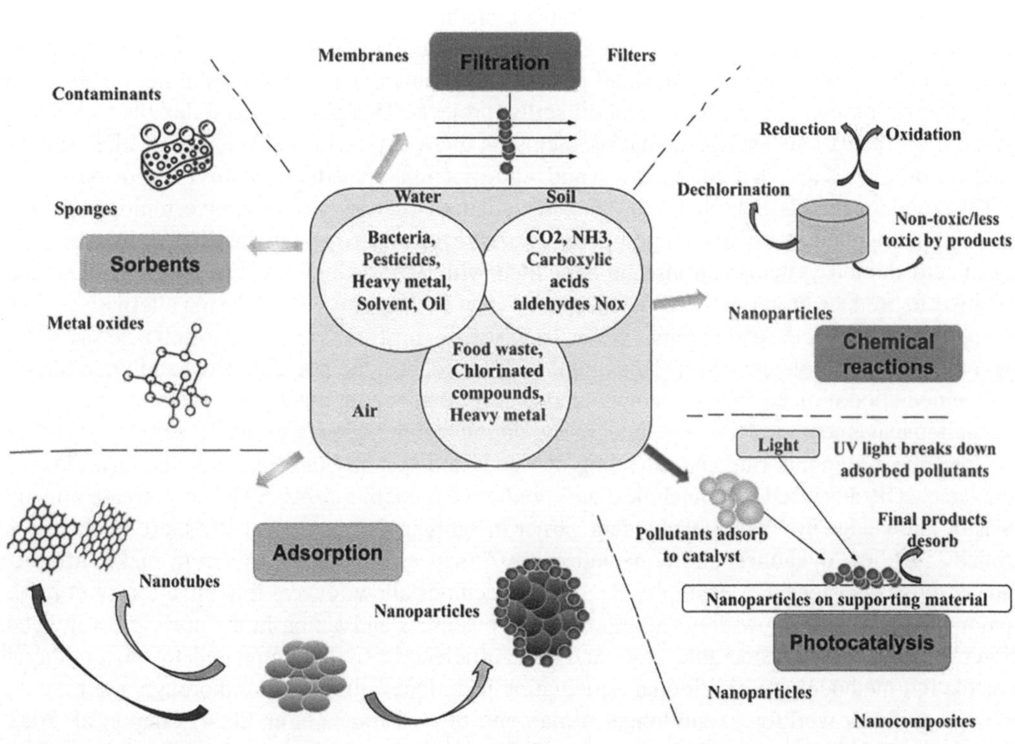

FIGURE 15.4 Applications in Environmental Remediation.

(Source: created by authors)

15.5 EFFICIENCY AND ACCURACY OF 3DP TECHNOLOGY

The importance of 3DP in these cleanup processes lies in its ability to offer practical, precise, and sustainable solutions for environmental remediation. The versatility and adaptability of 3DP technology is one of its main assets, as remediation solutions can be tailored and optimized to suit the contaminant, concentration, and site condition. Traditional remediation methods rely on generic (and often inappropriate) forms of remediation, whereas 3DPs build bespoke remediation systems and materials that can address a diverse range of pollutants, including heavy metals, organics, microplastics, and nutrients.

Conversely, customization is helpful in complex remediation situations with multiple contaminants or site circumstantial conditions requiring enhanced mitigation measures. On a somewhat more technical note, 3DP has significant resource-use efficiency and waste-generation advantages. Conventional soil remediation methods, such as excavation and washing, are resource-intensive and generate large quantities of secondary waste. It will lead to residuals that must be disposed of in landfills or incineration plants, which has an environmental impact and additional costs. Different materials can also be used less through 3DP as remedial materials can be situated directly where required, inducing minimum wastage generation and resource use. 3DP uses additive manufacturing methods to build results layer-by-layer, allowing for precise control of material deposition, which helps reduce material waste. The bio-remediation method of environmental pollution minimizes the eco-footprint of remediation procedures, making it an eco-friendly treatment for controlling ecological contamination. This provides many advantages as remediation approaches since 3DP is characterized by its high scalability and adaptability to site conditions and project specifications. Standard remediation strategies usually necessitate large-scale infrastructures or excavation works that might cause significant upheaval, costs and time (Vargas-Muñoz et al. 2023; Gonçalves et al. 2023; Prakash et al. 2023; Phang et al. 2021). In contrast, 3DP provides an adaptable process tailored to the particular requirements of individual remediation projects. Site-specific 3DP of soil can be completed at urban brown site locations or remote rural landscapes, targeting a wide range of environmental issues, ranging from small point source contamination hotspots to more substantial ecological restoration (i.e. mine site and oil spills) projects. This powerful and flexible technology can be used for various environmental problems. Is there any other applicability in 3DP, such as rapid prototyping & design iteration, to test and refine potential remediation paths for wide-scale use?

This functionality is helpful in complex remediation scenarios that involve multiple factors, including compound interactions, hydrogeology and regulatory requirements. The design and testing of remediation systems can also be expedited with 3DP technology, enabling engineers and scientists to produce prototypes and mock-ups. So, you can explore several design alternatives, look into performance details, and gravitate toward any strong constraints or issues to solve before building a full-fledged implementation. This dramatically mitigates the potential risks and uncertainties such remediation projects face. This may enable several novel approaches to weed detection, control, and management in 3DPs (e.g., integrating digital technologies (e.g. remote sensing, GIS, and AI) for improved monitoring and modeling of weeds and spatially targeted remediation). Through simulated 3DP, advanced data analytics, and predictive modeling tools, STKE can create a media basis of knowledge in which stakeholders garner insights into contaminant transport mechanisms, groundwater flow dynamics, and remediation performance. This enables them to make informed and proactive decisions. For example, data-driven methods allow for real-time monitoring of remediation strategies and allow directed allocation of resources and action in the areas most in need. Moreover, 3DP has considerable cost- and time-efficiency benefits compared to the traditional remediation methodology. Traditional remediation techniques often require thorough site preparation, a significant workforce, and longer turnaround times Farhana et al. 2024; Chen et al. 2022; Zhao et al. 2023; Zhang et al. 2023). Even so, 3DP technology can streamline remediation and keep project costs down. We achieve this by reducing material waste, optimizing resources, and fast-tracking the construction timelines.

Project stakeholders can optimize both time and cost efficiency in support of enhanced remediation results via the 3DP of remediation systems and components on-site or off-site. 3DP in remediation is a great benefit and future-oriented and innovative approach to treating environmental contaminants simply, correctly, and ecologically friendly. 3DP technology offers stakeholders agility, flexibility, and scalability to tailor remediation solutions to specific project requirements, ecological footprints, and resource consumption. Yet, 3DP can also impact the area of environmental remediation, as it can enable rapid prototyping and software interfacing, as well as significantly reduce remediation costs and lead times. On top of all the previous benefits, even more points specific to the place make 3DP ingenious and efficient in cleaning the environment. 3DP technology can also produce customized soil amendments and bioaugmentation products developed specifically for local contaminations and soil properties in soil remediation. Such materials can improve clean-up processes, depending on their capacity to decompose contaminants, form contaminant bounds, or seek land construction and fertilization. 3DP can create in situ treatment systems for groundwater remediation, including permeable reactive barriers (PRBs) and groundwater extraction wells. They have a high degree of complexity and specific geometries and dimensions and are effective for groundwater remediation. They work by blocking, trapping and remediating contaminated groundwater plumes before they can migrate and pollute groundwater.

Through enhancing flow regimes, surface coating, and catalyst composition, 3DP technology can be adopted in air pollution control to design advanced and ultrahigh-performance filtration systems (e.g., air filters and catalytic converters). That leads to better air quality and lower health risks from air pollution. Furthermore, implementing 3DP technology enables the manufacture of new engineered habitat restoration structures and artificial reefs with complex geometric forms and bioinspired designs. One example of this technology is used to improve biodiversity and combat ecological resilience in severed natural systems. These work as a nook for aquatic species, a habitat for coral reef formations, and aid in countering the effects of habitat loss and climate change in coastal and marine ecosystems. The 3DP technology can also advance the development of new materials and devices for environmental monitoring and pollution detection (Siqueira et al. 2023; Phang et al. 2021; Wu et al. 2021; Peng et al. 2022; Dong et al. 2022). They can be customized to have different functionalities and deployed in monitoring networks for real-time information on environmental conditions, contaminant concentrations, and remediation progress. This enables better management and decision support for ecological site remediation projects. Thereby, 3DP technology, advanced materials, digital design capabilities, and interdisciplinary collaboration constitute unprecedented opportunities for innovation and enhancement of environmental remediation. As a result, via 3DP technology, the respective main body (enterprises, governments, etc.) can provide more personalized and greener solutions to tackle complex environmental pollution and ecological deterioration. In turn, this helps to protect and restore natural resources and ecosystem services for the benefit of future generations. Its inherent accuracy and efficiency greatly enhance 3DP technology for environmental remediation. 3DP methods can enable the creation of remediation systems and components with extreme accuracy and reliability, ensuring their efficient operation and optimal contaminant removal or neutralization capability. Precision is key in environmental remediation; differences in design or construction or variableness, even at a small scale, can substantially impact the success of remediation efforts and the protection of human and ecosystem health.

Accurate control of material deposition and layer-by-layer construction is critical in enabling accuracy in 3DP. This feature ensures continuity and standardization when creating remediation materials and infrastructures. This exacting capability enables complex geometries, intricate designs, and tailored features that maximize remediation and address problems site-specific. What's more, 3DP also guarantees proper manufacturing of remediation systems to project specifications, be it physical dimensions (e.g., pore size for permeable reactive barriers) or structural performance (e.g., adsorbent materials with surface characteristics tuned for contaminant adsorption). Additionally, the efficacy of 3DP technology boosts the overall influence of environmental remediation projects by minimizing labor and material costs while also speeding up the process. Typically, manufacturing processes would be complex, labor- or time-intensive, and involve molding, casting, or machining so that 3DP can provide a more likely, meaningful approach to manufacturing remediation systems and their components. By eliminating tooling, fixtures, and manual assembly, 3DP reduces production lead times from months to weeks. It expedites the implementation of corrective actions, thereby enabling relevant stakeholders to promptly and efficiently respond to environmental contamination (Khan et al. 2022; Cui et al. 2023; Jiang et al. 2024; S. Kumar et al. 2022; Mohd Yusoff et al. 2023). In addition, the efficiency of 3DP is not limited to the advanced manufacturing phase; it includes the total life cycle of remediation projects, providing design, prototyping, monitoring, and maintenance services. Using 3DP technology, engineers and scientists can rapidly iterate and optimize remediation designs, while continuously incorporating learnings from field tests and monitoring data to help drive long-term optimization and effectiveness. In addition, 3D-printed components can be designed for ease of assembly, disassembly, and replacement, resulting in maintenance of routine maintenance being less complex and remediation systems being taken out of service for shorter periods. The well-established accuracy and precision of 3DP technologies are key features in the future evolution of remediation action. Fueled by academic and industrial research, the advancement of 3DP is likely to be fundamental to the innovation and optimization of environmental remediation. They design sustainable approaches to address threats of environmental pollution and ecological degradation. Thus, for large-scale contamination cases, the incorporation of 3DP technology into the repair of ecological elements could create a revolutionary window to increase the cleanup of its operation scale. As the 3DP technology advances, the technology evolves, and the cost decreases significantly. Consequently, leveraging this technology during environmental restoration brings unprecedented potential for effectively remediating polluted sites, restoring natural environments, and safeguarding human health. It is just one of many possibilities and a glimmer of hope for a future in which environmental remediation is efficient, effective, and enduring. A deep-dive examination of the potential for significantly expanding the availability of ecological remediation made possible through 3DP technology (Bharat Jivrakh, Kuppieddy et al. 2024 and Guo et al. 2023).

15.5.1 Tackling Sites of Significant Contamination

Pollution is mainly geographical, covering large land, water, and air areas. Remedial measures for conventional media, such as soil excavation and washing, can be expensive, time-consuming, and logistically challenging, mainly when the contaminated area is large, or a wide area needs to be decontaminated. 3DP technologies enable quick and flexible manufacture of remediation systems and components specific to large projects requiring clean-up. In the environmental field, 3DP can be quickly and accurately addressed for the large-scale reclamation of contaminated exogenous sites, including constructing permeable reactive barriers for groundwater remediation, forming sediment traps for river remediation, or establishing habitat structures for ecological restoration.

15.5.2 Enhancing Resource Efficiency

3DP use in environmental cleanup provides more efficient use of resources with less wastage of processed material and mitigated ecological impacts. Traditional remediation methods depend on large quantities of construction materials, energy, and labor that are often overutilized, resulting in resources, energy, and waste. Moreover, the remediation materials (adsorbents, sorbents, and structural components) can also be placed accurately with less waste of materials and better resource utilization with 3DP technology. On-location crafters and building remediation infrastructures utilizing bite-producing methods and administering partners can effectively tackle the inundated results of media cleaning, while staying within the required remediation goals. This is a much more sustainable approach to cleaning up the environment.

15.5.3 Enhancing Efficiency in Project Workflows

In terms of device design, fabrication, and deployment stages, these aspects can lead to streamlined project workflows. Conventional cleanup processes typically include site assessment, engineering design, material procurement, construction, and maintenance. These processes are time-consuming and labor-intensive. The 3DP method intertwines these steps into a more efficient workflow, facilitating faster design iteration, prototyping, and production of numerous systems and components for remediation. Pre-digital-era efforts in design in remediation included physical 3D models of the environment and illustrative drawings; now, with both conceptual design tools and more advanced design strategies, stakeholders have at their disposal digital design tools such as CAD software and GIS to help develop design models, model environmental scenarios, and optimize remediation strategies before undertaking full-scale implementation. All of this translates into reduced project cycle time and cost. This is a testament to the fact that one of the projects the invested technology provides is monitoring and evaluation processes to clean the environment and ensure the sand supports remediation interventions. Standard cleanup methods for direct measurement of contaminants rely on manual sampling, lab-based analysis, and field monitoring to assess contamination levels, remediation capabilities, and overall effectiveness (Mohd Yusoff et al. 2023; Sahragard et al. 2024; Farhana et al. 2024; Ogunsola et al. 2024; Li et al. 2024). Although such methods may show a more comprehensive view of the environmental conditions, they are laborious, costly, and have issues with space coverage increase. 3DP technology allows you to integrate data collection technologies, remote monitoring systems, and autonomous data collection platforms into your remediation systems. This connection enables real-time monitoring of contaminant levels, groundwater flow, and treatment efficacy. Real-time data collection and analytics will assist stakeholders in making informed decisions and fine-tuning remediation plans and resources to meet the objectives of the remedial program.

15.5.4 Promoting Interdisciplinary Collaboration

The growth in environmental remediation via 3DP technology is not a solo project but a team effort. It requires participation from different sectors (scientists, engineers, policymakers, stakeholders). 3DP also solves the collaboration challenge among disciplines; it provides a standard interface to create, fabricate, and implement remediation solutions. Using 3DP technology in this context enables stakeholders to work effectively across various disciplines and sectors. This technology allows them to create custom-made remediation systems, work on research and development initiatives, and exchange data and best practices. The advances in 3DP technology not only accelerate the rate of innovation and transfer of technology but foster mechanisms that ensure the global deployment of sustainable solutions for environmental remediation by promoting collaboration and knowledge exchange. The part you play in this process is valuable.

15.5.5 Enhancing Community Engagement

Community leaders stimulate social innovation, identify emerging issues, address local concerns and mindsets, find and establish local partnerships that mobilize activist networks in the community. 3DP technology supports civic engagement by enabling stakeholders to participate in every aspect of the clean-up process design, planning, and implementation. There is the opportunity to build trust as communities become partners in shaping remediation solutions that are responsive to community needs and priorities in transparent and accountable ways. This can be achieved through participatory design workshops, public outreach campaigns, and citizen science initiatives. This engagement facilitates the development of a sense of ownership or guardianship of environmental resources in stakeholders, leading to more sustainable outcomes and more capable circumstances.

15.5.6 Geographical Scaling

The geographical limitation of resource source and demand resource is also relaxed, as the enterprise can expand from a local neural network to a globular neural network, enter the international market, and solve global environmental damage problems, such as mounting marine silt pollution problems. 3DP technology can help address complex and interconnected ecological issues that span local to global scales in a highly effective and innovative way. 3DP addresses climate change, one of the most significant challenges the world faces today, through scalable solutions that advocate environmental sustainability through deploying 3D printed reef structures to revive coral ecosystems, creating barrier islands to act as buffers for coastlines from wave action, and 3D printing modular habitats to suit vulnerable species. Using the systematic knowledge, resources, and cleverness of men and women, communities, and nations, 3DP offers collaborative platforms for achieving a sustainable environmental impact and building a stronger world for future generations. The potential for scaling up 3DP technology for application in environmental remediation is vast and transformational, offering unmatched opportunities to address large-scale contamination problems quickly, effectively, and sustainably. By enhancing resource allocation, minimizing layers of project delivery, establishing more effective monitoring and evaluation, enabling cross-sectoral solution collaboration, fostering pro-active community engagement, and being infinitely scalable across geographic barriers, 3DP technology offers solutions to complex socio-environmental challenges, with the opportunity for widespread impact at the global level. The continuing evolution of 3DP technology suggests a future that expands the range of environmental remediation technologies. This change can help support a more sustainable, resilient future for the Earth and its life (Figure 15.5).

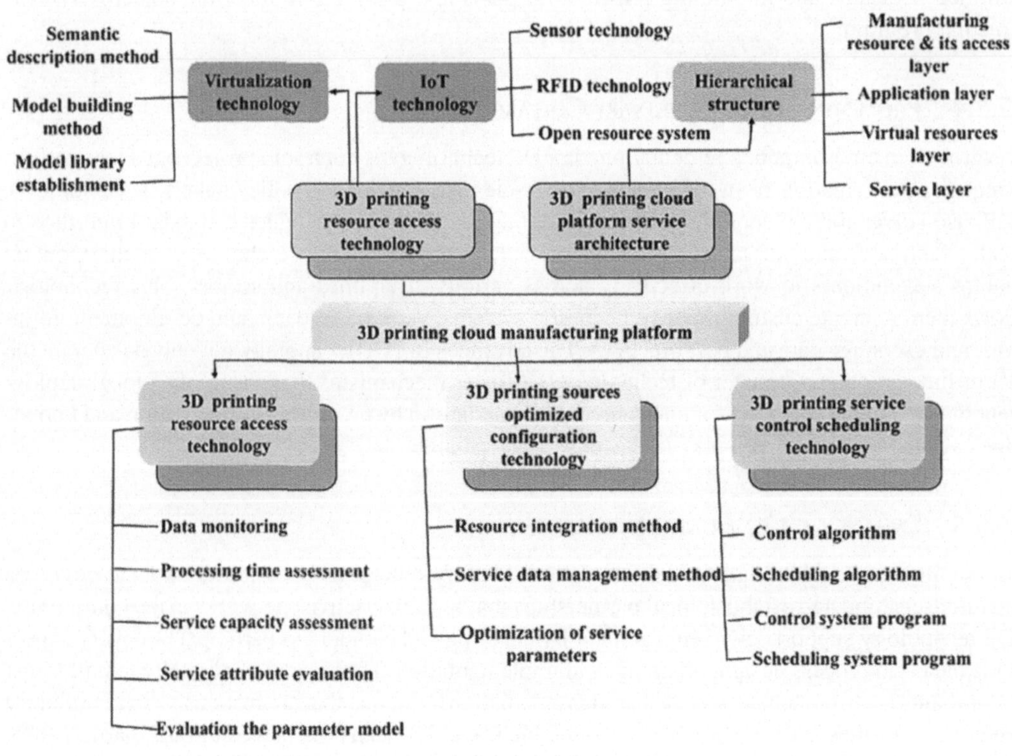

FIGURE 15.5 Efficiency and Accuracy of 3DP Technology.

(Source: created by authors)

15.6 CHALLENGES IN DEPLOYING 3D PRINTED MATERIALS

Adopting 3D-printed materials across various applications involves technological, environmental, and economic issues. There are great benefits available, both in terms of the innovation that 3DP has the potential to bring and the efficiency it can lend to productivity. These benefits can only be fully exploited where these questions are dealt with thoroughly.

15.6.1 Characteristics and Efficiency of Materials

It is a critical challenge to ensure the requisite properties and performance characteristics for the intended application of 3D printed materials. People love ink because it provides the necessary efficacy and durability for this highly eco-friendly craft product, composing an appropriate printing format with enough toughness to endorse innovative meter manufacturers. Additive manufacturing can easily handle complex combination parts. 3DP has high-performance requirements regarding its materials, strength, toughness, heat resistance, and chemical resistance. This disadvantage can significantly limit the technology's viability in more complex applications, such as aerospace. The materials must withstand harsh operational conditions that impose high-performance levels and highly stringent equipment specifications. A significant difficulty for 3DP is being consistently repeatable in processing and quality. Therefore, quality inspection is a must. Different printing parameters, raw materials, and environmental conditions can go all right, incorporating diversity in product performance. However, 3DP technology does not abandon the control practices or system software and monitoring used today by mature technologies, yielding a production process that can be reproduced or repeated. Except for all errors detected during processes, target variations should be remedied, whether big or small. Standardization and certification are essential for the chemistry industry's safety or regulated applications, which design and produce the raw powder materials that go into 3D-printed objects.

15.6.2 Flexibility and Effectiveness in Production

Despite its many design benefits and freedom of choice made possible by additive manufacturing, this may not always be the most efficient or cost-effective way for high-volume production. Meanwhile, addressing scalability problems means that new ways of culturing are needed in printing technology, automation, material handling and post-processing methods for fractures that can process increased throughput and reduce cycle time considerably while optimizing production workflows, adhering throughout all of these to standards essential for quality and up it quite neatly. Incorporating 3DP technology into conventional manufacturing poses technical obstacles to compatibility, interoperability, and process synchronization. Integrating 3D-printed parts with traditional manufacturing components is necessary in many cases. Implementing massive precision in this sort of manufacturing is a balance of coordinates between 3D and turned parts. Coordinating and aligning manufacturing processes, tooling, and quality assurance procedures are required to ensure that materials match between 3DP and turned parts accurately (Gonçalves et al. 2024; Dong et al. 2022; Peng et al. 2022).

15.6.3 Environmental Considerations and Impact

The environmental consequences of 3DP are dependant upon selecting materials, feedstock, and post-processing techniques that have been employed during the fabrication procedure. Certain 3DP materials (like biodegradable polymers and recycled plastics, etc.) are environmentally advantageous by minimizing waste and resource usage. However, other materials may be ecologically hazardous due to their capacity to release toxins, high consumption of energy, and improper disposal. To minimize the environmental impact of 3DP and promote circularity in material usage, it is crucial to choose environmentally sustainable materials and implement eco-friendly printing practices. The 3DP process has the potential to produce waste in the form of surplus material, unsuccessful prints, and unused powder or resin. Managing the waste produced by 3DP poses difficulties in gathering, segregating, recycling, and disposing of safely. While certain materials possess the potential for recycling, others may necessitate specific handling and treatment procedures owing to their composition or contamination.

New progress in 3DP technology has significantly influenced manufacturing, art, and several other industries. However, recent developments have expanded the use of 3DP and opened unique opportunities and prospects. These trends encompass advances in material science and printing techniques. That includes progress with the types of 3DP materials (metals, polymers, ceramics, and composites) and their properties in recent years. Material science is becoming richer, obtaining specific materials with optimal mechanical properties, thermal stability, and chemical resistance. Recent developments have enabled the production of functional prototypes, end-use parts, and complex geometries with high performance and reliability. Additive manufacturing technologies have become more rapid, precise, scalable and productive, with faster, higher-resolution and more productive printing. The ability of 3DP has grown immensely with the repeatability (within layers), creating volumes and complexities with different printing methods (VAT polymerization, Powder bed fusion, directed energy deposition) and, therefore, responding to various materials and potential applications on a vast scale. This advancement has enabled the design of microscale components and large-scale structures.

Recent developments in multi-material and multifunctional printing have enabled the simultaneous deposition of several materials, gradients, and functionalities on a single printed object, leading to unprecedented opportunities in designing and fabricating complex, heterogeneous architectures with tailored properties and functionalities. Typical such structures include intelligent sensors, actuators, biomedical implants, etc. The embedding of digital design and simulation tools at 3DP workflows has led to improvements in design optimization, performed validation, and behavior prediction before the print stage. Advanced software platforms, such as computer-aided design (CAD), finite element analysis (FEA), and computational fluid dynamics (CFD) can be used by engineers and designers to iterate quickly, explore design spaces, and optimize geometries for additive manufacturing, thus, allowing them to be designed for specific requirements and requirements, which is one of the most prominent advantages of 3D-printed materials. Manufacturers can design the material's properties, strength, flexibility, conductivity, etc., by altering printing parameters, material formulations and processing conditions. This allows the production of parts with better material properties. Recently, advances in 3DP technology have allowed the production of new functional, graded materials (FGMs) with spatially varying composition, microstructure and properties. FGMs provide substantial benefits in terms of structural stability, thermal regulations, and wear properties, which enable the production of engineered mechanical and thermal properties that match that of lighting FTC-rectified terminational components.

Biomimetic techniques consist of materials and structures whose properties and functionalities mimic those of biological systems, with researchers investigating using such methods for 3DP (Mallakpour, 2023; Prakash et al. 2023). For example, biomimetic materials, including hydrogels, scaffolds, and tissue-engineered constructs, can be significant for regenerative medicine, drug delivery, and biomedical engineering. This is difficult but important because it is a domain of biocompatibility and bioactivity. The field of material science and nanotechnology has advanced so that they can (1) sense, (2) respond, and (3) react to external stimuli (e.g., temperature, light, heat, and pH). Examples of stimuli-responsive or programmable materials are widely used in wearable electronics, soft robotics, adaptive structures, and environmental sensors. Combined with digital manufacturing technology characterized by AI, IoT, and cloud computing, 3DP propels the transfer to Industry 4.0. Smart factories can leverage platforms connecting many 3D printers or automated robotic systems with data analytics tools that allow real-time monitoring, predictive maintenance, and adaptive manufacturing, enabling novel ways to design, produce, and distribute products. 3DP technology provides for the mass customization and personalization of any product, allowing consumers to specify certain facets of the design, the materials used, and the functionalities that suit their needs and tastes.

Applications of mass customization range from custom-fit prosthetics and orthotics to personalized consumer goods and medical devices and can optimize the user experience, enhance product performance and reduce waste in manufacturing. The increasing focus on environmental sustainability and resource depletion has increasingly highlighted 3DP as a sector that could adopt both sustainable manufacturing and circular economy values. 3DP may play a role in minimizing manufacturing processes' environmental footprint and contribute to a more sustainable approach to production by replacing EUROs with recycled materials, bio-based polymers, and additive manufacturing techniques that minimize wasting raw materials and energy consumption. The 3DP technology is well-designed for bioprinting living tissues and organs and its applications in regenerative medicine, organ transplantation and tissue engineering. Improvements in bioprinting techniques, bio-inks and cell culture methods have overcome the difficulty of generating 3D structures of complex tissues and organs with vascular networks, cellular architecture, and biological function. These advancements have opened up possibilities for personalized medicine and tissue replacement therapies. Besides space exploration, where traditional methods are not feasible or even possible, 3DP technology is critical to off-world mass production.

Other uses of additive manufacturing technologies, such as Fused Deposition Modeling (FDM) and Selective Laser Sintering (SLS), manufacture spare parts, tools, and habitats on-site in space. This reduces dependence on terrestrial supply chains, enabling long-duration missions in space and constructing these outer worlds. It has changed the manufacturing landscape forever with all new capabilities, materials, and applications in numerous industries and sectors, due to recent advancements in the 3DP field. Exploring materials science and printing technology, 3D-printed materials are a progressive arena that provides new potential avenues for customization, sustainability, and innovation. 3DP is a core component of the future of manufacturing, healthcare, and society, building a more cohesive, sustainable, and technologically advanced world. It remains the focus of an inquiry into establishing new trends and potential opportunities (Zhao et al. 2023; Khan et al. 2022; McDonagh et al. 2023; Mohd Yusoff et al. 2023; Sahragard et al. 2024).

15.7 RECENT PROGRESS AND INNOVATIONS

3DP has seen significant advancements in recent years, with significant implications in manufacturing, design, and others. Within the last decade, exciting developments have emerged, enhancing the capabilities of 3DP and expanding its potential in interesting areas with great promise. Also, the advancements are related to material science and new printing technologies. Recent advances in 3DP materials that are relevant to metals, polymers, ceramics, and composites have been made, including well-known methodologies. Advancements in material science have led to the accelerated development of new materials with superior mechanical, thermal, and chemical properties. This, in turn, has driven the production of functional prototypes, end-use components, and complex geometries at high precision and reliability. Additive manufacturing technologies have increased velocity, the accuracy of parts, and expandability, which allow printing speeds to decrease and the resolution and production by volume to increase. Several relatively new technologies can now be used for production processes that may offer improved quality, utilize a wider range of materials and allow for more applications, such as vat polymerization, powder bed fusion, and directed energy deposition. The technologies have enabled the production of microscale components and construction structures at a much larger scale.

However, recent developments in multi-material and multifunctional printing have led to the simultaneous deposition of different materials, gradients, and functions onto the same printed objects. This opens new frontiers for designing and fabricating complex, heterogeneous structures with tailored properties and functionalities. These potentially include smart sensors, actuators, and biomedical implants. Integrating digital design and simulated tools allows performance validation and behavior prediction before printing, taking the 3DP processes to a higher level of design efficiency. A unique benefit of 3D-printed materials is their ability to be tuned and/or customized for specific applications and needs. Moreover, by varying the printing parameters, material compositions, and related processing factors, the characteristics of the solid plastic, such as mechanical strength, flexibility, and conductivity, can be further tailored. This enables the production of components with enhanced performance characteristics. Development in 3DP technology has led to the creation of FGMs with spatial gradients in composition, microstructure, and properties (Siddiqui et al. 2023; Chang et al. 2024; da Silva et al. 2023; Siqueira et al. 2023). The advantages of FGMs include structural strength and stability, management, and wear prevention. As a result, FGMs can form lightweight and durable components from a particular mechanical and thermal property.

Researchers are currently exploring biomimetic or 3DP methods developed from the properties and functions of living systems. Hydrogels, scaffolds, and tissue-engineered constructs hold immense potential in regenerative medicine, drug delivery, and biomedical engineering. Applications are essential due to the requirements for biocompatibility and bioactivity. Such progress in materials science and nanotechnology is significant. It has helped develop intelligent and responsive materials that sense, respond, and react to external stimuli (i.e., environmental factors such as temperature, light, and pH). Stimuli-responsive (or programmable) materials find applications across fields like wearable electronics, soft robotics, adaptive structures, and environmental sensors. The combination of 3DP with digital manufacturing technologies, including artificial intelligence (AI), Internet of Things (IoT) and cloud computing, is paving the way for the shift to Industry 4.0. Through smart factories that use interconnected 3D printers, automated robotics and data analytics platforms, these processes enable real-time monitoring, predictive maintenance and adaptive manufacturing, transforming how products are designed, produced, and distributed. 3D printing (3DP) technology allows for mass customization and personalization of various products, allowing consumers to customise design features, materials, and functionalities according to their preferences and requirements.

Mass customization has the potential to enhance user experience, optimize product functionality, and reduce waste in manufacturing. Still, its applications can differ widely, from custom-fit prosthetics and orthotics to tailored consumer products and medical devices. Stripped of environmental hysteria and based on finite resource consumption rationale, 3DP is marketed as a sustainable manufacturing path to a circular economy. By utilizing recycled materials, bio-based polymers and additive manufacturing techniques that minimize waste and lower energy footprint, 3DP may address the environmental impacts of manufacturing and contribute to a more sustainable mode of production. The applications of 3DP technology for bioprinting, which has potential future applications in regenerative medicine, tissue engineering, and even organ transplantation, are prerequisites for the 3DP of living tissues and organs. Recent advances in bioprinting technology and the composition of bio-ink and cell culture methods have led to the in vitro engineering of more complex and functional tissues and organs, including vascularization, tissue cellular organization, and biological functionality. These advances in organoid technology hold great promise for personalized medicine and tissue replacement therapies. In the field, 3DP technology has become critical, especially in space exploration and beyond world manufacturing; thereby, traditional manufacturing methods are unfeasible or unreachable. Space travelers needed to produce spare parts, tools, and habitats using additive manufacturing technologies, such as fused deposition modeling (FDM) and selective laser sintering (SLS). Reducing dependence on supply chains from Earth allows for long-duration space missions and off-planet colonies. Over the previous years, 3D printing (3DP) processes have experienced impressive evolutions and have opened novel possibilities and applications to nearly all industrial and non-industrial sectors. We are constantly exploring how these emerging changes and potential opportunities can impact our future. Nevertheless, a transformative impact of 3DP on the future of manufacturing, healthcare, and society, with stronger interconnectivity, sustainability, and technological sophistication is envisioned (Iqbal, Bilal, and Zdarta 2022; Dong et al. 2022).

15.8 ENVIRONMENTAL IMPACT AND SUSTAINABILITY ASSESSMENT

Lastly, when using 3D printed materials, legwork must be done to ensure their sustainability and eco-impact. This becomes even more critical for applications that target ecological remediation, which strives to minimize damage to ecosystems and human health. Still, unlike more established remediation materials, knowledge of the environmental ramifications of 3D printed materials across their lifecycle (from production to end-of-life) is crucial to quantifying the responsible applications of 3DP technology and informing potential consequences for sustainability practices. Here's a breakdown of those different components. LCA is a comprehensive assessment tool for determining the environmental impact of ecology-hazardous products or processes over their lifetime, from the extraction of virgin materials to end-of-life disposal. The 3D printed materials LCA method will tell where these environmental impacts emerge at the different stages: material manufacturing, printing, performance and end of life. LCA accounts for energy consumption, resource depletion, greenhouse gas emissions, air and water pollution, and waste production. Several LCAs have been performed on various 3D printed materials, such as polymers, metals, and ceramics, to compare their environmental impacts to traditional manufacturing techniques. The relationship of 3D printed ecological function with material properties (e.g., manufacturing technology, printing process parameters, and applications) is heterogeneous. Although 3DP has numerous advantages related to design flexibility, material utilization, and waste minimization, fundamental constraints exist for this manufacturing technology concerning energy usage, emissions, and resource availability compared to conventional, well-established manufacturing technologies.

Determining the environmental impact of 3D-printed materials for remediation requires some baseline comparisons to commonly available remediation methods. The performance characteristics and ecological profiles of conventional remediation materials like activated carbon, zeolites, and clay minerals have been well-characterized and validated by decades of academic research and field experience. Several factors should be considered when approaching the environmental impact of 3D printed materials versus conventional remediation materials, including acquiring materials, production methods, transportation, implementation, and junk (da Silva et al. 2023; Siqueira et al. 2023). Recent works have examined the environmental impacts of 3D printed materials and traditional remediation materials for various application domains, such as water treatment, soil stabilization, and air pollution control. It was a theory of 3D printed materials that provided unprecedented design freedom and the ability to design, customize, and fabricate anywhere. However, the implications of such materials' energy, emissions, and resource use are claimed to be similar or even higher than those associated with conventional materials. However, the environmental footprint from 3D-printed materials can be minimized through thoughtful material selection, printing processes, and wise end-of-life practices.

In environmental remediation or other applications, the users would be responsible for conducting life cycle assessments on the environmental impacts and sustainability of 3D printed materials, developing and implementing sustainable strategies. When comparing the life cycle environmental performance of 3D printed materials to traditional remediation materials, stakeholders can make pollution reduction and sustainability enhancing decisions - material selection, process optimization and technology adoption. Because of this, they must use eco-friendly materials and have many more ways to think about their decision, including only using sustainable suppliers, recyclable, biodegradable, or non-toxic materials! This integration can mitigate the overall environmental burden of 3DP processes by combining high-performance materials with lower environmental impacts and can lead to sustainable developments of environmental remediation practices. Energy efficiency, material usage, and waste reduction are prerequisites to the sustainability of 3DP processes and minimizing their impact on the environment and resource use. Technology developments related to printers, such as additive manufacturing and robotics, will supply better era efficiency and reduced cycle time and role opportunities. As such, this advance can push changes that integrate more sustainable practices into environmental remediation. To minimize waste and achieve circularity, considerations for end-of-life materials management and recycling, reuse, and upcycling practices of

3D printed materials should also be included. If 3D printing technologies must be used in disaster circumstances, from a systems approach to disaster management, designing the material and product with disassembly, recyclability, and bioactivity considerations should be included to conquer the 3DP environmental challenges and strive for sustainable material technologies that incorporate closed-loop recycling and resource recovery. Environmental consideration should be included in decision-making when developing policies, regulations, and standards for the appropriate application of 3DP technology. Material applications not only pose environmental risks in addition to micro but also guide the process and end-of-life management of the material application.

15.9 CONCLUDING REMARKS

Using 3D printed materials for environmental remediation is a promising approach to addressing ecological issues. Throughout this study, we have considered the adaptability, precision, and innovative capability of 3DP to conquer contamination challenges in broad application fields, including air, water, and soil pollution. This demonstrates the burgeoning potential of 3DP in replacing traditional clean-up technologies to aid degradation toward target pollutants and produce intelligent, lightweight composites. However, as a new technology, there are challenges for 3DP in areas such as technology application methods, environmental impact and economic sustainability. Overcoming such challenges and maximizing the potential of 3DP for environmental remediation require breakthroughs in materials science, new printing technologies, and digital design tools. In addition, the introduction of 3D-printed materials is to put greater emphasis on sustainability. Responsible decision-making and lifelong environmental caretaking are the key takeaways from life cycle evaluations, comparison with traditional remediation materials, and suggestions for continuing practice. By selecting green materials with the least economic impact possible, optimising the 3D printing process and planning for a materials lifecycle's end, we can reduce environmental burdens and promote recycling business ideas and developments. However, the lack of a single export market or ceiling price means the success of employing 3D printed materials in environmental remediation is huge in sustainability. Through collaboration at the international level for a green skyscraper future, we seek from within our respective domains to open up new frontiers and thoroughly comprehensively promote the prosperity of development. Technologies such as 3D printing may lead to constructive environmental results and enable the environment to rejuvenate and be maintained for the next generations by stressing harmony, encouraging creation, and having an ambition toward ecological awareness.

REFERENCES

Ahmed, Imteaz, and Sung Hwa Jhung. 2021. "Covalent Organic Framework-Based Materials: Synthesis, Modification, and Application in Environmental Remediation." *Coordination Chemistry Reviews* 441: 213989. https://doi.org/10.1016/j.ccr.2021.213989

Baigorria, E, S Souza dos Santos, M R de Moura, and L F Fraceto. 2023. "Nanocomposite Hydrogels 3D Printed for Application in Water Remediation." *Materials Today Chemistry* 30: 101559. https://doi.org/10.1016/j.mtchem.2023.101559

Bharat Jivrakh, Kedar, Suresh Kuppieddy, Somayya E Taher, Kyriaki Polychronopoulou, Rashid Abu Al-Rub, Nahla Alamoodi, and Georgios N Karanikolos. 2024. "Zeolite-Coated 3D-Printed Gyroid Scaffolds for Carbon Dioxide Adsorption." *Separation and Purification Technology*, 127523. https://doi.org/10.1016/j.seppur.2024.127523

Brown, Hilary M, Trevor J McDaniel, Christopher P West, Ebenezer H Bondzie, Matthew R Aldeman, Brian T Molnar, Christopher C Mulligan, and Patrick W Fedick. 2022. "Characterization and Optimization of a Rapid, Automated 3D-Printed Cone Spray Ionization-Mass Spectrometry (3D-PCSI-MS) Methodology." *International Journal of Mass Spectrometry* 474: 116781. https://doi.org/10.1016/j.ijms.2021.116781

Chang, Guorui, Sirui Li, Yunqi Wang, Qingxia Ran, Qing Tan, Shuli Gou, Haiying Du, and Shuxia Xu. 2024. "Cu-C3N4 Nanoenzyme-Based Freezing-Dried Bioactive Capsule Integrated with 3D-Printed Smartphone Platform for Visual Detection of Organophosphorus Pesticides Paraoxon in Scallion." *Sensors and Actuators B: Chemical* 398: 134584. https://doi.org/10.1016/j.snb.2023.134584

Chen, Zhijie, Wei Wei, Hong Chen, and Bing-Jie Ni. 2022. "Recent Advances in Waste-Derived Functional Materials for Wastewater Remediation." *Eco-Environment & Health* 1 (2): 86–104. https://doi.org/10.1016/j.eehl.2022.05.001

Cui, Yuanlong, Jie Zhu, Hui Tong, and Ran Zou. 2023. "Advanced Perspectives on MXene Composite Nanomaterials: Types Synthetic Methods, Thermal Energy Utilization and 3D-Printed Techniques." *IScience* 26 (1): 105824. https://doi.org/10.1016/j.isci.2022.105824

Devi, M Sindhu, S Srinivasan, and A Muthuvel. 2023. "Selenium Nanomaterial Is a Promising Nanotechnology for Biomedical and Environmental Remediation: A Detailed Review." *Biocatalysis and Agricultural Biotechnology* 51: 102766. https://doi.org/10.1016/j.bcab.2023.102766

Dong, Xu, Yiren Lu, Xianhua Liu, Lihong Zhang, and Yindong Tong. 2022. "Nanostructured Tungsten Oxide as Photochromic Material for Smart Devices, Energy Conversion, and Environmental Remediation." *Journal of Photochemistry and Photobiology C: Photochemistry Reviews* 53: 100555. https://doi.org/10.1016/j.jphotochemrev.2022.100555

Farhana, Kaniz, Kumaran Kadirgama, Abu Shadate Faisal Mahamude, and Rajan Jose. 2024. "Review of MXenes as a Component in Smart Textiles and an Adsorbent for Textile Wastewater Remediation." *Chinese Chemical Letters* 35 (2): 108533. https://doi.org/10.1016/j.cclet.2023.108533

Finny, Abraham Samuel, Nadia Cheng, Oluwatosin Popoola, and Silvana Andreescu. 2022. "3D Printable Polyethyleneimine Based Hydrogel Adsorbents for Heavy Metal Ions Removal." *Environmental Science Advances* 1 (4): 443–455. https://doi.org/10.1039/d2va00064d

Gonçalves, Nuno P F, Susana M Olhero, João A Labrincha, and Rui M Novais. 2023. "3D-Printed Red Mud/Metakaolin-Based Geopolymers as Water Pollutant Sorbents of Methylene Blue." *Journal of Cleaner Production* 383: 135315. https://doi.org/10.1016/j.jclepro.2022.135315

Gonçalves, Nuno P F, Eduardo Ferreira da Silva, Luís A C Tarelho, João A Labrincha, and Rui M Novais. 2024. "Simultaneous Removal of Multiple Metal(Loid)s and Neutralization of Acid Mine Drainage Using 3D-Printed Bauxite-Containing Geopolymers." *Journal of Hazardous Materials* 462: 132718. https://doi.org/10.1016/j.jhazmat.2023.132718

Guo, Sheng, Meng Chen, Liming You, Yu Wei, Chao Cai, Qingsong Wei, Huali Zhang, and Kun Zhou. 2023. "3D Printed Hierarchically Porous Zero-Valent Copper for Efficient Pollutant Degradation through Peroxymonosulfate Activation." *Separation and Purification Technology* 305: 122437. https://doi.org/10.1016/j.seppur.2022.122437

Iqbal, Hafiz M N, Muhammad Bilal, and Jakub Zdarta. 2022. "Strategies for Environmental Contaminants Monitoring and Remediation." *Environmental Technology & Innovation* 28: 102561. https://doi.org/10.1016/j.eti.2022.102561

Jatoi, Abdul Sattar, Nabisab Mujawar Mubarak, Zubair Hashmi, Nadeem Hussain Solangi, Rama Rao Karri, Yie Hua Tan, Shaukat Ali Mazari, Janardhan Reddy Koduru, and Akram Alfantazi. 2023. "New Insights into MXene Applications for Sustainable Environmental Remediation." *Chemosphere* 313: 137497. https://doi.org/10.1016/j.chemosphere.2022.137497

Jiang, Ru, Hua-Yue Zhu, Xiao Zang, Yong-Qian Fu, Sheng-Tao Jiang, Jian-Bing Li, and Qi Wang. 2024. "A Review on Chitosan/Metal Oxide Nanocomposites for Applications in Environmental Remediation." *International Journal of Biological Macromolecules* 254: 127887. https://doi.org/10.1016/j.ijbiomac.2023.127887

Khan, Mujeeb, Mohamed E Assal, Muhammad Nawaz Tahir, Majad Khan, Muhammad Ashraf, Mohammad Rafe Hatshan, Merajuddin Khan, Ravi Varala, Nujud Mohammed Badawi, and Syed Farooq Adil. 2022. "Graphene/Inorganic Nanocomposites: Evolving Photocatalysts for Solar Energy Conversion for Environmental Remediation." *Journal of Saudi Chemical Society* 26 (6): 101544. https://doi.org/10.1016/j.jscs.2022.101544

Kim, Yoogyeong, Sungsil Park, Sejun Park, Kosisochi Ibebunjo, Eveliina Repo, Tae-Hyun Kim, and Yuhoon Hwang. 2024. "Preparation of Three-Dimensional Layered Double Hydroxide Filters by 3D Printing: Proof-of-Concept." *Applied Clay Science* 250: 107283. https://doi.org/10.1016/j.clay.2024.107283

Kumar, Ashwani, Sunil Dutt, Raj Kumar, and Sonika. 2023. "Chapter 9 - Use of 3D Printed Techniques for Organic Pollutants Removal." In Additive Manufacturing Materials and Technologies, edited by Jitendra Kumar Pandey, Suvendu Manna, Ravi Kumar Patel, and Ma *3D Printing Technology for Water Treatment Applications Qian*, 193–211. Elsevier. https://doi.org/10.1016/B978-0-323-99861-1.00006-0

Kumar, Sandeep, Parminder Kaur, Ravinderdeep Singh Brar, and J Nagendra Babu. 2022. "Nanoscale Zerovalent Copper (NZVC) Catalyzed Environmental Remediation of Organic and Inorganic Contaminants: A Review." *Heliyon* 8 (8): e10140. https://doi.org/10.1016/j.heliyon.2022.e10140

Li, Jiaming, Mengli Li, J Justin Koh, John Wang, and Zhiyang Lyu. 2024. "3D-Printed Biomimetic Structures for Energy and Environmental Applications." *DeCarbon* 3: 100026. https://doi.org/10.1016/j.decarb.2023.100026

Mallakpour, Shadpour, Vajiheh Behranvand, and Elham Azadi. 2023. "10 - Environmental Remediation Utilization of Polyurethanes/Carbon Nanomaterial Nanocomposite Sponges." In edited by Chaudhery Mustansar Hussain and Nashaat N, *Nanoremediation Nassar*, 283–310. Elsevier. https://doi.org/10.1016/B978-0-12-823874-5.00004-8

McDonagh, Pádraig, Matthew McCole, Nathan Skillen, Peter K J Robertson, and Denis McCrudden. 2023. "3D Printed Carbon Black-Polylactic Acid/Copper (CB-PLA/Cu) Sensor for Improved Sustainability in Monitoring Photocatalytic Hydrogen Peroxide Evolution." *Journal of Environmental Chemical Engineering* 11 (6): 111123. https://doi.org/10.1016/j.jece.2023.111123

Mohd Yusoff, Nurul Husna, Chien Hwa Chong, Yoke Kin Wan, Kean How Cheah, and Voon-Loong Wong. 2023. "Optimization Strategies and Emerging Application of Functionalized 3D-Printed Materials in Water Treatment: A Review." *Journal of Water Process Engineering* 51: 103410. https://doi.org/10.1016/j.jwpe.2022.103410

Montaner, Mireia Benito, and Stephen T Hilton. 2024. "Recent Advances in 3D Printing for Continuous Flow Chemistry." *Current Opinion in Green and Sustainable Chemistry*, 100923. https://doi.org/10.1016/j.cogsc.2024.100923

Nawaz, Shahid, Andleeb Tabassum, Sara Muslim, Tayyaba Nasreen, Ausra Baradoke, Tak H Kim, Grzegorz Boczkaj, Teofil Jesionowski, and Muhammad Bilal. 2023. "Effective Assessment of Biopolymer-Based Multifunctional Sorbents for the Remediation of Environmentally Hazardous Contaminants from Aqueous Solutions." *Chemosphere* 329: 138552. https://doi.org/10.1016/j.chemosphere.2023.138552

Ogunsola, Samuel Sunday, Mayowa Ezekiel Oladipo, Peter Olusakin Oladoye, and Mohammed Kadhom. 2024. "Carbon Nanotubes for Sustainable Environmental Remediation: A Critical and Comprehensive Review." *Nano-Structures & Nano-Objects* 37: 101099. https://doi.org/10.1016/j.nanoso.2024.101099

Peng, Haihao, Weiping Xiong, Zhaohui Yang, Zhengyong Xu, Jiao Cao, Meiying Jia, and Yinping Xiang. 2022. "Advanced MOFs@aerogel Composites: Construction and Application towards Environmental Remediation." *Journal of Hazardous Materials* 432: 128684. https://doi.org/10.1016/j.jhazmat.2022.128684

Phang, Sue Jiun, Voon-Loong Wong, Kean How Cheah, and Lling-Lling Tan. 2021. "3D-Printed Photoreactor with Robust g-C3N4 Homojunction Based Thermoset Coating as a New and Sustainable Approach for Photocatalytic Wastewater Treatment." *Journal of Environmental Chemical Engineering* 9 (6): 106437. https://doi.org/10.1016/j.jece.2021.106437

Phang, Sue Jiun, Voon-Loong Wong, Kean How Cheah, and Lling-Lling Tan. 2023. "Robust Graphitic Carbon Nitrite-Based Thermoset Coating for Synergistic Adsorption–Photocatalysis Process in 3D-Printed Photoreactors." *Energy Reports* 9: 30–37. https://doi.org/10.1016/j.egyr.2022.11.159

Prakash, V R Arun, Mostefa Bourchak, Hassan Alshahrani, and Khalid A Juhany. 2023. "Development of Cashew Nut Shell Lignin-Acrylonitrile Butadiene Styrene 3D Printed Core and Industrial Hemp/Aluminized Glass Fiber Epoxy Biocomposite for Morphing Wing and Unmanned Aerial Vehicle Applications." *International Journal of Biological Macromolecules* 253: 127068. https://doi.org/10.1016/j.ijbiomac.2023.127068

Ranjan, Rahul, Rohit Rai, Kaustubh Naik, Avanish Singh Parmar, and Prodyut Dhar. 2024. "Scalable Phosphorylated Cellulose Production with Improved Environmental Sustainability, Crosslinkability and Processability Using 3D Bioprinting for Dye Remediation." *International Journal of Biological Macromolecules* 264: 130577. https://doi.org/10.1016/j.ijbiomac.2024.130577

Rashid, Ruhma, Iqrash Shafiq, Muhammad Rehan Hasan Shah Gilani, Muhammad Maaz, Parveen Akhter, Murid Hussain, Kwang-Eun Jeong, Eilhann E Kwon, Sungjun Bae, and Young-Kwon Park. 2024. "Advancements in TiO2-Based Photocatalysis for Environmental Remediation: Strategies for Enhancing Visible-Light-Driven Activity." *Chemosphere* 349: 140703. https://doi.org/10.1016/j.chemosphere.2023.140703

Rivas-Sanchez, Andrea, Angelica Cruz-Cruz, Guillermo Gallareta-Olivares, Reyna Berenice González-González, Roberto Parra-Saldívar, and Hafiz M N Iqbal. 2022. "Carbon-Based Nanocomposite Materials with Multifunctional Attributes for Environmental Remediation of Emerging Pollutants." *Chemosphere* 303: 135054. https://doi.org/10.1016/j.chemosphere.2022.135054

Sahragard, Ali, Miloš Dvořák, Carlos Pagan-Galbarro, Enrique Javier Carrasco-Correa, Pavel Kubáň, and Manuel Miró. 2024. "3D-Printed Stereolithographic Fluidic Devices for Automatic Nonsupported Microelectromembrane Extraction and Clean-up of Wastewater Samples." *Analytica Chimica Acta* 1297: 342362. https://doi.org/10.1016/j.aca.2024.342362

Shin, Seonghun, Seongsu Cho, Ryungeun Song, Hyejeong Kim, and Jinkee Lee. 2023. "Oil Pollution Remediation with Mass-Producible and Recyclable Amphiphilic Magnetic Janus Particles." *Chemical Engineering Journal* 471: 144734. https://doi.org/10.1016/j.cej.2023.144734

Siddiqui, Shabnam, Srinath Surananai, Krishnamurthy Sainath, Mohammed Zubair Khan, Raghu Raja Pandiyan Kuppusamy, and Yalachigere Kempaiah Suneetha. 2023. "Emerging Trends in Development and Application of 3D Printed Nanocomposite Polymers for Sustainable Environmental Solutions." *European Polymer Journal* 196: 112298. https://doi.org/10.1016/j.eurpolymj.2023.112298

Silva, Leticia Mirella da, Ismael F Mena, Miguel A Montiel, Cristina Saez, Artur J Motheo, and Manuel A Rodrigo. 2023. "Electrochemical Generation of Chlorine Dioxide for Use in Environmental Remediation." *Process Safety and Environmental Protection* 177: 1249–1259. https://doi.org/10.1016/j.psep.2023.07.088

Siqueira, Gilvana P, Diele A G Araújo, Lucas V de Faria, David L O Ramos, Tiago A Matias, Eduardo M Richter, Thiago R L C Paixão, and Rodrigo A A Muñoz. 2023. "A Novel 3D-Printed Graphite/Polylactic Acid Sensor for the Electrochemical Determination of 2,4,6-Trinitrotoluene Residues in Environmental Waters." *Chemosphere* 340: 139796. https://doi.org/10.1016/j.chemosphere.2023.139796

Sreenivasan, S, and Balasubramanian Kandasubramanian. 2023. "Novel Geomaterials for the Remediation of Toxic Pollutants: A Review." *Hybrid Advances* 3: 100057. https://doi.org/10.1016/j.hybadv.2023.100057

Vaishampayan, Vijay, Oinam Robita Chanu, Balasubramanian Sivasamy, Muthamilselvi Ponnuchamy, Varshini Karthik, Ambar Pendharkar, Lohith Srinivas Thotakura, Aryan Prabhu, Venkatesan Dhananjeyan, and Ashish Kapoor. 2023. "Microfluidic Paper-Based Device Coupled with 3D Printed Imaging Box for Colorimetric Detection in Resource-Limited Settings." *HardwareX* 15: e00456. https://doi.org/10.1016/j.ohx.2023.e00456

Vargas-Muñoz, M A, Carlos Palomino, Gemma Turnes, and Edwin Palacio. 2023. "A Sampling Platform Incorporating 3D-Printed Paddles-Stirrers Coated with Metal-Organic Framework MIL-100(Fe) for in-Situ Extraction of Phenolic Pollutants in Biodigester Supernatant and Wastewater Effluent Samples." *Journal of Environmental Chemical Engineering* 11 (5): 110503. https://doi.org/10.1016/j.jece.2023.110503

Vo, Hoang Nhat Phong, Chawalit Chaiwong, Lei Zheng, Thi Minh Hong Nguyen, Thammarat Koottatep, and Thanh Tin Nguyen. 2022. "Chapter9 - Algae-Based Biomaterials in 3D Printing for Applications in Medical, Environmental Remediation, and Commercial Products." In *Biomass, Biofuels, and Biochemicals* edited by Huu Ngo, Wenshan Guo, Ashok Pandey, Jo-Shu Chang, and Duu-Jong, 185–202. Elsevier. https://doi.org/10.1016/B978-0-323-96142-4.00011-7

Wan, Keming, Caiyu Liu, Yan Wang, Wenliang Zhang, Pengfei Qi, Lei Guo, Gang Wei, and Xiaomin Liu. 2022. "Recent Advances in the Synthesis and Regulation of 3D Metal Carbide-Based Hybrid Architectures for Water Environmental Remediation and Monitoring." *Journal of Environmental Chemical Engineering* 10 (6): 108994. https://doi.org/10.1016/j.jece.2022.108994

Wang, Y, Y Zhang, S Wang, Y Guan, and Y Zhang. 2021. "A 3D Printed Synergistic Aerogel Microreactor toward Stable and High-Efficiency Photocatalytic Degradation." *Materials Today Chemistry* 22: 100566. https://doi.org/10.1016/j.mtchem.2021.100566

Wu, You, Xiaoming Li, Hui Zhao, Fubing Yao, Jiao Cao, Zhuo Chen, Xiaoding Huang, Dongbo Wang, and Qi Yang. 2021. "Recent Advances in Transition Metal Carbides and Nitrides (MXenes): Characteristics, Environmental Remediation and Challenges." *Chemical Engineering Journal* 418: 129296. https://doi.org/10.1016/j.cej.2021.129296

Zhang, Yun-fei, Jia-le Gao, Shen-gui Chen, Lei Li, Jian-Hui Xu, Dan Li, Yu-feng Liu, et al. 2023. "Advanced Electrooxidation of Florfenicol Using 3D Printed Nb2O5/Ti Electrodes: Degradation Efficiency, Dehalogenation Performance, and Toxicity Reduction." *Chemical Engineering Journal* 474: 145561. https://doi.org/10.1016/j.cej.2023.145561

Zhao, Tianyang, Yinuo Liu, Yichen Wu, Minghao Zhao, and Yingxin Zhao. 2023. "Controllable and Biocompatible 3D Bioprinting Technology for Microorganisms: Fundamental, Environmental Applications and Challenges." *Biotechnology Advances* 69: 108243. https://doi.org/10.1016/j.biotechadv.2023.108243

Zou, Rusen, Babak Rezaei, Stephan Sylvest Keller, and Yifeng Zhang. 2024. "Additive Manufacturing-Derived Free-Standing 3D Pyrolytic Carbon Electrodes for Sustainable Microbial Electrochemical Production of H2O2." *Journal of Hazardous Materials* 467: 133681. https://doi.org/10.1016/j.jhazmat.2024.133681

16 Assessment of Potential Environmental Risks of Nano Remediation

Neha Rani Kumar

16.1 INTRODUCTION

Every action taken by humans affects the environment, changing and impairing natural ecosystems and creating a steadily larger ecological imprint on a global scale. There are major repercussions for the health of humans and the environment as a result of pollutants from both natural and artificial sources contaminating the air, water, and soil (Halder and Islam, 2015). This has led to a degradation of the entire ecosystem. A significantly large percentage of human deaths are attributed to pollution, which continues to be the major cause of premature deaths globally (Bhatt et al., 2022). Environmental restoration has become a major industry over the last 20 years, and most of the global expenditures have been made in the use of technology solutions in different environmental contexts, including landfills, oil fields, manufacturing and industrial sites, military installations, and private properties. Several efforts have been focused on the benefits of public health and the Earth's natural systems, demonstrated by the strong advocacy of pollution management, remediation, and prevention efforts as a means of minimizing environmental damage (Whitmee et al., 2015).

A potential technique for the effective detection and removal of contaminants from damaged environments is the use of manmade NMs (MNMs) in environmental remediation, or nano remediation. Though their safety for people and the environment has frequently been disregarded, MNMs are intended to immobilize, convert, and degrade chemical contaminants, as well as improve their biotic degradation. The process of nano remediation makes use of a material's qualities at the nanoscale (less than 100 nm), such as its chemical affinity, surface charge density, and electron transfer capacity. Besides being excellent for environmental applications, these features also bestow remarkable optical, quantum, electrical, catalytic, and adsorptive properties (Baig et al., 2021). Various strategies have been employed to produce and successfully apply both inorganic and organic MNMs in water purification, disinfection, pollution detection, and removal of pollutants from water, soil, and air (Gajanan and Tijare, 2018). MNMs can also travel across wide distances and to far-off locations with ease, which improves cost-efficiency, particularly for in situ applications (Marcon et al., 2021).

Although there is no denying the advantages of using nanotechnology to prevent, control, and improve public health while preserving natural ecosystems, nano remediation is hindered by worries about the safety of metal-forming materials. Their fate and possible risks are determined by how MNMs interact with biological systems in the environment after being released. After being released, MNMs experience a variety of chemical, physical, and biological changes that have a substantial impact on their behavior, ultimate fate, and general effects on ecosystems and living things (Mitrano et al., 2015; Hochella et al., 2019). Agglomeration or aggregation can change the size of substances through interactions with other man-made or natural compounds, resulting in processes like (photo) reduction/oxidation, dissolution, and sulfidation (Corsi et al., 2014). MNMs also come into touch with biological fluids and engage in redox reactions with microbes, which leads to the creation of eco- and bio-coronas that are either internalized or expelled in the biota (Corsi et al., 2021).

DOI: 10.1201/9781003484189-16

With its unique use of NPs for environmental cleanup, nano remediation has great promise for reducing pollution and reestablishing ecosystems. To guarantee its safe and long-term application, nano remediation's environmental concerns must be thoroughly assessed, just like those of any newly developed technology. This chapter delves into the negative impacts of nano remediation, including the impact on organisms in the soil, the impact on aquatic organisms, as well as the impact on human health. This is followed by detailed insights into monitoring and risk management strategies.

16.2 APPLICATION OF NMS

NMs are a novel class of materials with special qualities that have found uses in a wide range of industries, including electronics, energy storage, medical, and environmental remediation. There are several ways to synthesize NMs; they can be broadly divided into top-down and bottom-up approaches, each with its own set of benefits and drawbacks (Mekuye and Abera, 2023). These materials are designed at the nanoscale level to have improved chemical, biological, and physical characteristics, allowing for advances in industry and technology.

There are two main methods used to create NMs:

- Bottom-up Approach: This approach builds structures starting at the atomic level. It is affordable and appropriate for production on a wide scale.
- Top-down Approach: This approach scales down from macroscopic structures. It is not economical for large-scale production and is slower.

Several methods can be used to create NPs from chemical solutions, such as microemulsion-based synthesis, colloid-based synthesis, hydrothermal-based synthesis, solvothermal synthesis, chemical vapor deposition synthesis, thermal decomposition synthesis, pulsed laser ablation synthesis, template-based synthesis, combustion, gas phase methods, microwave synthesis, etc (Baig et al., 2021).

Researchers are looking at environmentally friendly and sustainable ways to synthesize NPs to reduce the risks to the environment that come from physico-chemical synthesis. These environmentally friendly techniques are termed green synthesis. In green synthesis, suitable concentrations of metal ions are combined, under specific conditions, with plant extracts or microorganisms (Ying et al., 2022). NM synthesis is a broad field with a variety of approaches, from cutting-edge green techniques to conventional chemical methods, all of which contribute to the creation of unique applications across a range of industries.

Many different types of NPs have been investigated for use in nano remediation; that provide distinct benefits for stability, compatibility with various environmental conditions, and efficacy in removing pollutants. Among the most studied NPs are dendrimers, carbon nanotubes (CNTs), titanium dioxide (TiO_2), and zero-valent iron (nZVI) (Saleh, 2020).

The unique qualities of NMs have transformed several industries and provided creative solutions for a wide range of applications (Figure 16.1). NMs are utilized in the field of water treatment to create advanced nano adsorbents, nanomembranes, nano filters, and nano photocatalytic materials (Santhosh et al., 2016). The efficiency of water purification procedures is improved by several significant examples, such as graphene-based NMs, titanium dioxide (TiO_2) NPs, zerovalent iron (nZVI), and silver NPs. NMs, including carbon NMs, Fe_3O_4 NPs, TiO_2 NPs, and ZnO NPs, are applied in soil remediation for Fenton and Fenton-like oxidation, reduction reactions, immobilization, and photocatalysis, which greatly enhances the decontamination of contaminated soils. Technologies for air purification profit from the use of materials such as carbon nanotubes, Al_2O_3, zeolites, fullerenes, nano adsorbents, sensors, filters, and catalysts (Saleem et al., 2022). The oil and gas sector employs NMs such as aerogels, nanodispersants, magnetic nanocomposites, and carbon nanostructures for improved extraction, separation, and purification processes (Singh et al., 2020).

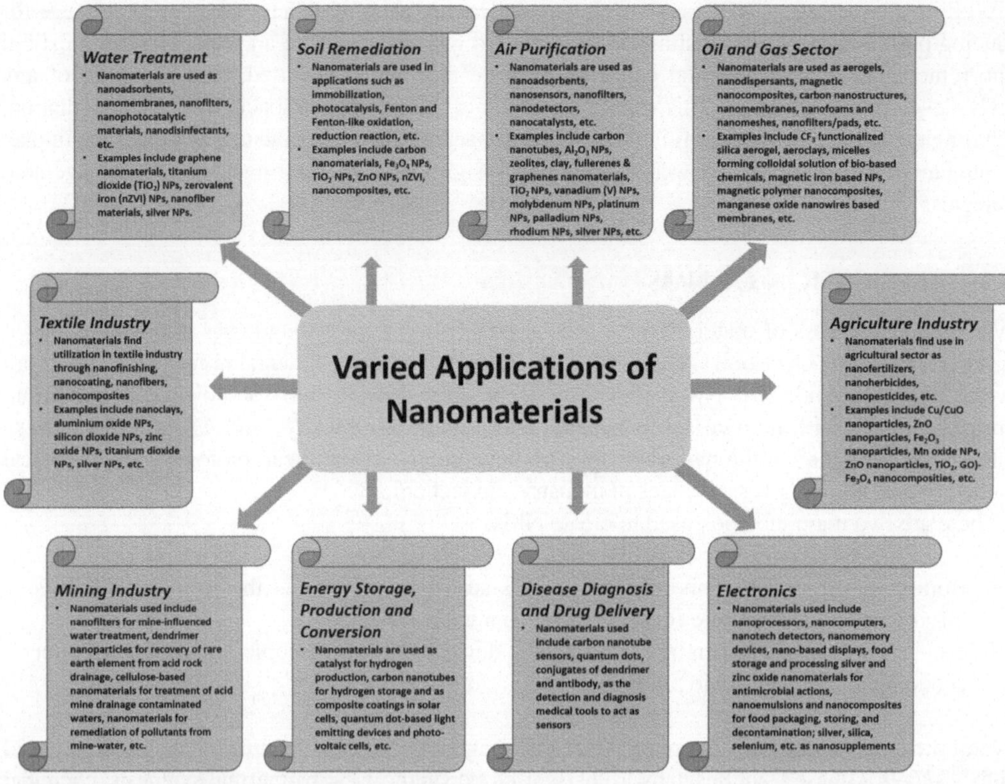

FIGURE 16.1 Applications of NMs in Various Domains.

Examples include CF_3 functionalized silica aerogel, micelle-forming colloidal solutions, and graphene-based structures. In the textile industry, NMs enable advancements in nano finishing, nanocoating, and the development of nanofibers and nanocomposites, with nano clays, aluminum oxide, silicon dioxide, and silver NPs enhancing fabric properties (Saleem et al., 2022).

With the integration of nano processors, nano computers, nanotech detectors, and nano memory devices, electronics have advanced significantly, resulting in greater performance and miniaturization. In food storage and processing, NMs such as silver and zinc oxide exhibit potent antimicrobial properties, while nanoemulsions and nanocomposites improve packaging, preservation, and decontamination processes. With their innovative uses, NMs are bringing forth a new era of environmental sustainability and technological innovation (Tiwari et al., 2021).

16.3 NEGATIVE IMPACT OF NMS

16.3.1 Impact on Organisms in Soil

Because soil has a far larger reactive sink than freshwater and marine settings, the amount of NMs that affect soil biota is frequently overestimated. NM contamination of soil can happen accidentally through uncontrolled spills during manufacture and transportation, or purposely for objectives like fertilization and remediation. Regardless of how they are meant to be used, these NMs may have negative effects on the creatures that live in or come into contact with the soil. Upon coming into contact with soil, NMs attach themselves to soil particles via a variety of interactions, such as electrostatic forces, hydrogen bonding, and hydrophobic interactions. These substances may undergo chemical changes due to the presence of organic matter, and they may spend a considerable amount of time trapped in the pores of soil particles. The retention and mobility of NMs in soil are impacted by variables such as the texture of the soil, pH, humic acid content, surface chemistry, as well as the size and coating of the NMs, alongside pH and humic acid playing a crucial part in their aggregation and stability.

Numerous species, such as nematodes, plants, and microbes, are found in soil and are vital for nitrogen fixation, material breakdown, and the cycling of nutrients. The health of the soil and the general quality of life might suffer severe, irreversible effects from any damage done to these vital elements of the ecosystem. More and more engineered NMs from different sources are being dumped into wastewater, where they end up in soils when biosolids are used. The complex balance of soil ecosystems is at risk due to the introduction of NMs, which could damage the organisms that keep soil healthy and fertile, as well as cause key processes to be disrupted.

16.3.1.1 Potential Toxicity to Plants

The presence of NMs in the soil has a major impact on plants. Different kinds of NMs can enter plants through their roots and spread throughout their structures in various forms, through translocation and biotransformation (Ma et al., 2015). Essential physiological processes, including photosynthesis, growth, and regeneration may be hampered by this presence (Yang et al., 2017). The ensuing disruptions have an impact on the germination of seeds and the growth of seedlings. They also increase the generation of reactive oxygen species (ROS), damage cell walls, and change hormones, proteins, carbohydrates, lipids, and pigments (Mazumdar et al., 2014). NMs frequently experience changes from the point of origin, such as a production facility, to the point of destination. The toxicity of NMs in their final location is influenced by a variety of environmental parameters, such as pH, surface charge, ionic strength, surface coatings, UV radiation exposure, humic acids, inorganic ligands, and the presence of mono- and divalent cations (McKee and Filser, 2016). A significant amount of research has been done on the detrimental effects of engineered NMs (ENMs) on plants. Inhibited seed germination, decreased root and shoot growth, hampered seedling development, decreased photosynthesis, and degraded reproductive and yield characteristics are just a few of the effects that have been seen during different phases of plant life (Du et al., 2017).

The hazardous effects of ENMs on plants are directly related to multiple variables, such as the size of the particles, solubility, bioaccessibility, test dose and time frame, and the specific physicochemical characteristics of the plant species being tested (Josko et al., 2017). Furthermore, the behavior of ENMs in the soil environment and the ensuing impacts on plants can be greatly influenced by environmental factors, such as contact time, light exposure, soil characteristics, and the presence of soil organisms (Dimkpa, 2018).

Metal oxide NMs, such as Al_2O_3, TiO_2, CeO_2, ZnO, CuO, and ZrO_2, are frequently employed as fertilizers in agriculture. Plants are most harmed by ZnO and CeO_2 NPs among them (Du et al., 2017). Furthermore, it has been discovered that carbon nanotubes (CNTs) and SiO_2 NPs have a detrimental effect on rice and tomato seed germination. The reaction of the plant is largely dependent on the concentration of these NMs, and different plants may respond differently to the same NM. NMs can change inside plants, which could provide new hazards or, in some situations, promote plant growth. In addition to interfering with plant growth processes, NMs can raise the generation of reactive oxygen species (ROS), which can cause lipid peroxidation, protein oxidation, DNA and cell membrane damage, electrolyte leakage, and finally cell death (Holden et al., 2014).

Hydrophobic engineered NMs (ENMs), such as carbon NPs, are taken up by natural organic matter in plants and are transported alongside nutrients and water through the xylem. For terrestrial plants, metal and metal oxide NPs are also dangerous. For example, it has been found that copper NPs encourage the production of lateral roots, due to oxidation induction, and diminish wheat root growth by around 60% (Zhang et al., 2018; Khalaj et al., 2018) Xun et al.'s (2017) study on the effects of zinc oxide (ZnO) on maize roots and shoots at the gene level demonstrated genotoxic consequences, with roots exhibiting a higher number of differentially expressed genes than shoots (Xun et al., 2017). It was discovered that the toxicity and bioaccumulation of ZnO in beans (Phaseolus vulgaris) were influenced by the presence of a root-associated bacterium, Pseudomonas chlororaphis, as bacterial siderophores might reduce metal uptake by plants.

Although nano pesticides are applied in agriculture to increase the sustainability and efficiency of pest control, they can also have a direct effect on the physiological activities of plant seeds. Research has demonstrated that the physiological and biochemical processes of green pea (Pisum sativum) seeds are altered by copper oxide NPs in combination with increased indole-3-acetic acid, which impacts yield and nutritional quality (Ochoa et al., 2018) Studies conducted by Majumdar and co-workers (2016) and Medino-Velo and co-workers (2017) have revealed that the active ingredients in nano pesticides modify the nutritional quality of soybean seeds (Medino-Velo et al., 2017; Adisa et al., 2019). Similar to traditional pesticides, nano pesticides cause a variety of physiological changes in plants, impacting the health and nutritional value of both seedlings and seeds. Experimental applications of nano pesticides on an extensive variety of crops, including forage crops, fruits, vegetables, and vegetation, have aimed to prevent fungal and bacterial diseases, but can unintentionally influence plant physiological responses, regardless of biomass and nutrition. There have been reports of physiological alterations in many plants, including lettuce, spinach, cucumber, maize, and basil, caused by nano pesticides (Simonin et al., 2018a). Because soil microorganisms are essential to plants' ability to convert chemicals into useful forms, using nano pesticides in fields may have unanticipated effects on soil microorganisms like nitrogen-fixing bacteria or mycorrhizal communities.

Thus, NMs significantly impact plant health by impacting physiological functions, growth processes, and biochemical processes. Their presence in the soil can impede photosynthesis, root elongation, and seed germination, which frequently results in oxidative stress and cellular damage. Even though certain NMs might be advantageous for agriculture, careful management and more research are required, due to their possible toxicity and long-term effects on plant ecosystems. It is essential to comprehend how plants and NMs interact to create sustainable farming methods. Our agricultural landscapes will be protected and productive if the benefits and risks of NMs are balanced.

16.3.1.2 Potential Toxicity to Soil Microorganisms

Soil microorganisms play an integral part in soil ecosystem functions, such as nutrient cycles and organic pollutants mitigation. By disrupting soil microbial communities, NMs can have an impact on the breakdown, cycling of nutrients, and general health of the soil (Asadishad et al., 2018; Chen et al., 2017). Their presence may change the activity and variety of microbes, causing imbalances in the way ecosystems work. Certain NMs may possess antibacterial characteristics, which could hinder advantageous soil microbes and cause disturbances to significant ecological processes. Furthermore, the buildup of NPs in soil has the potential to last over time, impacting soil fertility and microbial communities in addition. Comprehending the interplay between soil microorganisms and NMs is crucial in evaluating their ecological consequences and formulating enduring soil management strategies. The potential risks of ENMs to soil microorganisms include the genotoxicity on microbial cells, developmental and reproductive damage on individuals, the reduction or inhibition of soil enzyme activity, and the disruption of microbial community structure (Chen et al., 2017).

To comprehend the risk of NMs in the environment, both the dose-response effect of NMs and the channels of exposure assessing how NMs enter an organism must be considered. Previous studies have demonstrated that pure cultures of bacteria, such as Escherichia coli, Pseudomonas fluorescence, and Bacillus subtilis var. niger can be adversely affected by NPs, with oxidation of the nZVI resulting in the generation of reactive oxygen species in living cells. Furthermore, Fe^+ can penetrate cells, causing oxidative stress that damages their membranes, allowing intracellular components to leak out, ultimately resulting in cell death. In the presence of NMs like nZVI, a decrease in the population growth of earthworms, phyto- and zooplankton was also noted. The toxicity effect of NMs on soil microorganisms has been well established. For example, NMs such as CuO and Fe_3O_4 were found to cause changes in soil microbial communities caused by toxicity at 1% and 5% w/w dry soil. Thus toxicity is strongly related to bioavailability and solubility. ZnO is the most concerning of the several NMs, since it is 100% capable of causing the mortality of B. subtilis, E. coli, and P. fluorescens, and shows toxic effects on all species studied.

Because they regulate the turnover of organic matter and the cycle of mineral nutrients, microorganisms are essential to soil health and performance. They also have a significant impact on the physical properties of soil, which affects the soil's ability to retain water and its propensity to compaction or erosion. Therefore, any unfavorable impact that the released NMs may have on the community of microorganisms may have an indirect impact on the quality and functionality of soil. Unwanted NMs in the soil can have an impact on various aspects, including but not limited to soil respiration, microbial biomass, phospholipid fatty acid amount, methyl ester of fatty acids, enzymatic activity, colony forming unit, and DNA profile of bacterial community. Literature insights reveal that microorganisms that fix nitrogen and denitrify it are less susceptible to the harmful impacts of NMs than nitrifying bacteria. TiO_2 NPs, among other NMs, have demonstrated toxicity toward ammonia-oxidizing bacteria (Luo et al., 2015).

While the toxicity mechanisms of ENMs on soil microorganisms have not been thoroughly evaluated for most ENMs yet, potential pathways may be interpreted as disrupting cell membranes and cell walls, damaging organelles, generating reactive oxygen species (ROS) thus causing oxidative stress, disordering the metabolic system, inhibiting gene and protein expression, and so on (Xie et al., 2017). The breakdown of cell membranes requires close, direct contact; thus, if cell walls and membranes collapse, it may harm organelles like mitochondria. The primary cause of concern is ROS formation due to its potential to alter cell architecture and impair in vivo metabolism, both of which may lead to future cell death. For instance, it has been stated that nZVI's high reducibility may denature ion transporter proteins and lipopolysaccharides, reducing membrane permeability and encouraging the entry of Fe(II) into cells. According to Chen and co-workers, once within the cell, Fe(II) combines with H_2O_2 (generated from the mitochondria) to form ROS, which causes oxidative stress and ultimately cell death. An additional pathway for iron oxide precipitation could be the complexation that results from nZVI reacting with lipoteichoic acid, a component of Gram-positive bacteria's cell wall, or the anionic structure of the bacterial cell wall. These precipitants may obstruct the cell membrane's functional pores, preventing the cell from absorbing nutrients and ultimately leading to the cell's slow death (Sacca et al., 2014). A few studies have also examined the possible effects of ENMs on soil invertebrates, including earthworms, collembolans, and springtails, with an emphasis on endpoints related to avoidance performance, weight changes, mortality rate, and modifications to test species' capacity for reproduction (Tatsi et al., 2018; Bouguerra et al., 2016; Waalewijin-Kool et al., 2014).

Nano pesticides employed in the agricultural sector can lead to significant degradation of organic pollutants in soil. However, the type of soil also affects the toxicity level. For instance, soil with a low pH and little organic matter has increased toxicity levels for a copper-based nano pesticide, as these conditions increase copper's solubility and bioavailability. However, whether the effects of nano pesticides on non-target creatures are worse than those of traditional pesticides cannot be generalized.

NMs can significantly affect soil microorganisms, disrupting their functions and altering soil health. These effects may impair soil fertility and nutrient cycling, which may influence plant growth and ecosystem stability. To completely comprehend these effects and create mitigation techniques, more research is necessary. Encouraging the correct application of NMs in both agriculture and industry is essential for maintaining soil health and guaranteeing environmentally sound operations.

16.3.2 Impact on Aquatic Organisms

NMs infiltrate aquatic systems through diverse pathways. These sources include industrial effluents carrying NPs discharged during manufacturing processes in electronics, pharmaceuticals, and textiles. NPs from consumer products like cosmetics, sunscreens, paints, etc. are washed off by urban runoff and deposited into rivers, lakes, and coastal waters. Similarly, agricultural runoff transports nanopesticides and other agricultural NMs applied to crops or soil into nearby waterways through irrigation or rainfall (Zhenlan et al., 2022). Atmospheric deposition involves NPs settling onto water surfaces from industrial emissions and vehicle exhaust. When garbage containing NPs is improperly disposed of, it enters water bodies directly through streams or landfills, where it may eventually seep into surface and groundwater. Sustainable water resource management depends on controlling the discharge of NPs into aquatic environments and comprehending their effects on aquatic ecosystems. The NMs are released into different ecosystems, including the aquatic ones, either deliberately or accidentally.

NPs found in paints, textiles, and various healthcare products, including sunscreens and cosmetics, have a risk of being released into the environment. For example, due to their ability to successfully battle bacteria that cause odor, silver NPs are frequently employed in consumer goods like socks. Nevertheless, these NPs, if discharged into water streams, can also damage beneficial bacteria that are necessary for ammonia exclusion in wastewater treatment systems. Similar to this, there are several methods by which metal oxide NPs, such as those found in sunscreens, paints, coatings, and catalysts, can get into water systems. These NPs include TiO_2, ZnO, and CeO_2. After being properly surface treated, these released NPs settle in soil and surface water systems, where they float on water surfaces without aggregating. When they deposit onto land, they affect soil and can seep into groundwater, impacting existing biota while they travel through the environment. NPs from solid wastes, wastewater effluents, and direct or uncontrolled emissions into aquatic bodies can be further dispersed by wind and rainfall. While unintentional NM emissions can be managed within production facilities, spills that occur during transit to end users present serious difficulties.

Once released, most NPs tend to agglomerate upon hydration, resulting in sedimentation at different rates. The surface charge and magnitude of the NP, as well as the amount of mono- and divalent cations, natural organic matter (NOM), or other organic molecules covering it, all have an impact on the degree of aggregation. These properties determine which forces are more powerful, repulsive, or attractive, and whether or not NPs clump together or stick to other surfaces like watery colloids in the end.

The International Union of Pure and Applied Chemistry (IUPAC) defines aquatic colloids as materials with diameters less than 1 μm. These aquatic colloids may interact with NPs (NMs), causing aggregation and sedimentation. As such, the nature and amounts of aquatic colloids present have a significant impact on the fate of NPs. Furthermore, the concentration of NPs that are frequently found in natural waterways, such as those made of oxides of titanium, cerium, and zinc, as well as silver, normally varies between 1 and 10 μg/L, with cumulative concentrations as high as 100 μg/L. Humic materials frequently coat the surfaces of NPs in aquatic colloids, stabilizing their surface charge and decreasing the possibility of aggregation. This phenomenon has been observed, for example, with carbon nanotubes (CNTs). The behavior of NPs in naturally occurring water systems is greatly influenced by some important factors or properties, such as their chemical composition, mass, particle density, surface area, size distribution, surface charge, surface contamination (such as capping and shell agents), stability, and solubility.

Considering that discharges from industries predominantly impact marine environments, and freshwater streams and coastal runoff ultimately flow into the seas, it's important to tackle contamination in this medium. The marine environment is distinguished by its high ionic strength, alkalinity, and varied spectrum of colloids and natural organic materials (NOMs), which vary in type and concentration according to location, whether they are oceanic or coastal. NPs emitted through air deposition and coastal runoff have the potential to contaminate this environment. Aggregation and colloid formation are influenced by the physicochemical characteristics of water, which change with depth and include temperature, salinity, and type of NOM. Nanoparticulate pollutants create aggregates that gradually sink to the ocean floor, much like in freshwater systems. They might come into contact with warm and cold streams along the way, or they might get consumed by marine life already present. These NPs can affect the marine life that lives in these zones, whether they are deposited on the ocean floor or captured at the interface of two or more water masses. Moreover, birds, animals, and other species that live on the ocean's surface microlayer may be at risk from suspended and trapped NPs.

When living organisms uptake NPs, various toxicity effects can occur through numerous processes. NPs can attach to cell surfaces, penetrate cell membranes, and enter cells by endocytosis. After entering the cell, NPs begin to impair the body's normal activities by oxidizing proteins, causing genotoxicity, obstructing energy transmission, breaking down membrane structure, and producing reactive oxygen species (ROS) and other hazardous compounds. The organism's general health and well-being may suffer as a result of these interactions between NPs and biological components.

As NMs become increasingly utilized in the agricultural sector, there's increasing concern over the exposure of non-target aquatic creatures. Like other toxic substances, the number of nano pesticides that are introduced into aquatic habitats determines the environmental dangers. Several research works have examined the effects of NMs on aquatic life, providing insight into possible risks. For example, studies by Simonin and co-workers showed that long-term exposure to low concentrations of $Cu(OH)_2$ nano pesticide increased eutrophication in wetlands and caused algal blooms(Simonin et al., 2018b). Furthermore, morphological abnormalities in embryos and larvae, including tail malformations, aberrant head shapes, and missing body parts, were observed in response to greater doses (Aksakal and Arslan, 2020). Similarly, studies into the effects of a commercial nano pesticide, such as Kocide 3000, on gene expression related to detoxification and reproductive systems in Daphnia magna revealed significant alterations in gene regulation after short-term exposures (Aksakal and Sisman, 2020). In determining the detrimental effects of nano pesticides, it is important to consider the complex interplay between dosage, non-target organisms, and environmental variables. To enable the responsible use of nanotechnology in agriculture for sustainable practices, it will be essential going forward for environmental researchers and developers of nanopesticides to work closely together to develop a thorough understanding of the environmental risks associated with nanopesticides.

Thus, a variety of potentially hazardous effects of NMs on aquatic life highlight the necessity of cautious management and additional study. To maintain ecosystem health and biodiversity, cooperative efforts are essential for reducing hazards and ensuring the responsible application of nanotechnology in aquatic environments.

16.3.3 Impact on Human Health

The use of NMs in nano remediation poses several possible hazards to human health, mainly due to their distinctive features, which enable them to interact with biological structures in potentially harmful ways. NPs can cause several health problems when they are swallowed, absorbed via the skin, or inhaled. Once within the body, they can cross cellular membranes and enter delicate organs, where they may result in respiratory, cardiovascular, or even neurological complications (Exbrayat et al., 2015; Nallanthighal et al., 2017). NPs' small size and strong reactivity can also cause oxidative stress, inflammation, and cellular damage—all of which are connected to long-term illnesses like cancer and heart problems. Furthermore, there are questions regarding the safety of NMs because it is unclear what the long-term effects of exposure to them will be. Research findings have shown that some NPs can accumulate in the body, leading to prolonged exposure and potential toxicity. The accumulation raises the likelihood of developing chronic illnesses by interfering with biological processes and disrupting regular cellular activities. Furthermore, little is known about how NPs behave in the environment and how they might re-enter the human food chain and cause unintended health hazards. Even though nano remediation has great potential for cleaning up the environment, strict regulation and thorough research are needed to make sure that the advantages do not outweigh the risks to human health.

Human exposure to NPs may culminate in a variety of negative health outcomes, such as oxidative stress, lipid peroxidation, genetic damage, respiratory diseases, inflammation, and pathological changes in lung tissue. NPs can enter biological structures such as the cell nucleus and mitochondria.

This penetration has the potential to cause substantial structural damage to mitochondria and DNA mutations, which would ultimately result in cell death. The extensive use of NPs in environmental cleanup may be restricted as a result of these safety issues. Therefore, it's critical to quickly put monitoring and intervention procedures in place to make sure that the advantages of this technology outweigh the risks.

Engineered NPs represent a significant area of study due to their potential impact on human health. The two main ways that these particles reach the body are by inhalation and skin absorption. After entering the body, they often gather in the kidney, liver, spleen, and bone (Tavares et al., 2017). Given the extensive usage of artificial NMs and their possible health implications, current research efforts focus on understanding their toxicokinetics. Because manufactured NPs are frequently smaller than 100 nm, it is easier for them to enter cells, which raises questions regarding the potential for adverse reactions. Even while NPs have a wide range of characteristics, including size, shape, charge, chemical composition, and surface energy, their effects on living things can be unpredictable because of genetic factors and individual DNA covering ratios. The skin, with its rich blood supply and abundant macrophages, dendrites, lymph vessels, and sensory nerve endings, serves as a prime site for NP absorption (Filon et al., 2016).

Multiple investigations have highlighted the potentially harmful effects of NMs, including DNA damage, and this has garnered significant attention because of its link with cancer, neurological disorders, and aging (Xuan et al., 2023). The possible toxicity of designed NPs can be controlled by several factors, such as biodegradation, biodistribution, biocompatibility, inflammation, and disruption of cellular and organ functions. The reactivity, composition, shape, and size of the NPs are all closely related to these variables. A deeper comprehension of these variables is crucial for the creation of safer NMs and the reduction of any possible health hazards connected to their use, as research in this area progresses. To guarantee the ethical use of engineered NPs in a variety of industries, such as consumer products, medicine, and environmental remediation, more research into the toxicological impacts of these materials is essential.

Thus, the application of NMs in nano remediation presents both promising ecological advantages and potential risks to human health. Although environmental toxins can be effectively cleaned up using nano remediation, there are concerns regarding the specific properties of NPs and how they may interact with biological systems, potentially leading to negative health impacts. The diverse toxicological effects of NPs, which range from oxidative stress to DNA damage, highlight the significance of careful investigation and strict regulation of their use. To optimize the advantages of nano remediation while reducing the hazards to human health and the environment, we must comprehend the components that determine NP toxicity and swiftly put monitoring and intervention mechanisms into place. Sufficient research and cross-disciplinary cooperation are necessary for the conscientious development of nanotechnology in environmental restoration.

16.4 RISK OF NP RELEASE AND TRANSPORT

There are a lot of concerns and difficulties associated with the release of NPs from remediation sites and their subsequent movement through environmental routes. Unintentional ecological repercussions may arise from the dispersion of NPs outside the designated treatment zones, which can be facilitated by mechanisms such as leaching, erosion, groundwater movement, and bioaccumulation. In addition, research on the long-term behavior and fate of NPs in the environment is still underway, and this uncertainty makes it necessary to take proactive steps to reduce their release and associated dangers. During remediation operations, strategies like barrier technologies, containment protocols, and engineered NM design are used to minimize the release of NPs and stop them from migrating into nearby ecosystems. NPs used in remediation materials may leach out into the immediate surroundings, resulting in the spread of contaminants and adverse environmental consequences (Kurwadkar et al., 2014).

16.5 MONITORING AND RISK MANAGEMENT STRATEGIES

To identify and reduce any possible environmental concerns connected to nano remediation, effective monitoring techniques are crucial. A comprehensive strategy that integrates biomonitoring, remote sensing, fate and transport modeling, and environmental monitoring can yield important insights into the behavior and effects of NPs in the environment. The distribution, persistence, and possible effects of NPs on environmental receptors can all be evaluated in real-time, employing continuous monitoring of remediation sites. Additionally, the safe and responsible deployment of nano remediation technology is greatly aided by risk management strategies, such as regulatory frameworks, stakeholder involvement, and exposure controls. The effectiveness and sustainability of nano remediation initiatives can be improved by stakeholders by proactively identifying and addressing possible risks through the integration of monitoring and risk management into remediation procedures.

16.5.1 Green Nanoscience

Using the 12 principles of Green Chemistry, which increasingly influence the synthesis of new molecules and materials, Dahl, and co-authors proposed the "Greener Nanoscience and Nanotechnology Design Principles" in 2007 to offer helpful tools for a sustainable design of MNMs. Further research by Gilberson and co-workers, presented Green Nanoscience to the scientific community, to create highly effective MNMs while minimizing unfavorable effects (Figure 16.2) (Gilberson et al., 2015).

In 2016, Hutchinson and co-workers identified the unique challenges that have been encountered in the design of MNMs (Hutchinson, 2016). These challenges were:

(i) the lack of material definition, often neglecting important aspects that are strictly related to final reactivity, such as shape, size, and surface composition
(ii) the lack of general processes for efficient synthesis, the comprehension of mechanisms, and scalable productions
(iii) the need to define precise material complexity/structure-activity relationships and
(iv) the lack of measurable tools for predicting life cycle impacts/benefits

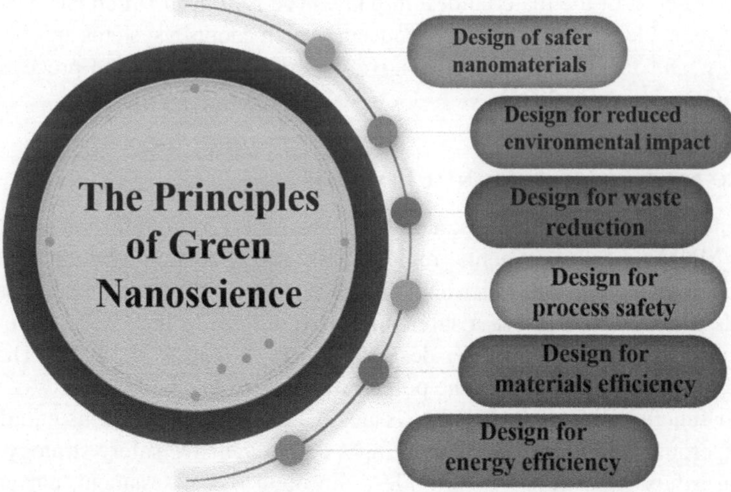

FIGURE 16.2 Six principles of green nanoscience were given by Dahl and co-workers.

All these aspects in numerous instances hinder the advancement of new nanotechnologies for nano remediation, due to the substantial ambiguity of the fate of MNMs and their interaction with the surrounding environment, which encourages international and local entities to adopt a conservative approach.

16.5.2 Life Cycle Assessment in Nano Remediation

To attain true long-term viability, a product or technique needs to be assessed in all its aspects, throughout its life cycle. These elements encompass its influence on the economy, ecology, and society. The methodology known as life cycle assessment or analysis, or LCA, is used to determine how a process or product affects the environment at every stage of its life cycle (Bartolozzi et al., 2019). The life cycle assessment (LCA) is a crucial tool for determining environmental hotspots and determining which area has the greatest environmental impact. When Life Cycle Assessment (LCA) is applied from the outset of product or process design, or eco-design, significant environmental and energy savings can be achieved. This is particularly true for new technologies that are now being scaled up, such as nano remediation, where suitable manufacturing best practices have not been developed.

A limitation of life cycle assessment (LCA) is its restricted consideration of the product or technique's social and economic aspects. Our comprehension of the true costs and advantages of nano remediation from a comprehensive standpoint will be improved with further clarification of these points. Besides also driving an appropriate and sustainable selection of sources, reagents, and solvents for the creation of safer ENMs in the context of Green Nanoscience, life cycle assessment (LCA) represents an essential tool for a proper evaluation and validation of emerging nanotechnologies (Nizam et al., 2021).

An issue that still hinders the avenues associated with the utilization of LCA to promote the eco-design and synthesis of MNMs is the necessity to define ecological impacts based on nano-enabled product design and characteristics. This is mostly because the models that are currently available to describe their fate and behavior in natural settings—that is, environmental matrices—are limited in their ability to precisely represent their specific properties. These properties include dimension, surface area/composition/charge, and interactions with other elements of the receiving environment, such as living things. This lack of knowledge becomes more critical at the early stages of development, which are those that should direct the eco-design of nanotechnology. In addition, there is still a lack of a methodical and comprehensive methodology that integrates nanosafety and sustainability concerns from the outset of the material design phase. The European Union has recently launched the Safe- and Sustainable-by-Design (SSbD) concept, which adopts a systems approach to integrate safety, circularity, and functionality of innovative materials, products, and processes throughout their life cycle (Gottardo et al., 2021).

16.5.3 Nano Safety in Environmental Remediation

Based on current nanosafety regulations, a new framework is required for the safer use of manufactured NMs (MNMs) in environmental remediation. The goal should be to create MNMs that are both environmentally safe and functional and do not pose a threat to human health. To lower the risk of NMs it is essential to use a safer innovative approach that assesses socio-economic analysis, life cycle assessment, and safer design framework (Lin et al., 2018). During research and development, it's critical to reduce the potentially dangerous characteristics of MNMs while preserving their functionality to meet safety issues (Chen et al., 2020). This entails eliminating potentially dangerous features and implementing a comprehensive safety strategy that incorporates ecological safety right away. To provide a thorough risk assessment, current nanosafety testing needs to be improved to take into account bioaccumulation, low dosages, and long-term impacts.

Predicting exposure, behavior, and dangers in nano remediation can be aided by understanding the precise quantity of MNMs utilized as well as their characteristics. MNM amounts, their physicochemical characteristics, and their interactions with biological systems are among the parameters that can be controlled to enable safer design without sacrificing effectiveness. To evaluate risks at all scales—from the cellular to the ecosystem—ecotoxicology should concentrate on bio-nano interactions (Pedersen et al., 2015). Understanding these dangers is aided by the creation of adverse outcome pathways in nanoecotoxicology. Research indicates that MNMs can alter biological interactions and toxicity in natural environments. Long-term impacts and realistic environmental scenarios are essential components of ecotoxicity testing. Before an MNM is released into the market, it should be tested using standardized protocols, and those that are already there should undergo further research. These tests are necessary for safe MNM development, even though they are expensive and time-consuming (Hernandez et al., 2024).

MNM advantages and hazards can be balanced by conducting comprehensive ecotoxicity tests, guaranteeing their safe usage in environmental treatment. To produce safer, more sustainable MNMs, the strategy involves implementing eco-design principles, minimum characterization standards, and standardized testing. Determining nanosafety for pollution cleanup is critical because we need to find ways to clean up contaminated regions without compromising ecosystem health or human health. Global support for nanosafety and environmental sustainability principles is necessary for nano remediation.

Environmentally friendly NMs can be designed by taking into account the following important points (Corsi et al., 2023)-

- Make use of green chemistry concepts and renewable resources.
- Evaluate the effects on the environment by applying Life Cycle Assessment (LCA).
- Create MNMs that are safer based on evaluations of ecological hazards, such as:
  - Comprehending the features of MNM in various environmental settings.
  - To predict MNM behavior in cells, biomolecular interactions should be studied.
  - Examining the trophic transfer and adsorption of pollutants.
  - Performing research on several trophic levels in authentic settings.
  - Assessing biological responses with the use of modern technologies.
  - Gaining knowledge about ecological interactions to protect ecosystem services and function.

Some sustainable solutions that can be adopted to combat existing problems associated with nano remediation are:

Synthesis of NPs from Plants: A greener approach can be utilized for the production of NMs, which leads to the significant reduction of the associated hazards. For example, recent years have seen the development of a green approach for the preparation of nZVI, using natural product extracts from green tea leaves, oak, etc (Jadoun et al., 2021).

Bio-NPs: Nanotechnology can utilize bio-NPs derived from microbes like bacteria, fungi, and algae (Chokriwal et al., 2014).

Emulsified Zero-Valent Iron: Encapsulated in a biodegradable oil membrane, E-ZVI can both absorb and degrade insoluble pollutants. This process promotes microbial growth while shielding iron particles from components of groundwater (Galdames et al., 2020).

Advanced Engineered Approach: Despite rapid progress in nanotechnology, there remains a gap in the development of suitable measurement tools for risk assessment. The quantitative analytical methods available today are insufficient and ineffective for quantifying NPs in environmental systems. Developing efficient analytical techniques to comprehend the fate and transport of NPs in various environmental environments is imperative.

Smarter Engineered NMs: Advanced engineered NMs with new coatings or functional groups can improve stability, mobility, and overall performance. Following restoration, these materials may turn benign, guaranteeing ecological safety.

Regulatory or Legislative Measures: To monitor and regulate the usage of NPs, strict local, national, and worldwide regulatory measures are required in addition to sustainable and sophisticated engineering systems. Under current rules, the United Kingdom considers NPs to be new chemicals, but underdeveloped countries must improve their regulatory systems. To tackle nanotechnology concerns proactively, a multifaceted strategy that includes cutting-edge research, public education, media coverage, and integrated legislation is necessary.

16.6 CONCLUSION

In summary, although nano remediation has great potential to solve environmental issues, careful evaluations of its possible hazards must be carried out to protect the ecosystem and public health. By exploiting multidisciplinary approaches, adopting proactive risk management strategies, and incorporating stakeholder engagement throughout the remediation process, the potential hazards of nano remediation can be effectively assessed and minimized. The safety and sustainability of nano remediation techniques can be further improved by ongoing research that aims to further our understanding of NP behavior, toxicity mechanisms, and environmental fate. This will open the door for their wider use in environmental remediation projects.

REFERENCES

Adisa, I. O., Pullagurala, V. L. R., Peralta-Videa, J. R., Dimkpa, C. O., Elmer, W. H., Gardea-Torresdeya, J. L., White, J. C., 2019. Recent Advances in Nano-Enabled Fertilizers and Pesticides: A Critical Review of Mechanisms of Action. *Environ. Sci. Nano* 6, 2002–2030. https://doi.org/10.1039/C9EN00265K

Aksakal, F. I., Arslan, H., 2020. Detoxification and Reproductive System-Related Gene Expression Following Exposure to $Cu(OH)_2$ Nanopesticide in Water Flea (Daphnai Magna Straus 1820). *Environ. Sci. Pollut. Res.* 27, 6103–6111. https://doi.org/10.1007/s11356-019-07414-x

Aksakal, F. I., Sisman, T., 2020. Developmental Toxicity Induced by $Cu(OH)_2$ Nanopesticide in Zebrafish Embryos. *Environ. Toxicol.* 35(12), 1289–1298. https://doi.org/10.1002/tox.22993

Asadishad, B., Chahal, S., Akbari, A., Cianciarelli, V., Azodi, M., Ghoshal, S., Tufenkji, N., 2018. "Amendment of Agricultural Soil with Metal NPs" Effects on Soil Enzyme Activity and Microbial Community Composition. *Environ. Sci. and Technol.* 52(4), 1908–1908. https://doi.org/10.1021/acs.est.7b05389

Baig, N., Kammakakam, I., Falath, W., 2021. Nanomaterials: A Review of Synthesis Methods, Properties, Recent Progress, and Challenges. *Mat. Adv.* 2, 1821–1871. https://doi.org/10.1039/D0MA00807A

Bartolozzi, I., Daddi, T., Punta, C., Fiorati, A., Iraldo, F., 2019. Life Cycle Assessment of Emerging EnvironmentalTechnologies in the Early Stage of Development: A Case Study on Nanostructured Materials. *J. Ind. Ecol.* 24(1), 101–115. https://doi.org/10.1111/jiec.12959

Bhatt, P., Pandey, C. S., Joshi, S., Chaudhary, P., Pathak, V. M., Huang, Y., Wu, X., Zhou, Z., Chen, S., 2022. Nanobioremediation: A Sustainable Approach for the Removal of Toxic Pollutants from the Environment. *J. Hazard. Mater.* 427, 128033. https://doi.org/10.1016/j.jhazmat.2021.128033

Bouguerra, S., Gavina, A., Ksibi, M., Rasteiro, M. G., Santos, T. R., Pereira, R., 2016. Ecotoxicity of Titanium Silicon Oxide ($TiSiO_4$) Nanomaterial for terrestrial Plants and Soil Invertebrates Species. *Ecotoxicol. Environ. Saf.* 129, 291–301. https://doi.org/10.1016/j.ecoenv.2016.03.038

Chen, C., Leong, D. T., Lynch, I., 2020. Rethinking Nanosafety: Harnessing Progress and Driiving Innovation. *Small.* 16(21), 2002053. https://doi.org/10.1002/smll.202002503

Chen, M., Qin, X., Zeng, G., 2017. Biodiversity Change behind Wide Application of Nanomaterials. *Nano Today.* 17, 11–13. https://doi.org/10.1016/j.nantod.2017.09.001

Chokriwal, A., Sharma, M. M., Singh, A., 2014. Biological Synthesis of NPs Using Bacteria and their Applications. *Am. J. for Pharmtech Res.* 4, 1–24.

Corsi, I., Bellingeri, A., Eliso, M. C., Grassi, G., Liberatori, G., Murano, C., Sturba, L. Vannuccini, M. L., Bergami, E., 2021. Eco-interactions of Engineered Nanomaterials in the Marine Environment: Towards an Eco-Design Framework. *Nanomater.* 11(8), 1903. https://doi.org/10.3390/nano11081903

Corsi, I., Cherr, G. N., Lenihan, H.S., Labille, J., Hassellov, M., Canesi, L., Dondero, F., Frenzilli, G., Hristozov, D., Puntes, V., Torre, C. D., Pinsino, A., Libralato, G., Marcomini, A., Sabbioni, E., Matranga, V., 2014. Common Strategies and Technologies for the Ecosafety Assessment and Design of Nanomaterials Entering the Marine Environment. *ACS Nano*. 8(10), 9694–9709. https://doi.org/10.1021/nn504684k

Corsi, I., Venditti, I., Trotta, F., Punta, C., 2023. Environmental Safety of Nanotechnologies: The Eco-Design of Manufactured Nanomateriald for Environmental Remediation. *Sci. Total Environt*. 854, 161181 https://doi.org/10.1016/j.scitotenv.2022.161181

Dimkpa, Christian. O., 2018. Soil Properties Influence the Response of Terrestrial Plants to Metallic NPs Exposure. *Curr. Opin. Environ. Sci. Health*. 6, 1–8. https://doi.org/10.1016/j.coesh.2018.06.007

Du, W., Tan, W., Peralta-Videa, J. R., Gardea-Torresday, J. L., Ji, R., Yin, Y., Guo, H., 2017. Interaction of Metal Oxide NPs with Higher Terrestrial Plants: Physiological and Biochemical Aspects Plant. *Physiol. Biochem*. 110, 210–225. https://doi.org/10.1016/j.plaphy.2016.04.024

Exbrayat, J. M., Moudilou, E. N., Lapied, E., 2015. Harmful Effects of NPs on Animals. *J. Nanotechnol*. 2015, 1–11 https://doi.org/10.1155/2015/861092

Filon, F. L., Bello, D., Cherrie, J. W., Sleeuwenhoek, A., Spaan, S., Brouwer, D. H., 2016. Occupational Dermal Exposure to NPs and Nano-Enabled Products: Part I-Factors Affecting Skin Absorption. *Int. J. Hyg. Environ. Health*. 219(6), 536–544. https://doi.org/10.1016/j.ijheh.2016.05.009

Gajanan, K., Tijare, S. N., 2018. Applications of Nanomaterials. *Mater. Today Proc*. 5(1), 1093–1096. https://doi.org/10.1016/j.matpr.2017.11.187

Galdames, A., Rubio, L. R., Orueta, M., Arzalluz, M. S., Vilela, J. L. V., 2020. Zero-Valent Iron NPs for Soil and Groundwater Remediation, *Int. J. Environ. Res. Public Health*. 17(16), 5817. https://doi.org/10.3390%2Fijerph17165817

Gilberson, L. M., Zimmerman, J. B., Plata, D. L., Hutchinson, J. E., Anastas, P., 2015. Designing Nanomaterials to Maximize Performance and Minimize Undesirable Implications Guided by the Principles of Green Chemistry. *Chem. Soc. Rev*. 44, 5758–5777. https://doi.org/10.1039/C4CS00445K

Gottardo, S., Mecha, A., Drbohlavova, D., Malyska, A., Bowadt, S., Sintes, J. R., Rauscher, H., 2021. Towards Safe and Sustainable Innovation in Nanotechnology: State-of-Play for Smart Nanomaterials. *NanoImact*. 21, 100297. https://doi.org/10.1016/j.impact.2021.100297

Halder, J. N., Islam, M. N., 2015. Water Pollution and its Impact on the Human Health *J. Environ. Human*. 2, 36–46. https://doi.org/10.15764/EH.2015.01005

Hernandez J. A. C., Salcedo, A. J. V., Tovar, G. N., Gonzalez, C., 2024. Safe Nanomaterials: From their Use, Application, and Disposal to Regulations. *Nanoscale Adv*. 6, 1583–1610. https://doi.org/10.1039/D3NA01097J

Hochella, M., Mogk, D. W., Ranville, J., Irving, I. C., Yang, Y., 2019. Natural, Incidental, and Engineered Nanomaterials and their Impacts on the Earth System. *Science* 363, 1414. https://doi.org/10.1126/science.aau8299

Holden, P. A., Klaessig, F., Turco, R. F., Priester, J. H., Rico, C. M., Arias, H. A., Mortimer, M., Pacpaco, K., Gardea-Torresdey, J. L., 2014. Evaluation of Exposure Concentrations Used in Assessing Manufactured Nanomaterial Environmental Hazards: Are They Relevant? *Environ. Sci. Technol*. 48(18), 10541–10551. https://doi.org/10.1021/es502440s

Hutchinson, James. E., 2016. The Road to Sustainable Nanotechnology: Challenges, Progress and Opportunities. *ACS Sustain. Chem. Eng*. 4(11), 5907–5914. https://doi.org/10.1021/acssuschemeng.6b02121

Jadoun, S., Arif, R., Jangid, N. K., Meena, R. K., 2021. Green Synthesis of NPs Using Plant Extracts: A Review. *Environm. Chem. Lett*. 19, 355–374. https://doi.org/10.1007/s10311-020-01074-x

Josko, I., Oleszczuk, P., Skwarek, E., 2017. Toxicity of Combined Mixtures of NPs to Plants. *J. Hazard. Mater*. 331, 200–209. https://doi.org/10.1016/j.jhazmat.2017.02.028

Khalaj, M., Kamali, M., Khodaparast, Z., Jahanshahi, A., 2018. Copper-Based Nanomaterials for Environmental Decontamination- An Overview on Technical and Toxicological Aspects. *Ecotoxicol. Environ. Saf*. 148, 813–824. https://doi.org/10.1016/j.ecoenv.2017.11.060

Kurwadkar S., Pugh, K., Gupta, A., Ingole, S., 2014. NPs in the Environment: Occurrence, Distribution, and Risks. *J. Hazard., Toxic, Radioact. Waste*. 19(3), 1233. https://doi.org/10.1061/(ASCE)HZ.2153-5515.0000258

Lin, S., Yu, T., Yu, Z., Hu, X., Yin, D., 2018. Nanomaterials Safer-By-Design: An Environmental Safety Perspective, *Adv. Mater*. 30(17), 1706591. https://doi.org/10.1002/adma.201705691

Luo, Z., Qiu, Z., Chen, Z., Du Laing, G., Liu, A., Yan, C., 2015. Impact of TiO_2 and ZnO NPs at Predicted Environmentally Relevant Concentrations on Ammonia-Oxidizing Bacteria Cultures Under Ammonia Oxidation. *Environ. Sci. Pollut. Res*. 22(4), 2891–2899. https://doi.org/10.1007/s11356-014-3545-9

Ma, Y., Zhang, P., Zhang, Z., He, X., Li, Y., Zhang, J., Zheng, L., Chu, S., Yang, K., Zhao, Y., Chai, Z., 2015. Origin of the Different Phytotoxicity and Biotransformation of Cerium and Lanthnum Oxide NPs in Cucucmber. *Nanotoxicology* 9, 262–270. https://doi.org/10.3109/17435390.2014.921344

Marcon, L., Oliveras, J., Puntes, V. F., 2021. In situ Nanoremediation of Soils and Groundwaters from the NP's Standpoint: A Review. *Sci. Total Environ.* 791, 148324. https://doi.org/10.1016/j.scitotenv.2021.148324

Mazumdar, S., Peralta-Videa, J. R., Bandyopadhyay, S., Castillo-Michel, H., Hernandez-Viezcas, J.-A., Sahi, S., Gardea-Torresdey, J. L., 2014. Exposure of Cerium Oxide NPs to Kidney Bean Shows Disturbance in the Plant Defence Mechanisms. *J. Hazard. Mater.* 278, 279–287. https://doi.org/10.1016/j.jhazmat.2014.06.009

McKee, Moira. S., Filser, Juliane, 2016. Impacts of Metal-Based Engineered Nanomaterials on Soil Communities. *Environ. Sci. Technol.* 3, 506–533. https://doi.org/10.1039/C6EN00007J

Medino-Velo, I. A., Dominguez, O. E., Ochoa, L., Barrios, A. C., Hernandez-Viezcas, J. A., White, J. C., Perlata-Videa, J. R., Gardea-Torresday, J. L., 2017. Nutritional Quality of Bean Seeds Harvested from Plants Grown in Different Soils Amended with Coated and Uncoated Zinc Oxide Nanomaterials. *Environ. Sci. Nano.* 4, 2336–2347. https://doi.org/10.1039/C7EN00495H

Mekuye, Bawoke, Abera, Birhanu, 2023. Nanomaterials: An Overview of Synthesis, Classification, Characterization, and Applications. *Nano Select.* 4(8), 486–501. https://doi.org/10.1002/nano.202300038

Mitrano, D. M., Motellier, S., Clavaguera, S., Nowack, B., 2015. Review of Nanomaterial Aging and Transformations Through the Life Cycle of Nano-Enhanced Products. *Environ. Int.* 77, 132–147. https://doi.org/10.1016/j.envint.2015.01.013

Nallanthighal, S., Chan, C., Murray, T. M., Mosier, A. P., Cady, N. C., Reliene, R., 2017. Different Effects of Silver NP on DNA Damage and DNA Repair Gene Expression in Ogg1-Deficient and Wild Type Mice. *Nanotoxicology.* 11(8), 996–1011. https://doi.org/10.1080/17435390.2017.1388863

Nizam, N. U. M., Hanafiah, M. M., Woon K. S. A., 2021. Content Review of Life Cycle Assessment of Nanomaterials: Current Practices, Challenges, and Future Prospects. *Nanomater.* 11(12), 3324. https://doi.org/10.3390/nano11123324

Ochoa, L., Zuverza-Mena, N., Medina-Velo, I. A., Flores-Margez, J. P., Peralta-Videa, J. R., Gardea-Torresdey, J. L., 2018. Copper Oxide NPs and Bulk Copper Oxide, Combined with Indole-3-acetic acid, Alter Aluminium, Boron, and Iron in Pisum Sativum Seeds, *Sci. Total Environ.* 634(5), 1238–1245. http://doi.org/10.1016/j.scitotenv.2018.04.003

Pedersen, E. J., Diamond, S. A., Kennedy, A. J., Goss, G. G., Ho, K., Lead, J., Hanna, S. K., Hartmann, N. B., Hund-Rinke, K., Mader, B., Manier, N., Pandard, P., Salinas, E. R., Sayre, P., 2015. Adapting OECD Aquatic toxicity Tests for Use with Manufactured Nanomaterials: Key Issues and Concencus Recommendations. *Environ. Sci. Technol.* 49(16), 9532–9547. https://doi.org/10.1021/acs.est.5b00997

Sacca, M. L., Fajardo, C., Costa, G., Lobo, C., Nande, M., Martin, M., 2014. Integrating Classical and Molecular Approaches to Evaluate the Impact of Nanosized Zero-Valent Iron (nZVI) on Soil Organisms. *Chemosphere.* 104, 184–189. https://doi.org/10.1016/j.chemosphere.2013.11.013

Saleem, H., Zaidi, S. J., Ismail, A. F., Goh, P. S., 2022. Advances of Nanomaterials for Air Pollution Remediation and their Impacts on the Environment. *Chemosphere.* 287(2), 132083. https://doi.org/10.1016/j.chemosphere.2021.132083

Saleh, Tawfik. A., 2020. Nanomaterials: Classification, Properties, and Environmental Toxicities. *Environ. Tech. Inno.* Vol 20, 101067. https://doi.org/10.1016/j.eti.2020.101067

Santhosh, C., Velmurugan, V., Jacob, G., Jeong, S. K., Grace, A. N., Bhatnagar, A., 2016. Role of Nanomaterials in Water Treatment Applications: A Review. *Chem. Eng. J.* 306, 1116–1137. https://doi.org/10.1016/j.cej.2016.08.053

Simonin, Marie, Colma, Benjamin. P., Tang, Weiyi, Judy, Jonathan. D., Anderson, Steven. M., Bergemann, Christina. M., Rocca, Jennifer. D., Unrine, Jason. M., Cassar, Nicolas, Bernhardt, Emily. S., 2018a. Plant and Microbial Responses to Repeated $Cu(OH)_2$ Nanopesticide Exposures Under Different Fertilization Levels in an Agro-Ecosystem. *Front. Biosci.* 9, 1769. https://doi.org/10.3389/fmicb.2018.01769

Simonin, M., Colman, B. P., Anderson, S. M., King, R. S., Ruis, M. T., Avellan, A., Bergemann, C. M., Perrotta, B. G., Geitner, N. K., Ho, M., De La Barrera, B., Unrine, J. M., Lowry, G. V., Richardson, C. J., Wiesner, M. R., Bernhardt, E. S., 2018b. Engineered NPs Interact With Nutrients to Intensify Eutrophicatopn in a Wetland Ecosystem Experiment. *Ecol. Appl.* 28(6), 1435–1449. https://doi.org/10.1002/eap.1742

Singh, H., Bhardwaj, N., Arya S. K., Khatri, M., 2020. Environmental Impacts of Oil Spills and their Remediation by Magnetic Nanomaterials. *Environ. Nanotechnol. Monit. Manag.* 14, 100305. https://doi.org/10.1016/j.enmm.2020.100305

Tatsi, K., Shaw, B. J., Hutchinson, T. H., Handy, R. D., 2018. Copper Accumulation and Toxicity in Earthworms Exposed to CuO Nanomaterials: Effects of Particle Coating and Soil Ageing. *Ecotoxicol. Environ. Saf.* 166, 462–473. https://doi.org/10.1016/j.ecoenv.2018.09.054

Tavares, A. J., Poon, W., Zhang, Y. N., Dai, Q., Besla, R., Ding, D., Ouyang, B., Li, A., Chen, J., Zheng, G., Robbins, C., Chan, W. C. W., 2017. Effect of Removing Kupffer Cells on NP Tumor Delivery. *Proc. Natl. Acad. Sci. USA*. 114(51), E10871–E10880. https://doi.org/10.1073/pnas.1713390114

Tiwari, K., Singh, R., Negi, P., Dani, R., Rawat, A., 2021. Application of Nanomaterials in Food Packaging Industry: A Review. *Mater. Tod. Proc.* 46(20), 10652–10655. https://doi.org/10.1016/j.matpr.2021.01.385

Waalewijin-Kool, P. L., Rupp, S., Lofts, S., Svendsen, C., Van Gestel, C. A. M., 2014. Effect of Soil Organic Matter Content and pH on the Toxicity of ZnO NPs to Folsomia Candida. *Ecotoxicol. Environ. Saf.* 108, 9–15. https://doi.org/10.1016/j.ecoenv.2014.06.031

Whitmee, S., Haines, A., Beyrer, C., Boltz, F., Capon, A. G., Dias, B. F. de S., Ezeh, A., Frumkin, H., Gong, P., Head, P., Horton, R., Mace, G. M., Marten, R., Myers, S. S., Nishtar, S., Osofsky, S. A., Pattanayak, S. K., Pongsiri, M. J., Romanelli, C., Soucat, A., Vega, J., Yach, D., 2015. Safeguarding Human Health in the Anthropocene Epoch: Report of the Rockefeller Foundation-Lancet Commission on Planetary Health. *Lancet* 386, 1973–2028. https://doi.org/10.1016/S0140-6736(15)60901-1

Xie, Y., Dong, H., Zeng, G., Tang, L., Jiang, Z., Zhang, C., Deng, J., Zhang, L., Zhang, Y., 2017. The interactions between Nanoscale Zero-Valent Iron and Microbes in the Subsurface Environment: A Review. *J. Hazard. Mater.* 321, 390–407. https://doi.org/10.1016/j.jhazmat.2016.09.028

Xuan, L., Ju, Z., Skonieczna, M., Zhou, P. K., Huang, R., 2023. NPs-Induced Potential Toxicity on Human Health: Applications, Toxicity Mechanisms, and Evaluation Models. *MedComm.* 4(4), e327 https://doi.org/10.1002/mco2.327

Xun, H., Ma, X., Chen, J., Yang, Z., Liu, B., Gao, X., Li, G., Yu, J., Wang, L., Pang, J., 2017. Zinc Oxide NP Exposure Triggers Different Gene Expression Patterns in Maize Shoots and Roots. *Environ. Pollut.* 229, 479–488. https://doi.org/10.1016/j.envpol.2017.05.066

Yang, J., Cao, W., Rui, Y., 2017. Interactions Between NPs and Plants: Phytotoxicity and Defense Mechanism. *J. Plant Interact.* 12(1), 158–169 https://doi.org/10.1080/17429145.2017.1310944

Ying, S., Guan, Z., Ofoegbu, P. C., Clubb, P., Rico, C., He, F., Hong, J., 2022 Green Synthesis of NPs: Current Developments and Limitations. *Environ. Tech. Inno.* 26, 102336. https://doi.org/10.1016/j.eti.2022.102336

Zhang, Z., Ke, M., Qu, Q., Peijnenburg, W. J. G. M., Lu, T., Zhang, Q., Ye, Y., Xu, P., Du, B., Sun, L., Qian, H., 2018. Impact of Copper NPs and Ionic Copper Exposure on Wheat (Triticum aestivum L.) Root Morphology and Antioxidant Response. *Environ. Pollut.* 239, 689–697. https://doi.org/10.1016/j.envpol.2018.04.066

Zhenlan, X., Tang, T., Lin, Q., Yu, J., Zhang, C., Zhao, X., Kah, M., Li, L., 2022. Environmental Risks and the Potential Benefits of Nanopesticides: A Review Environ. *Chem. Lett.* 20, 2097–2108. https://doi.org/10.1007/s10311-021-01338-0

17 Opportunities and Challenges of Using Nanomaterials in the Remediation of Pollutants

Saikat Banerjee, Uma Reddy Meka, Selvaraju Sivamani, and B.S. Naveen Prasad

17.1 INTRODUCTION

Nanotechnology indeed presents a promising avenue for tackling environmental pollution. Its potential lies in the unique properties and functionalities that emerge at the nanoscale - a realm where materials exhibit remarkable differences in behavior compared to their bulk counterparts. Nanotechnology holds great promise to contribute to the development of clean, green technologies with outstanding environmental and health benefits. Undoubtedly, nanotechnology holds great promise to manage and reduce pollution and enhance traditional environmental cleanup methods. Here are some of the methods for exploring and applying nanotechnology in this context, as well as pollution detection and monitoring. For instance; the development of nanoscale sensors can detect and monitor the presence of air, water, and soil pollutants with high sensitivity and accuracy. These sensors can provide real-time information, enabling rapid response to environmental damage events. Efficient water purification systems can be developed, using NMs such as NPs and nanofibers (Roy et al., 2021a; Pal et al., 2024; Akshay et al., 2024). These materials can absorb or degrade dirt, heavy metals, and pollutants, improving overall water quality. Nanotechnology is also used in the development of advanced air filtration systems. NM-based filters can capture particles, pollutants, and air better than conventional filters, improving air quality (Marlon et al., 2019). NPs can be engineered to better target and prevent polluted areas. This includes the degradation of organic pollutants and the immobilization or removal of heavy metals from soil and groundwater (Kaur and Roy, 2021). Nano-sponges, nano-gels, and other NMs are being investigated for their ability to adsorb and release oil and non-aqueous pollutants from the surface. These features can improve the efficiency of oil spill cleanup efforts. Nanotechnology can improve bioremediation processes by encapsulating microorganisms or enzymes and delivering them to target pollutants. This targeted approach can increase pollution emissions in a controlled and efficient manner. NMs can be incorporated into process wastes to enhance pollutant removal and facilitate the conversion of wastes into less harmful products. Nanotechnology itself contributes to the development of medical products that can repair environmental damage. NPs can be used as carriers for the targeted delivery of cleaning agents, such as enzymes and drugs, to specific contaminated sites. This targeted approach minimizes resource consumption and ensures that improvement efforts are focused on areas of greatest need. While these applications of nanotechnology in pollution management show great promise, it is important to continue investigating the potential risks and ethical considerations associated with the use of NMs in the environment (Khan, 2020).

17.1.1 Types of NMs

This refers to the constructing blocks of the NP, which includes metals (gold, silver), semiconductors (silicon, gallium nitride), polymers, or carbon-based total substances (graphene, nanotubes) (Hochella et al., 2015). Organic NMs are received from natural matter except for carbon-based or

DOI: 10.1201/9781003484189-17

inorganic-based resources; they are found in a bureaucracy like micelles, liposomes, and polymeric particles. Composite NPs are multiphasic in nature, such that certainly one of their phases can integrate with similar or distinct NPs in bulk (e.g., a gold NP in composition with ceramic or a polymer). Carbon-based NPs include fullerenes, nanotubes, and nanofibers that are observed in usually-dimensional morphological systems (Nagore et al., 2021).

Dimensions of NPs are typically defined as having at least one measurement between 1 and 100 nanometers. Their structure can be round, rod-like, cubic, or greater complex, influencing their behavior and properties (Bhatia, 2016). On the premise of structure, they may be classified into three classes: one dimension (1-D), two dimensions (2-D), and three dimensions (3-D). In zero dimension (0-D) the electrons can't move in any axis; they are dimensionless; in 1-D, they can circulate along a single axis followed via 2-D and 3-D shape: The inner arrangement of atoms or molecules within the NP can be amorphous (randomly packed) or crystalline (ordered lattice). This shape affects elements like power, conductivity, and reactivity (Jeevanandam et al., 2018).

The origin of NPs can be naturally happening, determined in minerals or volcanic ash, or engineered via numerous synthesis strategies. Synthetic NPs provide extra management over their houses for particular programs. Various techniques have been used for characterization, such as UV-vis spectroscopy, primarily based on the principle of floor plaque resonance (SPR). Another technique is X-ray diffraction (XRD) that is used to decide the crystalline section. Fourier transform infrared (FTIR) spectroscopy enables us to identify the practical organizations present within the sample. Characterization of NPs is completely based on size, morphology, and the surface area turned into the usage of atomic force microscopy (AFM), scanning electron microscopy, and transmission electron microscopy. Dynamic light scattering is used to measure NP length (Roy and Bharadvaja, 2017).

17.1.2 Overview of the Effect of Pollutants on the Environment

Pollutants can contaminate soil and water, as well as have major negative consequences on the environment. These contaminants can come from a variety of sources, including incorrect trash disposal, car emissions, industrial activity, and agricultural practices (Dey et al., 2024). Particulate particles, insecticides, heavy metals, and organic compounds are a few examples of frequent contaminants. Pollutants can damage ecosystems, taint water supplies, lower air quality, and endanger human health and endangered species. Pollutants have a significant effect on biodiversity, different ecosystems, and human health. Below is a summary of some typical contaminants and the harm they cause:

Particulate Matter (PM): Tiny airborne particles are able to enter the lungs deeply and cause cardiovascular disorders, pulmonary issues, and even early death. Sulfur oxides (SOx) and nitrogen oxides (NOx) harm plants, ecosystems, and structures by causing smog and acid rain. Volatile Organic Compounds (VOCs) which cause respiratory problems and damage to plants are released from a variety of sources, including vehicle exhaust and industrial activities. Carbon monoxide (CO), a colorless and odorless gas released by combustion processes and automobile exhaust; lowers oxygen transport capacity in the blood, resulting in illnesses and even death (Landrigan et al., 2024).

Chemical Contaminants That Pollute Water: Water bodies are contaminated by pesticides, heavy metals, medicines, and industrial pollutants, which disturb aquatic ecosystems, harm aquatic life, and pose health concerns to humans through contaminated drinking water. Overabundance of nutrients, such as nitrogen and phosphorus, from wastewater and agricultural runoff can result in eutrophication, which can kill aquatic life by depleting oxygen and causing algal blooms (Alidoust et al., 2024).

Soil Pollutants: Industrial waste, heavy metals, solvents, and pesticides are the examples of contaminants that build up in soil and have an impact on agricultural productivity, groundwater quality, and soil fertility. Landfills also cause soil pollution when solid waste is improperly disposed of or toxins seep into the groundwater and soil, endangering ecosystems and human health (Shi et al., 2024).

Noise Pollution and Light Pollution: Excessive noise from sources such as industry, urban growth, and transportation can harm wildlife, obstruct animal navigation and communication, and induce stress, hearing loss, and other health issues in people. Artificial nighttime light disturbs natural ecosystems, impacting the migratory and reproductive cycles of nocturnal species. In addition, it wastes energy and obstructs astronomy study and stargazing (Candolin, 2024).

17.2 APPLICATIONS OF NMS IN REMEDIATION OF POLLUTANTS

NMs offer unique properties, such as high surface area-to-volume ratio, tunable surface chemistry, and enhanced reactivity, making them effective in pollutant remediation. Here's an overview of how NMs can be used to control different types of pollutants (Figure 17.1).

Air Pollution Control: NMs like titanium dioxide (TiO2) and carbon-based NMs (such as graphene and carbon nanotubes) have shown promising results in removing air pollutants like nitrogen oxides (NOx), volatile organic compounds (VOCs), and particulate matter (PM) from the atmosphere. These materials can be incorporated into filters, coatings, or catalysts for air purification systems.

Water Pollution Control: NMs have been utilized for the removal of various contaminants from water bodies, including heavy metals, organic pollutants, and pathogens. Nanostructured materials like iron NPs, graphene oxide, and zeolites can adsorb or catalyze the degradation of pollutants, providing an efficient means of water treatment.

Soil Remediation: NMs can aid in the remediation of contaminated soils by facilitating the degradation or immobilization of pollutants. Nano zero-valent iron (nZVI), for example, can be used to reduce the concentration of heavy metals and organic compounds through processes like adsorption, precipitation, and chemical reactions.

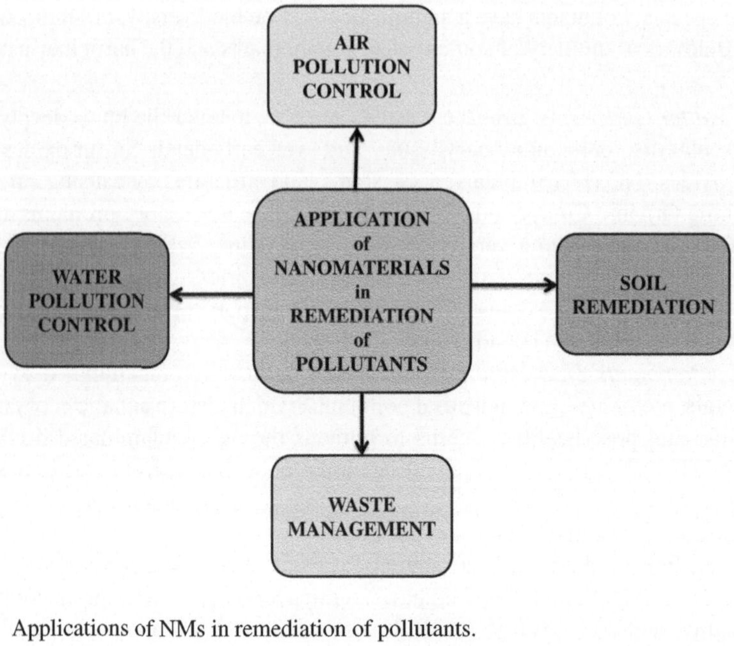

FIGURE 17.1 Applications of NMs in remediation of pollutants.

Waste Management: NMs can also contribute to more efficient waste management practices. Nano-enabled sensors and monitoring devices can detect and track pollutant levels in real-time, enabling better management and control of waste streams. Additionally, NMs can enhance the efficiency of waste treatment processes, such as landfill leachate treatment or wastewater sludge dewatering. The NM-based approaches hold promise for pollutant control, it's important to consider potential risks associated with their use, such as unintended environmental impacts and human health concerns. Proper regulation, risk assessment, and research into the long-term effects of NMs are essential to ensure their safe and effective implementation in environmental remediation efforts (Sathish et al., 2024).

17.3 FATE AND TRANSPORT OF NMS IN THE ENVIRONMENT

The fate and transport of NMs in the environment are crucial considerations, due to their potential impact on ecosystems and human health.

17.3.1 Fate of NMs

Deposition: NMs can deposit onto surfaces such as soil, sediments, and plants through various mechanisms like gravitational settling and electrostatic attraction (Lin et al., 2010).

Transformation: They may undergo transformations like aggregation, dissolution, or surface modifications, altering their properties and behavior.

Transport: NMs can be transported through air, water, and soil, influenced by factors such as particle size, shape, surface charge, and environmental conditions (Batley et al., 2013).

17.3.2 Transport Mechanisms

Air Transport: NPs can be carried by wind currents over long distances, impacting air quality and potentially reaching ecosystems far from their source (Westerhoff and Nowack, 2013).

Water Transport: NMs can be dispersed in water bodies, where they may settle, attach to sediments, or be carried downstream, affecting aquatic organisms and sediment quality (Rawat et al., 2018; Williams et al., 2019).

Soil Transport: NPs can enter soils through deposition or application of NM-containing products, potentially affecting soil organisms and groundwater quality (Baalousha et al., 2016).

17.3.3 Environmental Behavior

Bioavailability: NMs' bioavailability to organisms depends on factors like their size, surface properties, and interactions with biomolecules, influencing their potential ecological impacts (Lowry et al., 2010).

Ecological Interactions: NMs can interact with organisms at different trophic levels, affecting their physiology, behavior, and ecosystem dynamics (Scheringer et al., 2014).

Persistence: Some NMs may persist in the environment for extended periods, raising concerns about long-term exposure and accumulation in ecosystems (Shareena et al., 2022).

17.3.4 Risk Assessment and Management

Monitoring: Monitoring NMs in environmental compartments is essential for assessing their distribution, concentrations, and potential risks (Williams et al., 2019).

Risk Assessment: Evaluating the hazards and exposure pathways of NMs helps in assessing their environmental and human health risks (Anawar and Strezov, 2019).

Regulatory Frameworks: Developing and implementing regulatory frameworks for NMs' safe use and disposal is crucial for sustainable environmental management (Garner and Keller, 2014).

17.3.5 Research Challenges

Knowledge Gaps: There are still gaps in understanding the behavior, interactions, and long-term effects of NMs in different environmental matrices (Dale et al., 2015).

Standardization: Lack of standardized methods for assessing NMs' fate and transport hinders accurate risk assessment and management strategies (Peijnenburg et al., 2015).

Multifactorial Effects: NMs' behavior in the environment is influenced by complex interactions among physical, chemical, and biological factors, requiring integrated research approaches (Klaine et al., 2008).

Addressing these challenges and improving our understanding of NMs' fate and transport in the environment is essential for ensuring their safe and sustainable use while minimizing potential adverse impacts.

17.3.6 NM Mobility and Stability in Various Environmental Media

NM mobility and stability in different environmental media, such as air, water, soil, and sediment, are influenced by various factors. Here's an overview of these factors and how they affect NM behavior:

Size and Shape: NP size and shape significantly influence their mobility and stability. Smaller NPs generally exhibit higher mobility due to reduced settling rates, while certain shapes (e.g., rods, fibers) may have different transport behaviors than spherical particles (Darlington et al., 2009).

Surface Chemistry: Surface properties, including charge, functional groups, and coatings, affect NM interactions with environmental media. Positively charged NPs may adhere more strongly to negatively charged surfaces in soils and sediments, impacting their mobility (Farré et al., 2011).

Aggregation and Dispersion: NMs can aggregate or disperse based on environmental conditions. Aggregation can reduce mobility by increasing particle size, while dispersion enhances mobility but may also increase environmental exposure (Ju-Nam and Lead, 2016).

Environmental pH and Ionic Strength: pH and ionic strength influence NM stability and interactions. Changes in pH can affect surface charge and aggregation, while high ionic strength can promote NP aggregation and sedimentation (Mortimer and Holden, 2019).

Organic Matter: Presence of organic matter in environmental media can affect NM stability and mobility. Organic compounds can adsorb onto NP surfaces, altering their behavior in soils, sediments, and water (Anitha et al., 2015).

Temperature and Climate: Environmental temperature and climate conditions can impact NM stability and transport. Temperature variations can affect NP aggregation kinetics, while climate factors like rainfall and humidity can influence NP dispersion in water and air (Lead et al., 2018).

Redox Conditions: Redox potential plays a role in NM transformations and interactions. NPs may undergo redox reactions in different environmental compartments, affecting their mobility and fate (Jośko and Oleszczuk, 2013).

Biological Interactions: NMs can interact with biota in environmental media, influencing their mobility and behavior. Biological processes like microbial degradation and bioaccumulation can alter NM fate and transport pathways (Suhendra et al., 2020).

Understanding these factors is essential for predicting and managing NM behavior in the environment. It requires interdisciplinary research, integrating physical, chemical, biological, and environmental sciences to develop comprehensive models and guidelines for safe NM use and disposal.

17.4 OPPORTUNITIES AND CHALLENGES OF USING NMS IN THE REMEDIATION OF POLLUTANTS

17.4.1 Opportunities of NMs in Pollution Remediation

NMs offer several opportunities for the remediation of pollutants. These materials possess unique properties at the nanoscale that make them highly efficient in capturing and removing contaminants from various environmental matrices (Figure 17.2). Moreover, due to their distinctive features, NMs can be tailor-made and functionalized to exhibit enhanced adsorption capacities, which ultimately makes them promising candidates for pollution remediation (Aguilar-Pérez et al., 2021). By leveraging the exceptional properties of NMs, there emerges the potential for targeted remediation, allowing for the selective removal of specific contaminants while leaving the surrounding environment untouched and unharmed. Furthermore, the utilization of NMs in pollution remediation processes holds a myriad of advantages. For instance, NMs enable the minimization of environmental impact in comparison to conventional remediation methods (Ahmed et al., 2022). This stems from the fact that NMs require smaller quantities of materials to effectively remove pollutants from the environment. Additionally, NMs possess the ability to be designed and engineered for reuse, further reducing the overall environmental footprint. Hence, the utilization of NMs in pollution remediation endeavors presents an opportunity for the development of innovative and sustainable solutions for pollution control and remediation. These solutions have the potential to revolutionize the field and contribute to the preservation and protection of our ecosystems for generations to come (Al-Maqdi et al., 2021).

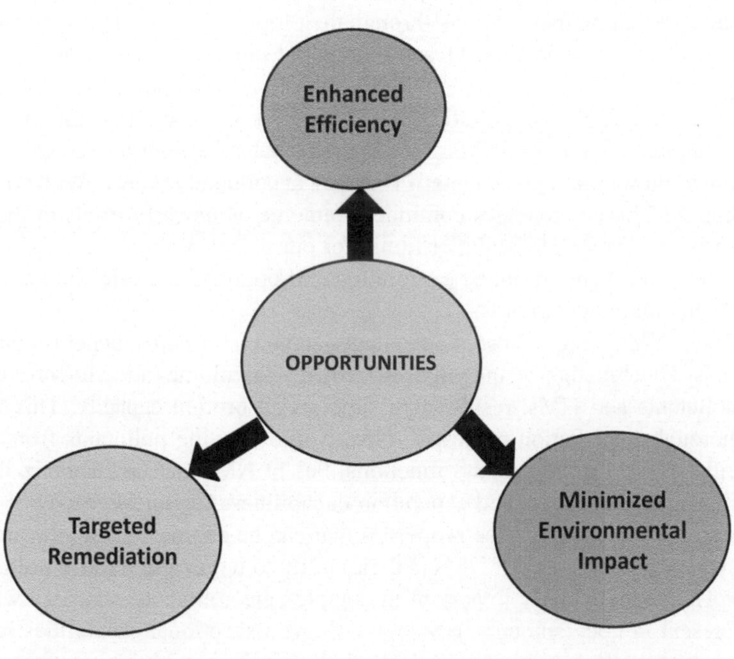

FIGURE 17.2 Opportunities of NMs in environmental remediation.

17.4.1.1 Enhanced Efficiency

The use of NMs in pollution remediation offers significantly enhanced efficiency compared to traditional remediation techniques. This is primarily due to their exceptionally small size and remarkable surface area-to-volume ratio, which allows for a significantly larger contact area for pollutant adsorption. This extended surface area allows for a substantially greater interaction between the NMs and contaminants, ultimately resulting in significantly improved pollutant removal efficiency. Furthermore, NMs can be meticulously engineered to possess specific surface properties and functionalities. These can include exceptional chemical reactivity or even magnetic responsiveness, both of which tremendously enhance their adsorption and separation capabilities. These enhanced properties enable NMs to effectively capture pollutants even at lower concentrations and in remarkably complex environmental matrices. Consequently, NMs have emerged as highly efficient tools in pollution remediation, allowing for the remediation of pollution on a scale which was not possible earlier. The distinctive capabilities possessed by NMs have been extensively explored and harnessed in pollution remediation. Researchers and scientists have continuously worked toward developing and optimizing the functionality of NMs, aiming to maximize their performance and efficacy. This ongoing research and development have resulted in a myriad of innovative NM-based strategies for pollution remediation (Bao et al., 2023). One such example is the utilization of NPs in the removal of heavy metals from contaminated sites. By carefully tailoring the composition and surface characteristics of NPs, researchers have achieved remarkable success in selectively adsorbing heavy metals, effectively detoxifying polluted areas. The high surface area-to-volume ratio of NMs greatly enhances their capacity for heavy metal adsorption, ensuring efficient and thorough remediation. Moreover, NMs have proven to be highly effective in the removal of organic contaminants from various environmental matrices. Their superior adsorption properties enable them to sequester organic pollutants, such as pesticides, industrial chemicals, and pharmaceutical residues, from soil, water, and air. The diverse surface functionalities of NMs can be precisely tuned to target specific organic contaminants, ensuring selective and efficient adsorption. Another cutting-edge application of NMs in pollution remediation is their integration into filtration systems. By incorporating NMs into membranes, researchers have created highly efficient filters capable of removing a wide range of pollutants from water and air. The densely packed and interconnected NM structures effectively trap contaminants, preventing their passage through the filtration system. This innovation has paved the way for the development of advanced water treatment technologies and air purification systems, contributing significantly to the preservation of the environment and public health.

In conclusion, the use of NMs in pollution remediation has revolutionized traditional techniques. Tapping into the unique properties of NMs, such as their small size, high surface area, and customizable functionalities, allows for unprecedented efficiency in pollutant removal. With ongoing research and development, NM-based strategies continue to emerge as powerful tools in the fight against pollution, promising a cleaner and healthier future for our planet (Bhatt et al., 2022). NMs possess incredible potential, and their continued exploration and optimization offer immense possibilities for the field of pollution remediation.

The utilization of NMs brings a host of advantages and transformative benefits compared to conventional methods. Their small size and large surface area-to-volume ratio allow for enhanced contact between pollutants and NMs, resulting in superior adsorption capacity. This leads to more efficient and thorough remediation processes, effectively removing pollutants from contaminated sites. The specific surface properties and functionalities of NMs can be customized and tailored, further enhancing their adsorption and separation capabilities. Chemical reactivity and magnetic responsiveness are examples of surface properties that can be engineered, significantly improving pollutant capture. A key advantage of NMs is their ability to target and remove pollutants at lower concentrations. This is particularly important in complex environmental matrices where contaminants may be present in trace amounts. NMs have shown exceptional capabilities in sequestering organic contaminants, such as pesticides, industrial chemicals, and pharmaceutical residues, from diverse matrices, including soil, water, and air (Brahmkhatri et al., 2021).

The versatile surface functionalities of NMs enable selective adsorption, ensuring efficient capture of specific organic pollutants. Integrating NMs into filtration systems is a cutting-edge approach to pollution remediation. The incorporation of NMs into membranes has led to the development of highly efficient filters capable of removing a wide range of pollutants from water and air. The densely packed and interconnected structures of NMs act as a barrier, effectively trapping contaminants and preventing their passage through the filtration system. This breakthrough in filtration technology has opened doors to advanced water treatment methods and air purification systems, significantly contributing to environmental preservation and public health. The use of NMs in pollution remediation continues to evolve as researchers work diligently to enhance their performance and further optimize their functionality (Chauhan et al., 2022).

Ongoing research and development efforts have resulted in a multitude of innovative NM-based strategies for pollution remediation. Scientists strive to maximize the efficiency and efficacy of NMs, ensuring their remarkable capabilities are fully utilized in the fight against pollution. The potential impact of NMs in pollution remediation cannot be overstated, as they promise a cleaner and healthier future for our planet. With their unique properties and continual advancements in the field, NMs are revolutionizing traditional techniques, bringing forth unprecedented efficiency in pollutant removal. In conclusion, NM-based strategies represent a powerful tool in combating pollution and creating a sustainable environment for future generations.

17.4.1.2 Targeted Remediation

NMs offer the potential for targeted remediation of pollutants, addressing the specific contaminants of concern while minimizing disruption to the surrounding environment. By modifying the surface properties and functionalizing NMs, they can be designed to selectively interact with and capture specific pollutants. This selectivity allows for the removal of targeted contaminants without affecting non-targeted substances, reducing the potential for unintended ecological consequences. Such targeted remediation approaches can be especially beneficial in sensitive ecosystems with limited tolerance for disturbance. In addition, NMs can be deployed in controlled release systems or integrated with monitoring technologies, enabling real-time tracking of pollutant levels and targeted remediation efforts when and where they are most needed. Furthermore, the versatility of NMs extends beyond their ability to target and capture pollutants. They can also serve as carriers for delivering drugs and therapeutic agents, revolutionizing the field of medicine.

The unique properties of NMs, such as their high surface area-to-volume ratio and tunable surface chemistry, allow for precise control over drug release kinetics and targeted delivery to specific sites within the body. By encapsulating therapeutic compounds within NMs, researchers have the potential to enhance drug stability, improve bioavailability, and reduce toxic side effects. This has led to advancements in personalized medicine, where treatments can be tailored to individual patients based on their unique genetic makeup and disease characteristics (Chen et al., 2023). Moreover, the applications of NMs are not limited to remediation and medicine. They have found utility in various industries, including electronics, energy, and catalysis. In electronics, NMs are used to create smaller, faster, and more efficient devices. They can be incorporated into transistors, sensors, and memory storage devices, allowing for high-performance electronics with improved energy efficiency (Ganie et al., 2021).

In the energy sector, NMs play a crucial role in the development of renewable energy sources. They can improve the efficiency of solar cells, facilitate hydrogen storage and fuel cell technologies, and improve energy storage systems. Additionally, NMs are widely used as catalysts in chemical reactions, enabling faster reaction rates, reduced energy consumption, and improved selectivity (Gottardo et al., 2021).

In conclusion, NMs offer immense potential for targeted remediation, personalized medicine, and technological advancements across various industries. Their unique properties and surface modifications enable selective interactions with pollutants, precise drug delivery, and enhanced performance in electronics, energy, and catalysis. As research continues to advance in the field of NMs, new applications and innovations are bound to emerge, further expanding their impact on society and the environment (Hussain et al., 2022).

17.4.1.3 Minimized Environmental Impact

The use of NMs in pollution remediation can lead to minimal environmental impact compared to traditional remediation methods. NMs can be designed to be highly efficient at pollutant capture, enabling the use of smaller quantities of materials for effective remediation (Intisar et al., 2022). This reduction in material usage can result in decreased operational costs and resource consumption, as well as reduced waste generation. Furthermore, the versatility and functionality of NMs allow for their potential reusability, further minimizing environmental impact and contributing to sustainable practices. Additionally, the targeted nature of NM-based remediation approaches can limit collateral damage to non-targeted organisms and ecosystems, ensuring a more sustainable and environmentally friendly remediation process that aligns with global conservation efforts (Joseph et al., 2023). By utilizing NMs, pollution remediation efforts can be optimized to achieve efficient and environmentally conscious outcomes, allowing for the preservation and conservation of natural resources for future generations. This advancement in technology presents opportunities for cleaner and healthier environments, where the detrimental effects of pollutants on human health and the ecosystem are significantly reduced or eliminated. Overall, the integration of NMs in pollution remediation represents a ground-breaking step toward a more sustainable, efficient, and environmentally conscious approach to combating pollution and restoring balance and harmony in our ecosystems (Khan et al., 2021).

17.4.2 Challenges of NMs in Pollution Remediation

The use of NMs in pollution remediation presents several challenges that need to be addressed. These challenges include toxicity concerns, regulatory issues, cost and scalability considerations, as well as ethical implications. Given the rapid advancements in nanotechnology, it becomes imperative to thoroughly understand the potential negative impacts on human health and the environment. Therefore, extensive research efforts are required to evaluate and mitigate any associated risks. The development and implementation of robust regulatory frameworks are crucial to ensure the safe and responsible use of NMs in pollution remediation practices. Moreover, the cost and scalability of NM-based remediation techniques need to be carefully evaluated to determine their feasibility and practicality for large-scale applications. While NMs have shown promise in enhancing the efficiency and targeted remediation of pollutants, it is essential to address economic constraints and intricacies (Liu et al., 2023).

Striking a balance between cost-effectiveness and environmental sustainability is of paramount importance. Additionally, the social and ethical implications surrounding the use of nanotechnology in pollution remediation cannot be overlooked. Stakeholder engagement, public awareness, and transparent communication are vital in fostering trust and understanding. Encouraging dialogue and addressing ethical concerns will help in navigating the complexities associated with the adoption of NMs for pollution remediation purposes (Manimegalai et al., 2023b). Furthermore, interdisciplinary collaborations involving scientists, policymakers, industry experts, and the public are necessary to ensure comprehensive decision-making and responsible technological development. In conclusion, the use of NMs in pollution remediation holds significant potential, but it is essential to address various challenges holistically. By addressing toxicity concerns, regulatory issues, cost and scalability considerations, as well as ethical implications, we can harness the transformative power of nanotechnology while ensuring the protection of human health and the environment. Through collaborative efforts and responsible innovation, we can pave the way for a sustainable and cleaner future (Mensah et al., 2021).

17.4.2.1 Toxicity Concerns

One of the key challenges associated with the utilization of NMs in pollution remediation is the concern over their toxicity. While NMs have undeniably illustrated the potential to efficiently eradicate pollutants, their diminutive size and substantial surface area can concurrently intensify their reactivity and conceivably endanger living organisms (Mofijur et al., 2023). It is of utmost importance to meticulously scrutinize the toxicity of NMs before their wide-ranging implementation, accounting for elements such as their physicochemical attributes, latent interactions with pollutants and the environment, and their potential to accumulate in biological systems (Mukhopadhyay et al., 2022). Comprehending the prospective hazards and implementing appropriate risk mitigation strategies plays a pivotal role in ensuring the secure and sustainable utilization of NMs in the field of pollution remediation (Roy et al., 2021b).

17.4.2.2 Regulatory Issues

The application of NMs in pollution remediation is accompanied by regulatory challenges that need to be addressed to ensure the safe and sustainable use of these materials. The use of NMs raises various questions and concerns regarding their potential impacts on human health and the environment. Regulatory agencies need to play a proactive role in assessing the risks associated with NMs and establishing robust guidelines and standards for their use. To effectively regulate the use of NMs, it is crucial to understand their fate and behavior in the environment. This includes studying their transport, transformation, and potential accumulation in different ecological systems. Regulatory agencies need to be equipped with the necessary tools and knowledge to evaluate the risks and benefits of these materials, taking into account their unique properties and behavior (Lowry and Casman, 2009).

Establishing comprehensive regulatory frameworks is essential for ensuring the responsible use of NMs in pollution remediation. These frameworks should not only focus on the safe handling and disposal of NMs but also address the monitoring and assessment of their long-term impacts. This requires collaboration between researchers, policymakers, and regulatory agencies to develop evidence-based regulations that can adapt to the evolving nature of NMs. Furthermore, regulatory agencies should consider the potential interactions between NMs and other substances in the environment. This includes exploring the potential synergistic effects or unintended consequences that may arise from the use of NMs in combination with other remediation techniques or pollutants (Saleem et al., 2022). By considering the broader context of pollution remediation, regulatory agencies can ensure that the use of NMs is harmonized with other strategies and does not create additional risks or challenges.

In conclusion, the application of NMs in pollution remediation holds great promise but also requires careful regulatory oversight. Collaboration between researchers, policymakers, and regulatory agencies is crucial to establishing comprehensive frameworks that consider the unique properties of NMs and provide clear guidance on their safe handling, disposal, and monitoring. By doing so, we can ensure the responsible and sustainable use of NMs in addressing the challenges of pollution remediation (Thakur and Kumar, 2023).

17.4.2.3 Cost and Scalability

The cost and scalability of using NMs in pollution remediation pose significant challenges that must be overcome. While NMs have shown promise in efficiently removing pollutants, their production, synthesis, and application can be quite expensive. Moreover, the scalability of NM-based remediation techniques is yet another concern that demands attention. The effectiveness of these techniques at small scales may not necessarily translate to large-scale applications, which raises further questions about their practicality. Therefore, the cost-effectiveness and practicality of NM-based remediation methods need to be meticulously evaluated (Thakur et al., 2024). This evaluation must take into account various factors, including the availability and cost of NMs, the manufacturing processes involved, and the intricate logistics required for implementing large-scale remediation projects. It is paramount to develop cost-effective and scalable approaches that can ensure the widespread adoption of NMs in pollution remediation. Only by doing so can we effectively tackle the pressing issue of pollution and its harmful consequences on our environment (Wang et al., 2022).

17.5 LIMITATIONS OF USING NMS IN THE REMEDIATION OF POLLUTANTS

Using NMs to clean up pollutants has great potential. However, several issues and restrictions must be resolved:

Cost: Producing NMs can be costly, particularly if they have particular qualities needed for pollutant remediation. It is difficult to increase production while maintaining control over costs.

Environmental Impact: There may be environmental repercussions from the production and disposal of NMs. It's critical to take into account these materials' entire life cycle, including any potential toxicity and ecological effects.

Risk of NP Release: NMs used to remediate pollutants may unintentionally release NPs into the environment, which could have negative effects on ecosystems (Kania et al., 2018).

Regulatory Obstacles: The use of NMs in environmental applications may not be sufficiently covered by current regulations. It is difficult but necessary to establish clear guidelines for their safe use and disposal (Rather et al., 2023).

Long-Term Stability: The efficacy of NMs used in remediation must be sustained over time. Their performance and stability can be impacted by elements like surface modification, aggregation, and degradation (Vashchenko et al., 2017).

Selectivity and Specificity: Achieving selectivity for target pollutants while avoiding interference with non-target compounds can be difficult, despite the high reactivity of many NMs (Elsaid et al., 2023).

Scale-up Challenges: Although studies conducted on a laboratory scale frequently show that NMs are effective at remediating pollutants, bringing these processes to a real-world setting presents some difficulties, including preserving cost-effectiveness and performance consistency (Rani et al., 2021).

Interactions with Environmental Elements: Organic matter, minerals, and microbial communities are just a few of the elements in the environment that NMs may interact with. Comprehending these interplays is essential to precisely projecting their conduct and possible consequences (Bhattu and Singh, 2023).

Ethical Concerns: The use of NMs in environmental remediation raises ethical questions about fair distribution of risks and benefits, as well as possible unforeseen consequences (Wang et al., 2019).

Public Perception and Acceptance: The use of NM-based remediation techniques may be influenced by public acceptance of nanotechnology in environmental applications (Dhanapal et al., 2024).

Variability in External Environments: Environmental factors, such as pH, temperature, and the existence of additional contaminants, can differ significantly between sites. One of the biggest challenges in nanotechnology is creating materials that function consistently in a variety of environments (Kaushik et al., 2010).

Limited Knowledge of Fate and Transport: Research is still being done to better understand how NMs behave once they are released into the environment, including their mobility, bioavailability, and potential for accumulation. Strategies for risk assessment and management are complicated by this knowledge gap (Kumar et al., 2023).

Nano-Bio Interactions: NMs can have intricate interactions with biological systems that could have unanticipated effects on the environment or human health. Comprehending the aforementioned nano-bio interactions is imperative in evaluating the safety and effectiveness of remediation strategies based on NMs (Rani et al., 2017).

Observation and Identification: Because of their small size and low concentrations, NMs are difficult to detect and quantify in environmental matrices. It is crucial to create sensitive and trustworthy detection techniques to keep an eye on their existence and any possible effects on the environment (Inobeme et al., 2023a).

Resource Restrictions: The synthesis of certain NMs necessitates expensive or rare resources, which presents problems for their widespread use. Although ongoing, research into substitute materials with similar qualities but fewer resource requirements is essential.

Energy Consumption: NMs' overall environmental footprint is influenced by the significant energy inputs required for their production and deployment. The sustainability of remediation techniques based on NMs can be enhanced by identifying strategies to reduce energy consumption, both in the synthesis and application phases (Inobeme et al., 2023b).

Lack of Standardization: Repeatability and cross-study comparability are hampered by the absence of standard protocols for the synthesis, characterization, and testing of NMs. Creating uniform standards and procedures can help the field advance and increase trust in the efficacy of remediation techniques based on NMs.

Social and Economic Equity: It's possible that the advantages of remediation technologies based on NMs won't be distributed equally, which could make already-existing social and economic inequalities worse. Promoting environmental justice requires making sure that these technologies are implemented and have equitable access.

Nanotoxicity: Although some NMs show promise in the remediation of pollutants, others may be hazardous to people, animals, and ecosystems. It's critical to comprehend the potentially harmful effects of NMs and to develop mitigation strategies (Manimegalai, et al., 2023a).

Limited Remediation Efficacy for Specific Pollutants: NMs may not be an effective remediation method for all pollutants. It's possible that some contaminants and NMs don't interact well, or that the remediation procedure is too laborious or slow to be useful.

Possibility of Secondary Pollution: Using NMs to remediate pollution may unintentionally cause secondary pollutants or by-products to be produced. To stop additional environmental harm, these unintended consequences must be evaluated and mitigated.

Environmental Systems' Complexity: Because they have many interrelated processes and feedback loops, environmental systems are by nature complex. Forecasting the long-term effects of remediation based on NMs on these systems thorough modeling and risk assessment techniques is needed.

Legacy Pollution: Remediation techniques based on NMs might not be appropriate for dealing with pollutants that have long since built up in the environment. Addressing such persistent contaminants may require the use of alternative remediation techniques (Sharma et al., 2023).

Public Perception and Trust: How the general public views nanotechnology in environmental remediation can have an impact on how widely it is used and accepted. The effective application of NM-based remediation technologies depends on fostering public trust via open communication, stakeholder involvement, and responsible innovation.

Commercialization and Intellectual Property Barriers: These two factors could prevent the widespread application of remediation technologies based on NMs. The government, business, and academic community must work together to remove these obstacles and hasten the adoption and transfer of new technologies (Ansari et al., 2023).

Global Issues and Equity: Reducing pollution is a worldwide issue that calls for cooperation across national and international borders. Realistic environmental impact requires global knowledge and resource sharing, as well as equitable access to remediation technologies based on NMs (Cheng et al., 2023).

Long-Term Monitoring and Maintenance: To guarantee the continued efficacy of NM-based remediation technologies and avoid unforeseen environmental repercussions, long-term monitoring and maintenance are required after they are implemented (Nzilu et al., 2023).

Gaining the public's trust requires addressing worries about efficacy, safety, and possible risks. To create safe, efficient, and long-lasting NM-based solutions for pollutant remediation, scientists, engineers, legislators, and stakeholders must work together across disciplinary boundaries. In addition, to get past these obstacles and realize the full promise of nanotechnology for environmental protection, more research and innovation are required. To overcome these obstacles, the scientific community, legislators, industry stakeholders, and the general public must work together to promote the ethical development and application of NM-based solutions for pollutant remediation that minimize potential risks and maximize benefits.

It is crucial to create sustainable management plans to keep these technologies in use. A comprehensive strategy that takes into account socioeconomic, ethical, technical, scientific, and regulatory aspects is needed to address these obstacles and constraints. To fully realize the potential of NMs in reducing pollution and safeguarding the environment, more research, creativity, and cooperation are needed.

17.6 REGULATORY AND ETHICAL CONSIDERATIONS

17.6.1 Current Regulations and Guidelines for the Use of NMs in Environmental Applications

It's fascinating to explore the distinctions between comprehensive and limited nano safety regulatory regimes in the context of international politics and the landscape of nanotechnology risk policies. Understanding these different approaches is crucial for navigating the complexities of nanotechnology regulation and its impact on global markets. Comprehensive nano safety regulatory regimes typically involve a more thorough and proactive approach to managing nanotechnology risks. These regimes often prioritize robust risk assessment, monitoring, and regulatory oversight to ensure the safe and responsible use of NMs. On the other hand, limited nano safety regulatory regimes may have a narrower focus and fewer regulatory requirements, potentially leading to gaps in oversight and risk management. However, as you mentioned, the limited approach has faced increasing pressure to evolve toward a more stringent, proactive, and precautionary stance in light of growing concerns about nanotechnology risks and uncertainties. The shift toward a more precautionary approach reflects the recognition of the importance of addressing potential risks associated with nanotechnology early on to prevent adverse impacts on human health and the environment. This evolution in regulatory approaches highlights the dynamic nature of nanotechnology governance and the need to adapt to emerging challenges and scientific knowledge.

Regulatory frameworks and standards for NMs and nanotechnologies can be challenging due to differences in conceptualization, characterization techniques, toxicity/ecotoxicity tests, and synthesis methods among countries. These variations can create obstacles to achieving consistent and effective regulations at the international level.

Conceptualization: Different countries may have varying definitions and classifications of NMs, leading to discrepancies in how these materials are regulated and managed.

Characterization techniques: The methods used to characterize NMs can vary, affecting how their properties and potential risks are assessed. Standardizing characterization techniques can help ensure more accurate and reliable data for regulatory decision-making.

Toxicity/ecotoxicity tests: There may be differences in the testing protocols and criteria used to assess the toxicity and ecotoxicity of NMs, making it challenging to compare results and establish consistent risk assessment frameworks.

Synthesis methods: Variations in the synthesis methods of NMs can impact their properties and behavior, influencing their potential risks and regulatory requirements. Standardizing synthesis methods can promote better understanding and control of NM properties.

Addressing these challenges requires international collaboration, coordination, and efforts to harmonize regulatory approaches and standards for NMs and nanotechnologies. By promoting information sharing, mutual recognition of testing data, and alignment of regulatory principles, countries can work together to overcome these obstacles and establish more coherent and effective regulatory frameworks (Chavez-Hernandez et al., 2024).

The use of Life Cycle Assessments (LCAs) and Risk Assessments (RAs) in regulatory frameworks and legislations for NMs is crucial for understanding the potential toxic effects on health, the environment, and the different stages of NMs' life cycle. It's important to conduct LCAs and RAs specifically on NMs themselves, rather than on the final manufactured products, as the life cycles of NMs can differ significantly. By focusing on NMs in their raw form, regulators can gain a more accurate understanding of the potential risks and impacts associated with these materials throughout their life cycle, from production to disposal. This targeted approach allows for more precise and tailored regulatory measures to be developed to address the specific characteristics and behaviors of NMs. Establishing clear and standardized methodologies for conducting LCAs and RAs on NMs is essential to ensure consistency and reliability in regulatory decision-making. Ambiguity must be minimized to avoid confusion and ensure that regulatory frameworks at both national and international levels are well-defined and effective. The integration of LCAs and RAs into regulatory frameworks provides a powerful tool for developing more informed and proactive legislation that can mitigate potential risks associated with NMs. When used in conjunction, LCAs and RAs can offer complementary insights into the environmental and health impacts of NMs, enabling regulators to make more precise and evidence-based decisions. Overall, the systematic inclusion of LCAs and RAs in regulatory processes worldwide is key to advancing the responsible governance of NMs and ensuring the development of robust and effective regulatory frameworks (Linkov et al., 2007).

Ethical considerations play a crucial role in ensuring that regulations for NMs are not only effective but also aligned with societal values and expectations. In addition to the technical knowledge, frameworks, and legislation governing NMs, it's essential to prioritize consumer awareness of the benefits and risks associated with these materials. Clear communication channels between scientific-academic institutions, governments, and society are vital to ensure that the public is well-informed and engaged in discussions about nanotechnology. Ethical regulation should extend beyond the research and development of NMs to include aspects such as patenting. Proper regulation of nanotechnology patents should encompass high-quality standards, risk assessment, communication strategies, and the evaluation of risks and benefits to human health and the environment. By applying ethical principles to patent regulation, we can ensure that innovations in nanotechnology are guided by considerations of safety, utility, and societal well-being. Ultimately, the goal of NM regulations should be to guarantee the safety and efficacy of nanotechnological products for consumers, while fostering public trust and supporting ongoing scientific advancements. By upholding ethical standards in the development and regulation of NMs, we can promote responsible innovation, improve quality of life, and safeguard environmental sustainability (Hodge et al., 2010).

By implementing effective regulatory measures for NMs, we can work toward minimizing environmental impacts and ensuring responsible management of these advanced materials. These frameworks can help reduce the negative effects of anthropogenic waste on ecosystems, biodiversity, and overall environmental health. Furthermore, continuous exposure to NMs, whether direct or indirect, can pose potential risks to human health. Regulatory frameworks play a key role in safeguarding human well-being by setting standards, guidelines, and protocols to minimize or prevent adverse health effects associated with NM exposure. Through proactive regulatory oversight, risk assessment, and monitoring, we can better understand and address the potential health and environmental implications of nanotechnology. By integrating ethical considerations, scientific knowledge, and stakeholder engagement, regulatory frameworks can support the safe and sustainable development of NMs while protecting both human health and the environment (Corsi, et al., 2018).

17.6.2 Ethical Considerations in Deploying NMs for Pollution Control

Ethical considerations are paramount when deploying NMs for pollution control. Here are some key ethical aspects to consider:

Ensuring the safety of both human health and the environment is crucial when using NMs for pollution control. Ethical practices involve conducting thorough risk assessments to understand the potential impacts of these materials on ecosystems and human populations.

Transparent communication about the risks and benefits of using NMs for pollution control is essential. Providing clear information to all stakeholders, including communities affected by pollution, fosters trust and informed decision-making. It is important to consider equity in the deployment of NMs for pollution control.

Ensuring that vulnerable populations are not disproportionately impacted and have a voice in decision-making processes is essential for ethical practice. Ethical deployment of NMs for pollution control involves considering long-term sustainability. This includes evaluating the lifecycle impacts of these materials, promoting circular economy principles, and minimizing unintended consequences on ecosystems.

Adhering to existing regulations and standards related to NM use is a fundamental ethical consideration.

Compliance with legal frameworks helps protect public health and the environment, as well as ensures responsible innovation. By incorporating ethical considerations into the deployment of nanomaterials (NMs) for pollution control, organizations can develop effective, sustainable, and socially responsible solutions that address environmental challenges while adhering to ethical principles.

Taking a balanced approach that considers ethical, environmental, and regulatory aspects is crucial when harnessing the benefits of nanotechnology while mitigating potential risks. Some key considerations for achieving this balance follow:

Ethical Considerations that uphold ethical principles, such as safety, transparency, equity, and sustainability, are essential in the development and deployment of nanotechnology.

Ensuring the maximum benefits of nanotechnology while minimizing harm to individuals and the environment, is a key ethical responsibility. Understanding and mitigating the environmental impact of nanotechnology is vital. This includes assessing the lifecycle impacts of NMs, promoting eco-friendly practices, and prioritizing pollution prevention and resource conservation.

Regulatory adhering to regulatory frameworks and standards is critical for ensuring the safe and responsible use of nanotechnology. Compliance with regulations helps protect public health and the environment. On the other hand, compliance also maintains accountability in the development and deployment of nanotechnological solutions.

Implementing robust risk management strategies is essential to identify, assess, and mitigate potential risks associated with nanotechnology. This includes conducting thorough risk assessments, monitoring NM exposure, and implementing precautionary measures to safeguard human health and the environment (Akshay et al., 2024).

By integrating ethical, environmental, and regulatory considerations into decision-making processes related to nanotechnology, we can strike a balance that maximizes the benefits of nanotechnology innovation, while minimizing potential risks. This holistic approach fosters responsible and sustainable development in the field of nanotechnology (Dada et al., 2024). Because of the possible effects on human health and socioeconomic equity, it is crucial to ensure the ethical application of nanotechnology applications.

17.7 FUTURE DIRECTIONS AND EMERGING TECHNOLOGIES

Future directions and emerging technologies in NM design for improved remediation, integration with other remediation technologies, and potential breakthroughs include:

Tailored NMs for Specific Contaminants: Designing NMs with specific properties (e.g., surface functionalization, size, morphology) tailored for the remediation of targeted contaminants (e.g., heavy metals, organic pollutants) to enhance removal efficiency and selectivity (Lai et al., 2018).

Multifunctional NMs: Developing multifunctional NMs that combine remediation capabilities (e.g., adsorption, catalysis, degradation) with sensing or monitoring functionalities for real-time environmental assessment and feedback-controlled remediation (Isibor et al., 2024).

Smart NMs with Stimuli-Responsive Properties: Engineering NMs with stimuli-responsive properties (e.g., pH-responsive, temperature-responsive) to enable controlled release of remediation agents or targeted responses to environmental conditions for enhanced efficiency and reduced environmental impact (Sohail et al., 2019).

NMs for In-Situ Remediation: Advancing in situ remediation technologies using NMs that can be delivered and activated directly at contaminated sites, reducing costs, minimizing environmental disturbance, and improving remediation effectiveness (Nnaji et al., 2018).

Integration of NMs with Biological Remediation: Integrating NMs with biological remediation approaches (e.g., bioremediation, phytoremediation) to synergistically enhance contaminant removal, microbial activity, and plant uptake, leading to more sustainable and comprehensive remediation strategies (Rao and Cheetham, 2001).

Nano-Biohybrid Systems: Developing nano-biohybrid systems that combine NMs with engineered microbes or enzymes for targeted and efficient degradation of contaminants, offering potential breakthroughs in complex pollutant remediation scenarios (Roco et al., 2011).

NMs for Emerging Contaminants: Researching and developing NM-based solutions for emerging contaminants (e.g., pharmaceuticals, microplastics, per-, and polyfluoroalkyl substances) to address evolving environmental challenges and ensure comprehensive remediation strategies (Palit, 2018).

NMs for Contaminated Water Treatment: Exploring advanced NM-based technologies for water treatment, including nanofiltration membranes, photocatalytic NPs, and hybrid NMs for simultaneous removal of multiple contaminants and improved water quality (Liu et al., 2016).

Sustainable NM Synthesis and Application: Emphasizing sustainable synthesis methods and life cycle assessments of NMs to ensure environmental compatibility, reduce resource consumption, and mitigate potential risks associated with their production and use (Roco et al., 2011).

Regulatory Frameworks and Risk Assessment: Developing robust regulatory frameworks, risk assessment methodologies, and guidelines for the safe and responsible use of NMs in remediation applications, addressing concerns related to environmental fate, toxicity, and long-term impacts.

These future directions and emerging technologies represent a proactive approach to advancing NM-based remediation strategies, fostering innovation, sustainability, and efficacy in addressing environmental contamination challenges.

17.8 CONCLUSIONS AND RECOMMENDATIONS

The utilization of NMs in pollutant remediation presents a dual scenario of both advantages and hurdles. Some significant findings and insights follow. NMs boast high surface area-to-volume ratios, thereby enhancing their efficacy in the adsorption and degradation of pollutants. They can be customized to target specific pollutants, thereby minimizing adverse impacts on the environment. Moreover, they exhibit versatility by being applicable in various domains, such as water and soil remediation, air purification, and waste treatment. NPs employ distinct mechanisms like photocatalysis, utilizing light to facilitate pollutant degradation. Despite the initial investment, the long-term benefits of NMs might outweigh those of traditional remediation methods, due to their heightened efficiency and durability. Nevertheless, concerns emerge regarding the potential toxicity and inadvertent environmental ramifications associated with NMs, including their potential release into ecosystems. Additionally, NPs may tend to agglomerate, lose efficacy, or degrade over time, thereby diminishing their effectiveness in pollutant remediation. The scalability of NM synthesis and deployment for large-scale remediation initiatives remains a significant challenge. Furthermore, the absence of comprehensive regulations and standardized testing protocols acts as a barrier to the widespread adoption of NMs in remediation efforts. Efforts to address these challenges while capitalizing on the opportunities offered by NMs can pave the way for the development of more efficient and sustainable strategies for pollutant remediation.

REFERENCES

Aguilar-Pérez, K. M., J. I. Avilés-Castrillo, Gustavo Ruiz-Pulido, Dora I. Medina, Roberto Parra-Saldivar, and Hafiz MN Iqbal. "Nanoadsorbents in focus for the remediation of environmentally-related contaminants with rising toxicity concerns." *Science of the Total Environment* 779 (2021): 146465.

Ahmed, Shams Forruque, M. Mofijur, Nazifa Rafa, Anika Tasnim Chowdhury, Sidratun Chowdhury, Muntasha Nahrin, ABM Saiful Islam, and Hwai Chyuan Ong. "Green approaches in synthesising NMs for environmental nanobioremediation: Technological advancements, applications, benefits and challenges." *Environmental Research* 204 (2022): 111967.

Akshay Mehta, Alkesh Yadav, Aman Kumar, Kanika and Manish. "Role of nanotechnology in horticulture: An overview." 8(1), (2024): 702–708.

Alidoust, Mona, Yu Saito, Hideshige Takada, Kaoruko Mizukawa, Therese Karlsson, Sara Brosché, Bjorn Beeler, and Hrissi K. Karapanagioti. "A Unique Monitoring Method for Fecal and Sewage-Derived Chemical Pollution Utilizing International Pellet Watch Approach." *Environmental Science & Technology* 58(10, (2024).

Al-Maqdi, Khadega A., Nada Elmerhi, Khawlah Athamneh, Muhammad Bilal, Ahmed Alzamly, Syed Salman Ashraf, and Iltaf Shah. "Challenges and recent advances in enzyme-mediated wastewater remediation: A review." *NMs* 11, no. 11 (2021): 3124.

Anawar, Hossain Md, and Vladimir Strezov. "Transport, fate, and toxicity of the emerging and NM contaminants in aquatic ecosystems: Removal by natural processes." In Ajay Kumar Mishra, Hossain M.D. Anawar, Nadjib Drouiche (eds.), *Emerging and NM Contaminants in Wastewater*, pp. 43–62. Elsevier, 2019. https://doi.org/10.1016/B978-0-12-814673-6.00002-4

Ansari, Amad Ashraf, Mohd Arham Shamim, Amjad Mumtaz Khan, Khursheed Anwar, and Ajaz Ahmad Wani. "NMs as a cutting edge in the removal of toxic contaminants from water." *Materials Chemistry and Physics* 295 (2023): 127092.

Baalousha, M., G. Cornelis, T. A. J. Kuhlbusch, I. Lynch, C. Nickel, W. J. G. M. Peijnenburg, and N. W. Van Den Brink. "Modeling NM fate and uptake in the environment: current knowledge and future trends." *Environmental Science: Nano* 3, no. 2 (2016): 323–345.

Bao, Jianfeng, Shuangshaung Guo, Dandan Fan, Jingliang Cheng, Yong Zhang, and Xin Pang. "Sonoactivated NMs: A potent armament for wastewater treatment." *Ultrasonics Sonochemistry* 99 (2023): 106569.

Batley, Graeme E., Jason K. Kirby, and Michael J. McLaughlin. "Fate and risks of NMs in aquatic and terrestrial environments." *Accounts of Chemical Research* 46, no. 3 (2013): 854–862.

Roy, Saurabh, and Saurabh Bhatia. "NPs types, classification, characterization, fabrication methods and drug delivery applications." *Natural Polymer Drug Delivery Systems: NPs, Plants, and Algae* (2016): 33–93.

Bhatt, Pankaj, Satish Chandra Pandey, Samiksha Joshi, Parul Chaudhary, Vinay Mohan Pathak, Yaohua Huang, Xiaozhen Wu, Zhe Zhou, and Shaohua Chen. "Nanobioremediation: A sustainable approach for the removal of toxic pollutants from the environment." *Journal of Hazardous Materials* 427 (2022): 128033.

Bhattu, Monika, and Jagpreet Singh. "Recent advances in NMs based sustainable approaches for mitigation of emerging organic pollutants." *Chemosphere* 321 (2023): 138072.

Brahmkhatri, Varsha, Parimal Pandit, Pranita Rananaware, Aviva D'Souza, and Mahaveer D. Kurkuri. "Recent progress in detection of chemical and biological toxins in Water using plasmonic nanosensors." *Trends in Environmental Analytical Chemistry* 30 (2021): e00117.

Candolin, Ulrika. "Coping with light pollution in urban environments: patterns and challenges." *iScience, Volume 27, Issue 3* (2024), 109244.

Chauhan, Gaurav, Reyna Berenice González-González, and Hafiz MN Iqbal. "Bioremediation and decontamination potentials of metallic NPs loaded nanohybrid matrices: A review." *Environmental Research* 204 (2022): 112407.

Chavez-Hernandez, Jorge Antonio, Aida Jimena Velarde-Salcedo, Gabriela Navarro-Tovar, and Carmen Gonzalez. "Safe NMs: From their use, application and disposal to regulations." *Nanoscale Advances*, vol 6, (2024), 1583–1610.

Chen, Zhongshan, Yang Li, Yawen Cai, Suhua Wang, Baowei Hu, Bingfeng Li, Xiaodong Ding, Li Zhuang, and Xiangke Wang. "Application of covalent organic frameworks and metal–organic frameworks NMs in organic/inorganic pollutants removal from solutions through sorption-catalysis strategies." *Carbon Research* 2, no. 1 (2023): 8.

Cheng, Yueqin, Changlei Xia, Hakim AL Garalleh, Mazen Garaleh, Nguyen Thuy Lan Chi, and Kathirvel Brindhadevi. "A review on optimistic development of polymeric nanocomposite membrane on environmental remediation." *Chemosphere* 315 (2023): 137706.

Corsi, I., M. Winther-Nielsen, R. Sethi, C. Punta, C. Della Torre, G. Libralato, G. Lofrano et al. "Ecofriendly nanotechnologies and NMs for environmental applications: Key issue and consensus recommendations for sustainable and ecosafe nanoremediation." *Ecotoxicology and Environmental Safety* 154 (2018): 237–244.

Dada, Michael Ayorinde, Johnson Sunday Oliha, Michael Tega Majemite, Alexander Obaigbena, Zamathula Queen Sikhakhane Nwokediegwu, and Onyeka Henry Daraojimba. "Review of nanotechnology in water treatment: Adoption in the USA and Prospects for Africa." *World Journal of Advanced Research and Reviews* 21, no. 1 (2024): 1412–1421.

Dale, Amy L., Elizabeth A. Casman, Gregory V. Lowry, Jamie R. Lead, Enrica Viparelli, and Mohammed Baalousha. "Modeling NM environmental fate in aquatic systems." *Environmental Science & Technology* 2015 *49*(5), 2587–2593. DOI: 10.1021/es505076w

Darlington, Thomas K., Arianne M. Neigh, Matthew T. Spencer, Oanh TN Guyen, and Steven J. Oldenburg. "NP characteristics affecting environmental fate and transport through soil." *Environmental Toxicology and Chemistry: An International Journal* 28, no. 6 (2009): 1191–1199.

Dey, Sudeshna, Akankshya Das, Krishnamayee Mallick, Aishwarya Sahu, and Alok Prasad Das. "Environmental petroleum waste: Pollution, toxicity, sustainable remediation." In Ipsita Diparmitra Behera and Alok Prasad Das (eds.), *Impact of Petroleum Waste on Environmental Pollution and its Sustainable Management Through Circular Economy*, pp. 159–175. Cham: Springer Nature Switzerland, 2024.

Dhanapal, Anand Raj, Muthu Thiruvengadam, Jayavarshini Vairavanathan, Baskar Venkidasamy, Maheswaran Easwaran, and Mansour Ghorbanpour. "Nanotechnology Approaches for the Remediation of Agricultural Polluted Soils." *ACS Omega* (2024) 9(12):13522–13533.

Elsaid, Khaled, A. G. Olabi, Ahmed Abdel-Wahab, Ali Elkamel, Abdul Hai Alami, Abrar Inayat, Kyu-Jung Chae, and Mohammad Ali Abdelkareem. "Membrane processes for environmental remediation of NMs: Potentials and challenges." *Science of the Total Environment* 879 (2023): 162569.

Farré, Marinella, Josep Sanchís, and Damià Barceló. "Analysis and assessment of the occurrence, the fate and the behavior of NMs in the environment." *TrAC Trends in Analytical Chemistry* 30, no. 3 (2011): 517–527.

Ganie, Adil Shafi, Sayfa Bano, Nishat Khan, Saima Sultana, Zubair Rehman, Mohammed M. Rahman, Suhail Sabir, Frederic Coulon, and Mohammad Zain Khan. "Nanoremediation technologies for sustainable remediation of contaminated environments: Recent advances and challenges." *Chemosphere* 275 (2021): 130065.

Garner, Kendra L., and Arturo A. Keller. "Emerging patterns for engineered NMs in the environment: a review of fate and toxicity studies." *Journal of NP Research* 16 (2014): 1–28.

Gottardo, Stefania, Agnieszka Mech, Jana Drbohlavová, Aleksandra Małyska, Søren Bøwadt, Juan Riego Sintes, and Hubert Rauscher. "Towards safe and sustainable innovation in nanotechnology: State-of-play for smart NMs." *NanoImpact* 21 (2021): 100297.

Hochella, Michael F., Michael G. Spencer, and Kimberly L. Jones. "Nanotechnology: Nature's gift or scientists' brainchild?." *Environmental Science: Nano* 2, no. 2 (2015): 114–119.

Hodge, Graeme A., Diana M. Bowman, and Andrew D. Maynard. In Edward Edgar (ed.), *International Handbook on Regulating Nanotechnologies*. Edward Elgar Publishing Ltd., 2010. USA.

Hussain, Asim, Fazeelat Rehman, Hamza Rafeeq, Muhammad Waqas, Asma Asghar, Nadia Afsheen, Abbas Rahdar, Muhammad Bilal, and Hafiz MN Iqbal. "In-situ, Ex-situ, and nano-remediation strategies to treat polluted soil, water, and air: A review." *Chemosphere* 289 (2022): 133252.

Inobeme, Abel, John Tsado Mathew, Charles Oluwaseun Adetunji, Alexander Ikechukwu Ajai, Jonathan Inobeme, Muniratu Maliki, Stanley Okonkwo et al. "Recent advances in nanotechnology for remediation of heavy metals." *Environmental Monitoring and Assessment* 195, no. 1 (2023a): 111.

Inobeme, Abel, John Tsado Mathew, Charles Oluwaseun Adetunji, Alexander Ikechukwu Ajai, Jonathan Inobeme, Muniratu Maliki, Stanley Okonkwo et al. "Recent advances in nanotechnology for remediation of heavy metals." *Environmental Monitoring and Assessment* 195, no. 1 (2023b): 111.

Intisar, Azeem, Arooj Ramzan, Tehzeeb Sawaira, Ama Tul Kareem, Nazim Hussain, Muhammad Imran Din, Muhammad Bilal, and Hafiz MN Iqbal. "Occurrence, toxic effects, and mitigation of pesticides as emerging environmental pollutants using robust NMs–A review." *Chemosphere* 293 (2022): 133538.

Isibor, Patrick Omoregie, Ifeoluwa Ihotu Kayode-Edwards, and Olugbenga Samson Taiwo."Emerging Technology and Future Directions in Environmental Nanotoxicology." In *Environmental Nanotoxicology: Combatting the Minute Contaminants*, pp. 325–346. Cham: Springer Nature Switzerland, 2024.

Jeevanandam, Jaison, Ahmed Barhoum, Yen S. Chan, Alain Dufresne, and Michael K. Danquah. "Review on NPs and nanostructured materials: history, sources, toxicity and regulations." *Beilstein Journal of Nanotechnology* 9, no. 1 (2018): 1050–1074.

Joseph, Tomy Muringayil, Hussein E. Al-Hazmi, Bogna Sniatała, Amin Esmaeili, and Sajjad Habibzadeh. "NPs and nanofiltration for wastewater treatment: From polluted to fresh water." *Environmental Research, 238, Part 1*, (2023): 117114.

Jośko, Izabela, and Patryk Oleszczuk. "Manufactured NMs: the connection between environmental fate and toxicity." *Critical Reviews in Environmental Science and Technology* 43, no. 23 (2013): 2581–2616.

Ju-Nam, Yon, and Jamie Lead. "Properties, sources, pathways, and fate of NPs in the environment." *Engineered NPs and the Environment Biophysicochemical Processes and Toxicity* 4 (2016): 95–117.

Kania, Gabriela, Magdalena Sternak, Agnieszka Jasztal, Stefan Chlopicki, Agnieszka Błażejczyk, Anna Nasulewicz-Goldeman, Joanna Wietrzyk et al. "Uptake and bioreactivity of charged chitosan-coated superparamagnetic NPs as promising contrast agents for magnetic resonance imaging." *Nanomedicine: Nanotechnology, Biology and Medicine* 14, no. 1 (2018): 131–140.

Kaur, Simran, and Arpita Roy. "Bioremediation of heavy metals from wastewater using NMs." *Environment, Development and Sustainability* 23 (2021): 9617–9640.

Kaushik, N. T. et al., "Biological synthesis of metallic NPs, Nano-med." *Bio-med, Nanotechnology* 6 (2010): 257–262.

Khan, Samreen Heena. "Green nanotechnology for the environment and sustainable development." *Green Materials for Wastewater Treatment* In: Naushad, M., Lichtfouse, E. (eds.) Environmental Chemistry for a Sustainable World, vol 38. Springer, Cham: (2020):13–46. https://doi.org/10.1007/978-3-030-17724-9_2

Khan, Shamshad, Mu Naushad, Adel Al-Gheethi, and Jibran Iqbal. "Engineered NPs for removal of pollutants from wastewater: Current status and future prospects of nanotechnology for remediation strategies." *Journal of Environmental Chemical Engineering* 9, no. 5 (2021): 106160.

Klaine, Stephen J., Pedro JJ Alvarez, Graeme E. Batley, Teresa F. Fernandes, Richard D. Handy, Delina Y. Lyon, Shaily Mahendra, Michael J. McLaughlin, and Jamie R. Lead. "NMs in the environment: Behavior, fate, bioavailability, and effects." *Environmental Toxicology and Chemistry: An International Journal* 27, no. 9 (2008): 1825–1851.

Kumar, Sandeep, Sangita Yadav, Navish Kataria, Amit Kumar Chauhan, Seema Joshi, Renuka Gupta, Parmod Kumar, Jun Wei Roy Chong, Kuan Shiong Khoo, and Pau Loke Show. "Recent advancement in nanotechnology for the treatment of pharmaceutical wastewater: sources, toxicity, and remediation technology." *Current Pollution Reports* 9, no. 2 (2023): 110–142.

Anitha, Kunhikrishnan, Ho Kyong Shon, Nanthi S. Bolan, Ibrahim El Saliby, and Saravanamuthu Vigneswaran. "Sources, distribution, environmental fate, and ecological effects of NMs in wastewater streams." *Critical Reviews in Environmental Science and Technology* 45, no. 4 (2015): 277–318.

Lai, Racliffe WS, Katie WY Yeung, Mana MN Yung, Aleksandra B. Djurišić, John P. Giesy, and Kenneth MY Leung. "Regulation of engineered NMs: Current challenges, insights and future directions." *Environmental Science and Pollution Research* 25 (2018): 3060–3077.

Landrigan, Philip J., Michael Britt, Samantha Fisher, Amelia Holmes, Manasi Kumar, Jenna Mu, Isabella Rizzo, Anna Sather, Aroub Yousuf, and Pushpam Kumar. "Assessing the human health benefits of climate mitigation, pollution prevention, and biodiversity preservation." *Annals of Global Health* 90, no. 1 (2024): 1–23.

Lead, Jamie R., Graeme E. Batley, Pedro JJ Alvarez, Marie-Noële Croteau, Richard D. Handy, Michael J. McLaughlin, Jonathan D. Judy, and Kristin Schirmer. "NMs in the environment: Behavior, fate, bioavailability, and effects: An updated review." *Environmental Toxicology and Chemistry* 37, no. 8 (2018): 2029–2063.

Lin, Daohui, Xiaoli Tian, Fengchang Wu, and Baoshan Xing. "Fate and transport of engineered NMs in the environment." *Journal of Environmental Quality* 39, no. 6 (2010): 1896–1908.

Linkov, Igor, F. Kyle Satterstrom, Jeffery Steevens, Elizabeth Ferguson, and Richard C. Pleus. "Multi-criteria decision analysis and environmental risk assessment for NMs." *Journal of NP Research* 9 (2007): 543–554.

Liu, Xiaolu, Yang Li, Zhongshan Chen, Hui Yang, Yawen Cai, Suhua Wang, Jianrong Chen et al. "Advanced porous NMs as superior adsorbents for environmental pollutants removal from aqueous solutions." *Critical Reviews in Environmental Science and Technology* 53, no. 13 (2023): 1289–1309.

Liu, Yanbiao, Choon Nam Ong, and Jianping Xie. "Emerging nanotechnology for environmental applications." *Nanotechnology Reviews* 5, no. 1 (2016): 1–2.

Lowry, G. V., and E. A. Casman. "NM transport, transformation, and fate in the environment." In: Linkov, I., Steevens, J. (eds) *NMs: Risks and Benefits* NATO Science for Peace and Security Series C: Environmental Security. Springer, Dordrecht. (2009): 125–137. https://doi.org/10.1007/978-1-4020-9491-0_9

Lowry, Gregory V., Ernest M. Hotze, Emily S. Bernhardt, Dionysios D. Dionysiou, Joel A. Pedersen, Mark R. Wiesner, and Baoshan Xing. "Environmental occurrences, behavior, fate, and ecological effects of NMs: An introduction to the special series." *Journal of Environmental Quality* 39, no. 6 (2010): 1867–1874.

Manimegalai, Sengani, Sundaram Vickram, Santhana Raj Deena, Karunakaran Rohini, Sundaram Thanigaivel, S. Manikandan, R. Subbaiya, N. Karmegam, Woong Kim, and M. Govarthanan. "Carbon-based NM intervention and efficient removal of various contaminants from effluents: A review." *Chemosphere* 312 (2023a): 137319.

Manimegalai, Sengani, Sundaram Vickram, Santhana Raj Deena, Karunakaran Rohini, Sundaram Thanigaivel, S. Manikandan, R. Subbaiya, N. Karmegam, Woong Kim, and M. Govarthanan. "Carbon-based NM intervention and efficient removal of various contaminants from effluents: A review." *Chemosphere* 312 (2023b): 137319.

Marlon, Jennifer R., Brittany Bloodhart, Matthew T. Ballew, Justin Rolfe-Redding, Connie Roser-Renouf, Anthony Leiserowitz, and Edward Maibach. "How hope and doubt affect climate change mobilization." *Frontiers in Communication* 4 (2019): 20.

Mensah, Michael B., David J. Lewis, Nathaniel O. Boadi, and Johannes AM Awudza. "Heavy metal pollution and the role of inorganic NMs in environmental remediation." *Royal Society open science* 8, no. 10 (2021): 201485.

Mofijur, M., M. M. Hasan, Shams Forruque Ahmed, F. Djavanroodi, I. M. R. Fattah, A. S. Silitonga, M. A. Kalam, John L. Zhou, and TM Yunus Khan. "Advances in identifying and managing emerging contaminants in aquatic ecosystems: Analytical approaches, toxicity assessment, transformation pathways, environmental fate, and remediation strategies." *Environmental Pollution* (2023): 122889.

Mortimer, Monika, and Patricia A. Holden. "Fate of engineered NMs in natural environments and impacts on ecosystems." In Nelson Marmiroli, Jason C. White, Jing Song (eds.),*Exposure to Engineered NMs in the Environment*, pp. 61–103. Elsevier, 2019, https://doi.org/10.1016/B978-0-12-814835-8.00003-0

Mukhopadhyay, Raj, Binoy Sarkar, Eakalak Khan, Daniel S. Alessi, Jayanta Kumar Biswas, K. M. Manjaiah, Miharu Eguchi, Kevin CW Wu, Yusuke Yamauchi, and Yong Sik Ok. "NMs for sustainable remediation of chemical contaminants in water and soil." *Critical Reviews in Environmental Science and Technology* 52, no. 15 (2022): 2611–2660.

Nagore, Pravinkumar, Suresh Ghotekar, Kanchan Mane, Amol Ghoti, Muhammad Bilal, and Arpita Roy. "Structural properties and antimicrobial activities of Polyalthia longifolia leaf extract-mediated CuO NPs." *BioNanoScience* 11 (2021): 579–589.

Nnaji, Chukwuemeka O., Jaison Jeevanandam, Yen S. Chan, Michael K. Danquah, Sharadwata Pan, and Ahmed Barhoum. "Engineered NMs for wastewater treatment: Current and future trends." Editor(s): Ahmed Barhoum, Abdel Salam Hamdy Makhlouf, In *Micro and Nano Technologies, Fundamentals of NPs*, Elsevier, (2018): 129–168.

Nzilu, Dennis Mwanza, Edwin Shigwenya Madivoli, David Sujee Makhanu, Brian Victor Otenda, Patrick Gachoki Kareru, Pius Kinoti Kairigo, and Tuula Tuhkanen. "Environmental remediation using NM as adsorbents for emerging micropollutants." *Environmental Nanotechnology, Monitoring & Management* 20 (2023): 100789.

Pal, Roopal, Lakhan Kumar, Shaubhik Anand, and Navneeta Bharadvaja. "Environmental pollutants remediation using Phyto-NPs: An overview on synthesis, characterization, and remediation potential." In: Shah, M. P., Bharadvaja, N., Kumar, L. (eds.), *Biogenic NMs for Environmental Sustainability: Principles, Practices, and Opportunities*, Environmental Science and Engineering. Springer, Cham. (2024): 111–145.

Palit, Sukanchan. "Recent advances in the application of engineered NMs in the environment industry: A critical overview and a vision for the future." In Chaudhery Mustansar Hussain (ed.), *Handbook of NMs for Industrial Applications* (2018): 883–893. Elsevier, United States.

Peijnenburg, Willie JGM, Mohammed Baalousha, Jingwen Chen, Qasim Chaudry, Frank Von der Kammer, Thomas AJ Kuhlbusch, Jamie Lead et al. "A review of the properties and processes determining the fate of engineered NMs in the aquatic environment." *Critical Reviews in Environmental Science and Technology* 45, no. 19 (2015): 2084–2134.

Rani, Manviri, Jyoti Yadav, and Uma Shanker. "Green synthesized Zn-based catalysts." In Kamel A. Abd-Elsalam (ed.), *Nanobiotechnology for Plant Protection, Zinc-Based Nanostructures for Environmental and Agricultural Applications*, pp. 93–121. Elsevier, 2021. https://doi.org/10.1016/B978-0-12-822836-4.00008-2

Rani, Manviri, Uma Shanker, and Amit K. Chaurasia. "Catalytic potential of laccase immobilized on transition metal oxides NMs: Degradation of alizarin red S dye." *Journal of Environmental Chemical Engineering* 5, no. 3 (2017): 2730–2739.

Rao, C. N. R., and A. K. Cheetham. "Science and technology of NMs: Current status and future prospects." *Journal of Materials Chemistry* 11, no. 12 (2001): 2887–2894.

Rather, Muzamil Ahmad, Shuvam Bhuyan, Ratan Chowdhury, Rahul Sarma, Subhm Roy, and Panchi Rani Neog. "Nanoremediation strategies to address environmental problems." *Science of the Total Environment, 886*, (2023): 163998.

Rawat, Swati, Venkata LR Pullagurala, Ishaq O. Adisa, Yi Wang, Jose R. Peralta-Videa, and Jorge L. Gardea-Torresdey. "Factors affecting fate and transport of engineered NMs in terrestrial environments." *Current Opinion in Environmental Science & Health* 6 (2018): 47–53.

Roco, Mihail C., Chad A. Mirkin, and Mark C. Hersam. "Nanotechnology research directions for societal needs in 2020: Summary of international study." *Journal of NP Research* 13 (2011): 897–919.

Roy, Arpita, and Navneeta Bharadvaja. "Qualitative analysis of phytocompounds and synthesis of silver NPs from Centella asiatica." *Innovative Techniques in Agriculture* 1, no. 2 (2017): 88–95.

Roy, Arpita, Apoorva Sharma, Saanya Yadav, Leta Tesfaye Jule, and Ramaswamy Krishnaraj. "Nanomaterials for remediation of environmental pollutants." *Bioinorganic Chemistry and Applications* (2021a), 1764647. doi: 10.1155/2021/1764647. PMID: 34992641; PMCID: PMC8727162.

Roy, Arpita, Apoorva Sharma, Saanya Yadav, Leta Tesfaye Jule, and Ramaswamy Krishnaraj. "NMs for remediation of environmental pollutants." *Bioinorganic Chemistry and Applications* (2021b):1–16: 1764647.

Saleem, Haleema, Syed Javaid Zaidi, Ahmad Fauzi Ismail, and Pei Sean Goh. "Advances of NMs for air pollution remediation and their impacts on the environment." *Chemosphere* 287 (2022): 132083.

Sathish, T., N. Ahalya, M. Thirunavukkarasu, T. S. Senthil, Zakir Hussain, Md Irfanul Haque Siddiqui, Hitesh Panchal, and Kishor Kumar Sadasivuni. "A comprehensive review on the novel approaches using NMs for the remediation of soil and water pollution." *Alexandria Engineering Journal* 86 (2024): 373–385.

Scheringer, Martin, Antonia Praetorius, and Eli S. Goldberg. "Environmental fate and exposure modeling of NMs." In *Frontiers of Nanoscience*, vol. 7, pp. 89–125. Elsevier, 2014.

Shareena, Thabitha P. Dasari, Asok K. Dasmahapatra, and Paul B. Tchounwou. "NMs in the environment: Sources, fate, transport and ecotoxicology." In Thabitha P. Dasari Shareena, Asok K Dasmahapatra, Paul B Tchounwou (eds.), *NMs for Environmental Applications* (2022): 317–341.

Sharma, Shashank, Priyanka Dhingra, and Saurabh Jain. "Advancements & challenges of nanotechnology in waste water treatment." *Materials Today: Proceedings* 80 (2023): 18–23.

Shi, Bin, Jing Meng, Tieyu Wang, Qianqian Li, Qifan Zhang, and Guijin Su. "The main strategies for soil pollution apportionment: A review of the numerical methods." *Journal of Environmental Sciences* 136 (2024): 95–109.

Sohail, Muhammad Irfan, Aisha A. Waris, Muhammad Ashar Ayub, Muhammad Usman, Muhammad Zia Ur Rehman, Muhammad Sabir, and Tehmina Faiz. "Environmental application of NMs: A promise to sustainable future." In Sandeep Kumar Verma, Ashok Kumar Das (eds.), *Comprehensive Analytical Chemistry*, vol. 87, pp. 1–54. Elsevier, 2019. https://doi.org/10.1016/bs.coac.2019.10.002

Suhendra, Edward, Chih-Hua Chang, Wen-Che Hou, and Yi-Chin Hsieh. "A review on the environmental fate models for predicting the distribution of engineered NMs in surface waters." *International Journal of Molecular Sciences* 21, no. 12 (2020): 4554.

Thakur, Abhinay, and Ashish Kumar. "Ecotoxicity analysis and risk assessment of NMs for the environmental remediation." In *Macromolecular Symposia*, vol. 410, no. 1, p. 2100438. 2023.

Thakur, Abhinay, Ashish Kumar, and Ambrish Singh. "Adsorptive removal of heavy metals, dyes, and pharmaceuticals: Carbon-based NMs in focus." *Carbon* 217 (2024): 118621.

Vashchenko, Olena, Jacek Gapiński, Barbara Peplińska, Łucja Przysiecka, Tomasz Zalewski, Grzegorz Nowaczyk, Marcin Jarek, Anna Marcinkowska-Gapińska, and Stefan Jurga. "Self-organizing silver and ultrasmall iron oxide NPs prepared with ginger rhizome extract: Characterization, biomedical potential and microstructure analysis of hydrocolloids." *Materials & Design* 133 (2017): 307–324.

Wang, Yinan, David O'Connor, Zhengtao Shen, Irene MC Lo, Daniel CW Tsang, Simo Pehkonen, Shengyan Pu, and Deyi Hou. "Green synthesis of NPs for the remediation of contaminated waters and soils: Constituents, synthesizing methods, and influencing factors." *Journal of Cleaner Production* 226 (2019): 540–549.

Wang, Zhou, Pingping Luo, Xianbao Zha, Chengyi Xu, Shuxin Kang, Meimei Zhou, Daniel Nover, and Yihe Wang. "Overview assessment of risk evaluation and treatment technologies for heavy metal pollution of water and soil." *Journal of Cleaner Production*, 379 part 2, (2022): 134043.

Westerhoff, Paul, and Bernd Nowack. "Searching for global descriptors of engineered NM fate and transport in the environment." *Accounts of Chemical Research* 46, no. 3 (2013): 844–853.

Williams, Richard J., Samuel Harrison, Virginie Keller, Jeroen Kuenen, Stephen Lofts, Antonia Praetorius, Claus Svendsen, Lucie C. Vermeulen, and Jikke van Wijnen. "Models for assessing engineered NM fate and behaviour in the aquatic environment." *Current Opinion in Environmental Sustainability* 36 (2019): 105–115.

18 Benefits, Risks, and Responsible Use of Nanomaterials in Remediation

Chitra Bhattacharya, Nidhiba Rayjada, and Bhawana Pandey

18.1 INTRODUCTION

Our environment is a complex web of life, with every living organism playing a crucial role in maintaining ecosystem balance. However, our actions, such as industrialization, deforestation, pollution, and over-exploitation of natural resources, have disrupted this delicate balance, leading to severe environmental degradation and climate change. Environmental pollutants, such as sulphur dioxide, carbon monoxide, nitrogen oxides, heavy metals, biological contaminants, ozone, and tobacco smoke, contribute to human disease, climate change, increased mortality, and morbidity. Exposure to soil with high levels of contaminants can cause acute and chronic health problems, and plant diseases that deteriorate crop production due to the high toxicity of soil. Similarly, water pollution, a major issue worldwide, is triggered by both organic and artificial human activities, with carbonic compounds being the most common component (Wasi et al., 2013). Industrial waste toxins can lead to severe poisoning, immunosuppression, and reproductive failure. Environmental pollution is a global issue with significant risks and health consequences. Recently, pesticide overuse and negative impacts on the environment have become a major concern. Pesticides, designed to kill, are often used to control insects, weeds, nematodes, rodents, and pests. The World Health Organization estimates that there are 3 million cases of pesticide poisoning every year and up to 220,000 deaths, primarily in developing countries. These pesticides produce reactive oxygen species, which decrease antioxidant levels and defense mechanism against oxidative damage in the cellular system. This imbalance affects lipids, proteins, and nucleic acids, leading to long-term health effects such as carcinogenesis, neurodegeneration, cardiovascular, respiratory, renal, endocrine, and reproductive problems. The application of pesticides is often not precise, and their use can cause damage to the environment and ecosystems. Pesticides are used to kill pests and insects that damage crops, yet they also have an adverse effect on the ecosystem. Excessive use of pesticides may lead to the destruction of biodiversity and threaten the survival of birds, aquatic organisms, and animals (Kaur et al., 2019). Therefore, the removal of pesticides from the environment is of utmost importance. There are numerous traditional approaches, each with its limitations; consequently, finding a different strategy that can effectively remove these toxins from the atmosphere is imperative. Microbes are used in bioremediation techniques to eradicate toxic substances; however, it has limitations, such as a slow response time and low specificity. Nanotechnology has emerged as a tool for sensing and remediating contaminants, particularly NPs with size-dependent properties. NPs can detect, degrade, and remove pollutants through adsorption or chemical modification, and can facilitate microbial remediation through immobilization or enzyme production (Benjamin et al., 2019). NMs are cost-effective and faster than conventional approaches, due to their enhanced surface area, transport properties, and sequestration characteristics. Recently, nanoscale zero-valent iron, carbon nanotubes, and nano-fibers have been used for various contaminants (Bhandari, 2018).

DOI: 10.1201/9781003484189-18

18.2 NANOTECHNOLOGY: NEW APPROACH

The field of "nanotechnology," which arose in the 1980s as a result of the union of K.E. Drexler's theoretical and practical work, is currently receiving widespread attention from the scientific and publicly universe. According to the National Nanotechnology Initiative (2009), nanotechnology is "the understanding and control of matter at dimensions between 1 and 100 nm, where unique phenomena enable novel applications." As per the calculation of 2012, over 1300 manufacturers identified, prepared, and marketized nanoproducts at a pace of 3–4 weeks (Bondarenko et al., 2013). Nanotechnology has been more prevalent in recent years in all fields of science and technology, including engineering, medicine, pharmacology, agriculture, environment, and many more. Among the various common uses of nanotechnology, successful applications include the manufacture of cosmetics, sunscreens, surface coatings, and some food items, using NMs like titanium dioxide and zinc oxide. Silver NPs are becoming increasingly employed in water purification systems, bandages, household appliances, food packaging, textiles, and disinfectants. More proof of the commercial influence of nanotechnology may be seen in the usage of carbon nanotubes in solar panels (Royal Society and Royal Academy of Engineering, 2004). According to the report of Roco, (2005) the worldwide money investment in the field of nanotechnology research and development increased around nine times over between 1997 and 2005, from \$432 million to \$4100 million. By 2015, \$US 1 trillion worth of products have been added as essential functioning components in nanotechnology. According to the high-end estimation, 17% of these items are utilized in soils, 21% in water, 2.5% in the air, and the remaining portion in landfills (Keller and Lazareva, 2013). The use of nanotechnology, also known as environmental nanotechnology or E-Nanoin, the environmental sector is rapidly gaining attention. E-Nano uses nanoscale particles to apply remediation processes more effectively. The advantages of nanotechnology for ecological cleaning have been shown in several research studies and business reports (Ghasemzadeh et al., 2014, Thatai et al., 2014; Zhang, 2003). However, the threats to the environment and human health that come with this next-generation technology have only recently come to light (Keller and Lazareva, 2013). This critical study will quickly look at nanotechnology's potential, especially in the environment, highlighting both the advantages, responsibilities, and risks of NMs in the use of environmental remediation.

18.3 TYPES OF NMS

Nanotechnology involves manipulating the shape and size of structures, electronics, and systems at the nanometer scale, ranging from 1 nm to 100 nm. NPs have gained attention due to their small size, significant surface areas, higher reactivity, and tuneable properties (Khan et al., 2022a). These properties have led to their application in various fields, such as biomedicine, cosmetics, electronics, food analysis, environmental remediation, and paints. Nanoscale science and engineering provide a new level of understanding and control of matter at atomic and molecular dimensions. NPs have remarkable electrical, optical, and magnetic properties, making them suitable candidates for nanoengineering. The desire for novel technology applications in data storage, biomedical sciences, and drug delivery has fueled NP research (Salem et al., 2022). Figure 18.1 depicts the types of NMs, with their subcategories, as follows:

18.3.1 Organic NMs

Organic molecules are transformed into organic NMs at the nanoscale. The organic NPs and polymers ferritin, liposomes, dendrimers, and micelles are a few examples of organic NMs. Biodegradable, non-toxic NPs with hollow interiors that are susceptible to light, heat, and electromagnetic radiation are called liposomes and nanocapsule micelles (Khan et al., 2022b). Dendrimers are covered with many chain ends that can carry out particular chemical reactions on their surface. Dendrimers find use in light harvesting, nanosensing, molecular recognition, and optoelectrochemical systems.

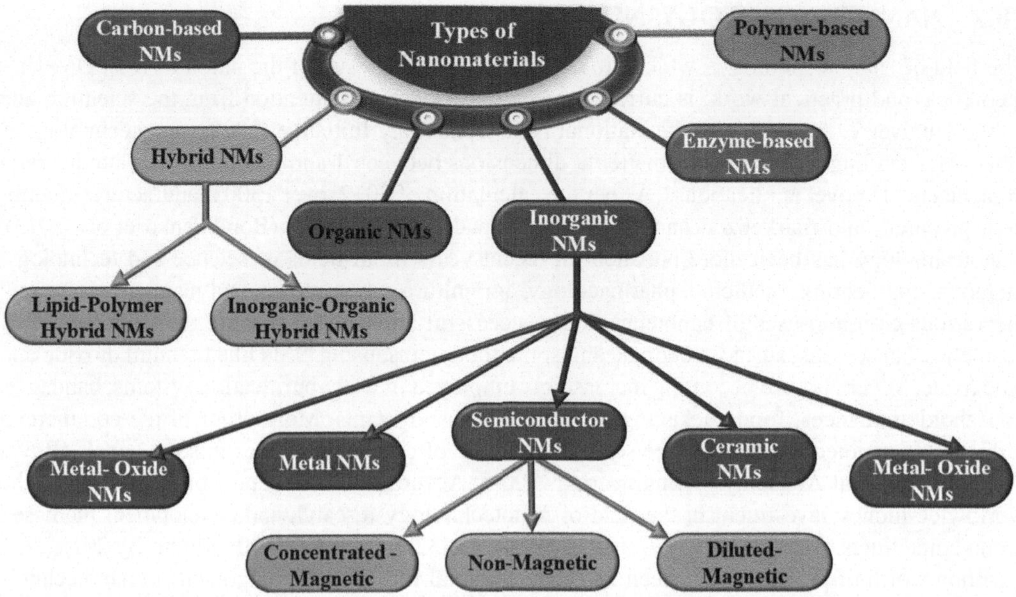

FIGURE 18.1 Types of NMs.

Moreover, three-dimensional (3D) dendrimers may be helpful for drug administration, since they have interior holes that can hold extra molecules (Mekuye and Abera, 2023).

18.3.2 Inorganic NMs

Inorganic NPs that lack carbon atoms are inorganic NMs. They are typically classified as those composed of metal-based or metal oxide-based NMs.

18.3.2.1 Metal-Based NMs

Destructive or constructive methods can be used to create metal-based NPs. In the synthesis of NPs, metals such as aluminium (Al), cadmium (Cd), cobalt (Co), copper (Cu), gold (Au), iron (Fe), lead (Pb), silver (Ag), and zinc (Zn) are commonly utilized. In addition to their exceptional UV-visible sensitivity, metal NPs possess electrical, catalytic, thermal, and antibacterial capabilities, due to their quantum effects and large surface-to-volume ratio. Because metal NPs offer exceptional optical qualities, they are used in many different types of study (Mekuye and Abera, 2023).

18.3.2.2 Metal Oxide-Based NMs

Fusion of positive metallic ions and negative oxygen ions can synthesize metal oxide NMs. Metal oxide NPs, such as silicon dioxide (SiO_2), titanium oxide (TiO_2), zinc oxide (ZnO), and aluminium oxide (Al_2O_3), are commonly synthesized. These NPs exhibit remarkable properties compared to their metal analogs (Khan et al., 2022c).

18.3.2.3 Semiconductor NMs

Semiconductor NPs share the same characteristics as metals and insulators. They are categorized into three sub categories (Mekuye and Abera, 2023).

Concentrated Magnetic Semiconductor NMs: It has spontaneous magnetic order and can be found in binary compounds such as EuTe (anti-ferromagnetic).
Nonmagnetic Semiconductor NMs: Nonmagnetic semiconductors, which lack magnetic ions and are utilized for information processing and communications, have had considerable success due to the

charge of electrons in semiconductors, but are not employed for mass storage of information in dispensable information technology.

Diluted Magnetic Semiconductor NMs: Semiconducting materials become magnetic by adding a few magnetic impurities to the host matrix, with some of the diamagnetic host cations replaced at random by magnetic (TM) cations. These materials maintain semiconducting characteristic features and possess magnetic properties that are a mixture of ordinary and magnetic semiconductors (Mekuye and Abera, 2023).

18.3.2.4 Ceramic NMs

Ceramic NPs are inorganic solids composed of carbides, carbonates, oxides, carbides, carbonates, and phosphates synthesized using heat and subsequent cooling. Ceramic NMs can be formulated in drug delivery systems, particularly for the treatment of tumors, glaucoma, and certain bacterial infections. NMs are also gaining popularity among researchers due to their use in catalysis, photocatalysis, dye degradation, and imaging applications (Khan et al., 2019).

18.3.2.5 Lipid-Based NMs

Lipid-based NPs are typically spherical, with sizes ranging from 10 to 100 nm. It is made up of a lipid-rich solid core and a soluble lipophilic matrix. Lipid-based NPs have medicinal applications, such as medication delivery and RNA release therapy in cancer treatment (Puri et al., 2009).

18.3.3 Carbon-Based NMs

Carbon-based NMs are constituted of five primary materials: carbon nanotubes, graphene, fullerenes, carbon nanofiber, and carbon black. Fullerenes, which have a spherical or ellipsoidal shape, are sometimes known as Buckyballs. Fullerenes are spherical structures with diameters of up to 8.2 nm for a single layer and 4–36 nm for multi-layered fullerenes, formed by 28–1500 carbon atoms. Graphene is a hexagonal network of honeycomb lattices composed of carbon atoms on a two-dimensional (2D) planar surface, with the sheet measuring approximately 1 nm, whereas cylindrical ones are referred to as nanotubes. Hollow cylinders are used to form nanotubes with diameters as small as 0.7 nm for a single-layered and 100 nm for a multi-layered carbon nanotube, and lengths ranging from a few micrometers to several millimeters. The same graphene nano fossils are used to produce carbon nano fiber, and carbon black is an amorphous material made of carbon, with diameters ranging from 20 to 70 nm (Kumar and Kumbhat, 2016). Carbon-based NMs are mostly employed for structural reinforcement because they are stronger than steel at times. Carbon NMs have infinite possibilities for modification and tailoring, due to their small size, which is similar to that of many fundamental biomolecules, large specific surface area, high electrical and thermal conductivity, unique optical properties, and superior mechanical properties, which have opened the door to a wide range of applications (Mekuye and Abera, 2023). Fullerene derivatives, in particular, have been used to scavenge solar energy; graphene has been widely used in flexible electronics; carbon nanotubes have been modified to have molecular recognition capabilities; graphene quantum dots have been extensively used for bio-imaging and sensing due to their photoluminescence properties; and nanodiamonds are useful in super-resolution imaging and nanoscale temperature sensing.

18.3.4 Enzyme-Based NMs

The synthesis of green NPs has lately gained popularity, due to its advantages over traditional chemical and physical processes, particularly in terms of cost-effectiveness, simplicity, environmental friendliness, and broad applicability. Enzyme-driven NMs synthesis is a promising approach to efficient NP manufacture. However, traditional NM manufacturing methods may involve the use of dangerous chemicals and energy-intensive techniques, resulting in environmental contamination and possible dangers to human health. Enzymes, which are found in all living creatures, act as biocatalysts, driving a wide range of chemical and biological processes. Microbial enzymes are being extensively investigated due to their low-cost synthesis, purification, and characterization approaches. Enzymes are also extremely important in regulating environmental health since they help detoxify or convert hazardous substances into beneficial products (Seth & Meena, 2024).

18.3.5 Polymer-Based NMs

Polymer-based NPs are nanospherical or nanocapsule-shaped particles, with a complex polymeric matrix that allows them to embed and/or encapsulate chemicals or additives. It can also be adsorbed on the particle's surface. Polymeric particles are primarily used in biomedical applications, such as drug delivery, due to their exceptional properties such as biodegradability and biocompatibility, the ability to release drugs in a controlled manner, and the ability to entrap drugs via various mechanisms, such as cross-linking, ionic cross-linking, and ionic complexation (Kumar et al., 2012). Some disadvantages of polymeric NPs include high cost, greater toxicity levels, low purity, due to the presence of solvent residue during the synthesis and purification process, and the inability to produce on a large scale, degradability as well as specific requirements for biodegradable polymers (e.g., high purity and quality) (Kahraman et al., 2017).

Polymeric NPs can be classified as natural or manufactured, with the latter being further separated into biodegradable and nonbiodegradable polymeric NPs. Natural, biodegradable polymeric NPs include chitosan, cellulose, gelatine, alginate, and others; synthetic biodegradable polymers, such as polylactic acid (PLA), poly-(lactide-co-glycolide) (PLGA), and polyanhydrides, are widely recognized. Additionally, nonbiodegradable synthetic polymers such as polymethacrylate and polymethylmethacrylate (PMMA) have been utilized in drug delivery and diagnostic imaging applications (Anboo et al., 2022).

18.3.6 Hybrid NMs

18.3.6.1 Lipid-Polymer Hybrid NPs (LPN)

The concept of lipid-polymer-based NPs is the combination of polymeric and lipid-based NPs. LPNs are made up of a hydrophobic polymeric core, a hydrophilic shell, and a lipid shell sandwiched between the two. The hydrophobic core entraps poorly water-soluble drugs, while the hydrophilic shell improves stability and half-life due to their anti-biofouling properties, and the interface lipid shell acts as a protective wall to ensure drug encapsulation efficiency, high loading yield, controlled release of drugs, and high immune-compatibility properties (e.g., complement and coagulation system activation and the ability to bind onto human plasma protein) (Zhang et al., 2008).

18.3.6.2 Inorganic–Organic Hybrid NPs

The hybrid NPs are formed by binding an inorganic compound to an organic compound, resulting in a hybrid NM. Hybrid NPs have several advantages over conventional materials, including the capacity to immobilize enzyme activity and much superior stability, physical, and reproducible properties. Hybrid NPs can be used to stabilize enzyme activity, transport drugs, create biosensors, image cells, and perform other biological functions (Lee et al., 2015).

18.4 APPLICATION OF NMS IN ENVIRONMENTAL REMEDIATION

18.4.1 Pollutant Removal by Nanoscale Sorbents

By developing and applying nanoscale sorbents, nanotechnology provides sophisticated ways to address environmental contamination. These materials effectively absorb and eliminate contaminants by utilizing the special properties of NPs and nanostructured materials. Their compact size allows them to have a vast surface area and reactive sites to interact with various pollutants (Usha Rani et al., 2024). NPs, typically less than 100 nanometers, are highly efficient in environmental cleanup, due to their small size and surface area. Metal oxide NPs, like iron oxide and titanium dioxide; attract heavy metals and persistent organic pollutants. Surface chemistry can be manipulated to enhance selectivity and affinity (Boulkhessaim et al., 2022).

Nanostructured materials like carbon nanotubes and nanofibers enhance the removal of contaminants by providing organized frameworks. Carbon nanotubes have cylindrical nanostructures that adsorb organic compounds and heavy metal ions, while nanofibers can be customized with functional groups to attract specific pollutants. These materials can be used in water treatment plants or air purification systems to trap and immobilize contaminants. Their flexibility and specific purpose make them essential in reducing pollution and improving ecological well-being (Abbas et al., 2016).

18.4.2 Pollutant Degradation by Employing Nanocatalysts

Nanocatalysts are advanced NPs used in environmental remediation, providing a precise and efficient method for breaking down contaminants. They have a large surface area, adjustable surface chemistry, and catalytic activity, allowing them to be engineered to target specific contaminants and adapt to specific environmental conditions. Nanocatalysts can be optimized by modifying their surface, introducing components, or immobilizing them onto support materials. They are used in water treatment, soil remediation, and air purification to enhance the breakdown of organic pollutants, promote the conversion of heavy metal contaminants into less harmful forms, and improve air quality. Research and development in nanocatalyst design and use show potential for tackling pollution problems and promoting environmental stewardship initiatives (Zhang et al.,2013).

18.4.3 Nano-Empowered Filtration Membranes and Adsorbents

Nano-enabled filtration membranes and adsorbents are innovatively used in environmental remediation. These materials, with their tiny pores, large surface area, and regulating surface chemistry, provide efficient and precise removal of pollutants from water, air, and other substances. They offer mechanical durability, chemical endurance, and resistance to fouling, making them suitable for desalination, wastewater reuse, and drinking water purification (Rasheed et al., 2019). Nanotechnology-enhanced membranes and adsorbents can be modified to selectively bind to specific pollutants, enhancing environmental quality and human health. Ongoing research and innovation in nanostructured materials could advance environmental remediation and address rising pollution concerns (Cashin et al., 2018).

18.4.4 Nanosensors and Monitoring Devices

Nanotechnology-enabled sensors and monitoring devices offer advanced methods for environmental monitoring and pollution control. These sensors utilize NMs' surface reactivity, electrical conductivity, and optical properties to detect pollutants in air, water, soil, and biological matter. They use optical, electrochemical, and mechanical sensing principles to identify changes in target analytes (Mirzaei et al., 2016). NM-based sensors are used in several environmental monitoring situations, including air quality, water quality, and soil pollution. They can be integrated into portable, wearable, or networked monitoring systems, and can be used in autonomous or remote sensing systems (Thakur et al., 2022).

18.4.5 Nanoremediation

Different worldwide events, such as industrial development and population growth, have resulted in the existence and permanence of organic and inorganic pollutants, posing a risk to the environment and human health. As a result, the development and implementation of novel solutions for addressing the issue of environmental contamination is critical. Nanotechnology is a rapidly developing science with applications in a variety of fields. Nanoremediation, in particular, is a promising method defined as the use of designed materials to clean up the environment. It is an authentic, fast, and efficient technology for dealing with persistent pollutants such as pesticides, chlorinated solvents, halogenated chemicals, or heavy metals. Furthermore, nanoremediation is a sustainable way to eradicate new pollutants including pharmaceuticals and personal care items. Because of their variety and versatility, NMs can be used in water, soil, or air media. Nanoremediation uses engineered NMs to clean up polluted media. This technique is less costly and more effective than most typical methods (Del Prado-Audelo et al., 2021).

18.4.5.1 Nanoremediation of Water

Over the past decade, the use of NMs in water and wastewater treatment has grown significantly. Contaminated groundwater poses significant threats to ecosystems, including heavy metals, pesticides, radioactive elements, and new contaminants like pharmaceuticals and personal care items (Zamora-Ledezma et al., 2021). Research on efficient water remediation methods using NMs like metallic NPs, biopolymeric membranes, and carbon-derived materials is crucial due to their selectivity and absorption capacity (Saikia et al., 2019). Metal oxide NPs, such as iron oxide, zinc oxide, and titanium dioxide are used for water purification, due to their high reactivity, photolytic properties, and adsorption properties (Aragaw et al., 2021). Nanoporous carbon-based materials like activated carbons, carbon nanotubes, and graphene oxide are suitable for water treatment, removing contaminants like heavy metals, fluorides, textile dyes, and pharmaceutical products (Mpouras et al., 2021). Polymer nanotechnology can be utilized in water treatment to create NPs, nanocomposites, and nanofiltration membranes. Chitosan is a popular polymer for NF membranes, used for textile wastewater treatment (Long et al., 2020). Synthetic and natural polymers like polyamide, cellulose, and chitosan enhance these membranes' stability and effectiveness (Yan et al., 2016; Lakhotia et al., 2018; Del Prado-Audelo et al., 2021).

18.4.5.2 Nanoremediation of Soil

The transition from hunter-gatherer to farmer has led to irreversible impacts on nature, including the disappearance of animal and plant species, river course diversion, and soil erosion. Industrialization and urbanization have accelerated soil deterioration, leading to the use of NMs for soil remediation (Kumar et al., 2021). NMs like metallic and magnetic NPs, carbon nanotubes, and nanoscale zero-valent iron have been used for soil contamination removal. Nanoscale zero-valent iron (nZVI) is effective in removing chlorinated organic solvents, polychlorinated biphenyls, and organochlorine pesticides (Stefaniuk et al., 2016). Carbonaceous NMs have unique characteristics, such as large surface area, high microporosity, and eco-friendliness, but their high surface area and surface-to-volume ratio can hinder nanoremediation properties. The use of surfactant poloxamer 407 stabilizes multi-wall carbon nanotubes. CNTs can remove heavy metal ions, but their immobilization depends on pH, organic matter content, and silt and clay particles. They can also remediate soil from petroleum hydrocarbons, crude oil, and other pollutants, increasing microbial population and plant growth. CNTs can be used in membrane filtration, separation columns, and aqueous dispersion (Del Prado-Audelo et al., 2021).

Polymer nanotechnology can be used in water treatment to create NPs, nanocomposites, and nanofiltration membranes. These membranes eliminate unwanted NPs in the aqueous phase by detouring particles and immobilizing pollutants. Chitosan is a widely used polymer for NF membranes, which are used to clean textile wastewater (Bassyouni et al., 2019). These membranes have lower rejection to electroneutral and negatively charged dyes than positively charged ones. The stability and effectiveness of these nanofiltration membranes can be enhanced by using them as a matrix or support for other materials. Synthetic and natural polymers, like polyamide, cellulose, and chitosan, have been used as membrane matrices, with carboxylated MWCNTs increasing salt rejection rates. Polyethersulfone membranes functionalized with MWCNTs and graphene show excellent heavy metals and dye rejection in aqueous media (Neeta et al., 2017; Del Prado-Audelo et al., 2021).

18.4.5.3 Nanoremediation of Air

Air pollution is a significant global issue affecting climate change and public health. Common outdoor air pollutants include PM10 and PM2.5, NOx, SO_2, CO, lead, and ground-level ozone. CO_2 is the most significant greenhouse gas emitted by human activities. To combat this, researchers have explored using graphene oxides, graphite oxides, and mesoporous silica materials. Researchers are exploring the use of graphene oxides, graphite oxides, and mesoporous silica materials to combat this issue. These materials offer high surface area, large pore volume, and thermal stability, making them promising solutions for air pollution control. Nanotechnology offers benefits in remediation, treatment, prevention, and detection and sensing. Graphite oxide can be controlled by changing the reaction temperature with water, while carbon-based NMs can form nanocomposites. Silicon-based NMs are versatile and have been used for catalysis and adsorption (Del Prado-Audelo et al., 2021). Nanotechnology offers numerous benefits in air pollution control, including remediation, treatment, prevention, and detection and sensing. Graphite oxide, a material with oxygen-containing functional groups, can be controlled by changing the reaction temperature with water. Carbon-based NMs can combine different properties to form nanocomposites, such as graphene oxide and zirconium hydroxide/graphene composites. GO has been used for environmental remediation, partially reduced via photoreduction, and used as a photocatalyst to degrade VOCs. Modified MWCNTs or SWCNTs have been used to detect various pollutants, including H_2S, SO_2, CO, NH_3, NO_2, VOCs, NOx, and CO_2.

Silica-based NMs have gained interest for their versatility, wide surface area, adjustable pore size, and adaptability. They have been used for catalysis and adsorption, leading to their use in remediating polluted air and eliminating gas phase contaminants. Surface modification can enhance their properties, such as incorporating hydroxyl groups on the surface for gas adsorption and wetting (Shukla et al., 2020). Studies have shown that modified mesoporous silica can effectively capture H_2S and CO_2 from natural gas, removing up to 80% of total H_2S and CO_2 in 35 minutes. Aminosilicates have the potential to eliminate CO_2 from ambient air, potentially mitigating climate change. These amine-modified silicates can also eliminate other organic contaminants, making them suitable for industrial pollutants. Silica NPs have been developed to address atmospheric lead contamination, making them attractive environmental agents against industrial pollution by Pb and other heavy metals (Del Prado-Audelo et al., 2021). Figure 18.2 depicts the summarized overview of sources of environmental contaminants, and also its possible remedies suggested, undergoing trial in field and lab study for the successful implementation of the same.

18.5 BENEFITS OF NMS IN ENVIRONMENTAL REMEDIATION

Natural causes of pollution are unavoidable; however, man-made pollution is on the rise, generating alarming health problems, dwindling access to clean water, and a scarcity of nutritious food. The limits of traditional technologies and materials used in remediation to remove organic, inorganic, and microbiological contaminants have prompted the research community to look at nanotechnological solutions for environmental problems (Durgalakshmi et al., 2019). Nanotechnology has significant potential for environmental remediation, offering several key benefits:

Enhanced Contaminant Removal: NPs can interact with pollutants at a molecular level, providing improved efficiency in removing contaminants from soil, water, and air. For example, nanoscale zero-valent iron (nZVI) can effectively degrade organic pollutants and heavy metals.

Increased Surface Area: NMs have a high surface-area-to-volume ratio, which means they can interact more effectively with contaminants. This property enhances their ability to adsorb and break down pollutants.

Targeted Remediation: Nanotechnology allows for the design of particles that can target specific contaminants. Functionalized NPs can be engineered to selectively bind to particular pollutants, making the remediation process more precise and efficient.

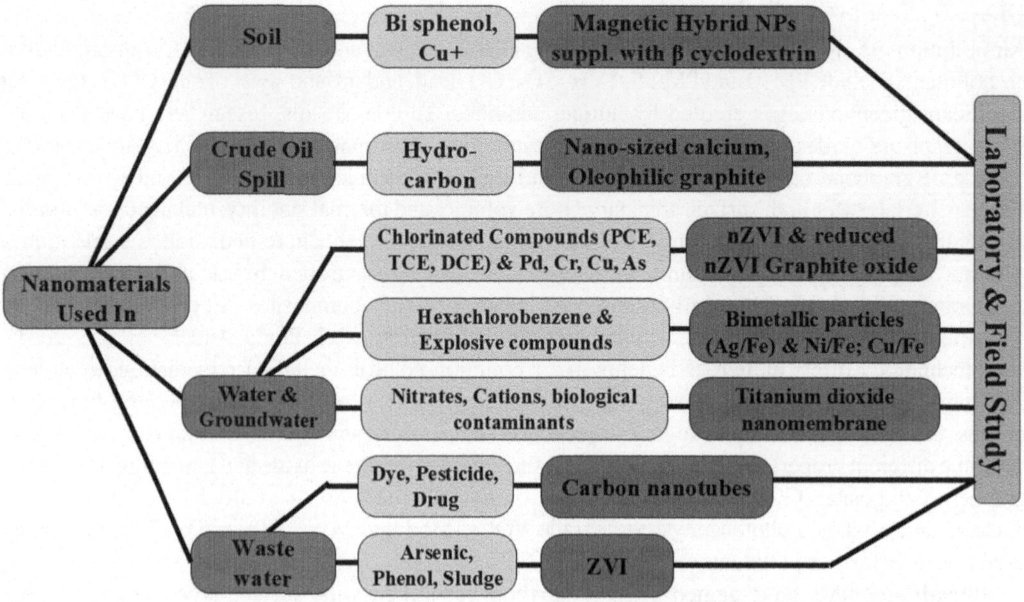

FIGURE 18.2 Application of various NMs (NMs) for environmental remediation: case studies in which gray color indicates the name of contaminants, blue shows treatment, and peach color shows the study area (Patil et al., 2016).

Improved Detection and Monitoring: Nanosensors can detect low concentrations of pollutants, with high sensitivity. These sensors can provide real-time data on environmental conditions, allowing for more effective monitoring and management of contamination.

Faster Remediation Processes: The high reactivity of NMs can speed up the degradation of contaminants, potentially reducing the time required for cleanup efforts, compared to conventional methods.

Reduced Secondary Pollution: NMs can often be used in ways that minimize the generation of secondary pollutants, a common issue with some traditional remediation techniques.

Versatility: Nanotechnology can be applied to a wide range of contaminants, including organic pollutants, heavy metals, and radioactive materials. This versatility makes it a valuable tool for various environmental challenges.

Cost-Effectiveness: While initial costs might be high, nanotechnology can lead to cost savings in the long term by reducing the need for extensive treatment and disposal processes, as well as by improving the efficiency of remediation efforts.

Enhanced Soil and Water Quality: By effectively removing pollutants, nanotechnology can help restore the quality of soil and water, leading to healthier ecosystems and improved public health.

Sustainable Solutions: Nanotechnology can contribute to more sustainable environmental practices by offering innovative ways to manage and mitigate pollution, supporting overall environmental stewardship (Hristozov et al., 2009).

There are myriad reports on possible remedial strategies employing NMs usages to mitigate environmental problems. Some of the representative data are well documented in **Table 18.1**:

Despite these benefits, it's important to carefully consider potential risks and regulatory challenges associated with the use of nanotechnology, including issues related to NP toxicity and environmental impact. Research and development in this field are ongoing to address these concerns and maximize the positive outcomes of nanotechnology in environmental remediation.

TABLE 18.1
List of Targeted Remediation NMs of Specific Contaminants

NM	Target Contaminant	References
Iron oxide	Textile dyes, paints, dyes of paper industries Chromium (VI)	Mashkoor and Nasar (2020)
Titanium dioxide	Pharmaceutical waste	Mahmoud et al. (2017)
Graphene and its oxide	Heavy metals Fluorides Textile dyes Pharmaceutical products	Mpouras et al. (2021)
Chitosan (polymer)	Textile wastewater	Long et al. (2020)
MWCNTs in polyamide membranes	Industrial salt from textile effluents	Al-Hobaib et al. (2017)
Polyethersulfone membranes + MWCNTs/graphene/other polymers	Heavy metals Dyes	Vatsha et al. (2014); Ma et al. (2017); Peydayesh et al. (2020)
nZVI	Chlorinated organic solvents Polychlorinated biphenyls Organochlorine pesticides Trichloroethene Hexavalent chromium Nitrate Lead Cadmium DDT	Stefaniuk et al. (2016); Guerra et al. (2018)
CNTs	Heavy metal ions (Pb^{2+}, Cu^{2+}, Ni^{2+}, & ZN^{2+}) Petroleum hydrocarbons Crude oil Cr (VI) Cd DDT Hexachlorocyclohexane	Shan et al. (2015)
Silica	H_2S, CO_2 Atmospheric Pb Organic contaminants-aldehyde & ketone	Huang et al. (2003); Choi et al. (2011); Nomura and Jones (2013, 2014).

18.6 USAGE OF NMS AND ITS POSSIBLE CHALLENGES (RISKS)

The risks accompanying exposure to NMs escalate with acquaintance length and concentrations of NPs in the atmosphere. Ingestion, cutaneous contact, inhalation, and unintentional injection are additional ways to expose people to NMs. The biggest risk of exposure is from inhalation only. Although there is reduced risk when NMs are suspended in a solution or slurry, inhalation exposure can still occur from sonication, shaking, stirring, pouring, or spraying. Fixed within a matrix NPs are the least dangerous, provided there is no mechanical disruption, including grinding, cutting, burning, etc. Figure 18.3 represents the basic NMs properties that mitigate the potential risks and challenges.

18.6.1 Toxicity and Ecological Consequences of NMs

The incorporation of NMs into environmental remediation systems raises significant concerns, particularly with their potential toxicity and environmental impact. Although NPs are extremely

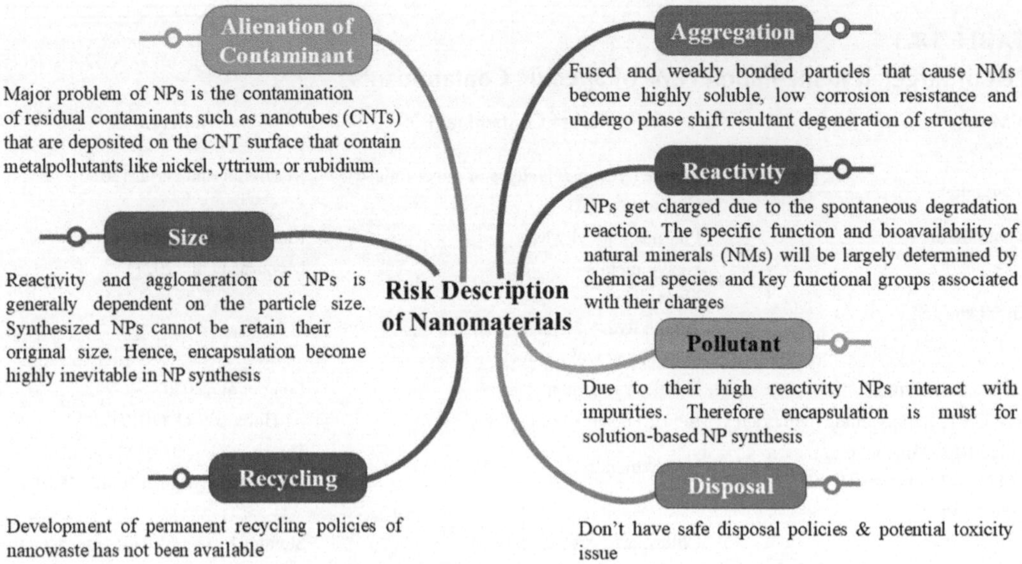

FIGURE 18.3 Basic NM properties and their potential risks and challenges (Jeevandam et al., 2018).

effective in removing and breaking down pollutants, their fundamental physicochemical properties may have harmful effects on ecosystems and human health. There have been reports of cytotoxicity, genotoxicity, and oxidative stress, raising serious concerns about their environmental repercussions. Furthermore, the complex changes and interactions of NMs in environmental contexts necessitate rigorous risk assessment processes to comprehend their future consequences and ecological impacts. To address these challenges, it is required to combine several fields of study to understand how NMs may be detrimental, assess how long they remain in the environment, and build plans to reduce hazards and ensure the safe use of nanotechnology for environmental cleaning (Donaldson et al., 2005).

18.6.2 Robustness and Lucrativeness of Nanotechnology-Based Remediation Technologies

The transition from small-scale laboratory research to the practical application of nanotechnology-based remediation solutions poses major challenges in terms of scalability and cost-efficiency. Although laboratory studies indicate NMs' efficacy in eliminating and degrading contaminants, the problem is to scale up production while maintaining uniform quality (Guerra et al., 2018). Furthermore, the large capital investment required for infrastructure, manufacturing techniques, and quality assurance standards adds to the economic barriers to increasing the use of nanotechnology-based remediation. To address these issues, research efforts should be directed toward developing effective methodologies for producing nanotechnology-based remediation solutions on a wide scale. This includes improving production processes and devising lucrative strategies to make these technologies economically viable and available.

18.6.3 Governing Hindrances and Amenability Deliberations

Dealing with the legal framework that oversees nanotechnology in environmental restoration presents significant challenges, characterized by fragmented regulation, imprecise standards, and changing norms. Because of the complexity of NMs, regulatory organizations have difficulty assessing their safety and effectiveness. They generally lack standardized methods for assessing risks and providing regulatory approvals. Furthermore, the lack of precise instructions for NPs' environmental discharge,

TABLE 18.2
Types, Application, Benefits, and Risks of NMs

Types of NMs	Name of NM	Application	Benefits	Risks	References
Metal-based	Iron oxide Titanium dioxide nZVI	Treatment of textile dyes, paints, dyes from paper industries, Chromium (VI), pharmaceutical waste, chlorinated organic solvents, polychlorinated biphenyls, organochlorine pesticides, trichloroethene, nitrate, lead, cadmium, DDT	High specific surface area Removal of diverse pollutants Compatibility with other treatment	NPs can have adverse effects on pure cultures of bacteria Research on the risks in human and environmental health is missing	Mashkoor and Nasar (2020), Mahmoud et al. (2017), Stefaniuk et al. (2016), Guerra et al. (2018).
Carbon based	Activated carbon Carbon Nanotubes MWCNTs SWCNTs Graphene and its oxide	To remove contaminants like Heavy metal ions (Pb^{2+}, Cu^{2+}, Ni^{2+}, & ZN^{2+}), petroleum hydrocarbons, crude oil, Cr(VI), Cd, DDT, Hexachlorocyclohexane, Fluorides, Textile dyes, Pharmaceutical products	High surface area Microporosity Sorption properties Ecofriendly nature Compatibility with other treatments	Different cell toxicity effects (reactive oxygen species production, lysosomal and DNA damage) Rapid saturation High cost	Mpouras et al. (2021), Shan et al. (2015).
Polymer based	Chitosan MWCNTs in polyamide membranes Polyethersulphone membranes + MWCNTs/ grapheme/other polymers	Clean textile wastewater, remove industrial salts from textile effluents, heavy metals and dyes	Employment of polymer derived from waste materials Compatibility with other treatments	Denaturation by extreme temperature	Long et al. (2020), Al-Hobaib et al. (2017), Vatsha et al. (2014), Ma et al. (2017), Peydayesh et al. (2020)
Silica based	Silica incorporated with hydroxyl group Aminosilicates	Removes H_2S, CO_2 Atmospheric Pb, Organic contaminants-aldehyde & ketone	Versatility on surface modification Adaptable pore size Compatible with other treatments	Performance depends on pH Scattered size distribution	Huang et al. (2003), Choi et al. (2011), Nomura and Jones (2014).

disposal, and monitoring adds to regulatory uncertainty and impedes the implementation of nanotechnology-based remedial solutions. To address these challenges, academics, lawmakers, industry stakeholders, and regulatory agencies must collaborate to develop solid regulatory frameworks, adopt standardized risk assessment systems, and ensure compliance with environmental standards. To lessen the risks associated with pollution and safeguard the health of the environment and people, people or organizations with an interest in the issue can work to address these challenges and encourage the thoughtful and long-term integration of nanotechnology into methods of environmental restoration (Usha Rani et al., 2024). In general, **Table 18.2** summarizes the uses, advantages, and risk factors associated with using NMs in this chapter.

18.7 RESPONSIBLE USAGE OF NMs

The appropriate use of nanotechnology in environmental restoration necessitates the creation and implementation of well-planned methods to reduce dangers and assure the long-term use of NMs. These solutions promote environmental integrity and human health by taking a cautious approach in the face of uncertainty about NM behavior and ramifications. Engaging in open talks with stakeholders and the general public is critical for encouraging transparency, accountability, and confidence in decision-making regarding nanotechnology applications. Furthermore, thorough monitoring and surveillance programs are required to accurately monitor and understand NMs' environmental fate and behaviour, evaluate their long-term consequences, and enable the deployment of adaptive management strategies. Incorporating green chemistry and sustainability themes into nanotechnology development and implementation can lessen environmental consequences while also improving overall remedial sustainability. By implementing these complex and carefully planned strategies, stakeholders will be able to successfully navigate the complexities of nanotechnology implementation, effectively reduce potential risks, and promote responsible and sustainable environmental cleanup practices based on scientific rigor and caution (Usha Rani et al., 2024).

18.8 CONCLUSION

The application of nanotechnology to environmental remediation represents a new era of inventiveness and possibility in the realm of environmental protection. By using the revolutionary capabilities of NMs and adopting a multidisciplinary approach, individuals involved may effectively navigate the complexities and take advantage of nanotechnology to tackle urgent environmental issues and lay the foundation for a more sustainable future. A robust plan for further research and action in harnessing nanotechnology to clean up the environment emerges. By acknowledging the benefits and risks, and adopting responsible use practices, we can harness the potential of NMs for effective and sustainable environmental remediation.

ACKNOWLEDGEMENT

We extend our sincere gratitude to the Department of Microbiology, Atmiya University, Rajkot for giving us this opportunity and providing all the required facilities to accomplish our task. We would also like to thank Krishnaba Zala, a student of B.Sc. Microbiology, for helping us in searching the review articles.

REFERENCES

Abbas, A., Al-Amer, A. M., Laoui, T., Al-Marri, M. J., Nasser, M. S., Khraisheh, M., & Atieh, M. A. (2016). Heavy metal removal from aqueous solution by advanced carbon nanotubes: Critical review of adsorption applications. *Separation and Purification Technology*, *157*, 141–161. https://doi.org/10.1016/j.seppur.2015.11.039

Al-Hobaib, A. S., Al-Sheetan, K. M., Shaik, M. R., & Al-Suhybani, M. S. (2017). Modification of thin-film polyamide membrane with multi-walled carbon nanotubes by interfacial polymerization. *Applied Water Science*, *7*, 4341–4350. https://doi.org/10.1007/s13201-017-0578-5

Anboo, S., Lau, S. Y., Kansedo, J., Yap, P. S., Hadibarata, T., Jeevanandam, J., & Kamaruddin, A. H. (2022). Recent advancements in enzyme-incorporated NMs: Synthesis, mechanistic formation, and applications. *Biotechnology and Bioengineering*, *119*(10), 2609–2638. https://doi.org/10.1002/bit.28185

Aragaw, T. A., Bogale, F. M., & Aragaw, B. A. (2021). Iron-based NPs in wastewater treatment: A review on synthesis methods, applications, and removal mechanisms. *Journal of Saudi Chemical Society*, *25*(8), 101280. https://doi.org/10.1016/j.jscs.2021.101280

Bassyouni, M., Abdel-Aziz, M. H., Zoromba, M. S., Abdel-Hamid, S. M. S., & Drioli, E. (2019). A review of polymeric nanocomposite membranes for water purification. *Journal of Industrial and Engineering Chemistry 73*, 19–46. https://doi.org/10.1016/j.jiec.2019.01.045

Benjamin, S. R., Lima, F. D., Florean, E. O. P. T., & Guedes, M. I. F. (2019). Current trends in nanotechnology for bioremediation. *International Journal of Environment and Pollution*, *66*(1–3), 19–40.

Bhandari, G. (2018). Environmental nanotechnology: Applications of NPs for bioremediation. *Approaches in Bioremediation: The New Era of Environmental Microbiology and Nanobiotechnology*, *1*, 301–315.

Bondarenko, O., Juganson, K., Ivask, A., Kasemets, K., Mortimer, M., & Kahru, A., 2013. Toxicity of Ag, CuO and ZnO NPs to selected environmentally relevant test organisms and mammalian cells in vitro: A critical review. *Archives of Toxicology*, *87*, 1181–1200.

Boulkhessaim, S., Gacem, A., Khan, S. H., Amari, A., Yadav, V. K., Harharah, H. N., … & Jeon, B. H. (2022). Emerging trends in the remediation of persistent organic pollutants using NMs and related processes: A review. *NMs*, *12*(13), 2148. https://doi.org/10.3390/nano12132148

Cashin, V. B., Eldridge, D. S., Yu, A., & Zhao, D. (2018). Surface functionalization and manipulation of mesoporous silica adsorbents for improved removal of pollutants: A review. *Environmental Science: Water Research & Technology*, *4*(2), 110–128.

Choi, S., Drese, J. H., Eisenberger, P. M., & Jones, C. W. (2011). Application of amine-tethered solid sorbents for direct CO_2 capture from the ambient air. *Environmental Science & Technology*, *45*(6), 2420–2427. https://doi.org/10.1021/es102797w

Del Prado-Audelo, M. L., García Kerdan, I., Escutia-Guadarrama, L., Reyna-González, J. M., Magaña, J. J., & Leyva-Gómez, G. (2021). Nanoremediation: NMs and nanotechnologies for environmental cleanup. *Frontiers in Environmental Science*, *9*, 793765. https://doi.org/10.3389/fenvs.2021.793765

Donaldson, K., Tran, L., Jimenez, L., Duffin, R., Newby, D. E., & Mills, N. (2005). Combustion-derived NPs: A review of their toxicology following inhalation exposure 1. *Particle and Fibre Toxicology*, *2*, 10.

Durgalakshmi, D., Rajendran, S., & Naushad, M. (2019). Current role of NMs in environmental remediation. *Advanced Nanostructured Materials for Environmental Remediation*, *30*, 1–20.

Ghasemzadeh, G., Momenpour, M., et al., 2014. Applications of NMs in water treatment and environmental remediation. *Frontiers of Environmental Science & Engineering*, *8*, 471–482.

Guerra, F. D., Attia, M. F., Whitehead, D. C., & Alexis, F. (2018). Nanotechnology for environmental remediation: Materials and applications. *Molecules*, *23*(7), 1760. https://doi.org/10.3390/molecules23071760

Hristozov, D., Ertel, J., & Techno Valuation, M. (2009). Nanotechnology and sustainability: Benefits and risks of nanotechnology for environmental sustainability. *Forum der Forschung*, *22*, 161–168. https://doi.org/10.33472/AFJBS.6.10.2024.4359-4380

Huang, H. Y., Yang, R. T., Chinn, D., & Munson, C. L. (2003). Amine-grafted MCM-48 and silica xerogel as superior sorbents for acidic gas removal from natural gas. *Industrial & Engineering Chemistry Research*, *42*(12), 2427–2433. https://doi.org/10.1021/ie020440u

Jeevanandam, Jaison, Barhoum Ahmed, Chan Yen S., Dufresne Alain, Danquah, Michael K., Review on nanoparticles and nanostructured materials: history, sources, toxicity and regulations, *Beilstein Journal of Nanotechnology*, 2018, 9, 1050–1074. https://doi.org/10.3762/bjnano.9.98

Kahraman, E., Gungor, S., & Ozsoy, Y. (2017). Potential enhancement and targeting strategies of polymeric and lipid-based nanocarriers in dermal drug delivery. *Therapeutic Delivery*, *8*(11), 967–985. https://doi.org/10.4155/tde-2017-0075

Kaur, R., Mavi, G. K., Raghav, S., & Khan, I. (2019). Pesticides classification and its impact on environment. *International Journal of Current Microbiology and Applied Sciences*, *8*(3), 1889–1897.

Keller, A.A., & Lazareva, A., 2013. Predicted release of engineered NMs: From global to regional to local. *Environmental Science & Technology Letters*, *1*(1), 65–70.

Khan, I., Saeed, K., & Khan, I. (2019). NPs: Properties, applications and toxicities. *Arabian Journal of Chemistry*, *12*(7), 908–931. https://doi.org/10.1016/j.arabjc.2017.05.011

Khan, S., Mansoor, S., Rafi, Z., Kumari, B., Shoaib, A., Saeed, M., … & Shakeel, F. (2022a). A review on nanotechnology: Properties, applications, and mechanistic insights of cellular uptake mechanisms. *Journal of Molecular Liquids*, *348*, 118008. https://doi.org/10.1016/j.molliq.2021.118008

Khan, Y., Sadia, H., Ali Shah, S. Z., Khan, M. N., Shah, A. A., Ullah, N., … & Khan, M. I. (2022b). Classification, synthetic, and characterization approaches to NPs, and their applications in various fields of nanotechnology: A review. *Catalysts*, *12*(11), 1386. https://doi.org/10.3390/catal12111386

Khan, Y., Sadia, H., Ali Shah, S. Z., Khan, M. N., Shah, A. A., Ullah, N., Ullah, M. F., Bibi, H., Bafakeeh, O. T., Khedher, N. B., Eldin, S. M., Fadhl, B. M., & Khan, M. I. Classification, synthetic, and characterization approaches to NPs, and their applications in various fields of nanotechnology: A review. *Catalysts* 2022c, 12 (11), 1386. https://doi.org/10.3390/catal12111386

Kumar, L., Ragunathan, V., Chugh, M., & Bharadvaja, N. (2021). NMs for remediation of contaminants: A review. *Environmental Chemistry Letters*, *19*(4), 3139–3163. https://doi.org/10.1007/s10311-021-01212-z

Kumar, N., & Kumbhat, S. Carbon-Based NMs. *Essentials in Nanoscience and Nanotechnology*; John Wiley&Sons,Inc.:Hoboken,NJ,U.S.A.,2016;pp189–236.https://doi.org/10.1002/9781119096122.ch5

Kumar, S., Dilbaghi, N., Saharan, R., & Bhanjana, G. (2012). Nanotechnology as emerging tool for enhancing solubility of poorly water-soluble drugs. *Bionanoscience*, 2, 227–250. https://doi.org/10.1002/bit.28185

Lakhotia, S. R., Mukhopadhyay, M., & Kumari, P. (2018). Cerium oxide NPs embedded thin-film nanocomposite nanofiltration membrane for water treatment. *Scientific Reports*, *8*(1), 4976. https://doi.org/10.1038/s41598-018-23188-7

Lee, S. W., Cheon, S. A., Kim, M. I., & Park, T. J. (2015). Organic–inorganic hybrid nanoflowers: Types, characteristics, and future prospects. *Journal of Nanobiotechnology*, *13*, 1–10. https://doi.org/10.1186/s12951-015-0118-0

Long, Q., Zhang, Z., Qi, G., Wang, Z., Chen, Y., & Liu, Z. Q. (2020). Fabrication of chitosan nanofiltration membranes by the film casting strategy for effective removal of dyes/salts in textile wastewater. *ACS Sustainable Chemistry & Engineering*, *8*(6), 2512–2522. https://doi.org/10.1021/acssuschemeng.9b07026

Ma, J., Guo, X., Ying, Y., Liu, D., & Zhong, C. (2017). Composite ultrafiltration membrane tailored by MOF@GO with highly improved water purification performance. *Chemical Engineering Journal*, *313*, 890–898. https://doi.org/10.1016/j.cej.2016.10.127

Mahmoud, W. M., Rastogi, T., & Kümmerer, K. (2017). Application of titanium dioxide NPs as a photocatalyst for the removal of micropollutants such as pharmaceuticals from water. *Current Opinion in Green and Sustainable Chemistry*, *6*, 1–10. https://doi.org/10.1016/j.cogsc.2017.04.001

Mashkoor, F., & Nasar, A. (2020). Magsorbents: Potential candidates in wastewater treatment technology: A review on the removal of methylene blue dye. *Journal of Magnetism and Magnetic Materials*, *500*, 166408. https://doi.org/10.1016/j.jmmm.2020.166408

Mekuye, B., & Abera, B. (2023). NMs: An overview of synthesis, classification, characterization, and applications. *Nano Select*, *4*(8), 486–501. https://doi.org/10.1002/nano.202300038

Mirzaei, A., Leonardi, S. G., & Neri, G. (2016). Detection of hazardous volatile organic compounds (VOCs) by metal oxide nanostructures-based gas sensors: A review. *Ceramics International*, *42*(14), 15119–15141. https://doi.org/10.1016/j.ceramint.2016.06.145

Mpouras, T., Polydera, A., Dermatas, D., Verdone, N., & Vilardi, G. (2021). Multi wall carbon nanotubes application for treatment of Cr (VI)-contaminated groundwater; Modeling of batch & column experiments. *Chemosphere*, *269*, 128749. https://doi.org/10.1016/j.chemosphere.2020.128749

National Nanotechnology Initiative. What is Nanotechnology? 2009. Available: https://www.nano.gov/html/facts/whatisNano.html (accessed 19.04.14).

Neeta, Pandey, Shukla, S. K., & Singh, N. B. 2017. Water purification by polymer nanocomposites: An overview *Nanocomposites*, *3*(2), 47–66. https://doi.org/10.1080/20550324.2016.1329983

Nomura, A., & Jones, C. W. (2013). Amine-functionalized porous silicas as adsorbents for aldehyde abatement. *ACS Applied Materials & Interfaces*, *5*(12), 5569–5577. https://doi.org/10.1021/am400810s

Nomura, A., & Jones, C. W. (2014). Enhanced formaldehyde-vapor adsorption capacity of polymeric amine-incorporated aminosilicas. *Chemistry: A European Journal*, *20*(21), 6381–6390. https://doi.org/10.1002/chem.201304954

Patil, S. S., Shedbalkar, U. U., Truskewycz, A., Chopade, B. A., & Ball, A. S. (2016). NPs for environmental clean-up: A review of potential risks and emerging solutions. *Environmental Technology & Innovation 5*, 10–21. https://doi.org/10.1016/j.eti.2015.11.001

Peydayesh, M., Mohammadi, T., & Nikouzad, S. K. (2020). A positively charged composite loose nanofiltration membrane for water purification from heavy metals. *Journal of Membrane Science*, *611*, 118205. https://doi.org/10.1016/j.memsci.2020.118205

Puri, A., Loomis, K., Smith, B., Lee, J. H., Yavlovich, A., Heldman, E., & Blumenthal, R. (2009). Lipid-based NPs as pharmaceutical drug carriers: From concepts to clinic. *Critical Reviews™ in Therapeutic Drug Carrier Systems*, *26*(6). https://doi.org/10.1615/CritRevTherDrugCarrierSyst.v26.i6.10

Rasheed, T., Adeel, M., Nabeel, F., Bilal, M., & Iqbal, H. M. (2019). TiO_2/SiO_2 decorated carbon nanostructured materials as a multifunctional platform for emerging pollutants removal. *Science of the Total Environment*, *688*, 299–311. https://doi.org/10.1016/j.scitotenv.2019.06.200

Roco, M.C., 2005. International perspective on government nanotechnology funding in 2005. *Journal of Nanoparticle Research*, *7*, 707–712.

Royal Society and Royal Academy of Engineering. Nanoscience and nanotechnologies: Opportunities and uncertainties, 2004. https://www.nanotec.org.uk/report/summary.pdf (accessed on 03.04.14).

Saikia, J., Gogoi, A., & Baruah, S. (2019). Nanotechnology for water remediation. *Environmental Nanotechnology* 2, 195–211. https://doi.org/10.1007/978-3-319-98708-8_7

Salem, S. S., Hammad, E. N., Mohamed, A. A., & El-Dougdoug, W. (2022). A comprehensive review of NMs: Types, synthesis, characterization, and applications. *Biointerface Research in Applied Chemistry, 13*(1), 41.

Seth, R., & Meena, A. (2024). Enzymes-based NM synthesis: An eco-friendly and green synthesis approach. *Clean Technologies and Environmental Policy*, 1–24. https://doi.org/10.1007/s10098-024-02854-7

Shan, J., Ji, R., Yu, Y., Xie, Z., & Yan, X. (2015). Biochar, activated carbon and carbon nanotubes have different effects on fate of 14C-catechol and microbial community in soil. *Scientific Reports, 5*(1), 16000. https://doi.org/10.1038/srep16000

Shukla, S., Khan, R., & Hussain, C. M. (2020). "Nanoremediation," in *The Handbook of Environmental Remediation*. Editor C. M. Hussain (London: The Royal Society of Chemistry).

Stefaniuk, M., Oleszczuk, P., & Ok, Y. S. (2016). Review on nano zerovalent iron (nZVI): From synthesis to environmental applications. *Chemical Engineering Journal, 287*, 618–632. https://doi.org/10.1016/j.cej.2015.11.046

Thakur, A., & Kumar, A. (2022). Recent advances on rapid detection and remediation of environmental pollutants utilizing NMs-based (bio) sensors. *Science of the Total Environment, 834*, 155219. https://doi.org/10.1016/j.scitotenv.2022.155219

Thatai, S., Khurana, P., Boken, J., et al., 2014. NPs and core–shell nanocomposite based new generation water remediation materials and analytical techniques: A review. *The Microchemical Journal, 116*, 62–76.

Usha Rani, N., Pratibha, S., Rajendra Kumar, S., Kavita, C., Shanmuga, S., & Shankar, S. P., The role of nanotechnology in environmental remediation opportunities and challenges. *African Journal of Biological Sciences, 6*(10) (2024), 4359–4380. https://doi.org/10.33472/AFJBS.6.10.2024.4359-4380

Vatsha, B., Ngila, J. C., & Moutloali, R. M. (2014). Preparation of antifouling polyvinylpyrrolidone (PVP 40K) modified polyethersulfone (PES) ultrafiltration (UF) membrane for water purification. *Physics and Chemistry of the Earth, Parts A/B/C, 67*, 125–131. https://doi.org/10.1016/j.pce.2013.09.021

Wasi, S., Tabrez, S., & Ahmad, M. (2013). Toxicological effects of major environmental pollutants: An overview. *Environmental Monitoring and Assessment, 185*, 2585–2593.

Yan, F., Chen, H., Lü, Y., Lü, Z., Yu, S., Liu, M., & Gao, C. (2016). Improving the water permeability and antifouling property of thin-film composite polyamide nanofiltration membrane by modifying the active layer with triethanolamine. *Journal of Membrane Science, 513*, 108–116. https://doi.org/10.1016/j.memsci.2016.04.049

Zamora-Ledezma, C., Negrete-Bolagay, D., Figueroa, F., Zamora-Ledezma, E., Ni, M., Alexis, F., & Guerrero, V. H. (2021). Heavy metal water pollution: A fresh look about hazards, novel and conventional remediation methods. *Environmental Technology & Innovation, 22*, 101504. https://doi.org/10.1016/j.eti.2021.101504

Zhang, L., Chan, J. M., Gu, F. X., Rhee, J. W., Wang, A. Z., Radovic-Moreno, A. F., … &Farokhzad, O. C. (2008). Self-assembled lipid– polymer hybrid NPs: A robust drug delivery platform. *ACS Nano, 2*(8), 1696–1702. https://doi.org/10.1021/nn800275r

Zhang, W.X., (2003). Nanoscale iron particles for environmental remediation: An overview. *Journal of Nanoparticle Research, 5*, 323–332.

Zhang, X., Wang, H., He, L., Lu, K., Sarmah, A., Li, J., & Huang, H. (2013). Using biochar for remediation of soils contaminated with heavy metals and organic pollutants. *Environmental Science and Pollution Research, 20*, 8472–8483. https://doi.org/10.1007/s11356-013-1659-0

19 Advancements in Nanoremediation

Sustaining the Natural Environment Through Innovative Techniques

Pankaj Kumar, Snigdha Singh, Apurva Sharma, Manoj Kumar, Virendra Kumar Yadav, Mohd Tariq, Nakul Kumar, and Yamee Shah

19.1 INTRODUCTION

19.1.1 Environmental Pollution

Environmental pollution is a critical global issue resulting from the introduction of harmful substances or energy into the environment through human activities, leading to a decline in environmental quality and functionality. This pollution can have adverse effects on ecosystems, human health, and overall environmental sustainability (Kumar et al. 2022). Environmental pollution has wide-ranging effects that go beyond just harming the environment. It also causes severe toxicity to ecosystems, poses a risk to biodiversity, and endangers human health. Research has shown that environmental pollution plays a significant role in causing global diseases, disability, and mortality. It has a greater impact on human health than the combined impact of infectious diseases like HIV/AIDS, malaria, and tuberculosis (Kumar et al. 2024). The severity of environmental contamination in developing nations is alarming, requiring corrective actions to mitigate its adverse effects on the environment and human populations (Kumar and Fulekar 2021).

The relationship between environmental pollution and its impact on various aspects of life is complex. For example, environmental pollution has been linked to respiratory diseases, genetic mutations, and cancer, emphasizing the critical need for effective pollution control measures (Kumar and Fulekar 2018). Moreover, the economic implications of environmental pollution are significant, with stringent environmental policies and regulations playing a crucial role in mitigating the adverse effects of pollution on economic growth and sustainability (Kumar, et al., 2023). The dynamics of infectious diseases are also influenced by environmental pollution, highlighting the interconnectedness of environmental factors and public health outcomes. Furthermore, the correlation between energy use, commercial development, and environmental contamination underscores the delicate balance required to achieve sustainable development without compromising environmental integrity. In China, for instance, the relationship between pollution, economic growth, and marine fishery culture underscores the need for comprehensive strategies to address pollution and its impact on diverse sectors. Additionally, the impact of pollution on urban areas, such as air pollution from vehicle emissions, underscores the importance of implementing effective environmental planning and management practices to safeguard public health and environmental quality (Kumar 2018).

 DOI: 10.1201/9781003484189-19

19.1.2 Role of Nanotechnology in Environmental Cleanup

Nanotechnology plays an important role in environmental cleanup by providing novel ways to meet various environmental concerns. NMs, with their unique properties, are being utilized to combat issues such as water and air pollution, as well as oil spill cleanup. The nanotechnological application has garnered significant attention and financial support from both service providers and the scientific community (Gebre and Sendeku 2019). This technology is not only limited to theoretical discussions but is also being practically implemented in various areas, such as pollution prevention, treatment, and cleanup. Nanotechnology has been recognized as a crucial factor in the effort to decrease environmental pollution and mitigate threats to humans.

One of the significant contributions of nanotechnology to environmental cleanup is the development of nano-sensors and nanoscale coatings that can replace traditional, thicker polymer coatings to prevent corrosion, as well as the use of nanoscale biopolymers for enhanced decontamination and recycling of heavy metals. Furthermore, nanostructured metals have been developed to break toxic organics at room temperature, and smart particles are being used for environmental purification. NPs have also shown promise as novel photocatalysts for environmental cleanup, highlighting the diverse applications of nanotechnology in addressing environmental challenges. In addition, nanoremediation technologies have evolved as highly effective and efficient solutions for cleaning up soil and groundwater that have been tainted with petroleum chemicals and metals (Alazaiza et al. 2021).

The role of nanotechnology in wastewater treatment has been extensively studied, with research highlighting the promising outcomes and recent advancements in nanotechnology-mediated remediation methods. Nanotechnology has enabled the development of nanoscale materials, such as carbon dots, which have found applications in environmental cleanup and monitoring (Khayal et al. 2021). Moreover, the use of biosynthetic NPs in bioremediation processes is emerging as a green and efficient method for environmental cleanup. Nanotechnology has also enabled the use of microbial cells to absorb heavy metals, providing a potential option for extracting heavy metals from industrial waste. The application of biosurfactants, due to their diversity, is being explored as a viable option for the environmental cleanup of pollutants. Nanotechnology has revolutionized environmental cleanup efforts by providing innovative solutions to address pollution, contamination, and remediation challenges. NMs possess distinct characteristics that have facilitated the advancement of advanced technologies, including nano-sensors, nanoremediation techniques, and biosynthetic NPs. These technologies are paramount in protecting the environment and promoting human well-being. The interdisciplinary nature of nanotechnology has allowed for the integration of biology and environmental science to create sustainable and efficient environmental cleanup strategies. As research in nanotechnology continues to advance, the potential for further applications in environmental cleanup remains promising, paving the way for a cleaner and healthier environment. **Table 19.1** shows the various nanomaterials for effective wastewater treatment.

19.1.3 Overview of Nanoremediation Techniques

Nanoremediation techniques have attracted considerable interest in recent years as new methods for conserving polluted environments. These methods utilize NMs to combat many forms of pollution, such as persistent organic chemicals like pesticides, chlorinated solvents, heavy metals, and perfluoroalkyl and polyfluoroalkyl substances (PFAS) (Ganie et al. 2021). Nanoremediation encompasses several applications like chemical degradation, Fenton oxidation, photocatalytic degradation and bioremediation techniques, making it a versatile and promising technology for mitigating soil contamination. By utilizing engineered NMs (ENMs), nanoremediation offers a novel and effective method to clean up polluted matrices, providing a sustainable cleanup technique to restore natural equilibrium (Guidi et al. 2023).

One of the main benefits of nanoremediation is its capacity to improve the adsorption and sequestration of contaminants; hence, it is a modern method with interesting in-situ and ex-situ

TABLE 19.1
Nanoscale Materials for Wastewater Treatment

S.No.	Type of NP	Treatment Mechanism	Removal Target
1	TiO_2 NPs	Photocatalysis	Organic pollutants
2	Iron NPs	Reduction, adsorption	Metals, organic pollutants
3	Bimetallic NPs	Reduction, adsorption	Dichlorination
4	Nanofiltration and nanomembranes	Nanofiltration	Organic and inorganic materials
5	Magnetite NPs	Adsorption	Metals, organic mixtures
6	Metal-sorbing vesicles	Adsorption	Metals
7	Micelles	Adsorption	Organic chemicals
8	Dendrimers	Encapsulation	Metals, organic chemicals
9	Nanotube	Adsorption	Metals, anions, organic pollutants
10	Nanoclay	Adsorption	Heavy metals and anions

uses for soil and groundwater treatment. Nanoremediation has been identified as a cost-effective and environmentally sustainable method compared to conventional in situ remediation techniques, offering better efficiency and economic viability. Nanoremediation techniques are effective for treating tainted aquifers, highlighting the potential of nanoremediation for addressing groundwater pollution. Furthermore, nanoremediation was developed as a highly effective technique for cleaning contaminated soil and groundwater, including heavy metal and petroleum pollution sources. It has demonstrated its ability to rapidly and effectively remediate such pollutants (Alazaiza et al. 2021).

In addition, researchers have investigated nanoremediation methods for addressing organic contaminants, metals, and pesticides. These methods include nano-bioremediation and nano-phytoremediation (Dhanapal et al. 2024). These techniques leverage NMs to remove pollutants, demonstrating their effectiveness in enhancing remediation processes (El-Ramady et al. 2020). Nanoremediation has also been proposed as a promising tool for the treatment of toxic organic pollutants, highlighting the potential of NPs in mitigating environmental contamination. The use of green leaf-produced nanoscale zero-valent iron (nZVI) particles in nanoremediation techniques has shown promising results in the cleanup of lead and nickel-contaminated soil. This demonstrates how well nanoremediation works to address soil contamination (Francy et al. 2020).

Moreover, nanoremediation techniques offer a sustainable and eco-friendly solution for soil and water treatment, particularly in the context of expanding environmental pollution issues (Chaudhari et al. 2023). The large surface area and high reactivity of engineered NMs (ENMs) make them efficient in removing pollutants from environmental matrices, outperforming conventional techniques in terms of performance. Nanoremediation has also been identified as a modern technology for immobilizing arsenic in polluted soils, showcasing its potential for reducing human health risks and improving soil properties. Additionally, nanoremediation techniques have been recommended for reducing the harmful effects of hazardous pesticides in wastewater, emphasizing the importance of adsorption techniques and nanoremediation in mitigating contamination (Ugwu et al. 2022).

Nanoremediation techniques represent a promising and innovative approach for addressing various environmental challenges, ranging from soil and groundwater contamination to the treatment of inorganic and organic contaminants. By leveraging NMs and advanced technologies, nanoremediation offers efficient, cost-effective, and sustainable solutions for cleaning up polluted environments and restoring natural equilibrium. Even though facing encounters such as high fabrication costs and ecological impacts, current research and advancements in nanoremediation continue to enhance its effectiveness and applicability in environmental remediation.

19.2 FUNDAMENTALS OF NMS IN REMEDIATION

19.2.1 Properties of NMs

NMs possess a diverse range of distinctive characteristics that make them exceptionally adaptable and helpful in multiple domains. These qualities arise from the nanoscale size and distinct traits that distinguish them from bulk materials. An important benefit of NMs is their elevated surface area-to-volume ratio, which increases their reactivity and ability to interact with the surroundings (Virendra K. Yadav et al. 2022). This property is crucial in applications where surface interactions play a significant role, such as in catalysis and sensing. NMs often hold greater mechanical properties due to their small size and unique structures, making them suitable for reinforcing composite materials. These improved mechanical properties can lead to enhanced durability and strength in the resulting materials.

Furthermore, the optical properties of NMs are of particular interest, especially in fields like optoelectronics and photonics. NMs can exhibit tunable optical properties, allowing for applications in wavelength-tunable lasers and broadband lasers (Yadav, et al. 2022). The ability to manipulate the optical properties of NMs opens possibilities for developing advanced optical devices and sensors. Moreover, the semiconducting properties of certain NMs make them suitable for applications in electronics and energy storage.

Another crucial aspect of NMs is the competence to relate with biological systems, leading to advancements in nanomedicine and theranostics. NMs can be created to have special properties that enable targeted drug delivery, imaging, and therapy. The unique physicochemical assets play a vital role in their efficacy and safety in medical applications. In addition, the surface characteristics of NMs, including their bio-corona, can influence their interactions with the immune system and living entities (Fadeel 2022).

Moreover, the thermal properties of NMs are of interest in various applications, including in improving the thermal and mechanical stability of materials. NMs can exhibit excellent thermal conductivity, which is beneficial in thermal management applications (S. Zhang et al. 2023). By leveraging the thermal properties of NMs, it is possible to enhance the performance of thermal interface materials and heat dissipation systems. Additionally, the piezoelectric properties of certain NMs make them valuable in sensors and actuators. These materials can convert mechanical stress into electrical signals, enabling their use in diverse sensing applications.

Additionally, NMs, particularly in remediation and catalysis, are affected by their chemical properties. NMs can serve as efficient catalysts due to their prominent surface area. These properties enable enhanced catalytic activity and selectivity, making NMs promising candidates for various catalytic reactions. Additionally, the chemical stability of NMs is essential in applications where materials are exposed to harsh environments or corrosive conditions (Bounegru and Apetrei 2020). NMs with high chemical stability can withstand challenging operational conditions, expanding their utility in demanding applications.

NMs exhibit a diverse range of properties that make them highly valuable in numerous fields, including materials science, nanotechnology, medicine, and environmental science. Their unique physicochemical, optical, mechanical, thermal, and chemical properties enable a wide array of applications, ranging from electronics and photonics to biomedicine and catalysis (Choudhary et al. 2021). Understanding and harnessing the distinct properties of NMs is essential for advancing technology and developing innovative solutions to complex challenges across various disciplines. Figure 19.1 demonstrates various nanoremediation approaches to address environmental problems.

19.2.2 Interaction of NPs with Contaminants

The interaction of NPs with contaminants is crucial in environmental remediation strategies, influencing the fate and transport of pollutants. Studies have indicated that NPs can absorb impurities

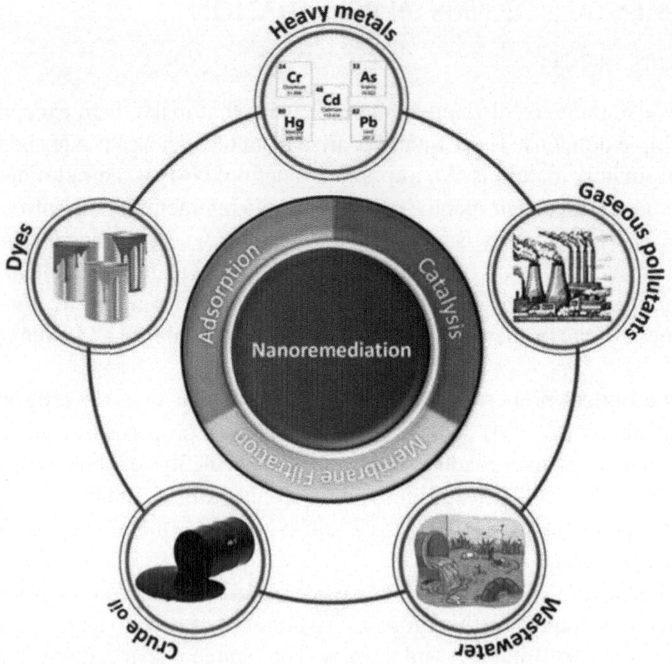

FIGURE 19.1 Nanoremediation approaches to address environmental problems. Adapted and modified (Rather et al. 2023).

through electrostatic interactions, π-π interactions, and van der Waals forces, which results in the destruction or sequestration of pollutants (Vinzant, Rashid, and Khodakovskaya 2023). These interactions are needed for understanding the efficiency and effectiveness of nanoremediation approaches, as they determine the ability of NPs to capture, transform, or immobilize contaminants in soil, water, or air environments. By elucidating the underlying mechanisms of NP-contaminant interactions, researchers can tailor NM properties to enhance remediation processes and mitigate environmental pollution.

Furthermore, the co-exposure effects of NPs and contaminants have been a subject of interest in environmental research, particularly in assessing the combined impacts on ecosystems and human health. NPs can interact with co-contaminants, altering their bioavailability, toxicity, and transport in the environment (Fellet, Conte, and Marchiol 2023). These interactions may lead to synergistic, antagonistic, or additive effects on biological systems, highlighting the importance of considering the complex interactions between NPs and contaminants in risk assessment and remediation strategies. Understanding the joint interactions of NPs with various contaminants is necessary for using sustainable and effective approaches to mitigate environmental pollution and protect ecosystem health.

Moreover, the reactivity of NPs with contaminants can result in the formation of new compounds that may exhibit different properties and behaviors compared to individual pollutants. The high reactivity of NPs can lead to the transformation or degradation of contaminants, influencing their environmental persistence and toxicity (Aslam and Mazhar 2023). NPs, through their interactions with contaminants, have the potential to mitigate the adverse effects of pollutants and contribute to the remediation of contaminated sites. By exploring the chemical and physical interactions between NPs and contaminants, researchers can design innovative nanotechnological solutions for addressing environmental challenges and promoting sustainable practices in pollution control and remediation efforts. Figure 19.2 shows the effects of NPs on human health.

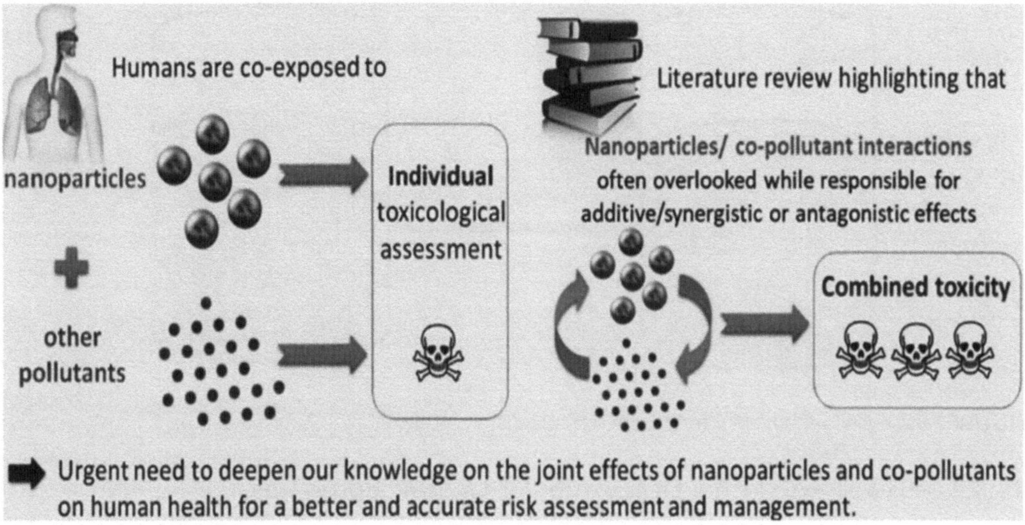

FIGURE 19.2 Consequences of NPs on human health. Adapted and Modified (Forest 2021).

19.3 TYPES OF NANOREMEDIATION APPROACHES

19.3.1 Nano-Adsorbents for Water and Soil Treatment

Nano-adsorbents are essential in water and soil treatment because they remove pollutants efficiently through adsorption mechanisms. NMs, characterized by their extensive surface area and reactivity, engage with contaminants in water and soil environments, resulting in their sequestration and immobilization. For instance, the process of adsorbing levofloxacin and antibiotic-resistance genes from hospital wastewater requires the chemical production of nano-zero-valent iron and nano-copper to create particular nano-adsorbents that can effectively target these contaminants (Hamad et al., 2023). The adsorption mechanism of stable iodine ^{127} and radioactive iodineI^{131} onto Nano ZnO/MWCNTs follows a multilayer physical adsorption process, as demonstrated by kinetic studies and response surface methodology. These studies highlight the diverse adsorption mechanisms employed by nano-adsorbents to capture and remove pollutants from water systems.

Case studies have exhibited the practical usage of nano-adsorbents in real-world scenarios, showcasing their efficacy in treating polluted water and soil. Additionally, biosynthesized, silica-supported iron oxide nanocomposites for the adsorptive removal of metal from aqueous solutions exemplify the economic and eco-friendly potential of nano-adsorbents in water treatment (Garg et al. 2023). These case studies provide insights into the practicality and effectiveness of nano-adsorbents in addressing contamination issues in water and soil treatment processes.

The process by which nano-adsorbents are used in water and soil treatment involves the attachment of pollutants onto the surface of these NMs by different types of interactions, including physical adsorption, chemical bonding, and ion exchange. Studies have demonstrated that nano-adsorbents can proficiently eliminate contaminants by attracting and attaching them to their surfaces. This process results in a decrease in the levels of these substances in water and soil systems (Garg et al., 2023). The adsorption process is affected by the surface properties of the nano-adsorbents, the composition of the pollutants, and the surrounding environmental conditions. These factors collectively determine the effectiveness of removing contaminants. Overall, the complex procedure by which nano-adsorbents are used in water and soil treatment highlights their capacity to be efficient and long-lasting solutions for environmental rehabilitation. Figure 19.3 indicates the nano-adsorbents for wastewater treatment.

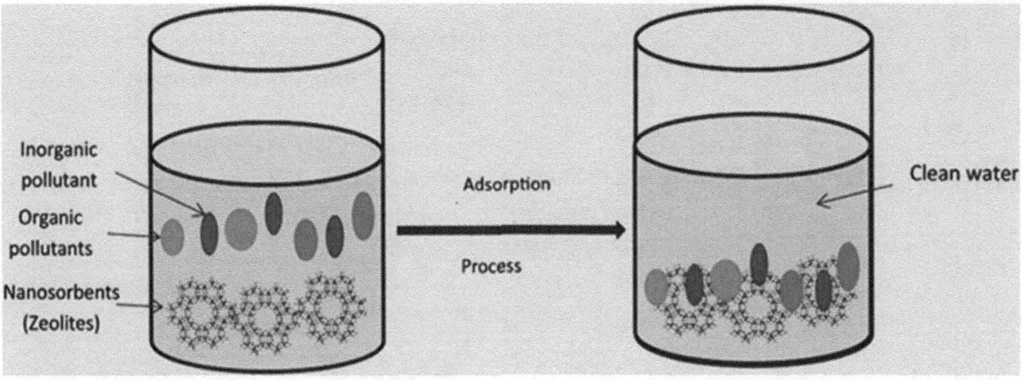

FIGURE 19.3 Nano-adsorbents for wastewater treatment. Adapted by (El-Sayed 2020).

19.3.2 Nano-Catalysts for Degradation of Organic Pollutants

Nano-catalysts play a crucial role in breaking down organic contaminants, offering efficient and durable options for environmental remediation. NMs possess a large surface area and exhibit catalytic activity, which facilitates the decomposition of organic pollutants into less dangerous substances through diverse catalytic mechanisms. For example, a study by Sasmal et al. (2021) showcased the use of porous covalent organic framework nanofilms as catalysts for C-H functionalization reactions in water, demonstrating the efficacy of nano-catalysts in driving organic transformations. The study highlighted the transformative potential of nano-catalysts in environmentally friendly catalytic reactions for organic pollutant degradation.

Real-world case studies have further illustrated the practical application of nano-catalysts in organic pollutant degradation. For instance, the extraction of iron from laterite soil and the green synthesis of laterite nano iron catalysts (GLaNICs) for Fenton's catalysis in the degradation of triclosan exemplify the use of nano-catalysts in advanced oxidation processes for pollutant removal. The efficient degradation of triclosan using nano-iron catalysts emphasizes the capability of nano-catalysts to address complex organic contaminants in water systems. Additionally, research by Yang et al. (2021) on metallic glass/copper composite catalysts for wastewater treatment demonstrates the development of hierarchical porous structures for effective organic pollutant degradation, showcasing the versatility of nano-catalysts in environmental applications. The mechanism of nano-catalysts in organic pollutant degradation involves catalytic reactions that facilitate the conversion of complex organic compounds into simpler, less toxic substances. Investigation has shown that nano-catalysts can improve degradation processes by offering active sites for catalytic reactions and facilitating the creation of reactive species that assist in the breakdown of pollutants. Nano-catalysts can effectively tackle environmental pollution concerns and encourage sustainable remediation methods in areas such as wastewater and soil treatment and organic pollutant degradation. Nano-catalysts are pivotal in advancing environmental catalysis, offering innovative solutions for the efficient deprivation of organic chemicals in water and soil systems. Table 19.2 shows the microbial-assisted nanoremediation of petroleum products.

19.3.3 Nano-Encapsulation for Controlled Release and Targeted Treatment

Nano-encapsulation is an advanced technology that involves enclosing active substances within nanoscale carriers to achieve controlled release and targeted treatment in drug delivery and agriculture. Recent advancements in nano-encapsulation have transformed the field of controlled release, enabling the precise delivery of therapeutic agents or chemicals to specific sites within the body or the environment. Nano-encapsulation offers several advantages, such as protecting the encapsulated

TABLE 19.2
Microbial-Assisted Nanoremediation of Petroleum Products

NMs	Microorganisms Involved	Petroleum Hydrocarbon Components	Removal Efficiency	References
Iron and zinc oxide	*Bacillus thuringiensis* S1, *P. aeruginosa* S2, and *Psychrobacter faecalis* S3	Crude oil	98% and 96% by Fe2O3 and ZnO NPs, respectively	Ehmedan et al. (2021)
Iron NPs	*Pseudomonas* stutzeri NA3 and *Acinetobacter baumannii* MN3	Polycyclic aromatic hydrocarbons (PAHs)	85%	Parthipan et al. (2022)
Fe_3O_4 NPs	*Rhodococcus erythropolis SY095*	Hexadecane and crude oil		X. Ma et al. (2021)
Fe_2O_3, and $Zn_5(OH)_8Cl_2$ NPs	*Bacillus licheniformis*	Crude oil (paraffin)	60%	El-Sheshtawy and Ahmed (2017)
Silver NPs	*Aspergillus flavus* AF15 and *Trichoderma harzianum* TH07	Crude oil	41.94% (*A. flavus* AF15) and 37.36% (*T. harzianum* TH07), respectively	Al-Zaban et al. (2020)
Silver	Ochrobactrum sp., *Stenotrophomonas maltophilia* and *P. aeruginosa*	Resin and asphaltenes	83.49%	Varjani et al. (2015)

Adapted and modified (Rather et al. 2023)

material from degradation, enhancing its bioavailability, and enabling sustained release over an extended period. By encapsulating active substances in nano-carriers, researchers can customize the release kinetics, improve efficacy, and minimize side effects, making nano-encapsulation a versatile and promising approach for targeted treatment (Yousefi et al. 2024).

Case studies have illustrated the effectiveness of nano-encapsulation for controlled release and targeted treatment in various contexts. For instance, Saleem et al. (2023) developed an adaptable magnetic mesoporous carbon-derived nano-adsorbent for synchronized metal removal from water. This nano-adsorbent demonstrated high adsorption capacity for toxic thallium ions, effective disinfection against bacteria, rapid disinfection kinetics, and notable regeneration capability, highlighting the potential of nano-encapsulation for multifunctional applications in water treatment. Additionally, they investigated the remediation of metals in soil using nano-bentonite, nano-hydroxyapatite, and nanocomposite, emphasizing the role of nano-encapsulation in enhancing the efficiency of pollutant removal. These case studies emphasize the versatility and efficacy of nano-encapsulation strategies for addressing complex environmental challenges and targeted treatment needs.

The mechanism of nano-encapsulation involves encapsulating active substances within nano-carriers, such as liposomes, polymeric NPs, or mesoporous materials, to achieve controlled release and targeted delivery. Nano-encapsulation protects the encapsulated material from degradation, aids in its transport to specific sites, and enables sustained release through diffusion or stimuli-responsive mechanisms. The release kinetics of the encapsulated material can be adjusted by modifying the properties of the nano-carrier, such as its composition, size, surface charge, and porosity. Studies have indicated that nano-encapsulation can increase the stability, solubility, and bioavailability of encapsulated substances, leading to improved therapeutic efficacy and reduced toxicity (Saleem et al. 2023). Overall, nano-encapsulation shows significant promise for advancing controlled release and targeted treatment strategies in various fields, offering innovative solutions for precise and efficient delivery of active substances.

19.3.4 Nano-Sensors for Monitoring and Detection

Nano-sensors are essential for monitoring and detecting various substances and conditions due to their high sensitivity, small size, and efficiency. These sensors, often based on nanotechnology, offer unique advantages in environmental monitoring, healthcare, agriculture, and industrial applications. The application of one-dimensional metal oxide nanostructures in chemical and gas sensors has been thoroughly studied for the aim of environmental monitoring. These nanostructures provide superior surface-to-volume ratios, stability, and cost-effectiveness, making them ideal for detecting the presence of different chemical species. Additionally, nanostructured sensors like the ZnO/RTV hemispherical shell structure have been developed for monitoring human breathing patterns, enabling early detection of respiratory diseases such as asthma, bronchitis, and sleep apnea (Liu et al. 2024).

In the realm of water pathogen detection, CRISPR-Cas-based nano-sensors have shown promise in quickly identifying pathogen contamination, thereby enhancing water quality using molecular-recognition probes. Furthermore, advancements in electrochemical sensors, such as Bi2Te3-based sensors, have demonstrated high sensitivity and low detection limits for detecting substances like hydrogen peroxide. These sensors exhibit superior performance compared to traditional noble metal-based electrochemical sensors. In addition, the advancement of portable data-gathering systems for nano and ultra-micro-scale electrochemical sensors has facilitated the ability to monitor and collect data in real-time at the location of utilization (Murphy et al. 2021).

Nanomaterial film-based gas sensors integrated into automated monitoring and control systems have been utilized in greenhouse crop production to measure the levels of oxygen and CO_2, showcasing the versatility of nano-sensors in diverse applications. Wireless sensor systems based on hierarchical CuO microspheres have been designed for volatile organic compound detection in restricted Internet of Things environments, highlighting the potential of nano-sensors in wireless and remote monitoring applications. Additionally, MEMS resistive pressure sensors and temperature sensors have been proposed for monitoring pressure and temperature in high-voltage substation intelligent perception resilience technologies (B. Chen et al. 2024).

Graphene-based strain and pressure sensors have shown high sensitivity and probable applications in human health. NP-based quartz crystal microbalance (QCM) sensors have attracted interest due to their cost-effectiveness, selectivity, sensitivity, and quick response time in the detection of hazardous metal ions, such as cadmium (Tarek et al. 2024). Multi-sensor data fusion approaches have been employed to monitor tool conditions in real-time during machining processes, showcasing the integration of sensors for industrial applications. Plasmonic bridge sensors aided by carbon nanotubes and Au-Ag nano-rambutans have demonstrated improved stress sensing capabilities, enhancing detection sensitivity for various applications (D. Chen et al. 2023).

Gold nanomesh platforms have been developed for cost-efficient multiplex sensing, providing complementary information for high-performance sensing platforms. Carbon quantum dots have been utilized as nano-sensors for the quality and safety of food, demonstrating the potential of nanotechnology in ensuring food security. Nano biosensors, consisting of immobilized bio-receptor probes, have been employed for detecting target analyte molecules in agriculture and other fields. Furthermore, solid-state nano-electrochemical sensors have been designed for Agri 4.0 applications, showcasing the integration of nano-sensors in agricultural technologies (Seymour et al. 2021).

In the medical field, nano-nodes integrated with nano-sensors circulating in the bloodstream have been proposed for detecting biomarkers, filling the gap where larger devices are not feasible. Additionally, nano-material-coupled aptamer sensors have been developed for detecting body blood glucose levels, offering real-time monitoring and analysis for healthcare applications. Humidity sensors based on porous PDMS/AgNWs and GO have been designed for humidity detection, highlighting the importance of wearable sensors for health monitoring. Moreover, nano-sensors have been utilized in the detection of acetaldehyde, showcasing their potential in gas-sensing applications (R. K. Zhang et al. 2020).

In general, nano-sensors have completely transformed the ability to monitor and detect in different areas, providing exceptional sensitivity, quick reaction times, and cost-efficient solutions for a diverse set of uses. Nano-sensors are becoming increasingly important in many industries, including healthcare, agriculture, and environmental monitoring. They are driving innovation and developments in sensor technology, leading to greater sensing capabilities and more effective data collection methods.

19.4 CASE STUDIES AND APPLICATIONS

19.4.1 Application of Nanoremediation in Groundwater and Soil Cleanup

Nanoremediation, an innovative technology that utilizes NPs for soil and groundwater cleanup, has gained noteworthy consideration due to its potential to address environmental contamination effectively. Recent case studies have highlighted the application of nanoremediation in soil and groundwater cleanup, demonstrating its effectiveness in remediation processes. Alazaiza et al. (2021) have emphasized the advantages of nanoremediation technologies in reducing costs and cleanup time. These advancements have paved the way for more sustainable and efficient remediation practices in contaminated sites.

Zero-valent iron NPs have emerged as a prominent NM for soil and groundwater cleanup, as highlighted in studies by Semerad et al. (2020). The use of zero-valent iron NPs in nanoremediation technologies has been widely recognized for its effectiveness in treating contaminated sites. These findings have exhibited the successful application of zero-valent iron NPs in remediating soil and groundwater, showcasing their potential for widespread environmental cleanup initiatives.

Furthermore, the integration of nanoremediation with other remediation techniques, such as phytoremediation, has shown promising results in enhancing soil and groundwater cleanup processes. Silva et al. (2023) have highlighted the synergistic effects of combining nanoremediation with phytoremediation to achieve rapid and effective remediation outcomes. This integrated approach offers a comprehensive solution for addressing soil and groundwater contamination, emphasizing the importance of utilizing multiple remediation strategies for optimal results.

Moreover, the use of nanoremediation in agriculture and environmental sustainability has been explored by Ur Rahim et al. (2021). These studies have emphasized the significance of NMs in enhancing sustainability in advanced agriculture, as well as in environmentally acceptable methods for the remediation of contaminated soils. Nanoremediation technologies offer innovative solutions for addressing environmental challenges in agriculture, highlighting their potential for enhancing food security and environmental protection.

In the context of groundwater quality and contamination, Adadzi et al. (2022) have focused on assessing groundwater quality, pollution patterns, and risk factors associated with groundwater contamination. These studies have utilized advanced methodologies, such as geostatistical analyses and numerical modeling, to evaluate groundwater pollution risks and design effective management and remediation strategies. By incorporating high-resolution data and modeling techniques, researchers can gain valuable insights into groundwater contamination dynamics and develop targeted remediation approaches.

Overall, recent case studies on the application of nanoremediation in soil and groundwater cleanup have demonstrated the effectiveness of nanotechnologies in addressing environmental contamination challenges. By leveraging the unique properties of NPs, researchers and environmental practitioners can develop innovative remediation strategies that offer sustainable, cost-effective, and efficient solutions for restoring contaminated sites. The integration of nanoremediation with other remediation techniques and the exploration of NMs in agriculture and environmental sustainability further highlight the diverse applications and potential benefits of nanoremediation in addressing complex environmental issues.

19.4.2 Nanotechnology in Oil Spill Cleanup

Recent studies have demonstrated significant progress in utilizing nanotechnology for oil spill cleanup applications. Zhang et al. (2020) explored the use of graphene-based aerogels (GAs) as super-absorbents for oil and chemical cleaning. Faqihi et al. (2023) highlighted the effectiveness of functionalized magnetite NPs (MNPs) for oil cleanup due to high oil elimination efficacy. Additionally, Musa et al. (2021) explored polymer-coated magnetic NPs for cleaning crude oil.

Li (2023) developed open-cell long-chain branched polypropylene foams with excellent oil absorption capacity, cyclic performance, and reusability for oil adsorption in spill cleanup efforts. Another study by Li et al. (2021) introduced Joule-heated superhydrophobic foam for fast cleanup of viscous crude oil spills, emphasizing its porous structure and low operational costs. Lu et al. (2021) created a light-triggered thermo-responsive material that can absorb ultra-heavy oil spills. This study demonstrates the use of nanotechnology for clearing up the environment in a novel way.

Furthermore, Song et al. (2022) investigated highly branched polyethylene as an effective sorbent for oil spill cleanup. Amar et al. (2023) synthesized zinc ferrite magnetic NPs for oil spill cleanup, demonstrating the multifunctionality of NMs in environmental remediation. Subramoniapillai & Thilagavathi (2022) reviewed the use of natural fibers for oil spill cleanup, showcasing diverse approaches in utilizing environmentally friendly materials for remediation efforts.

Chen et al. (2021) explored graphene-based techniques for oil-water separation and cleanup of oil, highlighting the versatility of graphene in addressing environmental challenges. These studies collectively illustrate the various applications of nanotechnology in oil spill cleanup, from super-absorbent materials to magnetic NPs and environmentally friendly sorbents, showcasing the potential for advanced technologies to enhance the proficiency of oil spill cleanup operations. Figure 19.4 indicates the effects of oil spills and their remedy by magnetic NPs.

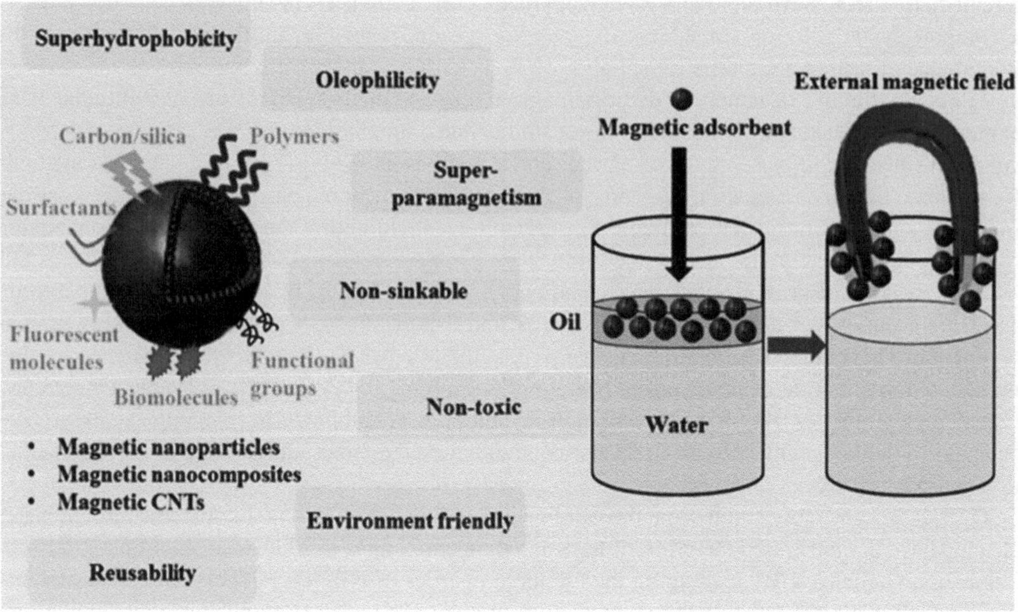

FIGURE 19.4 Ecological influences of oil spills and application of magnetic NPs for treatment. (Adapted and modified) Singh et al. 2020).

19.5 ENVIRONMENTAL IMPACTS AND SAFETY CONSIDERATIONS

19.5.1 Ecotoxicological Effects of NMs

NMs have attracted significant attention in diverse disciplines due to their distinctive characteristics, although apprehensions have been expressed regarding their possible ecotoxicological impacts on numerous creatures. Research investigations have demonstrated that manufactured NMs can have diverse ecotoxicological impacts on bacteria, plants, invertebrates, and fish. The proposed measure of the ecotoxicological impacts of NPs on biological systems is the total surface area of the NMs. Moreover, investigations have highlighted the role of oxidative stress in contributing to the ecotoxicological effects of NMs, with the generation of reactive oxygen species (ROS) by NPs' surfaces playing a crucial role in inducing toxicity (Girigoswami, Pallavi, and Girigoswami 2023).

Environmental corona-coated engineered NMs (ENMs) have been studied for their ecotoxicological effects on aquatic organisms at different trophic levels compared to pristine ENMs. NMs, such as silver NPs and nanocomposites, have restrained exact ecotoxicological evidence, impacting ecosystems (Thakur and Kumar 2023). Additionally, research has described the ecotoxicological effects of NMs and the testing methods used to assess their impact on the environment. Another study has examined the impact of NMs on environmental microbial communities, which has revealed specific harmful effects on various organisms in these environments.

Research has raised worries about the health effects of NM toxicity, underscoring the importance of understanding nanosafety. The underlying mechanisms of ecotoxicological effects of fungal-mediated selenium NPs on Snails and Daphnia magna have been explored (Ibrahim et al. 2023). Furthermore, the potential ecotoxicological impacts of NMs on aquatic organisms have been examined. The ecotoxicological effects of NMs represent a critical area of research that necessitates further investigation to comprehend the potential risks in different applications. Studies across different disciplines have provided valuable insights into the mechanisms of NM toxicity and the significance of assessing their impact on people and ecosystems (Jennings, Goodhead, and Tyler 2015).

19.5.2 Risk Assessment and Mitigation Strategies

Risk estimation and mitigation approaches are essential for the safe handling and management of NMs, due to their potential hazards and uncertainties. Different studies have focused on creating efficient instruments and methods to warrant the safe application of NMs for several uses.

A recent study unveiled NanoSafe III, a safety control mechanism designed for small labs and other institutions using NMs (Buitrago et al. 2021). This system utilizes a three-step control banding approach and the precautionary principle to assess and mitigate risks associated with NM handling, enhancing safety protocols, and minimizing potential hazards.

Another study stressed the importance of comprehensive risk assessment and health and safety when dealing with carbon NMs. Their work emphasized the significance of identifying, evaluating, eliminating, mitigating, monitoring, and communicating risks associated with NMs to ensure safe handling (Silva et al. 2020).

In the realm of nanomedicine and nanotechnology, Singh et al. (2020) discussed the role of artificial intelligence and machine learning in advancing biomedical material design and toxicity prediction. Their study highlighted the importance of using predictive models for quantitative risk assessment and safety evaluation of NMs to enhance accuracy and develop targeted mitigation strategies.

Halil et al. (2022) proposed the NanoRisk framework, an integrated health risk assessment approach that combines Bayesian Network modeling, control banding, and process parameters to evaluate NM exposure hazards. This approach focuses on factors like humidity, NM mass, and operating temperatures to assess potential biological effects on human health, providing a holistic understanding of NM risks.

Furthermore, ongoing research initiatives, as emphasized by Festus-Ikhuoria (2023), play a crucial role in understanding NM toxicity and developing effective risk assessment strategies. Advancements in nano-toxicology contribute to refining risk assessment methodologies and enhancing the development of targeted mitigation approaches for NM-related risks.

When it comes to managing and handling NMs safely, these studies show how important it is to have solid risk assessment and mitigation plans. By adopting systematic approaches, leveraging advanced technologies, and integrating multiple risk assessment methodologies, organizations can effectively identify, evaluate, and mitigate risks associated with NMs, safeguarding human health and the environment.

19.6 ADVANCEMENTS AND FUTURE PERSPECTIVES

19.6.1 Recent Innovations in Nanoremediation Techniques

Recent advancements in nanoremediation techniques have significantly progressed the field of environmental remediation, providing sustainable and effective solutions for cleaning up contaminated environments. Nanoremediation is the use of NPs to clean up contaminants in soil, water, and air. It is a potential method for addressing environmental problems. Ganie et al. (2021) highlighted the latest developments in nano-remediation technologies, which are used to clean up polluted environments in a sustainable manner. The focus is on removing persistent organic chemicals and heavy metals. Nanoremediation techniques present a sustainable approach to remediating various environmental contaminants, aiding in the restoration of contaminated sites.

Guidi et al. (2023) introduced eco-friendly engineered NMs combined with filtering fine-mesh nets as a favorable tool for remediating sludges. This novel method employs tailored NPs to remediate contaminated substances efficiently. By integrating NMs with filtering technologies, this method shows potential in addressing contamination in freshwater environments, underscoring the promise of nanoremediation in environmental cleanup endeavors.

Erkoç (2024) discussed the dye removal by a bacterial consortium immobilized on cellulose acetate nanofiber mats. This study emphasized the fast and effective nature of nanoremediation, which involves engineered or biogenic NMs in treatment processes. By combining NMs with bacterial consortia, this approach provides a sustainable and efficient method for remediating dye-contaminated environments.

Boregowda et al. (2022) discussed recent advances in nanoremediation for cleaning up polluted agricultural soils, emphasizing the importance of sustainable solutions in environmental cleanup efforts. This study showcased how nanoremediation techniques can effectively address soil pollutants in agricultural settings, offering a sustainable approach to restoring contaminated agricultural soils and ensuring environmental health and sustainability.

Recent innovations in nanoremediation techniques have transformed environmental remediation practices, providing sustainable and efficient solutions for cleaning up contaminated environments. By leveraging NMs in remediation processes, researchers and environmental practitioners can effectively address a wide range of pollutants, contributing to the restoration and preservation of environmental quality.

19.6.2 Emerging Trends and Technologies

Nanoremediation techniques have emerged as innovative approaches for the sustainable cleanup of polluted environmental domains. These techniques leverage the unique properties of NMs to address various environmental challenges effectively. Several recent studies have highlighted the trends and advancements in nanoremediation techniques, showcasing their potential for environmental cleanup.

Dhanapal (2024) explored nanotechnology for agricultural polluted soils, emphasizing the use of biologically produced NPs in nano bioremediation. This cutting-edge technique offers a novel way to remove pollutants from the environment efficiently, showcasing the potential of NMs in addressing agricultural pollution and soil contamination.

Boulkhessaim et al. (2022) reviewed current developments in the treatment of persistent organic pollutants by the utilization of NMs and associated techniques. The study emphasized nanoremediation as a cutting-edge method for effectively eliminating organic contaminants, including pesticides, PCBs, and brominated chemicals, sustainably. The utilization of NMs in environmental remediation procedures presents a hopeful resolution for tackling intricate pollution predicaments.

Shaibu et al. (2023) explored the prospects of nanosorption and photocatalysis in the remediation of oil spills, emphasizing the advantages of NMs in treating a wide spectrum of contaminants. The study highlighted the high surface area of NMs, their ability to treat various pollutants, and their efficiency in contaminant destruction, positioning nanoremediation as a superior technology for oil spill cleanup.

Mukhopadhyay et al. (2022) focused on the use of NMs for the sustainable remediation of chemical contaminants in water and soil. The study highlighted the high reactivity of NMs in remediation processes, showcasing the effectiveness of nanoremediation techniques in environmental cleanup efforts.

Overall, these studies underscore the significant role of nanoremediation techniques in addressing environmental pollution and contamination challenges. By leveraging the unique properties of NMs, nanoremediation offers sustainable and effective solutions for remediation efforts, paving the way for a cleaner and healthier environment.

19.6.3 Potential Challenges and Future Directions

Nanoremediation techniques offer promising solutions for water treatment, addressing various contaminants and pollutants to ensure water quality and environmental sustainability. However, these techniques come with potential challenges that need to be addressed for their effective implementation. Future directions in nanoremediation techniques for water treatment involve overcoming these challenges and advancing the field towards more efficient and sustainable practices.

One of the key challenges in nanoremediation techniques for water treatment is the issue of fouling in nanofiltration processes. Fouling can reduce the efficiency of membrane-based water treatment systems, impacting their overall performance. Addressing fouling through improved membrane design, surface modifications, and fouling-resistant materials is crucial for enhancing the longevity and effectiveness of nanofiltration processes in water treatment. Another challenge in nanoremediation techniques is the potential leaching of NMs into treated water. Leaching can lead to unintended consequences, such as secondary contamination or reduced treatment efficiency. Developing strategies to prevent or minimize leaching, such as encapsulation of NMs or using stable nanocomposites, is essential for ensuring the safety and efficacy of nanoremediation processes in water treatment.

Furthermore, the selection of appropriate NMs for water treatment applications is critical to ensure both effectiveness and environmental safety. Girigoswami and Girigoswami (2023) emphasized the importance of using non-toxic and low-cost NMs, such as iron-based NMs and chitosan, for industrial wastewater treatment. Future directions in nanoremediation techniques involve the exploration of novel NMs with enhanced adsorption capacities, selectivity, and environmental compatibility for diverse water treatment challenges.

Covalent Organic Frameworks (COFs) are favourable for water treatment applications, offering unique properties for various remediation processes, as highlighted by Xia et al. (2021). Future directions in utilizing COFs for water treatment involve exploring their potential in solar steam generation, photocatalysis, and capacitive deionization, expanding their applications to address a wide range of water quality issues effectively.

Incorporating advanced polymeric nanocomposites in water treatment, as discussed by Adeola & Nomngongo (2022), presents opportunities and challenges for enhancing treatment efficiency and pollutant removal. Future research directions involve optimizing the synthesis and performance of polymeric nanocomposites, understanding their interactions with contaminants, and developing tailored nanocomposite materials for specific water treatment applications.

Furthermore, the use of NMs to remediate persistent organic contaminants necessitates the development of novel strategies to tackle complicated environmental issues, as highlighted by Boulkhessaim et al. (2022). Future directions in nanoremediation techniques involve exploring the synergistic effects of NMs in removing persistent pollutants, optimizing treatment processes, and ensuring the long-term sustainability of remediation efforts.

Addressing challenges, such as fouling, leaching, material selection, and pollutant removal efficiency, is crucial for advancing nanoremediation techniques in water treatment. Future directions in nanoremediation involve the development of innovative NMs, optimization of treatment processes, and integration of sustainable practices to ensure effective and environmentally friendly solutions for water quality management.

19.7 CONCLUSION

Advancements in nanoremediation techniques are pivotal for sustaining the natural environment through innovative and sustainable practices. Nanoremediation technologies present an encouraging method for remediating polluted ecosystems by targeting a diverse array of pollutants, which includes persistent organic compounds, heavy metals, pesticides, and other environmental contaminants. The development of functional nanohybrids and nanocomposites has enabled the removal of environmental pollutants and facilitated bioremediation processes, showcasing the potential of nanotechnology in environmental cleanup efforts.

Nanoremediation techniques employ the distinctive characteristics of NMs, like their large surface area, catalytic ability, and reactivity, to create sophisticated and environmentally friendly solutions for environmental remediation. These novel methods show significant potential for creating effective and environmentally friendly strategies to purify polluted water, soil, and air. By utilizing NMs in bioremediation processes and employing green synthesis methods, researchers are paving the way for sustainable solutions to clean up polluted environments and protect natural ecosystems.

The integration of nanotechnology in remediation strategies, such as nano sorption, photocatalysis, and nano bioremediation, offers effective means to address environmental challenges and promote sustainable development. Nanoremediation technologies possess the capacity to revolutionize the realm of environmental cleanup through the provision of intelligent, secure, and economical methods for addressing pollution and reinstating ecological equilibrium. By harnessing the power of NMs and innovative techniques, the future of nanoremediation holds immense promise for creating a cleaner and healthier environment for current and future generations. Overall, the continuous advancements in nanoremediation techniques, coupled with a focus on sustainability and environmental protection, are essential for preserving the natural environment and ensuring a greener future. A more sustainable and resilient Planet can be achieved with the continued use of nanotechnology, which is being developed through research, collaborations, and environmentally friendly actions.

ACKNOWLEDGEMENT

The authors are thankful to the Department of Earth and Environmental Science, Parul Institute of Applied Sciences, Parul University, Vadodara, for providing the necessary facilities.

REFERENCES

Adadzi, Patrick, Amy Allwright, and Francois Fourie. 2022. "Multivariate and Geostatistical Analyses of Groundwater Quality for Acid Rock Drainage at Waste Rock and Tailings Storage Site." *Journal of Ecological Engineering* 23 (12): 203–16. doi:10.12911/22998993/153091

Adeola, Adedapo Oluwasanu, and Philiswa Nosizo Nomngongo. 2022. "Advanced Polymeric Nanocomposites for Water Treatment Applications: A Holistic Perspective." *Polymers* 14 (12): 2462. doi:10.3390/polym14122462

Alazaiza, Motasem Y.D., Ahmed Albahnasawi, Gomaa A.M. Ali, Mohammed J.K. Bashir, Nadim K. Copty, Salem S. Abu Amr, Mohammed F.M. Abushammala, and Tahra Al Maskari. 2021. "Recent Advances of Nanoremediation Technologies for Soil and Groundwater Remediation: A Review." *Water (Switzerland)*. doi:10.3390/w13162186

Al-Zaban, Mayasar I., Mohamed A. Mahmoud, Maha A. Alharbi, and Aisha M. Bahatheq. 2020. "Bioremediation of Crude Oil by Rhizosphere Fungal Isolates in the Presence of Silver Nanoparticles." *International Journal of Environmental Research and Public Health*. doi:10.3390/ijerph17186564

Amar, Ibrahim A., Sarah S. Kanah, Hibah A. Hijaz, Mabroukah A. Abdulqadir, Shamsi A. Shamsi, Ihssin A. Abdalsamed, and Mohammed A. Samba. 2023. "Surfactant-Assisted Sol-Gel Synthesis of Zinc Ferrite Magnetic Nanoparticles for Oil Spills Cleanup from Seawater and Antibacterial Activity Applications." *World Journal of Engineering* 20 (4): 713–21. doi:10.1108/WJE-10-2021-0605

Aslam, Farah, and Sumaira Mazhar. 2023. "Nano-Bioremediation of Heavy Metals from Environment Using a Green Synthesis Approach." *International Journal of Advances in Applied Sciences* 12 (1): 7–14. doi:10.11591/ijaas.v12.i1.pp7-14

Boregowda, Nandini, Sanjay C. Jogigowda, Gurulingaiah Bhavya, Channarayapatna Ramesh Sunilkumar, Nagaraja Geetha, Shashikant Shiddappa Udikeri, Srinivas Chowdappa, Muthusamy Govarthanan, Sudisha Jogaiah. 2022. Recent advances in nanoremediation: Carving sustainable solution to clean-up polluted agriculture soils. *Environmental Pollution*, 297, 118728. doi:10.1016/j.envpol.2021.118728

Boulkhessaim, Salim, Amel Gacem, Samreen Heena Khan, Abdelfattah Amari, Virendra Kumar Yadav, Hamed N. Harharah, Abubakr M. Elkhaleefa, et al. 2022. "Emerging Trends in the Remediation of Persistent Organic Pollutants Using Nanomaterials and Related Processes: A Review." *Nanomaterials* 12 (13): 2148. doi:10.3390/nano12132148

Bounegru, Alexandra Virginia, and Constantin Apetrei. 2020. "Carbonaceous Nanomaterials Employed in the Development of Electrochemical Sensors Based on Screen-Printing Technique: A Review." *Catalysts* 10 (6): 680. doi:10.3390/catal10060680

Buitrago, Elina, Anna Maria Novello, Alke Fink, Michael Riediker, Barbara Rothen-Rutishauser, and Thierry Meyer. 2021. "Nanosafe III: A User Friendly Safety Management System for Nanomaterials in Laboratories and Small Facilities." *Nanomaterials* 11 (10): 2768. doi:10.3390/nano11102768

Chaudhari, Yogesh S., Pankaj Kumar, Sunil Soni, Amel Gacem, Vinay Kumar, Snigdha Singh, Virendra Kumar Yadav, et al. 2023. "An Inclusive Outlook on the Fate and Persistence of Pesticides in the Environment and Integrated Eco-Technologies for Their Degradation." *Toxicology and Applied Pharmacology* 466. doi:10.1016/j.taap.2023.116449

Chen, Baiquan, Zilan He, Daopin Chena, and Feng Zhang. 2024. "High-Voltage Substation Intelligent Perception Resilience Technology Research," 45. doi:10.1117/12.3029153

Chen, Dongzhen, Jianzhao Gao, Dan He, Jingshun He, Yang Li, Meng Zhang, Wenya Li, Xin Chen, Xinhai He, and Tao Fu. 2023. "Plasmonic Bridge Sensor Enabled by Carbon Nanotubes and Au-Ag Nano-Rambutan for Multifunctional Detection of Biomechanics and Bio/Chemical Molecules." *ACS Applied Materials and Interfaces* 15 (6): 8783–93. doi:10.1021/acsami.2c22634

Chen, Honglei, Shao Lin Wu, Hua Li Wang, Qing Yun Wu, and Hao Cheng Yang. 2021. "Photothermal Devices for Sustainable Uses Beyond Desalination." *Advanced Energy and Sustainability Research* 2 (3). doi:10.1002/aesr.202000056

Choudhary, Nisha, Virendra Kumar Yadav, Krishna Kumar Yadav, Abdulaziz Ibrahim Almohana, Sattam Fahad Almojil, Govhindhan Gnanamoorthy, Do Hyeon Kim, Saiful Islam, Pankaj Kumar, and Byong Hun Jeon. 2021. "Application of Green Synthesized Mmt/Ag Nanocomposite for Removal of Methylene Blue from Aqueous Solution." *Water (Switzerland)* 13 (22). doi:10.3390/w13223206

Dhanapal, Anand Raj, Muthu Thiruvengadam, Jayavarshini Vairavanathan, Baskar Venkidasamy, Maheswaran Easwaran, and Mansour Ghorbanpour. 2024. "Nanotechnology Approaches for the Remediation of Agricultural Polluted Soils." *ACS Omega* 9 (12): 13522–33. doi:10.1021/acsomega.3c09776

Ehmedan, Sally Said, Mohamed Khaled Ibrahim, Ahmed M. Azzam, Hend Abdulhameed Hamedo, and Ali M. Saeed. 2021. "Acceleration the Bacterial Biodegradation of Crude Oil Pollution Using Fe2O3 and ZnO Nanoparticles." *Environmental Nanotechnology, Monitoring and Management* 16: 100613. doi:10.1016/j.enmm.2021.100613

El-Ramady, Hassan, Ahmed El-Henawy, Megahed Amer, Alaa El-Dein Omara, Tamer Elsakhawy, Abdel-Moety Salama, Ahmed Ezzat, Aly Elsherif, Mohamed Elmahrouk, and Tarek Shalaby. 2020. "Agro-Pollutants and Their Nano-Remediation from Soil and Water: A Mini-Review." *Environment, Biodiversity and Soil Security*. doi:10.21608/jenvbs.2020.47751.1111

El-Sayed, Mohamed E.A. 2020. "Nanoadsorbents for Water and Wastewater Remediation." *Science of the Total Environment* 739: 139903. doi:10.1016/j.scitotenv.2020.139903

El-Sheshtawy, H. S., and W. Ahmed. 2017. "Bioremediation of Crude Oil by Bacillus Licheniformis in the Presence of Different Concentration Nanoparticles and Produced Biosurfactant." *International Journal of Environmental Science and Technology* 14 (8): 1603–14. doi:10.1007/s13762-016-1190-1

Erkoç, Esra, İmre Tüzün, Filiz Korkmaz, Nalan Oya San Keskin, and Nur Koçberber Kiliç. 2024. "Nanoremediation of Toxic Dyes Using a Bacterial Consortium Immobilized on Cellulose Acetate Nanofiber Mats." *Polymer Engineering and Science* 64 (1): 339–49. doi:10.1002/pen.26551

Fadeel, Bengt. 2022. "Understanding the Immunological Interactions of Engineered Nanomaterials: Role of the Bio-Corona." *Wiley Interdisciplinary Reviews: Nanomedicine and Nanobiotechnology* 14 (6). doi:10.1002/wnan.1798

Faqihi, Noorah A., Mahmood M.S. Abdullah, Mohd Sajid Ali, Hamad A. Al-Lohedan, and Zainab M. Almarhoon. 2023. "Functionalized Magnetite Nanoparticles Using Two New Ionic Liquids for Efficient Oil Spill Cleanup." *Journal of Chemistry* 2023: 1–12. doi:10.1155/2023/7148988

Fellet, Guido, Pellegrino Conte, and Luca Marchiol. 2023. "Biochar Effects on Ce Leaching and Plant Uptake in Lepidium Sativum L. Grown on a Ceria Nanoparticle Spiked Soil." *Applied Sciences (Switzerland)* 13 (11): 6846. doi:10.3390/app13116846

Festus-Ikhuoria, Igberaese Clinton, Nwankwo Constance Obiuto, Riliwan Adekola Adebayo, and Oladiran Kayode Olajiga. 2023. "Nanotechnology in Consumer Products: A Review of Applications and Safety Considerations." *World Journal of Advanced Research and Reviews* 21 (3): 2050–59. doi:10.30574/wjarr.2024.21.3.0923

Forest, Valérie. 2021. "Combined Effects of Nanoparticles and Other Environmental Contaminants on Human Health - an Issue Often Overlooked." *NanoImpact* 23: 100344. doi:10.1016/j.impact.2021.100344

Francy, Nimita, Subramanian Shanthakumar, Fulvia Chiampo, and Yendaluru Raja Sekhar. 2020. "Remediation of Lead and Nickel Contaminated Soil Using Nanoscale Zero-Valent Iron (Nzvi) Particles Synthesized Using Green Leaves: First Results." *Processes* 8 (11): 1–12. doi:10.3390/pr8111453

Ganie, Adil Shafi, Sayfa Bano, Nishat Khan, Saima Sultana, Zubair Rehman, Mohammed M. Rahman, Suhail Sabir, Frederic Coulon, and Mohammad Zain Khan. 2021. "Nanoremediation Technologies for Sustainable Remediation of Contaminated Environments: Recent Advances and Challenges." *Chemosphere* 275: 130065. doi:10.1016/j.chemosphere.2021.130065

Garg, Rishav, Rajni Garg, Md Amir Khan, Manjeet Bansal, and Vinod Kumar Garg. 2023. "Utilization of Biosynthesized Silica-Supported Iron Oxide Nanocomposites for the Adsorptive Removal of Heavy Metal Ions from Aqueous Solutions." *Environmental Science and Pollution Research* 30 (34): 81319–32. doi:10.1007/s11356-022-21111-2

Gebre, Shushay Hagos, and Marshet Getaye Sendeku. 2019. "New Frontiers in the Biosynthesis of Metal Oxide Nanoparticles and Their Environmental Applications: An Overview." *SN Applied Sciences* 1 (8). doi:10.1007/s42452-019-0931-4

Girigoswami, Agnishwar, and Koyeli Girigoswami. "Potential applications of nanoparticles in improving the outcome of lung cancer treatment." *Genes* 14.7 (2023): 1370.

Girigoswami, Koyeli, Pragya Pallavi, and Agnishwar Girigoswami. 2023. "Intricate Subcellular Journey of Nanoparticles to the Enigmatic Domains of Endoplasmic Reticulum." *Drug Delivery* 30 (1). doi:10.1080/10717544.2023.2284684

Guidi, Patrizia, Margherita Bernardeschi, Mara Palumbo, Isabella Buttino, Valentina Vitiello, Vittoria Scarcelli, Gianluca Chiaretti, et al. 2023. "Eco-Friendly Engineered Nanomaterials Coupled with Filtering Fine-Mesh Net as a Promising Tool to Remediate Contaminated Freshwater Sludges: An Ecotoxicity Investigation." *Nanomaterials* 13 (3): 396. doi:10.3390/nano13030396

Halil, Norsuzieanah, Risza Rusli, Mardhati Zainal Abidin, Shahrulnizam Jamen, and Faisal Khan. 2022. "An Integrated Health Risk Assessment with Control Banding for Nanomaterials Exposure." *Process Safety Progress* 41 (S1): S84–97. doi:10.1002/prs.12327

Hamad, Mohammed Taha Moustafa Hussien, and Marwa E. El-Sesy. 2023. "Adsorptive Removal of Levofloxacin and Antibiotic Resistance Genes from Hospital Wastewater by Nano-Zero-Valent Iron and Nano-Copper Using Kinetic Studies and Response Surface Methodology." *Bioresources and Bioprocessing* 10 (1). doi:10.1186/s40643-022-00616-1

Ibrahim, Amina M., Mahassen Ghazy, Heba El-Sayed, Rehab M. Abd El-Hameed, Rehab G. Khalil, Shereen M. Korany, Abeer S. Aloufi, Olfat A. Hammam, and Mostafa Y. Morad. 2023. "Histopathological, Immunohistochemical, Biochemical, and In Silico Molecular Docking Study of Fungal-Mediated Selenium Oxide Nanoparticles on Biomphalaria Alexandrina (Ehrenberg, 1831) Snails." *Microorganisms* 11 (3): 811. doi:10.3390/microorganisms11030811

Jennings, Victoria, Rhys Goodhead, and Charles R. Tyler. 2015. "Ecotoxicology of Nanomaterials in Aquatic Systems." In *Frontiers of Nanoscience*, edited by Mohammed Baalousha and Jamie R B T - Frontiers of Nanoscience Lead, 8:3–45. Elsevier. doi:10.1016/B978-0-08-099948-7.00001-4

Khayal, Areeba, Vinars Dawane, Mohammed A. Amin, Vineet Tirth, Virendra Kumar Yadav, Ali Algahtani, Samreen Heena Khan, Saiful Islam, Krishna Kumar Yadav, and Byong Hun Jeon. 2021. "Advances in the Methods for the Synthesis of Carbon Dots and Their Emerging Applications." *Polymers* 13 (18): 3190. doi:10.3390/polym13183190

Kumar, Pankaj. 2018. "Electronic Waste - Hazards, Management and Available Green Technologies for Remediation: A Review." *International Research Journal of Environmental Sciences* 7 (5): 57–68.

Kumar, Pankaj, Muhammad Arshad, Amel Gacem, Sunil Soni, Snigdha Singh, Manoj Kumar, Virendra Kumar Yadav, et al. 2023. "Insight into the Environmental Fate, Hazard, Detection, and Sustainable Degradation Technologies of Chlorpyrifos: An Organophosphorus Pesticide." *Environmental Science and Pollution Research* 30 (50): 108347–69. doi:10.1007/s11356-023-30049-y

Kumar, Pankaj, and Madhusudan Hiraman Fulekar. 2018. "Rhizosphere Bioremediation of Heavy Metals (Copper and Lead) by Cenchrus Ciliaris." *Research Journal of Environmental Sciences* 12 (4): 166–76. doi:10.3923/rjes.2018.166.176

Kumar, Pankaj, and Madhusudan Hiraman Fulekar 2021. "Cadmium Phytoremediation Potential of Deenanath Grass (Pennisetum Pedicellatum) and the Assessment of Bacterial Communities in the Rhizospheric Soil." *Environmental Science and Pollution Research.* doi:10.1007/s11356-021-15667-8

Kumar, Pankaj, Amel Gacem, Mohammad Tauheed Ahmad, Virendra Kumar Yadav, Snigdha Singh, Krishna Kumar Yadav, Md Mottahir Alam, et al. 2022. "Environmental and Human Health Implications of Metal(Loid)s: Source Identification, Contamination, Toxicity, and Sustainable Cleanup Technologies." *Frontiers in Environmental Science.* https://doi.org/10.3389/fenvs.2022.949581

Kumar, Pankaj, Manoj Kumar, Abdulwasa Bakr Barnawi, Parul Maurya, Snigdha Singh, Deepankshi Shah, Virendra Kumar Yadav, et al. 2024. "A Review on Fluoride Contamination in Groundwater and Human Health Implications and Its Remediation: A Sustainable Approaches." *Environmental Toxicology and Pharmacology* 106: 104356. doi:10.1016/j.etap.2023.104356

Li, Chenhui, Jirun Hu, Haikuo Yan, Yuyuan Yao, Li Zhang, and Jinbiao Bao. 2023. "Preparation of Open-Cell Long-Chain Branched Polypropylene Foams for Oil Absorption." *ACS Omega* 8 (51): 49372–82. doi:10.1021/acsomega.3c07915

Li, Zhangdi, Qiong Tian, Jicheng Xu, Shouzhen Sun, Ying Cheng, Fengxian Qiu, and Tao Zhang. 2021. "Easily Fabricated Low-Energy Consumption Joule-Heated Superhydrophobic Foam for Fast Cleanup of Viscous Crude Oil Spills." *ACS Applied Materials and Interfaces* 13 (43): 51652–60. doi:10.1021/acsami.1c13574

Liu, Caixia, Shuai Li, Qichang Sun, Fei Ma, Yifan Sun, Yugang Zhang, Lisha Xie, Ying Huang, Xiaoming Yang, and Xingtong Hao. 2024. "High-Sensitivity Capacitive Flexible Tactile Sensor Based on Hemispherical Shell Microstructure and ZnO/RTV Composite Dielectric Layer." *ACS Applied Electronic Materials* 6 (5): 3440–53. doi:10.1021/acsaelm.4c00249

Lu, Yi, Yeling Zhu, Fan Yang, Zhenghe Xu, and Qingxia Liu. 2021. "Advanced Switchable Molecules and Materials for Oil Recovery and Oily Waste Cleanup." *Advanced Science* 8 (15). doi:10.1002/advs.202004082

Ma, Xiaolei, Duomo Duan, Xunliang Wang, Junrui Cao, Jinquan Qiu, Baolong Xie, Xiaolei Ma, and Baolong Xie. 2021. "Degradation of Rhodococcus Erythropolis SY095 Modified with Functional Magnetic Fe3O4 nanoparticles." *Royal Society Open Science* 8 (12). Royal Society: 211172. doi:10.1098/rsos.211172

Mukhopadhyay, Raj, Binoy Sarkar, Eakalak Khan, Daniel S. Alessi, Jayanta Kumar Biswas, K. M. Manjaiah, Miharu Eguchi, Kevin C.W. Wu, Yusuke Yamauchi, and Yong Sik Ok. 2022. "Nanomaterials for Sustainable Remediation of Chemical Contaminants in Water and Soil." *Critical Reviews in Environmental Science and Technology* 52 (15): 2611–60. doi:10.1080/10643389.2021.1886891

Murphy, Aidan, Ian Seymour, James Rohan, Alan O'Riordan, and Ivan O'Connell. 2021. "Portable Data Acquisition System for Nano and Ultra-Micro Scale Electrochemical Sensors." *IEEE Sensors Journal* 21 (3): 3210–15. doi:10.1109/JSEN.2020.3021941

Musa, A., M. A. Muhammad, N. M. Maharaz, and A. S. Gidado. 2021. "Magnetic Characterization of Co and Zn Nano Ferrites: Property Consideration for Oil-Spill Clean Up." *Asian Journal of Research and Reviews in Physics*, 5(3): 14–25. doi:10.9734/ajr2p/2021/v5i330164

Parthipan, Punniyakotti, Liang Cheng, Perumal Dhandapani, Punniyakotti Elumalai, Mingzhi Huang, and Aruliah Rajasekar. 2022. "Impact of Biosurfactant and Iron Nanoparticles on Biodegradation of Polyaromatic Hydrocarbons (PAHs)." *Environmental Pollution* 306: 119384. doi:10.1016/j.envpol.2022.119384

Ur Rahim, Hafeez, Muhammad Qaswar, Misbah Uddin, Cinzia Giannini, Maria Lidia Herrera, and Giuseppina Rea. 2021. "Nano-Enable Materials Promoting Sustainability and Resilience in Modern Agriculture." *Nanomaterials* 11 (8): 2068. doi:10.3390/nano11082068

Rather, Muzamil Ahmad, Shuvam Bhuyan, Ratan Chowdhury, Rahul Sarma, Subham Roy, and Panchi Rani Neog. 2023. "Nanoremediation Strategies to Address Environmental Problems." *Science of the Total Environment* 886: 163998. doi:10.1016/j.scitotenv.2023.163998

Saleem, Atif, Jingjie Chen, Meng Liu, Nian Liu, Muhammad Usman, Ke Wang, Muhammad Haris, Yuezhou Zhang, and Peng Li. 2023. "Versatile Magnetic Mesoporous Carbon Derived Nano-Adsorbent for Synchronized Toxic Metal Removal and Bacterial Disinfection from Water Matrices." *Small* 19 (15). doi:10.1002/smll.202207348

Sasmal, Himadri Sekhar, Saikat Bag, Bittu Chandra, Poulami Majumder, Himangshu Kuiry, Suvendu Karak, Sayam Sen Gupta, and Rahul Banerjee. 2021. "Heterogeneous C-H Functionalization in Water via Porous Covalent Organic Framework Nanofilms: A Case of Catalytic Sphere Transmutation." *Journal of the American Chemical Society* 143 (22): 8426–36. doi:10.1021/jacs.1c02425

Semerad, J.; N.I.N. Pacheco, A. Grasserova, P. Prochazkova, M. Pivokonsky; L. Pivokonska, T. Cajthaml. 2020. In Vitro Study of the Toxicity Mechanisms of Nanoscale Zero-Valent Iron (nZVI) and Released Iron Ions Using Earthworm Cells. *Nanomaterials*, 10, 2189. doi:10.3390/nano10112189

Seymour, Ian, Tarun Narayan, Niamh Creedon, Kathleen Kennedy, Aidan Murphy, Riona Sayers, Emer Kennedy, Ivan O'Connell, James F. Rohan, and Alan O'Riordan. 2021. "Advanced Solid State Nano-Electrochemical Sensors and System for Agri 4.0 Applications." *Sensors* 21 (9): 3149. doi:10.3390/s21093149

Shaibu, S. E., E. J. Inam, E. A. Moses, U. A. Ofon, O. K. Fatunla, C. O. Obadimu, N. D. Ibuotenang, et al. 2023. "Prospects of Nanosorption and Photocatalysis in Remediation of Oil Spills." *Journal of the Nigerian Society of Physical Sciences* 5 (1): 1043. doi:10.46481/jnsps.2023.1043

Silva, Guilherme Lenz, Camila Viana, Danieli Domingues, and Fernanda Vieira. 2020. "Risk Assessment and Health, Safety, and Environmental Management of Carbon Nanomaterials." *Nanomaterials: Toxicity, Human Health and Environment*. doi:10.5772/intechopen.85485

Silva, Jhonatan da, Gabriel Bachega Rosa, William Gustavo Sganzerla, Jocleita Peruzzo Ferrareze, Flávio José Simioni, and Mari Lucia Campos. 2023. "Strategies and Prospects in the Recovery of Contaminated Soils by Phytoremediation: An Updated Overview." *Communications in Plant Sciences* 13, doi: 10.26814/cps2023001

Singh, Ajay Vikram, Daniel Rosenkranz, Mohammad Hasan Dad Ansari, Rishabh Singh, Anurag Kanase, Shubham Pratap Singh, Blair Johnston, Jutta Tentschert, Peter Laux, and Andreas Luch. 2020. "Artificial Intelligence and Machine Learning Empower Advanced Biomedical Material Design to Toxicity Prediction." *Advanced Intelligent Systems* 2(12). doi:10.1002/aisy.202000084

Singh, Harpreet, Neha Bhardwaj, Shailendra Kumar Arya, and Madhu Khatri. 2020. "Environmental Impacts of Oil Spills and Their Remediation by Magnetic Nanomaterials." *Environmental Nanotechnology, Monitoring and Management* 14: 100305. doi:10.1016/j.enmm.2020.100305

Song, Shao Fei, Cheng Wang, Zhi Sheng Fu, and Zhi Qiang Fan. 2022. "Highly Branched Polyethylene Used as Sorbents for Oil-Spill Cleanup and Separation." *Journal of Polymer Science* 60 (15): 2418–27. doi:10.1002/pol.20210950

Subramoniapillai, Viju, and Govindharajan Thilagavathi. 2022. "Oil Spill Cleanup by Natural Fibers: A Review." *Research Journal of Textile and Apparel* 26 (4): 390–404. doi:10.1108/RJTA-03-2021-0031

Tarek, Mohamed, Walaa Mahmoud, Mohamed Elmosallamy, and Ahmed El-Sherif. 2024. "Cadmium Detection with Nickel Complex Nanoparticles Sensor Using QCM Technique Mohamed T. Radwan 1, Walaa H. Mahmoud 1, Mohamed A.F. ElMosallamy 2 and Ahmed A. El-Sherif 1." *Bulletin of Faculty of Science, Zagazig University* 2024 (1): 132–42. doi:10.21608/bfszu.2023.233505.1303

Thakur, Abhinay, and Ashish Kumar. 2023. "Ecotoxicity Analysis and Risk Assessment of Nanomaterials for the Environmental Remediation." *Macromolecular Symposia* 410 (1). doi:10.1002/masy.202100438

Ugwu, B. I., J. C. Nnaji, H. O. Chukwuemeka-Okorie, M. J. Siyaka, F. J. Amaku, C. Ngwu, and S. A. Odoemelam. 2022. "Discharge of Emerging Contaminant Laden Effluents By Industries in Nigeria -A Review." *Journal of Chemical Society of Nigeria* 47 (3). doi:10.46602/jcsn.v47i3.760

Varjani, Sunita J., Dolly P. Rana, Ajay K. Jain, Surendra Bateja, and Vivek N. Upasani. 2015. "Synergistic Ex-Situ Biodegradation of Crude Oil by Halotolerant Bacterial Consortium of Indigenous Strains Isolated from on Shore Sites of Gujarat, India." *International Biodeterioration and Biodegradation* 103: 116–24. doi:10.1016/j.ibiod.2015.03.030

Vinzant, Kari, Mohammad Rashid, and Mariya V. Khodakovskaya. "Advanced applications of sustainable and biological nano-polymers in agricultural production." *Frontiers in Plant Science* 13 (2023): 1081165.

Xia, Zijing, Yusen Zhao, and Seth B. Darling. 2021. "Covalent Organic Frameworks for Water Treatment." *Advanced Materials Interfaces* 8 (1). doi:10.1002/admi.202001507

Yadav, Virendra K., Samreen H. Khan, Nisha Choudhary, Vineet Tirth, Pankaj Kumar, Raman K. Ravi, Shreya Modi, et al. 2022. "Nanobioremediation: A Sustainable Approach towards the Degradation of Sodium Dodecyl Sulfate in the Environment and Simulated Conditions." *Journal of Basic Microbiology* 62 (3–4): 348–60. doi:10.1002/jobm.202100217

Yang, Chong, Cheng Zhang, Zheng Jie Chen, Yu Li, Wen Yuan Yan, Hai Bin Yu, and Lin Liu. 2021. "Three-Dimensional Hierarchical Porous Structures of Metallic Glass/Copper Composite Catalysts by 3D Printing for Efficient Wastewater Treatments." *ACS Applied Materials and Interfaces* 13 (6): 7227–37. doi:10.1021/acsami.0c20832

Yousefi, Shima, Weria Weisany, Seyed Ebrahim Hosseini, and Mehran Ghasemlou. 2024. "Mechanisms of Nanoencapsulation to Boost the Antimicrobial Efficacy of Essential Oils: A Review." *Food Hydrocolloids* 150: 109655. doi:10.1016/j.foodhyd.2023.109655

Zhang, Chenxi, Congyu Zhang, Yiwei Li, Yaran Shi, Jiale Chao, Yifan Zhao, He Yang, and Bo Fu. 2023. "Wavelength-Tunable Broadband Lasers Based on Nanomaterials." *Nanotechnology* 34 (49): 492001. doi:10.1088/1361-6528/acf66d

Zhang, Chunmei, Jiazhan Zhu, Ziting Cao, Ping Song, and Tianliang Zhai. 2020. "Flow-Induced Microchannel Structure of the Graphene-Based Aerogel Microspheres and Their Use as Superabsorbents." *Polymers for Advanced Technologies* 31 (11): 2789–96. doi:10.1002/pat.5005

Zhang, Run Kun, Die Wang, Yan Jun Wu, Yi Han Hu, Jian Yu Chen, Jin Can He, and Jing Xin Wang. 2020. "A Cataluminescence Sensor Based on NiO Nanoparticles for Sensitive Detection of Acetaldehyde." *Molecules* 25 (5): 1097. doi:10.3390/molecules25051097

20 Current Status and Future Prospects of Application of Engineered Nanomaterials

Ashwini Kumar Mishra, Rahul Maurya, Chandrashekhar Y Jagtap, Rajesh Singh, and Vijay Kumar

20.1 INTRODUCTION

Over the past few years, nanotechnology has emerged as a game-changing technology that has paved the way for creating innovative and advanced applications in nanomedicine. This technology has opened up new avenues for developing potential medical treatments and therapies that can revolutionize how we approach healthcare. From a commercial standpoint, the nanomedicine sector is expected to reach a valuation of USD 293.2 billion by 2022 and USD 350.7 billion by 2025 (Albalawi et al. 2021). ENs have greater potential for biomedical use in nanomedicine development, due to their unique physiochemical property in sensitivity, functionality, specificity, and efficiency. By focusing on the surfaces of the particles, one can alter the adsorptive physio-chemical characteristics of the NPs and investigate the particular interactions between NPs and cells. (Navya and Daima 2016). Photosensitizer molecules can be used to achieve nano-scaled photothermal treatment in combined immunotherapy treatment of malignant tumors. To achieve targeted irradiation, these molecules are injected into the tumor site (Conde et al. 2015; Dickerson et al. 2008). In addition to the instances described above, NMs possess other distinctive characteristics, such as the ability to control nanocluster size, high surface dispersity, and adjustable morphology. These traits allow for the emergence of new possibilities and advancements in various applications within the field of nanomedicine (Conde et al. 2014). As NPs become more common in commercial articles, a public discussion arises regarding whether nanotechnology's environmental and societal drawbacks outweigh its numerous advantages. There have been limited research efforts to examine the toxicological and ecological impacts of both direct and indirect contact with NMs. Furthermore, there is a lack of established criteria to measure these effects accurately (Colvin 2003). ENs are substances created by intentionally manipulating matter at the nanoscale to generate new materials, structures, and electronics. These substances are artificially made and have at least one dimension measuring between 1 and 100 nanometers. ENs, or engineered NMs, are a subset of NMs that can be formed through natural, anthropogenic, and engineered processes. Currently, nanotechnology is generating extraordinary scientific marvels and surpassing visionary boundaries. ENs have been developed to fight against microorganisms and restrict their growth. They have shown exceptional efficacy in various applications, such as healthcare, water treatment, alternative food packaging, facility substrates, and antifouling coatings (Palit 2018; Ogunsona et al. 2020). Due to their unique characteristics, ENs are widely used in electronics, chemical engineering, solar cells, catalysis, space technology, and other fields. It also has great potential to revolutionize modern pharmaceuticals (Gajewicz et al. 2012). The agricultural industry is now focused on balancing the growing need for food crops with limited resources, such as water and land, by effectively utilizing these resources. To mitigate the negative impacts of climate change, it is imperative to implement strategies that involve adopting contemporary methods, such as using ENs in agricultural practices and advocating

 DOI: 10.1201/9781003484189-20

for the intelligent exploitation of natural resources. Many projects and investments are required to meet the increasing demand for agricultural products while embracing sustainable methodologies (FAO 2017). The process of encapsulating ENs with other substances improves plants' absorption of nutrients, leading to a reduction in the amount of agricultural chemicals that are wasted. Nano-encapsulated materials exhibit improved stability and solubility, making them suitable for precise and efficient delivery of nutrients (Shang et al. 2019). Researchers are focused on developing sustainable techniques for creating agricultural materials using plant extracts. NMs can be synthesized with the help of natural extracts, which act as stabilizing and reducing agents and can be customized to fit specific needs. The agricultural industry is concerned with improving crop output and yield while resisting environmental pressures. Nano-fertilizers, which are ENs, can be utilized in the agricultural industry to deliver vital nutrients for the growth of plants. Furthermore, they can be used for foliar spraying and seed priming in order to mitigate environmental stresses and fight plant diseases. The growing need for food, medicine, and other crops, coupled with the shifting environment, necessitates the development of innovative technology to bridge the gap between supply and demand (Bora et al. 2022). Common examples of ENs include oxides of metals such as titanium dioxide, gold (Au), iron (Fe), silver (Ag), and zinc oxide. Other forms of ENs include polymers, carbon-based ENs (CENs) like fullerenes, carbon nanotubes, graphene-based nano delivery platforms, and hybrid particles and biopolymer-based nanocomposites. The primary factors that significantly impact the binding and uptake mechanism of ENs are their particle size and shape. These features also play an important role in determining the kind and distribution of reactive sites involved in chemical reactions. The reactivity of NM in air, soil, water, or biota is influenced by changes in surface structure due to the high surface area-to-volume ratio (Dontsova, Nahirniak, and Astrelin 2019; Coman et al. 2019). The ENs interact with environmental chemical molecules, including natural and manufactured contaminants. When ENs are released into the aquatic environment, they interact with various biological and organic molecules in the water. These interactions lead to the formation of eco- and bio-coronas, essentially a layer of organic and inorganic substances surrounding the ENs. These corona structures can significantly impact the uptake and toxicity of ENs by marine species and play a crucial role in determining their environmental fate and potential risks. An eco-design strategy was created for designing environmentally friendly ENs for applications like nano-remediation, while balancing their effectiveness for the safety of marine species and other organisms (Corsi et al. 2021). ENs are made for various purposes according to the needs of advanced healthcare systems, such as imaging of cancer and related diseases, drug delivery platforms, and targeted curative agents. By utilizing these multifunctional ENs, theragnostic therapy can be provided to patients for cancer diagnosis and treatment (Malik, Muhammad, and Waheed 2023). In this chapter, utilization of various ENs for environmental remediation, opportunities and challenges in the synthesis of ENs, ENs as an aid to environmental pollution monitoring systems, and the impact of ENs on a variety of areas like chemical, energy, electronics, and space industries and practical implication of ENs in healthcare with their future perspectives are discussed.

20.2 CLASSIFICATION OF ENGINEERED NMS (ENS)

ENs can be classified in several ways, depending on the specific property or characteristic of interest. Here are two common classification methods:

20.2.1 DIMENSIONALITY

This classification system (depicted in Figure 20.1) categorizes NMs based on the number of dimensions and size of the nanoscale (1-100 nanometers). A diagram illustrates the proportional sizes of NPs across various categories, with each exemplified. NMs (NMs) exhibit diverse dimensionalities, with classification schemes delineating zero-dimensional (0D), one-dimensional (1D), two-dimensional (2D), and three-dimensional (3D) nano-structured materials (NSM). When a number of

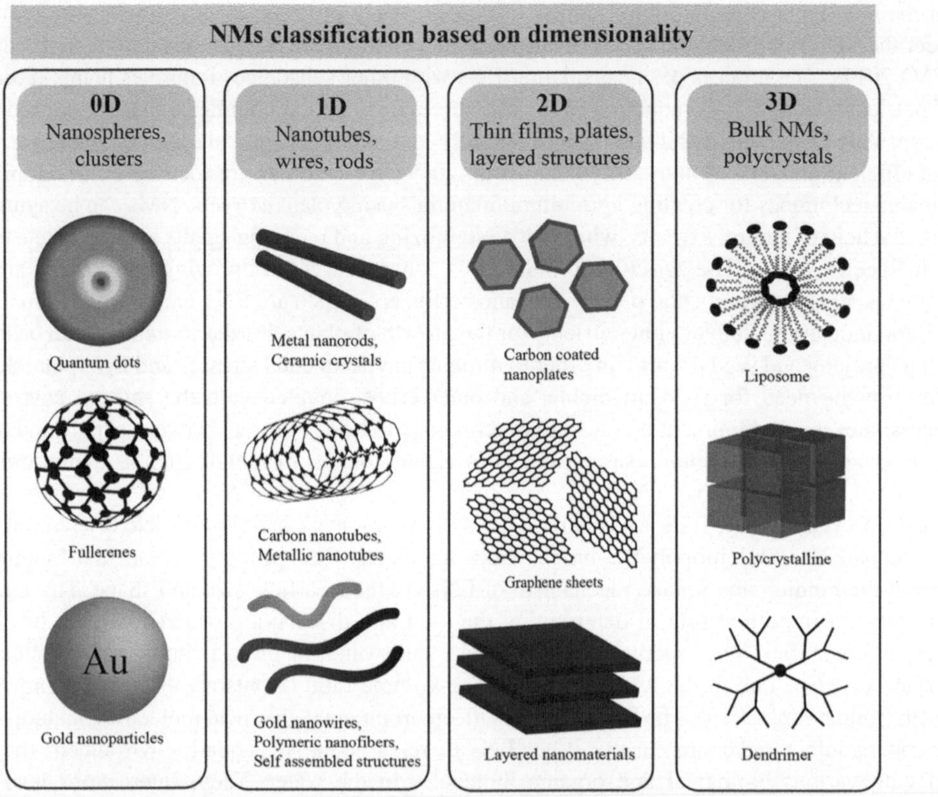

FIGURE 20.1 Classification of NMs based on dimension. Image reprint from Poh et al. 2018. Open access article is distributed under the terms of the Creative Commons Attribution 4.0 International License.

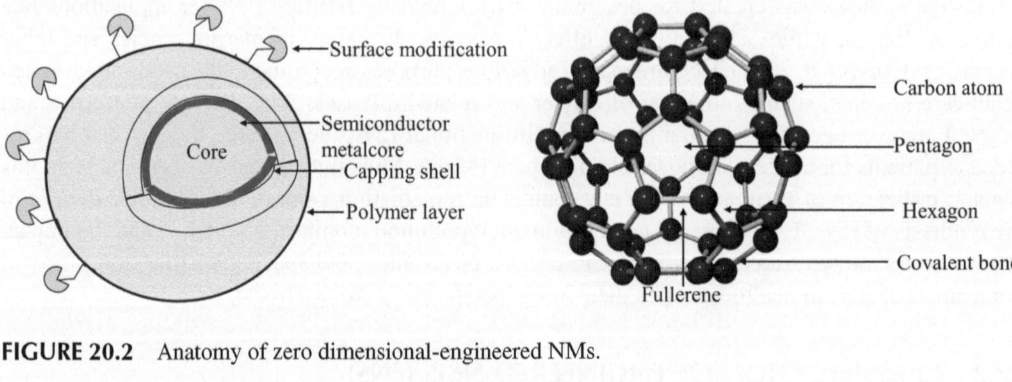

FIGURE 20.2 Anatomy of zero dimensional-engineered NMs.

materials with different properties work together harmoniously, they form 3D nanocomposites, which are composites with unique properties (Poh et al. 2018).

0-dimensional (0D) ENs refer to NPs with a uniform dimension in all directions, resulting in a structure with zero-dimensional characteristics. These NMs are extremely adjustable for a variety of applications since they are usually synthesized with exact control over their size, shape, and content. Fullerene and quantum dots are two instances of zero-dimensional NMs (Sudha et al. 2018). For example, semiconductor nanoparticles (NPs) known as quantum dots have special optical and electrical characteristics, like size-dependent fluorescence emission and high quantum yield. Carbon atoms are organized in a sphere cage-like shape to form fullerenes, as shown in Figure 20.2. These hollow carbon molecules exhibit unique properties due to their 0D nature, such as high electron mobility and good chemical stability. They find applications in diverse fields, including drug delivery, electronics, and materials science, owing to their potential as versatile building blocks for nanotechnology (Gaur et al. 2014).

Two main strategies shown in Table 20.1 for crafting 0-dimensional NMs are breaking down bulk materials (top-down) or building them from scratch (bottom-up). The bottom-up methodology involves constructing smaller entities that are gradually built up to match a material's specifications. Graphene can be synthesized with atomically precise control from smaller carbon precursors, resulting in zigzag or armchair configurations that affect their electron behavior. The top-down approach is more scalable, cost-effective, and reproducible than the bottom-up approach, as it does not require substrate transfer (Zhang et al. 2019).

Top-down methods like mechanical milling use intense grinding to pulverize materials (see Figure 20.3a) (de Lima Barizão et al. 2021). At the same time, lithography employs light or electron beams to create the desired pattern on a light-sensitive material called photoresist. It employs etching to shape a nanostructure by eliminating material from beneath the pattern (see Figure 20.3b).

TABLE 20.1
Outlook for the Synthesis of NMs (Majumder et al. 2007)

Top-Down Approach	Bottom-Up Approach
Breaking down more extensive materials into NPs	Builds NPs from individual atoms or molecules
Mechanical Milling: This technique uses high-energy grinding to pulverize bulk materials into NPs	*Chemical Vapor Deposition (CVD)*: This technique uses light or electron beams to create patterns on a surface.
Lithography: This technique uses light or electron beams to create patterns on a surface, which are then used to etch NPs	*Molecular Precursor Synthesis*: This method involves synthesizing molecules that can self-assemble into NPs under specific conditions
	Sol-gel synthesis: Condensation-derived precursor molecules and hydrolysis-derived precursor molecules
	Wet Chemical Methods--
	Hydrothermal/Solvothermal Synthesis: Precursor materials are heated in a sealed container with water or organic solvents at high temperatures and pressures to induce reactions that form NPs.
	Precipitation: By rapidly changing the chemical conditions of a solution, precursor materials can be induced to precipitate out of the solution as NPs
	Reduction: NPs are created when a reducing agent is used to convert a metal salt to its elemental form.

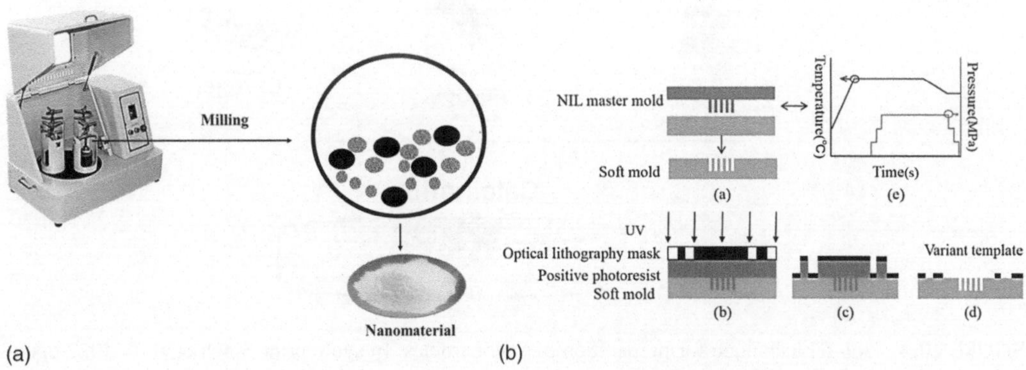

FIGURE 20.3 (a) NMs are mechanically milled using a planetary milling machine. (b) Illustration of the variant template fabrication process.

Variant templates are created using polymer films that are transparent to UV light. This templet includes nanoscale structures and larger metal light-blocking layers (MBLs). MBLs can be used as imprint molds for nanopatterns and lithographic masks for micronscale. The nanoscale features of the variant template are fabricated via the Nanoimprint lithography (NIL) process, followed by deposition and lift-off processes to fabricate the light-blocking layers on the variant template as a lithographic mask (Savouchkina et al. 2010).

(a) Nanostructure replication utilizing T-NIL based on the variation template.
(b) Optical lithography can be used to make micropatterns based on the variation template.
(c) Cover the variation template with deposited metal to create a light-blocking coating.
(d) Variant template after lift-off procedure.
(e) T-NIL parameters for the fabrication process (Liu et al. 2022). Image reprint from Liu et al. 2022

On the other hand, bottom-up approaches meticulously construct NPs. Figure 20.4 illustrates sol-gel synthesis, a solid macromolecule (gel) scattered in a continuous media or a colloidal solution (sol) with solid particles embedded in a liquid is used. This technique relies on two reaction pathways: condensation-derived precursor molecules and hydrolysis-derived precursor molecules. It has proven effective in producing various bimetallic NPs with high homogeneity, such as high-quality Au-Ag NPs (Singh et al. 2022).

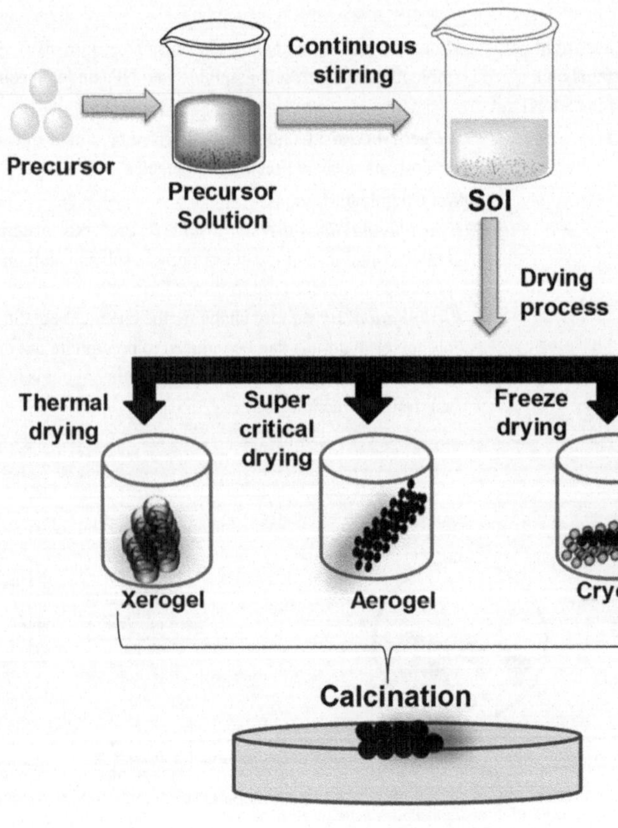

FIGURE 20.4 Sol-gel technique for preparation of nanoparticles. Image reprint Singh et al. © 2022 open-access articles distributed under the terms and conditions of the Creative Commons Attribution (CC BY 4.0) license).

However, Chemical Vapor Deposition (CVD) introduces precursor molecules that decompose onto a surface to form NPs. The schematic diagram in Figure 20.5 illustrates the chemical vapor deposition (CVD) process for synthesizing graphene from hydrogen and methane, with each step numbered for clarity. Initially, reactants are transported by convection in the gas flow (step 1) and then thermally activated (step 2). The reactants then diffuse across the stationary boundary layer (step 3), and depending on the substrate's characteristics and carbon solubility, they are either adsorbed on the substrate surface (step 4) or diffused through the substrate bulk (step 5). A graphene film is formed on low-solubility substrates by the catalytic breakdown of methane, which leads to graphene development on the surface. On the other hand, carbon diffuses into the bulk via high solubility substrates, while graphene grows as a result of carbon precipitating to the surface after cooling. It is recommended to cool quickly to avoid the production of multi-layer graphene. In addition to additional heterogeneous reactions, surface processes (step 6) include the migration of reactive molecules to attachment sites and catalytic breakdown. Step 7 of the film growth process involves the desorption of by-products from the substrate. Step 8 involves their diffusion through the boundary layer to the main gas stream, and step 9 involves their convection away to the exhaust system. This detailed understanding aids in optimizing CVD methods for efficient graphene production (Saeed et al. 2020).

Wet chemical methods are another bottom-up approach, encompassing hydrothermal/solvothermal synthesis, where precursors react in hot, pressurized solutions to form NPs. Precipitation involves rapid changes in a solution's chemistry, forcing precursor materials to form NPs (Ndlwana et al. 2021).

Using a co-precipitation method, Milichko et al. synthesized spherical Fe_3O_4 NPs that exhibit a broad nonlinear absorption band in visible light. The findings reveal that under low-intensity visible radiation (with intensities ≤ 0.2 kW/cm^2 and wavelengths of 442 nm and 561 nm), the electric polarizability of Fe_3O_4 NPs changes, reaching a value of 107 Å3. This change is attributed to the intra-band photo transition of charge carriers. This optical effect could potentially enhance the drug uptake properties of Fe_3O_4 NPs (see Figure 20.6) (Milichko et al. 2013).

Finally, reduction uses a reducing agent to transform metal salts into elemental NPs. The chosen method hinges on the desired properties of the final product, such as size, shape, and surface characteristics (Abid et al. 2022). ENs can be built in two ways: top-down, where bulk materials are

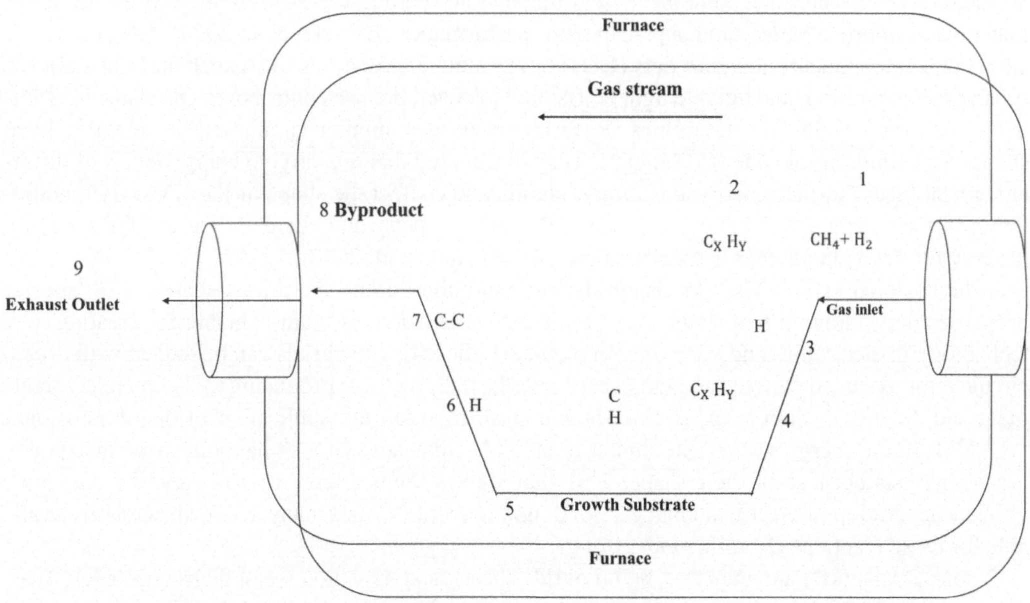

FIGURE 20.5 Schematic diagram of thermal CVD growth of graphene.

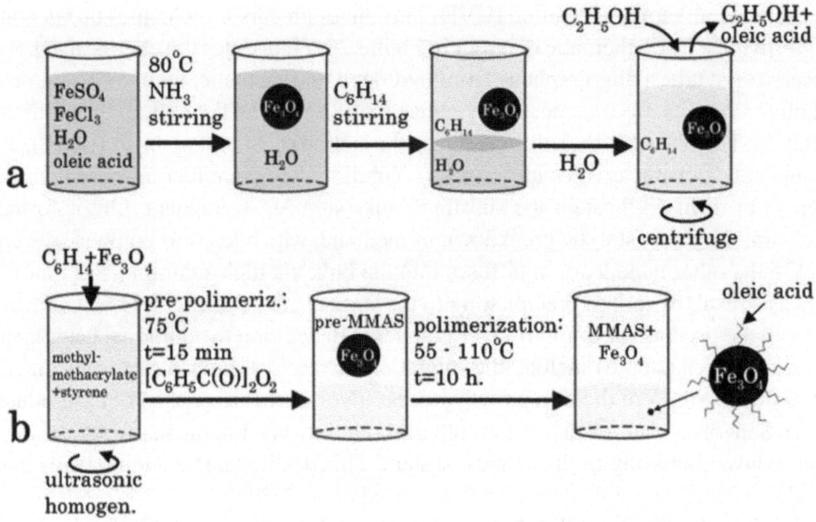

FIGURE 20.6 The developed co-precipitation method. (a) The synthesis of Fe_3O_4 NPs with a monolayer of oleic acid by the developed co-precipitation method and (b) the composite MMAS + Fe_3O_4 preparation (Milichko et al. 2013). Image reprint from Milichko et al., This article is distributed under the terms of the Creative Commons Attribution 2.0 International License.

broken down into tiny pieces, or bottom-up, where individual atoms are meticulously assembled into complex structures. Each approach offers size, shape, and production advantages, making the choice dependent on the desired outcome (see Table 20.1) (Majumder et al. 2007). Breczko et al. describe 0-dimensional carbon NMs like fullerenes and quantum dots for use in batteries and supercapacitors. These 1-10 nanometers particles have properties that make them ideal for electrochemical energy storage due to their small and controllable size and unique structure. The study highlights fullerenes for their precise structure and rich chemical properties, while carbon quantum dots and small carbon nano-onions bridge the gap between fullerenes and larger carbon structures. Researchers are creating composites that combine these 0-dimensional carbon structures with other electroactive materials to improve battery and supercapacitor performance (Breczko et al. 2024). Yang et al. discuss that semiconductor quantum dots (QDs) have valuable properties like excellent light stability, broad light absorption, and tunable light emission. However, precise control over QD shape is essential to achieve specific functionalities. Researchers used computer simulations to identify ideal shapes for cadmium selenide (CdSe) QDs. They discovered that the varying energy levels of different crystal faces (surface energy anisotropy) significantly affect the shape of the NMs. By manipulating surface energy, they believe they can tailor the geometry and characteristics of QDs, paving the way for designing them for specific applications (Yang et al. 2018).

1-dimensional (1D) NMs, like nanorods and nanotubes, come in various shapes, with special properties depending on that shape (see Figure 20.7). This makes them valuable for creating new technology in electronics and medicine. In medicine, these tiny materials can be loaded with drugs, circulate for extended periods, and target cancer cells, making them promising tools for cancer treatment and diagnosis (Shin et al. 2023). Machin et al. reported the application of one-dimensional (1D) NMs in the energy sector. One-dimensional NMs (like nanowires or nanotubes) are also popular in catalysis because of their shape. This shape gives them a large surface area for reactions, efficient separation of electrical charges, good light absorption, and many active sites readily available for those reactions (Machín et al. 2021).

2-dimensional (2D) are graphene, boron nitride sheets, and transition metal dichalcogenides. Two-dimensional (2D) NMs are being explored for their potential in various biological applications, particularly drug delivery. Researchers are especially interested in layered materials like layered double salts

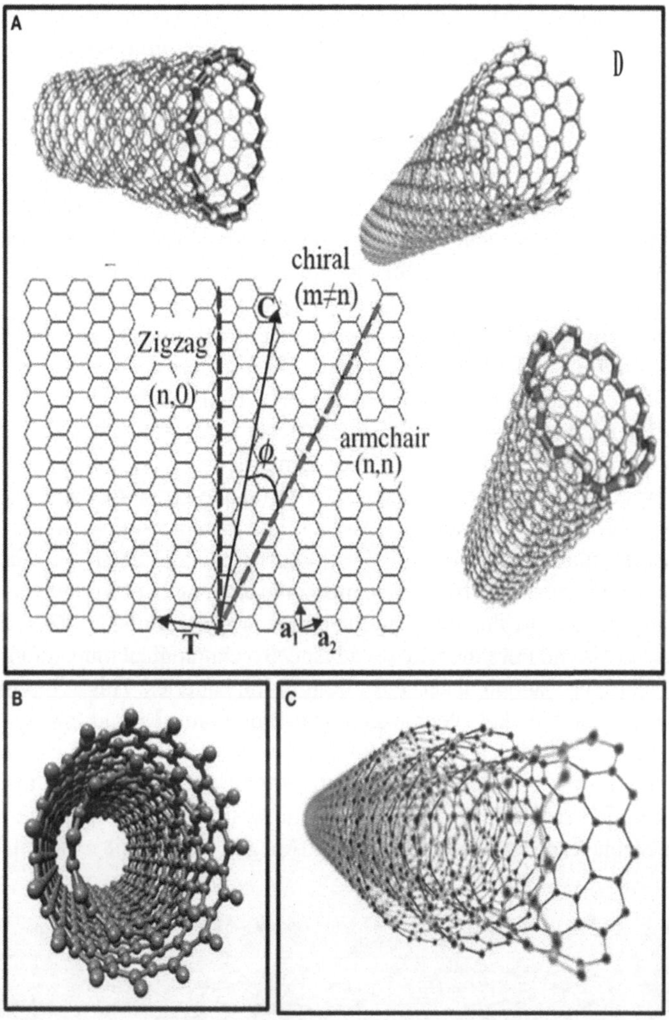

FIGURE 20.7 Anatomy of one-dimensional NMs A: "variations in carbon nanotube structures based on their number of walls. (A) Distinct configurations of single-walled carbon nanotubes (SWNTs) are determined by their chirality, including zigzag, armchair, and chiral orientations. (B) Schematic representation illustrating double-walled carbon nanotubes (DWNTs). (C) Multi-walled carbon nanotubes (MWNTs) exhibit a layered structure composed of multiple concentric shells.". Image reprint from Copyright © 2015 Tîlmaciu and Morris (Tîlmaciu and Morris 2015). This open-access article is distributed under the terms of the Creative Commons Attribution License (CC BY).

(LDSs), layered rear-earth hydroxides (LRHs), and layered double hydroxides (LDHs) because these 2D structures offer several advantages: minimal toxicity, good dissolvability within the body, efficient targeting of tumors, and the ability to carry large amounts of drugs. This makes them promising candidates for the development of new drug delivery systems (Eom et al. 2020). Researchers addressed the limitations of using graphene (NG) to improve metal oxide performance in water purification by creating ZnO/NG nanohybrids with a new one-step method. This method uses a special solvent to disperse graphene and fabricate the nanohybrids with uniformly distributed ZnO NPs. The resulting material exhibits excellent photocatalytic activity under both UV and visible light. This is due to a combination of factors: graphene's ability to absorb more visible light and efficiently transport electrons and forming a bond between ZnO and graphene that narrows the bandgap of ZnO and allows visible light activation. Overall, this study provides a simple and green approach to creating highly effective photocatalysts for degrading pollutants in water (Jin et al. 2020). Figure 20.8 shows that the graphene-based NMs (GNs) encompass a spectrum of structures defined by layer count, sheet dimensions, surface modifications, and defect density. Notable variants include single-layer graphene, few-layered graphene, graphene oxide (GO), and reduced graphene oxide (rGO). Single-layer graphene exhibits a pristine hexagonal lattice of carbon atoms, while few-layered graphene comprises stacked single layers. Despite their exceptional properties, challenges in bulk production, suspension in solution, and isolation impede their application in neural tissue engineering. GO features oxygen-functionalized groups like carboxylic acid, epoxide, and hydroxyl, enhancing colloidal stability and surface interactions with neurocytes. However, these functional groups decrease electrical conductivity and mechanical strength. rGO, an intermediary between graphene and GO, boasts reduced oxygen content and improved properties, making it a cost-effective alternative for neural tissue engineering. Furthermore, variations in GN forms influence stiffness, flexibility, and cellular interactions. For instance, uptake mechanisms shift with GN size, and cytotoxicity correlates with lateral dimensions. Understanding these nuances is crucial for harnessing GNs' potential in biomedical applications (Bei et al. 2019).

3-dimensional (3D) are bulk NMs with all three dimensions in the nanoscale. Three-dimensional (3D) NMs are a new frontier in materials science, with unique properties and applications. Various techniques, such as self-assembly, templating, and additive manufacturing, can produce these structures. Their complex structures make them suitable for energy storage, catalysis, sensing, and biomedical devices, by providing increased surface area, mechanical strength, and conductivity (Papadas et al. 2016). By harnessing the design flexibility and superior properties of 3D NMs, researchers aim to revolutionize diverse fields and pave the way for advanced technological innovations. Scientists developed a new high-performance anode material for lithium-ion batteries. This composite material, CF@ Co-Co3O4/CNT, combines three key elements: a 3D carbon foam scaffold for high surface area, carbon nanotubes for improved electrical conductivity, and cobalt-cobalt oxide NPs for enhanced lithium storage capacity. They achieved this structure through a simple two-step process, making it suitable for large-scale production. The resulting material demonstrates superior performance to traditional graphite anodes, with a stable capacity exceeding 500 mAh g-1, making it a promising candidate for

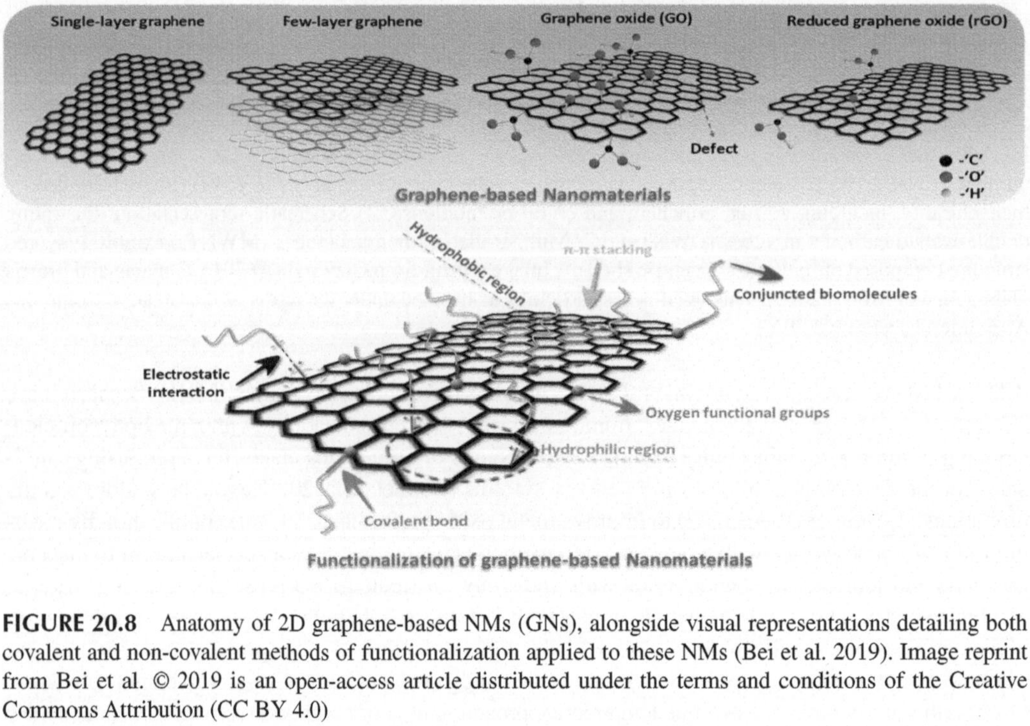

FIGURE 20.8 Anatomy of 2D graphene-based NMs (GNs), alongside visual representations detailing both covalent and non-covalent methods of functionalization applied to these NMs (Bei et al. 2019). Image reprint from Bei et al. © 2019 is an open-access article distributed under the terms and conditions of the Creative Commons Attribution (CC BY 4.0)

next-generation lithium-ion batteries (Liu et al. 2019). Figure 20.9 exhibits a Three-dimensional liposomal delivery system that offers versatility through four primary types, distinguished by their surface characteristics: conventional liposomes, PEGylated liposomes, ligand-targeted liposomes, and theranostic liposomes. Ligand-targeted liposomes enable precise drug delivery to specific tissues expressing targeted ligands, enhancing therapeutic efficacy. PEGylated liposomes, by contrast, extend circulation half-life through steric hindrance following systemic administration, improving overall bioavailability. Furthermore, combining different delivery platforms within liposomes can synergistically enhance drug delivery, tailored to the route of administration and the disease site, offering promising avenues for improved therapeutic outcomes (Sercombe et al. 2015).

20.2.2 Composition

This classification system categorizes NMs based on their chemical makeup. Inorganic metal-based NMs include gold, silver, iron oxide, and titanium dioxide NPs. Metal-based engineered NPs, often called nanometals, are tailored structures with dimensions typically ranging from 1 to 100 nanometers.

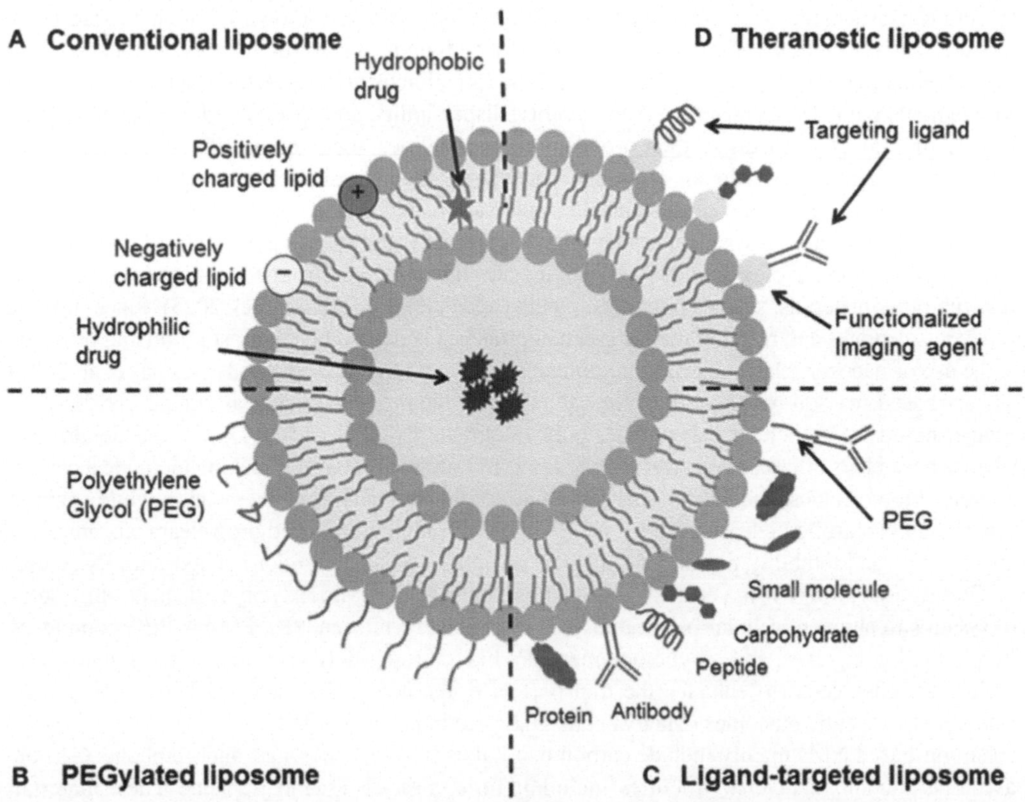

FIGURE 20.9 Schematic overview of 3D NMs liposomal based drug delivery systems. (a) Conventional liposomes comprise a lipid bilayer enclosing an aqueous core capable of encapsulating hydrophobic and hydrophilic compounds. (b) PEGylated liposomes feature a hydrophilic polymer coating, polyethene glycol (PEG), on their surface, stabilizing steric and altering their behavior ***in vivo***. (c) Theranostic liposomes symbolize a multifaceted system that combines a nanoparticle, an imaging component, a targeting element, and a medicinal component. (d) Liposomes target ligands bind ligands, such as antibodies or peptides, to the liposome surface or the terminal end of PEG chains to allow for particular targeting. (Sercombe et al. 2015). Image reprint from © 2015 Sercombe, Veerati, Moheimani, Wu, Sood and Hua open-access article distributed under the Creative Commons Attribution License (CC BY).

These NPs exhibit distinct properties due to their shape, size, and composition, offering numerous advantages over bulk materials. They can be synthesized through various techniques, such as chemical reduction, thermal decomposition, and electrodeposition, allowing precise control over their size and morphology. Metal NPs have extensive applications in catalysis, sensing, imaging, drug delivery, and environmental remediation, due to their high surface-to-volume ratio, enhanced reactivity, and tunable optical and electronic properties (Shnoudeh et al. 2019). Additionally, surface functionalization techniques enable the modification of their surface chemistry, enhancing stability and enabling targeted interactions with biological systems (Ahangaran and Navarchian 2020). Despite their promising potential, the synthesis, characterization, and safe handling of metal-based NPs remain active research areas to ensure their effective and responsible utilization in diverse technological and biomedical applications. Metal oxide-based NPs, including alumina (Al_2O_3), silica (SiO_2), and ceria (CeO_2), represent a versatile class of ENs with wide-ranging applications. These NPs are typically synthesized through sol-gel, hydrothermal, or precipitation techniques, allowing precise control over their size, morphology, and surface properties. Alumina NPs exhibit exceptional mechanical strength, high thermal stability, and chemical inertness, making them suitable for applications in catalyst supports, abrasives, and biomedical implants (Congreve, Quezada, and Kokkarachedu 2024). Silica NPs possess excellent biocompatibility, optical transparency, and tunable porosity, making them valuable in drug delivery, imaging, and sensor technologies (Murugan et al. 2020). Ceria NPs are known for their unique redox properties and oxygen storage capacity, which have been found to be useful in catalysis, fuel cells, and environmental remediation. Additionally, surface modification techniques enable the functionalization of these metal oxide NPs, enhancing their stability, dispersibility, and specificity for targeted applications (Lord et al. 2021). Despite their promising functionalities, metal oxide NPs' potential toxicity and environmental impact remain important considerations, necessitating thorough evaluation and regulation to ensure their safe and sustainable use in various industries.

Semiconductor-based NMs include silicon, gallium nitride, and indium phosphide NPs. Using various synthetic methods, semiconductor nanocomposites have emerged from combining semiconductor NPs with other materials, such as polymers, metals, and cellulose (Prasad et al. 2023). Since they use non-toxic substrates and reaction media, green approaches appear to be the most promising protocols for the use of nanoparticles in biomedical domains among all the other methods (Kumar et al. 2021). Polymer-based, carbon-based, and biological NMs are commonly known as organic NMs. First, polymer-based NMs include polystyrene, poly (lactic-co-glycolic acid) (PLGA), and dendrimers. Polymer-based NPs offer tailored size, morphology, and surface properties, making them ideal carriers for drugs, biomolecules, and genes. Their ability to shield payloads from degradation while enabling sustained and controlled release of drugs is highlighted, leading to reduced drug clearance, enhanced cargo stability and solubility, prolonged half-life, and optimized concentration at target sites (Niculescu and Grumezescu 2021; Mishra et al. 2023). Scientists have devised various polymeric nano-drug delivery systems to augment cellular permeability. These include Chitosan NPs, PMAA-PEG complexes, phospholipid complexes, and acrylic polymer NPs like chitosan-poly (γ-glutamic acid) NPs. These systems are engineered to enhance the transport of drugs across the intestinal barrier, potentially improving therapeutic outcomes (Maurya et al. 2022; Rudrapal et al. 2022).

Carbon-based NMs mainly include carbon nanotubes (CNT), fullerenes, and graphene. Carbon-based ENs encompass diverse structures, including fullerenes, CNTs, and graphene. These materials exhibit unique properties due to their nanoscale dimensions, making them highly attractive for various applications across numerous fields. Fullerenes, often called buckyballs, are spherical carbon molecules with a cage-like structure (Ghosh et al. 2023). They possess exceptional mechanical, electrical, and optical properties, leading to their use in electronics, photonics, and medicine. Fullerenes have shown potential as drug-delivery vehicles, antioxidants, and photovoltaic devices. CNTs are cylindrical structures composed of rolled-up graphene sheets, exhibiting remarkable strength, flexibility, and electrical conductivity. They come in two main types: single-walled nanotubes (SWCNTs) and multi-walled nanotubes (MWCNTs). CNTs find applications in nanoelectronics, composite materials, sensors, and biomedical devices (Manikandan et al. 2021). Their high aspect ratio and tunable properties make them promising candidates for reinforcing materials and drug delivery systems. Graphene, a single layer of carbon atoms arranged in a two-dimensional honeycomb lattice, possesses extraordinary mechanical, thermal, and electrical properties. Its large surface area, high conductivity, and biocompatibility have led to its exploration in various fields, including electronics, sensors, energy storage, and biomedical applications. Graphene-based materials have been investigated for use in biosensors, tissue engineering, drug delivery systems, and water purification, among other areas. Carbon-based ENs hold immense promise for advancing

TABLE 20.2
Function, Structure, and Production of the Most Commonly Used Protein Based Nanoparticles (PNPs) (Diaz, Care, and Sunna 2018)

PNP		Native Organism	Biological Function	Geometry	Number of Subunits	Size (Diameter)	Heterologous Production
Virus-like particles (VLPs)	CCMV[a]	Cowpea chlorotic mottle virus capsid protein	Plant virus	Icosahedral	182	28 nm	Plants; yeast; Escherichia coli; Pseudomonas fluorescens
	CPMV[b]	Cowpea mosaic virus capsid protein	Plant virus	Pseudo icosahedral	120 Large (L) and 120 Small (S)	28 nm	Insect cells; plants
	HBc[c]	Hepatitis B virus capsid protein	Human virus	Icosahedral	180 or 240	30 nm or 34 nm	Mammalian cells; insect cells: plants; yeast; E. coli, cell-free
	MS2	Enterobacteriaceae	Bacteriophage	Icosahedral	180	26 nm	Yeast; E coli; cell-free
	P22	Salmonella typhimurium	Bacteriophage	Icosahedral	420	60 nm	E. coll
	QB	E coli	Bacteriophage	Icosahedral	180	28 nm	Yeast; E coli; cell-free
Non-viral PNPs	Dps[d] (mini-ferritin)	Archaea; Bacteria (e.g., Listeria innocua)	Involved in oxidative and starvation responses	Tetrahedral	12	9 nm	E coli
	E2	Bacillus stearothermophilus	Core of the pyruvate dehydrogenase multienzyme complex	Dodecahedral	60	24 nm	E coll
	Encapsulin	Archaea; Bacteria	Involved in oxidative stress response	Icosahedral	60 or 180	20-40 nm	Mammalian cells; yeast; E coli
	Ferritin (maxi-ferritin)	Archaea; Bacteria; Eukarya	Iron storage	Octahedral	24	12 nm	Mammalian cells; insect cells; yeast; E coli
	Hsp[e]	Archaea; Bacteria: Eukarya (e.g., Methanococcus jannaschii)	Chaperone	Octahedral	24	12 nm	E coli

continued

TABLE 20.2 (Continued)

PNP	Native Organism	Biological Function	Geometry	Number of Subunits	Size (Diameter)	Heterologous Production
Lumazine synthase	Archaea; Bacteria: Eukarya (e.g., Aquifex aeolicus)	Mediates the biosynthesis of riboflavin	Icosahedral	60	15.4 nm	E coli
Vault protein	Eukarya	Involved in signaling and immune responses	39-fold dihedral	78 Major vault protein	Diameter 40 nm Length: 67 nm	Insect cells; cell-free

Table adapted from reference Diaz et al.,

[a] Cow pea chlorotic mottle virus

[b] Cow pea mosaic virus

[c] Hepatitis virus

[d] DNA binding protein from starved cells

[e] Heat shock proteins.

technology and addressing challenges across diverse domains, owing to their exceptional properties and versatility. Continued research and development in this field will unlock further potential for innovative applications and solutions (Levchenko et al. 2023).

Protein-based ENs encompass a wide range of structures and functionalities, leveraging the unique properties of proteins for various applications (Kaltbeitzel and Wich 2023). Three prominent categories within this domain are enzymes, antibodies, and virus-like particles (VLPs). Biological catalysts called enzymes quicken chemical reactions in living things. When engineered at the nanoscale, enzymes can be utilized for targeted and efficient catalysis in various industrial processes, environmental remediation, and biomedical applications (Elegbede and Lateef 2021). Their specificity, efficiency, and biocompatibility make them invaluable tools in biocatalysis, biosensing, and drug delivery. The immune system creates specialized proteins called antibodies, or immunoglobulins, to recognize and neutralize foreign molecules, or antigens, such bacteria, viruses, and poisons. Engineered antibody-based NMs, often called antibody conjugates or antibody-drug conjugates (ADCs), combine the antigen-recognition capability of antibodies with therapeutic or diagnostic payloads. These constructs are utilized in targeted drug delivery, cancer therapy, and in vitro diagnostics, enabling precise targeting of diseased cells while minimizing off-target effects (Joubert et al. 2020). VLPs are self-assembled structures derived from viral proteins but lack the viral genome, rendering them non-infectious. Engineered VLPs mimic the structure and surface characteristics of native viruses, making them ideal platforms for vaccine development, drug delivery, and imaging applications. VLP-based vaccines have been successful against various infectious diseases, including hepatitis B virus (HBV) and human papillomavirus (HPV), owing to their ability to elicit strong immune responses without causing infection (Joubert et al. 2020). Protein-based ENs (PBENMs) offer versatile platforms with tailored functionalities for diverse biotechnology, medicine, and materials science applications. Continued advancements in protein engineering and nanotechnology are expected to drive further innovation in this field, leading to novel solutions for addressing complex challenges in healthcare, industry, and beyond. Nonviral PBENMs such as ferritin, heat shock proteins (Hsp), DNA-binding proteins from starved cells (Dps), lumazine synthase, the E2 protein of pyruvate dehydrogenase, vault proteins, encapsulin, and virus-like particles (VLPs) are among the PBENMs that have been thoroughly investigated. These VLPs have a virus-like appearance, but they are not contagious since they do not contain viral DNA. Table 20.2 presents the natural functions and structural attributes of the most researched PBENMs (Diaz, Care, and Sunna 2018).

20.3 UTILIZATION OF VARIOUS ENS FOR ENVIRONMENTAL REMEDIATION

The pristine nanoparticles (PNPs) are currently used for environmental remediation. Still, their utilization is limited due to their natural tendency to aggregate, limited stability, suboptimal effectiveness, inconsistent size distribution, and uneven toxicological profile. ENs have great potential in improving nanoscale environmental remediation methods by reducing the aggregation of NPs using repulsion forces between the particles. Solubility-enhancing compounds found in ENs can improve metal bioavailability and degrade environmental contaminants. Therefore, modification of NMs is crucial for fully utilizing their capabilities and expanding their range of use (Ahmad et al. 2023). The impact of NMs on the food chain is significant when they come in contact with land and water systems. They undergo changes such as disintegration, dissolution, and reaggregation in the environment. The toxicity of NMs depends on various elements, including their surrounding conditions and the Physicochemical properties of these materials. Understanding the impact of bioaccumulation, biotransformation, and biomagnification mechanisms is crucial in determining the environmental fate of NMs. The pathway by which NPs reach living cells offers valuable insights into their toxicity characteristics (Vineeth Kumar et al. 2022).

20.3.1 ENs in Water Pollution Remediation

Water pollution is a major issue today, Finding effective solutions to treat and remediate it is crucial. Fortunately, nanotechnology provides a promising strategy for addressing this challenge. Researchers can explore new ways to purify and restore our precious water resources with advanced methods and techniques. Various NMs are employed in water purification via several methods, such as the adsorption of heavy metals and other pollutants, the inactivation of pathogens, and the conversion of dangerous contaminants into less harmful molecules. Nanotechnology has revolutionized science, technology, and engineering. The environment is exposed to NMs from natural or accidental sources. ENs are produced for engineering, imaging, and medication (Hendricks et al. 2023). The use of NPs/NMs has seen significant growth worldwide, with applications ranging from medicinal implants to sensing equipment. They have revolutionized the way different techniques are carried out. The NPs synthesized by green methods, such as silver NPs, have antibacterial properties. These NMs are useful in numerous medical applications, such as silicone replacement heart valves, dental sets, catheters, bone prostheses, and bandages (Rafique et al. 2017). These NPs/NMs can be engineered and modified for special purposes to offer unique functionalities. ENs exhibit high reactivity, due to their enormous surface-area-to-volume ratio and the abundance of reactive sites. These attributes facilitate increased interaction with pollutants, rapidly decreasing pollutant concentration. The improper disposal of wastewater from commercial and industrial establishments, untreated residential sludge, and the introduction of synthetic toxins into aquatic systems pose significant risks to human health and the environment. The main water pollutants, in this case, are heavy metal ions, organic compounds, and oils. These substances can make any water source unsuitable for drinking (Lu et al. 2016). NMs, functioning as adsorbent systems, possess a significantly large surface area for reactions with environmental pollutants while having a relatively low weight. They are produced at a lower cost than activated carbon and can efficiently remove impurities. Extensive research has been conducted on various adsorbents, based on NMs, which come in many shapes and dimensions. These include NPs with zero dimensions, nanofibers and nanotubes with one dimension, nanosheets with two dimensions, and nanoflowers with three dimensions (Homaeigohar 2020). There is great variability in terms of composition, and various types of NMs, both carbon-based and non-carbon-based, can help clean up contaminated water. The physical and chemical interaction between the pollutant and the NM's surface is upheld by the segregation process. One important manufacturing component for the permeable segregator unit can be NMs. They can function as a thin film addition to enhance the hydrophilicity and thermomechanical qualities of the membrane or as a nanocomposite ultimate filtering membrane (Bansal, Chaudhary, and Mehta 2015). There are several techniques available for the remediation of contaminated substances. These techniques involve using carbon nanotubes, polymeric membranes, ceramics, submicron particles, nanowires, and titanium oxides. The remediation methods include lysis, which involves breaking down the contaminant into smaller, less harmful parts. Another method is filtration, which involves separating the contaminant from the solution; adsorption, which involves using a material to attract and capture the contaminant; and oxidation, which involves using chemicals to break down the contaminant into non-harmful compounds (Singh et al. 2012). Certain dyes (containing the azo group) in the textile, painting, and pigment industries may have lethal and cancerous effects. Almost 0.1 million distinct synthetic colors are produced in today's commercial world. On an annual basis, over 1.6 million metric tons of dyes are consumed. Of the total, 15% is lost during use (Maheshwari, Agrawal, and Gupta 2021). The adsorption efficiency of chitosan sugar-coated magnetite NPs was higher for Crocein orange G and acid green 25 dyes. Unusually, a magnetic field can effortlessly retrieve the adsorbent NPs. Electro-spun polyether sulfone nanofibers embedded with vanadium NPs are employed to eliminate methylene blue dye from water. Due to their low isoelectric point, the nanofibers exhibit a significant increase in their surface area when exposed to an alkaline environment. This increase is accompanied by a high degree of hydroxylation, which facilitates the adsorption of cationic methylene blue molecules (Homaeigohar et al. 2016). Adsorption has limitations when it comes to fully removing or neutralizing color pollutants. Instead, it only accumulates specific color components by transporting them through various molecular stages. Implementing this process could be difficult due to the complex nature of removing dye-based sewage. These dyes tend to be adsorbed on the membrane and are not frequently reused (Bansal, Chaudhary, and Mehta 2015). Several advanced oxidation techniques that generate free radicals show potential for removing colors from textile effluents. The hydroxyl radical is extremely reactive and oxidizes stubborn organic pollutants because of its unpaired electrons. Photocatalysis, the photoreaction process, is a highly studied technology for advanced oxidative remediation. It is a practical and cost-effective approach to degrading organic dyes and pesticides. Zinc oxide and titanium oxide, metal oxides with semiconductor capabilities, exhibited remarkable photodecomposition capacity when employed to degrade dye pollutants (Li et al. 2019). While organic dyes can cause semiconductors to decompose, visible light photocatalysts in various industries exhibit low stability and efficiency when exposed to semiconductors with energy gaps in the visible light spectrum belonging to groups II-VI. As a solution, these semiconductors have been proposed as promising options to address these challenges (Ghadimi, Zangenehtabar, and Homaeigohar 2020).

20.3.2 ENs in Soil Pollution Remediation

Soil contamination and degradation persist as a substantial environmental issue, requiring global efforts for remediation. The degradation of soil reservoirs worldwide significantly impacts agricultural production and food safety. Consequently, immediate action is necessary to solve this issue (Singh et al. 2019). Soil contamination is impaired by the presence of heavy metals (like As, Cd, Hg, Pb), herbicides (sodium chlorate ($NaClO_3$), sodium arsenate (Na_3AsO_3), etc.), and persistent organic pollutants, which are toxic substances. Due to biomagnification, soil contaminated with these substances has a higher risk of intoxicating the food chain. The pressing need to increase food production and the challenge of preventing further land degradation substantially influence agricultural output. The immobilization or adsorption approach is a widely used therapy technique. One of the major advantages of this remediation technology is its efficacy, cheap cost, and ecologically friendly method for eliminating metal contaminants from soils. Various ENs, including metal oxides like ferric oxide and titanium oxide, carbon-based NMs such as single-walled carbon nanotubes, multi-walled carbon nanotubes, and different advanced composite materials, have been employed for soil contaminant immobilization (Singh et al. 2019; Baby, Saifullah, and Hussein 2019). Ferric oxide NPs can adsorb and immobilize heavy metals from diverse samples in different mediums (Gaur et al. 2014). The discharge of oily sewage is a significant environmental concern that can harm the aquatic ecosystem in the surrounding area (Peng et al. 2020). Iron NPs effectively remove Total Petroleum Hydrocarbons from water, achieving an impressive 88.34% improvement in findings

(Murgueitio et al. 2018). ENs-based treatments can effectively provide purified water with reduced pollutants and hazardous chemicals, as well as the elimination of heavy metals. Nanotechnology can be used with biological treatments to mitigate the risks associated with environmental contaminants. Industrial techniques supported by NPs can effectively decolorize organic colors. Gold and silver NPs are highly effective catalysts for dye degradation. The process involves two essential steps to initiate the reduction reaction. First, the concentration of electrons takes place on the surface of the particles, and then the diffusion of the dye occurs. This entire process is crucial for the reduction reaction to take place successfully (Bastús et al. 2014). Immobilizing pollutants on-site has garnered significant attention as a viable and cost-effective approach to remediating contaminated soils. Several studies have examined several modifications of nanotechnology-based materials to identify a suitable substance that considers factors such as affordability, effectiveness, stability, environmental impact, and reusability. Nanohydroxyapatite particles effectively immobilized metal concentrations in pore water by decreasing the exchangeable percentage of metals in polluted sediments and soils (Wang et al. 2014). This study demonstrated that by using sodium carboxymethyl cellulose-stabilized nano zero-valent iron (nZVI), they achieved an 80% reduction in chromium (VI) bonded to the soil. Pollutants containing redox-sensitive elements can be treated through abiotic reductive processes. These reactions directly affect the movement and toxicity of the elements in soil, either on their own or in conjunction with immobilization procedures. Adding nZVI to soil contaminated with pyrene significantly reduced the soil's oxidizing conditions, facilitating the removal of pyrene through a reduction process. ENs are utilized in advanced oxidation processes, which utilize diverse oxidants to degrade organic pollutants in contaminated soil (Usman et al. 2020). When mixed with soluble Fe (II), hydrogen peroxide or persulfate generates highly reactive radicals such as water and sulfur dioxide radicals. These radicals can efficiently dissociate organic pollutants by oxidation. However, the need for an acidic pH to prevent Fe (II) precipitation and its associated problems, such as early acidification expenses and harmful effects on soil quality and microorganisms, limit such operations. Consequently, there have been suggestions to use solid iron phases instead of soluble Fe (II) to enable chemical oxidation without changing the pH. Iron particles, combined with chelating ligands, augment the process. Green NPs provide significant potential for addressing environmental contaminants and treating localized ecosystems contaminated with hazardous metal ions. These products are beneficial for remediating and decontaminating waste areas. Ferrous NPs can be used to cleanse water contaminated with heavy metals and soil containing heavy metals. They can serve as effective and environmentally acceptable fertilizers. Nano fertilizers have the ability to coordinate the absorption of nutrients by plants, while also reducing the polluting of groundwater. This is because the nano fertilizers remain fixed in place and do not convert into long-lasting chemical or gaseous forms that cannot be accessed by the growing plant (Prasad, Jain, and Varma 2010). In the field of solar energy, there has been a recent interest in using magnetite (Fe3O4) and andgreigite (Fe3S4), along with siliceous material. These materials are created through bacterial processes and have been found to be highly effective in the production of optical coatings specifically designed for solar energy applications. The combination of these materials has shown promising results in increasing the efficiency of solar energy systems (Hussain et al. 2016). However, biogenic fabrications utilizing materials produced from biological components are expected to be widely implemented in several fields, including tissue engineering, environmental remediation, pharmaceuticals, nutrition, and ornamental sectors. The bio-fabrication of metals and their oxide NPs employing algae, fungi, bacteria, and medicinal plants is yet largely unexplored (Chouke et al. 2022). Therefore, there is still much potential to explore novel environmentally friendly preparation procedures based on biogenic synthesis.

20.3.3 ENs in Air Pollution Remediation

Nanotechnology has the potential to mitigate air pollution through many means. One approach for gaseous reactions involves employing nanocatalysts with increased surface area. Catalysts expedite chemical reactions that transform noxious vapors emitted by vehicles and industrial establishments into harmless gases. The process of removing airborne waste from the ground is not efficient; as a result, disposing of the filtered chemicals becomes challenging (Jain and Singh 2012). Filters are advantageous in mitigating air pollution. A filter comprises a compact, porous structure that permits gas passage while retaining particles. The filter membrane employs three techniques to address toxins: surface contact with the filter structure, application of inertial force to redirect gas flow, and generation of electric charge through particle and filter construction. The main drawback of filters is that they decrease pressure, which requires a significant amount of energy consumption. Nano-filters are more efficient than conventional filters due to their narrow pore size distribution, typically ranging from 1 to 10 nm. This characteristic enables them to effectively eliminate a wide spectrum of microbes and organic contaminants. Membranes made from carbon nanotubes exhibit superior efficacy in separating CO2 from other gases (Skoulidas, Sholl, and Johnson 2006). Carbon nanotubes have the ability to absorb gases at a rate that is 100 times faster than existing gas separation techniques. Consequently, they are well-suited for extensive applications. The effectiveness of the gas separation process appears to be inversely related to the volume passing through standard layers. However, this issue is resolved by using carbon-based nanotube (single-walled or multi-walled) membranes. Nanomembrane technology can be employed for the separation and purification of gases and polluting vapors across several industries, with the added benefit of preventing their release into the environment (Zhang et al. 2015). Nanosensors are advanced devices capable of detecting and responding to minute physical changes at the nanometer level. 3D NPs serve as sensors and relay information about the presence of toxins in a certain area. The potential for toxic and dangerous spills is an inherent danger in the commercial industry. Recently, sensors mostly utilized in business settings have detected trace concentrations of these gases. This necessitates the utilization of technological innovations in developing sensors that are faster and more precise. This biosensor variant has individual or multiple membrane nanotubes capable of capturing hazardous gas molecules. Smart dust refers to a novel type of nanosensor. The primary objective is to create a collection of sophisticated sensors in the shape of ultralight nanocomputers. These tiny particles made of silicon have a structure that allows them to transmit data wirelessly to the main server. The essential data transmission rate associated with this type of sensor is around one kbps or higher. Fueled by solar energy, these nanosensors can remain suspended in the air for extended periods. Smart dust can transmit data on climate conditions, humidity levels, and pollutant levels for distances of up to 20 kilometers. This technology can consistently monitor and control air pollution in a specific area (Steinegger, Wolfbeis, and Borisov 2020).

20.4 OPPORTUNITIES AND CHALLENGES IN THE SYNTHESIS OF ENS

The studies explore the transformative impact of nanotechnology on nanomedicine, highlighting its emergence as a pivotal enabling technology for innovative applications. Significant opportunities have arisen, with the nanomedicine market estimated to be worth USD 293.1 billion in 2022 and projected to grow to USD 350.8 billion by 2025. However, challenges persist, particularly regarding the future integration of ENs(ENs) into clinical practice (Albalawi et al. 2021). Key obstacles include the need for unbiased methods to assess the toxicity of ENMs, as existing detection and characterization techniques are often inadequate and inconsistent. Engineered NPs (ENPs) represent a fascinating frontier in science and technology. These microscopic marvels, meticulously crafted at the nanoscale, possess unique properties that open doors to a world of possibilities across various sectors (Albalawi et al. 2021). Let us delve deeper into some of the exciting opportunities that ENP synthesis presents:

20.4.1 Revolutionizing Medicine

ENPs hold immense promise for targeted drug delivery in healthcare. Imagine NPs acting as microscopic surgeons, precisely delivering potent medications directly to diseased cells while minimizing side effects on healthy tissues. This targeted approach could revolutionize cancer treatment and other diseases. ENPs are also being explored for disease diagnosis. NPs that bind to specific biomarkers could enable early detection and monitoring of various illnesses. Furthermore, the potential of ENPs in regenerative medicine is truly groundbreaking. Researchers are exploring the use of ENPs to help repair damaged tissues and organs, offering hope for patients with conditions like heart disease and neurodegenerative disorders (Cui et al. 2021).

20.4.2 Transforming Materials Science

The world of materials is poised for a significant transformation with the advent of ENPs. Scientists can create novel materials with exceptional properties by precisely engineering NPs' size, shape, and composition. These advanced materials could be significantly stronger and lighter than their conventional counterparts. Additionally, ENPs can enhance the conductivity of materials, paving the way for more efficient electronics and energy transmission. The possibilities are endless, with applications ranging from self-cleaning surfaces to next-generation solar cells (Padmanabhan and John 2020).

20.4.3 Championing Clean Technologies

In the fight against environmental pollution and the quest for sustainable solutions, ENPs are emerging as powerful allies. NPs can be used to develop efficient methods for environmental remediation, cleaning up contaminated soil and water. Their unique properties can also be harnessed for water purification, removing harmful contaminants, and ensuring access to clean water for all. Furthermore, ENPs can play a crucial role in developing clean energy sources. Researchers are exploring using NPs to improve solar cells' efficiency, capturing more sunlight and generating cleaner energy (Alkhalayfeh, Aziz, and Pakhuruddin 2021). Despite the exciting possibilities, significant challenges remain in synthesizing ENs. Scaling up production methods from the lab to an industrial level remains a hurdle, often requiring expensive and complex processes. Precise control over the size, shape, and surface properties of ENPs is crucial for their function, yet current techniques often lack the desired level of control. Perhaps the most pressing challenge lies in ensuring the safety of ENPs. Their unique properties raise concerns about potential toxicity and robust methods for assessing and mitigating risks are essential before widespread adoption (Abbas et al. 2020). Researchers are actively developing solutions to these challenges, but navigating this complex landscape is crucial for unlocking the full potential of ENs.

20.5 UTILIZING ENS FOR ENVIRONMENTAL POLLUTION MONITORING SYSTEMS

ENs have emerged as promising tools for revolutionizing environmental pollution monitoring systems, due to their unique properties and versatile applications. ENMs, nano-sized materials engineered to exhibit specific properties, offer several advantages in sensitivity, selectivity, and versatility, making them well-suited for detecting and monitoring various environmental pollutants. One of the key advantages of utilizing ENs in pollution monitoring systems is their exceptional sensitivity to environmental pollutants. Various organic, inorganic, and organometallic compounds are classified as NMs because of their distinct size, shape, and chemical, physical, and electrical characteristics. Typically, these properties become pronounced at the nanoscale, ranging from 1 nanometer to several tens of nanometers in at least one dimension. NMs exhibit a high surface-to-volume ratio, leading to distinctive surface-dominated behaviors. Quantum confinement effects further contribute to size-dependent variations in their properties. Processes such as growth, dissolution, surface reactivity, and aggregation dynamics significantly influence the behavior and interactions of NMs in natural and engineered environments (Gottschalk, Sun, and Nowack 2013). These interactions have implications at local and global scales, impacting environmental processes and systems. ENMs can be functionalized or tailored to interact selectively with specific pollutants, such as heavy metals, organic contaminants, and NPs (Mu et al.

2014). This high sensitivity enables the detection of trace levels of pollutants in environmental samples, even at concentrations below regulatory limits, thereby enhancing the early detection and prevention of pollution-related risks. Moreover, ENs offer rapid and real-time monitoring capabilities, facilitating timely intervention and response to pollution incidents. By incorporating ENMs into sensor devices or monitoring platforms, researchers and environmental agencies can achieve continuous monitoring of pollution parameters, such as air quality, water purity, and soil contamination. ENMs have been widely applied in the remediation of soil and groundwater contaminated with organic and inorganic pollutants through advanced oxidation and chemical reduction, sorption, complexation, precipitation, membrane filtration, and their widespread use in removing contaminants from surface water. The characteristics of ENMs cannot be extrapolated from their bulk counterparts because the properties fundamentally change when bulk materials are downsized to the nanoscale range (1–100 nm) (Park, Wang, and Su 2018). This real-time monitoring capability enables proactive management of environmental resources and minimizes the potential impact of pollution on ecosystems and human health. Additionally, ENMs can be engineered to exhibit multifunctional properties, allowing for the simultaneous detection of multiple pollutants and environmental parameters. Better NM-based sensors have been developed due to cutting-edge nanotechnology methods, especially for real-time pollution analysis. These sensors demonstrate exceptional sensitivity and rapid detection capabilities, offering significant improvements over traditional methods. For instance, utilizing techniques such as nano bioconjugation and nanofabrication, researchers have achieved ultra-sensitive detection of pollutants like Hg2+ ions using AgNP-MD, with detection limits as low as 5–45 μM. This represents a remarkable advancement in environmental sensing technology (Thakur and Kumar 2022). These strategies encompass a range of approaches, including nano biosensors, enzyme-based biosensors, electrochemical-based biosensors, carbon-based biosensors, and optical biosensors. In recent times, there has been a growing emphasis on research concerning nucleic acid biosensors. The Figure 20.10 illustrates the biosensors procedures employed for detecting specific waterborne pathogens in environmental samples. These sensors include immunosensors, DNA-based sensors, and other types. Preprocessing steps are highlighted as essential prerequisites, involving filtration and concentration, followed by an immunosuppression step, such as immunomagnetic separation, across various assay types (Koedrith et al. 2015).

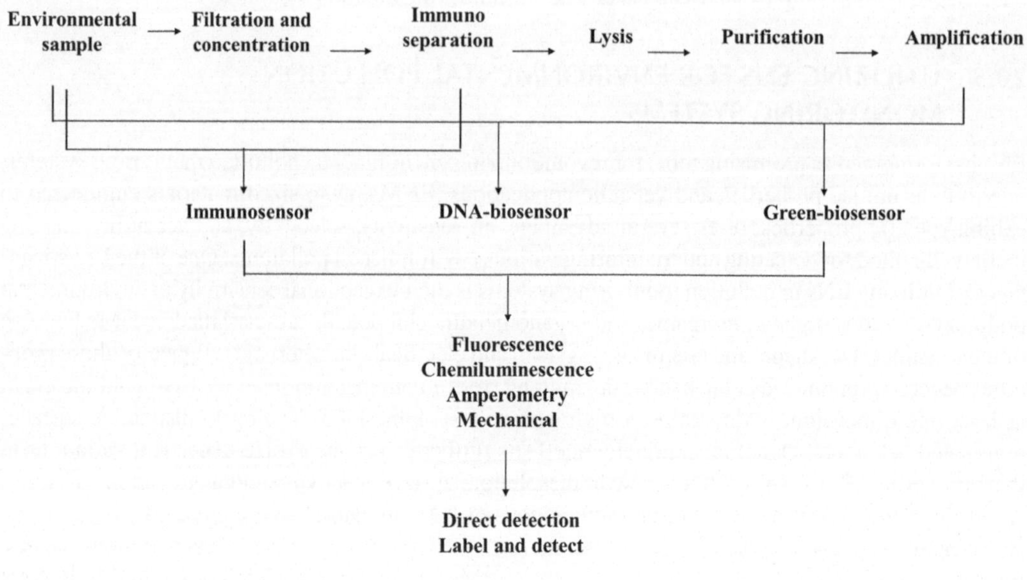

FIGURE 20.10 Mechanism of biosensors for detecting specific waterborne pathogens in environmental samples.

These biosensors are designed for on-site identification and detection of target analytes, offering superior sensitivity, reliability, and selectivity for environmental monitoring applications. For instance, nanocomposite materials incorporating various NMs can be designed to detect and differentiate between different types of pollutants in complex environmental matrices. This multifunctionality enhances the efficiency and reliability of pollution monitoring systems, enabling comprehensive assessment of environmental quality and pollutant sources (Zeng et al. 2021). Furthermore, ENs enable the development of miniaturized and portable monitoring devices, offering greater flexibility and accessibility in environmental monitoring applications. NM-based sensors and detectors can be integrated into handheld devices or wearable technologies, allowing on-site measurements and rapid deployment in remote or inaccessible areas. This portability and versatility empower communities, researchers, and environmental practitioners to conduct real-time monitoring and surveillance of pollution hotspots with minimal infrastructure requirements (Sadik et al. 2009). In conclusion, the utilization of ENs holds significant promise for advancing environmental pollution monitoring systems. Through their high sensitivity, real-time monitoring capabilities, multifunctionality, and portability, ENMs offer innovative solutions for detecting, assessing, and mitigating environmental pollution. As research and development in nanotechnology continue to progress, the integration of ENs into pollution monitoring systems is poised to contribute to more effective and sustainable management of environmental resources and the protection of ecosystems.

20.6 IMPACT OF ENS ON THE CHEMICAL, ENERGY, ELECTRONICS, AND SPACE INDUSTRIES

ENs have significantly impacted various industries, including chemical, energy, electronics, and space. Use and effects of ENs in some of the sectors are discussed here in this section.

20.6.1 Chemical Industries

20.6.1.1 Catalysis

NMs offer high surface area-to-volume ratios and unique surface properties, making them highly effective catalysts for chemical reactions. They enhance reaction rates, selectivity, and efficiency in hydrogenation, oxidation, and synthesis of fine chemicals. Nanoscale catalysts possess exceptional properties, including high specific surface area and surface energy, contributing to their remarkable catalytic activity. Their unique characteristics enable reactions at lower temperatures, reducing side reactions and improving selectivity. Additionally, nano-catalysts facilitate higher recycling rates and energy recovery, making them integral to various fields such as green chemistry, environmental remediation, biomass conversion, and renewable energy development. Their widespread application underscores their pivotal role in advancing sustainable practices and addressing global challenges (Sharma et al. 2015).

20.6.1.2 Sensors

NM-based sensors are revolutionizing chemical detection and monitoring. They offer increased sensitivity, selectivity, and responsiveness, enabling rapid and accurate analysis of chemical compositions, contaminants, and pollutants. Nanoscale materials exhibit unique physical and chemical properties, such as size-dependent behavior, quantum effects, and high surface-to-volume ratios, contributing to their superior catalytic activity. These attributes make them ideal for creating sensors with heightened sensitivity, durability, and rapid response times. With the rise of the Internet of Things (IoT), there's a growing demand for sensors, prompting intensified research into NM-based sensing technologies. Additive manufacturing (AM) is crucial in scaling sensor production from laboratory prototypes to cost-effective, industrial-scale manufacturing, ensuring high reproducibility. Leveraging the design flexibility and cost efficiencies of AM techniques, integrating NM-based sensing platforms with IoT devices becomes more feasible, ushering in a new era of advanced sensing solutions for consumer applications (Sui et al. 2021).

20.6.2 Energy Industry

20.6.2.1 Solar Cells

NMs, such as quantum dots and nanowires, are used to improve the efficiency and cost-effectiveness of solar cells. They enhance light absorption, charge separation, and electron transport, leading to higher power conversion efficiencies. Solar cells offer a promising solution to address global energy crises sustainably. Photovoltaic technology focuses on improving cost-effectiveness and increasing power conversion efficiency. Through the integration of nanotechnology, solar cells are poised to become more efficient, affordable, and reliable, thus advancing the global transition to renewable energy sources (Deshmukh et al. 2021).

20.6.2.2 Battery Technology

NMs enable advancements in battery technology by enhancing electrode performance, energy density, and cycling stability. They are used in lithium-ion batteries, supercapacitors, and emerging technologies, like solid-state and lithium-sulfur batteries. The most promising technology for cost-effectively storing significant amounts of electrical energy is rechargeable lithium-sulfur (Li–S) batteries. However, fundamental issues like the low electrical conductivity of sulfur and sulfides and the intricate solution chemistry of lithiated sulfur compounds in electrolytes must be resolved in order to fully realize the promise of Li–S batteries. Nanotechnology and NMs provide creative answers to these problems (Chen et al. 2021).

20.6.3 Electronics Industry

20.6.3.1 Miniaturization

NMs facilitate the continued miniaturization of electronic devices, enabling the development of faster, smaller, and more efficient components. They are used in transistors, displays, memory devices, and sensors.

20.6.3.2 Conductive Materials

NMs, such as carbon nanotubes and graphene, possess exceptional electrical conductivity and mechanical strength. They are utilized in transparent conductive films, flexible electronics, and high-performance interconnects. Kumar et al. described promising photovoltaic (PV) technology for designing quantum dot-sensitized solar cells (QDSSCs) for energy conversion. Significant advancements have been made in improving the energy conversion efficiency (ECE) of QDSSCs. This is a competitive alternative to traditional crystalline and thin film PV technologies. Advancements in components such as photo-electrodes, electrolytes, sensitizers, and counter electrodes are enhancing the efficiency of green QDSSCs, with their ECE nearing that of dye-sensitized solar cells at around 13%. Consequently, QDSSCs demonstrate considerable potential for further research aimed at developing highly efficient solar cells (Kumar et al. 2017).

20.6.4 Space Industry

20.6.4.1 Lightweight Materials

NMs offer high strength-to-weight ratios, thermal stability, and resistance to radiation, making them ideal for space applications. They are used in spacecraft components, structural materials, and protective coatings. Polymer nanocomposites have become the forefront of high-performance materials, offering significant advancements over conventional aerospace materials like fiber-reinforced composites. While traditional composites already boast high strength-to-weight and modulus-to-weight ratios compared to metallic materials, the past two decades have seen a shift toward NM-based composites. These nanocomposites demonstrate the potential for enhancing properties while also offering multifunctional performance. This shift underscores the recognition of NMs' ability to improve material characteristics and functionality, making them increasingly attractive for aerospace and other high-performance applications (Joshi and Chatterjee 2016).

20.6.4.2 Sensors and Imaging

NM-based sensors and imaging technologies play crucial roles in space exploration, enabling remote sensing, environmental monitoring, and detecting extraterrestrial phenomena. They enhance the sensitivity, resolution, and functionality of spaceborne instruments. Overall, ENs have revolutionized various industries by offering unique properties and capabilities that enable innovation, efficiency, and sustainability in chemical processes, energy generation and storage, electronics manufacturing, and space exploration. However, it is essential to address potential environmental, health, and safety concerns associated with NMs' production, use, and disposal to ensure their responsible application and sustainable development (Rowland et al. 2016).

20.7 PRACTICAL IMPLICATION OF ENS IN HEALTHCARE

In previous years, the US and EU food and drug administrations (FDA) have approved a significant number of nanotechnology-based medications for cancer diagnosis and treatment. Additionally, several other drugs are now undergoing evaluation for potential release in the market. Polymeric NPs are widely used in nanomedicine due to their versatility, ease of production, and extensive utilization in several fields. Among the nanomedicines that have been approved by the FDA, a significant portion comprises polymeric micelles. Polymeric micelles are ideal for the controlled release of hydrophobic medicinal agents because of their self-assembling polymeric amphiphile structures (Bobo et al. 2016). The term "liposome" has also seen extensive usage in medicine that has been approved by the FDA, in addition to polymeric micelles. Liposomes are structures that can integrate themselves and have amphipathic domains all around them. Meanwhile, different molecular species with different physicochemical features might undergo immediate chemical hybridization in the hydrophilic core. Liposomes appear to be the most easily manufactured class of NPs because of their distinctive blend of structural characteristics. Doxorubicin and Amphotericin B, two antifungal drugs used to treat severe fungal infections and leishmaniasis, were among the first liposomal NPs to receive FDA approval (Taylor and Gercel-Taylor 2008; Allen and Cullis 2013). Nanocarrier technologies play a crucial role in improving the effectiveness of DNA and RNA-based therapeutics during the process of medication selection. ENMs are extensively used as a foundational tool to support the creation of various nanomedicines that are influenced by the genome. This includes RNA-based vaccinations that are designed to treat infections and specific types of cancer. The RNA encoded in the vaccinations can serve as a biological template for producing proteins when transported to hepatocytes in the liver (Sahin and Türeci 2018). In addition to choosing the appropriate drugs, effective delivery of the active components to tumor locations is a crucial aspect of cancer treatment. An effective approach is to utilize a synergistic treatment method that combines the use of nanomedicine with the systematic administration of curative agents. The integration of many therapies can effectively surmount obstacles in the delivery of drugs specifically targeted to tumors. Biological barriers, such as the dense extracellular matrix (ECM), high interstitial fluid pressure (IFP), and high metabolism in carcinomas, pose significant challenges in delivery of drugs to target sites. However, by employing a combination therapy approaches that involves the use of prescribed drugs along with the angiotensin II receptor inhibitor losartan, it is possible to reduce the levels of ECM and IFP in tumors. Consequently, it enhances the clinical effectiveness and performance of nanomedicine (Chauhan et al. 2013). Several types of ENs have been recognized for the purpose of delivering drugs and treating eye-related disorders by experts who focus on the eyes. Eleraky et al. studied that surfactants are incorporated into ENs composites to enhance the dispersion of nanoclusters, hence decreasing the size of ocular and pharmacological emulsions. Although ENs have shown potential benefits in ocular treatment, ionic surfactants are often hazardous. It is crucial to carefully adjust the quantity of the surfactant, as larger concentrations might lead to irritation and damage to the cornea. Additional approaches to address the issue of toxicity in ionic surfactants involve the utilization of non-ionic surfactants, such as poloxamers, polyethylene glycol, polysorbates, and tyloxapol (Eleraky et al. 2020). In addition to the aforementioned uses, promising applications of ENs include their use in combating microbial infections. ENs have demonstrated remarkable antifungal, antibacterial, and antiviral properties, making them a potential solution for addressing various infections. Multiple studies have documented the utilization of ENs as a treatment for infectious diseases. For example, elements such as silver (Ag), iron (Fe), titanium dioxide (TiO2), copper (Cu), and zinc oxide nanoparticles (ZnONPs) have demonstrated notable antibacterial properties against bacteria that are resistant to conventional drugs. The primary antibacterial action of these ENs is ionization, which induces oxidative stress in bacteria through the generation of free radicals. Alavi and colleagues have confirmed that Fe3O4 NPs possess the ability to control the spread of multidrug-resistant bacterial growth when utilized in iron-based ENs. The antimicrobial properties of Fe3O4 NPs were tested on E. coli ATCC 25922, *S. marcescens* ATCC 13880, and *S. aureus* ATCC 43300 in the study. A higher concentration of Fe3O4 NPs effectively reduced the growth of the *S. aureus* ATCC 43300 colony, according to the study. In addition to Fe-based NPs, nanoformulations designed to combat bacterial growth have been developed by combining Ag-based particles with various bioactive molecules. As previously noted in a related study, there are multiple factors that can impact the bioactivity of Ag-based nanoformulations, such as the type of formulation (films, foams, or hydrogels) and the concentration or volume ratio of each ingredient (Alavi and Karimi 2019). Aside from nanomedicine, the antibacterial characteristics of ENs have potential uses in food packaging, water disinfection, antimicrobial coatings, and more. In addition to ENs based on metals, carbon-based ENs like carbon-nanotubes have also demonstrated antibacterial properties. The bactericidal effects of the antibacterial agent can be enhanced by encapsulating it in the open-ended structures of carbon-nanotubes (Alavi, Jabari, and Jabbari 2021). ENs have proved to be highly effective in the screening of patients, selecting medications, immunomodulation, and combination therapies for cancer, ocular, and infectious diseases, as discussed in this section. There have been several reports on the use of various NMs in eye therapy. However, a significant hindrance in implementing these advancements in current nanomedicine-based therapy is the toxicity of most ENs. Therefore, several initiatives will be launched to address the concerns of nanotoxicology in nanomedicine.

20.8 POTENTIAL AREAS FOR FUTURE RESEARCH

ENs have great potential to enhance different aspects of the environment and healthcare systems. However, it is crucial to comprehend the safety of these NMs. One prevalent toxicological consequence observed across numerous types of NPs is the initiation of oxidative stress. Research has shown that exposure to ENs can cause oxidative stress, which in turn can result in inflammation, cell death, and altered genomic profiles. However, many of these studies have used high doses of ENs for short periods of time, which may not truly reflect the levels of exposure that occur through the use of consumer items or nanotherapeutics (Mendoza and Brown 2019). Research studies consistently demonstrates the potential applications of ENs in the treatment and purification of wastewater. Nanosorbents, antimicrobial conmponds, nanocatalysts, sensors, and molecularly imprinted polymers are examples of nanotechnology advancements that can efficiently eliminate contaminants from water. Magnetic NPs have the ability to attract magnetic and metallic substances when present in water. Currently under investigation are novel ENs that function as both filters and photocatalysts. These are membrane filters. Future studies on nanotechnology for water purification and wastewater treatment will concentrate on multifunctional nanoparticles (NPs) that have some of the previously mentioned characteristics (Nnaji et al. 2018). The area of ENs is still emerging, and as a result, the monitoring methods used are typically derived from existing approaches for detecting particle pollution in air and water settings. The primary difficulty in monitoring the concentration of ENMs in any given environment is in the ability to differentiate them from the NPs that naturally occur in that environment. Various techniques exist for quantifying the quantity of particles and determining their properties. Particle concentration and size distribution can be quantified using several instruments, including DLS (Dynamic Light Scattering), CPC (Condensation Particle Counter), DMPS (Differential Mobility Particle Sizer), FMPS (Fast

Mobility Particle Sizer), ELPI (Electrical Low-Pressure Impactor), and SMPS (Scanning Mobility Particle Sizer). The surface properties of ENs are often analyzsed using several spectroscopic techniques such as scanning electron microscopy (SEM), transmission electron microscopy (TEM), energy-dispersive X-ray spectroscopy (EDX), Brunauer-Emmett-Teller (BET) analysis, thermogravimetric analysis (TG), Fourier-transform infrared spectroscopy (FTIR), near-infrared spectroscopy, and Raman spectroscopy (Kumar et al. 2014). These approaches are typically intricate, labor-intensive, and costly, rendering them unsuitable for regular monitoring of ENMs in various environmental media. Although there has been a rise in public awareness and recent reporting on the presence of ENs in the ecological system, there is still a lack of sufficient data to effectively regulate them, primarily due to ongoing technical challenges. The suggested future actions could facilitate the distribution of information regarding the monitoring of ENs and the determination of dose-response parameters that accurately reflect environmental exposure circumstances. These initiatives aim to advance the targeting of cancer and other infectious diseases by developing quantitative nanotoxicology research (QNTR).

20.9 CONCLUSION

The chapter highlighted significant trends in the global market for ENs. It specifically emphasizes the chemical, energy, electronics, and space industries sector, drawing from recent analytical reports and research papers on the subject. The authors provided evidence for the rapid expansion of the energy-related ENs by considering the available proof of their distinct electrical, thermal, mechanical, optical, and catalytic capabilities, as well as the increasing demands in this sector. The application possibilities of energy-related ENs in the generation, saving, storage, conversion, and transmission of energy in both conventional and renewable energy systems are greatly enhanced by these aspects. The text characterizes the current and future applications and growth prospects of commonly used ENs in global healthcare applications. These include graphene, carbon nanotubes, metal oxide NPs, and others.

REFERENCES

Abbas, Qumber, Balal Yousaf, Muhammad Ubaid Ali, Mehr Ahmed Mujtaba Munir, Ali El-Naggar, Jörg Rinklebe, and Mu Naushad. 2020. “Transformation pathways and fate of engineered nanoparticles (ENPs) in distinct interactive environmental compartments: A review.” *Environment International* 138: 105646.

Abid, Namra, Aqib Muhammad Khan, Sara Shujait, Kainat Chaudhary, Muhammad Ikram, Muhammad Imran, Junaid Haider, Maaz Khan, Qasim Khan, and Muhammad Maqbool. 2022. “Synthesis of nanomaterials using various top-down and bottom-up approaches, influencing factors, advantages, and disadvantages: A review.” *Advances in Colloid and Interface Science* 300:102597.

Ahangaran, Fatemeh, and Amir H Navarchian. 2020. “Recent advances in chemical surface modification of metal oxide nanoparticles with silane coupling agents: A review.” *Advances in Colloid and Interface Science* 286:102298.

Ahmad, Muhammad Arslan, Muhammad Adeel, Noman Shakoor, Rabia Javed, Muhammad Ishfaq, Yutao Peng, Muhammad Zain, Imran Azeem, Ilyas Ali, Muhammad Usman, Zihao Wu, Gholamreza Gohari, Ming Xu, Yukui Rui, Zhiyong Zhang, Jason C. White, and Xu Deng. 2023. “Modifying engineered nanomaterials to produce next generation agents for environmental remediation.” *Science of The Total Environment* 894:164861. https://doi.org/10.1016/j.scitotenv.2023.164861

Alavi, Mehran, Erfan Jabari, and Esmaiel Jabbari. 2021. “Functionalized carbon-based nanomaterials and quantum dots with antibacterial activity: a review.” *Expert Review of Anti-infective Therapy* 19 (1):35–44.

Alavi, Mehran, and Naser Karimi. 2019. “Ultrasound assisted-phytofabricated Fe3O4 NPs with antioxidant properties and antibacterial effects on growth, biofilm formation, and spreading ability of multidrug resistant bacteria.” *Artificial Cells, Nanomedicine, and Biotechnology* 47 (1):2405–2423.

Albalawi, Fahad, Mohd Zobir Hussein, Sharida Fakurazi, and Mas Jaffri Masarudin. 2021. “Engineered nanomaterials: The challenges and opportunities for nanomedicines.” *International Journal of Nanomedicine* 30:161–184.

Alkhalayfeh, Muheeb Ahmad, Azlan Abdul Aziz, and Mohd Zamir Pakhuruddin. 2021. "An overview of enhanced polymer solar cells with embedded plasmonic nanoparticles." *Renewable and Sustainable Energy Reviews* 141:110726.

Allen, Theresa M, and Pieter R Cullis. 2013. "Liposomal drug delivery systems: From concept to clinical applications." *Advanced Drug Delivery Reviews* 65 (1):36–48.

Baby, Rabia, Bullo Saifullah, and Mohd Zobir Hussein. 2019. "Carbon nanomaterials for the treatment of heavy metal-contaminated water and environmental remediation." *Nanoscale Research Letters* 14 (1):341.

Bansal, Pratibha, Ganga Ram Chaudhary, and SK Mehta. 2015. "Comparative study of catalytic activity of ZrO2 nanoparticles for sonocatalytic and photocatalytic degradation of cationic and anionic dyes." *Chemical Engineering Journal* 280:475–485.

Bastús, Neus G, Florind Merkoçi, Jordi Piella, and Victor Puntes. 2014. "Synthesis of highly monodisperse citrate-stabilized silver nanoparticles of up to 200 nm: Kinetic control and catalytic properties." *Chemistry of Materials* 26 (9):2836–2846.

Bei, Ho Pan, Yuhe Yang, Qiang Zhang, Yu Tian, Xiaoming Luo, Mo Yang, and Xin Zhao. 2019. "Graphene-based nanocomposites for neural tissue engineering." *Molecules* 24 (4):658.

Bobo, Daniel, Kye J Robinson, Jiaul Islam, Kristofer J Thurecht, and Simon R Corrie. 2016. "Nanoparticle-based medicines: A review of FDA-approved materials and clinical trials to date." *Pharmaceutical Research* 33:2373–2387.

Bora, Kainat Amin, Saud Hashmi, Faisal Zulfiqar, Zainul Abideen, Haibat Ali, Zamin Shaheed Siddiqui, and Kadambot H. M. Siddique. 2022. "Recent progress in bio-mediated synthesis and applications of engineered nanomaterials for sustainable agriculture." *Frontiers in Plant Science* 13. https://doi.org/10.3389/fpls.2022.999505

Breczko, Joanna, Monika Wysocka-Zołopa, Emilia Grądzka, and Krzysztof Winkler. 2024. "Zero-Dimensional Carbon Nanomaterials for Electrochemical Energy Storage." *ChemElectroChem* 11e202300752.

Chauhan, Vikash P, John D Martin, Hao Liu, Delphine A Lacorre, Saloni R Jain, Sergey V Kozin, Triantafyllos Stylianopoulos, Ahmed S Mousa, Xiaoxing Han, and Pichet Adstamongkonkul. 2013. "Angiotensin inhibition enhances drug delivery and potentiates chemotherapy by decompressing tumour blood vessels." *Nature Communications* 4 (1):2516.

Chen, Yi, Tianyi Wang, Huajun Tian, Dawei Su, Qiang Zhang, and Guoxiu Wang. 2021. "Advances in lithium–sulfur batteries: From academic research to commercial viability." *Advanced Materials* 33 (29):2003666.

Chouke, Prashant B., Trupti Shrirame, Ajay K. Potbhare, Aniruddha Mondal, Ankita R. Chaudhary, Sudip Mondal, Sanjay R. Thakare, Eugenie Nepovimova, Martin Valis, Kamil Kuca, Rohit Sharma, and Ratiram Gomaji Chaudhary. 2022. "Bioinspired metal/metal oxide nanoparticles: A road map to potential applications." *Materials Today Advances* 16:100314. https://doi.org/10.1016/j.mtadv.2022.100314

Colvin, Vicki L. 2003. "The potential environmental impact of engineered nanomaterials." *Nature Biotechnology* 21 (10):1166–1170. https://doi.org/10.1038/nbt875

Coman, Vasile, Ioana Oprea, Loredana Florina Leopold, Dan Cristian Vodnar, and Cristina Coman. 2019. "Soybean interaction with engineered nanomaterials: a literature review of recent data." *Nanomaterials* 9 (9):1248.

Conde, João, Chenchen Bao, Yeqi Tan, Daxiang Cui, Elazer R Edelman, Helena S Azevedo, Hugh J Byrne, Natalie Artzi, and Furong Tian. 2015. "Dual targeted immunotherapy via in vivo delivery of biohybrid RNAi-peptide nanoparticles to tumor-associated macrophages and cancer cells." *Advanced Functional Materials* 25 (27):4183–4194.

Conde, João, Jorge T Dias, Valeria Grazú, Maria Moros, Pedro V Baptista, and Jesus M de la Fuente. 2014. "Revisiting 30 years of biofunctionalization and surface chemistry of inorganic nanoparticles for nanomedicine." *Frontiers in Chemistry* 2:48.

Congreve, Rodrigo Cáceres, Carolina Paz Quezada, and Varaprasad Kokkarachedu. 2024. "Aluminum Oxide Nanoparticles: Properties and Applications Overview." In *Nanoparticles in Modern Antimicrobial and Antiviral Applications*, 265–288. Springer.

Corsi, Ilaria, Arianna Bellingeri, Maria C. Eliso, Giacomo Grassi, Giulia Liberatori, Carola Murano, Lucrezia Sturba, Maria L. Vannuccini, and Elisa Bergami. 2021. Eco-Interactions of Engineered Nanomaterials in the Marine Environment: Towards an Eco-Design Framework. *Nanomaterials* 11 (8). https://doi.org/10.3390/nano11081903

Cui, Weitong, Wei Fu, Yunfeng Lin, and Tianxu Zhang. 2021. "Application of nanomaterials in neurodegenerative diseases." *Current Stem Cell Research & Therapy* 16 (1):83–94.

de Lima Barizão, Ana Carolina, Jairo Pinto de Oliveira, Ricardo Franci Gonçalves, and Sérvio Túlio Cassini. 2021. "Nanomagnetic approach applied to microalgae biomass harvesting: Advances, gaps, and perspectives." *Environmental Science and Pollution Research* 28:44795–44811.

Deshmukh, Megha A, Sang-Joon Park, Bhavna S Hedau, and Tae-Jun Ha. 2021. "Recent progress in solar cells based on carbon nanomaterials." *Solar Energy* 220:953–990.

Diaz, Dennis, Andrew Care, and Anwar Sunna. 2018. "Bioengineering strategies for protein-based nanoparticles." *Genes* 9 (7):370.

Dickerson, Erin B, Erik C Dreaden, Xiaohua Huang, Ivan H El-Sayed, Hunghao Chu, Sujatha Pushpanketh, John F McDonald, and Mostafa A El-Sayed. 2008. "Gold nanorod assisted near-infrared plasmonic photothermal therapy (PPTT) of squamous cell carcinoma in mice." *Cancer Letters* 269 (1):57–66.

Dontsova, Tetiana A, Svitlana V Nahirniak, and Ihor M Astrelin. 2019. Metaloxide nanomaterials and nanocomposites of ecological purpose. *Journal of Nanomaterials* 2019: 1–31. https://doi.org/10.1155/2019/5942194

Elegbede, JA, and A Lateef. 2021. "Microbial enzymes in nanotechnology and fabrication of nanozymes: A perspective." In Lateef, A., Gueguim-Kana, E.B., Dasgupta, N., Ranjan, S. (eds), *Microbial Nanobiotechnology: Principles and Applications*: Springer, Singapore: 185–232. https://doi.org/10.1007/978-981-33-4777-9_7

Eleraky, Nermin E, Ayat Allam, Sahar B Hassan, and Mahmoud M Omar. 2020. "Nanomedicine fight against antibacterial resistance: An overview of the recent pharmaceutical innovations." *Pharmaceutics* 12 (2):142.

Eom, Sairan, Goeun Choi, Hiroyuki Nakamura, and Jin-Ho Choy. 2020. "2-Dimensional nanomaterials with imaging and diagnostic functions for nanomedicine: review." *Bulletin of the Chemical Society of Japan* 93 (1):1–12.

FAO. 2017. *Water for Sustainable Food and Agriculture: A report produced for the G20 Presidency of Germany*. Food and Agriculture Organization of the United Nations Rome.

Gajewicz, Agnieszka, Bakhtiyor Rasulev, Tandabany C. Dinadayalane, Piotr Urbaszek, Tomasz Puzyn, Danuta Leszczynska, and Jerzy Leszczynski. 2012. "Advancing risk assessment of engineered nanomaterials: Application of computational approaches." *Advanced Drug Delivery Reviews* 64 (15):1663–1693. https://doi.org/10.1016/j.addr.2012.05.014

Gaur, Nisha, Gagan Flora, Mahavir Yadav, and Archana Tiwari. 2014. "A review with recent advancements on bioremediation-based abolition of heavy metals." *Environmental Science: Processes & Impacts* 16 (2):180–193.

Ghadimi, Mehrnoosh, Sasan Zangenehtabar, and Shahin Homaeigohar. 2020. "An overview of the water remediation potential of nanomaterials and their ecotoxicological impacts." *Water* 12 (4):1150.

Ghosh, Dipanjan, Gouranga Dutta, Abimanyu Sugumaran, Gopal Chakrabarti, and Biplab Debnath. 2023. "Fullerenes: Bucky balls in the therapeutic application." In *Carbon Nanostructures in Biomedical Applications*, 1–25. Springer.

Gottschalk, Fadri, TianYin Sun, and Bernd Nowack. 2013. "Environmental concentrations of engineered nanomaterials: Review of modeling and analytical studies." *Environmental Pollution* 181:287–300.

Hendricks, N., O. Olatunji, I. Ofori, and B. P. Gumbi. 2023. "Occurrence, fate, and impact of engineered metal/carbonaceous nanomaterials in the environment, detection, and quantitation methods." *International Journal of Environmental Science and Technology* 20 (11):12937–12954. https://doi.org/10.1007/s13762-023-04977-8

Homaeigohar, Shahin. 2020. "The nanosized dye adsorbents for water treatment." *Nanomaterials* 10 (2):295.

Homaeigohar, Shahin, Ahnaf Usman Zillohu, Ramzy Abdelaziz, Mehdi Keshavarz Hedayati, and Mady Elbahri. 2016. "A novel nanohybrid nanofibrous adsorbent for water purification from dye pollutants." *Materials* 9 (10):848.

Hussain, Imtiyaz, N.B. Singh, Ajey Singh, Himani Singh, and SC Singh. 2016. "Green synthesis of nanoparticles and its potential application." *Biotechnology letters* 38:545–560.

Jain, CK, and RD Singh. 2012. "Technological options for the removal of arsenic with special reference to South East Asia." *Journal of Environmental Management* 107:1–18. https://doi.org/10.1016/j.jenvman.2012.04.016

Jin, Xin, Zijie Ma, Guangrong Liu, Dongwen Hu, Chuangfu Song, and Qiang Huang. 2020. "In-situ ionothermal precipitation of well-dispersed ZnO nanoparticles onto 2-dimension neat graphene sheets with excellent photocatalytic activity." *Journal of Environmental Chemical Engineering* 8 (4):104030.

Joshi, M., and U. Chatterjee. 2016. "Polymer nanocomposite: An advanced material for aerospace applications." In Sohel Rana, Raul Fangueiro (eds), *Advanced Composite Materials for Aerospace Engineering*, Woodhead Publishing: 241–264. Elsevier. https://doi.org/10.1016/B978-0-08-100037-3.00008-0

Joubert, Nicolas, Alain Beck, Charles Dumontet, and Caroline Denevault-Sabourin. 2020. "Antibody–drug conjugates: The last decade." *Pharmaceuticals* 13 (9):245.

Kaltbeitzel, Jonas, and Peter R Wich. 2023. "Protein-based nanoparticles: From drug delivery to imaging, nanocatalysis and protein therapy." *Angewandte Chemie International Edition* 62 (44):e202216097.

Koedrith, Preeyaporn, Thalisa Thasiphu, Jong-Il Weon, Rattana Boonprasert, Kooranee Tuitemwong, and Pravate Tuitemwong. 2015. "Recent trends in rapid environmental monitoring of pathogens and toxicants: Potential of nanoparticle-based biosensor and applications." *The Scientific World Journal* 2015 (1), 510982.

Kumar, Arun, Prashant Kumar, Ananthitha Anandan, Teresa F Fernandes, Godwin A Ayoko, and George Biskos. 2014. "Engineered nanomaterials: knowledge gaps in fate, exposure, toxicity, and future directions." *Journal of Nanomaterials* 2014:5–5.

Kumar, J.A., Krithiga, T., Manigandan, S., Sathish, S., Renita, A.A., Prakash, P., Prasad, B.N., Kumar, T.P., Rajasimman, M., Hosseini-Bandegharaei, A. and Prabu, D. 2021. "A focus to green synthesis of metal/metal based oxide nanoparticles: Various mechanisms and applications towards ecological approach." *Journal of Cleaner Production* 324:129198.

Kumar, Sandeep, Monika Nehra, Akash Deep, Deepak Kedia, Neeraj Dilbaghi, and Ki-Hyun Kim. 2017. "Quantum-sized nanomaterials for solar cell applications." *Renewable and Sustainable Energy Reviews* 73:821–839.

Levchenko, Igor, Oleg Baranov, Claudia Riccardi, H Eduardo Roman, Uroš Cvelbar, Elena P Ivanova, Mandhakini Mohandas, Patrik Ščajev, Tadas Malinauskas, and Shuyan Xu. 2023. "Nanoengineered carbon-based interfaces for advanced energy and photonics applications: A recent progress and innovations." *Advanced Materials Interfaces* 10 (1):2201739.

Li, Qin, Tingting Zhao, Mei Li, Wenting Li, Bin Yang, Dongran Qin, Kangle Lv, Xian Wang, Lamei Wu, and Xiaofeng Wu. 2019. "One-step construction of Pickering emulsion via commercial TiO_2 nanoparticles for photocatalytic dye degradation." *Applied Catalysis B: Environmental* 249:1–8.

Liu, Wenbin, Yuanyuan Fu, Yanfei Li, Shouhui Chen, Yonghai Song, and Li Wang. 2019. "Three-dimensional carbon foam surrounded by carbon nanotubes and Co-Co3O4 nanoparticles for stable lithium-ion batteries." *Composites Part B: Engineering* 163:464–470.

Liu, Ze, Jinkui Chu, Ran Zhang, Chuanlong Guan, and Yuanyi Fan. 2022. "Preparation of an integrated polarization navigation sensor via a nanoimprint photolithography process." *Photonics* 9, (11): 806. https://doi.org/10.3390/photonics9110806

Lord, Megan S, Jean Francois Berret, Sanjay Singh, Ajayan Vinu, and Ajay S Karakoti. 2021. "Redox active cerium oxide nanoparticles: Current status and burning issues." *Small* 17 (51):2102342.

Lu, Haijiao, Jingkang Wang, Marco Stoller, Ting Wang, Ying Bao, and Hongxun Hao. 2016. "An overview of nanomaterials for water and wastewater treatment." *Advances in Materials Science and Engineering* 2016. 1–10. https://doi.org/10.1155/2016/4964828

Machín, Abniel, Kenneth Fontánez, Juan C Arango, Dayna Ortiz, Jimmy De León, Sergio Pinilla, Valeria Nicolosi, Florian I Petrescu, Carmen Morant, and Francisco Márquez. 2021. "One-dimensional (1D) nanostructured materials for energy applications." *Materials* 14 (10):2609.

Maheshwari, Karishma, Madhu Agrawal, and Akhilendra Bhushan Gupta. 2021. "Dye pollution in water and wastewater." in Muthu, S. S., Khadir, A. (eds.), *Novel Materials for Dye-Containing Wastewater Treatment*:1–25. Sustainable Textiles: Production, Processing, Manufacturing & Chemistry. Springer, Singapore. https://doi.org/10.1007/978-981-16-2892-4_1

Majumder, D Dutta, Rajat Banerjee, CH Ulrichs, Inga Mewis, and Arunava Goswami. 2007. "Nano-materials: Science of bottom-up and top-down." *IETE Technical Review* 24 (1):9–25.

Malik, Shiza, Khalid Muhammad, and Yasir Waheed. 2023. "Emerging applications of nanotechnology in healthcare and medicine." *Molecules* 28 (18). https://doi.org/10.3390/molecules28186624

Manikandan, N, Suresh Kumar VP, G Rathis, and TK Shabariganesh. 2021. "Carbon nanotubes and their properties: The review." *Materials Today: Proceedings* 47:4682–4685.

Maurya, Rahul, Suman Ramteke, Pranay Guru, and Narendra Kumar Jain. 2022. "Oral glucose-responsive nanocarrier system for management of diabetes." *Journal of Endocrinology and Metabolism* 12 (4–5):146–160.

Mendoza, Ryan P., and Jared M. Brown. 2019. "Engineered nanomaterials and oxidative stress: Current understanding and future challenges." *Current Opinion in Toxicology* 13:74–80. https://doi.org/10.1016/j.cotox.2018.09.001

Milichko, Valentin A, Anton I Nechaev, Viktor A Valtsifer, Vladimir N Strelnikov, Yurii N Kulchin, and Vladimir P Dzyuba. 2013. "Photo-induced electric polarizability of Fe 3 O 4 nanoparticles in weak optical fields." *Nanoscale Research Letters* 8:1–7.

Mishra, Ashwini Kumar, S. L. Neha, Laxmi Rani, Ankit Jain, Hitesh Kumar Dewangan, and Pravat Kumar Sahoo. 2023. "Rationally designed nanoparticulate delivery approach for silymarin with natural bio-enhancer: In vitro characterization and in vivo evaluations of hepatoprotective effects in a mouse model." *Journal of Drug Delivery Science and Technology* 86:104580. https://doi.org/10.1016/j.jddst.2023.104580

Mu, Qingxin, Guibin Jiang, Lingxin Chen, Hongyu Zhou, Denis Fourches, Alexander Tropsha, and Bing Yan. 2014. "Chemical basis of interactions between engineered nanoparticles and biological systems." *Chemical Reviews* 114 (15):7740–7781.

Murgueitio, Erika, Luis Cumbal, Mayra Abril, Andrés Izquierdo, Alexis Debut, and Oscar Tinoco. 2018. "Green synthesis of iron nanoparticles: Application on the removal of petroleum oil from contaminated water and soils." *Journal of Nanotechnology*. 1–8, https://doi.org/10.1155/2018/4184769

Murugan, Baranya, Suresh Sagadevan, Anita Lett, Is Fatimah, Kamrun Nahar Fatema, Won-Chun Oh, Faruq Mohammad, and Mohd Rafie Johan. 2020. "Role of mesoporous silica nanoparticles for the drug delivery applications." *Materials Research Express* 7 (10):102002.

Navya, PN, and Hemant Kumar Daima. 2016. "Rational engineering of physicochemical properties of nanomaterials for biomedical applications with nanotoxicological perspectives." *Nano Convergence* 3:1–14.

Ndlwana, Lwazi, Naledi Raleie, Kgogobi M Dimpe, Hezron F Ogutu, Ekemena O Oseghe, Mxolisi M Motsa, Titus AM Msagati, and Bhekie B Mamba. 2021. "Sustainable hydrothermal and solvothermal synthesis of advanced carbon materials in multidimensional applications: A review." *Materials* 14 (17):5094.

Niculescu, Adelina-Gabriela, and Alexandru Mihai Grumezescu. 2021. "Polymer-based nanosystems: A versatile delivery approach." *Materials* 14 (22):6812.

Nnaji, Chukwuemeka O., Jaison Jeevanandam, Yen S. Chan, Michael K. Danquah, Sharadwata Pan, and Ahmed Barhoum. 2018. "Chapter 6 - Engineered nanomaterials for wastewater treatment: current and future trends." In *Fundamentals of Nanoparticles*, edited by Ahmed Barhoum and Abdel Salam Hamdy Makhlouf, 129–168. Elsevier.

Ogunsona, Emmanuel O., Rajendran Muthuraj, Ewomazino Ojogbo, Oscar Valerio, and Tizazu H. Mekonnen. 2020. "Engineered nanomaterials for antimicrobial applications: A review." *Applied Materials Today* 18:100473. https://doi.org/10.1016/j.apmt.2019.100473

Padmanabhan, Nisha T, and Honey John. 2020. "Titanium dioxide based self-cleaning smart surfaces: A short review." *Journal of Environmental Chemical Engineering* 8 (5):104211.

Palit, Sukanchan. 2018. "Chapter 47 - Recent Advances in the Application of Engineered Nanomaterials in the Environment Industry—A Critical Overview and a Vision for the Future." In *Handbook of Nanomaterials for Industrial Applications*, edited by Chaudhery Mustansar Hussain, 883–893. Elsevier.

Papadas, Ioannis T, Ioannis Vamvasakis, Ioannis Tamiolakis, and Gerasimos S Armatas. 2016. "Templated self-assembly of colloidal nanocrystals into three-dimensional mesoscopic structures: A perspective on synthesis and catalytic prospects." *Chemistry of Materials* 28 (9):2886–2896.

Park, Chang Min, Dengjun Wang, and Chunming Su. 2018. "Recent developments in engineered nanomaterials for water treatment and environmental remediation." *Handbook of Nanomaterials for Industrial Applications*:Elsevier, Amsterdam, Holland, 849–882, https://doi.org/10.1016/B978-0-12-813351-4.00048-1

Peng, Yunyan, Yuan Liu, Jinyue Dai, Lijun Cao, and Xiaoqing Liu. 2020. "A sustainable strategy for remediation of oily sewage: Clean and safe." *Separation and Purification Technology* 240:116592.

Poh, T. Y., N. A. T. B. M. Ali, M. Mac Aogáin, M. H. Kathawala, M. I. Setyawati, K. W. Ng, & S. H. Chotirmall 2018. "Inhaled nanomaterials and the respiratory microbiome: Clinical, immunological and toxicological perspectives." *Particle and Fibre Toxicology* 15:1–16.

Prasad, Cheera, Nawal Madkhali, Seong-Geun Jeong, Kuruma Malkappa, Hyeong Yeol Choi, and V Govinda. 2023. "Recent advances in the hybridization of cellulose and semiconductors: Design, fabrication and emerging multidimensional applications: A review." *International Journal of Biological Macromolecules* 233:123551.

Prasad, Ram, VK Jain, and Ajit Varma. 2010. "Role of nanomaterials in symbiotic fungus growth enhancement." *Current Science* 99 (9):1189–1191.

Rafique, Muhammad, M. Iqra Sadaf Shahid Rafique, and M Bilal Tahir. 2017. "A review on green synthesis of silver nanoparticles and their applications." *Artificial Cells, Nanomedicine, and Biotechnology* 45 (7):1272–1291.

Rowland, Clare E, Carl W Brown, James B Delehanty, and Igor L Medintz. 2016. "Nanomaterial-based sensors for the detection of biological threat agents." *Materials Today* 19 (8):464–477.

Rudrapal, Mithun, Ashwini K. Mishra, Laxmi Rani, Khomendra K. Sarwa, James H. Zothantluanga, Johra Khan, Mehnaz Kamal, Santwana Palai, Atul R. Bendale, Swati G. Talele, Vasim T. Pathan, Laxmikant B. Borse, Vishnu S. Neharkar, and Pravat K. Sahoo. 2022. "Nanodelivery of dietary polyphenols for therapeutic applications." *Molecules* 27 (24):8706.

Sadik, Omowunmi (Wunmi) A, AL Zhou, S Kikandi, N Du, Q Wang, and K Varner. 2009. "Sensors as tools for quantitation, nanotoxicity and nanomonitoring assessment of engineered nanomaterials." *Journal of Environmental Monitoring* 11 (10):1782–1800.

Saeed, Maryam, Yousef Alshammari, Shereen A Majeed, and Eissa Al-Nasrallah. 2020. "Chemical vapour deposition of graphene: Synthesis, characterisation, and applications: A review." *Molecules* 25 (17):3856.

Sahin, Ugur, and Özlem Türeci. 2018. "Personalized vaccines for cancer immunotherapy." *Science* 359 (6382):1355–1360.

Savouchkina, Anastasia, Annette Foelske-Schmitz, Rudiger Kötz, A Wokaun, Guenther Georg Scherer, Celestino Padeste, Jorg Ziegler, Vaida Auzelyte, and Harun H. Solak. 2010. "Extreme ultraviolet interference lithography for generation of platinum nanoparticles on glassy carbon." *ECS Transactions* 25 (24):175.

Sercombe, Lisa, Tejaswi Veerati, Fatemeh Moheimani, Sherry Y Wu, Anil K Sood, and Susan Hua. 2015. "Advances and challenges of liposome assisted drug delivery." *Frontiers in Pharmacology* 6:163819.

Shang, Yifen, Md Kamrul Hasan, Golam Jalal Ahammed, Mengqi Li, Hanqin Yin, and Jie Zhou. 2019. "Applications of nanotechnology in plant growth and crop protection: A review." *Molecules* 24 (14):2558.

Sharma, Navneet, Himanshu Ojha, Ambika Bharadwaj, Dharam Pal Pathak, and Rakesh Kumar Sharma. 2015. "Preparation and catalytic applications of nanomaterials: A review." *Rsc Advances* 5 (66):53381–53403.

Shin, Jinwoo, Nayeon Kang, Byungkook Kim, Hyunsik Hong, Le Yu, Jungryun Kim, Heemin Kang, and Jong Seung Kim. 2023. "One-dimensional nanomaterials for cancer therapy and diagnosis." *Chemical Society Reviews* 52, (13), 4488–4514.

Shnoudeh, Abeer Jabra, Islam Hamad, Ruwaida W Abdo, Lana Qadumii, Abdulmutallab Yousef Jaber, Hiba Salim Surchi, and Shahd Z Alkelany. 2019. "Synthesis, characterization, and applications of metal nanoparticles." In *Biomaterials and Bionanotechnology*, 527–612. Elsevier.

Singh, Chandrashekhar, Abhishesh Kumar Mehata, Vishnu Priya, Ankit Kumar Malik, Aseem Setia, M Nikitha Lakshmi Suseela, Vikas, Patharaj Gokul, Samridhi, and Sanjeev K Singh. 2022. "Bimetallic Au–Ag nanoparticles: Advanced nanotechnology for tackling antimicrobial resistance." *Molecules* 27 (20):7059.

Singh, Pardeep, Anwesha Borthakur, PK Mishra, and Dhanesh Tiwary. 2019. "Nano-materials as photocatalysts for degradation of environmental pollutants: Challenges and possibilities."

Singh, PKTR, Smriti Vats, Dharmendra Kumar, and Shruti Tyagi. 2012. "Nanomaterials use in wastewater treatment." *Proceedings of the International Conference on Nanotechnology and Chemical Engineering*. (ICNCS'2012) December 21-22, vol. 2012, Bangkok (Thailand)

Skoulidas, Anastasios I, David S Sholl, and J Karl Johnson. (2006, March). "Adsorption and diffusion of carbon dioxide and nitrogen through single-walled carbon nanotube membranes." *The Journal of Chemical Physics* 124 (5): 054708. https://doi.org/10.1063/1.2151173

Steinegger, Andreas, Otto S Wolfbeis, and Sergey M Borisov. 2020. "Optical sensing and imaging of pH values: Spectroscopies, materials, and applications." *Chemical Reviews* 120 (22):12357–12489.

Sudha, Parappurath N, Kirubanandam Sangeetha, Kumar Vijayalakshmi, and Ahmed Barhoum. 2018. "Nanomaterials history, classification, unique properties, production and market." In *Emerging Applications of Nanoparticles and Architecture Nanostructures*, 341–384. Elsevier.

Sui, Xiaoyu, Julia R Downing, Mark C Hersam, and Junhong Chen. 2021. "Additive manufacturing and applications of nanomaterial-based sensors." *Materials Today* 48:135–154.

Taylor, Douglas D, and Cicek Gercel-Taylor. 2008. *Retracted: MicroRNA Signatures of Tumor-derived Exosomes as Diagnostic Biomarkers of Ovarian Cancer*. Elsevier.

Thakur, Abhinay, and Ashish Kumar. 2022. "Recent advances on rapid detection and remediation of environmental pollutants utilizing nanomaterials-based (bio) sensors." *Science of the Total Environment* 834:155219.

Tîlmaciu, Carmen-Mihaela, and May C Morris. 2015. "Carbon nanotube biosensors." *Frontiers in Chemistry* 3:59.

Usman, Muhammad, Muhammad Farooq, Abdul Wakeel, Ahmad Nawaz, Sardar Alam Cheema, Hafeez ur Rehman, Imran Ashraf, and Muhammad Sanaullah. 2020. "Nanotechnology in agriculture: Current status, challenges and future opportunities." *Science of the Total Environment* 721:137778.

Vineeth Kumar, C. M., V. Karthick, V. Ganesh Kumar, D. Inbakandan, Eldon R. Rene, K. S. Uma Suganya, Asha Embrandiri, T. Stalin Dhas, M. Ravi, and P. Sowmiya. 2022. "The impact of engineered nanomaterials on the environment: Release mechanism, toxicity, transformation, and remediation." *Environmental Research* 212:113202. https://doi.org/10.1016/j.envres.2022.113202

Wang, Yu, Zhanqiang Fang, Bin Liang, and Eric Pokeung Tsang. 2014. "Remediation of hexavalent chromium contaminated soil by stabilized nanoscale zero-valent iron prepared from steel pickling waste liquor." *Chemical Engineering Journal* 247:283–290.

Yang, Lan-Hee, Sangil Hyun, Eunhae Koo, and Dong June Ahn. 2018. "Structure optimization of low-dimensional quantum dots via anisotropic surface energy." *Journal of the Korean Physical Society* 72:582–587.

Zeng, Minxiang, Mingfeng Chen, Dali Huang, Shijun Lei, Xuan Zhang, Ling Wang, and Zhengdong Cheng. 2021. "Engineered two-dimensional nanomaterials: An emerging paradigm for water purification and monitoring." *Materials Horizons* 8 (3):758–802.

Zhang, Lei, Bin Zhao, Chuan Jiang, Junhe Yang, and Guangping Zheng. 2015. "Preparation and transport performances of high-density, aligned carbon nanotube membranes." *Nanoscale Research Letters* 10:1–8.

Zhang, Zishuai, Alison Fraser, Siyu Ye, Geraldine Merle, and Jake Barralet. 2019. "Top-down bottom-up graphene synthesis." *Nano Futures* 3 (4):042003.

Index

Pages in **bold** refer to tables.